DISCRETE OPTIMIZATION
The State of the Art

TOPICS IN DISCRETE MATHEMATICS

Reprints from the journals Discrete Mathematics and
Discrete Applied Mathematics

General Editor: Peter L. Hammer
Rutgers University, Piscataway, NJ, USA

ELSEVIER
Amsterdam – Boston – London – New York – Oxford – Paris – San Diego
San Francisco – Singapore – Sydney – Tokyo

DISCRETE OPTIMIZATION
The State of the Art

Edited by

Endre Boros
Rutgers University, Piscataway, NJ, USA

Peter L. Hammer
Rutgers University, Piscataway, NJ, USA

Reprinted from the journal
Discrete Applied Mathematics, Volume 123, Numbers 1-3 (2002)

N·H

2003
ELSEVIER
Amsterdam – Boston – London – New York – Oxford – Paris – San Diego
San Francisco – Singapore – Sydney – Tokyo

ELSEVIER SCIENCE B.V.
Radarweg 29, PO Box 211, 1000 AE Amsterdam, The Netherlands

Reprinted from: Discrete Applied Mathematics, Volume 123, Numbers 1-3 (2002)

Library of Congress Cataloging in Publication Data
A catalog record from the Library of Congress has been applied for.

British Library Cataloguing in Publication Data
A catalogue record from the British Library has been applied for.

ISBN: 0 444 51295 0

Transferred to digital printing 2007

Contents

Discrete Optimization

N·H

ELSEVIER

Discrete Applied Mathematics 123 (2002) 1–4

DISCRETE
APPLIED
MATHEMATICS

Preface

We have collected in this volume ($2^{2^2}=$) 16 surveys presenting the state of the art in the rapidly growing area of discrete optimization. The surveys were presented by some of the most prominent researchers in this field at the Workshop on Discrete Optimization (DO'99), which was held at RUTCOR—Rutgers University Center for Operations Research, in the summer of 1999, and was attended by a large group of researchers representing the various facets of discrete optimization. We are extremely grateful to the authors of the expository lectures, which represent a comprehensive overview of the theoretical foundations, the methodology and the applications of discrete optimization.

DO'99 was held 22 years after DO'77, which had very similar goals, and whose collection of surveys gives even to this day a valuable image of the state of the art of our area a quarter of a century ago. It is interesting to compare the topics of the surveys presented in 1977 to those of 1999. Two striking characteristics emerge immediately, one being the continuity of topics, and the other being the maturing of new ideas which grow organically on the foundations present for more than 20 years. Among the noticeable research trends reflected in the collection we mention the evolution of large scale local search techniques, the substantial presence of applications geared to VLSI and telecommunications, and the impressive algebraization of the field.

We append to this preface, the table of contents of the two volumes of the state-of-the-art surveys of DO'77. We are happy to remark that those old surveys are still young, being used by many researchers, and still being quoted frequently in their papers. We do hope that the present collection of the state-of-the-art surveys in D.O., appearing here as a new volume of an ongoing series, will provide a similarly useful source of information and inspiration to the community as the first two volumes did for 25 years.

We are all looking forward to the next DO, and to many forthcoming volumes.

Endre Boros, Peter L. Hammer
Rutgers University, RUTCOR, 640 Bartholomew Road,
Piscataway, NJ 08854-8003, USA

E-mail address: hammer@rutcor.rutgers.edu (P.L. Hammer).

DO'77 THE STATE of the ART 2^{-2} CENTURIES AGO
ANNALS of DISCRETE MATHEMATICS, VOLUMES 4 and 5, 1979

1. Combinatorial and polyhedral aspects of discrete optimization

Surveys

C. Berge, Packing problems and hypergraph theory: a survey

J. Edmonds, Matroid intersection

P.L. Hammer, Boolean elements in combinatorial optimization

A.J. Hoffman, The role of unimodularity in applying linear inequalities to combinatorial theorems

B. Korte, Approximative algorithms for discrete optimization problems

J.K. Lenstra and A.H.G. Rinnooy Kan, Computational complexity of discrete optimization problems

L. Lovász, Graph theory and integer programming

J. Tind, Blocking and antiblocking polyhedra

Reports

Complexity of combinatorial problems (R.L. Graham)

Structural aspects of discrete problems (P. Hansen)

Polyhedral aspects of discrete optimization (A.J. Hoffman)

2. Some fundamental classes of problems

Surveys

R.E. Burkard, Travelling salesman and assignment problems: a survey

P.C. Gilmore, Cutting stock, linear programming, knapsacking, dynamic programming and integer programming, some interconnections

V. Klee and D. Larman, Use of Floyd's algorithm to find shortest restricted paths

E.L. Lawler, Shortest path and network flow algorithms

M.W. Padberg, Covering, Packing and Knapsack problems

Reports

Network flow, assignment and travelling salesman problems (D. De Ghellinck)

Algorithms for special classes of combinatorial optimization problems (J.K. Lenstra)

3. Methodology

Surveys

E. Balas Disjunctive programming
P. Hansen, Methods of nonlinear 0–1 programming
R.G. Jeroslow, An introduction to the theory of cutting planes
E.L. Johnson, On the group problem and a subadditive approach to integer programming
J.F. Shapiro, A survey of lagrangian techniques for discrete optimization
K. Spielberg, Enumerative methods in integer programming

Reports

Branch and bound/implicit enumeration (E. Balas)
Cutting planes (M. Gondran)
Group theoretic and lagrangean methods (M.W. Padberg)

4. Computer codes

Surveys

E.M.L. Beale, Branch and bound methods for mathematical programming systems
A. Land and S. Powell, Computer codes for problems of integer programming

Reports

Current state of computer codes for discrete optimization (J.J.H. Forrest)
Codes for special problems (F. Giannessi)
Current computer codes (S. Powell)

5. Applications

Surveys

R.L. Graham, E.L. Lawler, J.K. Lenstra and A.H.G. Rinnooy Kan, Optimization and approximation in deterministic sequencing and scheduling: a survey
J. Krarup and P. Pruzan, Selected families of location problems
S. Zionts, A survey of multiple criteria integer programming methods

Reports

Industrial applications (E.M.L. Beale)

Modelling (D. Klingman)
Location and distribution problems (J. Krarup)
Communication and electrical networks (M. Segal)
Scheduling (A.H.G. Rinnooy Kan)
Conclusive remarks

N·H

ELSEVIER

Discrete Applied Mathematics 123 (2002) 5–74

DISCRETE
APPLIED
MATHEMATICS

Non-standard approaches to integer programming ☆

Karen Aardal[a,b,1], Robert Weismantel[a,c,2], Laurence A. Wolsey[a,*]

[a]CORE and INMA, Universite Catholique de Louvain, 34 Voie du Roman Pays,
1348 Louvain-la-Neuve, Belgium
[b]Department of Mathematics, Universiteit Utrecht, Budapestlaan 6, 3584 CH Utrecht, The Netherlands
[c]Fakultät für Mathematik, IMO Otto-van-Guericke Universität Magdeburg, Universitätsplatz 2,
D-39106 Magdeburg, Germany

Received 6 December 1999; received in revised form 27 June 2001; accepted 2 July 2001

Abstract

In this survey we address three of the principal algebraic approaches to integer programming. After introducing lattices and basis reduction, we first survey their use in integer programming, presenting among others Lenstra's algorithm that is polynomial in fixed dimension, and the solution of diophanine equations using basis reduction. The second topic concerns augmentation algorithms and test sets, including the role played by Hilbert and Gröbner bases in the development of a primal approach to solve a family of problems for all right-hand sides. Thirdly we survey the group approach of Gomory, showing the importance of subadditivity in integer programming and the generation of valid inequalities, as well the relation to the parametric problem cited above of solving for all right-hand sides. © 2002 Elsevier Science B.V. All rights reserved.

MSC: ; 90C10; 90C11

Keywords: Integer programming; Lattice basis reduction; Lenstra's algorithm; Test sets; Augmentation algorithms; Gröbner basis; Test sets; Asymptotic group problem; Subadditivity; Corner polyhedron

☆ Work carried out as part of DONET (Discrete Optimization Network) TMR project No. ERB FMRX-CT98-0202 of the EU.
* Corresponding author. Tel.: 32-10-47-4307; fax: 32-10-47-4301.
  E-mail address: wolsey@core.ucl.ac.be (L.A. Wolsey).
[1] Research partially supported by the ESPRIT Long Term Research Project No. 20244 (Project ALCOM-IT: *Algorithms* and Complexity in Information Technology of the EU and by NSF through the Center for Research on Parallel Computation, Rice University, under Cooperative Agreement No. CCR-9120008.
[2] Supported by a Gerhard–Hess-Forschungsförderpreis and grant WE1462 of the German Science Foundation (DFG), and grants FKZ 0037KD0099 and FKZ 2945A/0028G of the Kultusministerium Sachsen-Anhalt.

0. Introduction

The standard approach to the integer programming optimization problem

$$(IP) \quad \min\{c^{\mathrm{T}}x\colon x \in X\} \quad \text{where } X = \{x \in \mathbb{Z}_+^n\colon Ax = b\}$$

or the equivalent integer programming feasibility problem

$$(FP) \quad \text{is } X \neq \emptyset?$$

is to use linear programming within a branch-and-bound or branch-and-cut framework, using whenever possible polyhedral results about the structure of conv(X) or approximations to conv(X). Here we examine alternative approaches depending explicitly on the *discrete* nature of the set X.

Given a specific point $x^0 \in X$ and a generic point $x \in X$, the vector $y = x - x^0$ lies in

$$L = \{x \in \mathbb{Z}^n\colon Ax = 0\},$$

the set of integer points in a subspace. Every such set can be shown to form an *integer lattice*, namely it can be rewritten in the form

$$L = \{x\colon x = B\lambda,\ \lambda \in \mathbb{Z}^p\}.$$

When the columns of B are linearly independent, B is known as a *lattice basis*.

In Section 1 we introduce lattices and the basis reduction algorithms of Lovász [70] and of Lovász and Scarf [76]. A *reduced* basis is a basis in which the vectors are short and nearly orthogonal. The basis reduction algorithm of Lovász runs in polynomial time and starting from a lattice basis produces another basis with basis vectors of short Euclidean length. The algorithm of Lovász and Scarf works with a more general norm and runs in polynomial time for fixed n.

Lattice basis reduction has played an important role in the theory of integer programming. It was first used in this area by Lenstra in 1983 [71] in proving that the integer programming problem can be solved in polynomial time for a fixed number of variables. The proof was constructive and consisted of two main steps: a linear transformation, and Lovász' basis reduction algorithm [70]. Later, Grötschel et al. [48], Kannan [64], and Lovász and Scarf [76] developed algorithms using similar principles to Lenstra's algorithm. In computational integer programming, however, basis reduction has received less attention. One of the few implementations that we are aware is that of Cook et al. [23] in which some difficult, not previously solved, network design problems were solved using the generalized basis reduction algorithm of Lovász and Scarf. Recently Aardal et al. [2,3] have developed an algorithm for solving a system of diophantine equations with bounds on the variables. They used basis reduction to reformulate a certain integer relaxation of the problem, and were able to solve several integer programming instances that proved hard, or even unsolvable, for several other algorithms. Their algorithm was partly inspired by algorithms used in cryptography to solve subset sum problems that occur in knapsack public-key cryptosystems. In the area of cryptography, basis reduction has been used successfully to solve such subset sum problems, see for instance the survey article by Joux and Stern [60]. These lattice based algorithms are presented in Section 2.

Alternatively given a point $x^0 \in X$, suppose that there exists a point $x \in X$ having a smaller objective function value $c^{\mathrm{T}}x < c^{\mathrm{T}}x^0$, and also satisfying the condition $x \geqslant x^0$. Now $y = x - x^0$ lies in the set

$$X^0 = \{x \in \mathbb{Z}^n : Ax = 0, \ x \geqslant 0\},$$

the set of non-negative integer points in a cone. Here again such sets can be finitely generated, and rewritten in the form

$$X^0 = \{x : x = H\lambda, \ \lambda \in \mathbb{Z}_+^p\},$$

where the minimal set of columns H is known as a *Hilbert basis*. Note that it follows that there is some column h of H for which h is an improving vector for x^0, i.e. $x' = x^0 + h \in X$ with $x' \geqslant x^0$ and $c^{\mathrm{T}}x' < c^{\mathrm{T}}x^0$.

Generalizing this idea leads to test sets for families of integer programs. Test sets are collections of integral vectors with the property that every feasible non-optimal point of any integer program in the family can be improved by a vector in the test set. Given a test set T explicitly, there is a straightforward algorithm for solving (IP). Starting with a feasible point x, one searches for an element $t \in T$ such that $x + t$ is feasible and has a smaller objective function value, replaces x by $x + t$ and iterates until no such t exists. In general one cannot expect that a test set is polynomial in the size of a given integer program. This raises the question of designing an efficient augmentation algorithm for the problem:

Let x be any feasible point of the linear integer program. While x is not optimal, determine an integral vector z and a non-negative integer λ such that (i) $x + \lambda z$ is feasible and (ii) $x + \lambda z$ attains a smaller objective function value than x. Set $x := x + \lambda z$.

Augmentation algorithms have been designed for and applied to a range of linear integer programming problems: augmenting path methods for solving maximum flow problems or algorithms for solving the minimum cost flow problem via augmentation along negative cycles are of this type. Other examples include the greedy algorithm for solving the matroid optimization problem, alternating path algorithms for solving the maximum weighted matching problem or primal methods for optimizing over the intersection of two matroids. In Section 3 we investigate these primal approaches to solving (IP).

We next consider a family of relaxations for (IP) in which we drop the nonnegativity constraints on a subset of the variables, namely relaxations of the form:

$$\min\{c^{\mathrm{T}}x : Ax = b, \ x_j \in \mathbb{Z}_+^1 \text{ for } j \in V \setminus S, \ x_j \in \mathbb{Z}^1 \text{ for } j \in S\}.$$

where V is the index set of the variables, and $S \subseteq V$. Note that when $S = V$, this leads us back to the lattice viewpoint.

Gomory [40–42] in the 1960s studied the "asymptotic group" relaxation in which S is the index set of an optimal linear programming basis A_B. The resulting solution set $X^G = \{x = (x_B, x_N) \in \mathbb{Z}^n : A_B x_B + A_N x_N = b, \ x_N \geqslant 0\}$ can be reformulated in the space of the nonbasic variables as the set $\tilde{X}^G = \{x_N \in \mathbb{Z}_+^{n-m} : A_B^{-1} A_N x_N \equiv A_B^{-1} b \,(\mathrm{mod}\,1)\}$. Optimization over $\tilde{X}^G$ reduces to a shortest path problem in a graph with $|\det(A_B)|$ nodes, known as the *group problem*. Gomory's study of the convex hull of solutions of $\tilde{X}^G$, known as the "corner polyhedron", showed the crucial role of subadditivity in describing strong valid inequalities. These developments are discussed in Section 4.

Notation. For $z \in \mathbb{R}^n$, $||z||$, $||z||_1$ and $||z||_\infty$ denote the Euclidean, L_1 and maximum norms of z respectively. x^{T} denotes the transpose of the vector x such that $x^{\mathrm{T}}y$ is the inner product in $\mathbb{R}^n$ of the vectors x and y. The symbol e^j represents the unit vector in the corresponding Euclidean space having a one in position j and zeros everywhere else.

For $x \in \mathbb{R}$, $\lfloor x \rfloor$ is the largest integer not greater than x, $\lceil x \rceil$ is the smallest integer not less than x, and $[x] = \lceil x - \frac{1}{2} \rceil$, i.e., the nearest integer to x, where we round up if the fraction is equal to one half.

If $\lambda_1, \ldots, \lambda_n$ are integers, $\gcd(\lambda_1, \ldots, \lambda_n)$ denotes the greatest common divisor of these integers. For x, y non-zero integers, $x|y$ means that y is an integer multiple of x. A square integral matrix C is *unimodular* if $|\det C| = 1$.

Finally for $v \in \mathbb{R}^n$ we denote by $\mathrm{supp}(v) := \{i \in \{1, \ldots, n\}: v_i \neq 0\}$ the *support* of v. v^+ is the vector such that $v_i^+ = v_i$ if $v_i > 0$ and $v_i^+ = 0$, otherwise. Similarly, v^- is the vector with $v_i^- = -v_i$ if $v_i < 0$ and $v_i^- = 0$, otherwise. So $v = v^+ - v^-$.

1. Lattices and basis reduction

Here we define a lattice and a lattice basis. In the lattice approaches to integer programming that will be discussed in Section 2 we need lattice representations using bases with short, almost orthogonal basis vectors. Such bases are called reduced. We describe two algorithms for finding a reduced basis. The algorithm of Lovász, as presented by Lenstra et al. [70], is described in Section 1.2, and the algorithm of Lovász and Scarf [76] is presented in Section 1.3. We also discuss some recent implementations. Before starting the more formal presentation we give a brief sketch of the principle behind the integer programming algorithms discussed in Section 2, such as the algorithm of Lenstra [71], so as to motivate the study of reduced bases in this context.

Note that in Sections 1 and 2 we use b_j, b_j^*, b_j' to denote distinct vectors associated with the basis of a lattice, and not the jth component of the vectors b, b^* and b', respectively.

1.1. Preliminaries

Given a set of l linearly independent vectors $b_1, \ldots, b_l \in \mathbb{R}^n$ with $l \leqslant n$, let B be the matrix with column vectors $b_1, \ldots, b_l$.

Definition 1.1. The *lattice* L spanned by $b_1, \ldots, b_l$ is the set of vectors that can be obtained by taking integer linear combinations of the vectors $b_1, \ldots, b_l$,

$$L = \left\{ x : x = \sum_{j=1}^{l} \lambda_j b_j, \ \lambda_j \in \mathbb{Z}, \ 1 \leqslant j \leqslant l \right\}. \tag{1}$$

The set of vectors $\{b_1, \ldots, b_l\}$ is called a *basis* of the lattice.

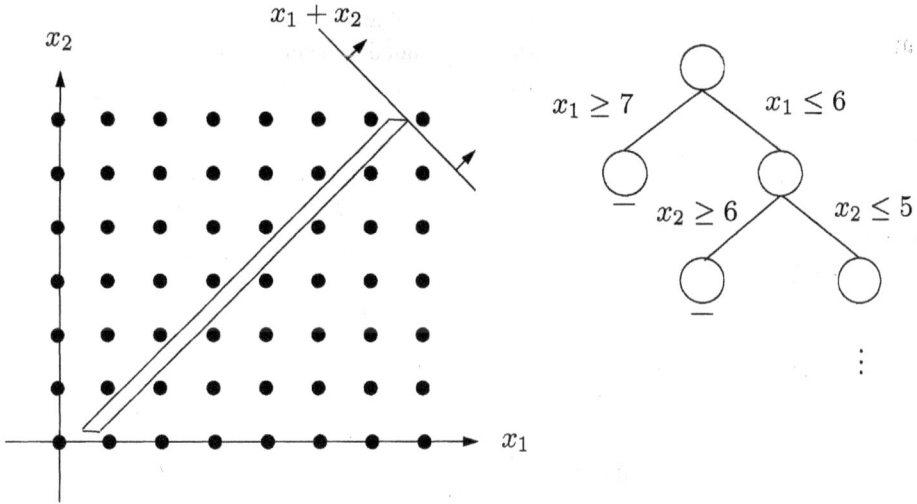

Fig. 1. A difficult type of instance for branch-and-bound.

The following operations on a matrix are called *elementary column operations*:

- exchanging two columns,
- multiplying a column by -1,
- adding an integral multiple of one column to another column.

Theorem 1.1. *An integral matrix U is unimodular if and only if U can be derived from the identity matrix by elementary column operations.*

A lattice may have several bases.

Observation 1.1. *If B and B' are bases for the same lattice L, then $B' = BU$ for some $l \times l$ unimodular matrix U.*

Consider the integer feasibility problem "Is $\{x \in \mathbb{Z}^n : Ax \leqslant b\} \neq \emptyset$?". Branch-and-bound applied to this problem may give rise to arbitrarily deep search trees even in dimension 2, which shows that branch-and-bound is not a polynomial time algorithm for integer programming in fixed dimension.

Example 1.1. Consider the two-dimensional polytope in Fig. 1. If we use linear programming based branch-and-bound on this instance with objective function max $x_1 + x_2$, then we see that the variables x_1 and x_2 alternately take fractional values, which forces us to branch. If we extend the polytope arbitrarily far, then the branch-and-bound tree will become arbitrarily deep. It is easy to construct an example that is equally bad for branch-and-bound in which the polytope contains an integer vector.   □

The ideas behind Lenstra's [71] integer programming algorithm are as follows. For simplicity we consider a full-dimensional bounded integer programming problem in the standard lattice $\mathbb{Z}^n$. Notice that the unit vectors in $\mathbb{R}^n$ form a basis for the lattice $\mathbb{Z}^n$. Since a "thin" polytope $X = \{x \in \mathbb{R}^n : Ax \leqslant b\}$ is bad for branch-and-bound enumeration, Lenstra first transforms the polytope using a linear transformation τ such that the transformed polytope τX appears "round". In order not to change the problem one has to apply the same linear transformation to $\mathbb{Z}^n$. This yields the equivalent problem "Is $\tau X \cup \tau \mathbb{Z}^n \neq \emptyset$?". A basis for the lattice $\tau \mathbb{Z}^n$ is formed by the vectors $\tau e^j, 1 \leqslant j \leqslant n$.

We now consider a real vector x reasonably in the middle of the polytope τX. It is easy to find a vector $y \in \tau \mathbb{Z}^n$ that is close to x, see Lemma 2.1 in Section 2.1. Either $y \in \tau X$, which means that we are done since we have found a feasible point, or y is outside of τX. If y does not belong to τX, then we still do not know the answer to our integer feasibility problem, but we do know that the polytope τX cannot be too large since the point y is quite close to x. The next observation is that each lattice is contained in countably many parallel hyperplanes. In order to determine whether or not τX contains a lattice point we need to enumerate hyperplanes intersecting τX. To avoid enumerating too many hyperplanes we want a representation of the lattice such that the distance h between each pair of consecutive hyperplanes is not too small. This is the step where basis reduction is used. In Fig. 2 we illustrate three different representations of the same lattice. In Fig. 2(a) the lattice is represented by non-orthogonal basis vectors yielding a small value of h, and therefore relatively many hyperplanes intersecting τX. In Fig. 2(b) an almost orthogonal basis is used, but the enumeration of hyperplanes is done along the shortest basis vector. Finally, in Fig. 2(c) we have again a nearly orthogonal basis, but now the enumeration of hyperplanes is done along the longest of the basis vectors, which yields a relatively large value of h. This is the desired situation.

In Lenstra's algorithm basis reduction is applied to the basis $\tau e_j, 1 \leqslant j \leqslant n$ to obtain an almost orthogonal basis. Next, the basis vectors are reordered such that the longest vector is the last one (cf. Fig. 2(c)). Then we consider the hyperplane H spanned by the first $n - 1$ basis vectors. The lattice $\tau \mathbb{Z}^n$ is now contained in the union of H plus an integer multiple of the last basis vector, i.e., countably many parallel hyperplanes. Once the hyperplane representation is determined we fix a hyperplane intersecting τX, which gives rise to a problem in dimension at least one lower than the original problem, and the whole process is repeated again.

To get an algorithm that is polynomial in fixed dimension n, one needs a basis reduction algorithm that is polynomial for fixed n. Lenstra uses Lovász' basis reduction algorithm, which is more efficient than that since it is polynomial for arbitrary n, see Theorem 1.2 in Section 1.2. Moreover, one needs to prove that the number of hyperplanes that are enumerated at each level is bounded by a constant depending only on the dimension at that level. To do this Lenstra determines an appropriate linear transformation in polynomial time such that τX is indeed "round" (for the technical details, see Section 2.1.1), and he shows that the distance h between consecutive lattice hyperplanes is bounded from below by $c \cdot ||b_n||$, where c is a constant depending only

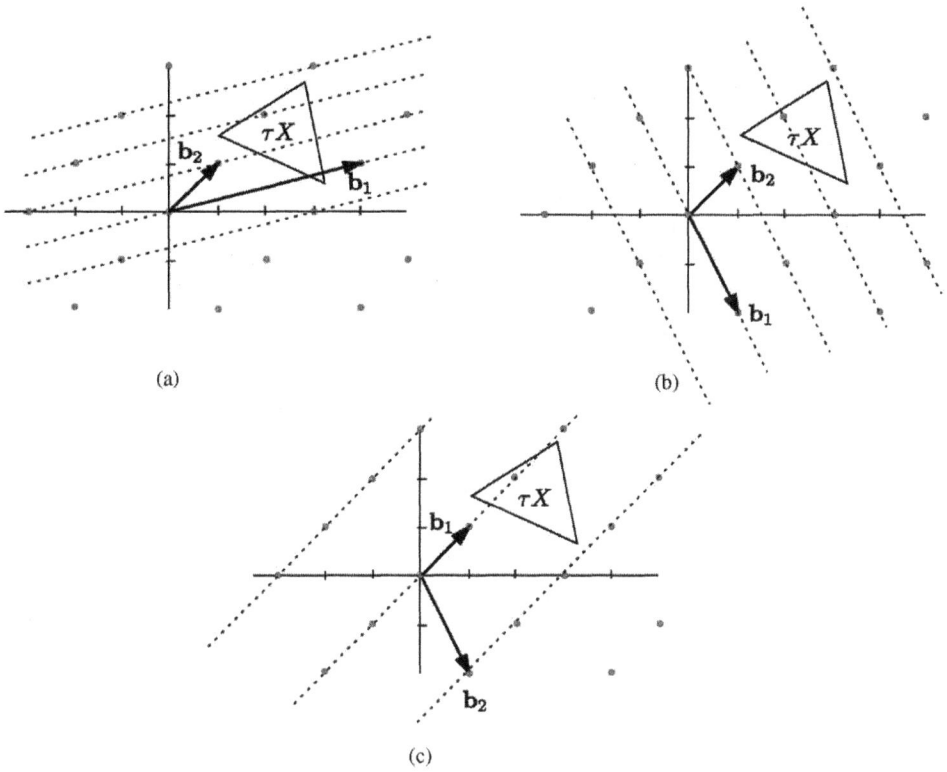

Fig. 2. (a) Non-orthogonal basis. (b) Nearly orthogonal basis. (c) Nearly orthogonal basis, branching on the longest vector.

on the dimension, and b_n is the longest basis vector in a reduced basis (see Proposition 1.3, and Corollary 1.1 in Section 1.2).

1.2. Lovász' basis reduction algorithm

Lovász' basis reduction algorithm [70] consists of a series of elementary column operations on an initial basis B for a given lattice and produces a so-called *reduced basis* B' such that the basis vectors $b'_1, \ldots, b'_l$ are short and nearly orthogonal, and such that b'_1 is an approximation of the shortest vector in the lattice. So, B' is obtained as $B' = BU$ for some unimodular matrix U. Given a basis B one can obtain orthogonal vectors by applying Gram–Schmidt orthogonalization. The Gram–Schmidt vectors, however, do not necessarily belong to the lattice, but they do span the same real vector space as $b_1, \ldots, b_l$, so they are used as a "reference" for the basis reduction algorithm.

Definition 1.2. The Gram–Schmidt process derives orthogonal vectors b_j^*, $1 \leqslant j \leqslant l$, from linearly independent vectors b_j, $1 \leqslant j \leqslant l$. The vectors b_j^*, $1 \leqslant j \leqslant l$, and the

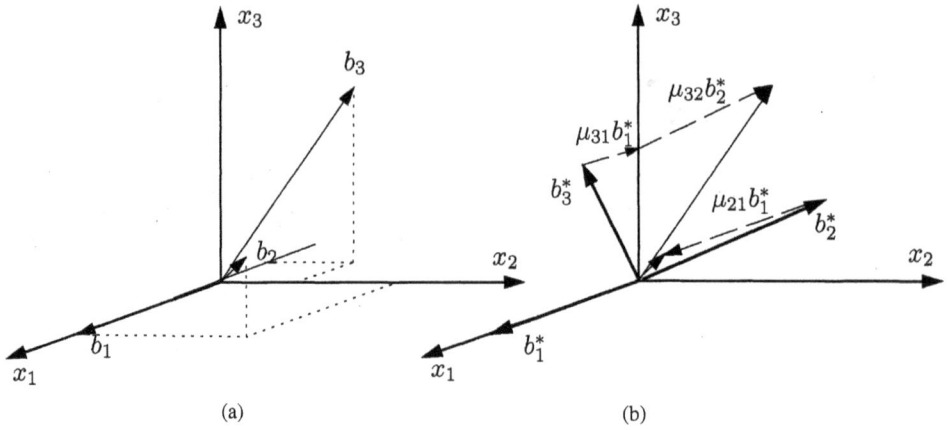

Fig. 3.

real numbers μ_{jk}, $1 \leqslant k < j \leqslant l$, are determined from b_j, $1 \leqslant j \leqslant l$, by the recursion

$$b_1^* = b_1, \tag{2}$$

$$b_j^* = b_j - \sum_{k=1}^{j-1} \mu_{jk} b_k^*, \quad 2 \leqslant j \leqslant l, \tag{3}$$

$$\mu_{jk} = \frac{b_j^\mathrm{T} b_k^*}{\|b_k^*\|^2}, \quad 1 \leqslant k < j \leqslant l. \tag{4}$$

Example 1.2. Here we illustrate the Gram–Schmidt vectors obtained by applying the orthogonalization procedure given in Definition 1.2 to the vectors

$$b_1 = \begin{pmatrix} 2 \\ 0 \\ 0 \end{pmatrix}, \quad b_2 = \begin{pmatrix} 2 \\ 2 \\ 1 \end{pmatrix}, \quad b_3 = \begin{pmatrix} -1 \\ 1 \\ 2 \end{pmatrix},$$

shown in Fig. 3(a).

We obtain $\mu_{21} = 1$, $\mu_{31} = -\frac{1}{2}$, $\mu_{32} = \frac{4}{5}$, and

$$b_1^* = b_1 = \begin{pmatrix} 2 \\ 0 \\ 0 \end{pmatrix}, \quad b_2^* = b_2 - \mu_{21} b_1^* = \begin{pmatrix} 0 \\ 2 \\ 1 \end{pmatrix},$$

$$b_3^* = b_3 - \mu_{31} b_1^* - \mu_{32} b_2^* = \begin{pmatrix} 0 \\ -\frac{3}{5} \\ \frac{6}{5} \end{pmatrix}.$$

The Gram–Schmidt vectors are shown in Fig. 3(b). □

As mentioned above, the vectors $b_1^*, \ldots, b_j^*$, span the same real vector space as the vectors $b_1, \ldots, b_j$, $1 \leqslant j \leqslant l$. The vector b_j^* is the projection of b_j on the orthogonal complement of $\sum_{k=1}^{j-1} \mathbb{R}b_k = \{\sum_{k=1}^{j-1} m_k b_k : m_k \in \mathbb{R}, \ 1 \leqslant k \leqslant j - 1\}$, i.e., b_j^* is the component of b_j orthogonal to the real subspace spanned by $b_1, \ldots, b_{j-1}$. Thus, any pair b_i^*, b_k^* of the Gram–Schmidt vectors are mutually orthogonal. The multiplier μ_{jk} gives the length, relative to b_k^*, of the component of the vector b_j in direction b_k^*. The multiplier μ_{jk} is equal to zero if and only if b_j is orthogonal to b_k^*.

Definition 1.3 (Lenstra et al. [71]). A basis $b_1, b_2, \ldots, b_l$ is called *reduced* if

$$|\mu_{jk}| \leqslant \tfrac{1}{2} \quad \text{for } 1 \leqslant k < j \leqslant l, \tag{5}$$

$$||b_j^* + \mu_{j,j-1} b_{j-1}^*||^2 \geqslant \frac{3}{4} ||b_{j-1}^*||^2 \quad \text{for } 1 < j \leqslant l. \tag{6}$$

A reduced basis according to Lovász is a basis in which the vectors are short and nearly orthogonal. Below we explain why vectors satisfying Conditions (5) and (6) have these characteristics.

Condition (5) is satisfied if the component of vector b_j in direction b_k^* is short relative to b_k^*. This is the case if b_j and b_k^* are nearly orthogonal, or if b_j is short relative to b_k^*. If condition (5) is violated, i.e., the component of vector b_j in direction b_k^* is *relatively long*, then Lovász' basis reduction algorithm will replace b_j by $b_j - \lceil \mu_{jk} \rfloor b_k$. Such a step is called *size reduction* and will ensure relatively short basis vectors. Next, suppose that (5) is satisfied because b_j is short relative to b_k^*, $k < j$. Then we may end up with a basis where the vectors are not at all orthogonal, and where the first vector is very long, the next one relatively short compared to the first one, and so on. To prevent this from happening we enforce Condition (6). Here we relate to the interpretation of the Gram–Schmidt vectors above, and notice that the vectors $b_j^* + \mu_{j,j-1} b_{j-1}^*$ and b_{j-1}^* are the projections of b_j and b_{j-1} on the orthogonal complement of $\sum_{k=1}^{j-2} \mathbb{R}b_k$. Consider the case where $k = j - 1$, i.e., suppose that b_j is short compared to b_{j-1}^*, which implies that b_j^* is short compared to b_{j-1}^* as $||b_j^*|| \leqslant ||b_j||$. Suppose we *interchange* b_j and b_{j-1}. Then the new b_{j-1}^* will be the vector $b_j^* + \mu_{j,j-1} b_{j-1}^*$, which will be short compared to the old b_{j-1}^*, i.e., Condition (6) will be violated. To summarize, Conditions (5) and (6) ensure that we obtain a basis in which the vectors are short and nearly orthogonal. To achieve such a basis, Lovász' algorithm applies a sequence of *size reductions* and *interchanges* in order to reduce the length of the vectors, and to prevent us from obtaining non-orthogonal basis vectors of decreasing length, where the first basis vector may be arbitrarily long. The constant $\frac{3}{4}$ in inequality (6) is arbitrarily chosen and can be replaced by any fixed real number $\frac{1}{4} < y < 1$. In a practical implementation one chooses a constant close to one.

A brief outline of Lovász' basis reduction algorithm is as follows. For precise details we refer to [70]. First compute the Gram–Schmidt vectors b_j^*, $1 \leqslant j \leqslant l$ and the numbers μ_{jk}, $1 \leqslant k < j \leqslant l$. Initialize $i := 2$. Perform, if necessary, a size reduction to obtain $|\mu_{i,i-1}| \leqslant 1/2$. Update $\mu_{i,i-1}$. Then check whether Condition (6) holds for $j = i$. If Condition (6) is violated, then exchange b_i and b_{i-1}, and update the relevant

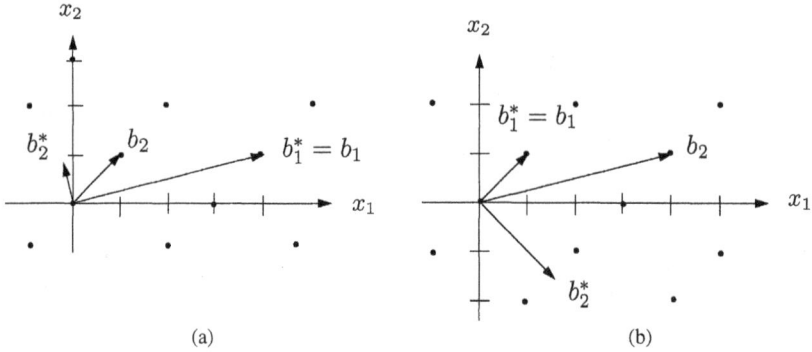

Fig. 4.

Gram–Schmidt vectors and numbers μ_{jk}. If $i > 2$, then let $i:=i-1$. Next, achieve $|\mu_{im}| \leqslant 1/2$ for $m = i-2, i-3, \ldots, 1$. If $i = l$, stop. Otherwise, let $i:=i+1$.

From this short description, it is not obvious that the algorithm is efficient, but as the following theorem states, Lovász' basis reduction algorithm runs in polynomial time.

Theorem 1.2 (Lenstra et al. [70]). *Let $L \subseteq \mathbb{Z}^n$ be a lattice with basis $b_1, \ldots, b_n$, and let $\beta \in \mathbb{R}$, $\beta \geqslant 2$, be such that $\|b_j\|^2 \leqslant \beta$ for $1 \leqslant j \leqslant n$. Then the number of arithmetic operations needed by the basis reduction algorithm as described in [70] is $O(n^4 \log \beta)$, and the integers on which these operations are performed each have binary length $O(n \log \beta)$.*

In terms of bit operations, Theorem 1.2 implies that Lovász' basis reduction algorithm has a running time of $O(n^6 (\log \beta)^3)$ using classical algorithms for addition and multiplication. There are reasons to believe that it is possible in practice to find a reduced basis in $O(n(\log \beta)^3)$ bit operations, see Section 4 of [61] and [82].

Example 1.3. Here we give an example of an initial and a reduced basis for a given lattice. Let L be the lattice generated by the vectors

$$b_1 = \begin{pmatrix} 4 \\ 1 \end{pmatrix}, \quad b_2 = \begin{pmatrix} 1 \\ 1 \end{pmatrix}.$$

The Gram–Schmidt vectors are $b_1^* = b_1$ and $b_2^* = b_2 - \mu_{21} b_1^* = (1,1)^{\mathrm{T}} - \frac{5}{17} b_1^* = \frac{1}{17}(-3, 12)^{\mathrm{T}}$, see Fig. 4(a). Condition (5) is satisfied since b_2 is short relative to b_1^*. However, Condition (6) is violated, so we exchange b_1 and b_2, giving

$$b_1 = \begin{pmatrix} 1 \\ 1 \end{pmatrix}, \quad b_2 = \begin{pmatrix} 4 \\ 1 \end{pmatrix}.$$

We now have $b_1^* = b_1$, $\mu_{21} = \frac{5}{2}$ and $b_2^* = \frac{1}{2}(3, -3)^{\mathrm{T}}$, see Fig. 4(b).

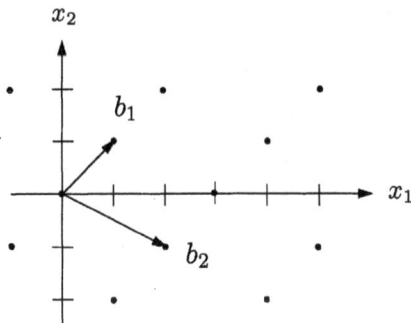

Fig. 5. The reduced basis.

Condition (5) is now violated, so we replace b_2 by $b_2 - 2b_1 = (2,-1)^T$. Conditions (5) and (6) are satisfied for the resulting basis

$$b_1 = \begin{pmatrix} 1 \\ 1 \end{pmatrix}, \quad b_2 = \begin{pmatrix} 2 \\ -1 \end{pmatrix},$$

and hence this basis is reduced, see Fig. 5. □

Let W be the vector space spanned by the lattice L, and let B_W be an orthonormal basis for W. The determinant of the lattice L, $d(L)$, is defined as the absolute value of the determinant of any non-singular linear transformation $W \to W$ that maps B_W onto a basis of L. Below we give three different formulae for computing $d(L)$. Let $B = (b_1, \ldots, b_m)$ be a basis for the lattice $L \subset \mathbb{R}^n$, with $m \leq n$, and let $b_1^*, \ldots, b_m^*$ be the vectors obtained from applying the Gram–Schmidt orthogonalization procedure, see Definition 1.2, to $b_1, \ldots, b_m$.

$$d(L) = ||b_1^*|| \cdot ||b_2^*|| \cdot \; \cdots \; \cdot ||b_m^*||, \tag{7}$$

$$d(L) = \sqrt{\det(B^T B)}, \tag{8}$$

$$d(L) = \lim_{r \to \infty} \frac{|\{x \in L: ||x|| < r\}|}{\text{vol}(B_m(r))}, \tag{9}$$

where $\text{vol}(B_m(r))$ is the volume of the m-dimensional ball with radius r. If L is full-dimensional, $d(L)$ can be interpreted as the volume of the parallelepiped $\sum_{j=1}^{n} [0,1)b_j$. In this case the determinant of the lattice can be computed straightforwardly as $d(L) = |\det(b_1, \ldots, b_n)|$. Note that the determinant of a lattice depends only on the lattice and not on the choice of basis (cf. Observation 1.1, and expression (9)). The determinant of $\mathbb{Z}^n$ is equal to one.

In Propositions 1.3 and 1.4 we assume that the lattice L is full-dimensional.

Proposition 1.3 (Lenstra et al. [71]). *Let $b_1, \ldots, b_n$ be a reduced basis for the lattice $L \subset \mathbb{R}^n$. Then,*

$$d(L) \leqslant \prod_{j=1}^{n} ||b_j|| \leqslant c_1 \cdot d(L), \tag{10}$$

where $c_1 = 2^{n(n-1)/4}$.

The first inequality in (10) is the so called *inequality of Hadamard* that holds for any basis of L. Hadamard's inequality holds with equality if and only if the basis is orthogonal. Hermite [56] proved that each lattice $L \subset \mathbb{R}^n$ has a basis $b_1, \ldots, b_n$ such that $\prod_{j=1}^{n} ||b_j|| \leqslant c \cdot d(L)$, where c is a constant depending only on n. The basis produced by Lovász' basis reduction algorithm yields the constant $c = c_1$ in Proposition 1.3. Better constants than c_1 are possible, but the question is then whether the basis can be obtained in polynomial time.

A consequence of Proposition 1.3 is that if we consider a basis that satisfies (10), then the distance of the basis vector b_n to the hyperplane generated by the reduced basis vectors $b_1, \ldots, b_{n-1}$ is not too small as stated in the following corollary.

Corollary 1.1 (Lenstra [71]). *Assume that $b_1, \ldots, b_n$ is a basis such that (10) holds, and that, after possible reordering, $||b_n|| = \max_{1 \leqslant j \leqslant n} \{||b_j||\}$. Let $H = \sum_{j=1}^{n-1} \mathbb{R} b_j$ and let h be the distance of basis vector b_n to H. Then*

$$c_1^{-1} \cdot ||b_n|| \leqslant h \leqslant ||b_n||, \tag{11}$$

where $c_1 = 2^{n(n-1)/4}$.

Proof. Let $L' = \sum_{j=1}^{n-1} \mathbb{Z} b_j$. We have

$$d(L) = h \cdot d(L'). \tag{12}$$

Expressions (10) and (12) give

$$\prod_{j=1}^{n} ||b_j|| \leqslant c_1 \cdot d(L) = c_1 \cdot h \cdot d(L') \leqslant c_1 \cdot h \cdot \prod_{j=1}^{n-1} ||b_j||, \tag{13}$$

where the first inequality follows from the second inequality of (10), and where the last inequality follows from the inequality of Hadamard (first inequality of (10)). From (13) we obtain $h \geqslant c_1^{-1} ||b_n||$. From the definition of h we have $h \leqslant ||b_n||$, and this bound holds with equality if and only if the vector b_n is perpendicular to H. □

The lower bound on h given in Corollary 1.1 plays a crucial role in the algorithm of Lenstra described in Section 2.1.1.

Proposition 1.4 (Lenstra et al. [70]). *Let* $L \subset \mathbb{R}^n$ *be a lattice with reduced basis* $b_1, \ldots, b_n \in \mathbb{R}^n$. *Let* $x^1, \ldots, x^t \in L$ *be linearly independent. Then we have*

$$||b_1||^2 \leqslant 2^{n-1} ||x||^2 \quad \text{for all } x \in L, \ x \neq 0, \tag{14}$$

$$||b_j||^2 \leqslant 2^{n-1} \max\{||x^1||^2, ||x^2||^2, \ldots, ||x^t||^2\} \quad \text{for } 1 \leqslant j \leqslant t. \tag{15}$$

Inequality (14) implies that the first reduced basis vector b_1 is an approximation of the shortest non-zero vector in L. Kannan [64] presents an algorithm based on Lovász' basis reduction algorithm that computes the shortest non-zero lattice vector in polynomial time for fixed n. The *shortest vector problem* with respect to a given l_p norm is the problem of finding the shortest (with respect to the given norm) non-zero vector in a given lattice, whereas the *closest vector problem* with respect to a given l_p norm is the problem, given a lattice L and a target vector $y \in L$, to find a vector $x \in L$ such that $||x - y||_p$ is minimal. Van Emde Boas [33] proved that the shortest vector problem with respect to the l_∞ norm is NP-hard, and he conjectured that it is NP-hard with respect to the Euclidean norm. In the same paper he proved that the closest vector problem is NP-hard for any norm. Recently substantial progress has been made in gaining more information about the complexity status of the two problems. Ajtai [5] proved that the shortest vector problem is NP-hard for randomized problem reductions. This means that the reduction makes use of results of a probabilistic algorithm. These results are true with probability arbitrarily close to one. Ajtai also showed that approximating the length of a shortest vector in a given lattice within a factor $1 + 1/2^{n^c}$ is NP-hard for some constant c. The non-approximability factor was improved to $(1 + 1/n^\epsilon)$ by Cai and Nerurkar [16]. Micciancio [79] improved this factor substantially by showing that it is NP-hard to approximate the shortest vector in a given lattice within any constant factor less that $\sqrt{2}$ for randomized problem reductions, and that the same result holds for deterministic problem reductions (the "normal" type of reductions used in an NP-hardness proof) under the condition that a certain number theoretic conjecture holds. Micciancio's results hold for any l_p norm. Goldreich et al. [38] show that given oracle access to a subroutine that returns approximate closest vectors in a given lattice, one can find in polynomial time approximate shortest vectors in the same lattice. This implies that the shortest vector problem is not harder than the closest vector problem.

Just as the first basis vector is an approximation of the shortest vector of the lattice (14), the other basis vectors are approximations of the *successive minima* of the lattice. The jth successive minimum of $|| \ ||$ on L is the smallest positive value v_j such that there exists j linearly independent elements of the lattice L in the ball of radius v_j centered at the origin.

Proposition 1.5 (Lenstra et al. [70]). *Let* $v_1, \ldots, v_l$ *denote the successive minima of* $|| \ ||$ *on* L, *and let* $b_1, \ldots, b_l$ *be a reduced basis for* L. *Then*

$$2^{(1-j)/2} v_j \leqslant ||b_j|| \leqslant 2^{(l-1)/2} v_j \quad \text{for } 1 \leqslant j \leqslant l. \tag{16}$$

In recent years several new variants of Lovász' basis reduction algorithm have been developed and a number of variants for implementation have been suggested. We mention a few below, and recommend the paper by Schnorr and Euchner [89] for a more detailed overview. Schnorr [87] extended Lovász' algorithm to a family of polynomial time algorithms that, given $\epsilon > 0$, finds a non-zero vector in an n-dimensional lattice that is no longer than $(1 + \epsilon)^n$ times the length of the shortest vector in the lattice. The degree of the polynomial that bounds the running time of the family of algorithms increases as ϵ goes to zero. Seysen [95] developed an algorithm in which the intermediate integers that are produced are no larger than the input integers. Seysen's algorithm performs well particularly on lower-dimensional lattices. Schnorr and Euchner [89] discuss the possibility of computing the Gram–Schmidt vectors using floating point arithmetic while keeping the basis vectors in exact arithmetic in order to improve the practical performance of the algorithm. The drawback of this approach is that the basis reduction algorithm tends to become unstable. They propose a floating point version with good stability, but cannot prove that the algorithm always terminates. Empirical studies indicate that their version is stable on instances of dimension up to 125 having input numbers of bit length as large as 300. Our experience is that one can use basis reduction for problems of larger dimensions if the input numbers are smaller, but once the dimension reaches about 300–400 basis reduction will be slow. Another version considered by Schnorr and Euchner is basis reduction *with deep insertions*. Here, they allow for a vector b_k to be swapped with a vector with lower index than $k - 1$. Schnorr [87,88] also developed a variant of Lovász' algorithm in which not only two vectors are interchanged during the reduction process, but where blocks $b_j, b_{j+1}, \ldots, b_{j+\beta-1}$ of β consecutive vectors are transformed so as to minimize the jth Gram Schmidt vector b_j^*. This so-called block reduction produces shorter basis vectors but needs more computing time. The shortest vector b_j^* in a block of size β is determined by complete enumeration of all short lattice vectors. Schnorr and Hörner [90] develop and analyze a rule for pruning this enumeration process.

For the reader interested in using a version of Lovász' basis reduction algorithm there are some useful libraries available on the Internet. Two of them are LiDIA—a C++ Library for Computational Number Theory [72], developed at TH Darmstadt, and NTL—a Library for doing Number Theory [96], developed by V. Shoup, IBM, Zürich.

1.3. The generalized basis reduction algorithm

In the generalized basis reduction algorithm a norm related to a full-dimensional compact convex set C is used, instead of the Euclidean norm as in Lovász' algorithm. A compact convex set $C \in \mathbb{R}^n$ that is symmetric about the origin gives rise to a norm $F(c) = \inf\{t \geq 0: c/t \in C\}$. Lovász and Scarf [76] call the function F the *distance function* with respect to C. As in Lovász' basis reduction algorithm, the generalized basis reduction algorithm finds short basis vectors with respect to the chosen norm. Moreover, the first basis vector is an approximation of the shortest non-zero lattice vector.

Given the convex set C we define a dual set $C^* = \{y: y^{\mathrm{T}}c \leq 1 \text{ for all } c \in C\}$. We also define a distance function associated with a projection of C. Let $b_1, \ldots, b_n$ be a basis for $\mathbb{Z}^n$, and let C_j be the projection of C onto the orthogonal complement of

$b_1, \ldots, b_{j-1}$. We have that $c = \beta_j b_j + \cdots + \beta_n b_n \in C_j$ if and only if there exist $\alpha_1, \ldots, \alpha_{j-1}$ such that $c + \alpha_1 b_1 + \cdots + \alpha_{j-1} b_{j-1} \in C$. The distance function associated with C_j is defined as

$$F_j(c) = \min_{\alpha_1, \ldots, \alpha_{j-1}} F(c + \alpha_1 b_1 + \cdots + \alpha_{j-1} b_{j-1}). \tag{17}$$

Using duality, one can show that $F_j(c)$ is also the optimal value of the maximization problem:

$$F_j(c) = \max\{c^{\mathrm{T}} z \colon z \in C^*, \; b_1^{\mathrm{T}} z = 0, \ldots, b_{j-1}^{\mathrm{T}} z = 0\}. \tag{18}$$

In Expression (18), note that only vectors z that are orthogonal to the basis vectors $b_1, \ldots, b_{j-1}$ are considered. This is similar to the role played by the Gram–Schmidt basis in Lovász' basis reduction algorithm. Also, notice that if C is a polytope, then (18) is a linear program, which can be solved in polynomial time. The distance function F has the following properties:

- F can be computed in polynomial time,
- F is convex,
- $F(-x) = F(x)$,
- $F(tx) = tF(x)$ for $t > 0$.

Lovász and Scarf use the following definition of a reduced basis.

Definition 1.4. A basis $b_1, \ldots, b_n$ is called *reduced* if

$$F_j(b_{j+1} + \mu b_j) \geqslant F_j(b_{j+1}) \quad \text{for } 1 \leqslant j \leqslant n - 1 \text{ and all integers } \mu, \tag{19}$$

$$F_j(b_{j+1}) \geqslant (1 - \epsilon) F_j(b_j) \quad \text{for } 1 \leqslant j \leqslant n - 1, \tag{20}$$

where ϵ satisfies $0 < \epsilon < \frac{1}{2}$.

Definition 1.5. A basis $b_1, \ldots, b_n$, not necessarily reduced, is called *proper* if

$$F_k(b_j + \mu b_k) \geqslant F_k(b_j) \quad \text{for } 1 \leqslant k < j \leqslant n. \tag{21}$$

Remark 1.1. The algorithm is called *generalized* basis reduction since it generalizes Lovász' basis reduction algorithm in the following sense. If the convex set C is an ellipsoid, then a proper reduced basis is precisely a reduced basis according to Lenstra et al. [70] (cf. Definition 1.3).

An important question is how to check whether Condition (19) is satisfied for all integers μ. Here we make use of the dual relationship between Formulations (17) and (18). We have the following equality: $\min_{\alpha \in \mathbb{R}} F_j(b_{j+1} + \alpha b_j) = F_{j+1}(b_{j+1})$. Let α^* denote the optimal α in the minimization. The function F_j is convex, and hence the integer μ that minimizes $F_j(b_{j+1} + \mu b_j)$ is either $\lfloor \alpha^* \rfloor$ or $\lceil \alpha^* \rceil$. If the convex set C is a rational polytope, then $\alpha^* \in \mathbb{Q}$ is the optimal dual variable corresponding to the constraint $b_j^T z = 0$, which implies that the integral μ that minimizes $F_j(b_{j+1} + \mu b_j)$ can be determined by solving two additional linear programs, unless α^* is integral.

Condition (21) is analogous to Condition (5) of Lovász' basis reduction algorithm, and is violated if adding an integer multiple of b_k to b_j yields a distance function value $F_k(b_j + \mu b_k)$ that is smaller than $F_k(b_j)$. In the generalized basis reduction algorithm we only check whether the condition is satisfied for $k = j - 1$ (cf. Condition (19)), and we use the value of μ that minimizes $F_j(b_{j+1} + \mu b_j)$ as mentioned above. If Condition (19) is violated, we do a *size reduction*, i.e., we replace b_{j+1} by $b_{j+1} + \mu b_j$.

Condition (20) corresponds to Condition (6) in Lovász' algorithm, and ensures that the basis vectors are in the order of increasing distance function value, aside from the factor $(1 - \epsilon)$. Recall that we want the first basis vector to be an approximation of the shortest lattice vector. If Condition (20) is violated we interchange vectors b_j and b_{j+1}.

The algorithm works as follows. Let $b_1, \ldots, b_n$ be an initial basis for $\mathbb{Z}^n$. Typically $b_j = e^j$, where e^j is the jth unit vector in $\mathbb{R}^n$. Let j be the first index for which Conditions (19) or (20) are not satisfied. If (19) is violated, we replace b_{j+1} by $b_{j+1} + \mu b_j$ with the appropriate value of μ. If Condition (20) is satisfied after the replacement, we let $j := j + 1$. If Condition (20) is violated, we interchange b_j and b_{j+1}, and let $j := j - 1$ if $j \geq 2$. If $j = 1$, we remain at this level. The operations that the algorithm performs on the basis vectors are elementary column operations as in Lovász' algorithm. The vectors that we obtain as output from the generalized basis reduction algorithm can therefore be written as the product of the initial basis matrix and a unimodular matrix, which implies that the output vectors form a basis for the lattice $\mathbb{Z}^n$. The question is how efficient the algorithm is.

Theorem 1.6 (Lovász and Scarf [76]). *Let ϵ be chosen as in (20), let $\gamma = 2 + 1/\log(1/(1 - \epsilon))$, and let $B(R)$ be a ball with radius R containing C. Moreover, let $U = \max_{1 \leq j \leq n}\{F_j(b_j)\}$, where $b_1, \ldots, b_n$ is the initial basis, and let $V = 1/(R(nRU)^{n-1})$.*

The generalized basis reduction algorithm runs in polynomial time for fixed n. The maximum number of interchanges performed during the execution of the algorithm is

$$\left(\frac{\gamma^n - 1}{\gamma - 1}\right)\left(\frac{\log(U/V)}{\log(1/(1 - \epsilon))}\right). \tag{22}$$

It is important to notice that, so far, the generalized basis reduction algorithm has been proved to run in polynomial time for *fixed* n only, whereas Lovász' basis reduction algorithm runs in polynomial time for arbitrary n (cf. Theorem 1.2).

We now give a few properties of a Lovász–Scarf reduced basis. If one can obtain a basis $b_1, \ldots, b_n$ such that $F_1(b_1) \leq F_2(b_2) \leq \cdots \leq F_n(b_n)$, then one can prove that b_1 is the shortest integral vector with respect to the distance function. The generalized basis reduction algorithm does not produce a basis with the above property, but it gives a basis that satisfies the following weaker condition.

Theorem 1.7 (Lovász and Scarf [76]). *Let* $0 < \epsilon < \frac{1}{2}$, *and let* $b_1,\ldots,b_n$ *be a Lovász–Scarf reduced basis. Then*

$$F_{j+1}(b_{j+1}) \geq (\tfrac{1}{2} - \epsilon)F_j(b_j) \quad \text{for } 1 \leq j \leq n-1. \tag{23}$$

We can use this theorem to obtain a result analogous to (14) of Proposition 1.4.

Proposition 1.8 (Lovász and Scarf [76]). *Let* $0 < \epsilon < \frac{1}{2}$, *and let* $b_1,\ldots,b_n$ *be a Lovász–Scarf reduced basis. Then*

$$F(b_1) \leq (\tfrac{1}{2} - \epsilon)^{1-n}F(x) \quad \text{for all } x \in \mathbb{Z}^n,\ x \neq 0. \tag{24}$$

We can also relate the distance function $F_j(b_j)$ to the jth *successive minimum* of F on the lattice $\mathbb{Z}^n$ (cf. Proposition 1.5). $v_1,\ldots,v_n$ are the *successive minima* of F on $\mathbb{Z}^n$ if there are vectors $x^1,\ldots,x^n \in \mathbb{Z}^n$ with $v_j = F(x^j)$, such that for each $1 \leq j \leq n$, x^j is the shortest lattice vector (with respect to F) that is linearly independent of $x^1,\ldots,x^{j-1}$.

Proposition 1.9. *Let* $v_1,\ldots,v_n$ *denote the successive minima of F on the lattice $\mathbb{Z}^n$, let* $0 < \epsilon < \frac{1}{2}$, *and let* $b_1,\ldots,b_n$ *be a Lovász–Scarf reduced basis. Then*

$$(\tfrac{1}{2} - \epsilon)^{j-1}v_j \leq F_j(b_j) \leq (\tfrac{1}{2} - \epsilon)^{j-n}v_j \quad \text{for } 1 \leq j \leq n. \tag{25}$$

The first reduced basis vector is an approximation of the shortest lattice vector (Proposition 1.8). In fact the generalized basis reduction algorithm can be used to find the shortest vector in the lattice in polynomial time for fixed n. This algorithm is used as a subroutine of Lovász and Scarf's algorithm for solving the integer programming problem "Is $X \cap \mathbb{Z}^n \neq \emptyset$?" described in Section 2.1.3. To find the shortest lattice vector we proceed as follows. If the basis $b_1,\ldots,b_n$ is Lovász–Scarf reduced, we can obtain a bound on the coordinates of lattice vectors c that satisfy $F_1(c) \leq F_1(b_1)$. We express the vector c as an integer linear combination of the basis vectors, i.e., $c = \lambda_1 b_1 + \cdots + \lambda_n b_n$, where $\lambda_j \in \mathbb{Z}$. We have

$$F_1(b_1) \geq F_1(c) \geq F_n(c) = F_n(\lambda_n b_n) = |\lambda_n|F_n(b_n), \tag{26}$$

where the second inequality holds since $F_n(c)$ is more constrained than $F_1(c)$, the first equality holds due to the constraints $b_i^{\mathrm{T}}z = 0$, $1 \leq i \leq n-1$, and the second equality holds as $F(tx) = tF(x)$ for $t > 0$. We can now use (26) to obtain the following bound on $|\lambda_n|$:

$$|\lambda_n| \leq \frac{F_1(b_1)}{F_n(b_n)} \leq \frac{1}{(\tfrac{1}{2} - \epsilon)^{n-1}}, \tag{27}$$

where the last inequality is obtained by applying Theorem 1.7 iteratively. Notice that the bound on λ_n is polynomial for fixed n. In a similar fashion we can obtain a bound on λ_j for $n-1 \geq j \geq 1$. Suppose that we have chosen multipliers $\lambda_n,\ldots,\lambda_{j+1}$ and that we want to determine a bound on λ_j. Let γ^* be the value of γ that minimizes $F_j(\lambda_n b_n + \cdots + \lambda_{j+1}b_{j+1} + \gamma b_j)$. If this minimum is greater than $F_1(b_1)$, then there does not exist a vector c, with $\lambda_n, \ldots, \lambda_{j+1}$ fixed such that $F_1(c) \leqslant F_1(b_1)$, since in that case $F_1(b_1) < F_j(\lambda_n b_n + \cdots + \lambda_{j+1} b_{j+1} + \gamma^* b_j) \leqslant F_j(\lambda_n b_n + \cdots + \lambda_j b_j) = F_j(c) \leqslant F_1(c)$, which yields a contradiction. If the minimum is less than or equal to $F_1(b_1)$, then we can obtain the bound

$$|\lambda_j - \gamma^*| \leqslant 2 \frac{F_1(b_1)}{F_j(b_j)} \leqslant \frac{2}{(\frac{1}{2} - \epsilon)^{j-1}}. \tag{28}$$

Hence, we obtain a search tree that has at most n levels, and, given the bounds on the multipliers λ_j, each level consists of a number of nodes that is polynomial if n is fixed.

The generalized basis reduction algorithm was implemented by Cook et al. [23], and by Wang [105]. Cook et al. used generalized basis reduction to derive a heuristic version of the integer programming algorithm by Lovász and Scarf (see Section 2.1.3) to solve difficult integer network design instances. Wang solved both linear and non-linear integer programming problems using the generalized basis reduction algorithm as a subroutine.

An example illustrating a few iterations of the generalized basis reduction algorithm is given in Section 2.1.3.

2. Basis reduction in integer programming

The main ideas behind the integer programming algorithms by Lenstra [71], Grötschel et al. [48], Kannan [64], and Lovász and Scarf [76] described in Section 2.1 are as follows. A lattice is contained in countably many parallel hyperplanes. If one wants to decide whether or not a certain polyhedron contains an integral vector, then one can enumerate some of these lattice hyperplanes. To avoid an unnecessarily large enumeration tree one wants to find a representation of the lattice hyperplanes such that the distance between them is not too small. In particular, for given dimension n one should only need to enumerate a polynomial number of hyperplanes. To find such a lattice representation basis reduction is used.

The use of basis reduction in cryptography will be briefly discussed in Section 2.2 since several interesting theoretical and computational results have been obtained in this area using basis reduction, and since the lattices and the bases that have been used in attacking knapsack cryptosystems are related to the lattice used by Aardal et al. [2,3]. Their algorithm is outlined in Section 2.3. The basic idea behind the algorithms discussed in Sections 2.2 and 2.3 is to reformulate the problem as a problem of finding a short vector in a certain lattice. One therefore needs to construct a lattice in which any feasible vector to the considered problem is provably short.

For the reader wishing to study this topic in more detail we refer to the articles mentioned in this introduction, to the survey article by Kannan [63], and to the textbooks by Lovász [75], Schrijver [84], Grötschel et al. [49], Nemhauser and Wolsey [80], and Cohen [17]. In these references, and in the article by Lenstra et al. [70], several applications of basis reduction are mentioned, other than integer programming, such as finding a short non-zero vector in a lattice, finding the Hermite normal form of a matrix, simultaneous diophantine approximation, factoring polynomials with rational coefficients, and finding $\mathbb{Q}$-linear relations among real numbers $\alpha_1, \alpha_1, \ldots, \alpha_n$. Reviewing these other topics is outside the scope of this section.

2.1. Integer programming in fixed dimension

Let A be a rational $m \times n$-matrix and let d be a rational m-vector. We consider the integer programming problem in the following form:

$$\text{Does there exist an integral vector } x \text{ such that } Ax \leqslant d? \tag{29}$$

Karp [67] showed that the zero-one integer programming problem is NP-complete, and Borosh and Treybig [10] proved that the integer programming problem (29) belongs to NP. Combining these results implies that (29) is NP-complete. The NP-completeness of the zero-one version is a fairly straightforward consequence of the proof by Cook [19] that the satisfiability problem is NP-complete. An important open question was still: Can the integer programming problem be solved in polynomial time in bounded dimension? If the dimension $n = 1$, the affirmative answer is trivial. Some special cases of $n = 2$ were proven to be polynomially solvable by Hirschberg and Wong [57], and by Kannan [62]. Scarf [85] showed that (29), for the general case $n = 2$, is polynomially solvable. Both Hirschberg and Wong, and Scarf conjectured that the integer programming problem could be solved in polynomial time if the dimension is fixed. The proof of this conjecture was given by Lenstra [71]. We describe three algorithms for solving the integer programming problem in fixed dimension: Lenstra's algorithm [71] and the algorithm of Grötschel et al. [48], which are both based on Lovász' basis reduction algorithm [70], and, finally, the algorithm of Lovász and Scarf [76], which is based on the generalized basis reduction algorithm.

It is worthwhile pointing out here that Barvinok [7] showed that there exists a polynomial time algorithm for *counting* the number of integral points in a polyhedron if the dimension is fixed. Barvinok's result therefore generalizes the result of Lenstra. Barvinok, however, based his algorithm on an identity by Brion for exponential sums over polytopes. Later, Dyer and Kannan [30] developed a simpler algorithm for counting the number of integral points in fixed dimension. Their algorithm uses only elementary properties of exponential sums. To describe Barvinok's result and the improvement by Dyer and Kannan is outside the scope of this chapter.

2.1.1. Lenstra's algorithm

Let $X = \{x \in \mathbb{R}^n : Ax \leqslant d\}$. The question we consider is

$$\text{Is } X \cap \mathbb{Z}^n \neq \emptyset? \tag{30}$$

An observation made by Lenstra was that "thin" polytopes as in Example 1.1 were "bad" from the worst-case perspective. He therefore suggested to transform the polytope using a linear transformation τ such that the polytope τX becomes "round" according to a certain measure. Assume without loss of generality that the polytope X is full-dimensional and bounded, and let $B(p,z) = \{x \in \mathbb{R}^n : \|x - p\| \leqslant z\}$ be the closed

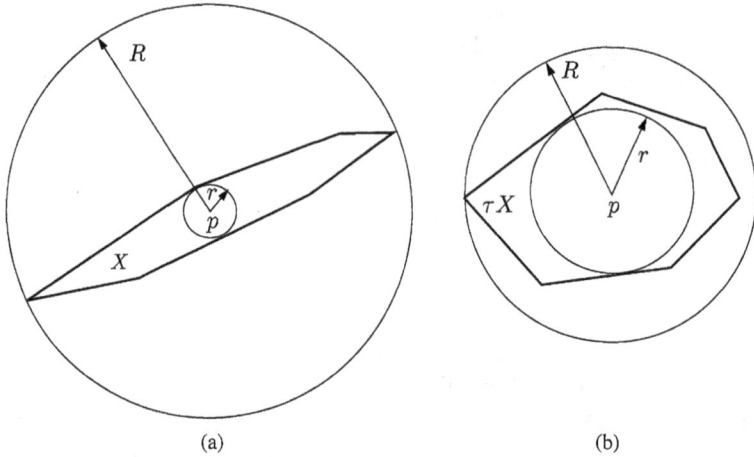

Fig. 6. (a) The original polytope X is thin, and the ratio R/r is large. (b) The transformed polytope τX is "round", and R/r is relatively small.

ball with center p and radius z. The transformation τ that we apply to the polytope is constructed such that $B(p,r) \subset \tau X \subset B(p,R)$ for some $p \in \tau X$ and such that

$$\frac{R}{r} \leqslant c_2, \tag{31}$$

where c_2 is a constant that depends only on the dimension n. Relation (31) is the measure of "roundness" that Lenstra uses. For an illustration, see Fig. 6. Once we have transformed the polytope, we need to apply the same transformation to the lattice, which gives us the following problem:

$$\text{Is } \tau \mathbb{Z}^n \cap \tau X \neq \emptyset? \tag{32}$$

Note that problems (30) and (32) are equivalent. The vectors τe^j, $1 \leqslant j \leqslant n$, where e^j is the jth unit vector in $\mathbb{R}^n$, form a basis for the lattice $\tau \mathbb{Z}^n$. If the polytope X is thin, then this will translate to the lattice basis vectors τe^j, $1 \leqslant j \leqslant n$ in the sense that these vectors are long and non-orthogonal. This is where lattice basis reduction becomes useful. Once we have the transformed polytope τX, Lenstra uses the following lemma to find a lattice point quickly.

Lemma 2.1 (Lenstra [71]). *Let $b_1, \ldots, b_n$ be any basis for L. Then for all $x \in \mathbb{R}^n$ there exists a vector $y \in L$ such that*

$$\|x - y\|^2 \leqslant \tfrac{1}{4}(\|b_1\|^2 + \cdots + \|b_n\|^2). \tag{33}$$

The proof of this lemma suggests a fast construction of the vector $y \in L$ given the vector x.

Next, let $L = \tau \mathbb{Z}^n$, and let $b_1, \ldots, b_n$ be a basis for L such that (10) holds. Notice that (10) holds if the basis is reduced. Also, reorder the vectors such that $\|b_n\| =$

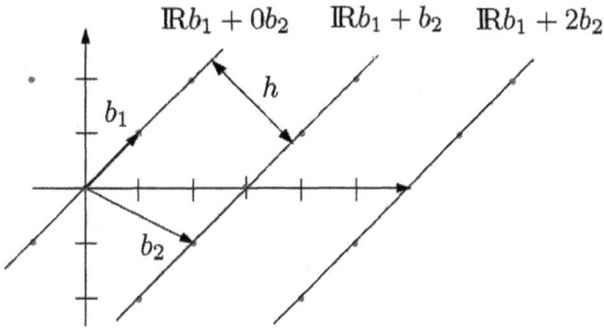

Fig. 7.

$\max_{1 \leqslant j \leqslant n}\{||b_j||\}$. Let $x = p$ where p is the center of the closed balls $B(p,r)$ and $B(p,R)$. Apply Lemma 2.1 to the given x. This gives a lattice vector $y \in \tau \mathbb{Z}^n$ such that

$$||p - y||^2 \leqslant \tfrac{1}{4}(||b_1||^2 + \cdots + ||b_n||^2) \leqslant \tfrac{1}{4} \cdot n \cdot ||b_n||^2 \qquad (34)$$

in polynomial time. We now distinguish two cases. Either $y \in \tau X$ or $y \notin \tau X$. The first case implies that τX is relatively large, and if we are in this case, then we are done, so we assume we are in the second case. Since $y \notin \tau X$ we know that y is not inside the ball $B(p,r)$ as $B(p,r)$ is completely contained in τX. Hence we know that $||p - y|| > r$, or using (34), that

$$r < \tfrac{1}{2} \cdot \sqrt{n} \cdot ||b_n||. \qquad (35)$$

Below we will demonstrate that the lattice L is contained in countably many parallel hyperplanes. The distance between any two consecutive hyperplanes is equal to a certain constant. We now create t subproblems by considering intersections between the polytope τX with t of these parallel hyperplanes. Each of the subproblems has dimension at least one lower than the parent problem and they are solved recursively. The procedure of splitting the problem into subproblems of lower dimension is called "branching", and each subproblem is represented by a node in the enumeration tree. In each node we repeat the whole process of transformation, basis reduction and, if necessary, branching. The enumeration tree created by this recursive process is of depth at most n, and the number of nodes at each level is polynomially bounded by a constant that depends only on the dimension. The value of t will be computed below.

Let H, h and L' be defined as in Corollary 1.1 and its proof. We can write L as

$$L = L' + \mathbb{Z}b_n \subset H + \mathbb{Z}b_n = \bigcup_{k \in \mathbb{Z}} (H + kb_n). \qquad (36)$$

So the lattice L is contained in countably many parallel hyperplanes. For an example we refer to Fig. 7. The distance between two consecutive hyperplanes is h, and Corollary 1.1 says that h is bounded from below by $c_1^{-1}||b_n||$, which implies that not too many hyperplanes intersect τX. To determine precisely how many hyperplanes intersect τX, we approximate τX by the ball $B(p,R)$. If t is the number of hyperplanes intersecting $B(p,R)$ we have

$$t - 1 \leqslant \frac{2R}{h}. \tag{37}$$

Using relationship (31) between the radii R and r we have $2R \leqslant 2rc_2 < c_2\sqrt{n}\|b_n\|$, where the last inequality follows from (35). Since $h \geqslant c_1^{-1}\|b_n\|$ (cf. Corollary 1.1), we get the following bound on the number of hyperplanes that we need to consider:

$$t - 1 \leqslant \frac{2R}{h} < c_1 c_2 \sqrt{n}, \tag{38}$$

which depends on the dimension only. The values of the constants c_1 and c_2 that are used by Lenstra are: $c_1 = 2^{n(n-1)/4}$ and $c_2 = 2n^{3/2}$. Lenstra [71] discusses ways of improving these values. To determine the values of k in expression (36), we express p as a linear combination of the basis vectors $b_1, \ldots, b_n$. Recall that p is the center of the ball $B(p,R)$ that was used to approximate τX.

So far we have not mentioned how to determine the transformation τ and hence the balls $B(p,r)$ and $B(p,R)$. We give the general idea here without going into detail. First, determine an n-simplex contained in X. This can be done by repeated calls to the ellipsoid algorithm. The resulting simplex is described by its extreme points $v_0, \ldots, v_n$. By applying the ellipsoid algorithm repeatedly we can decide whether there exists an extreme point x of X such that if we replace v_j by x we obtain a new simplex whose volume is at least a factor of $\frac{3}{2}$ larger than the current simplex. We stop the procedure if we cannot find such a new simplex. The factor $\frac{3}{2}$ can be modified, but the choice will affect the value of the constant c_2, see [71] for further details. We now map the extreme points of the simplex to the unit vectors of $\mathbb{R}^{n+1}$ so as to obtain a regular n-simplex, and we denote this transformation by τ. Lenstra [71] shows that τ has the property that if we let $p = 1/(n+1)\sum_{j=0}^{n} e^j$, where e^j is the jth unit vector of $\mathbb{R}^{n+1}$ (i.e., p is the center of the regular simplex), then there exists closed balls $B(p,r)$ and $B(p,R)$ such that $B(p,r) \subset \tau X \subset B(p,R)$ for some $p \in \tau X$, and such that $R/r \leqslant c_2$.

Kannan [64] developed a variant of Lenstra's algorithm. The algorithm follows Lenstra's algorithm up to the point where he has applied a linear transformation to the polytope X and obtained a polytope τX such that $B(p,r) \subset \tau X \subset B(p,R)$ for some $p \in \tau X$. Here Kannan proceeds as follows. He applies a reduction algorithm to a basis of the lattice $\tau \mathbb{Z}^n$ that produces a "reduced" basis defined differently to a Lovász' reduced basis. In particular, in Kannan's reduced basis the first basis vector is the shortest non-zero lattice vector. As in Lenstra's algorithm two cases are considered. Either τX is relatively large which implies that τX contains a lattice vector, or τX is small, which means that not too many lattice hyperplanes can intersect τX. Each such intersection gives rise to a subproblem of at least one dimension lower. Kannan's reduced basis makes it possible to improve the bound on the number of hyperplanes that has to be considered to $O(n^{5/2})$. As far as we know, no implementation of Lenstra's or Kannan's algorithms has been reported on in the literature.

2.1.2. The algorithm of Grötschel et al.

Grötschel et al. [48] used ellipsoidal approximations of the feasible set X and derived an algorithm based on the same principles as Lenstra's algorithm. Here we will give a sketch of their approach. Assume without loss of generality that $X = \{x \in \mathbb{R}^n : Ax \leqslant d\}$ is bounded and full-dimensional. The key idea is to rapidly find a vector $y \in \mathbb{Z}^n$, as Lenstra does through Lemma 2.1, and if y does not belong to X, to find a non-zero integral direction c such that the width of the polytope X in this direction is bounded by a constant depending only on n. This is expressed in the following theorem.

Theorem 2.2 (Grötschel et al. [48]). *Let $Ax \leqslant d$ be a system of m rational inequalities in n variables, and let $X = \{x : Ax \leqslant d\}$. There exists a polynomial algorithm that finds either an integral vector $y \in X$, or a vector $c \in \mathbb{Z}^n \setminus 0$ such that*

$$\max\{c^\mathrm{T}x : x \in X\} - \min\{c^\mathrm{T}x : x \in X\} \leqslant 2n(n+1)2^{n(n-1)/4}. \tag{39}$$

Remark 2.1. Grötschel et al. in fact gave the polytope $\{x : Ax \leqslant d\}$ in terms of a separation oracle, and not by an explicit description. This gives rise to a slightly more involved proof. Here we follow the presentation of Schrijver [91]. Notice that the algorithm referred to in Theorem 2.2 is polynomial for *arbitrary* n.

Here we will not make a transformation to a lattice $\tau \mathbb{Z}^n$, but remain in the lattice $\mathbb{Z}^n$. The first step is to find two ellipsoids; one contained in X, and one containing X. Let D be a positive semidefinite $n \times n$-matrix, and let $p \in \mathbb{R}^n$. The *ellipsoid* associated with p and D is defined as $E(p, D) = \{x \in \mathbb{R}^n : (x - p)^\mathrm{T}D^{-1}(x - p) \leqslant 1\}$. The vector p is called the *center* of the ellipsoid $E(p, D)$. Goffin [37] showed that it is possible to find ellipsoids $E(p, (1/(n+1)^2)D)$, $E(p, D)$ in polynomial time such that

$$E\left(p, \frac{1}{(n+1)^2}D\right) \subseteq X \subseteq E(p, D). \tag{40}$$

Next, we apply basis reduction, but instead of using the Euclidean norm to measure the length of the basis vectors, as described in Section 2.2, we use a norm defined by the positive definite matrix D^{-1} describing the ellipsoids, see [91, Chapters 6 and 18]. The norm $\|\,\|$ defined by the matrix D^{-1} is given by $\|x\| = \sqrt{x^\mathrm{T}D^{-1}x}$. Given a positive definite rational matrix D^{-1}, we can apply basis reduction to the unit basis to obtain a basis $b_1, \ldots, b_n$ for the lattice $\mathbb{Z}^n$ in polynomial time that satisfies (cf. the second inequality of (10))

$$\prod_{j=1}^{n} \|b_j\| \leqslant 2^{n(n-1)/4} \sqrt{\det(D^{-1})}. \tag{41}$$

Next, reorder the basis vectors such that $\|b_n\| = \max_{1 \leqslant j \leqslant n}\{\|b_j\|\}$. After reordering, inequality (41) still holds. Suppose that the vector $y \in \mathbb{Z}^n$, which can be found by applying Lemma 2.1 with $x = p$, does not belong to X. We then have that $y \notin E(p, (1/(n+1)^2)D)$ as this ellipsoid is contained in X, which implies that $\|p - y\| > 1/(n+1)$. Using (34) we obtain $1/2 \cdot \sqrt{n} \cdot \|b_n\| \geqslant \|p - y\| > 1/(n+1)$ which gives the following bound on the length of the nth basis vector:

$$\|b_n\| > \frac{2}{\sqrt{n(n+1)}} > \frac{1}{n(n+1)}. \tag{42}$$

Choose a direction c such that the components of c are relatively prime integers, and such that c is orthogonal to the subspace generated by the basis vectors $b_1, \ldots, b_{n-1}$. One can show, see [91, pp. 257–258], that if we consider a vector z such that $z^T D^{-1} z \leqslant 1$, then

$$|c^T z| \leqslant \sqrt{\det(D)} \|b_1\| \cdot \cdots \cdot \|b_{n-1}\| \leqslant 2^{n(n-1)/4} \|b_n\|^{-1} < n(n+1) 2^{n(n-1)/4}, \tag{43}$$

where the second inequality follows from inequality (41), and the last inequality follows from (42). If a vector z satisfies $z^T D^{-1} z \leqslant 1$, then $z \in E(p, D)$, which implies that $|c^T(z - p)| \leqslant n(n+1) 2^{n(n-1)/4}$. We then obtain

$$\max\{c^T x \colon x \in X\} - \min\{c^T x \colon x \in X\}$$

$$\leqslant \max\{c^T x \colon x \in E(p, D)\} - \min\{c^T x \colon x \in E(p, D)\}$$

$$\leqslant 2n(n+1) 2^{n(n-1)/4}, \tag{44}$$

which gives the desired result.

Lenstra's result that the integer programming problem can be solved in polynomial time for fixed n follows from Theorem 2.2. If we apply the algorithm implied by Theorem 2.2, we either find an integral point $y \in X$ or a thin direction c, i.e., a direction c such that Eq. (44) holds. Assume that the direction c is the outcome of the algorithm. Let $\mu = \lceil \min\{c^T x \colon x \in X\} \rceil$. All points in $X \cap \mathbb{Z}^n$ are contained in the parallel hyperplanes $c^T x = t$ where $t = \mu, \ldots, \mu + 2n(n+1) 2^{n(n-1)/4}$, so, if n is fixed we get polynomially many hyperplanes, each giving rise to a subproblem of dimension less than or equal to $n - 1$: does there exist an integral vector $x \in \{X \colon c^T x = t\}$? For each of these lower-dimensional problems we repeat the algorithm of Theorem 2.2. The search tree has at most n levels and each level has polynomially many nodes if the dimension is fixed.

2.1.3. The algorithm of Lovász and Scarf

The integer programming algorithm of Lovász and Scarf [76] determines, in polynomial time for fixed n, whether there exists a thin direction for the polytope X. If X is not thin in any direction, then X has to contain an integral vector. If a thin direction is found, then one needs to branch, i.e., divide the problem into lower-dimensional subproblems, in order to determine whether or not a feasible vector exists, but then the number of branches is polynomially bounded for fixed n. If the algorithm indicates that X contains an integral vector, then one needs to determine a so-called Korkine–Zolotarev basis in order to construct a feasible vector. The Lovász–Scarf algorithm avoids the approximations by balls as in Lenstra's algorithm, or by ellipsoids as in the algorithm by Grötschel et al. Again, we assume that $X = \{x \in \mathbb{R}^n \colon Ax \leqslant d\}$ is bounded, rational, and full-dimensional.

Definition 2.1. The *width* of the polytope X in the non-zero direction c is determined as

$$\max\{c^{\mathrm{T}}x: x \in X\} - \min\{c^{\mathrm{T}}x: x \in X\} = \max\{c^{\mathrm{T}}(x - y): x \in X, \ y \in X\}. \tag{45}$$

Let $(X - X) = \{(x - y): x \in X, \ y \in X)\}$ be the difference set corresponding to X. Recall that $(X - X)^*$ denotes the dual set corresponding to $(X - X)$, and notice that $(X - X)^*$ is symmetric about the origin. The distance functions associated with $(X - X)^*$ are:

$$F_j(c) = \min_{\alpha_1,\dots,\alpha_{j-1} \in \mathbb{Q}} F(c + \alpha_1 b_1 + \cdots + \alpha_{j-1} b_{j-1}) \tag{46}$$

$$= \max\{c^{\mathrm{T}}(x - y): x \in X, \ y \in X, \ b_1^{\mathrm{T}}(x - y) = 0, \dots, b_{j-1}^{\mathrm{T}}(x - y) = 0\}, \tag{47}$$

(cf. expressions (17) and (18)). Here, we notice that $F(c) = F_1(c)$ is the width of X in the direction c. From the above we see that a lattice vector c that minimizes the width of the polytope X is a *shortest lattice vector* for the polytope $(X - X)^*$.

To outline the algorithm by Lovász and Scarf we need the results given in Theorems 2.3 and 2.4 below, and the definition of a so-called *generalized Korkine–Zolotarev basis*. Let b_j, $1 \leqslant j \leqslant n$ be defined recursively as follows. Given $b_1,\dots,b_{j-1}$, the vector b_j minimizes $F_j(x)$ over all lattice vectors that are linearly independent of $b_1,\dots,b_{j-1}$. A generalized Korkine–Zolotarev (KZ) basis is defined to be any proper basis $b'_1,\dots,b'_n$ associated with b_j, $1 \leqslant j \leqslant n$. (See Definition 1.5 for the definition of a proper basis.) The notion of a generalized KZ basis was introduced by Kannan and Lovász [65,66]. Kannan and Lovász [65] gave an algorithm for computing a generalized KZ basis in polynomial time for fixed n.

Theorem 2.3 (Kannan and Lovász [66]). *Let $F(c)$ be the length of the shortest lattice vector c with respect to the set $(X - X)^*$, and let $\rho_{\mathrm{KZ}} = \sum_{j=1}^{n} F_j(b'_j)$, where b'_j, $1 \leqslant j \leqslant n$ is a generalized Korkine–Zolotarev basis. There exists a universal constant c_0 such that*

$$F(c)\rho_{\mathrm{KZ}} \leqslant c_0 \cdot n \cdot (n+1)/2. \tag{48}$$

To derive their result, Kannan and Lovász used a lower bound on the product of the volume of a convex set $C \subset \mathbb{R}^n$ that is symmetric about the origin, and the volume of its dual C^*. The bound, due to Bourgain and Milman [11], is equal to c_{BM}^n/n^n, where c_{BM} is a constant depending only on n. In Theorem 2.3 we have $c_0 = 4/c_{\mathrm{BM}}$. See also the remark below.

Theorem 2.4 (Kannan and Lovász [66]). *Let $b_1,\dots,b_n$ be any basis for $\mathbb{Z}^n$, and let X be a bounded convex set that is symmetric about the origin. If $\rho = \sum_{j=1}^{n} F_j(b_j) \leqslant 1$, then X contains an integral vector.*

The first step of the Lovász–Scarf algorithm is to compute the shortest vector c with respect to $(X - X)^*$ using the algorithm described in Section 2.3. If $F(c) \geqslant c_0 \cdot n \cdot (n + 1)/2$, then $\rho_{KZ} \leqslant 1$, which by Theorem 2.4 implies that X contains an integral vector. If $F(c) < c_0 \cdot n \cdot (n+1)/2$, then we need to branch. Due to the definition of $F(c)$ we have in this case that $\max\{c^\mathrm{T} x : x \in X\} - \min\{c^\mathrm{T} x : x \in X\} < c_0 \cdot n \cdot (n+1)/2$, which implies that the polytope X in the direction c is "thin". As in the algorithm by Grötschel et al. we create one subproblem for every hyperplane $c^\mathrm{T} x = \mu, \ldots, c^\mathrm{T} x = \mu + c_0 \cdot n \cdot (n+1)/2$, where $\mu = \lceil \min\{c^\mathrm{T} x : x \in X\} \rceil$. Once we have fixed a hyperplane $c^\mathrm{T} x = t$, we have obtained a problem in dimension less than or equal to $n - 1$, and we repeat the process. This procedure creates a search tree that is at most n deep, and that has a polynomial number of branches at each level. The algorithm called in each branch is, however, polynomial for fixed dimension only. First, the generalized basis reduction algorithm runs in polynomial time for fixed dimension, and second, computing the shortest vector c is done in polynomial time for fixed dimension. An alternative would be to use the first reduced basis vector with respect to $(X - X)^*$, instead of the shortest vector c. According to Proposition 1.8, $F(b_1) \leqslant (\frac{1}{2} - \epsilon)^{1-n} F(c)$. In this version of the algorithm we would first check whether $F(b_1) \geqslant c_0 \cdot n \cdot (n + 1)/(2(\frac{1}{2} - \epsilon)^{1-n})$. If yes, then X contains an integral vector, and if no, we need to branch, and we create at most $c_0 \cdot n \cdot (n + 1)/(2(\frac{1}{2} - \epsilon)^{n-1})$ hyperplanes. We again obtain a search tree of at most n levels, but in this version the number of branches created at each level is polynomially bounded for fixed n only.

If the algorithm terminates with the result that X contains an integral vector, then Lovász and Scarf describe how such a vector can be constructed by using the Korkine–Zolotarev basis (see [76, proof of Theorem 10]).

Remark 2.2. Lagarias et al. [68] derive bounds on the Euclidean length of Korkine–Zolotarev reduced basis vectors of a lattice and its dual lattice. Let W be the vector space spanned by the lattice L. The lattice L^* *dual* to L is defined as $L^* = \{w \in W : w^\mathrm{T} v$ is an integer for all $v \in L\}$. The bounds are given in terms of the successive minima of L and L^*. These bounds, in turn, imply bounds on the product of successive minima of L and L^*. Later, Kannan and Lovász [65,66] introduced the generalized Korkine–Zolotarev basis, as defined above, and derived bounds such as developed in the paper by Lagarias et al. These bounds were used to study covering minima of a convex set with respect to a lattice, such as the covering radius, and the lattice width. An important result by Kannan and Lovász is that the product of the first successive minima of the lattices L and L^* is bounded from above by $c_0 \cdot n$. This improves on a similar result of Lagarias et al. and implies Theorem 2.3 above. There are many interesting results on properties of various lattice constants. Many of them are described in the survey by Kannan [63], and will not be discussed further here.

Example 2.1. The following example demonstrates a few iterations with the generalized basis reduction algorithm. Consider the polytope $X = \{x \in \mathbb{R}^2_{\geqslant 0} : x_1 + 7x_2 \geqslant 7, \ 2x_1 + 7x_2 \leqslant 14, \ -5x_1 + 4x_2 \leqslant 4\}$. Let $j = 1$ and $\epsilon = \frac{1}{4}$. Assume we want to use the generalized basis reduction algorithm to find a direction in which the width of X is small. Recall

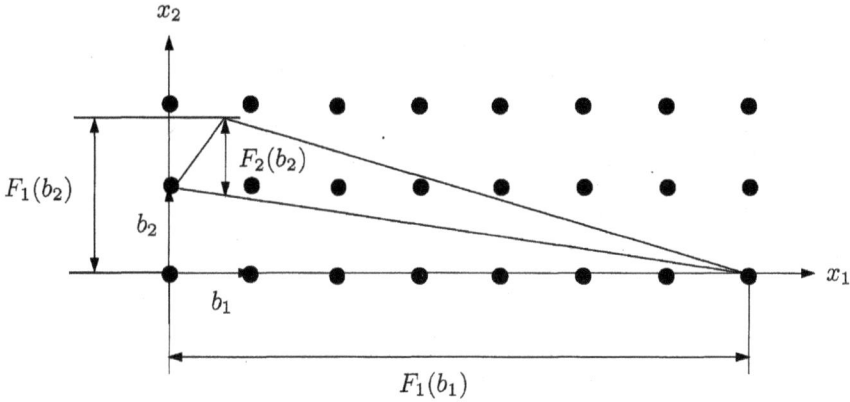

Fig. 8.

that a lattice vector c that minimizes the width of X is a shortest lattice vector with respect to the set $(X - X)^*$. The first reduced basis vector is an approximation of the shortest vector for $(X - X)^*$ and hence an approximation of the thinnest direction for X. The distance functions associated with $(X - X)^*$ are

$$F_j(c) = \max\{c^T(x - y): x \in X,\ y \in X,\ b_i^T(x - y) = 0,\ 1 \leqslant i \leqslant j - 1\}.$$

The initial basis is

$$b_1 = \begin{pmatrix} 1 \\ 0 \end{pmatrix}, \quad b_2 = \begin{pmatrix} 0 \\ 1 \end{pmatrix}.$$

We obtain $F_1(b_1) = 7.0$, $F_1(b_2) = 1.8$, $F_2(b_2) = 0.9$, $\mu = 0$, and $F_1(b_2 + 0b_1) = 1.8$, see Fig. 8. Notice that the widths F_j are not the geometric widths, but the widths with respect to the indicated directions.

Checking Conditions (19) and (20) shows that Condition (19) is satisfied as $F_1(b_2 + 0b_1) \geqslant F_1(b_2)$, but that Condition (20) is violated as $F_1(b_2) \not\geqslant (3/4)F_1(b_1)$, so we interchange b_1 and b_2 and remain at $j = 1$.

Now we have $j = 1$ and

$$b_1 = \begin{pmatrix} 0 \\ 1 \end{pmatrix}, \quad b_2 = \begin{pmatrix} 1 \\ 0 \end{pmatrix}.$$

$F_1(b_1) = 1.8$, $F_1(b_2) = 7.0$, $F_2(b_2) = 3.5$, $\mu = 4$, and $F_1(b_2 + 4b_1) = 3.9$, see Fig. 9. Condition (19) is violated as $F_1(b_2 + 4b_1) \not\geqslant F_1(b_2)$, so we replace b_2 by $b_2 + 4b_1 = (1, 4)^T$. Given the new basis vector b_2 we check Condition (20) and we conclude that this condition is satisfied. Hence the basis

$$b_1 = \begin{pmatrix} 0 \\ 1 \end{pmatrix}, \quad b_2 = \begin{pmatrix} 1 \\ 4 \end{pmatrix}$$

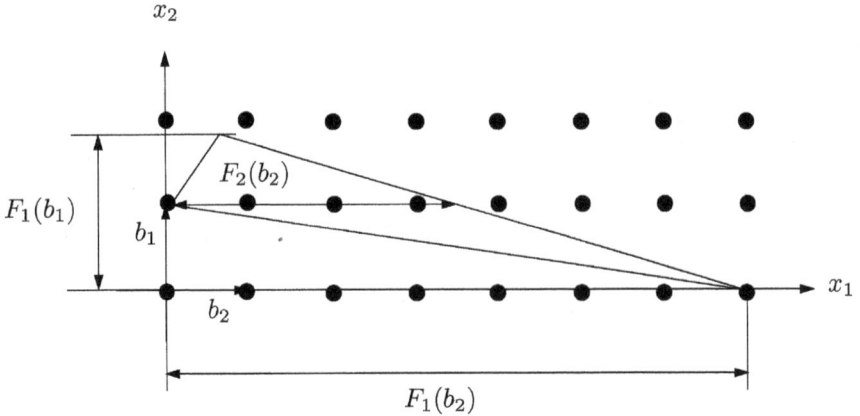

Fig. 9.

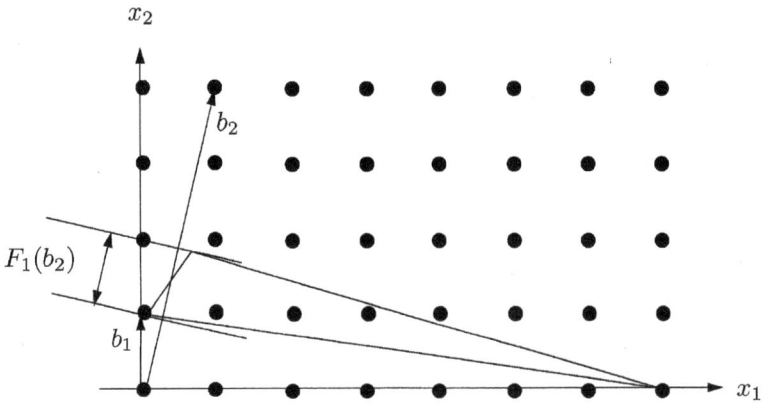

Fig. 10.

is Lovász–Scarf reduced, see Fig. 10. The vectors b_1 and b_2 indicate directions in which the polytope X is thin. □

2.2. Basis reduction and knapsack cryptosystems

Basis reduction has been used successfully to find solutions to subset sum problems arising in knapsack cryptosystems. For a recent excellent overview we refer to Joux and Stern [60].

A sender wants to transmit a message to a receiver. The plaintext message of the sender consists of a 0–1 vector $x_1, \ldots, x_n$, and this message is encrypted by using integral weights $a_1, \ldots, a_n$ leading to an encrypted message $a_0 = \sum_{j=1}^{n} a_j x_j$. The coefficients a_j, $1 \leqslant j \leqslant n$, are known to the public, but there is a hidden structure in the relation between these coefficients, called a trapdoor, which only the receiver knows. If the trapdoor is known, then the subset sum problem:

Determine a 0–1 vector x such that

$$\sum_{j=1}^{n} a_j x_j = a_0 \qquad (49)$$

can be solved easily. For an eavesdropper that does not know the trapdoor, however, the subset sum problem should be hard to solve in order to obtain a secure transmission.

The *density* of a set of coefficients a_j, $1 \leqslant j \leqslant n$ is defined as

$$d(a) = d(\{a_1, \ldots, a_n\}) = \frac{n}{\log_2(\max_{1 \leqslant j \leqslant n}\{a_j\})}. \qquad (50)$$

The density, as defined above, is an approximation of the information rate at which bits are transmitted. The interesting case is $d(a) \leqslant 1$, since for $d(a) > 1$ the subset sum problem (49) will in general have several solutions, which makes it unsuitable for generating encrypted messages. Lagarias and Odlyzko [69] proposed an algorithm based on basis reduction that often finds a solution to the subset sum problem (49) for instances having relatively low density. Earlier research had found methods based on recovering trapdoor information. If the information rate is high, i.e., $d(a)$ is high, then the trapdoor information is relatively hard to conceal. The result of Lagarias and Odlyzko therefore complements the earlier results by providing a method that is successful for low-density instances. In their algorithm Lagarias and Odlyzko consider a lattice in $\mathbb{Z}^{n+1}$ consisting of vectors of the following form:

$$L_{a,a_0} = \{(x_1, \ldots, x_n, (ax - a_0\xi))^{\mathrm{T}}\}, \qquad (51)$$

where ξ is a variable associated with the right-hand side of $ax = a_0$. Notice that the lattice vectors that are interesting for the subset sum problem all have $\xi = 1$ and $ax - a_0\xi = 0$. It is easy to write down an initial basis B for L_{a,a_0}:

$$B = \begin{pmatrix} I^{(n)} & 0^{(n \times 1)} \\ a & -a_0 \end{pmatrix}, \qquad (52)$$

where $I^{(n)}$ denotes the n-dimensional identity matrix, and where $0^{(n \times 1)}$ denotes an $(n \times 1)$ matrix (i.e. a column vector) consisting only of zeros. To see that B is a basis for L_{a,a_0}, we note that taking integer linear combinations of the column vectors of B generates vectors of type (51). Let $x \in \mathbb{Z}^n$ and $\xi \in \mathbb{Z}$. We obtain

$$\begin{pmatrix} x \\ ax - a_0\xi \end{pmatrix} = B \begin{pmatrix} x \\ \xi \end{pmatrix}. \qquad (53)$$

The algorithm SV (short vector) by Lagarias and Odlyzko consists of the following steps:

1. Apply Lovász' basis reduction algorithm to the basis B (52), which yields a reduced basis B'.

2. Check if any of the columns $b'_k = (b'_{1k}, \ldots, b'_{n+1,k})$ has all $b'_{jk} = 0$ or γ for some fixed constant γ, for $1 \leqslant j \leqslant n$. If such a reduced basis vector is found, check if the vector $x_j = b'_{jk}/\gamma$, $1 \leqslant j \leqslant n$ is a solution to $\sum_{j=1}^{n} a_j x_j = a_0$, and if yes, stop. Otherwise go to Step 3.

3. Repeat Steps 1 and 2 for the basis B with $a_0 = \sum_{j=1}^{n} a_j - a_0$, which corresponds to complementing all x_j-variables, i.e., considering $1 - x_j$ instead of x_j.

Algorithm SV runs in polynomial time as Lovász' basis reduction algorithm runs in polynomial time. It is not certain, however, that algorithm SV actually produces a solution to the subset sum problem. As Theorem 2.5 below shows, however, we can expect algorithm SV to work well on instances of (49) having low density. Consider a 0–1 vector x, which we will consider as fixed. We assume that $\sum_{j=1}^{n} x_j \leqslant n/2$. The reason for this assumption is that either $\sum_{j=1}^{n} x_j \leqslant n/2$, or $\sum_{j=1}^{n} x'_j \leqslant n/2$, where $x'_j = (1 - x_j)$, and since algorithm SV is run for both cases, one can perform the analysis for the vector that does satisfy the assumption. Let $\bar{x} = (x_1, \ldots, x_n, 0)$. Let the sample space $\Lambda(A, \bar{x})$ of lattices be defined to consist of all lattices L_{a,a_0} generated by basis (52) such that

$$1 \leqslant a_j \leqslant A \quad \text{for } 1 \leqslant j \leqslant n, \tag{54}$$

and

$$a_0 = \sum_{j=1}^{n} a_j \bar{x}_j. \tag{55}$$

There is precisely one lattice in the sample space for each vector a satisfying (54). Therefore the sample space consists of A^n lattices.

Theorem 2.5 (Lagarias and Odlyzko [69]). *Let $\bar{x}$ be a 0–1 vector for which $\sum_{j=1}^{n} \bar{x}_j \leqslant n/2$. If $A = 2^{\beta n}$ for any constant $\beta > 1.54725$, then the number of lattices L_{a,a_0} in $\Lambda(A, \bar{x})$ that contain a vector v such that $v \neq k\bar{x}$ for all $k \in \mathbb{Z}$, and such that $\|v\|^2 \leqslant n/2$ is*

$$O(A^{n - c_1(\beta)} (\log A)^2), \tag{56}$$

where $c_1(\beta) = 1 - 1.54725/\beta > 0$.

For $A = 2^{\beta n}$, the density of the subset sum problems associated with the lattices in the sample space can be proved to be equal to β^{-1}. This implies that Theorem 2.5 applies to lattices having density $d(a) < (1.54725)^{-1} \approx 0.6464$. Expression (56) gives a bound on the number of lattices we need to subtract from the total number of lattices in the sample space, A^n, in order to obtain the number of lattices in $\Lambda(A, \bar{x})$ for which $\bar{x}$ is the *shortest* non-zero vector. Here we notice that term (56) grows slower than the term A^n as n goes to infinity, and hence we can conclude that "almost all" lattices in the sample space $\Lambda(A, \bar{x})$ have $\bar{x}$ as the shortest vector. So, the subset sum problems (49) with density $d(a) < 0.6464$ could be solved in polynomial time if we had an oracle that could compute the shortest vector in the lattice L_{a,a_0}. Lagarias and Odlyzko also prove that the algorithm SV actually finds a solution to "almost all" feasible subset sum problems (49) having density $d(a) < (2 - \epsilon)(\log(\frac{4}{3}))^{-1} n^{-1}$ for any fixed $\epsilon > 0$.

Coster et al. [27] proposed two ways of improving Theorem 2.5. They showed that "almost all" subset sum problems (49) having density $d(a) < 0.9408$ can be solved in polynomial time in presence of an oracle that finds the shortest vector in certain lattices. Both ways of improving the bound on the density involve some changes in the lattice considered by Lagarias and Odlyzko. The first lattice $L'_{a,a_0} \in \mathbb{Q}^{n+1}$ considered by Coster et al. is defined as

$$L'_{a,a_0} = \{(x_1 - \tfrac{1}{2}\xi, \dots, x_n - \tfrac{1}{2}\xi, \; N(ax - a_0\xi))^{\mathrm{T}}\}, \tag{57}$$

where N is a natural number. The following basis $\bar{B}$ spans L':

$$\bar{B} = \begin{pmatrix} I^{(n)} & (-\tfrac{1}{2})^{(n\times 1)} \\ Na & -Na_0 \end{pmatrix}. \tag{58}$$

Here $(-\tfrac{1}{2})^{(n\times 1)}$ denotes the $(n \times 1)$-matrix consisting of elements $-\tfrac{1}{2}$ only. As in the analysis by Lagarias and Odlyzko, we consider a fixed vector $x \in \{0,1\}^n$, and we let $\bar{x} = (x_1, \dots, x_n, 0)$. The vector $\bar{x}$ does not belong to the lattice L', but the vector $w = (w_1, \dots, w_n, 0)$, where $w_j = x_j - \tfrac{1}{2}$, $1 \leqslant j \leqslant n$ does. So, if Lovász' basis reduction algorithm is applied to $\bar{B}$ and if the reduced basis $\bar{B}'$ contains a vector $(w_1, \dots, w_n, 0)$ with $w_j = \{-\tfrac{1}{2}, \tfrac{1}{2}\}$, $1 \leqslant j \leqslant n$, then the vector $(w_j + \tfrac{1}{2})$, $1 \leqslant j \leqslant n$ solves the subset sum problem (49). By shifting the feasible region to be symmetric about the origin we now look for vectors of shorter Euclidean length. Coster et al. prove the following theorem that is analogous to Theorem 2.5.

Theorem 2.6 (Coster et al. [27]). *Let A be a natural number, and let $a_1, \dots, a_n$ be random integers such that $1 \leqslant a_j \leqslant A$, for $1 \leqslant j \leqslant n$. Let $x = (x_1, \dots, x_n)$, $x_j \in \{0,1\}$, be fixed, and let $a_0 = \sum_{j=1}^{n} a_j x_j$. If the density $d(a) < 0.9408$, then the subset sum problem (49) defined by $a_1, \dots, a_n$ can "almost always" be solved in polynomial time by a single call to an oracle that finds the shortest vector in the lattice L'_{a,a_0}.*

Coster et al. prove Theorem 2.6 by showing that the probability that the lattice L'_{a,a_0} contains a vector $v = (v_1, \dots, v_{n+1})$ satisfying

$$v \neq kw \quad \text{for all } k \in \mathbb{Z}, \text{ and } ||v||^2 \leqslant ||w||^2 \tag{59}$$

is bounded by

$$n(4n\sqrt{n} + 1)\frac{2^{c_0 n}}{A} \tag{60}$$

for $c_0 = 1.0628$. Using the lattice L', note that $||w||^2 \leqslant n/4$. The number N in basis (58) is used in the following sense. Any vector in the lattice L' is an integer linear combination of the basis vectors. Hence, the $(n+1)$th element of a such a lattice vector is an integer multiple of N. If N is chosen large enough, then a lattice vector can be "short" only if the $(n+1)$th element is equal to zero. Since it is known that the length of w is bounded by $\tfrac{1}{2}\sqrt{n}$, then it suffices to choose $N > \tfrac{1}{2}\sqrt{n}$ in order to conclude that for a vector v to be shorter than w it should satisfy $v_{n+1} = 0$. Hence, Coster et al. only need to consider lattice vectors v in their proof that satisfy $v_{n+1} = 0$. In the theorem we assume that the density $d(a)$ of the subset sum problems is less than 0.9408.

Using the definition of $d(a)$ we obtain $d(a) = n/\log_2(\max_{1 \leqslant j \leqslant n}\{a_j\}) < 0.9408$, which implies that $\max_{1 \leqslant j \leqslant n}\{a_j\} > 2^{n/0.9408}$, giving $A > 2^{c_0 n}$. For $A > 2^{c_0 n}$, the bound (60) goes to zero as n goes to infinity, which shows that "almost all" subset sum problems having density $d(a) < 0.9408$ can be solved in polynomial time given the existence of a shortest vector oracle. Coster et al. also gave another lattice $L''(a, a_0) \in \mathbb{Z}^{n+2}$ that could be used to obtain the result given in Theorem 2.6. The lattice $L''(a, a_0)$ consists of vectors

$$
L''(a, a_0)
$$

$$
= \left\{ \left((n+1)x_1 - \sum_{\substack{k=1 \\ k \neq 1}}^{n} x_k - \xi, \ldots, (n+1)x_n - \sum_{\substack{k=1 \\ k \neq n}}^{n} x_k - \xi, \right.\right.
$$

$$
\left.\left. (n+1)\xi - \sum_{j=1}^{n} x_j, \ N(ax - a_0\xi) \right)^{\mathrm{T}} \right\}, \tag{61}
$$

and is spanned by the basis

$$
\begin{pmatrix}
(n+1) & -1 & -1 & \cdots & & -1 \\
-1 & (n+1) & -1 & \cdots & & -1 \\
\vdots & & \ddots & & & \vdots \\
-1 & \cdots & & -1 & (n+1) & -1 \\
-1 & \cdots & & & -1 & (n+1) \\
Na_1 & Na_2 & \cdots & & Na_n & -Na_0
\end{pmatrix}. \tag{62}
$$

Note that the lattice $L''(a, a_0)$ is not full-dimensional as the basis consists of $n+1$ vectors. Given a reduced basis vector $w = (w_1, \ldots, w_{n+1}, 0)$, we solve the system of equations

$$
w_j = (n+1)x_j - \sum_{k=1, k \neq j}^{n} x_k - \xi, \ 1 \leqslant j \leqslant n, \quad w_{n+1} = (n+1)\xi - \sum_{j=1}^{n} x_j
$$

and check whether $\xi = 1$, and the vector $x \in \{0, 1\}^n$. If so, x solves the subset sum problem (49). Coster et al. show that for $x \in \{0, 1\}^n$, $\xi = 1$, we obtain $\|w\|^2 \leqslant n^3/4$, and they indicate how to show that most of the time there will be no shorter vectors in $L''(a, a_0)$.

2.3. Solving diophantine equations using basis reduction

Aardal et al. [2,3] considered the following integer feasibility problem:

Does there exist a vector $x \in \mathbb{Z}^n$ such that $Ax = d$, $l \leqslant x \leqslant u$? \tag{63}

Here A is an $m \times n$-matrix, with $m \leqslant n$, and the vectors d, l, and u are of compatible dimensions. We assume that all input data is integral. Problem (63) is NP-complete, but if we remove the bound constraints $l \leqslant x \leqslant u$, it is polynomially solvable. A standard way of tackling problem (63) is by branch-and-bound, but for the applications considered by Aardal et al. this method did not work well. Let $X = \{x \in \mathbb{Z}^n : Ax = d, \ l \leqslant x \leqslant u\}$. Instead of using a method based on the linear relaxation of the problem, they considered the following integer relaxation of X, $X_{IR} = \{x \in \mathbb{Z}^n : Ax = d\}$. Determining whether X_{IR} is empty can be carried out in polynomial time for instance by generating the Hermite normal form of the matrix A. Let x^0 be an integral vector satisfying $Ax^0 = d$, and let Y be an $n \times (n - m)$-matrix consisting of integer, linearly independent column vectors y^j, $1 \leqslant j \leqslant n - m$, such that $Ay^j = 0$ for $1 \leqslant j \leqslant n - m$. We can now rewrite X_{IR} as

$$X_{IR} = \{x \in \mathbb{Z}^n : x = x^0 + Y\lambda, \ \lambda \in \mathbb{Z}^{n-m}\}, \tag{64}$$

that is, we express any vector x that satisfies $Ax = d$ as a vector x^0, satisfying $Ax^0 = d$, plus an integer linear combination of vectors that form a basis of the lattice $L_0 = \{x \in \mathbb{Z}^n : Ax = 0\}$. Since a lattice may have several bases, reformulation (64) is not unique.

The intuition behind the approach of Aardal et al. is as follows. Suppose that we are able to obtain a vector x^0 that is short with respect to the bounds. Then, we may hope that x^0 satisfies $l \leqslant x^0 \leqslant u$, in which case we are done. If x^0 does not satisfy the bounds, then we observe that $A(x^0 + \lambda y) = d$ for any integer multiplier λ and any vector y satisfying $Ay = 0$. Hence, we can derive an enumeration scheme in which we branch on integer linear combinations of vectors y satisfying $Ay = 0$, which explains the reformulation (64) of X_{IR}. Similar to Lagarias and Odlyzko, we choose a lattice, different from the standard lattice $\mathbb{Z}^n$, in which solutions to our problem (63) are relatively short vectors, and then apply basis reduction to the initial basis of the chosen lattice.

Aardal et al. [3] suggested a lattice $L_{A,d} \in \mathbb{Z}^{n+m+1}$ that contains vectors of the following form:

$$(x^T, \ N_1\xi, \ N_2(a_1 x - d_1 \xi), \ldots, N_2(a_m x - d_m \xi))^T, \tag{65}$$

where a_i is the ith row of the matrix A, where N_1 and N_2 are natural numbers, and where ξ, as in Section 2.2, is a variable associated with the right-hand side vector d. The basis B given below spans the lattice $L_{A,d}$:

$$B = \begin{pmatrix} I^{(n)} & 0^{(n \times 1)} \\ 0^{(1 \times n)} & N_1 \\ N_2 A & -N_2 d \end{pmatrix}. \tag{66}$$

The lattice $L_{A,d} \subset \mathbb{Z}^{m+n+1}$ is not full-dimensional as B only contains $n + 1$ columns. The numbers N_1 and N_2 are chosen so as to *guarantee* that certain elements of the reduced basis are equal to zero (cf. the different role of the number N used in the bases (58) and (62)). The following proposition states precisely which type of vectors we wish to obtain.

Proposition 2.7. *The integer vector x^0 satisfies $Ax^0 = d$ if and only if the vector*

$$((x^0)^{\mathrm{T}}, N_1, 0^{(1 \times m)})^{\mathrm{T}} = B \begin{pmatrix} x^0 \\ 1 \end{pmatrix} \tag{67}$$

belongs to the lattice L, and the integer vector y satisfies $Ay = 0$ if and only if the vector

$$(y^{\mathrm{T}}, 0, 0^{(1 \times m)})^{\mathrm{T}} = B \begin{pmatrix} y \\ 0 \end{pmatrix} \tag{68}$$

belongs to the lattice L.

Let $\hat{B}$ be the basis obtained by applying Lovász' basis reduction algorithm to the basis B, and let $\hat{b}_j = (\hat{b}_{1j}, \ldots, \hat{b}_{n+m+1,j})$ be the jth column vector of $\hat{B}$. Aardal et al. [3] prove that if the numbers N_1 and N_2 are chosen appropriately, then the $(n - m + 1)$th column of $\hat{B}$ is of type (67), and the first $n - m$ columns of $\hat{B}$ are of type (68), i.e., the first $n - m + 1$ columns of $\hat{B}$ are of the following form:

$$\begin{pmatrix} Y^{(n \times (n-m))} & x^0 \\ 0^{(1 \times (n-m))} & N_1 \\ 0^{(m \times (n-m))} & 0 \end{pmatrix}. \tag{69}$$

This result is stated in the following theorem.

Theorem 2.8 (Aardal et al. [3]). *Assume that there exists an integral vector x satisfying the system $Ax = d$. There exist numbers N_{01} and N_{02} such that if $N_1 > N_{01}$, and if $N_2 > 2^{n+m}N_1^2 + N_{02}$, then the vectors $\hat{b}_j \in \mathbb{Z}^{n+m+1}$ of the reduced basis $\hat{B}$ have the following properties:*
(1) $\hat{b}_{n+1,j} = 0$ for $1 \leqslant j \leqslant n - m$,
(2) $\hat{b}_{ij} = 0$ for $n + 2 \leqslant i \leqslant n + m + 1$ and $1 \leqslant j \leqslant n - m + 1$,
(3) $|\hat{b}_{n+1,n-m+1}| = N_1$.
Moreover, the sizes of N_{01} and N_{02} are polynomially bounded in the sizes of A and d.

In the proof of Properties 1 and 2 of Theorem 2.8, Aardal et al. make use of inequality (15) of Proposition 1.4.

Once we have obtained the matrix Y and the vector x^0, we can derive the following equivalent formulation of problem (63):

Does there exist a vector $\lambda \in \mathbb{Z}^{n-m}$ such that $l \leqslant x^0 + Y\lambda \leqslant u$? (70)

Aardal et al. [3] and Aardal et al. [1] investigated the effect of the reformulation on the number of nodes of a linear programming based branch-and-bound algorithm. They considered three sets of instances: instances obtained from Philips Research Labs, the Frobenius instances of Cornuéjols et al. [25], and the market split instances of Cornuéjols and Dawande [24]. The results were encouraging. For instance, after transforming problem (63) to problem (70), the size of the market split instances that

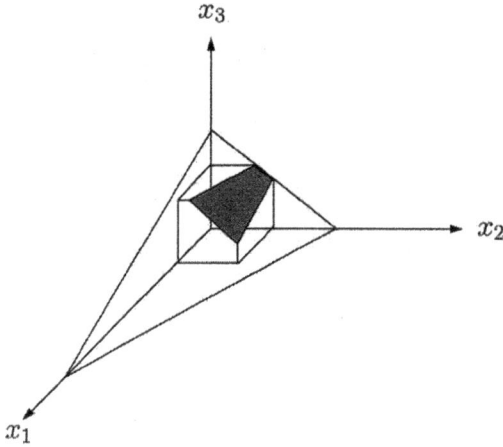

Fig. 11.

could be solved doubled. Aardal et al. [1] also investigated the performance of integer branching. Let $P = \{\lambda \in \mathbb{Z}^{n-m}: l \leqslant x^0 + Y\lambda \leqslant u\}$. At node k of the enumeration tree they choose a unit vector e^j, $1 \leqslant j \leqslant n - m$ that has not yet been chosen at any of the predecessors of node k. Then, they compute $\mu_k = \lceil \min\{(e^j)^{\mathrm{T}}\lambda: \lambda \in P \cap \{\lambda_j\text{'s fixed at predecessors of } k\}\}\rceil$ and $\gamma_k = \lfloor \max\{(e^j)^{\mathrm{T}}\lambda: \lambda \in P \cap \{\lambda_j\text{'s fixed at predecessors of } k\}\}\rfloor$. At node k, $\gamma_k - \mu_k + 1$ subproblems, or branches, are created by fixing λ_j to $\mu_k, \mu_k + 1, \ldots, \gamma_k$. Different strategies for choosing a unit direction e^j were considered. This branching scheme can be viewed as a heuristic version of the integer programming algorithms described in Section 2.1. Instead of using vectors that give provably thin directions, only unit vectors were used. The experiments indicated that the unit vectors yield good directions, i.e., only few nodes were created at each branch, and typically, at a modest depth of the search tree only one branch was created. One way of explaining why the reformulated problem is so much easier to solve is that the index of the lattice $L_0 = \{x \in \mathbb{Z}^n: Ax = 0\}$ in $\mathbb{Z}^n$ is, in general, larger than one. Let Λ be a sublattice of the lattice M. The index I of Λ in M is defined as $I = d(\Lambda)/d(M)$. If the index of Λ in M is large, then M contains a large number of vectors that are different from the vectors in Λ, which means that a certain "scaling effect" is obtained. We illustrate this effect in the following example.

Example 2.2. Consider the polytope $X = \{x \in \mathbb{R}^3: 2x_1 + 4x_2 + 5x_3 = 8, \ 0 \leqslant x_j \leqslant 1, \ 1 \leqslant j \leqslant 3\}$. The set X is illustrated in grey in Fig. 11. The question is: does X contain an integral vector? To use branch-and-bound we need to introduce an objective function. Here we have chosen $\min(x_1 + x_2 + x_3)$. The optimal solution to the linear relaxation of this instance is $x = (0, \frac{3}{4}, 1)^{\mathrm{T}}$. Two branch-and-bound nodes are created by adding the constraints $x_2 = 0$ and 1. The subproblem implied by $x_2 = 0$ is infeasible, but if we impose $x_2 = 1$ we obtain the solution $x = (0, 1, \frac{4}{5})$, and we need to branch on variable x_3.

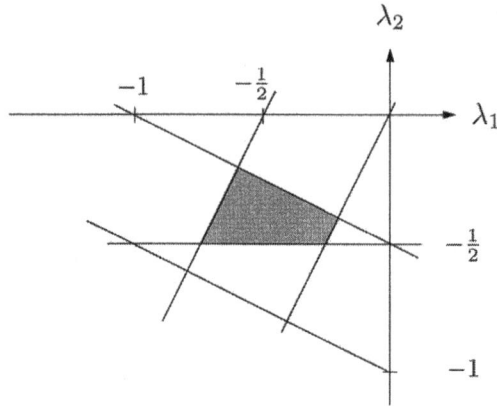

Fig. 12.

If we reformulate the integer feasibility problem according to (70) we obtain, through basis reduction, the vector $x^0 = (0,2,0)^T$ and the matrix

$$Y = \begin{pmatrix} -2 & 1 \\ 1 & 2 \\ 0 & -2 \end{pmatrix}.$$

The question is: Does there exist a vector $\lambda \in \mathbb{Z}^2$ such that $\lambda \in P$, where $P = \{\lambda \in \mathbb{Z}^2 : 0 \leqslant -2\lambda_1 + \lambda_2 \leqslant 1, -2 \leqslant \lambda_1 + 2\lambda_2 \leqslant -1, 0 \leqslant -2\lambda_2 \leqslant 1\}$. The linear relaxation of P is given in Fig. 12. If we use $\min(\lambda_1 + \lambda_2)$ as objective function, we obtain the fractional point $\lambda = (-\frac{3}{4}, -\frac{1}{2})^T$, but, the subproblems created by branching on λ_1 as well as on λ_2 are infeasible. In fact, regardless of the objective function that is used, integer infeasibility is detected at the root node. This example is of course so small that it is hard to draw any conclusions, but if we draw the coordinate system corresponding to the formulation in λ-variables in the coordinate system of the x-variables, we can observe the scaling effect discussed above. This is done by translating the lattice $L_0 = \{x \in \mathbb{Z}^3 : 2x_1 + 4x_2 + 5x_3 = 0\}$ to the point x^0, i.e., the origin of the λ-coordinate system is located at the vector x^0. The unit vector $\lambda = (-1,0)^T$ corresponds to the vector $x = (2,1,0)^T$, and the vector $\lambda = (0,-1)^T$ corresponds to the vector $x = (-1,0,2)^T$, see Fig. 13. The determinant of the lattice L_0 is equal to $\sqrt{45}$, whereas the determinant of $\mathbb{Z}^3$ is equal to 1. ☐

The computational study by Aardal et al. [1] indicated that integer branching on the unit vectors in the space of the λ-variables taken in the order $j = n - m, \ldots, 1$ was quite effective, and in general much better than the order $1, \ldots, n - m$. This can be explained as follows. Due to Lovász' algorithm, the vectors of Y are more or less in order of increasing length, so typically, the $(n - m)$th vector of Y is the longest one. Branching on this vector first should generate relatively few hyperplanes intersecting the linear relaxation of X, if this set has a regular shape. Note, that to branch on the jth vector of Y corresponds to branching on the jth unit vector in the space of the λ-variables.

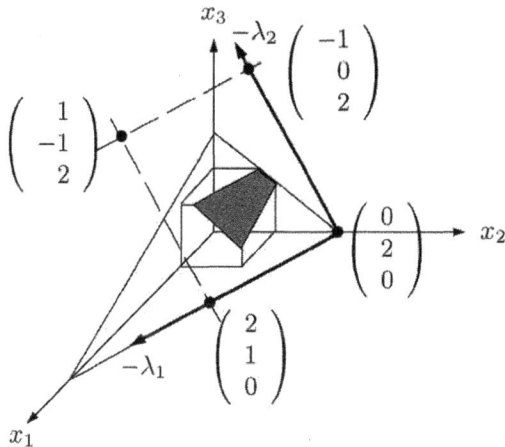

Fig. 13.

Recently, Louveaux and Wolsey [77] considered the problem: "Does there exist a matrix $X \in \mathbb{Z}^{m \times n}$ such that $XA = C$, and $BX = D$?", where $A \in \mathbb{Z}^{n \times p}$ and $B \in \mathbb{Z}^{q \times m}$. Their study was motivated by a portfolio planning problem, where variable x_{ij} denotes the number of shares of type j included in portfolio i. This problem can be written in the same form as problem (63), so in principle the approach discussed in this section could be applied. For reasonable problem sizes Louveaux and Wolsey observed that the basis reduction step became too time consuming. Instead they determine reduced bases for the lattices $L_0^A = \{y \in \mathbb{Z}^n : y^T A = 0\}$, and $L_0^B = \{z \in \mathbb{Z}^m : Bz = 0\}$. Let B_A be a basis for the lattice L_0^A, and let B_B be a basis for the lattice L_0^B. They showed that taking the so-called *Kronecker product* of the matrices B_A^T and B_B yields a basis for the lattice $L_0 = \{X \in \mathbb{Z}^{m \times n} : XA = 0, \ BX = 0\}$. The Kronecker product of two matrices $M \in \mathbb{R}^{m \times n}$, and $N \in \mathbb{R}^{p \times q}$ is defined as

$$M \otimes N = \begin{pmatrix} m_{11}N & \cdots & m_{1n}N \\ \cdots & \ddots & \cdots \\ m_{m1}N & \cdots & m_{mn}N \end{pmatrix}.$$

Moreover, they showed that the basis of L_0 obtained by taking the Kronecker product between B_A^T and B_B is reduced, up to a reordering of the basis vectors, if the bases B_A and B_B are reduced. Computational experience is reported.

2.4. Discussion

One important question is whether there exist versions of the integer programming algorithms presented in Section 2.1 that can be used with good results in practice. It should be noted that the main purpose of the algorithms by Lenstra [71], and by Lovász and Scarf [76], was to prove a theorem. No particular care was taken to ensure good performance in practice. However there is some evidence that the concepts discussed in Sections 1 and 2 can be used to design effective practical integer programming algorithms, namely the studies by Cook et al. [23], by Aardal et al. [1,3], and by Louveaux and Wolsey [77]. Two of these concepts appear to be worth emphasizing: branching on hyperplanes, and considering sublattices.

Branching on hyperplanes, or "integer branching", in directions in which the polytope is thin may reduce the number of nodes that one needs to evaluate in an enumeration tree quite drastically. One problem that needs to be dealt with is the amount of effort spent in each node. To compute search directions that are provably thin is quite time consuming, so heuristic algorithms are needed.

One of the features of the approach by Aardal et al. [3] is to consider a sublattice of $\mathbb{Z}^n$. Combining this idea with integer branching led to a decrease in the number of enumeration nodes of up to a factor of 10^4, compared with the number of nodes needed using branch-and-bound on the original formulation [1]. Similar results were obtained by Louveaux and Wolsey [77].

The instances tackled by Cook et al. [23] and by Aardal et al. [1], were relatively small. If one applies Lovász' algorithm to such instances to obtain a reformulation such as (70), then the basis reduction only takes a couple of seconds. Therefore, the branching phase is the bottleneck. If one wants to solve medium size instances, then the reduction phase will be time consuming using the current versions of Lovász' algorithm. A faster basis reduction algorithm and further studies on how to construct composite bases would be extremely useful.

3. Augmentation algorithms and test sets

A natural approach to attack a linear integer program with constraint set

$$X = \{x \in \mathbb{Z}^n \colon Ax = b,\ 0 \leqslant x \leqslant u\},$$

such that $A \in \mathbb{Z}^{m \times n}$, $b \in \mathbb{Z}^m$, $u \in \mathbb{Z}_+^n$, is via the following augmentation algorithm.

Algorithm 3.1. An Augmentation Algorithm for a minimization problem.
Let x be any feasible point of the linear integer program.

While x is not optimal, determine an integral vector z and a positive integer number $\lambda > 0$ such that (i) $x + \lambda z$ is feasible and (ii) $x + \lambda z$ attains a smaller objective function value than x. Set $x := x + \lambda z$.

One question that arises immediately is whether this algorithm can be made effective in terms of the number of augmentations that one needs to find an optimal solution. This topic is addressed in Section 3.1. We will see that one can solve the optimization problem with a polynomial number of calls of a *directed augmentation oracle*.

From a mathematical point of view a study of the augmentation problem leads naturally to an investigation of Hilbert bases of pointed polyhedral rational cones, and of test sets. This approach is discussed in Section 3.2. Test sets for families of integer programs are collections of integral vectors with the property that every feasible non-optimal point of any integer program in the family can be improved by a vector in the test set. They can be designed from various mathematical viewpoints. One of these approaches is based on Hilbert bases, another one comes from Gröbner bases associated with toric ideals. Test sets are the central topic in Section 3.3.

3.1. From augmentation to optimization

There are two elementary questions that arise in the analysis of an augmentation algorithm for a linear integer program: how can one solve the subproblem of detecting an improving direction and secondly what is a bound on the number of improvement steps required in order to reach an optimal point. This subsection is dedicated to the latter question.

To be more formal, let $A \in \mathbb{Z}^{m \times n}$, $b \in \mathbb{Z}^m$, $u \in \mathbb{Z}_+^n$. Throughout this section we assume that

$$X = \{x \in \mathbb{Z}^n : Ax = b, \ 0 \leqslant x \leqslant u\}, \tag{71}$$

is the set of all feasible solutions of the integer program. Our goal is to solve the optimization problem (OPT) by an augmentation algorithm, i.e., by repeated calls to an oracle that solves the augmentation problem.

The optimization problem (OPT)
Given a vector $c \in \mathbb{Z}^n$ and a point $x \in X$, find a vector $x^* \in X$
that minimizes c over X.

The augmentation problem (AUG)
Given a vector $c \in \mathbb{Z}^n$ and a point $x \in X$, find a point $y \in X$ such
that $c^T y < c^T x$, or assert that no such y exists.

A classical example of an augmentation algorithm for solving the minimum cost flow problem in digraphs is a cycle cancelling algorithm that improves feasible flow along negative cycles. Such negative cycles can be detected efficiently in an augmentation network that one constructs from a feasible solution. In this network each original arc, on which the value of a feasible flow can increase or decrease without exceeding the corresponding lower and upper bound requirement, is replaced by a forward arc and a backward arc. Note that this replacement operation allows one in particular to evaluate forward arcs and backward arcs differently. A generalization of this directed augmentation network to general integer programs is the directed augmentation problem.

The directed augmentation problem (DIR-AUG)
Given vectors $c, d \in \mathbb{Z}^n$ and a point $x \in X$, find vectors $z^1, z^2 \in \mathbb{Z}_+^n$
such that $c^T z^1 - d^T z^2 < 0$ and $x + z^1 - z^2$ is feasible, or assert
that no such vectors z^1, z^2 exist.

In the case of the minimum cost flow problem in digraphs it is well known that a cycle cancelling algorithm that augments along any negative cycle does not necessarily converge to an optimal solution in polynomial time in the encoding length of the input data. Indeed, a more sophisticated strategy for augmenting is required. In the min-cost-flow application it is for instance the augmentation of feasible flows along maximum mean ratio cycles that makes the primal algorithm work efficiently. Maximum mean ratio cycles are very special objects and there is no obvious counterpart in the case of general integer programs. Indeed, to show that a polynomial number of calls to the directed augmentation oracle suffices to solve the optimization problem, we need a combination of an interior point philosophy by using a barrier function and a maximum mean ratio augmentation.

Definition 3.1.

(a) For $x \in X$ and $j \in \{1, \ldots, n\}$, let

$$p(x)_j := 1/(u_j - x_j) \quad \text{if } x_j < u_j \text{ and } p(x)_j := 0, \text{ otherwise.}$$

$$n(x)_j := 1/x_j \quad \text{if } x_j > 0 \text{ and } n(x)_j := 0, \text{ otherwise.}$$

(b) A vector $z \in \mathbb{Z}^n$ is called *exhaustive* w.r.t. a point $x \in X$ if $x + z \in X$ and for all $\lambda \in \mathbb{Z}^+$, $\lambda \geqslant 2$ we have that $x + \lambda z \notin X$.

(c) For the integer program (IP), let $C := \max\{|c_i| : i = 1, \ldots, n\}$ and $U := \max\{|u_i| : i = 1, \ldots, n\}$.

The maximum ratio augmentation problem (MRA)

Given a vector $c \in \mathbb{Z}^n$ and a point $x \in X$, find vectors $z^1, z^2 \in \mathbb{Z}^n_+$ such that $c^{\mathrm{T}}(z^1 - z^2) < 0$, $x + z^1 - z^2$ is feasible and the objective $|c^{\mathrm{T}}(z^1 - z^2)|/(p(x)^{\mathrm{T}}z^1 + n(x)^{\mathrm{T}}z^2)$ is maximum.

An important relation between the oracles (MRA) and (DIR-AUG) is stated below.

Lemma 3.1 (Schulz and Weismantel [94]). *(MRA) can be solved with* $O(n \log(nCU))$ *calls to an oracle that solves (DIR-AUG).*

We are now ready to analyze a specific augmentation algorithm that we call MMA-augmentation-algorithm. This algorithm has been invented for the minimum cost flow problem by Wallacher [104]. Later McCormick and Shioura extended one of Wallacher's algorithms to linear programming over unimodular spaces [78]. For its analysis we resort to

Lemma 3.2 (Geometric improvement Ahuja et al. [4]). *Let* $x^1, x^2, \ldots$ *be a sequence of feasible points in X produced by some algorithm* $\mathscr{A}$ *such that* $c^{\mathrm{T}}x^1 > c^{\mathrm{T}}x^2 > \cdots$. *Let* x^* *be a solution of* $\min\{c^{\mathrm{T}}x : x \in X\}$. *If there exists a constant* $0 < \alpha < 1$ *such that for all k*

$$|c^{\mathrm{T}}(x^{k+1} - x^k)| \geqslant \alpha|c^{\mathrm{T}}(x^* - x^k)|,$$

then $\mathscr{A}$ *terminates after* $O(\log(nCU)/\alpha)$ *steps with an optimal solution.*

Proof. Consider a consecutive sequence of $\beta := 2/\alpha$ iterations starting with iteration k. If each of these iterations improves the objective function value by at least $\alpha/2|c^{\mathrm{T}}(x^* - x^k)|$, then $x^{k+\beta}$ is an optimal solution. Otherwise, there exists q such that

$$\alpha(|c^{\mathrm{T}}(x^* - x^q)|) \leqslant |c^{\mathrm{T}}(x^{q+1} - x^q)| \leqslant \frac{\alpha}{2}|c^{\mathrm{T}}(x^* - x^k)|$$

$$\Leftrightarrow |c^{\mathrm{T}}(x^* - x^q)| \leqslant \frac{1}{2}|c^{\mathrm{T}}(x^* - x^k)|,$$

i.e., after β iterations we have halved the gap between $c^{\mathrm{T}}x^*$ and $c^{\mathrm{T}}x^k$. $\square$

Algorithm 3.2. *Algorithm [MMA]*

1. Let $x \in X$.
2. Call (MRA) with input x and objective function c
3. If (MRA) does not return vectors z^1, z^2, then STOP. Otherwise let z^1, z^2 be the output of (MRA).
4. Using binary search determine a maximum step length, i.e., a number $\lambda \in \mathbb{Z}_+$ such that $\lambda(z^1 - z^2)$ is exhaustive.
5. Set $x := x + \lambda(z^1 - z^2)$ and return to Step 2.

Theorem 3.3 (Schulz and Weismantel [93]). *For any $x \in X$ and $c \in \mathbb{Z}^n$, Algorithm [MMA] solves (OPT) with at most $\mathrm{O}(n\log(nCU))$ calls of (MRA).*

Proof. Let $x \in X$ and $c \in \mathbb{Z}^n$. Assuming that x is not minimal w.r.t. c, let z^1, z^2 be the output of (MRA) and $\lambda \in \mathbb{Z}_+$ such that $\lambda(z^1 - z^2)$ is exhaustive. Let $z := \lambda(z^1 - z^2)$ and x^* be an optimal solution. We set $z^* := x^* - x$. Since z is exhaustive, there exists $j \in \{1,\ldots,n\}$ such that $x_j + 2z_j > u_j$ or $x_j + 2z_j < 0$. This situation occurs if and only if $z_j^+ > (u_j - x_j)/2$ or $z_j^- > x_j/2$. Therefore, $p(x)^{\mathrm{T}}z^+ + n(x)^{\mathrm{T}}z^- \geqslant \frac{1}{2}$. Moreover, $p(x)^{\mathrm{T}}(z^*)^+ + n(x)^{\mathrm{T}}(z^*)^- \leqslant n$. On account of the condition that

$$|c^{\mathrm{T}}z|/(p(x)^{\mathrm{T}}z^+ + n(x)^{\mathrm{T}}z^-) \geqslant |c^{\mathrm{T}}z^*|/(p(x)^{\mathrm{T}}(z^*)^+ + n(x)^{\mathrm{T}}(z^*)^-),$$

we obtain that $|c^{\mathrm{T}}z| \geqslant |c^{\mathrm{T}}z^*|/(2n)$. Applying Lemma 3.2 yields the result. $\square$

A consequence of Theorem 3.3 is

Theorem 3.4 (Schulz and Weismantel [93]). *Let X be given by an oracle that solves (DIR-AUG). Then for every $c \in \mathbb{Z}^n$, the optimization problem can be solved in oracle polynomial time.*

We remark that one can also use the method of *bit-scaling* (see [31]) in order to show that for a class of 0/1-integer programming problems the optimization problem can be solved by a polynomial number of calls of the (directed) augmentation oracle. This is discussed in Schulz et al. [92] and in Grötschel and Lovász [47]. For a thorough introduction to oracles and oracle-polynomial time algorithms, we refer to Grötschel et al. [49].

3.2. From augmentation to Hilbert bases

In this section we summarize elementary links between the study of test sets for families of integer programs and the augmentation problem. The augmentation problem is the task of determining an improving integral direction for a specific non-optimal point. Test sets for families of integer programs are collections of integral vectors with the property that every feasible non-optimal point of the integer program can be improved by a vector in the test set. We will see that test sets can be derived from Hilbert bases of rational polyhedral cones. The following definitions about cones are used.

Definition 3.2. A subset C of $\mathbb{R}^n$ is called a *cone* if for all $x, y \in C$ and $\lambda, \mu \geqslant 0$ we have that $\lambda x + \mu y \in C$. A cone C is called *polyhedral* if it is finitely generated, i.e., there exists a finite set $V = \{v^1, \ldots, v^k\} \subseteq \mathbb{R}^n$ such that

$$C = \left\{ y \colon y = \sum_{i=1}^{k} \lambda_i v^i \colon \lambda_1, \ldots, \lambda_k \geqslant 0 \right\}.$$

If C is generated by V, we write $C = C(V)$. If $V \subseteq \mathbb{Q}^n$, $C(V)$ is called *rational*. A cone C is *pointed* if there exists a hyperplane $a^T x \leqslant 0$ such that $\{0\} = \{x \in C \colon a^T x \leqslant 0\}$.

In the following we always consider rational polyhedral cones and call them cones for short. We are interested in a special subset of integral vectors in a cone, namely an integral generating set for all the integer points in the cone.

Definition 3.3 (Giles and Pulleyblank [36]). Let C be a rational polyhedral cone. A finite set $H \subseteq C \cap \mathbb{Z}^n$ is a *Hilbert basis* of C if every integral vector in C can be represented as a non-negative integral combination of the elements of H.

Example 3.1. Let $C = \{y \in \mathbb{R}^2 \colon y_1 = \lambda_1 + 3\lambda_2, \ y_2 = 3\lambda_1 + \lambda_2 \colon \lambda_1, \lambda_2 \geqslant 0\}$. The set $\{(1,3)^T, (3,1)^T\}$ does not form a Hilbert basis of C because $(2,2)^T \in C$ cannot be represented as a non-negative integral combination of the vectors $(1,3)^T$ and $(3,1)^T$. However the set

$$\{(1,1)^T, (2,1)^T, (1,2)^T, (1,3)^T, (3,1)^T\}$$

is a Hilbert basis of C. $\square$

Theorem 3.6 tells us that Hilbert bases of rational cones exist. This result is fundamental. Its proof may be derived from the Gordan Lemma [44], but can also be given directly.

Theorem 3.5 (Gordan's Lemma [44]). *Let $P \neq \emptyset \subseteq \mathbb{Z}^n_+$. There exists a unique minimal and finite subset $\{p_1, \ldots, p_m\}$ of P such that $p \in P$ implies that $p^j \leqslant p$ for at least one $j \in \{1, \ldots, m\}$.*

Theorem 3.6 (Gordan [44] see also Schrijver [91]). *Every rational polyhedral cone possesses a Hilbert basis.*

Proof. Let $C(p^1, \ldots, p^m)$ be a rational polyhedral cone generated by the vectors $p^1, \ldots, p^m \in \mathbb{Z}^n$. A Hilbert basis H of C is always contained in the zonotope. More precisely,

$$H \subseteq \mathscr{Z} = \{p^1, \ldots, p^m\}$$

$$\cup \left\{ p \in C \setminus \{0\}: p = \sum_{i=1}^{m} \lambda_i p^i, \ 0 \leqslant \lambda_i < 1, \ 1 \leqslant i \leqslant m \right\}. \tag{72}$$

Since $|\mathscr{Z} \cap \mathbb{Z}^n|$ is finite, the claim follows. $\square$

Not every rational cone has however a unique Hilbert basis that is minimal with respect to inclusion.

Example 3.2. Let $C = \{x \in \mathbb{R}^2: x_1 + 2x_2 = 0\}$. It may be checked that the set $(2, -1)^T$, $(-2, 1)^T$ is a Hilbert basis that is minimal with respect to taking subsets. The cone C also possesses a second Hilbert basis that is minimal with respect to inclusion consisting of the vectors $(4, -2)^T, (-2, 1)^T$. $\square$

If $C = C(p^1, \ldots, p^m) \subseteq \mathbb{R}^n$ is a pointed cone, then a Hilbert basis H of C that is minimal with respect to inclusion is uniquely determined (cf. [26,91]),

$$H = \{z \in C \cap \mathbb{Z}^n \setminus \{0\}: z \text{ is not the sum of two other vectors in}$$

$$C \cap \mathbb{Z}^n \setminus \{0\}\}. \tag{73}$$

Theorem 3.7 (van der Corput [26]). *If a rational polyhedral cone C is pointed, then there exists a unique Hilbert basis that is minimal with respect to inclusion. This minimal Hilbert basis is denoted $H(C)$.*

Let O_j denote the jth orthant in $\mathbb{R}^n$. We denote by

$$IP(b, c, u) \quad \min\{c^T x: Ax = b, \ 0 \leqslant x \leqslant u, \ x \in \mathbb{Z}^n\},$$

the family of integer programs associated with a fixed matrix A and varying $b \in \mathbb{Z}^m$, $u \in \mathbb{Z}^n_+$, $c \in \mathbb{R}^n$. Then $C_j := O_j \cap \{x \in \mathbb{R}^n: Ax = 0\}$ is a pointed polyhedral cone in $\mathbb{R}^n$. On account of Theorems 3.6 and 3.7, C_j possesses a unique and finite Hilbert basis $H_j := H(C_j)$. The set

$$\mathscr{H} := \bigcup_j H_j \setminus \{0\}$$

is called the *Graver test set* for the family of integer programs $IP(b, c, u)$. In particular, the Graver test set is a finite set. Moreover, it contains a test set for every member of the family of integer programs $IP(b, c, u)$.

Theorem 3.8 (Graver [46]). *The Graver test set $\mathcal{H}$ contains a test set for all integer programs of the family IP(b,c,u) with varying $b \in \mathbb{Z}^m$, $u \in \mathbb{Z}_+^n$, $c \in \mathbb{R}^n$.*

Proof. Let $b \in \mathbb{Z}^m$, $u \in \mathbb{Z}_+^n$ and $c \in \mathbb{R}^n$ and consider the integer program $\min c^\mathsf{T} x$: $Ax = b$, $0 \leqslant x \leqslant u$, $x \in \mathbb{Z}^n$. Let x be a feasible point for this program that is not minimal with respect to c and let y be an optimal solution. On account of $Ay = b = Ax$, it follows that $A(y-x) = 0$, $y - x \in \mathbb{Z}^n$ and $c^\mathsf{T}(y-x) < 0$. Let O_j denote the orthant that contains $y - x$. As $y - x$ is an integral point in C_j, there exist multipliers $\varepsilon_h \in \mathbb{Z}_+$ for all $h \in H_j$ such that $y - x = \sum_{h \in H_j} \varepsilon_h h$. As $c^\mathsf{T}(y-x) < 0$ and $\varepsilon_h \geqslant 0$ for all $h \in H_j$, there exists a vector $h^* \in H_j$ such that $c^\mathsf{T} h^* < 0$ and $\varepsilon_{h^*} > 0$. h^* lies in the same orthant as $y - x$, i.e., if $y_j - x_j > 0$, then $y_j - x_j \geqslant h_j^* \geqslant 0$ and if $y_j - x_j < 0$, then $y_j - x_j \leqslant h_j^* \leqslant 0$. Since x and y are feasible for the same integer program, we obtain that h^* is an "augmenting vector" and that $x + h^*$ is feasible. □

Example 3.3. Consider the family of equality knapsack problems,

$$\min\{c_1 x_1 + c_2 x_2 + c_3 x_3: x_1 + 2x_2 + 3x_3 = b, \ 0 \leqslant x \leqslant u, \ x \in \mathbb{Z}^3\}$$

with varying $b \in \mathbb{Z}$ and c, u. Since the Graver test set is symmetric about the origin, it suffices to analyze four orthants:

$$O_1 = \{x_1 \geqslant 0, x_2 \geqslant 0, x_3 \geqslant 0\}, \qquad O_2 = \{x_1 \geqslant 0, x_2 \geqslant 0, x_3 \leqslant 0\},$$

$$O_3 = \{x_1 \geqslant 0, x_2 \leqslant 0, x_3 \geqslant 0\}, \qquad O_4 = \{x_1 \leqslant 0, x_2 \geqslant 0, x_3 \geqslant 0\}.$$

For $j \in \{1,2,3,4\}$ we need to determine a Hilbert basis H_j of the cone

$$C_j = \{x \in O_j: x_1 + 2x_2 + 3x_3 = 0\}.$$

We obtain

$$H_1 = \emptyset,$$
$$H_2 = \{(3,0,-1),(0,3,-2),(1,1,-1)\},$$
$$H_3 = \{(2,-1,0),(0,-3,2),(1,-2,1)\},$$
$$H_4 = \{(-3,0,1),(-2,1,0)\}. \tag{74}$$

Therefore, the Graver test set for this family of integer programs is the set

$$\mathcal{H} = \{\pm(3,0,-1),\pm(0,3,-2),\pm(1,1,-1),\pm(2,-1,0),\pm(1,-2,1)\}. \quad □$$

Using the notion of irreducibility of vectors, there is a second equivalent characterization of the Graver test set.

Definition 3.4. Let $A \in \mathbb{Z}^{m \times n}$. We say that a vector $v \in \mathbb{Z}^n \setminus \{0\}$ *reduces* $w \in \mathbb{Z}^n \setminus \{0,v\}$ w.r.t. A if the following properties hold:

$$v^+ \leqslant w^+, \quad v^- \leqslant w^-, \quad (Av)^+ \leqslant (Aw)^+, \quad (Av)^- \leqslant (Aw)^-. \tag{75}$$

If such a v exists, w is called *reducible*. Otherwise, w is *irreducible*.

If v reduces w, then $w - v$ also reduces w, and we have that

$$v^+ + (w - v)^+ = w^+, \quad v^- + (w - v)^- = w^-,$$

$$(Av)^+ + (A(w - v))^+ = (Aw)^+, \quad (Av)^- + (A(w - v))^- = (Aw)^-. \tag{76}$$

For a cone C_j of the form $\{x \in O_j: Ax = 0\}$, these conditions ensure that the set of all irreducible integral points that lie in C_j define a Hilbert basis of this cone. Moreover, $w \in C_j \cap \mathbb{Z}^n$ implies that w is an element of the lattice

$$L = \{x \in \mathbb{Z}^n: Ax = 0\}.$$

This yields

Remark 3.1. The set of all irreducible lattice vectors $v \in L$ is the Graver test set for the family of integer programs $IP(b, c, u)$.

Proof. Let $\mathscr{H}$ be the Graver test set for the family of integer programs $IP(b, c, u)$. Let H_j denote the unique Hilbert basis H_j of the pointed cone $C_j = \{x \in O_j: Ax = 0\}$. A vector $v \in L \cap C_j$ is irreducible if and only if v cannot be written as the sum of other lattice vectors in C_j. This is true if and only if v is contained in H_j. $\square$

We will see in the next section that the notion of reducibility provides a way to determine Hilbert bases algorithmically. In this context a question arises concerning the complexity of deciding whether a vector is reducible.

Theorem 3.9 (Sebö [94]). *Given a pointed cone $C \subset \mathbb{R}^n$ and a vector $z \in C \cap \mathbb{Z}^n$, it is $co\mathcal{NP}$-complete to decide whether z is contained in the minimal Hilbert basis of C.*

Theorem 3.9 asserts the difficulty of deciding whether an integral vector is reducible. On the other hand, every augmentation vector can be decomposed into irreducible ones. In fact, we can write every integral vector in a pointed cone as a non-negative integer combination of at most $2n - 2$ irreducible vectors.

Theorem 3.10 (Sebö [94]). *Let C be a pointed cone in $\mathbb{R}^n$ and $H(C)$ its minimal Hilbert basis. Every integral point in C can be written as the non-negative integral combination of at most $2n - 2$ elements from $H(C)$.*

Theorem 3.10 improves a result of Cook et al. [21] who showed that every integral vector in a pointed n-dimensional cone is the non-negative integral combination of at most $2n - 1$ vectors from the minimal Hilbert basis.

From a result of Sebö [94] it follows that in dimensions $n = 2$ and 3 every integral vector in a pointed n-dimensional cone is the non-negative integral combination of at most n vectors from the Hilbert basis. This also holds for cones arising from perfect graphs [21] and a class of cones described in [54]. However, in general at least

$n + \lfloor 1/6 \cdot n \rfloor$ elements of the Hilbert basis are needed to represent any integral vector in the cone.

Theorem 3.11 (Bruns et al. [14]). *Let C be a pointed cone in $\mathbb{R}^n$ and $H(C)$ its minimal Hilbert basis. In general at least $n + \lfloor 1/6 \cdot n \rfloor$ elements from $H(C)$ are needed in order to represent any vector in $C \cap \mathbb{Z}^n$ as a non-negative integral combination of elements of $H(C)$.*

We have seen that Hilbert bases of rational polyhedral cones are central in the design of a test set. In fact, Hilbert bases play a central role in the theory of integer programming in general. Of major importance is their link to the integrality of polyhedra, i.e., to totally dual integral systems of inequalities.

Definition 3.5. Let $A \in \mathbb{Q}^{m \times n}$, $b \in \mathbb{Q}^m$. The system of inequalities $Ax \leqslant b$ is called *totally dual integral (TDI)* if for every $c \in \mathbb{Z}^m$ such that $|\min\{b^{\mathrm{T}} y : A^{\mathrm{T}} y = c,\ y \geqslant 0\}| < \infty$, there exists an integral vector $y^* \in \mathbb{Z}^m$, $A^{\mathrm{T}} y^* = c$, $y^* \geqslant 0$ with $b^{\mathrm{T}} y^* = \min\{b^{\mathrm{T}} y : A^{\mathrm{T}} y = c,\ y \geqslant 0\}$.

The TDIness of the system $Ax \leqslant b$ has an important consequence for polyhedra and a geometric meaning.

Theorem 3.12 (Edmonds and Giles [32]). *If $Ax \leqslant b$ is TDI and b is integral, then $\{x \in \mathbb{R}^n : Ax \leqslant b\}$ is integral.*

Let C be the cone generated by all the row vectors of A. Among all possible ways of writing any integral $c \in C$ as a conical combination of the row vectors of A, let S_c be the set of minimum weight combinations with respect to the function b, i.e.,

$$S_c = \{y \geqslant 0 : A^{\mathrm{T}} y = c \text{ and } b^{\mathrm{T}} y \text{ is minimal}\}.$$

Then $Ax \leqslant b$ is TDI if for every $c \in C \cap \mathbb{Z}^n$ there exists an integral vector in S_c. This geometric property can be expressed using Hilbert bases.

Theorem 3.13 (Giles and Pulleyblank [36]). *Let $A \in \mathbb{Q}^{m \times n}$ and $b \in \mathbb{Q}^m$. The system $Ax \leqslant b$ is TDI if and only if for every face F of $P = \{x \in \mathbb{R}^n : Ax \leqslant b\}$ the set of row vectors that determine F is a Hilbert basis of the cone generated by these row vectors. In fact, the converse of Theorem 3.12 is also true.*

Theorem 3.14 (Giles and Pulleyblank [36]). *If a rational polyhedron P is integral, then there exists a TDI system $Ax \leqslant b$ such that b is integral and $P = \{x \in \mathbb{R}^n : Ax \leqslant b\}$.*

Hilbert bases can also be used to estimate the distance between feasible solutions of an integer program.

Theorem 3.15 (Henk and Weismantel [53]). *Let $A \in \mathbb{Z}^{m \times n}$ with all subdeterminants at most α in absolute value and $b \in \mathbb{Z}^m$, $c \in \mathbb{Z}^n$. If $\tilde{x}$ is a feasible, non-optimal solution of the program $\min\{c^{\mathrm{T}}x : Ax \leqslant b, x \in \mathbb{Z}^n\}$, then there exists a feasible solution $\hat{x}$ such that $c^{\mathrm{T}}\hat{x} < c^{\mathrm{T}}\tilde{x}$ and $\|\hat{x} - \tilde{x}\|_\infty \leqslant (n-1)\alpha - (n-2)/(n^{n/2}\alpha^{n-2})$.*

The bound of Theorem 3.15 strengthens the bound of $n\alpha$ given in [22]. Its proof is based on an analysis of the *height* of a Hilbert basis, see [53] and also [74]. For further results about the structure and applications of Hilbert bases to combinatorial convexity, toric varieties and polynomial rings and ideals, we refer the reader to the papers mentioned above and to Schrijver [91], Liu [73], Bruns and Gubeladze [12], Firla and Ziegler [35], Henk and Weismantel [55], Dais et al. [29], Ewald [34], Oda [81], Sturmfels [98] and Bruns et al. [13]).

3.3. Hilbert bases versus Gröbner bases

We have seen that the Graver test set is naturally derived from a study of Hilbert bases of cones. There are two other ways of defining test sets that rely on a different mathematical approach. The *neighbors of the origin* define a test set that was introduced by Scarf [85,86]. It is based on a study of lattice point free convex bodies and establishes a beautiful link between the area of test sets and the geometry of numbers that we do not discuss here. The *reduced Gröbner basis* of an integer program is a test set obtained from a study of generators of polynomial ideals. The latter topic is a classical field of algebra. The reduced Gröbner basis of a toric ideal that one associates with an integral matrix A and a term order induced by c yields a test set for the family of integer programs

$$IP(b) \quad \min\{c^{\mathrm{T}}x : Ax = b, x \in \mathbb{Z}^n_+\}$$

associated with a fixed matrix $A \in \mathbb{Z}^{m \times n}$ and varying $b \in \mathbb{Z}^m$. The connection between test sets for integer programming and Gröbner bases of toric ideals was first established by Conti and Traverso [18]. This "algebraic view of test sets" is important from an algorithmic point of view. Reduced Gröbner bases of toric ideals can be computed by the Buchberger algorithm [15]. Reinterpreting the steps of this algorithm as operations on lattice vectors yields a combinatorial algorithm for computing test sets, see [101,103]. We consider here a geometric interpretation of Gröbner bases for integer programs. We refer to Cox et al. [28] and Becker and Weispfenning [8] for basics on Gröbner bases and on Buchberger's algorithm for polynomial ideals that motivated these constructions. As in the previous section, let L denote the lattice $\{x \in \mathbb{Z}^n : Ax = 0\}$.

In order to avoid technical difficulties we make the following two assumptions:

Assumption 3.1. c is generic, i.e., $c^{\mathrm{T}}x = 0$ for $x \in L$ if and only if $x = 0$. Moreover, A is a matrix in $\mathbb{Z}^{m \times n}$ such that $\{x \in \mathbb{Z}^n_+ : Ax = 0\} = \{0\}$. The latter assumption ensures that the integer program $IP(b)$ is bounded for every $b \in \mathbb{Z}^m$.

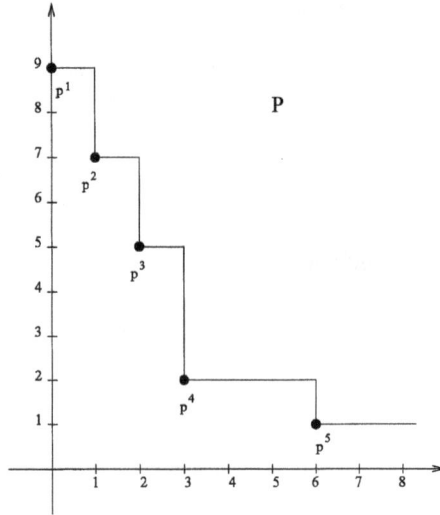

Fig. 14. The geometric structure of the set P.

Let P be the set of all non-negative integer points that are not optimal in $IP(b)$ for any value of b. More formally,

$$P = \{x \in \mathbb{Z}_+^n: \exists y \in \mathbb{Z}_+^n \text{ such that } Ay = Ax, \ c^T y < c^T x\}.$$

Note that the set P is well defined, because c is generic. The geometric structure of the set P can be nicely characterized, see Fig. 14.

Lemma 3.16 (Thomas [101]). *There exists a unique minimal finite set of vectors $p^1, \ldots, p^t$ in P such that*

$$P = \bigcup_{i=1}^{t} (p^i + \mathbb{Z}_+^n),$$

where $v + \mathbb{Z}_+^n := \{w \in \mathbb{Z}_+^n: w \geq v\}$. *Moreover, for any* $x \in \{p^1, \ldots, p^t\}$ *and* $y \in \mathbb{Z}_+^n$ *such that* $c^T y < c^T x$ *and* $Ay = Ax$, $\operatorname{supp}(y) \cap \operatorname{supp}(x) = \emptyset$.

Proof. Let $x \in P$. Then x does not attain the minimal objective function value with respect to c in the integer program with right hand side vector Ax. Let $y \in \mathbb{Z}_+^n$, be the unique minimal solution of $IP(Ax)$. Then for every $v \in \mathbb{Z}_+^n$ the vector $x + v$ is not minimal w.r.t. c because $c^T(y + v) < c^T(x + v)$ and $A(y + v) = Ay + Av = A(x + v)$. Therefore $x \in P$ implies that $(x + \mathbb{Z}_+^n) \subseteq P$. From Gordan's Lemma 3.5 we conclude that there exists a unique minimal and finite set of vectors $p^1, \ldots, p^t \in P$ such that $P \subseteq \bigcup_{i=1}^{t} (p^i + \mathbb{Z}_+^n)$. This shows that $P = \bigcup_{i=1}^{t} (p^i + \mathbb{Z}_+^n)$.

Let $x \in \{p^1, \ldots, p^t\}$ and $y \in \mathbb{Z}_+^n$ such that $c^T y < c^T x$ and $Ay = Ax$. Assuming that $k \in (\operatorname{supp}(y) \cap \operatorname{supp}(x))$ we have that $x - e^k \in \mathbb{Z}_+^n$ and $y - e^k \in \mathbb{Z}_+^n$. This implies that $x - e^k \in P$, a contradiction to the definition of $\{p^1, \ldots, p^t\}$. □

Taking into account the structure of the set P, we are ready to define a test set for the family of integer programs $IP(b)$ with varying $b \in \mathbb{Z}^m$.

Definition 3.6. Let $p^1, \ldots, p^t \in \mathbb{Z}^n_+$ be the set of vectors defined in Lemma 3.16 such that $P = \bigcup_{i=1}^t (p^i + \mathbb{Z}^n_+)$. For each $i \in \{1, \ldots, t\}$ let y^i denote the optimal solution with respect to c of the program $IP(A p^i)$. The set

$$G_c := \{y^i - p^i : i = 1, \ldots, t\}$$

is called the *reduced Gröbner basis* of the family of integer programs $IP(b)$.

Theorem 3.17 (Thomas [101]). *The reduced Gröbner basis G_c contains a test set for $IP(b)$ for every $b \in \mathbb{Z}^m$.*

Proof. Let $x \in P$. By Lemma 3.16 there exists $i \in \{1, \ldots, t\}$ and $v \in \mathbb{Z}^n_+$ such that $x = p^i + v$. So, $x' := x + (y^i - p^i)$ satisfies $Ax' = Ax$ and $c^T x' < c^T x$ since $c^T y^i < c^T p^i$. Moreover, $x' = y^i + v \in \mathbb{Z}^n_+$. It follows that x is improved by a vector from the set G_c. $\square$

Next, we show that the reduced Gröbner basis G_c is contained in the Graver test set $\mathcal{H}$.

Theorem 3.18 (Thomas [103]). *The reduced Gröbner basis G_c is contained in the Graver test set $\mathcal{H}$.*

Proof. Let $z \in G_c$ and $P = \bigcup_{i=1}^t (p^i + \mathbb{Z}^n_+)$. From Lemma 3.16 we have that $z = z^+ - z^-$ with $z^- \in \{p^1, \ldots, p^t\}$. Let O_j denote the orthant that contains z, $C_j = \{x \in O_j : Ax = 0\}$ and $H_j = H(C_j)$ be the minimal Hilbert basis of C_j. We conclude that $z \in C_j$. Suppose that $z \notin H_j$. Then $z = v + w$ where $v, w \in C_j \cap \mathbb{Z}^n$. As $c^T z < 0$, we can assume w.l.o.g. $c^T v < 0$. We obtain $v^+, v^- \in \mathbb{Z}^n_+$, $Av^+ = Av^-$ and $v^- \in P$. We have $z^- = v^- + w^-$. But $w^- \in \mathbb{Z}^n_+$ and $z^- \in \{p^1, \ldots, p^t\}$. This contradicts the definition of the set $\{p^1, \ldots, p^t\}$. Therefore $z \in H_j$. $\square$

Definition 3.7. The set

$$\mathscr{G} := \bigcup_{c \in \mathbb{R}^n, c \text{ generic}} G_c$$

is called the *universal Gröbner basis* associated with a matrix A.

Lemma 3.18 implies that $\mathscr{G}$ is contained in the Graver test set $\mathcal{H}$. In particular, $\mathscr{G}$ is finite. In fact, this containment relation is not always strict. Note, however, that the Graver test set is designed for a family of integer programs $\min \{c^T x; \ Ax = b, \ 0 \leqslant x \leqslant u, \ x \in \mathbb{Z}^n\}$ with varying upper bounds on the variables, whereas the universal Gröbner basis applies to a family of integer programs with no upper bounds on the variables $\min\{c^T x: Ax = b,\ 0 \leqslant x,\ x \in \mathbb{Z}^n\}$. To make a comparison between the two objects possible, one may transform a program $\min\{c^T x: Ax = b,\ 0 \leqslant x \leqslant u,\ x \in \mathbb{Z}^n\}$ into the form $\min\{c^T x + 0^T y: Ax + 0y = b,\ x + y = u,\ x, y \in \mathbb{Z}^n_+\}$. Then the universal Gröbner basis setting applies to the latter integer program in dimension $2n$. We obtain

Theorem 3.19 (Sturmfels and Thomas [99]). *Let $\tilde{G}$ be the universal Gröbner basis associated with the family of integer programs*

$$\min\{\tilde{c}^T(x, y): Ax + 0y = b,\ x + y = u,\ (x, y) \in \mathbb{Z}^{2n}_+\}$$

with varying $u \in \mathbb{Z}^n$, $b \in \mathbb{Z}^n$ and generic $\tilde{c} \in \mathbb{R}^{2n}$. Let $\mathcal{H}$ be the Graver test set associated with the family of integer programs

$$\min\{c^T x: Ax = b,\ 0 \leqslant x \leqslant u,\ x \in \mathbb{Z}^n\}$$

with varying $u \in \mathbb{Z}^n$, $b \in \mathbb{Z}^n$ and $c \in \mathbb{R}^n$. Then $(x, -x) \in \tilde{G}$ if and only if $x \in \mathcal{H}$.

Example 3.4. Consider the family of integer programs with varying $c \in \mathbb{R}^3$, c generic and $b \in \mathbb{Z}$ of the form

$$\max\{c_1 x_1 + c_2 x_2 + c_3 x_3: x_1 + x_2 + 2x_3 = b,\ x \in \mathbb{Z}^3_+\}.$$

In this example the Graver test set $\mathcal{H}$ is the set

$$\mathcal{H} = \{\pm(1, -1, 0,\ ), \pm(2, 0, -1), \pm(0, 2, -1), \pm(1, 1, -1)\}.$$

The universal Gröbner basis $\mathcal{G}$ is the set

$$\mathcal{G} = \{\pm(1, -1, 0,\ ), \pm(0, 2, -1), \pm(2, 0, -1)\}.$$

To see this, note that $\pm(1, 1, -1) = \pm\frac{1}{2}(0, 2, -1) \pm \frac{1}{2}(2, 0, -1)$. In fact, $(1, 1, 0) = \frac{1}{2}(0, 2, 0) + \frac{1}{2}(2, 0, 0)$, i.e., $(1, 1, 0)$ cannot be a vertex of $\text{conv}\{x \in \mathbb{Z}^3_+: x_1 + x_2 + 2x_3 = 2\}$. Accordingly, $(-1, -1, 1) \notin \mathcal{G}$, because for any objective function $c \in \mathbb{R}^3$, c generic such that $(1, 1, 0) \in P$, we have that $(1, 0, 0) \in P$ or $(0, 1, 0) \in P$. Therefore, $(1, 1, 0) \notin \{p^1, \ldots, p^t\}$, see Lemma 3.16. The six remaining vectors in $\mathcal{H}$ are also contained in $\mathcal{G}$, because there exist objective functions c for which these vectors define differences of a point $p^i \in P$ and the optimal solution y^i, see Definition 3.6. The example demonstrates that $\mathcal{G} \subseteq \mathcal{H}$, but $\mathcal{H}$ can be strictly bigger than $\mathcal{G}$. $\square$

The universal Gröbner basis $\mathcal{G}$ can be characterized geometrically. This is made precise in Theorem 3.20. For its proof in various versions we refer to the papers of Sturmfels and Thomas [99], Thomas and Weismantel [102], Sturmfels et al. [100].

Theorem 3.20 (Sturmfels and Thomas [99]). *Let $A \in \mathbb{Z}^{m \times n}$ of rank m.*

(i) *Let $z = z^+ - z^- \in \mathcal{G}$. Then z^+ and z^- are vertices of the polyhedron $\text{conv}\{x \in \mathbb{Z}^n_+: Ax = Az^+\}$. Moreover, the line joining z^+ and z^- is an edge of the polyhedron $\text{conv}\{x \in \mathbb{Z}^n_+: Ax = Az^+\}$.*

(ii) *Let z^1 and z^2 be two adjacent vertices of $\text{conv}\{x \in \mathbb{Z}^n_+: Ax = Az^1\}$, then $(z^1 - z^2)/\gcd(z^1 - z^2) \in \mathcal{G}$.*

We mentioned that Conti and Traverso [18] established the connection between test sets of integer programs and Gröbner bases of toric ideals. The latter objects can be computed by the Buchberger algorithm [15]. We discuss below a combinatorial variant of this procedure that allows us to determine a superset of the Graver test set and therefore a superset of the universal Gröbner basis for the family of integer programs $IP(b, c, u)$ and $IP(b)$, respectively. Starting with input $T := \{\pm e^i : i = 1, \ldots, n\}$, we repeatedly take all the sums of two vectors in T, reduce each of these vectors as much as possible by the elements of T and add all the reduced vectors that are different from the origin to the set T. On termination the set T contains the set of all irreducible vectors w.r.t. the matrix A.

Algorithm 3.3. *Input*: $A \in \mathbb{Z}^{m \times n}$.
 Output: A finite set T containing all the irreducible vectors w.r.t. A.
(1) Set $T_{\text{old}} := \emptyset$ and $T := \{\pm e^i : i = 1, \ldots, n\}$.
(2) While $T_{\text{old}} \neq T$ repeat the following steps:
  (2.1) Set $T_{\text{old}} := T$.
  (2.2) For all pairs of vectors $v, w \in T_{\text{old}}$, set $z := v + w$:
    (2.2.1) While there exists $y \in T \setminus \{z\}$ reducing z, set $z := z - y$.
    (2.2.2) If $z \neq 0$, update $T := T \cup \{z\}$.

Algorithm 3.3 is a simple combinatorial variant of a Buchberger type algorithm. We refer to [103,25] for earlier versions of this algorithm as well as other proofs of their correctness. We first illustrate the performance of Algorithm 3.3 on a small example.

Example 3.5. Consider the family of integer programs

$$\min\{c_1 x_1 + c_2 x_2 + c_3 x_3 : x_1 + 2x_2 + 3x_3 = b,\ 0 \leqslant x \leqslant u,\ x \in \mathbb{Z}_+^3\},$$

with varying $b \in \mathbb{Z}_+$, $u \in \mathbb{Z}_+^3$, $c \in \mathbb{R}^3$. Algorithm 3.3 starts with all the unit vectors,

$$T = \{\pm e^1, \pm e^2, \pm e^3\}.$$

Taking all sums of vectors of G gives rise after reduction to a new set

$$T = \{\pm e^1, \pm e^2, \pm e^3, \pm(e^1 - e^2), \pm(e^1 - e^3), \pm(e^2 - e^3)\}.$$

Note that for $i = 1, 2, 3$ the vectors $2e^i$ reduce to 0. Also $(e^1 + e^2)$ can be reduced by e^1, and a vector of the form $(e^i + e^3)$ is reducible by both e^i and by e^3. With the updated set T we again perform the operation of taking all the sums of vectors of T and checking for reducibility. This yields after reduction an updated set

$$T = T_{\text{old}} \cup \{\pm(2e^1 - e^2), \pm(2e^1 - e^3), \pm(2e^2 - e^3), \pm(e^1 + e^2 - e^3)\}.$$

Denoting this set T by T_{old}, taking the sums of vectors $\pm(e^1 + [2e^2 - e^3])$, $\pm(e^1 + [-2e^2 + e^3])$ and $\pm([e^2 - e^3] + [2e^2 - e^3])$ yields three additional vectors that are irreducible and added to T. All other sums of vectors of T_{old} can be reduced by T to 0. Algorithm 3.3 terminates with the following set:

$$
\begin{aligned}
T = \{ \quad &\pm e^1, \pm e^2, \pm e^3, \\
&\pm(e^1 - e^2), \pm(e^1 - e^3), \pm(e^2 - e^3), \\
&\pm(2e^1 - e^2), \pm(2e^1 - e^3), \pm(2e^2 - e^3), \pm(e^1 + e^2 - e^3), \\
&\pm(3e^1 - e^3), \pm(3e^2 - 2e^3), \pm(e^1 + e^3 - 2e^2) \quad \}.
\end{aligned}
\tag{77}
$$

The Graver test set for this family of knapsack problems is

$$
\{\pm(2e^1 - e^2), \pm(3e^1 - e^3), \pm(e^1 + e^3 - 2e^2), \pm(3e^2 - 2e^3), \pm(e^1 + e^2 - e^3)\},
$$

namely a subset of the vectors in T satisfying $x_1 + 2x_2 + 3x_3 = 0$.   □

Theorem 3.21. *Algorithm 3.3 is finite. The set T that is returned by the algorithm contains the set of all irreducible vectors w.r.t. the matrix A.*

Proof. Let G denote the set of all irreducible elements w.r.t. A. Let $t \in G$. Let T^u denote the current set T of Algorithm 3.3 before uth performance of Step (2). We remark that $\{\pm e^i : i = 1, \ldots, n\} \subseteq T^u$. Therefore, there exists a multiset $S = \{t^1, \ldots, t^k\} \subseteq T^u$ such that

$$
t = t^1 + \cdots + t^k.
$$

For every multiset $S_t = \{t^1, \ldots, t^k\} \subseteq T^u$ with $t = \sum_{i=1}^{k} t^i$, let

$$
\phi(S_t) := \sum_{i=1}^{k} (\|At^i\|_1 + \|t^i\|_1).
$$

Let S_t^* denote a multiset such that $\phi(S_t^*)$ is minimal. Note that $t \in G$ if and only if t is irreducible. On account of Definition 3.4, t is irreducible if and only if for all decompositions of the form $t = \sum_{i=1}^{k} t^i$ the following condition holds:

$$
\|At\|_1 + \|t\|_1 < \sum_{i=1}^{k} (\|At^i\|_1 + \|t^i\|_1).
$$

We conclude that

$$
\phi(S_t^*) > \|At\|_1 + \|t\|_1 \quad \Leftrightarrow \quad t \notin T^u.
$$

However, if $t \notin T^u$, then there exist indices $i, j \in \{1, \ldots, k\}$ such that the vectors (t^i, At^i) and (t^j, At^j) lie in different orthants of $\mathbb{R}^{n+m}$. On account of the minimality of $\phi(S_t^*)$, $t^i + t^j$ is neither in T^u, nor can $t^i + t^j$ be written as the sum of elements from T^u all of which reduce $t^i + t^j$. However $z = t^i + t^j$ will be considered in the subsequent performance of Step (2.2.1) of the algorithm. Then z will be added to T^u and the value $\phi(S_t^*)$ will decrease by at least one. Since $\phi(S_t^*) > \|At\|_1 + \|t\|_1$ for all iterations of Step (2) in which $t \notin T^u$, the algorithm will detect t in a finite number of steps. These arguments apply to any irreducible vector. There is only a finite number of irreducible vectors, and hence, the algorithm is finite. □

We have seen that at least from a theoretical point of view a primal approach for integer programming can be designed using test sets.

Although test sets are finite under modest assumptions, they are usually huge. This fact is not surprising because minimal test sets w.r.t. inclusion can be obtained as the union of Hilbert bases of rational polyhedral cones. In particular, the union is taken over all orthants of the space of variables. This is of course already by definition an exponential construction.

On the other hand, the primal approach for combinatorial optimization problems has been the starting point for many deep algorithmic results related to the theory about flows, matroids, and matchings. The practical applicability of this theoretical knowledge for general combinatorial or integer programs has however never been proved. Primal cutting plane algorithms for general integer programs were developed by Gomory [39] and Young [108,109] in the 1960s. A modified version of Young's algorithm with combinatorial cuts instead of Gomory cuts has been applied to the TSP [83]. The computational performance of the latter algorithm is interesting, and—from our point of view—there is no obvious reason why primal cutting plane methods should not be further pursued.

Today's commercial integer programming algorithms based on branch-and-cut are essentially dual algorithms, and little theory enters the design of the primal phase. As far as we know, they use mainly simple rounding heuristics to generate feasible solutions. This indicates that there is a significant gap between theory and practice regarding primal algorithms for integer and combinatorial programming.

Very recently, an algorithm for solving general integer programs has been proposed that is based on integral generating sets [50–52]. It is an exact and finite algorithm that generalizes the primal algorithm for set partitioning [6]. Starting off with a feasible solution, the algorithm replaces iteratively one column by other columns that correspond to irreducible solutions of a system of linear inequalities. The algorithm terminates by either returning an augmenting direction or providing a proof that the current point is optimal. Preliminary computational results on various instances of the MIPLIB show that this primal approach is able to tackle certain medium-sized integer programs.

This suggests that in the future augmentation algorithms and test sets may be able to contribute to the solution of general integer programs.

4. Group relaxations, corner polyhedra and subadditivity

Here we look at relaxations in which the nonnegativity constraints are dropped on a subset of the variables. We then present the Gomory group relaxation [40]. Though the subject of groups may be new to some, all the reader needs to understand is linear equations in integers under addition modulo integers. After discussing briefly algorithms based on the group relaxation, we study the corner polyhedron [41,42], which is the convex hull of solutions to the group problem. The study of the structure of valid inequalities for the corner polyhedron indicates the importance of subadditivity. Finally we briefly consider solving (IP) for all possible vectors b, and the question of how to choose the set of variables on which the nonnegativity constraints should be relaxed.

4.1. A family of relaxations and a canonical form

Consider the integer program

$$IP(b) \quad z = \min\{c^{\mathrm{T}}x\colon Ax = b,\ x \in \mathbb{Z}_+^{m+n}\},$$

where A is an $m \times (m+n)$ integer matrix, b is an integer m vector and $V = \{1, \ldots, m+n\}$ is the index set of the variables. One idea is to drop the nonnegativity constraints on a subset $S \subseteq V$ of the variables, which leads to the relaxation

$$IP_S(b) \quad z_S = \min\{c^{\mathrm{T}}x\colon Ax = b,\ x_j \in \mathbb{Z}_+^1 \text{ for } j \in V \setminus S,\ x_j \in \mathbb{Z}^1 \text{ for } j \in S\}.$$

This can be viewed as a special case of a more general family of relaxations

$$IP_{K,\beta}(b) \quad z_{K,\beta} = \min\{c^{\mathrm{T}}x + \beta^{\mathrm{T}}w\colon Ax + Kw = b,\ x \in \mathbb{Z}_+^{m+n},\ w \in \mathbb{Z}^p\}$$

where K is an integral $m \times p$ matrix. Two special cases of potential interest are the case where $K = A_S$, the submatrix of A indexed by the set S, $\beta = c_S$ and $IP_{K,\beta}(b)$ reduces to $IP_S(b)$, and the case in which $K = \binom{0}{I^{(m-1)}}$, $\beta = 0$ and $IP_{K,\beta}(b)$ reduces to a knapsack relaxation

$$z^{KN} = \min\{c^{\mathrm{T}}x\colon a^1 x = b_1,\ x \in \mathbb{Z}_+^{m+n}\},$$

where a^1 denotes the first row of A.

Instead of working with $IP_{K,\beta}(b)$, it is possible to work with its projection $X_K(b)$ on the space of x variables. Suppose that $p \leqslant m$ and that there exists a dual feasible vector $u \in R^m$ with $u^{\mathrm{T}}A \leqslant c^{\mathrm{T}}$ and $u^{\mathrm{T}}K = \beta^{\mathrm{T}}$. Now $IP_{K,\beta}(b)$ can be rewritten as

$$z_{K,\beta}(b) = u^{\mathrm{T}}b + \min\{(c^{\mathrm{T}} - u^{\mathrm{T}}A)x\colon x \in X_K(b)\},$$

where

$$X_K(b) = \{x \in Z_+^{m+n}\colon Ax + Kw = b \text{ for some } w \in Z^p\}.$$

To see the structure of $X_K(b)$ we make use of the Smith normal form of a matrix. Remember that a square integral matrix C is unimodular if $|\det C| = 1$, and if x, $y \in \mathbb{Z}_+^1 \setminus \{0\}$, and $x|y$ means that y is an integer multiple of x.

Theorem 4.1 (Smith [97]). *Given an $m \times p$ integer matrix K of rank $p \leqslant m$, there exist unimodular integer matrices R and C with R an $m \times m$ matrix, C a $p \times p$ matrix such that $RKC = \binom{\Delta}{0}$ where the diagonal matrix Δ has diagonal elements $\delta_i \in \mathbb{Z}_+^1$ for $i = 1, \ldots, p$ with $\delta_1|\delta_2 \cdots |\delta_p$, and Δ is unique.*

Now we can derive a canonical representation of $X_K(b)$. Here a^j denotes the jth column of A and $(\rho)_i$ is the ith coordinate of the vector ρ.

Theorem 4.2.

$$X_K(b) = \left\{ x \in \mathbb{Z}_+^{m+n}: \sum_{j=1}^{m+n} (Ra^j)_i x_j \equiv (Rb)_i \,(\mathrm{mod}\,\delta_i) \; for \; i = 1, \ldots, p, \right.$$

$$\left. \sum_{j=1}^{m+n} (Ra^j)_i x_j = (Rb)_i \; for \; i = p+1, \ldots, m \right\}$$

with $Ra^j, Rb \in \mathbb{Z}^m$ for $j = 1, \ldots, m+n$.

Proof. Observe that

$$RAx + RKw = Rb, \quad x \in \mathbb{Z}_+^{m+n}, \; w \in \mathbb{Z}^p$$

can be rewritten as

$$RAx + RKCC^{-1}w = Rb, \quad x \in \mathbb{Z}_+^{m+n}, \; C^{-1}w \in \mathbb{Z}^p$$

since C is unimodular. Now setting $v = -C^{-1}w$ and $\Delta' = \binom{\Delta}{0}$, this becomes

$$RAx = Rb + \Delta'v, \quad x \in \mathbb{Z}_+^{m+n}, \; v \in \mathbb{Z}^p,$$

where RA and Rb are integer as R is unimodular. $\square$

Note that when $K = A_S$, it is more natural to look at the feasible region in the space of the variables $x_{V \setminus S}$. Now

$$X_{V \setminus S} = \{ x_{V \setminus S} \in \mathbb{Z}_+^{|V \setminus S|}: A_{V \setminus S} x_{V \setminus S} + A_S x_S = b \text{ for some } x_S \in \mathbb{Z}^{|S|} \}$$

$$= \{ x_{V \setminus S} \in \mathbb{Z}_+^{|V \setminus S|}: RA_{V \setminus S} x_{V \setminus S} \equiv Rb \,(\mathrm{mod}\,\Delta') \},$$

where $(\mathrm{mod}\,\Delta')$ means $(\mathrm{mod}\,\delta_i)$ for rows $i = 1, \ldots, p$, and equality in rows $p+1, \ldots, m$.

4.2. Gomory's asymptotic group relaxation

Taking the integer program $IP(b)$, let $A = (A_B, A_N)$ with A_B an optimal *LP* basis. Now $IP(b)$ can be rewritten as

$$z = c_B^{\mathrm{T}} A_B^{-1} b + \min \sum_{j \in N} (c_j - c_B^{\mathrm{T}} A_B^{-1} a_j) x_j,$$

$$x_B + A_B^{-1} A_N x_N = A_B^{-1} b, \quad x_B \in \mathbb{Z}_+^m, \; x_N \in \mathbb{Z}_+^n,$$

where $N = \{1, \ldots, n\}$ is the set of nonbasic variables. The Gomory group relaxation [40] is obtained by dropping the nonnegativity constraint on x_B. It can be written as

$$z^G = c_B^{\mathrm{T}} A_B^{-1} b + \min \sum_{j \in N} \bar{c}_j x_j,$$

$$\sum_{j \in N} (A_B^{-1} a^j) x_j \equiv A_B^{-1} b \,(\mathrm{mod}\,1), \quad x \in \mathbb{Z}_+^n,$$

where $\bar{c}_j = (c_j - c_B^T A_B^{-1} a^j) \geqslant 0$ for $j \in N$ as the LP basis is optimal. Equivalently using the description following Theorem 4.2, the feasible region can be written in canonical form

$$X_N = \{x_N \in \mathbb{Z}_+^n : RA_N x_N \equiv Rb \,(\mathrm{mod}\,\Delta)\}.$$

Example 4.1. Consider the IP

$$
\begin{aligned}
z = \min \; & - 6x_5 - 4x_6 - x_1 + 4x_2 \qquad\quad + 2x_4, \\
& 5x_5 + 3x_6 + 2x_1 - 4x_2 + x_3 \qquad\quad = 5, \\
& x_5 + 2x_6 - 3x_1 + 5x_2 \qquad\;\; + x_4 = 2, \\
& x \in \mathbb{Z}_+^6
\end{aligned}
$$

$x_B = (x_5, x_6)$ are optimal basic variables to the LP relaxation, so we obtain

$$
\begin{aligned}
z = \min - \tfrac{44}{7} + \tfrac{3}{7}x_1 + \tfrac{6}{7}x_2 + \tfrac{8}{7}x_3 + \tfrac{16}{7}x_4, \\
x_5 + \tfrac{13}{7}x_1 - \tfrac{23}{7}x_2 + \tfrac{2}{7}x_3 - \tfrac{3}{7}x_4 = \tfrac{4}{7}, \\
x_6 - \tfrac{17}{7}x_1 + \tfrac{29}{7}x_2 - \tfrac{1}{7}x_3 + \tfrac{5}{7}x_4 = \tfrac{5}{7}, \quad x \in \mathbb{Z}_+^6.
\end{aligned}
$$

The Gomory group relaxation is thus

$$
\begin{aligned}
z^G = -\tfrac{44}{7} + \min \quad & \tfrac{3}{7}x_1 + \tfrac{6}{7}x_2 + \tfrac{8}{7}x_3 + \tfrac{16}{7}x_4, \\
& \tfrac{13}{7}x_1 - \tfrac{23}{7}x_2 + \tfrac{2}{7}x_3 - \tfrac{3}{7}x_4 = \tfrac{4}{7} \,(\mathrm{mod}\,1), \\
& - \tfrac{17}{7}x_1 + \tfrac{29}{7}x_2 - \tfrac{1}{7}x_3 + \tfrac{5}{7}x_4 = \tfrac{5}{7} \,(\mathrm{mod}\,1), \\
& x \in \mathbb{Z}_+^4,
\end{aligned}
$$

which simplifies to

$$
\begin{aligned}
z^G = -\tfrac{44}{7} + \min \; & \tfrac{3}{7}x_1 + \tfrac{6}{7}x_2 + \tfrac{8}{7}x_3 + \tfrac{16}{7}x_4, \\
& \tfrac{6}{7}x_1 + \tfrac{5}{7}x_2 + \tfrac{2}{7}x_3 + \tfrac{4}{7}x_4 = \tfrac{4}{7} \,(\mathrm{mod}\,1), \\
& \tfrac{4}{7}x_1 + \tfrac{1}{7}x_2 + \tfrac{6}{7}x_3 + \tfrac{5}{7}x_4 = \tfrac{5}{7} \,(\mathrm{mod}\,1), \quad x \in \mathbb{Z}_+^4.
\end{aligned}
$$

Now consider the Gomory group relaxation in canonical form. With

$$
R = \begin{pmatrix} 1 & 0 \\ 3 & -1 \end{pmatrix} \quad \text{and} \quad C = \begin{pmatrix} -1 & 3 \\ 2 & -5 \end{pmatrix}, \quad \Delta = RA_B C = \begin{pmatrix} 1 & 0 \\ 0 & 7 \end{pmatrix},
$$

so we obtain

$$
\begin{aligned}
z^G = -\tfrac{44}{7} + \min \; & \tfrac{3}{7}x_1 + \tfrac{6}{7}x_2 + \tfrac{8}{7}x_3 + \tfrac{16}{7}x_4 \\
& 2x_1 - 4x_2 + 1x_3 + 0x_4 = 5 \,(\mathrm{mod}\,1), \\
& 9x_1 - 17x_2 + 3x_3 - 1x_4 = 13 \,(\mathrm{mod}\,7), \quad x \in \mathbb{Z}_+^4.
\end{aligned}
$$

As all $x \in \mathbb{Z}^4$ satisfy $\sum_j a_j x_j = b \,(\mathrm{mod}\,1)$ when all the coefficients a_j and b are integers, the first equation is redundant, so the problem can finally be written as

$$
\begin{aligned}
z^G = -\tfrac{44}{7} + \min \; & \tfrac{3}{7}x_1 + \tfrac{6}{7}x_2 + \tfrac{8}{7}x_3 + \tfrac{16}{7}x_4, \\
& 2x_1 + 4x_2 + 3x_3 + 6x_4 = 6 \,(\mathrm{mod}\,7), \quad x \in \mathbb{Z}_+^4. \qquad \square
\end{aligned}
$$

The canonical form that we have derived for the group problem is not surprising given that all finite commutative groups reduce to sets of integer vectors under addition modulo some given integer vector. More precisely the following classical theorem gives a complete classification of all finite abelian groups.

Theorem 4.3. *Every finite abelian* (*commutative*) *group G is isomorphic to the group consisting of integer p vectors under addition modulo* $(\delta_1, \delta_2, \ldots, \delta_p)$ *for some p with* $\delta_i \in \mathbb{Z}_+^1 \setminus \{0, 1\}$ *and* $\delta_1 | \delta_2 |, \ldots, | \delta_p$. *Such a group is denoted by* $\mathbb{Z}_{\delta_1} \times \cdots \times \mathbb{Z}_{\delta_p}$.

Thus we see that the canonical form provides an explicit description of the group, and hence Gomory chose to speak of the "group relaxation" which has the general form

$$IP_G(g_0) \qquad \begin{aligned} z^G &= \min \sum_{j \in N} \bar{c}_j x_j, \\ &\sum_{j \in N} g_j x_j = g_0 \quad \text{in } G, \quad x \in \mathbb{Z}_+^n. \end{aligned}$$

where $g_j \in G$ for $j \in N$ and $g_j x_j$ denotes $g_j + \cdots + g_j$ added x_j times with addition in the group.

4.3. Algorithms based on group relaxations

The group relaxation $IP_G(g_0)$ can be viewed as a shortest path (or minimum cost) problem in a graph with $|G| = \prod_{i=1}^p \delta_i = |\det A_B|$ nodes. For each $g \in G$, there is an arc $(g, g + g_j)$ with length (cost) $\bar{c}_j \geq 0$, and the problem is to find a shortest path from $0 \in G$ to $g_0 \in G$. In Fig. 15, the shortest path from 0 to 2 gives an optimal solution to the group problem:

$$\min\{3x_1 + 7x_2 : 1x_1 + 3x_2 \equiv 2 \,(\mathrm{mod}\, 5), \ x \in \mathbb{Z}_+^2\}.$$

Proposition 4.4. *Using a shortest path algorithm, $IP_G(g_0)$ can be solved for all $g_0 \in G$ with* $O(n|\det A_B|)$ *operations.*

Note that the original problem $IP(b)$ and the relaxations $IP_{K,\beta}(b)$ can also be viewed, a priori, as shortest path problems on an infinite graph whose nodes are the vectors $d \in \mathbb{Z}^m$. For $IP(b)$, there is an arc $(d, d + a^j)$ with cost c_j for all $d \in \mathbb{Z}^m$ and $j = 1, \ldots, m + n$, and the problem is to find a shortest path from 0 to b. To obtain $IP_{K,\beta}$, the original problem $IP(b)$ is relaxed by adding additional arcs $(d, d \pm k^i)$ with cost $\pm \beta_i$ for each column k^i of K and for all $d \in \mathbb{Z}^m$.

Something can also be said about the magnitude of solutions to the group problems.

Observation 4.1. *There exists an optimal solution $\bar{x}$ to the group problem $IP_G(g_0)$ with* $\prod_{j \in N}(1 + \bar{x}_j) \leq |\det A_B|$.

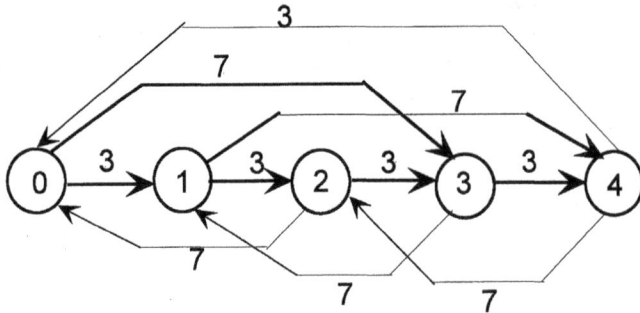

Fig. 15. Shortest path group problem.

Obviously one hopes that a solution to the group problem $IP_G(g_0)$ provides a solution of the original problem $IP(b)$.

Observation 4.2. *If $\bar{x}_N$ solves the group problem $IP_G(g_0)$, then by construction $\bar{x}_B = A_B^{-1}b - A_B^{-1}A_N\bar{x}_N \in \mathbb{Z}^m$. If in addition $\bar{x}_B \geqslant 0$, then $(\bar{x}_B, \bar{x}_N)$ solves $IP(b)$.*

From this, we see that a solution $\bar{x}_N$ to $IP_G(g_0)$ leads to a solution of the original problem $IP(b)$ for an infinity of values of b.

Observation 4.3. *Let $D = \{d \in \mathbb{Z}^m: A_B^{-1}d \geqslant A_B^{-1}N\bar{x}_N\}$. Then for all $b' \in D$ for which $Rb' \,(\mathrm{mod}\,\Delta) \equiv Rb \,(\mathrm{mod}\,\Delta)$, $(x_B, x_N) = (A_B^{-1}b' - A_B^{-1}A_N\bar{x}_N, \bar{x}_N)$ is optimal in $IP(b')$.*

When the original problem is not solved, $IP(b)$ has been reduced to finding the least cost group solution x^* with $A_B^{-1}b - A_B^{-1}A_N x^* \geqslant 0$. Based on this, and using the optimal value of the group problem for all right hand sides to obtain bounds, a branch-and-bound algorithm for $IP(b)$ was developed in [45]. A best bound variant of this approach based on finding the kth best solution to the group problem appears in [106].

Another natural approach is to systematically tighten the relaxation every time that the solution to the relaxed problem is infeasible in $IP(b)$. Specifically if $(\bar{x}_B)_u = (A_B^{-1}b)_u - (A_B^{-1}A_N\bar{x})_u < 0$, one can drop column u from A_B, and create a new relaxation in which the non-negativity of x_{B_u} is taken into account. However repeating this, we may be unlucky and return to the original problem $IP(b)$. An alternative that maintains finite group relaxations of the form $P_{\bar{G}}$ is just to increase the size of the group $\bar{G}$ so that $\bar{x}$ is no longer a feasible solution. One way to do this systematically is described in [9], see also [80].

Example 4.1 (continued). An optimal solution of the group relaxation is

$$x_1 = x_2 = 1, \quad x_3 = x_4 = 0.$$

Note that

$$\begin{pmatrix} x_5 \\ x_6 \end{pmatrix} = \begin{pmatrix} \frac{4}{7} \\ \frac{5}{7} \end{pmatrix} - A_B^{-1}A_N x_N = \begin{pmatrix} \frac{4}{7} \\ \frac{5}{7} \end{pmatrix} - \begin{pmatrix} \frac{10}{7} \\ \frac{12}{7} \end{pmatrix} = \begin{pmatrix} 2 \\ -1 \end{pmatrix}.$$

Thus the solution of the group relaxation is not feasible in the original problem, and thus $z_G = -\frac{44}{7} + \frac{9}{7} = -5$ is only a lower bound on $z(b)$. $\square$

4.4. The corner polyhedron

Given the basis A_B, the feasible region of the original problem $IP(b)$ can be viewed in the space of non-basic variables as

$$\tilde{X}(b) = \{x_N \in \mathbb{Z}_+^n : x_B = A_B^{-1}b - A_B^{-1}A_N x_N \in \mathbb{Z}_+^m\}$$

whereas the feasible region of the group problem is

$$\tilde{X}_G(g_0) = \{x_N \in \mathbb{Z}_+^n : x_B = A_B^{-1}b - A_B^{-1}A_N x_N \in \mathbb{Z}^m\}$$

$$= \left\{x_N \in \mathbb{Z}_+^n : \sum_{j\in N} g_j x_j = g_0 \text{ in } G\right\}$$

with $\tilde{X}(b) \subseteq \tilde{X}_G(g_0)$, and thus $\operatorname{conv}(\tilde{X}(b)) \subseteq \operatorname{conv}(\tilde{X}_G(g_0))$. This suggests the study of the *corner polyhedron* $\tilde{P}_G(g_0) = \operatorname{conv}(\tilde{X}_G(g_0))$, first introduced in [41]. Then in a remarkable paper [42], Gomory studied the valid inequalities and facet-defining inequalities of the corner polyhedron, introducing several of the ideas that have now become standard in polyhedral combinatorics, projection onto faces, subadditivity, master polytopes, using automorphisms to generate one facet from another, lifting, etc. For basic definitions and properties of valid inequalities, polyhedra, etc., see [80,91].

A first result about the form of the facet-defining inequalities is obtained by noting that if $g \in \tilde{X}_G(g_0)$, then $g + |G|e_j \in \tilde{X}_G(g_0)$ for all $j \in N$, and so the unit vectors e_j for $j \in N$ are extreme rays of the corner polyhedron.

Observation 4.4. *Except for non-negativity constraints $x_j \geqslant 0$ for $j \in N$, all facet-defining inequalities of $\tilde{P}_G(g_0)$ are of the form $\sum_{j\in N} \pi_j x_j \geqslant \pi_0$ with $\pi_j \geqslant 0$ for $j \in N$ and $\pi_0 > 0$.*

Gomory also introduced the idea of Master Polytopes for a given group G. Specifically let

$$X_G(g_0) = \left\{y \in \mathbb{Z}_+^{|G|} : \sum_{g\in G} g y_g = g_0 \text{ in } G\right\}$$

be a group problem in which every group element appears. Its convex hull

$$P_G(g_0) = \operatorname{conv}(X_G(g_0))$$

is called the *Master polytope*. The following theorem says that the Master polytope for G with right hand side g_0 provides the convex hull $\tilde{P}_G(g_0)$ for all instances of a group problem over the group G with right hand side g_0. Specifically this follows because $\tilde{p}_G(g_0) = p_G(g_0) \cap \{y : y_g = 0 \text{ for } g \notin \{g_1, \ldots, g_n\}\}$, defines a face of $P_G(g_0)$, and is therefore integral.

Theorem 4.5. *If* $P_G(g_0) = \{y \in \mathbb{R}_+^{|G|} : \sum_{g \in G} \pi_g^k y_g \geq \pi_0^k \text{ for } k = 1, \ldots, K\}$, *then* $\tilde{P}_G(g_0) = \{x \in \mathbb{R}_+^n : \sum_{j \in N} \pi_{g_j}^k x_j \geq \pi_0^k \text{ for } k = 1, \ldots, K\}$.

Gomory also derived a characterization for the facets of the Master polytope.

Theorem 4.6. *Let* $\{t^q\}_{q=1}^Q$ *be the vertices of* $P_G(g_0)$, *then* $\sum_{g \in G} \pi_g y_g \geq 1$ *is facet-defining if and only if* $\pi \in \mathbb{R}^{|G|}$ *is a basic feasible solution (vertex) of the polyhedron*

$$\sum_{g \in G} \pi_g t_g^q \geq 1, \quad q = 1, \ldots, Q,$$

$$\pi_g \geq 0 \quad \text{for } g \in G \setminus \{0\}, \quad \pi_0 = 0.$$

Theorem 4.7. *The inequality* $\sum_{g \in G} \pi_g y_g \geq 1$ *is facet defining for* $P_G(g_0)$ *with* $g_0 \neq 0$ *if and only if* $\pi \in \mathbb{R}^{|G|}$ *is an extreme point of the polyhedron*:

$$\pi_{g_1} + \pi_{g_2} \geq \pi_{g_1 + g_2} \quad \text{for } g_1, g_2 \in G \setminus \{0, g_0\},$$

$$\pi_g + \pi_{g_0 - g} = 1 \quad \text{for } g \in G \setminus \{0\},$$

$$\pi_g \geq 0 \quad \text{for } g \in G \setminus \{0\}, \quad \pi_0 = 0, \quad \pi_{g_0} = 1.$$

Example 4.2. Take $G = \mathbb{Z}_6$ and the master set $X_G(3)$:

$$0y_0 + 1y_1 + 2y_2 + 3y_3 + 4y_4 + 5y_5 \equiv 3 \,(\mathrm{mod}\, 6), \quad y \in \mathbb{Z}_+^6.$$

The polyhedron of Theorem 4.7 takes the form

$$
\begin{aligned}
2\pi_1 & & & & & \geq \pi_2, \\
\pi_1 &+ \pi_2 & & & & = 1, \\
\pi_1 &+ & \pi_4 & & & \geq \pi_5, \\
&2\pi_2 & & & & \geq \pi_4, \\
&\pi_2 & & &+ \pi_5 & \geq \pi_1, \\
& & 2\pi_4 & & & \geq \pi_2, \\
& & \pi_4 &+ \pi_5 & & = 1, \\
& & \pi & \geq 0, & \pi_0 = 0, & \pi_3 = 1.
\end{aligned}
$$

One extreme point is $(\pi_0, \pi_1, \pi_2, \pi_3, \pi_4, \pi_5) = (0, \frac{1}{3}, \frac{2}{3}, 1, \frac{1}{3}, \frac{2}{3})$ giving the facet-defining inequality

$$0y_0 + \tfrac{1}{3}y_1 + \tfrac{2}{3}y_2 + 1y_3 + \tfrac{1}{3}y_4 + \tfrac{2}{3}y_5 \geqslant 1. \qquad \square$$

Gomory also showed that because of the group structure, one vertex/facet could be used to obtain several vertices/facets for the same, or related group polyhedra. An automorphism is a one-to-one transformation from a group to itself preserving the addition structure of the group.

Proposition 4.8. *Suppose ϕ is an automorphism of G, then*
(i) *if $\sum_{g \in G} \pi_g y_g \geqslant 1$ defines a facet of $P_G(g_0)$,*
   then $\sum_{g \in G} \pi_{\phi^{-1}(g)} y_g \geqslant 1$ defines a facet for $P_G(\phi(g_0))$.
(ii) *if $t = (t_g)$ is a vertex of $P_G(g_0)$,*
   then $\bar{t} = (t_{\phi^{-1}(g)})$ is a vertex of $P_G(\phi(g_0))$.

Example 4.3. Consider the group $G = \mathbb{Z}_5$ of order 5 with $g_0 = 4$, and the associated corner polyhedron

$$P_{\mathbb{Z}_5}(4) = \text{conv}\{y \in Z_+^5 : 0y_0 + 1y_1 + 2y_2 + 3y_3 + 4y_4 \equiv 4\,(\text{mod } 5)\}.$$

It is not difficult to show that $0y_0 + \frac{1}{4}y_1 + \frac{2}{4}y_2 + \frac{3}{4}y_3 + 1y_4 \geqslant 1$ is a facet-defining inequality, and that $t = (0, 1, 0, 1, 0)$ is a vertex.
   Consider the automorphism $\phi : G \to G$ with $\phi(g) = 2g\,(\text{mod } 5)$. Then the inverse $\phi^{-1} : G \to G$ is given by $\phi^{-1}(g) = 3g\,(\text{mod } 5)$, so

$$(\phi^{-1}(0), \phi^{-1}(1), \phi^{-1}(2), \phi^{-1}(3), \phi^{-1}(4)) = (0, 3, 1, 4, 2) \quad \text{and} \quad \phi(g_0) = 3.$$

   Now by (i) of Proposition 4.8, the inequality $\pi_0 y_0 + \pi_3 y_1 + \pi_1 y_2 + \pi_4 y_3 + \pi_2 y_4 = 0y_0 + \frac{3}{4}y_1 + \frac{1}{4}y_2 + 1y_3 + \frac{1}{2}y_4 \geqslant 1$ defines a facet of $P_{\mathbb{Z}_5}(3)$.
   Also by (ii), $\bar{t} = (t_0, t_3, t_1, t_4, t_2) = (0, 1, 1, 0, 0)$ is a vertex of $P_{\mathbb{Z}_5}(3)$. $\quad \square$

Facet defining inequalities for subgroups can also be lifted into facets for larger groups. Specifically a subgroup H of a group G is a subset of the elements of G that is closed under addition within the group (if $h_1, h_2 \in H$, then $h_1 + h_2\,(\text{mod } \Delta) \in H$), and a homomorphism ϕ from a group G to a subgroup H is a transformation preserving the addition structure of the group ($\phi(h_1 + h_2)(\text{mod } \Delta) \equiv \phi(h_1) + \phi(h_2)(\text{mod } \Delta)$).

Proposition 4.9. *If $\phi : G \to H$ is a homomorphism into a subgroup H of G, and $\sum_{h \in H} \pi_h y_h \geqslant 1$ defines a facet of $\text{conv}(X_H(\beta_0))$, then if $\phi(g_0) = \beta_0 \neq 0$, $\sum_{g \in G} \pi_{\phi(g)} y_g \geqslant 1$ defines a facet of $\text{conv}(X_G(g_0))$.*

Example 4.4. Consider a group $G = \mathbb{Z}_2 \times \mathbb{Z}_4$ with subgroup $H = \mathbb{Z}_4$, and homomorphism $\phi : G \to H$ given by $\phi(\alpha, \beta) = \beta$. If $(\alpha, \beta) \in G$, $\phi(\alpha, \beta) = \beta \in H$. Now as $y_1 + y_3 \geqslant 1$ defines a facet of $P_{\mathbb{Z}_4}(3) = \text{conv}(X_H(g_0))$, we obtain from Proposition 4.9 that $y_{(0,1)} + y_{(1,1)} + y_{(0,3)} + y_{(1,3)} \geqslant 1$ defines a facet of both $P_{\mathbb{Z}_2 \times \mathbb{Z}_4}(1, 3)$ and of $P_{\mathbb{Z}_2 \times \mathbb{Z}_4}(0, 3)$.
$\qquad \square$

4.5. Subadditivity and duality

Here we limit our attention to cyclic (one-dimensional) groups $G = \mathbb{Z}_\delta$ for simplicity. We are still interested in the Master set

$$X_{\mathbb{Z}_\delta}(g_0) = \left\{ y \in \mathbb{Z}_+^\delta : \sum_{g=0}^{\delta-1} g y_g \equiv g_0 \,(\mathrm{mod}\,\delta) \right\}$$

and its convex hull $P_{\mathbb{Z}_\delta}(g_0)$.

Definition 4.1. $\pi : \{0, 1, \ldots, \delta - 1\} \to \mathbb{R}$ is *subadditive* on $\mathbb{Z}_\delta$ if $\pi(0) = 0$ and $\pi(u) + \pi(v) \geqslant \pi(u + v \,(\mathrm{mod}\,\delta))$ for all $u, v \in \mathbb{Z}_\delta$.

Theorem 4.7 can now be interpreted as saying that all the facet-defining inequalities arise from such a subadditive function. More generally, every subadditive function on $\mathbb{Z}_\delta$ leads to a valid inequality for $X_{\mathbb{Z}_\delta}(g_0)$.

Proposition 4.10. *If π is subadditive on $\mathbb{Z}_\delta$,*

$$\sum_{g=0}^{\delta-1} \pi(g) y_g \geqslant \pi(g_0)$$

is a valid inequality for $P_{\mathbb{Z}_\delta}(g_0)$.

Such functions also provide duals for the group problem.

Theorem 4.11. *Consider the (primal) group problem*

$$z_G(g_0) = \min \left\{ \sum_{j \in N} \bar{c}_j x_j : \sum_{j \in N} g_j x_j = g_0 \text{ in } \mathbb{Z}_\delta, \ x \in \mathbb{Z}_+^n \right\}.$$

The problem

$$w = \max \pi(g_0),$$

$$\pi(g_j) \leqslant \bar{c}_j \ for \ j \in N,$$

$$\pi \ subadditive \ on \ \mathbb{Z}_\delta$$

is a strong dual with $z_G(g_0) = w$.

Proof. Let $\sum_{j \in N} \pi^k(g_j) x_j \geqslant \pi^k(g_0)$ for $k = 1, \ldots, K$ be the facet-defining inequalities of the corner polyhedron $P_G(g_0)$ with π^k subadditive on $\mathbb{Z}_\delta$. Now

$$z_G(g_0) = \min \left\{ \sum_{j \in N} \bar{c}_j x_j : \sum_{j \in N} \pi^k(g_j) x_j \geqslant \pi^k(g_0) \text{ for } k = 1, \ldots, K, \ x \in \mathbb{R}_+^n \right\}.$$

Let $(u^1, \ldots, u^K)$ be a vector of optimal dual variables. Then clearly $\pi = \sum_{k=1}^K u^k \pi^k$ is a subadditive function on $\mathbb{Z}_\delta$ and is dual feasible with $\pi(g_0) = z_G(g_0)$. □

Now we examine how to generate a valid inequality from any constraint, and not just for groups corresponding to the integers modulo δ for some fixed δ. The results are from Gomory and Johnson [43].

Let I denote the unit interval $[0, 1)$ with addition modulo 1. Thinking of $\mathbb{Z}_\delta$ as a group with elements $\{0, 1/\delta, 2/\delta, \ldots, (\delta - 1)/\delta\}$ under addition modulo 1, we can let δ tend to $+\infty$, and then we obtain an infinite group I whose elements lie in $[0, 1)$ with addition modulo 1.

Here we are interested in generating valid inequalities for the set

$$X_I(u_0) = \left\{ x \colon \sum_{u \in I} u x(u) \equiv u_0 \, (\mathrm{mod}\ 1), \ x(u) \geqslant 0 \text{ and integer, } x \text{ has finite support} \right\}.$$

We first extend our definition of a subadditive function.

Definition 4.2. $\pi \colon [0, 1) \to \mathbb{R}$ is *subadditive on I* if $\pi(0) = 0$ and

$$\pi(u) + \pi(v) \geqslant \pi(u + v \, (\mathrm{mod}\ 1)) \quad \text{for all } u, v \in [0, 1).$$

Now we will use subadditive functions for finite groups $\mathbb{Z}_\delta$ to obtain subadditive functions on I. In analogy with Proposition 4.10, we have

Proposition 4.12. *If π is subadditive on I,*

$$\sum_{u \in I} \pi(u) x(u) \geqslant \pi(u_0)$$

is a valid inequality for $X_I(u_0)$.

Proposition 4.13 (Direct fill-in). *Let π be a subadditive function on $\mathbb{Z}_\delta$. Define $\pi(u)$ for $u \in [0, 1) \setminus \{0, 1/\delta, \ldots, (\delta - 1)/\delta\}$ by*

$$\pi(u) = \delta[(u - L(u)\pi(R(u)) + (R(u) - u)\pi(L(u))]$$

where $L(u) = (1/\delta)\lfloor \delta u \rfloor$ and $R(u) = (1/\delta)\lceil \delta u \rceil$. Then π is subadditive on I.

Example 4.5. Take $\delta = 6$, and consider the subadditive function π derived for $\mathbb{Z}_6$ in Example 4.2 with $(\pi(0), \pi(\frac{1}{6}), \pi(\frac{2}{6}), \pi(\frac{3}{6}), \pi(\frac{4}{6}), \pi(\frac{5}{6})) = (0, \frac{1}{3}, \frac{2}{3}, 1, \frac{1}{3}, \frac{2}{3})$. Direct fill in immediately gives the function π where

$$\begin{aligned}
\pi(u) &= 2u & \text{for } 0 \leqslant u \leqslant \tfrac{1}{2}, \\
\pi(u) &= 3 - 4u & \tfrac{1}{2} \leqslant u < \tfrac{2}{3}, \\
\pi(u) &= -1 + 2u & \tfrac{2}{3} \leqslant u < \tfrac{5}{6}, \\
\pi(u) &= 4 - 4u & \tfrac{5}{6} \leqslant u < 1.
\end{aligned}$$

In Fig. 16 we show the values of the original function π on $\mathbb{Z}_5$, and the function π on I obtained by direct fill-in.

Now consider the constraint set

$$x_0 + 0.76 x_1 - 0.35 x_2 + 2.41 x_3 = 4.49, \quad x_0, x_1, x_2, x_3 \in \mathbb{Z}_+^1$$

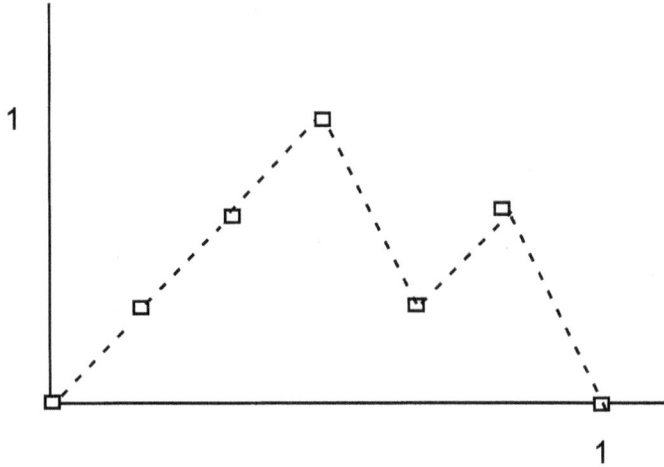

Fig. 16. Constructing a subadditive function by direct fill-in.

coming from some arbitrary integer program. Relaxing the nonnegativity on x_0 leads to the set

$$X_I(0.49) = \{(x_1, x_2, x_3) \in Z_+^3 : 0.76x_1 + 0.65x_2 + 0.41x_3 \equiv 0.49 \,(\text{mod } 1)\}$$

and the subadditive function π just constructed immediately gives the valid inequality

$$\pi(0.76)x_1 + \pi(0.65)x_2 + \pi(0.41)x_3 \geqslant \pi(0.49)$$

or

$$0.52x_1 + 0.30x_2 + 0.82x_3 \geqslant 0.98. \qquad \square$$

4.6. Solving IP(b) for all values of $b \in \mathbb{Z}^m$

Consider again the integer program $IP(b)$. For a given $b \in \mathbb{Z}^m$, let A_B be an optimal LP basis. We have seen in Observation 4.3 that the solution x_N^* of the group relaxation

$$IP_B(b) \quad \min\{c_B^T A_B^{-1} b + \bar{c}_N^T x_N : A_B^{-1} A_N x_N = A_B^{-1} b \,(\text{mod } 1), \ x_N \in \mathbb{Z}_+^n\}$$

solves $IP(d)$ for an infinity of values of $d \in \mathbb{Z}^m$, i.e. for all d such that $A_B^{-1}d = A_B^{-1}b \,(\text{mod } 1)$ and $x_B^* = A_B^{-1}d - A_B^{-1}A_N x_N^* \geqslant 0$. More generally, it is natural to ask what is the largest subset of columns $A_S \subset A_B$ for which the relaxation

$$IP_S(b) \quad z_S(b) = \min\{c^T x : Ax = b, \ x_j \in \mathbb{Z}_+^1 \text{ for } j \in V \setminus S, \ x_j \in \mathbb{Z}^1 \text{ for } j \in S\}$$

solves $IP(b)$. Solving $IP_S(b)$ will in turn provide a correction vector $x_{V \setminus S}^*$ that solves $IP(b)$ for many values of b.

In fact it can be shown that it suffices to solve a finite number of such problems $IP_{S_1}(b^1), \ldots, IP_{S_t}(b^t)$ in order to have a solution to $IP(b)$ for all b. Furthermore if for $u = 1, \ldots, t$, we take the subadditive functions $\{\pi^{k,u}\}_{k=1}^{K_u}$ describing the facets of the associated convex hulls for $u = 1, \ldots, t$, these suffice to describe the convex hulls of $IP(b)$ for all b, specifically

$$\sum_{j=1}^{m+n} \pi^{k,u}(a^j)x_j \geqslant \pi^{k,u}(b) \quad \text{for } k = 1, \ldots, K_u, \ u = 1, \ldots, t, \ x \in \mathbb{R}_+^{m+n}.$$

In other words

Theorem 4.14 (Cook et al. [20] and Wolsey [107]). *For each integral $m \times (m + n)$ matrix A, there exists an integral $m' \times (m + n)$ matrix Q such that for any $b \in \mathbb{Z}^m$, there exists a function $q : \mathbb{Z}^m \to \mathbb{R}^{m'}$ such that*

$$\mathrm{conv}(\{x \in \mathbb{Z}_+^m : Ax = b\}) = \{x \in \mathbb{R}^{m+n} : Qx \geqslant q(b)\}.$$

In addition the size of the coefficients in Q is bounded by $(m + n)^{2(m+n)} f(A)$ where $f(A)$ is the maximum absolute value of the subdeterminants of A.

In [20], it is also shown that the difference between the optimal values $z(b) - z(b')$ cannot be too large. See also Theorem 3.15.

Finally one might ask, given A and c fixed, but all possible b, for which sets S, is it necessary to solve $IP_S(b)$? Following Hosten and Thomas [58], a set $S \subset V$ is called *minimal* if for some b, the relaxation $IP_S(b)$ solves $IP(b)$, but $IP_{S'}(b)$ does not for any $S' \supset S$. Note that different vectors b and b' may imply that S is minimal, and, alternatively, different sets S and S' may be minimal due to the same vector b.

Theorem 4.15 (Hosten and Thomas [58]). *Given A, c with $\mathrm{rank}(A) = m$, if $S \subset V$ is a minimal set with $|S| < m$ for the family of problems $IP(b)$, there exists $v \in V \setminus S$ such that $S \cup \{v\}$ is minimal.*

4.7. Computational possibilities

The use of subadditive functions to generate valid inequalities for an arbitrary row of an integer program can be extended to mixed integer progams [43]. Given the recent computational interest in using Gomory's fractional cuts, mixed integer rounding inequalities and Gomory's mixed integer cuts, this reopens questions about the possible use of alternative subadditive functions to generate practically effective cutting planes. It is also natural to ask whether interesting higher dimensional functions can be found and put to use, see [80] for one two-dimensional function.

The subadditive duality on $\mathbb{Z}_\delta$ has been generalized to a subadditive duality theory for general integer and mixed integer programs [59]. This also raises the question of the conversion of subadditive functions on I into nondecreasing subadditive functions on $\mathbb{R}^n$ for use in arbitrary integer programs.

Acknowledgements

The authors would like to thank Bill Cook, Ravi Kannan, Arjen Lenstra, Hendrik Lenstra, Herb Scarf, Alexander Schrijver, and Rekha Thomas for enlightening discussions, comments on various versions of the manuscript, and for providing references, and to the referees.

References

[1] K. Aardal, R.E. Bixby, C.A.J. Hurkens, A.K. Lenstra, J.W. Smeltink, Market split and basis reduction: towards a solution of the Cornuéjols–Dawande instances, in: G. Cornuéjols, R.E. Burkard, G.J. Woeginger (Eds.), Integer Programming and Combinatorial Optimization, 7th International IPCO Conference, Lecture Notes in Computer Science, Vol. 1610, Springer, Berlin, Heidelberg, 1999, pp. 1–16.

[2] K. Aardal, C. Hurkens, A.K. Lenstra, Solving a linear diophantine equation with lower and upper bounds on the variables, in: R.E. Bixby, E.A. Boyd, R.Z. Ríos-Mercado (Eds.), Integer Programming and Combinatorial Optimization, 6th International IPCO Conference, Lecture Notes in Computer Science, Vol. 1412, Springer, Berlin, Heidelberg, 1998, pp. 229–242.

[3] K. Aardal, C. Hurkens, A.K. Lenstra, Solving a system of diophantine equations with lower and upper bounds on the variables, Math. Oper. Res. 25 (2000) 427–442.

[4] R.K. Ahuja, T.L. Magnanti, J.B. Orlin, Network flows: theory, algorithms, and applications, Prentice-Hall, Englewood Cliffs, NJ, 1993.

[5] M. Ajtai, The shortest vector problem in L_2 is NP-hard for randomized reductions, in Proceedings of the 30th Annual ACM Symposium on the Theory of Computing, ACM, New York, NY, 1998, pp. 10–19.

[6] E. Balas, M.W. Padberg, On the set covering problem II. An algorithm for set partitioning, Oper. Res. 23 (1975) 74–90.

[7] A.I. Barvinok, A polynomial time algorithm for counting integral points in polyhedra when the dimension is fixed, Math. Oper. Res. 19 (1994) 769–779.

[8] T. Becker, V. Weispfenning, Gröbner bases: a computational approach to commutative algebra, Springer, Verlag, New York, 1993.

[9] D.E. Bell, J.F. Shapiro, A convergent duality theory for integer programming, Oper. Res. 25 (1977) 419–434.

[10] I. Borosh, L.B. Treybig, Bounds on positive integral solutions of linear diophantine equations, Proc. Amer. Math. Soc. 55 (1976) 299–304.

[11] J. Bourgain, V.D. Milman, Sections euclidiennes et volume des corps symétriques convexes dans $\mathbb{R}^n$, C. R. Acad. Sc. Paris t. 300, Série I, (13) (1985) 435–438.

[12] W. Bruns, J. Gubeladze, Normality and covering properties, preprint, University of Osnabrück, 1998.

[13] W. Bruns, J. Gubeladze, N.V. Trung, Normal polytopes, triangulations, and Koszul algebras, J. Reine Angew. Math. 485 (1997) 123–160.

[14] W. Bruns, J. Gubeladze, M. Henk, A. Martin, R. Weismantel, A counterexample to an integer analogue of Carathéodory's Theorem, Preprint No. 32, Universität Magdeburg, 1998.

[15] B. Buchberger, Gröbner bases: an algorithmic method in Polynomial Ideal Theory, in: N.K. Bose (Ed.), Multidimensional Systems Theory, D. Reidel Publications, Dordrecht, 1985, pp. 184–232.

[16] J.-Y. Cai, A.P. Nerurkar, Approximating the svp to within a factor $(1 + 1/dim^\varepsilon)$ is NP-hard under randomized reductions, Manuscript, 1997.

[17] H. Cohen, A Course in Computational Algebraic Number Theory, Springer, Berlin, Heidelberg, 1996.

[18] P. Conti, C. Traverso, Buchberger algorithm and integer programming, Proceedings AAECC-9 (New Orleans), Lecture Notes in Computer Science, Vol. 539, Springer, Berlin 1991, pp. 130–139.

[19] S.A. Cook, The complexity of theorem-proving procedures, in Proceedings of Third Annual ACM Symposium on Theory of Computing, ACM, New York, 1991, pp. 151–158.

[20] W. Cook, A.M.H. Gerards, A. Schrijver, É. Tardos, Sensitivity results in integer programming, Math. Programming 34 (1986) 251–264.

[21] W. Cook, J. Fonlupt, A. Schrijver, An integer analogue of Carathéodory's theorem, J. Combin. Theory (B) 40 (1986) 63–70.

[22] W. Cook, A.M.H. Gerards, A. Schrijver, É. Tardos, Sensitivity theorems in integer programming problems, Math. Programming 34 (1986) 63–70.

[23] W. Cook, T. Rutherford, H.E. Scarf, D. Shallcross, An implementation of the generalized basis reduction algorithm for integer programming, ORSA J. Comput. 5 (1993) 206–212.

[24] G. Cornuéjols, M. Dawande, A class of hard small 0–1 programs, in: R.E. Bixby, E.A. Boyd, R.Z. Ríos-Mercado (Eds.), Integer Programming and Combinatorial Optimization, 6th International IPCO Conference, Lecture Notes in Computer Science, Vol. 1412, Springer, Berlin, Heidelberg, 1998, pp. 284–293.

[25] G. Cornuéjols, R. Urbaniak, R. Weismantel, L.A. Wolsey, Decomposition of integer programs and of generating sets, in: R. Burkard, G. Woeginger (Eds.), Algorithms—ESA '97, Lecture Notes in Computer Science, Vol. 1284, Springer, Berlin, Heidelberg, 1997, pp. 92–103.

[26] J.G. van der Corput, Über Systeme von linear-homogenen Gleichungen und Ungleichungen, Proceedings Koninklijke Akademie van Wetenschappen te Amsterdam 34 (1931) 368–371.

[27] M.J. Coster, A. Joux, B.A. LaMacchia, A.M. Odlyzko, C.P. Schnorr, J. Stern, Improved low-density subset sum algorithms, Comput. Complexity 2 (1992) 111–128.

[28] D.A. Cox, J.B. Little, D. O'Shea, Ideals, varieties, and algorithms: an introduction to computational algebraic geometry and commutative algebra, Springer, New York, 1992.

[29] D. Dais, U.U. Haus, M. Henk, On crepant resolutions of 2-parameter series of Gorenstein cyclic quotient singularities, Results Math. 33 (1998) 208–265.

[30] M. Dyer, R. Kannan, On Barvinok's algorithm for counting lattice points in fixed dimension, Math. Oper. Res. 22 (1997) 545–549.

[31] J. Edmonds, R.M. Karp, Theoretical improvements in algorithmic efficiency for network flow problems, J. Assoc. Comput. Mach. 19 (1972) 248–264.

[32] J. Edmonds, R. Giles, A min-max relation for submodular functions on graphs, in Studies in Integer Programming, P.L. Hammer et al. eds, Ann. Discrete Math. 1 (1977) 185–204.

[33] P. van Emde Boas, Another NP-complete partition problem and the complexity of computing short vectors in a lattice, Report 81-04, Mathematical Institute, University of Amsterdam, Amsterdam, 1981.

[34] G. Ewald, Combinatorial Convexity and Algebraic Geometry, Graduate Texts in Mathematics, Vol. 168, Springer-Verlag, Berlin, 1996.

[35] R.T. Firla, G.M. Ziegler, Hilbert bases, unimodular triangulations, and binary covers of rational polyhedral cones, Discrete Computational Geometry 21 (1999) 205–216.

[36] F.R. Giles, W.R. Pulleyblank, Total dual integrality and integer polyhedra, Linear Algebra Appl. 25 (1979) 191–196.

[37] J.-L. Goffin, Variable metric relaxation methods, Part II: the ellipsoid method, Math. Programming 30 (1984) 147–162.

[38] O. Goldreich, D. Micciancio, S. Safra, J.-P. Seifert, Approximating shortest lattice vectors is not harder than approximating closest lattice vectors, Inform. Process. Lett. 71 (1999) 55–61.

[39] R.E. Gomory, An all-integer integer programming algorithm, in: J.F. Muth, G.L. Thompson (Eds.), Industrial Scheduling, Prentice-Hall, Englewood Cliffs, NJ, 1963, pp. 193–206.

[40] R.E. Gomory, On the relation between integer and non-integer solutions to linear programs, Proc. Nat. Academy Sci. 53 (1965) 260–265.

[41] R.E. Gomory, Faces of an integer polyhedron, Proc. Nat. Academy Sci. 57 (1967) 16–18.

[42] R.E. Gomory, Some polyhedra related to combinatorial problems, Linear Algeb. Appl. 2 (1969) 451–558.

[43] R.E. Gomory, E.L. Johnson, Some continuous functions related to corner polyhedra, Math. Programming 3 (1972) 23–85.

[44] P. Gordan, Über die Auflösung linearer Gleichungen mit reellen Coefficienten, Math. Ann. 6 (1873) 23–28.

[45] G.A. Gorry, W.D. Northup, J.F. Shapiro, Computational experience with a group theoretic integer programming algorithm, Math. Programming 4 (1973) 171–192.

[46] J.E. Graver, On the foundations of linear and integer programming I, Math. Programming 8 (1975) 207–226.

[47] M. Grötschel, L. Lovász, Combinatorial optimization, in: R. Graham, M. Grötschel, L. Lovász (Eds.), Handbook of Combinatorics, North-Holland, Amsterdam, 1995.

[48] M. Grötschel, L. Lovász, A. Schrijver, Geometric methods in combinatorial optimization, in: W.R. Pulleyblank (Ed.), Progress in Combinatorial Optimization, Academic Press, Toronto, 1984, pp. 167–183.

[49] M. Grötschel, L. Lovász, A. Schrijver, Geometric Algorithms and Combinatorial Optimization, Springer-Verlag, Berlin, 1988.
[50] U. Haus, M. Köppe, R. Weismantel, The integral basis method for integer programming, Math. Methods Oper. Res. 53 (3) (2001).
[51] U. Haus, M. Köppe, R. Weismantel, A primal all-integer algorithm based on irreducible solutions, Manuscript, 2001.
[52] U. Haus, M. Köppe, R. Weismantel, The integral basis method in an augmentation framework for integer programming, Manuscript, 2001.
[53] M. Henk, R. Weismantel, The height of minimal Hilbert bases, Results Math. 32 (1997) 298–303.
[54] M. Henk, R. Weismantel, On minimal solutions of Diophantine equations, Contributions to Algebra and Geometry 41 (2000) 49–55.
[55] M. Henk, R. Weismantel, A theorem about minimal solutions of linear Diophantine equations, Contributions to Algebra and Geometry, 1998, to appear.
[56] Ch. Hermite, Extraits de lettres de M. Ch. Hermite à M. Jacobii sur différents objets de la théorie des nombres. J. angew. Math. 40 (1850) 261–278, 279–290, 291–307, 308–315, [reprinted in: É. Picard (Ed.), Oevres de Charles Hermite, Tome I, Gauthier-Villars, Paris, 1905, pp. 100–121, 122–135, 136–155, 155–163.]
[57] D.S. Hirschberg, C.K. Wong, A polynomial-time algorithm for the knapsack problem with two variables, J. ACM 23 (1976) 147–154.
[58] S. Hosten, R.R. Thomas, Standard pairs and group relaxations in integer programming, Technical Report, Department of Mathematics, Texas A& M University, February 1998.
[59] E.L. Johnson, Integer Programming—Facets, Subadditivity and Duality for Group and Semi-Group Problems, SIAM Publications, Philadelphia, 1980.
[60] A. Joux, J. Stern, Lattice reduction: a toolbox for the cryptanalyst, J. Cryptol. 11 (1998) 161–185.
[61] E. Kaltofen, On the complexity of finding short vectors in integer lattices, in: J.A. VanHulzen (Ed.), Computer Algebra: Proceedings of EUROCAL '83, European Computer Algebra Conference, Lecture Notes in Computer Science, Vol. 162, Springer-Verlag, New York, 1983, pp. 236–244.
[62] R. Kannan, A polynomial algorithm for the two-variable integer programming problem, J. ACM 27 (1980) 118–122.
[63] R. Kannan, Algorithmic geometry of numbers, Ann. Rev. Comput. Sci. 2 (1987) 231–267.
[64] R. Kannan, Minkowski's convex body theorem and integer programming, Math. Oper. Res. 12 (1987) 415–440.
[65] R. Kannan, L. Lovász, Covering minima and lattice point free convex bodies, in: K.V. Nori (Ed.), Foundations of software technology and theoretical computer science, Lecture Notes in Computer Science, Vol. 241, Springer, Berlin, 1986, pp. 193–213.
[66] R. Kannan, L. Lovász, Covering minima and lattice-point-free convex bodies, Ann. Math. 128 (1988) 577–602.
[67] R.M. Karp, Reducibility among combinatorial problems, in: R.E. Miller, J.W. Thatcher (Eds.), Complexity of Computer Computations, Plenum Press, New York, 1972, pp. 85–103.
[68] J.C. Lagarias, H.W. Lenstra Jr., C.P. Schnorr, Korkine-Zolotarev bases and successive minima of a lattice and its reciprocal lattice, Combinatorica 10 (1990) 333–348.
[69] J.C. Lagarias, A.M. Odlyzko, Solving low-density subset sum problems, J. Assoc. Comput. Mach. 32 (1985) 229–246.
[70] A.K. Lenstra, H.W. Lenstra Jr., L. Lovász, Factoring polynomials with rational coefficients, Mathematische Annalen 261 (1982) 515–534.
[71] H.W. Lenstra Jr., Integer programming with a fixed number of variables, Math. Oper. Res. 8 (1983) 538–548.
[72] LiDIA—A library for computational number theory. TH Darmstadt/Universität des Saarlandes, Fachbereich Informatik, Institut für Theoretische Informatik. http://www.informatik.th-darmstadt.de/pub/TI/LiDIA.
[73] J. Liu, Hilbert bases with the Carathéodory property, PhD. Thesis, Cornell University, 1991.
[74] J. Liu, L.E. Trotter Jr., G.M. Ziegler, On the height of the minimal Hilbert basis, Results Math. 23 (1993) 374–376.

[75] L. Lovász, An Algorithmic Theory of Numbers, Graphs and Convexity, CBMS-NSF Regional Conference Series in Applied Mathematics, Vol. 50, SIAM, Philadelphia, 1986.

[76] L. Lovász, H.E. Scarf, The generalized basis reduction algorithm, Math. Oper. Res. 17 (1992) 751–764.

[77] Q. Louveaux, L.A. Wolsey, Combining problem structure with basis reduction to solve a class of hard integer programs, CORE Discussion Paper 2000/51, CORE, Université Catholique de Louvain, Louvain-la-Neuve, Belgium, 2000.

[78] T. McCormick, A. Shioura, A minimum ratio cycle canceling algorithm for linear programming problems with application to network optimization, Manuscript, 1996.

[79] D. Micciancio, The shortest vector in a lattice is hard to approximate to within some constant, Proceedings of the 39th IEEE Symposium on Foundations of Computer Science, IEEE, Los Alamitos, CA, 1998, pp. 92–98.

[80] G.L. Nemhauser, L.A. Wolsey, Integer and Combinatorial Optimization, Wiley, New York, 1988.

[81] T. Oda, Convex bodies and algebraic geometry. An introduction to the theory of toric varieties, Ergebnisse der Mathematik und ihrer Grenzgebiete 3. Folge, Bd. 15, Springer, Berlin, 1988.

[82] A.M. Odlyzko, Cryptanalytic attacks on the multivariate knapsack cryptosystem and on Shamir's fast signature scheme, IEEE Trans. Inform. Theory IT-30 4 (1984) 584–601.

[83] M.W. Padberg, S. Hong, On the symmetric travelling salesman problem: a computational study, Math. Programming Study 12 (1980) 78–107.

[84] H. Röck, Scaling techniques for minimal cost network flows, in Discrete Structures and Algorithms, Carl Hanser, München, 1980, pp. 181–191.

[85] H.E. Scarf, Production sets with indivisibilities—Part I: Generalities, Econometrica 49 (1981) 1–32. Part II: The case of two activities, Econometrica 49 (1981) 395–423.

[86] H.E. Scarf, Neighborhood systems for production sets with indivisibilities, Econometrica 54 (1986) 507–532.

[87] C.P. Schnorr, A hierarchy of polynomial time lattice basis reduction algorithms, Theoret. Comput. Sci. 53 (1987) 201–224.

[88] C.P. Schnorr, Block reduced lattice bases and successive minima, Combin. Probab. Comput. 3 (1994) 507–522.

[89] C.P. Schnorr, M. Euchner, Lattice basis reduction: improved practical algorithms and solving subset sum problems, Math. Programming 66 (1994) 181–199.

[90] C.P. Schnorr, H.H. Hörner, Attacking the Chor–Rivest cryptosystem by improved lattice reduction, in: L.C. Guillou, J.-J Quisquater (Eds.), Advances in Cryptology—EUROCRYPT '95, Lecture Notes in Computer Science, Vol. 921, Springer, Berlin, 1995, pp. 1–12.

[91] A. Schrijver, Theory of Linear and Integer Programming, Wiley, Chichester, 1986.

[92] A.S. Schulz, R. Weismantel, G. Ziegler, 0/1 integer programming: optimization and augmentation are equivalent, in: Lecture Notes in Computer Science, Vol. 979, Springer, Berlin, 1995, pp. 473–483.

[93] A. Schulz, R. Weismantel, An oracle-polynomial time augmentation algorithm for integer programming, in Proceedings of the 10th ACM-SIAM Symposium on Discrete Algorithms, Baltimore, USA, 1999, pp. 967–968.

[94] A. Sebö, Hilbert bases, Carathéodory's theorem and combinatorial optimization, in Proceedings of the IPCO Conference, Waterloo, Canada, 1990, pp. 431–455.

[95] M. Seysen, Simultaneous reduction of a lattice basis and its reciprocal basis, Combinatorica 13 (1993) 363–376.

[96] V. Shoup, NTL: A Library for doing Number Theory, Department of Computer Science, University of Wisconsin-Madison. http://www.shoup.net/.

[97] H.J.S. Smith, On systems of indeterminate equations and congruences, Philosophical Transactions of the Royal Society of London (A) 151 (1861) 293–326.

[98] B. Sturmfels, Gröbner Bases and Convex Polytopes, University Lecture Series, Vol. 8, AMS, Providence, RI, 1996.

[99] B. Sturmfels, R. Thomas, Variation of cost functions in integer programming, Math. Programming 77 (1997) 357–388.

[100] B. Sturmfels, R. Weismantel, G. Ziegler, Gröbner bases of lattices, corner polyhedra and integer programming, Beiträge zur Geometrie und Algebra 36 (1995) 281–298.

[101] R. Thomas, A geometric Buchberger algorithm for integer programming, Math. Oper. Res. 20 (1995) 864–884.

[102] R. Thomas, R. Weismantel, Truncated Gröbner bases for integer programming, Appl. Algeb. Eng. Commun. Comput. 8 (1997) 241–257.

[103] R. Urbaniak, R. Weismantel, G. Ziegler, A variant of Buchberger's algorithm for integer programming, SIAM J. Discrete Math. 1 (1997) 96–108.

[104] C. Wallacher, Kombinatorische Algorithmen für Flußprobleme und submodulare Flußprobleme, Ph.D. Thesis, Technische Universität zu Braunschweig, 1992.

[105] X. Wang, A new implementation of the generalized basis reduction algorithm for convex integer programming, Ph.D. Thesis, Yale University, 1997.

[106] L.A. Wolsey, Generalized dynamic programming methods in integer programming, Math. Programming 4 (1973) 222–232.

[107] L.A. Wolsey, On the b-hull of an integer program, Discrete Appl. Math. 3 (1981) 193–201.

[108] R.D. Young, A primal (all integer), integer programming algorithm, J. Res. Nat. Bureau Standards 69b (1965).

[109] R.D. Young, A simplified primal (all integer) integer programming algorithm, Oper. Res. 16 (1968) 750–782, 213–250.

For further reading

H.N. Gabow, Scaling algorithms for network problems, J. Comput. System Sci. 31 (1985) 148–168.

L. Pottier, Minimal solutions of linear diophantine systems: bounds and algorithms, Proceedings RTA (Como), Lecture Notes in Computer Science, Vol. 488, Springer, Berlin, 1991.

R. Thomas, Gröbner basis methods for integer programming, Ph.D. Thesis, Cornell University, 1994.

N·H

ELSEVIER

Discrete Applied Mathematics 123 (2002) 75–102

DISCRETE
APPLIED
MATHEMATICS

A survey of very large-scale neighborhood search techniques

Ravindra K. Ahuja[a,*], Özlem Ergun[b], James B. Orlin[c],
Abraham P. Punnen[d]

[a] Department of Industrial & Systems Engineering, University of Florida, P.O. Box 116595, Gainesville,
FL 32611, USA
[b] School of Industrial and Systems Engineering, Georgia Institute of Technology, Atlanta,
GA 30332, USA
[c] Sloan School of Management, Massachusetts Institute of Technology, Cambridge, MA 02139, USA
[d] Department of Mathematics, Statistics and Computer Science, University of New Brunswick,
Saint John, N.B., Canada E2L 4L5

Received 9 September 1999; received in revised form 3 November 2000; accepted 8 January 2001

Abstract

Many optimization problems of practical interest are computationally intractable. Therefore, a practical approach for solving such problems is to employ heuristic (approximation) algorithms that can find nearly optimal solutions within a reasonable amount of computation time. An *improvement algorithm* is a heuristic algorithm that generally starts with a feasible solution and iteratively tries to obtain a better solution. Neighborhood search algorithms (alternatively called *local search algorithms*) are a wide class of improvement algorithms where at each iteration an improving solution is found by searching the "neighborhood" of the current solution. A critical issue in the design of a neighborhood search algorithm is the choice of the neighborhood structure, that is, the manner in which the neighborhood is defined. As a rule of thumb, the larger the neighborhood, the better is the quality of the locally optimal solutions, and the greater is the accuracy of the final solution that is obtained. At the same time, the larger the neighborhood, the longer it takes to search the neighborhood at each iteration. For this reason, a larger neighborhood does not necessarily produce a more effective heuristic unless one can search the larger neighborhood in a very efficient manner. This paper concentrates on neighborhood search algorithms where the size of the neighborhood is "*very large*" with respect to the size of the input data and in which the neighborhood is searched in an efficient manner. We survey three broad classes of very large-scale neighborhood search (VLSN) algorithms: (1) variable-depth methods in which large neighborhoods are searched heuristically, (2) large neighborhoods in which the neighborhoods are searched using network flow techniques or dynamic

---

*Corresponding author.

E-mail addresses: ahuja@ufl.edu (R.K. Ahuja), ozlem.ergun@isye.gatech.edu (Ö. Ergun), jorlin@mit.edu
(J.B. Orlin), punnen@unbsj.ca (A.P. Punnen).

programming, and (3) large neighborhoods induced by restrictions of the original problem that are solvable in polynomial time.

1. Introduction

Many optimization problems of practical interest are computationally intractable. Therefore, a practical approach for solving such problems is to employ heuristic (approximation) algorithms that can find nearly optimal solutions within a reasonable amount of computational time. The literature devoted to heuristic algorithms often distinguishes between two broad classes: constructive algorithms and improvement algorithms. A constructive algorithm builds a solution from scratch by assigning values to one or more decision variables at a time. An improvement algorithm generally starts with a feasible solution and iteratively tries to obtain a better solution. Neighborhood search algorithms (alternatively called *local search algorithms*) are a wide class of improvement algorithms where at each iteration an improving solution is found by searching the "neighborhood" of the current solution. This paper concentrates on neighborhood search algorithms where the size of the neighborhood is "*very large*" with respect to the size of the input data. For large problem instances, it is impractical to search these neighborhoods explicitly, and one must either search a small portion of the neighborhood or else develop efficient algorithms for searching the neighborhood implicitly.

A critical issue in the design of a neighborhood search approach is the choice of the neighborhood structure, that is, the manner in which the neighborhood is defined. This choice largely determines whether the neighborhood search will develop solutions that are highly accurate or whether they will develop solutions with very poor local optima. As a rule of thumb, the larger the neighborhood, the better is the quality of the locally optimal solutions, and the greater is the accuracy of the final solution that is obtained. At the same time, the larger the neighborhood, the longer it takes to search the neighborhood at each iteration. Since one generally performs many runs of a neighborhood search algorithm with different starting points, longer execution times per iteration lead to fewer runs per unit time. For this reason a larger neighborhood does not necessarily produce a more effective heuristic unless one can search the larger neighborhood in a very efficient manner.

Some very successful and widely used methods in operations research can be viewed as very large-scale neighborhood search techniques. For example, if the simplex algorithm for solving linear programs is viewed as a neighborhood search algorithm, then column generation is a very large-scale neighborhood search method. Also, the augmentation techniques used for solving many network flows problems can be categorized as very large-scale neighborhood search methods. The negative cost cycle canceling algorithm for solving the min cost flow problem and the augmenting path algorithm for solving matching problems are two such examples.

In this survey, we categorize very large-scale neighborhood methods into three possibly overlapping classes. The first category of neighborhood search algorithms we study are variable-depth methods. These algorithms focus on exponentially large neighborhoods and partially search these neighborhoods using heuristics. The second category contains network flow based improvement algorithms. These neighborhood search methods use network flow techniques to identify improving neighbors. Finally, in the third category we discuss neighborhoods for NP-hard problems induced by subclasses or restrictions of the problems that are solvable in polynomial time. Although we introduced the concept of large-scale neighborhood search by mentioning column generation techniques for linear programs and augmentation techniques for network flows, we will not address linear programs again. Rather, our survey will focus on applying very large-scale neighborhood search techniques to NP-hard optimization problems.

This paper is organized as follows. In Section 2, we give a brief overview of local search. We discuss variable-depth methods in Section 3. Very large-scale neighborhood search algorithms based on network flow techniques are considered in Section 4. In Section 5, efficiently solvable special cases of NP-hard combinatorial optimization problems and very large-scale neighborhoods based on these special cases are presented. We describe neighborhood metrics that might be a guide to the performance of local search algorithms with respect to the given neighborhoods in Section 6. Finally, in Section 7 we discuss the computational performance of some of the algorithms mentioned in the earlier sections.

2. Local search: an overview

We first formally introduce a combinatorial optimization problem and the concept of a neighborhood. There are alternative ways of representing combinatorial optimization problems, all relying on some method for representing the set of feasible solutions. Here, we will let the set of *feasible solutions* be represented as subsets of a finite set. We formalize this as follows:

Let $E = \{1, 2, \ldots, m\}$ be a finite set. In general, for a set S, we let $|S|$ denote its cardinality. Let $F \subseteq 2^E$, where 2^E denotes the set of all the subsets of E. The elements of F are called *feasible solutions*. Let $f : F \rightarrow \Re$. The function f is called the *objective function*. Then an instance of a *combinatorial optimization problem* (COP) is represented as follows:

Minimize $\{f(S): S \in F\}$.

We assume that the family F is not given explicitly by listing all its elements; instead, it is represented in a compact form of size polynomial in m. An instance of a combinatorial optimization problem is denoted by the pair (F, f). For most of the problems we consider, the cost function is linear, that is, there is a vector $f_1, f_2, \ldots, f_m$ such that for all feasible sets S, $f(S) = \sum_{i \in S} f_i$.

Suppose that (F, f) is an instance of a combinatorial optimization problem. A *neighborhood function* is a point to set map $N : F \rightarrow 2^E$. Under this function, each $S \in F$ has an associated subset $N(S)$ of E. The set $N(S)$ is called the *neighborhood* of the solution S, and we assume without loss of generality that $S \in N(S)$. A solution $S^* \in F$ is said to be *locally optimal* with respect to a neighborhood function N if $f(S^*) \leqslant f(S)$ for all $S \in N(S^*)$. The neighborhood $N(S)$ is said to be *exponential* if $|N(S)|$ grows exponentially in m as m increases. Throughout most of this survey, we will address exponential size neighborhoods, but we will also consider neighborhoods that are too large to search explicitly in practice. For example, a neighborhood with m^3 elements is too large to search in practice if m is large (say greater than a million).

We will refer to neighborhood search techniques using such neighborhoods as *very large-scale neighborhood search algorithms* or VLSN search algorithms.

For two solutions S and T, we let $S - T$ denote the set of elements that appear in S but not in T. We define the *distance* $d(S, T)$ as $|S - T| + |T - S|$, that is, the number of elements of E that appear in S or T but not both. Occasionally, we will permit neighborhoods to include infeasible solutions as well. For example, for the TSP we may permit the neighborhood of a tour to include each of the paths obtained by deleting an edge of the tour. To emphasize that the neighborhood contains more than tours, we normally give a combinatorial description of the non-feasible solutions permitted in the search. We refer to these non-feasible combinatorial structures as reference structures. For example, a Hamiltonian path may be a reference structure.

A neighborhood search algorithm (for a cost minimization problem) can be conceptualized as consisting of three parts:

(i) A neighborhood graph NG defined with respect to a specific problem instance, where NG is a directed graph with one node for each feasible solution (and/or instance of a non-feasible reference structure) created, and with an arc (S, T) whenever $T \in N(S)$.

(ii) A method for searching the neighborhood graph at each iteration.

(iii) A method for determining what is the next node of the neighborhood graph that the search in Step (ii) will choose. We will refer to this node as the *BaseSolution*.

The algorithm terminates when S is a locally optimal solution with respect to the given neighborhood. (See [1] for an extensive survey.)

We next define two neighborhoods based on the distance. The first neighborhood is $N_k(S) = \{T \in F: d(S, T) \leqslant k\}$. We will refer to this neighborhood as the *distance-k neighborhood*.

For some problem instances, any two feasible solutions have the same cardinality. This is true for the traveling salesman problem (TSP), where each feasible solution S represents a tour in a complete graph on n cities, and therefore has n arcs (see [46] for details on the TSP). In general, we say that T can be obtained by a single *exchange* from S if $|S - T| = |T - S| = 1$; we say T can be obtained by a *k-exchange* if $|T - S| = |S - T| = k$. We define the *k-exchange neighborhood* of S to be $\{T: |S - T| = |T - S| \leqslant k\}$. If any two feasible solutions have the same cardinality, then the *k*-exchange neighborhood of S is equal to $N_{2k}(S)$. A standard example for the *k*-exchange neighborhood for the TSP is the 2-exchange neighborhood, also called the 2-opt neighborhood. Each node in the 2-opt neighborhood graph for the TSP is a tour, and two tours are neighbors if one can be obtained from the other by a 2-exchange. The method for searching the neighborhood is exhaustive (or some shortcut), and the next BaseSolution will be an improving solution.

Since $N_m(S)=F$, it follows that searching the distance-k neighborhood can be difficult as k grows large. It is typically the case that this neighborhood grows exponentially if k is not fixed, and that finding the best solution (or even an improved solution) in the neighborhood is NP-hard if the original problem is NP-hard.

3. Variable-depth methods

For $k=1$ or 2, the k-exchange (or similarly k-distance) neighborhoods can often be efficiently searched, but on average the resulting local optima may be poor. For larger values of k, the k-exchange neighborhoods yield better local optima but the effort spent to search the neighborhood might be too large. Variable-depth search methods are techniques that search the k-exchange neighborhood partially. The goal in this partial search is to find solutions that are close in objective function value to the global optima while dramatically reducing the time to search the neighborhood. Typically, they do not guarantee to be local optima. In VLSN search algorithms, we are interested in several types of algorithms for searching a portion of the k-exchange neighborhood. In this section, we describe the Lin–Kernighan [48] algorithm for the traveling salesman problem as well as other variable-depth heuristics for searching the k-exchange neighborhood for different combinatorial optimization problems. In the next section, we describe other approaches that in polynomial time implicitly search an exponential size subset of the k-exchange neighborhood when k is not fixed.

Before describing the Lin–Kernighan approach, we introduce some notation. Subsequently, we will show how to generalize the Lin–Kernighan approach to variable-depth methods (and ejection chains) for heuristically solving combinatorial optimization problems. Suppose that T and T' are both subsets of E, but not necessarily feasible. A *path* from T to T' is a sequence $T=T_1,\ldots,T_K=T'$ such that $d(T_j,T_{j+1})=1$ for $j=1$ to $K-1$.

The variable-depth methods rely on a subroutine *Move* with the following features:
1. At each iteration, the subroutine Move creates a subset T_j and possibly also a feasible subset S_j from the input pair (S_{j-1}, T_{j-1}) according to some search criteria. The subset T_j may or may not be feasible. We represent this operation as $\text{Move}(S_{j-1}, T_{j-1}) = (S_j, T_j)$
2. $d(T_j, T_{j+1}) = 1$ for all $j=1$ to $K-1$.
3. T_j typically satisfies additional properties, depending on the variable-depth approach.

Let T be the current TSP tour and assume without loss of generality that T visits the cities in order $1,2,3,\ldots,n,1$. A *2-exchange neighborhood* for T can be defined as replacing two edges (i,j), and (k,l) by two other edges (i,k) and (j,l) or (i,l) and (j,k) to form another tour T'. Note that $d(T,T')=4$. The 2-exchange neighborhood may be described more formally by applying 4 Move operations where the first Move deletes the edge (i,j) from the tour, the second Move inserts the edge (i,k), the third Move deletes the edge (k,l) and the last Move inserts the edge (j,l).

Let $G=(N,A)$ be an undirected graph on n nodes. Let $P=v_1,\ldots,v_n$ be an n node Hamiltonian path of G. A *stem and cycle* (this terminology is introduced by Glover

[26]) is an n-arc spanning subgraph that can be obtained by adding an arc (i, j) to a Hamiltonian path, where i is an end node of the path. Note that if node i is an end node of the path, and if j is the other end node of the path, then the stem and cycle structure is a Hamiltonian cycle, or equivalently is a tour. If T is a path or a stem and cycle structure, we let $f(T)$ denote its total length.

The Lin–Kernighan heuristic allows the replacement of as many as n edges in moving from a tour S to a tour T, that is, $d(S, T)$ is equal to some arbitrary $k \leqslant 2n$. The algorithm starts by deleting an edge from the original tour T_1 constructing a Hamiltonian path T_2. Henceforth one of the end points of T_2 is fixed and stays fixed until the end of the iteration. The other end point is selected to initiate the search. The even moves insert an edge into the Hamiltonian path T_{2j} incident to the end point that is not fixed to obtain a stem and cycle T_{2j+1}. The odd moves in the iteration delete an edge from the current stem and cycle T_{2j-1} to obtain a Hamiltonian path T_{2j}. From each Hamiltonian path T_{2j}, one implicitly constructs a feasible tour S_{2j} by joining the two end nodes. At the end of the Lin–Kernighan iteration we obtain the new BaseSolution tour S_i such that $f(S_i) \leqslant f(S_{2j})$ for all j.

Now we will describe the steps of the Lin–Kernighan algorithm in more detail. During an even move the edge to be added is the minimum length edge incident to the unfixed end point, and this is added to the Hamiltonian path T_{2j} if and only if $f(S) - f(T_{2j+1}) > 0$. Lin–Kernighan [48] also describe a look-ahead refinement to the way this edge is chosen. The choice for the edge to be added to the Hamiltonian path T_{2j} is made by trying to maximize $f(T_{2j}) - f(T_{2j+2})$. On the other hand, the edges to be deleted during the odd moves are uniquely determined by the stem and cycle structure, T_{2j-1} created in the previous move so that T_{2j} will be a Hamiltonian path. Additional restrictions may be considered when choosing the edges to be added. Researchers have considered different combinations of restrictions such as an edge previously deleted cannot be added again or an edge previously added cannot be deleted again in a later move.

Finally, the Lin–Kernighan algorithm terminates with a local optimum when no improving tour can be constructed after considering all nodes as the original fixed node.

We now illustrate the Lin–Kernighan's algorithm using a numerical example. Consider the tour on 10 nodes shown in Fig. 1(a). The algorithm first deletes the arc (1, 2) creating the Hamiltonian path shown in Fig. 1(b). Then the arc (2, 6) is added giving the stem and cycle illustrated in Fig. 1(c).

Deleting edge (6, 5) from this structure yields a Hamiltonian path and adding edge (5, 8), we get another stem and cycle. The edge insertion moves in the Lin–Kernighan heuristic are guided by a cost criterion based on 'cumulative gain' and the edge deletion moves are defined uniquely to generate the path structure. The Lin–Kernighan algorithm reaches a local optimum when after considering all nodes as the starting node no improving solutions can be produced.

There are several variations of the Lin–Kernighan algorithm that have produced high quality heuristic solutions. These algorithms use several enhancements such as 2-opt, 3-opt, and special 4-opt moves [29,39,40,48,49,50,62] to obtain a tour that cannot be constructed via the basic Lin–Kernighan moves. Also, efficient data structures are used to update the tours to achieve computational efficiency and solution quality [24,40].

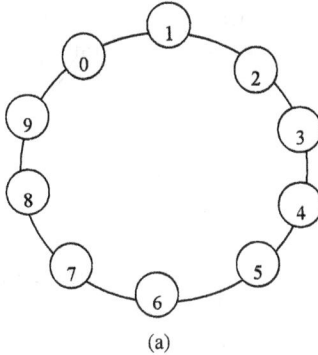

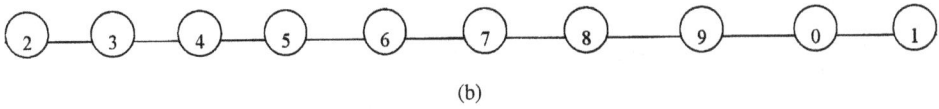

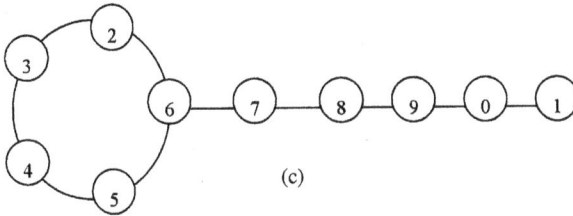

Fig. 1. Illustrating the Lin–Kernighan algorithm: (a) a tour on 10 nodes; (b) a Hamiltonian path; (c) a stem and cycle.

Papadimitriou [51] showed that the problem of determining a local optimum with respect to one version of the Lin–Kernighan algorithm is PLS-complete.

Now we can define the variable-depth methods for the TSP with the following procedure as a generalization of the Lin–Kernighan heuristic. This procedure takes as input a feasible tour S and then utilizes the function Move defined earlier. At each iteration the function Move creates the pair (T_j, S_j) where subset T_j is either a feasible solution or else an infeasible instance of a reference structure. The subset S_j is feasible. The function Move is called for r iterations where the value of r depends on an appropriate guidance rule. Finally, the procedure Variable-Depth-Search returns the feasible subset S_k that has the best objective function value found so far.

procedure Variable-Depth-Search(S);
begin
    $S_1 := T_1 := S$;
    for $j := 2$ to r **do** $(T_j, S_j) = \text{Move}(S_{j-1}, T_{j-1})$;
    select the set S_k that minimizes $(f(S_j): 1 \leqslant j \leqslant r)$;
end;

This particular type of variable-depth search relies on a heuristic called "Move" that systematically creates a path of solutions starting from the initial solution. This framework is quite flexible, and there are a variety of ways of designing the procedure Move. The details of how Move is designed can be the difference between a successful heuristic and one that is not so successful.

In the procedure described above, we assumed that Move creates a single feasible solution at each stage. In fact, it is possible for move to create multiple feasible solutions at each stage [26,59] or no feasible solution at a stage [26,57].

Many variable-depth methods require the intermediate solutions T_j to satisfy certain topological (or structural) properties. For example, in the Lin–Kernighan algorithm we required that for each odd value of j, T_j is a stem and cycle. We also required that for each even value of j, T_j is a Hamiltonian path. We will also see examples in the next section where T_j satisfies additional properties that are not structural. For example, additional properties may depend on the ordering of the indices.

Glover [26] considered a structured class of variable-depth methods called *ejection chains* based on classical alternating path methods, extending and generalizing the ideas of Lin–Kernighan. Glover writes "In rough overview, an ejection chain is initiated by selecting a set of elements to undergo a change of state (e.g., to occupy new positions and or receive new values). The result of this change leads to identifying a collection of other sets, with the property that the elements of at least one must be "ejected from" their current state. State change steps and ejection steps typically alternate, and the options for each depend on the cumulative effect of previous steps (usually, but not necessarily, being influenced by the step immediately preceding). In some cases, a cascading sequence of operations may be representing a domino effect. *The ejection chain terminology is intended to be suggestive rather than restrictive, providing a unifying thread that links a collection of useful procedures for exploiting structure, without establishing a narrow membership that excludes other forms of classification*".

In this paper, we will use the following more restrictive definition of ejection chains. We refer to the variable-depth method as an *ejection chain* if
 (i) $|T_1| = |T_3| = |T_5| = \cdots = n$, and
 (ii) $|T_2| = |T_4| = |T_6| = \cdots = n+1$ (or $n-1$).
For each even value of j, if $|T_j| = |S| - 1$, then T_j was obtained from T_{j-1} by ejecting an element. Otherwise, $|T_j| = |S| + 1$, and T_{j+1} is obtained from T_j by ejecting an element. Many of the variable-depth methods developed in the literature may be viewed as ejection chains. Typically these methods involve the construction of different reference structures along with a set of rules to obtain several different feasible solutions from them. To our knowledge, all the variable-depth methods for the traveling salesman problem considered in the literature may be viewed as ejection chains.

We can envision the nodes of the neighborhood graph with respect to the Lin–Kernighan neighborhood to consist of paths and stem-and-cycles. (Recall that tours are also instances of stem-and-cycles.) The endpoints of each edge of the neighborhood graph would link a path to a stem and cycle. The search technique would be the one proposed by Lin and Kernighan [42], and the selection procedure would be to select the best of the tours discovered along the way. Thus the reference structures described in ejection chains would be the nodes of the neighborhood graph in the ejection chain techniques. A technique in which the next BaseSolution is much more than a distance one from the current BaseSolution is a "variable-depth method". An ejection chain is a variable-depth method in which neighbors have the property that one is a subset of the other (and thus an element has been ejected in going from the larger to the smaller).

In the method described above, variable-depth methods rely on the function Move. One can also create exponential size subsets of N_k that are searched using network flows. In these neighborhoods, any neighbor can also be reached by a sequence of Moves in an appropriately defined neighborhood graph. Note that for the Lin–Kernighan algorithm the neighborhood size is polynomial, and it is the search that leads to finding solutions that are much different than the BaseSolution. We describe these network flows based techniques in the next section. Some of these techniques can also be viewed as ejection chain techniques if there is a natural way of associating an ejection chain (an alternating sequence of additions and deletions) with elements of the neighborhood [26,27,57,22].

Variable-depth and ejection chain based algorithms have been successfully applied in getting good solutions for a variety of combinatorial optimization problems. Glover [26], Rego [59], Zachariasen and Dum [81], Johnson and McGeoch [40], Mak and Morton [49], Pesch and Glover [53] considered such algorithms for the TSP. The vehicle routing problem is studied by Rego and Roucairol [61] and Rego [60]. Clustering algorithms using ejection chains are suggested by Dondorf and Pesch [13]. Variable-depth methods for the generalized assignment problem have been considered by Yagiura et al. [77] and ejection chain variations are considered by Yagiura et al. [76]. In addition, short ejection chain algorithms are applied to the multilevel generalized assignment problem by Laguna et al. [45]. These techniques are also applied to the uniform graph partitioning problem [16,20,42,52], categorized assignment problem [3], channel assignment problem [17], and nurse scheduling [14]. Sourd [68] apply a very general class of large neighborhood improvement procedures where the distance between two neighbors is variable to scheduling tasks on unrelated machines. These neighborhoods are developed by generating partial but still large enumeration trees based on the current solutions and searched heuristically.

4. Network flows based improvement algorithms

In this section, we study local improvement algorithms where the neighborhoods are searched using network flow based algorithms. The network flow techniques used to identify improving neighbors can be grouped into three categories: (i) minimum cost cycle finding methods; (ii) shortest path or dynamic programming based methods; and (iii) methods based on finding minimum cost assignments and matchings. The neighborhoods defined by cycles may be viewed as generalizations of 2-exchange neighborhoods. Neighborhoods based on assignments may be viewed as generalizations of insertion-based neighborhoods. In the following three subsections, we give general definitions of these exponential neighborhoods and describe the network flow algorithms used for finding an improving neighbor. For many problems, one determines an improving neighbor by applying a network flow algorithm to a related graph, which we refer to as an *improvement graph*.

4.1. Neighborhoods defined by cycles

In this subsection, we first define a generic partitioning problem. We then define the 2-exchange neighborhood and the cyclic exchange neighborhood.

Let $A = \{a_1, a_2, a_3, \ldots, a_n\}$ be a set of n elements. The collection $\{S_1, S_2, S_3, \ldots, S_K\}$ defines *a K-partition of A* if each set S_j is non-empty, the sets are pairwise disjoint, and their union is A. For any subset S of A, let $d[S]$ denote the cost of S. Then the set partitioning problem is to find a partition of A into at most K subsets so as to minimize $\Sigma_k d[S_k]$.

Let $\{S_1, S_2, S_3, \ldots, S_K\}$ be any feasible partition. We say that $\{T_1, T_2, T_3, \ldots, T_K\}$ is a *2-neighbor* of $\{S_1, S_2, S_3, \ldots, S_K\}$ if it can be obtained by swapping two elements that are in different subsets. The 2-exchange neighborhood of $\{S_1, S_2, S_3, \ldots, S_K\}$ consists of all 2-neighbors of $\{S_1, S_2, S_3, \ldots, S_K\}$. We say that $\{T_1, T_2, T_3, \ldots, T_K\}$ is a *cyclic-neighbor* of $\{S_1, S_2, S_3, \ldots, S_K\}$ if it can be obtained by transferring single elements among a sequence of $k \leqslant K$ subsets in S. Let $(S_h^1, S_m^2, S_n^3, \ldots, S_p^k)$ be such a sequence of k subsets, then we also require that $h = p$, that is, the last subset of the sequence is identical to S_h^1. We refer to the transferring of elements as a *cyclic exchange*. We illustrate a cyclic exchange using Fig. 2. In this example, node 9 is transferred from subset S_1 to subset S_2. Node 2 is transferred from subset S_2 to subset S_3. Node 3 is transferred from subset S_3 to subset S_3. Finally, the cycle exchange is completed by transferring node 14 from subset S_3 to subset S_4. Finally, the exchange is completed by transferring node 14 from subset S_4 to subset S_1. One may also define a *path neighbor* in an analogous way. From a mathematical perspective, it is easy to transform a path exchange into a cyclic exchange by adding appropriate dummy nodes.

In general, the number of cyclic neighbors is substantially greater than the number of 2-neighbors. Whereas there are $O(n^2)$ 2-neighbors, for fixed value of K, there are $O(n^K)$ cyclic neighbors. If K is allowed to vary with n, there may be an exponential number of cyclic neighbors.

Thompson [73], Thompson and Orlin [74], and Thompson and Psaraftis [75] show how to find an improving neighbor in the cyclic exchange neighborhood by finding a negative cost "subset-disjoint" cycle in an improvement graph. Here we will describe how to construct the improvement graph. Let $A = \{a_1, a_2, \ldots, a_n\}$ be the set of elements for the original set partitioning problem and let $S[i]$ denote the subset containing element a_i. The improvement graph is a graph $G = (V, E)$ where $V = \{1, 2, \ldots, n\}$ is a set of nodes corresponding to the indices of the elements of A of the original problem. Let $E = \{(i, j): S[i] \neq S[j]\}$, where an arc (i, j) corresponds to the transfer of node i from $S[i]$ to $S[j]$ and the removal of j from $S[j]$. For each arc $(i, j) \in E$, we let $c[i, j] = d[\{i\} \cup S[j] \setminus \{j\}] - d[S[j]]$, that is, the increase in the cost of $S[j]$ when i is added to the set and j is deleted. We say that a cycle W in G is *subset-disjoint* if for every pair i and j of nodes of W, $S[i] \neq S[j]$, that is, the elements of A corresponding to the nodes of W are all in different subsets. There is a one-to-one cost-preserving correspondence between cyclic exchanges for the partitioning problem and subset-disjoint

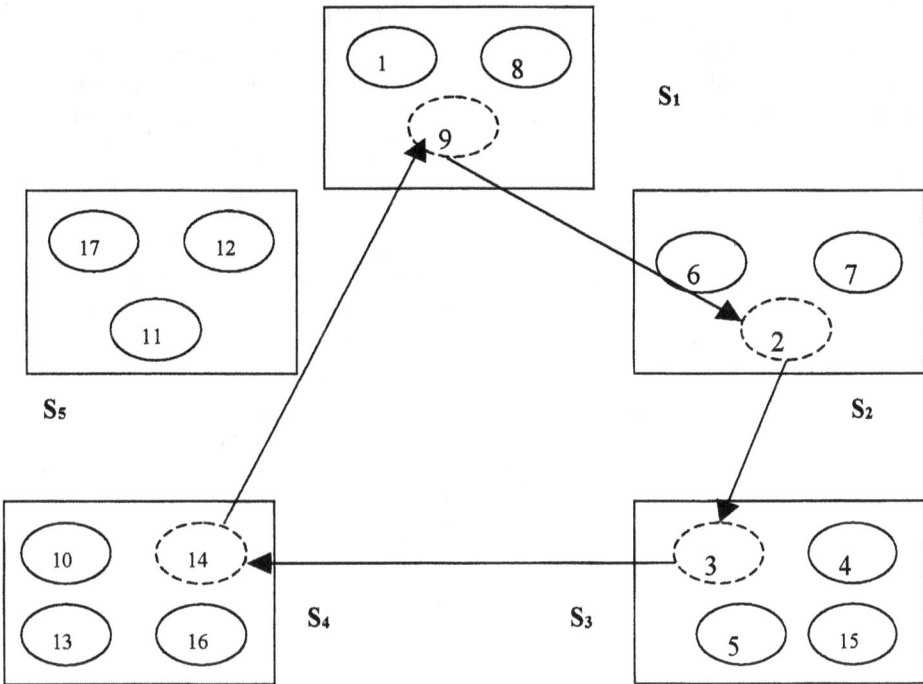

Fig. 2. Illustrating a cyclic exchange.

cycles in the improvement graph. In particular, for every negative cost cyclic exchange, there is a negative cost subset-disjoint cycle in the improvement graph. Unfortunately, the problem of determining whether there is a subset-disjoint cycle in the improvement graph is NP-complete, and the problem of finding a negative cost subset-disjoint cycle is NP-hard. (See, for example, Thompson [73], Thompson and Orlin [74], and Thompson and Psaraftis [75].)

Even though the problem of determining a negative cost subset-disjoint cycle in the improvement graph is NP-hard, there are effective heuristics for searching the graph. (See, for example, Thompson and Psaraftis [75] and Ahuja et al. [2].)

The cyclic exchange neighborhood search is successfully applied to several combinatorial optimization problems that can be characterized as specific partitioning problems. Thompson and Psaraftis [75], Gendreau et al. [25], and Fahrion and Wrede [19] solve the vehicle routing problem with the cyclic exchange neighborhood search. Frangioni et al. [23] apply cyclic exchanges to minimum makespan machine scheduling. Thompson and Psaraftis [75] also demonstrate its application to some scheduling problems. Ahuja et al. [2] developed the best available solutions for a widely used set of benchmark instances for the capacitated minimum spanning tree problems using cyclic exchanges.

The idea of finding improving solutions by determining negative cost cycles in improvement graphs has been used in several other contexts. Talluri [72] identifies cost saving exchanges of equipment type between flight legs for the daily airline fleet assignment problem by finding negative cost cycles in a related network. The fleet assignment problem can be modeled as an integer multicommodity flow problem subject to side constraints where each commodity refers to a fleet type. Talluri considers a given solution as restricted to two fleet types only, and looks for improvements that can be obtained by swapping a number of flights between the two fleet types. He develops an associated improvement graph and shows that improving neighbors correspond to negative cost cycles in the improvement graph. Schneur and Orlin [66] and Rockafellar [63] solve the linear multicommodity flow problem by iteratively detecting and sending flows around negative cost cycles. Their technique readily extends to a cycle-based improvement heuristic for the integer multicommodity flow problem. Wayne in [79] gives a cycle canceling algorithm for solving the generalized minimum cost flow problem. Firla et al. [21] introduce an improvement graph for the intersection of any two integer programs. Paths and cycles in this network correspond to candidates for improving feasible solutions. Furthermore, this network gives rise to an algorithmic characterization of the weighted bipartite b-matching problem. The algorithms discussed by Glover and Punnen [30] and Yeo [78] construct traveling salesman tours that are better than an exponential number of tours. These algorithms can also be viewed as computing a minimum cost cycle in an implicitly considered special layered network. We will discuss this heuristic in more detail in Section 5.

4.2. Neighborhoods defined by paths (or dynamic programming)

We will discuss three different types of neighborhood search algorithms based on shortest paths or dynamic programming. We discuss these approaches in the context of the traveling salesman problem. We can view these neighborhood search approaches when applied to the TSP as: (i) adding and deleting edges sequentially, (ii) accepting in parallel multiple swaps where a swap is defined by interchanging the current order of two cities in a tour, and (iii) cyclic shifts on the current tour. We next discuss these neighborhoods in more detail.

4.2.1. Creating a new neighbor by adding and deleting arcs sequentially

We first discuss a class of shortest path based methods that consider neighbors obtained by alternately adding and deleting edges from the current tour. These methods exhaustively search a subset of the ejection chain neighborhood of Section 3 with additional restrictions on the edges to be added. We assume for simplicity that the tour S visits the cities in the order $1, 2, 3, \ldots, n, 1$. Utilizing the terminology introduced in Section 3, let tour T be a k-exchange neighbor of S and let the path from S to T be the sequence $S = T_1, \ldots, T_K = T$. These neighborhoods correspond to trial solutions created by odd and even paths described in Punnen and Glover [57] among several other neighborhoods and trial solutions constructed from different kinds of path structures. The trial solution generated by odd paths was developed independently by Firla et al. [22]. To illustrate, this type of trial solution generated by either odd or even paths

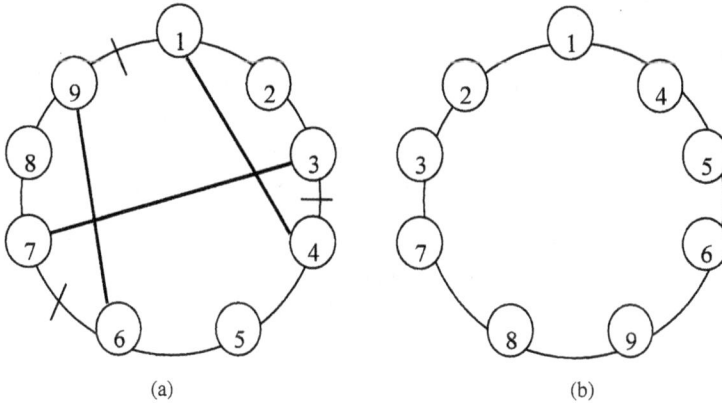

Fig. 3. Illustrating the alternate path exchange.

consider the following algorithm:

(i) Drop the edge $(n, 1)$ to obtain a Hamiltonian path T_2 and add an edge $(1, i)$ from node 1 to node i (where $i > 2$) to obtain a stem and cycle structure T_3.

(ii) The current terminal node of the path is node i. Drop edge $(i, i - 1)$ and add edge $(i - 1, j)$ for $i < j < n$, creating a Hamiltonian path first and then a stem and cycle structure.

(iii) Check if a termination criterion is reached. If yes, go to Step (iv). If no, let $i = j$ and go to step (ii).

(iv) Drop edge $(j, j - 1)$ and add the final edge $(j - 1, n)$ to complete the tour.

Fig. 3 illustrates this process on a 9 node tour $S = (1, 2, \ldots, 9, 1)$. The path exchange procedure initiates by deleting edge $(n, 1)$ and adding edge $(1, 4)$. Then edge $(3, 4)$ is dropped and edge $(3, 7)$ is added. Finally, the new tour T is created by deleting edge $(6, 7)$ and adding edge $(6, 9)$. In Fig. 3(a), we represent the edges that are added by bold lines and the edges that are dropped by a dash on the edge. Fig. 3(b) illustrates the new tour obtained after the path exchange.

Firla et al. [22], Glover [26] and Punnen and Glover [57] show that an improving solution in this neighborhood can be found in $O(n^2)$ time by finding an odd or even length shortest path in an improvement graph. Here, we describe an improvement graph that can be constructed to identify the best neighbor of a tour by finding a shortest path with even or odd number of nodes. Recall that $S = (1, 2, 3, \ldots, n, 1)$ is the current n node traveling salesman tour. The improvement graph is a graph $G = (V, E)$ where $V = \{1, 2, \ldots, n\}$ corresponding to nodes of the original problem, and $E = \{(i, j): 1 = i < j - 1 < n\}$ is a set of directed arcs. Arcs $(1, j) \in E$ (such that $2 < j = n$) correspond to the deletion of edge $(n, 1)$ and the addition of edge $(1, j)$, and arcs $(i, j) \in E$ (such that $1 < i < j - 1 < n$) correspond to the deletion of edge $(i - 1, i)$ and the addition of edge $(i - 1, j)$ on the original tour S. If $d[i, j]$ is the cost of going from city i to j then we associate a cost $c[1, j] = -d[n, 1] + d[1, j]$ for each arc $(1, j) \in E$ such that $2 < j = n$ and a cost $c[i, j] = -d[i - 1, i] + d[i - 1, j]$ for each arc $(i, j) \in E$ such that

$1 < i < j - 1 < n$. Finally, finding a negative cost path in G from node 1 to n identifies a profitable k-exchange.

In addition, trial solutions created from even and odd paths, new path structures such as broken paths and reverse paths leading to different trial solutions and reference structures are studied in [57]. The size of the neighborhood generated by even and odd paths alone is $\Omega(n2^n)$. Speed up techniques for searching neighborhoods is important even when the size of the neighborhood is not exponential. For example, using a shortest path algorithm on a directed acyclic improvement graph, Glover [27] obtained a class of best 4-opt moves in $O(n^2)$ time.

4.2.2. Creating a new neighbor by compounded swaps

The second class of local search algorithms defined by path exchanges is a generalization of the swap neighborhood. Given an n node traveling salesman tour $T = (1, 2, 3, \ldots, n, 1)$, the swap neighborhood generates solutions by interchanging the positions of nodes i and j for $1 \leqslant i < j \leqslant n$. For example letting $i = 3$ and $j = 6$, $T' = (1, 2, 6, 4, 5, 3, 7, \ldots, n, 1)$ is a neighbor of T under the swap operation. Two swap operations switching node i with j, and node k with l are said to be *independent* if $\max\{i, j\} < \min\{k, l\}$, or $\min\{i, j\} > \max\{k, l\}$. Then a large-scale neighborhood on the tour T can be defined by compounding (taking the union of) an arbitrary number of independent swap operations.

Congram et al. [9] and Potts and van de Velde [55] applied this compounded swap neighborhood to the single machine total weighted tardiness scheduling problem and the TSP, respectively. They refer to this approach as *dynasearch*. In their paper, Congram et al. [9] show that the size of the neighborhood is $O(2^{n-1})$ and give a dynamic programming recursion that finds the best neighbor in $O(n^3)$ time. Hurink [38] apply a special case of the compounded swap neighborhood where only adjacent pairs are allowed to switch in the context of one machine batching problems and show that an improving neighbor can be obtained in $O(n^2)$ time by finding a shortest path in the appropriate improvement graph.

Now we describe an improvement graph to aid in searching the compounded swap neighborhood. Let $T = (1, 2, 3, \ldots, n, 1)$ be an n node traveling salesman tour. The improvement graph is a graph $G = (V, E)$, where (i) $V = \{1, 2, \ldots, n, 1', 2', \ldots, n'\}$ is a set of nodes corresponding to the nodes of the original problem and a copy of them, and (ii) E is a set of directed arcs $(i, j') \cup (j', k)$, where an arc (i, j') corresponds to the swap of the nodes i and j, and an arc (j', k) indicates that node k will be the first node of the next swap. For example, a path of three arcs $(i, j'), (j', k), (k, l')$ in G represents two swap operations switching node i with j, and node k with l. To construct the arc set E, every pair (i, j') and (j', k) of nodes in V is considered, and arc (i, j') is added to E if and only if $j > i > 1$. Arc (j', k) is added to E if and only if $j = 1$ and $k > j$ or $j > 1$ and $k > j + 1$. For each arc $(i, j') \in E$, we associate a cost $c[i, j']$ that is equal to the net increase in the optimal cost of the TSP tour after deleting the edges $(i - 1, i)$, $(i, i + 1)$, $(j - 1, j)$ and $(j, j + 1)$ and adding the edges $(i - 1, j)$, $(j, i + 1)$, $(j - 1, i)$ and $(i, j + 1)$. In other words, if $d[i, j]$ is the cost of going from node i to node j in the original problem and $d[n, n + 1] = d[n, 1]$, then

$$c[i, j'] = (-d[i-1, i] - d[i, j] - d[j, j+1])$$

$$+ (d[i-1, j] + d[j, i] + d[i, j+1])$$

$$\text{for } j' = i+1,$$

and

$$c[i, j'] = (-d[i-1, i] - d[i, i+1] - d[j-1, j] - d[j, j+1])$$

$$+ (d[i-1, j] + d[j, i+1] + d[j-1, i] + d[i, j+1]) \quad \text{for } j' > i+1.$$

The cost $c[j', k]$ of all edges (j', k) are set equal to 0.

Now finding the best neighbor of a TSP tour for the compounded swap neighborhood is equivalent to finding a shortest path on this improvement graph, and hence takes $O(n^2)$ time. Note that since TSP is a cyclic problem, one of the nodes is held fixed during the exchange. In the above construction of the improvement graph, without loss of generality, we assumed that node 1 is not allowed to move, and hence the neighborhood is searched by finding a shortest path from node $1'$ to either node n or n'. The dynamic programming recursion given in Congram et al. [9] for searching the neighborhood will also take $O(n^2)$ time when applied to the TSP. The shortest path algorithm given above takes $O(n^3)$ time when applied to the total weighted tardiness scheduling problem because it takes $O(n^3)$ time to compute the arc costs.

4.2.3. Creating a new neighbor by a cyclical shift

The final class of local search algorithms in this section is based on a kind of cyclic shift of pyramidal tours [8]. A tour is called *pyramidal* if it starts in city 1, then visits cities in increasing order until it reaches city n, and finally returns through the remaining cities in decreasing order back to city 1. Let $T(i)$ represent the city in the ith position of tour T. A tour T' is a *pyramidal neighbor* of a tour T if there exists an integer p such that:

(i) $0 \leqslant p \leqslant n$,

(ii) $T'(1) = T(i_1), T'(2) = T(i_2), \ldots, T'(p) = T(i_p)$ with $i_1 < i_2 < \cdots < i_p$ and

(iii) $T'(p+1) = T(j_1), T'(p+2) = T(j_2), \ldots, T'(n) = T(j_{n-p})$ with $j_1 > j_2 > \cdots > j_{n-p}$.

For example, if tour $T = (1, 2, 3, 4, 5, 1)$ then tour $T' = (1, 3, 5, 4, 2, 1)$ is a pyramidal neighbor. Note that a drawback of this neighborhood is that edges $(1,2)$ and $(1,n)$ belong to all tours. To avoid this Carlier and Villon [8] consider the n rotations associated with a given tour. The size of this neighborhood is $\theta(n2^{n-1})$ and it can be searched in $O(n^3)$ time using n iterations of a shortest path algorithm in the improvement graph.

Now, we describe an improvement graph where for the TSP a best pyramidal neighbor can be found by solving for a shortest path on this graph. Let $T = (1, 2, 3, \ldots, n, 1)$ be an n node traveling salesman tour. The improvement graph is a graph $G = (V, E)$ where (i) $V = \{1, 2, \ldots, n, 1', 2', \ldots, n'\}$ corresponding to the nodes of the original problem and a copy of them, and (ii) E is a set of directed arcs $(i, j') \cup (j', k)$, where an arc (i, j') corresponds to having the nodes i to j in a consecutive order, and an arc

(j',k) corresponds to skipping the nodes $j+1$ to $k-1$ and appending them in reverse order to the end of the tour. To construct the arc set E, every pair (i,j') and (j',k) of nodes in V is considered. Arc (i,j') is added to E if and only if $i \leqslant j$, and arc (j',k) is added to E if and only if $j < k+1$. For each arc $(i,j') \in E$, we associate a cost $c[i,j']$ that is equal to the net increase in the cost of the pyramidal neighbor of the TSP tour after adding the edge $(j+1, i-1,)$ when the tour is visiting the previously skipped cities in reverse order. For each arc $(j',k) \in E$, we associate a cost $c[j',k]$ that is equal to the net increase in the optimal cost of the TSP tour after adding the edge (j,k) and deleting the edges $(j,j+1)$ and $(k-1,k)$. In other words, if $d[i,j]$ is the cost of going from city i to j in the original problem, then for $i < j$ and $j < n-1$

$$c[i,j'] = d[j+1, i-1],$$

$$c[j',k] = -d[j,j+1] - d[k-l,k] + d[k-1,k].$$

Note that some extra care must be taken when calculating the cost of the edges that have 1, 1', n, and n' as one of the end points. Now the neighborhood can be searched by finding a shortest path from node 1 to either node n or n'. Carlier and Villon [8] also show that if a tour is a local optimum for the above neighborhood, then it is a local optimum for the 2-exchange neighborhood.

In addition to these three classes of neighborhoods, dynamic programming has been used to determine optimal solutions for some special cases of the traveling salesman problem. Simonetti and Balas [67] solve the TSP with time windows under certain kinds of precedence constraints with a dynamic programming approach. Burkard et al. [7] prove that over a set of special structured tours that can be represented via PQ-trees, the TSP can be solved in polynomial time using a dynamic programming approach. They also show that the set of pyramidal tours can be represented by PQ-trees and there is an $O(n^2)$ algorithm for computing the shortest pyramidal tour. These results will be discussed in more detail in Section 5.

4.3. Neighborhoods defined by assignments and matchings

In this section, we discuss an exponential neighborhood structure defined by finding minimum cost assignments in an improvement graph. We illustrate this neighborhood in the context of the traveling salesman problem. Also, we show that the assignment neighborhood can be generalized to a neighborhood defined by finding a minimum cost matching on a non-bipartite improvement graph. We demonstrate this generalization on the set partitioning problem.

The assignment neighborhood for the traveling salesman problem can be viewed as a generalization of the simple neighborhood defined by ejecting a node from the tour and reinserting it optimally. Given an n node tour $T = (1, 2, 3, \ldots, n, 1)$, if the cost of going from city i to j is $d[i,j]$, then the first step in searching the assignment neighborhood is to create a bipartite improvement graph as follows:

(i) For some $k \leqslant \lfloor n/2 \rfloor$, choose and eject k nodes from the current tour T. Let the set of these ejected nodes be $V = \{v_1, v_2, \ldots, v_k\}$ and the set of remaining nodes be $U = \{u_1, u_2, \ldots, u_{n-k}\}$.

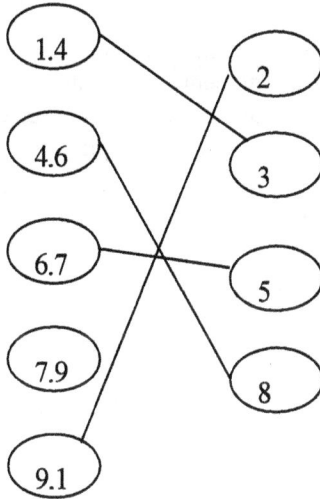

Fig. 4. Illustrating the matching neighborhood.

(ii) Construct a sub-tour $T' = (u_1, u_2, \ldots, u_{n-k}, u_1)$. Let q_i denote the edge for every (u_i, u_{i+1}) for $i = 1$ to $n - k - 1$, and let q_{n-k} denote the edge (u_{n-k}, u_1).

(iii) Now construct a complete bipartite graph $G = (N, N', E)$ such that $N = \{q_i: i = 1$ to $n - k\}$, $N' = V$, and the weight on each edge (q_i, v_j) is $c[q_i, v_j] = d[u_i, v_j] + d[v_j, u_{i+1}] - d[u_i, u_{i+1}]$.

A neighbor of T corresponds to a tour T^* obtained by inserting the nodes of V into the subtour T' with at most one node inserted between adjacent nodes of T'. The minimum cost assignment of k arcs corresponds to the minimum cost neighbor of T.

Using Fig. 4, we will demonstrate the assignment neighborhood on the 9 node tour $T = (1, 2, 3, 4, 5, 6, 7, 8, 9, 1)$. Let $V = \{2, 3, 5, 8\}$, then we can construct the sub-tour on the nodes in U as $T' = (1, 4, 6, 7, 9, 1)$. Fig. 4 illustrates the bipartite graph G with only the edges of the matching for simplicity. The new tour obtained is $T'' = (1, 3, 4, 8, 6, 5, 7, 9, 2, 1)$. Note that when $k = \lfloor n/2 \rfloor$, the size of the assignment neighborhood is equal to $\Omega(\lfloor n/2 \rfloor!)$.

The assignment neighborhood was first introduced in the context of the TSP by Sarvanov and Doroshko [64] for the case $k = n/2$ and n even. Gutin [31] gives a theoretical comparison of the assignment neighborhood search algorithm with the local steepest descent algorithms for $k = n/2$. Punnen [56] considered the general assignment neighborhood for arbitrary k and n. An extension of the neighborhood where paths instead of nodes are ejected and reinserted optimally by solving a minimum weight matching problem is also given in [56]. Gutin [32] shows that for certain values of k the neighborhood size can be maximized along with some low complexity algorithms searching related neighborhoods. Gutin and Yeo [35] constructed a neighborhood based on the assignment neighborhood and showed that moving from any tour T to another tour

T' using this neighborhood takes at most 4 steps. Deineko and Woeginger [12] study several exponential neighborhoods for the traveling salesman problem and the quadratic assignment problem based on assignments and matchings in bipartite graphs as well as neighborhoods based on partial orders, trees and other combinatorial structures. The matching based neighborhood heuristic is also applied to the inventory routing problem by Dror and Levy [15].

Another class of matching based neighborhoods can be obtained by packing subtours, where the subtours are generated by solving a bipartite minimum weight matching problem [41,58]. It is NP-hard to find the best such tour. Efficient heuristics can be used to search this neighborhood [28,41,58]. Neighborhood search algorithms using this neighborhood can be developed by utilizing cost modifications or other means to control the matching generated.

We next consider a neighborhood structure based on non-bipartite matchings. We will discuss this neighborhood in the context of the general set partitioning problem considered in Section 3. Let $S = \{S_1, S_2, S_3, \ldots, S_K\}$ be a partition of the set $A = \{a_1, a_2, a_3, \ldots, a_n\}$. Then construct a complete graph $G = (N, E)$ such that every node i for $1 \leqslant i \leqslant K$ represents the subset S_i in S. Now the weights $c[i,j]$ on the edge (i,j) of G can be constructed separately with respect to a variety of rules. One such rule can be as follows:

 (i) Let the cost contribution of subset S_i to the partitioning problem be $d[S_i]$.
 (ii) For each edge (i,j) in E, combine the elements in S_i and S_j, and repartition it into two subsets optimally. Let the new subsets be S_i' and S_j'.
(iii) Then $c[i,j] = (d[S_i'] + d[S_j']) - (d[S_i] + d[S_j])$.

Note that if the edges with non-negative weights are eliminated from G, then any negative cost matching on this graph will define a cost improving neighbor of S. Tailard apply these types of ideas to a general class of clustering problems in [70], and to the vehicle routing problem in [69].

5. Solvable special cases and related neighborhoods

There is a vast literature on efficiently solvable special cases of NP-hard combinatorial optimization problems. Of particular interest for our purposes are those special cases that can be obtained from the original NP-hard problem by restricting the problem topology, or by adding constraints to the original problem, or by a combination of these two factors. By basing neighborhoods on these efficiently solvable special cases, one can often develop exponential sized neighborhoods that may be searched in polynomial time. We point out that most of these techniques have not been tested experimentally and might yield poor local optima. Note that the cyclical shift neighborhood discussed in Section 4 is based on an $O(n^2)$ algorithm for finding the minimum cost pyramidal tour.

Our next illustration deals with Halin graphs. A *Halin graph* is a graph that may be obtained by embedding a tree that has no nodes of degree 2 in the plane, and joining the leaf nodes by a cycle so that the resulting graph is planar. Cornuejols et al. [10] gave an $O(n)$ algorithm for solving the traveling salesman problem on a Halin graph.

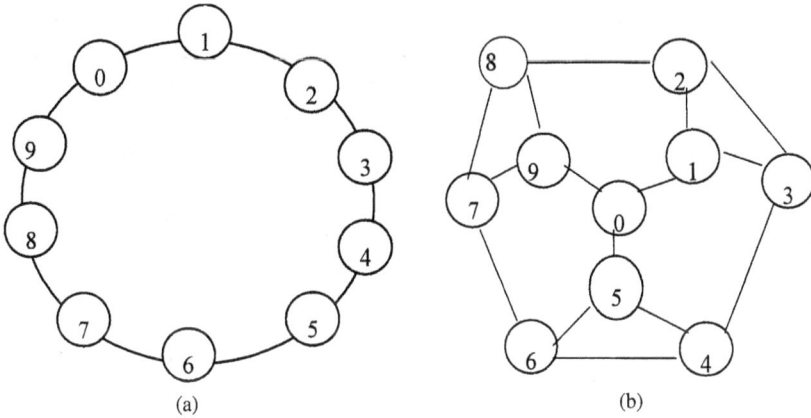

Fig. 5. A Halin graph: (a) a tour on 10 nodes; (b) a Halin extension.

Note that Halin graphs may have an exponential number of TSP tours, such as in the example of Fig. 3 (taken from [10]).

We now show how Halin graphs can be used to construct a very large neighborhood for the traveling salesman problem. Suppose that T is a tour. We say that H is a *Halin extension* of T if H is a Halin graph and if T is a subgraph of H. Fig. 5 illustrates a Halin extension for tour $T = (0, 1, 2, \ldots, 9, 0)$. Suppose that there is an efficient procedure HalinExtend(T), that creates some Halin extension of T. To create the neighborhood $N(T)$, one would let $H(T) = $ HalinExtend(T), and then let $N(T) = \{T': T'$ is a tour in $H(T)\}$. To find the best tour in this neighborhood, one would find the best tour in $H(T)$. In principle, one could define a much larger neighborhood: $N(T) = \{T':$ there exists a Halin extension of T containing in $T'\}$. Unfortunately, this neighborhood may be too difficult to search efficiently since it would involve simultaneously optimizing over all Halin extensions of T. Similar Halin graph based schemes for the bottleneck traveling salesman problem and the steiner tree problem could be developed using the linear time algorithm of Philips et al. [54], and Winter [80], respectively.

In our previous examples, we considered pyramidal tours, which may be viewed as the traveling salesman problem with additional constraints. We also considered the TSP as restricted to Halin graphs. In the next example, we consider (cf. [30]) a neighborhood that simultaneously relies on a restricted class of graphs plus additional side constraints. Glover and Punnen [30] identified the following class of tours among which the best member can be identified in linear time. Let $C_1, C_2, \ldots, C_k$ be k vertex-disjoint cycles each with at least three nodes, and such that each node is in one of the cycles. A *single edge ejection tour* is a tour T with the following properties:

1. $|T \cap C_i| = |C_i| - 1$ for $i = 1$ to k, that is, T has $|C_i| - 1$ arcs in common with C_i.
2. There is one arc of T directed from C_i to C_{i+1} for $i = 1$ to $k - 1$ and from C_k to C_1.

The number of ways of deleting arcs from the cycles is at least $\prod_i |C_i|$, which may be exponentially large, and so the number of single edge ejection tours is exponentially large. To find the optimum single edge ejection tour, one can solve a related shortest path problem on an improvement graph. The running time is linear in the number of arcs. Given a tour, one can delete $k + 1$ arcs of the tour, creating k paths, and then transform these k paths into the union of k cycles described above. In principle, the above neighborhood search approach is easy to implement; however, the quality of the search technique is likely to be sensitive to the details of the implementation. Glover and Punnen [30] consider broader neighborhoods as well, including something that they refer to as "double ejection tours". They also provide an efficient algorithm for optimizing over this class of tours.

Yeo [78] considered another neighborhood for the asymmetric version of the TSP. His neighborhood is related to that of Glover and Punnen [30], but with dramatically increased size. He showed that the search time for this neighborhood is $O(n^3)$. Burkard and Deineko [6] identified another class of exponential neighborhood such that the best member can be identified in quadratic time. Each of these algorithms can be used to develop VLSN search algorithms. However, to the best of our knowledge, no such algorithms have been implemented yet.

We summarize the results of this section with a general method for turning a solution method for a restricted problem into a very large-scale neighborhood search technique. Let X be a class of NP-hard combinatorial optimization problem. Suppose that X' is a restriction of X that is solvable in polynomial time. Further suppose that for a particular instance (F, f) of X, and for every feasible subset S in F, there is a subroutine "CreateNeighborhood(S)" that creates a well-structured instance (F', f) of X' such that

1. S is an element of F'.
2. F' is a subset of F.
3. (F', f) is an instance of X'.

We refer to F' as the X'-*induced neighborhood* of S. The neighborhood search approach consists of calling CreateNeighborhood(S) at each iteration, and then optimizing over (F', f) using the polynomial time algorithm. Then S is replaced with the optimum of (F', f), and the algorithm is iterated. Of particular interest is the case when the subroutine CreateNeighborhood runs in polynomial time, and the size of F' is exponential. The neighborhood would not be created explicitly. While we believe that this approach has enormous potential in the context of neighborhood search, this potential is largely unrealized. Moreover, it is not clear in many situations how to realize this potential. For example, many NP-hard combinatorial optimization problems are solvable in polynomial time when restricted to series-parallel graphs. Examples include network reliability problem [65], optimum communication spanning tree problem [18], vertex cover problem [5,71], the feedback vertex set problem [5,71], etc. Similar results are obtained for job shop scheduling problems with specific precedence restrictions [44]. It is an interesting open question how to best exploit these efficient algorithms for series-parallel graphs in the context of neighborhood search.

6. Neighborhood metrics

In this section we describe neighborhood metrics that might be a guide as to the performance of local search algorithms with respect to the given neighborhoods. As mentioned earlier, a critical issue in the design of neighborhood search heuristics is the balance between the size of the neighborhood and the time required to search it. Hence an important neighborhood metric is the *size of the neighborhood*. In terms of the neighborhood graph, the size of the neighborhood for a given solution S can be viewed as the number of directed arcs leaving S, or equivalently the outdegree of S. For the variable-depth methods the neighborhood size is not necessarily exponential, and it is the search that leads to finding solutions that are much different than the BaseSolution. On the other hand, for network based approaches and methods induced by solvable cases the neighborhood size is typically exponential. (See Table 1)

Table 1 summarizes the size of the neighborhoods discussed previously for the TSP (this table is mostly taken from [7]):

Another neighborhood metric that is studied in the literature is the *diameter of the neighborhood graph*. The *distance* from a node S to a node T in the neighborhood graph is the length of a shortest path from S to T. The *diameter* of the neighborhood graph NG is the least positive integer d such that $d(S,T) = d$ for all nodes S and T of NG. Gutin and Yeo [35] construct polynomially searchable neighborhoods of exponential size for the TSP where the corresponding neighborhood graphs has diameter four; that is, for any pair of tours T_1 and T_5, there exists tours T_2, T_3, and T_4 such that $T_i \in N(T_{i-1})$ for all $i = 2, 3, 4, 5$. The neighborhood graph for the cyclical-shift based neighborhood considered by Carlier and Villon [8] has diameter $\theta(\log n)$.

For a given neighborhood graph, we say that $P = i_1, i_2, \ldots, i_K$ is *monotone* if the objective value $f(i_j) < f(i_{j-1})$ for $j = 2$ to K. We let $d_m(S)$ denote the length of the shortest monotone path from S to a local optimum solution. If $d_m(S)$ is exponentially large for

Table 1

Neighborhood	Size	Log (size)	Time to search	Reference
2-Opt	$\Omega(n^2)$	$\Theta(\log n)$	$O(n^2)$	Croes [11]
k-Opt	$\Omega(n^k)$	$\Theta(\log n)$	$O(n^k)$	Lin [47]
Pyramidal	$\Omega(2^n)$	$\Theta(n)$	$O(n^2)$	Klyaus [43]
Cyclical shift	$\Omega(n2^n)$	$\Theta(n)$	$O(n^3)$	Carlier and Villion [8]
Edge ejection	$\Omega((12^n)^{1/3})$	$\Theta(n)$	$O(n)$	Glover and Punnen [30]
Shortest path based edge ejection	$\Omega(n2^n)$	$\Omega(n)$	$O(n^2)$	Punnen and Glover [57], Glover [26]
Cyclic-exchange (fixed-k subsets)	$\Theta(n^k)$	$\Theta(\log n)$	$O(n^2)$	Ahuja et al. [2]
Compound swaps	$\Theta(2^{n-1})$	$\Theta(n)$	$O(n^2)$	Potts and van de Velde [55]
Matching based[a]	$\Theta(n!/2)$	$\Theta(n \log n)$	$O(n^3)$	Sarvanov and Doroshko [64]
Halin graphs	$\Omega(2^n)$	$\Theta(n)$	$O(n)$	Cornuejols et al. [10]
PQ-Trees	$2^{\Theta(n \log \log n)}$	$\Theta(n \log \log n)$	$O(n^3)$	Burkard et al. [7]

[a] This assumes that $k = \lfloor n/2 \rfloor$, nodes are deleted. Better time bounds are available if fewer nodes are deleted, and the size of the neighborhood accordingly is decreased.

any S, this ensures that a neighborhood search approach starting at S will be exponentially long. In the converse direction, let $d_m^p(S,T)$ denote the longest monotone path from S to T. If $d_m^p(S,T)$ is guaranteed to be polynomial, than any neighborhood search technique based on this neighborhood will have a polynomial number of iterations.

Finally, we consider *domination analysis* of neighborhood search algorithms. The domination analysis of a heuristic analyzes the number of solutions that are 'dominated' by the solution produced. Let α be a heuristic algorithm for a combinatorial optimization problem which produces a solution S^* in F. The domination number of α, denoted by $\mathrm{dom}(\alpha)$, is the cardinality of the set $F(S^*)$, where $F(S^*) = \{S \in F: f(S) \geqslant f(S^*)\}$. If $\mathrm{dom}(\alpha) = |F|$, then S^* is an optimal solution. Domination analysis of various algorithms for the TSP has been studied in [28–30,33,34,36,58,56]

7. Computational performance of VLSN search algorithms

In this section, we briefly discuss the computational performance of some of the VLSN algorithms mentioned in the earlier sections. We first consider the traveling salesman problem. The Lin–Kernighan algorithm and its variants are widely believed to be the best heuristics for the TSP. In an extensive computational study, Johnson and McGeoch [40] substantiate this belief by providing a detailed comparative performance analysis. Rego [59] implemented a class of ejection chain algorithms for the TSP with very good experimental results indicating superiority of his method over the original Lin–Kernighan algorithm. Punnen and Glover [57] implemented a shortest path based ejection chain algorithm. The implementations of Rego [59] and Punnen and Glover [57] are relatively straightforward and use simple data structures.

Recently, Helsgaun [37] reported very impressive computational results based on a complex implementation of the Lin–Kernighan algorithm. Although this algorithm uses the core Lin–Kernighan variable-depth search, it differs from previous implementations in several key aspects. The superior computational result is achieved by efficient data handling, special 5-opt moves, new non-sequential moves, effective candidate lists, cost computations, effective use of upper bounds on element costs, information from the Held and Karp 1-tree algorithm and sensitivity analysis, among others. Helsgaun reports that his algorithm produced optimal solutions for all test problems for which an optimal solution is known, including the 7397 city problem and the 13,509 city problem considered by Applegate et al. [4]. Helsgaun estimated the average running time for his algorithm as $O(n^{2.2})$. To put this achievement in perspective, it may be noted that the 13,509 city problem solved to optimality by an exact branch and cut algorithm by Applegate et al. [4] used a cluster of three Digital Alpha 4100 servers (with 12 processors) and a cluster of 32 Pentium-II PCs and consumed three months of computation time. Helsgaun, on the other hand, used a 300 MHz G3 Power Macintosh. For the 85,900 city problem, pla85900 of the TSPLIB, Helsgaun obtained an improved solution using two weeks of CPU time. (We also note that the vast majority of time spent by Applegate et al. [4] is in proving that the optimal solution is indeed optimal.)

Table 2
Small TSP Problems: best % deviation from the optimum

Problem	REGO	HLK	MLK	SPG
bier127	0.12	0.00	0.42	0.10
u159	0.00	0.00	0.00	—
ch130	—	0.00	—	0.23
ch150	—	0.00	—	0.40
d198	0.30	0.00	0.53	0.33
d493	0.97	0.00	—	2.65
eil101	0.00	0.00	0.00	0.79
fl417	0.78	0.00	—	0.41
gil262	0.13	0.00	1.30	1.81
kroA150	0.00	0.00	0.00	0.00
kroA200	0.27	0.00	0.41	0.68
kroB150	0.02	0.00	0.01	0.07
kroB200	0.11	0.00	0.87	0.42
kroC100	0.00	0.00	0.00	0.00
kroD100	0.00	0.00	0.00	0.00
kroE100	0.00	0.00	0.21	0.02
lin105	0.00	0.00	0.00	0.00
lin318	0.00	0.00	0.57	1.03
pcb442	0.22	0.00	1.06	2.41
pr107	0.05	0.00	0.00	0.00
pr124	0.10	0.00	0.08	0.00
pr136	0.15	0.00	0.15	0.00
pr144	0.00	0.00	0.39	0.00
pr152	0.90	0.00	4.73	0.00
pr226	0.22	0.00	0.09	0.11
pr264	0.00	0.00	0.59	0.20
pr299	0.22	0.00	0.44	1.30
pr439	0.55	0.00	0.54	1.29
rd100	0.00	0.00	—	0.00
rd400	0.29	0.00	—	2.45
ts225	0.25	0.00	—	0.00
gr137	0.20	0.00	0.00	—
gr202	1.02	0.00	0.81	—
gr229	0.23	0.00	0.20	—
gr431	0.91	0.00	1.41	—

In Table 2, we summarize the performance of Rego's algorithm (REGO) [59], Helsgaun–Lin–Kernighan algorithm (HLK) [37], the modified Lin–Kernighan algorithm of Mak and Morton [49] and the shortest path algorithm of Punnen and Glover (SPG) [57] on small instances of the TSP problem. In the table we use the best case for each of these algorithms.

In Table 3, we summarize the performance of Rego's algorithm (REGO) [59], Helsgaun–Lin–Kernighan algorithm (HLK) [37], and the Lin–Kernighan implementation (JM-LK) of Johnson and McGeoch [40]. Reference source not found for large

Table 3
Large TSP: average % deviation over several runs

Problem	Rego	JM-LK	HLK
dsj1000	1.10	3.08	0.035
pr1002	0.86	2.61	0.00
pr2392	0.79	2.85	0.00
pcb3038	0.97	2.04	0.00
fl3795	7.16	8.41	—
fl4461	1.06	1.66	0.001
pla7397	1.57	2.19	0.001

instances of the TSP problem. In the table we use the average % deviation for each of these algorithms.

Next, we consider the capacitated minimum spanning tree problem. This is a special case of the partitioning problem discussed in Section 4. Exploiting the problem structure, Ahuja et al. [2] developed a VLSN search algorithm based on cyclic exchange neighborhood. This algorithm is highly efficient and obtained improved solutions for many benchmark problems. It currently has the best available solution for every instance listed in the set of benchmarks, accessible at http://www.ms.ic.ac.uk/info.html.

As our last example, we consider VLSN search algorithms for the generalized assignment problems (GAP). Yagiura et al. [77] developed an ejection chain based tabu search algorithm for the GAP. They report that in reasonable amount of computation time they obtained solutions that are superior or comparable with that of existing algorithms. Based on computational experiments and comparisons, it is reported that on benchmark instances the solutions produced by their algorithm are within 16% optimal.

Many of the references cited throughout this paper also report computational results based on their algorithms. For details we refer to these papers. Several neighborhoods of very large size discussed in the paper have not been tested experimentally within a VLSN search framework. Effective implementations of these and related neighborhoods are topics for further investigation.

Acknowledgements

The research of the first author was supported by the NSF Grant DMI-9900087. The second and third authors were in part supported by the NSF Grants DMI-9810359 and DMI-9820998. The fourth author was supported by NSERC Grant OPG0170381.

References

[1] E. Aarts, J.K. Lenstra, Local Search in Combinatorial Optimization, Wiley, New York, 1997.

[2] R.K. Ahuja, J.B. Orlin, D. Sharma, New neighborhood search structures for the capacitated minimum spanning tree problem, Research Report 99-2, Department of Industrial & Systems Engineering, University of Florida, 1999.

[3] V. Aggarwal, V.G. Tikekar, Lie-Fer Hsu, Bottleneck assignment problem under categorization, Comput. Oper. Res. 13 (1986) 11–26.

[4] D. Applegate, R. Bixby, V. Chvatal, W. Cook, On the solution of traveling salesman problems, Documenta Math. ICM (1998) 645–656.

[5] S. Arnborg, A. Proskurowski, Linear time algorithms for NP-hard problems restricted to partial k-trees, Discrete Appl. Math. 23 (1989) 11–24.

[6] R.E. Burkard, V.G. Deineko, Polynomially solvable cases of the traveling salesman problem and a new exponential neighborhood, Computing 54 (1995) 191–211.

[7] R.E. Burkard, V.G. Deineko, G.J. Woeginger, The travelling salesman problem and the PQ-tree, Mathematics of Operations Reserach 23 (1998) 613–623.

[8] J. Carlier, P. Villon, A new heuristic for the traveling salesman problem, RAIRO Oper. Res. 24 (1990) 245–253.

[9] R.K. Congram, C.N. Potts, S.L. van de Velde, An iterated dynasearch algorithm for the single machine total weighted tardiness scheduling problem, 1998, paper in preparation.

[10] G. Cornuejols, D. Naddef, W.R. Pulleyblank, Halin graphs and the traveling salesman problem, Math. Programming 26 (1983) 287–294.

[11] G.A. Croes, A method for solving traveling-salesman problems, Oper. Res. 6 (1958) 791–812.

[12] V. Deineko, G.J. Woeginger, A study of exponential neighborhoods for the traveling salesman problem and the quadratic assignment problem, Report Woe-05, Technical University Graz, 1997.

[13] U. Dorndorf, E. Pesch, Fast clustering algorithms, ORSA J. Comput. 6 (1994) 141–153.

[14] K.A. Dowsland, Nurse scheduling with tabu search and strategic oscillation, European J. Oper. Res. 106 (1998) 393–407.

[15] M. Dror, L. Levy, A vehicle routing improvement algorithm comparison of a "greedy" and a "matching" implementation for inventory routing, Comput. Oper. Res. 13 (1986) 33–45.

[16] A.E. Dunlop, B.W. Kernighan, A procedure for placement of standard cell VLSI circuits, IEEE Trans. Comput.-Aided Design 4 (1985) 92–98.

[17] M. Duque-Anton, Constructing efficient simulated annealing algorithms, Discrete Appl. Math. 77 (1997) 139–159.

[18] E.S. El-Mallah, C.J. Colbourn, Optimum communication spanning trees in series parallel networks, SIAM J. Comput. 14 (1985) 915–925.

[19] R. Fahrion, M. Wrede, On a principle of chain exchange for vehicle routing problems (I-VRP), J. Oper. Res. Soc. (1990) 821–827.

[20] C.M. Fiduccia, R.M. Mattheyses, A linear time heuristic for improving network partitions, in: ACM IEEE Nineteenth Design Automation Conference Proceedings, IEEE Computer Society, Los Alamitos, CA, 1982, pp. 175–181.

[21] R.T. Firla, B. Spille, R. Weismantel, A primal analogue of cutting plane algorithms, Department of Mathematics, Otto-von-Guericke-University Magdeburg, 1999.

[22] R.T. Firla, B. Spille, R. Weismantel, personal communication.

[23] A. Frangioni, E. Necciari, M.G. Scutella, Multi-exchange algorithms for the minimum makespan machine scheduling problem, Dipartimento di Informatica, University of Pisa, 2000, paper in preparation.

[24] M.L. Fredman, D.S. Johnson, L.A. McGeoch, Data structures for traveling salesman, J. Algorithms 16 (1995) 432–479.

[25] M. Gendreau, F. Guertin, J.Y. Potvin, R. Seguin, Neighborhood search heuristics for a dynamic vehicle dispatching problem with pick-ups and deliveries, CRT-98-10, 1998.

[26] F. Glover, Ejection chains, reference structures, and alternating path algorithms for the traveling salesman problem, Research report, University of Colorado-Boulder, Graduate School of Business, 1992. {A short version appeared in Discrete Appl. Math. 65 (1996) 223–253.}

[27] F. Glover, Finding the best traveling salesman 4-opt move in the same time as a best 2-opt move, J. Heuristics 2 (1996) 169–179.

[28] F. Glover, G.M. Gutin, A. Yeo, Zverovich, Construction heuristics and domination analysis for the asymmetric TSP, Research Report, Brunel University, 1999.

[29] F. Glover, M. Laguna, Tabu Search, Kluwer Academic Publishers, Dordrecht, 1997.
[30] F. Glover, A.P. Punnen, The traveling salesman problem: new solvable cases and linkages with the development of approximation algorithms, J. Oper. Res. Soc. 48 (1997) 502–510.
[31] G.M. Gutin, On the efficiency of a local algorithm for solving the traveling salesman problem, Automat. Remote Control 11 (part 2) (1988) 1514–1519.
[32] G.M. Gutin, Exponential neighborhood local search for the traveling salesman problem, Comput. Oper. Res. 26 (1999) 313–320.
[33] G.M. Gutin, A. Yeo, Polynomial algorithms for the TSP and the QAP with a factorial domination number, Manuscript, Brunel University, UK, 1998.
[34] G.M. Gutin, A. Yeo, TSP heuristics with large domination number, Report 12/98, Department of Mathematics and Statistics, Brunel University, UK, 1998.
[35] G.M. Gutin, A. Yeo, Small diameter neighborhood graphs for the traveling salesman problem, Comput. Oper. Res. 26 (1999) 321–327.
[36] G.M. Gutin, A. Yeo, TSP tour domination and Hamiltonian cycle decomposition of regular digraphs, Manuscript, Brunel University, UK, 1999.
[37] K. Helsgaun, An effective implementation of the Lin–Kernighan traveling salesman heuristic, Manuscript, Roskilde University, Denmark, 1999.
[38] J. Hurink, An exponential neighborhood for a one machine batching problem, OR Spektrum 21 (1999) 461–476.
[39] D.S. Johnson, Local search and the traveling salesman problem, in: Proceedings of 17th International Colloquium on Automata Languages and Programming, Lecture Notes in Computer Science, Springer, Berlin, 1990, pp. 443–460.
[40] D.S. Johnson, L.A. McGeoch, The travelling salesman problem: a case study in local optimization, in: E.H.L. Aarts, J.K. Lenstra (Eds.), Local Search in Combinatorial Optimization, Wiley, New York, 1997, pp. 215–310.
[41] R.M. Karp, A patching algorithm for the non-symmetric traveling salesman problem, SIAM J. Comput. 8 (1979) 561–573.
[42] B.W. Kernighan, S. Lin, An efficient heuristic procedure for partitioning graphs, Bell System Tech. J. 49 (1970) 291–307.
[43] P.S. Klyaus, The structure of the optimal solution of certain classes of the traveling salesman problems, Vestsi Akad. Nauk BSSR, Phys. Math. Sci., Minsk, (1976) 95–98 (in Russian).
[44] S. Knust, Optimality conditions and exact neighborhoods for sequencing problems, Universitat Osnabruck, Fachbereich Mathematik/Informatik, Osnabruck, Germany, 1997.
[45] M Laguna, J. Kelly, J.L. Gonzales-Velarde, F. Glover, Tabu search for multilevel generalized assignment problem, European J. Oper. Res. 82 (1995) 176–189.
[46] E.L. Lawler, J.K. Lenstra, A.H.G. Rinnooy Kan, D.B. Shmoys, The Traveling Salesman Problem, Wiley, New York, 1985.
[47] S. Lin, Computer solutions to the traveling salesman problem, Bell System Tech. J. 44 (1965) 2245–2269.
[48] S. Lin, B. Kernighan, An effective heuristic algorithm for the traveling salesman problem, Oper. Res. 21 (1973) 498–516.
[49] K. Mak, A. Morton, A modified Lin–Kernighan traveling salesman heuristic, ORSA J. Comput. 13 (1992) 127–132.
[50] I.I. Melamed, S.I. Sergeev, I.K. Sigal, The traveling salesman problem: approximation algorithms, Avtomati Telemekh 11 (1989) 3–26.
[51] C.H. Papadimitriou, The complexity of Lin–Kernighan algorithm, SIAM J. Comput. (1992) 450–465.
[52] C.H. Papadimitriou, K. Steiglitz, Combinatorial Optimization: Algorithms and Complexity, Prentice-Hall, Englewood Cliffs, NJ, 1982.
[53] E. Pesch, F. Glover, TSP ejection chains, Discrete Appl. Math. 76 (1997) 165–181.
[54] J.M. Phillips, A.P. Punnen, S.N. Kabadi, A linear time algorithm for the bottleneck traveling salesman problem on a Halin graph, Inform. Process. Lett. 67 (1998) 105–110.
[55] C.N. Potts, S.L. van de Velde, Dynasearch—Iterative local improvement by dynamic programming. part I. The traveling salesman problem, Technical Report, University of Twente, The Netherlands, 1995.

[56] A.P. Punnen, The traveling salesman problem: new polynomial approximation algorithms and domination analysis, J. Inform. Optim. Sci., to appear.

[57] A.P. Punnen, F. Glover, Ejection chains with combinatorial leverage for the TSP, Research Report, University of Colorado-Boulder, 1996.

[58] A.P. Punnen, S.N. Kabadi, Domination analysis of heuristics for the asymmetric traveling salesman problem, Manuscript, University of New Brunswick, 1998.

[59] C. Rego, Relaxed tours and path ejections for the traveling salesman problem, European J. Oper. Res. 106 (1998a) 522–538.

[60] C. Rego, A subpath ejection method for the vehicle routing problem, Management Sci. 44 (1998b) 1447–1459.

[61] C. Rego, C. Roucairol, A parallel tabu search algorithm using ejection chains for the vehicle routing problem, in: I.H. Osman, J.P. Kelly (Eds.), Meta-Heuristics: Theory and Applications, Kluwer Academic Publishers, Dordrecht, 1996.

[62] G. Reinelt, The traveling salesman computational solutions for TSP application, Lecture Notes in Computer Science, Vol. 840, Springer, Berlin, 1994.

[63] R.T. Rockafellar, Network Flows and Monotropic Optimization, Wiley, New York, 1984.

[64] V.I. Sarvanov, N.N. Doroshko, Approximate solution of the traveling salesman problem by a local algorithm with scanning neighborhoods of factorial cardinality in cubic time, Software: Algorithms and Programs, Vol. 31, Mathematics Institute of the Belorussia Academy of Science, Minsk, 1981, pp. 11–13 (in Russian).

[65] A. Satyanarayana, R.K. Wood, A linear time algorithm for computing k-terminal reliability in series parallel networks, SIAM J. Comput. 14 (1985) 818–832.

[66] R.R. Schneur, J.B. Orlin, A scaling algorithm for multicommodity flow problems, Oper. Res. 46 (1998).

[67] N. Simonetti, E. Balas, Implementation of a linear time algorithm for certain generalized traveling salesman problems, in: Proceedings of the IPCO V, Lecture Notes in Computer Science, Vol. 1084, Springer, Berlin, 1996, pp. 316–329.

[68] F. Sourd, Scheduling tasks on unrelated machines: large neighborhood improvement procedures, J. Heuristics, submitted for publication.

[69] E.D. Taillard, Parallel iterative search methods for vehicle routing problems, Network 23 (1993) 661–673.

[70] E.D. Taillard, Heuristic methods for large centroid clustering problems, Technical Report IDSIA-96-96, Lugano, 1996.

[71] T. Takamizawa, T. Nishizeki, N. Sato, Linear time computability of combinatorial problems on series-parallel graphs, J. ACM 29 (1982) 623–641.

[72] K.T. Talluri, Swapping applications in a daily airline fleet assignment, Transportation Sci. 30 (1996) 237–248.

[73] P.M. Thompson, Local search algorithms for vehicle routing and other combinatorial problems, Ph.D. Thesis, Operations Research Center, MIT, Cambridge, MA, 1988.

[74] P.M. Thompson, J.B. Orlin, The theory of cyclic transfers, Operations Research Center Working Paper, MIT, Cambridge MA, August 1989.

[75] P.M. Thompson, H.N. Psaraftis, Cyclic transfer algorithms for multivehicle routing and scheduling problems, Oper. Res. 41 (1993).

[76] M. Yagiura, T. Ibaraki, F. Glover, An ejection chain approach for the generalized assignment problem, Technical Report #99013, Department of Applied Mathematics and Physics, Kyoto University, 1999.

[77] M. Yagiura, T. Yamaguchi, T. Ibaraki, A variable-depth search algorithm for the generalized assignment problem, in: S. Voss, S. Martello, I.H. Osman (Eds.), Metaheuristics: Advances and Trends in Local Search Paradigms for Optimization, Kluwer Academic Publishers, Boston, 1999, pp. 459–471.

[78] A. Yeo, Large exponential neighborhoods for the traveling salesman problem, Preprint no. 47, Department of Mathematics and Computer Science, Odense University, 1997.

[79] K. Wayne, A polynomial combinatorial algorithm for generalized minimum cost flow, STOC 1999.

[80] P. Winter, Steiner problem in Halin networks, Discrete Appl. Math. 17 (1987) 281–294.

[81] M. Zachariasen, M. Dam, Tabu search on the geometric traveling salesman problem, in: I.H. Osman, J.P. Kelly (Eds.), Meta-Heuristics: Theory and Applications, Kluwer Academic Publishers, Dordrecht, 1996.

For further reading

P.C. Gilmore, E.L. Lawler, D.B. Shmoys, Well-solved special cases, in: E.L. Lawler, J.K. Lenstra, A.H.G. Rinnooy Kan, D.B. Shmoys (Eds.), The Traveling Salesman Problem, Wiley, New York, 1985, pp. 87–143.
G.M. Gutin, On approach to solving the traveling salesman problem, in: Theory, Methodology, and Practice of System Research, Mathematical Methods of Systems Analysis, VNIIST, Moscow, 1984, pp. 184–186.

Discrete Applied Mathematics 123 (2002) 103–127

DISCRETE
APPLIED
MATHEMATICS

Maximum mean weight cycle in a digraph and minimizing cycle time of a logic chip

Christoph Albrecht, Bernhard Korte*, Jürgen Schietke, Jens Vygen

Research Institute for Discrete Mathematics, University of Bonn, Lennéstraße 2, D-53113 Bonn, Germany

Received 13 December 1999; received in revised form 12 March 2001; accepted 19 March 2001

Abstract

The maximum mean weight cycle problem is well-known: given a digraph G with weights $c : E(G) \to \mathbb{R}$, find a directed circuit in G whose mean weight is maximum. Closely related is the minimum balance problem: Find a potential $\pi : V(G) \to \mathbb{R}$ such that the numbers $slack(e) := \pi(w) - \pi(v) - c((v, w))$ ($e = (v, w) \in E(G)$) are optimally balanced: for any subset of vertices, the minimum slack on an entering edge should equal the minimum slack on a leaving edge. Both problems can be solved by a parametric shortest path algorithm.

We describe an application of these problems to the design of logic chips. In the simplest model, optimizing the clock schedule of a chip to minimize the cycle time is equivalent to a maximum mean weight cycle problem. It is very important to find a solution with well-balanced slacks; this problem, in the simple model, is a minimum balance problem.

However, in practical situations many constraints have to be taken into account. Therefore minimizing the cycle time and finding the optimum slack distribution are more general problems. We show how a parametric shortest path algorithm can be extended to solve these problems efficiently.

Computational results with recent IBM processor chips show that the cycle time reduces considerably. Moreover, the number of critical paths (with small slack) decreases dramatically. As a result we obtain significantly faster chips. The running time of our algorithm is reasonable even for very large designs. © 2002 Elsevier Science B.V. All rights reserved.

1. Introduction

In this paper we consider the maximum mean cycle problem in digraphs, the minimum balance problem, and generalizations. We show how these problems apply to

---

* Corresponding author.
  E-mail address: dm@or.uni-bonn.de (B. Korte).

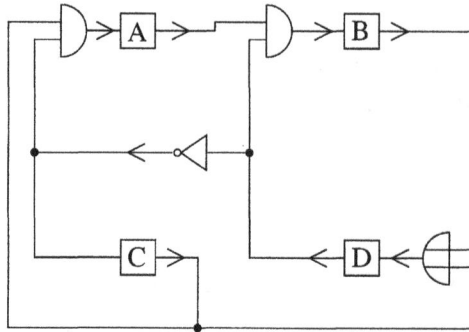

Fig. 1. A sample chip.

a major problem in the design of very large scale integrated (VLSI) circuits. We encounter the rare case that the original practical problem (with all constraints and without any simplification!) can be solved optimally by efficient algorithms.

A main goal in the design of logic chips is maximizing its frequency. All computations on a chip are synchronized by certain storage elements which receive a clock signal periodically. The period is called the cycle time; its inverse is the frequency of the chip.

In each period each storage element (latch) stores one bit. In the next period a fixed logical function is evaluated; the inputs are the bits stored in the previous period and some external inputs. Some of the output bits of the logical function are forwarded to the exterior, the others are stored in the latches; they will be input to the function in the next period.

The begin and end of a period for a certain latch is determined by a periodical clock signal arriving at that latch. The output bit of the function must arrive at this latch before the clock signal, and the propagation of this bit for computation in the next period begins when the clock signal has arrived.

The computation is done by a network of logical gates through which signals are propagated. If the clock signals arrive at all latches simultaneously, then the minimum possible cycle time is determined by the slowest computation, i.e. the longest path in the network. Here the length of a path is its propagation delay; this depends, among other things, on the number and type of gates and their positions. In this paper we assume these lengths to be fixed (already optimized).

Consider the very primitive example of Fig. 1: we have four latches (A, B, C, D) and seven paths between pairs of latches, each containing one or two gates. In this example there are no external inputs and outputs. For our purposes, the relevant information is shown in Fig. 2: the latches, the paths, and their lengths (which reflect the propagation time along the paths). From these numbers we see that the cycle time must not be shorter than 14 time units: this is the time the signal from D to A needs. So we have simultaneous clock signals at time 0, 14, 28, 42, and so on.

However, if we allow individual clock arrival times at the latches, we can do better: By having clock signal arrival times at latch B one unit earlier and at latch D four

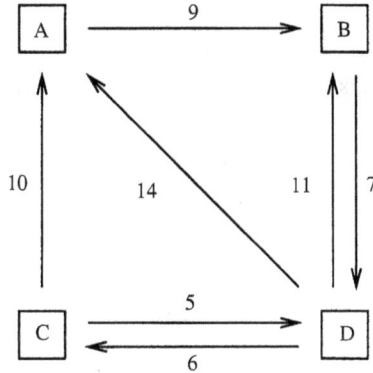

Fig. 2. Latch graph of the sample chip.

units earlier (than at A and C), we can achieve a cycle time of 10:
- Clock signal at latch A: $0, 10, 20, \ldots$
- Clock signal at latch B: $-1, 9, 19, \ldots$
- Clock signal at latch C: $0, 10, 20, \ldots$
- Clock signal at latch D: $-4, 6, 16, \ldots$

One easily checks that all data signals arrive in time: for example the path from D to B has length 11, and the allowed time is $9 - (-4) = 13$ units.

Observe that there exists a circuit in the "latch graph" of Fig. 2 which has mean weight 10. Indeed, the maximum mean weight of a circuit equals the optimum cycle time.

So the problem reduces to a maximum mean cycle problem: given a digraph G with edge weights, find a directed circuit in G whose mean weight (total weight divided by number of edges) is maximum. Karp [6] showed that this problem can be solved in $O(nm)$ time, where n and m denote the number of vertices and edges, respectively. Although this is still the best theoretical bound, other algorithms such as the $O(nm + n^2 \log n)$ parametric shortest path algorithm of Young et al. [22] are faster in practice.

However, the above solution has a serious disadvantage: four out of seven paths have zero slack: if any of these propagation times is larger than estimated, the chip will not function correctly anymore. To make the design more robust we prefer to have as large slacks as possible on as many paths as possible. For example if we change the clock signal arrival times at latch C to $-4, 6, 16, \ldots$, the solution remains valid, but the paths incident to C now all have slack at least 4. We of course prefer such a solution

There are two ways of formalizing this concept: Let G be a digraph and $c : E(G) \to \mathbb{R}$. First, one could consider the vector of all slacks in nondecreasing order. We prefer one solution to another if this vector is lexicographically greater. In other words, we look for a potential $\pi : V(G) \to \mathbb{R}$ such that the vector of slacks $slack(e) := \pi(w) - \pi(v) - c((v, w))$ $(e = (v, w) \in E(G))$ in nondecreasing order is lexicographically maximum.

The second way of formalizing the concept is to consider the following necessary condition for a solution to be called optimal: For any latch, and in fact for any group of latches, the minimum slack on an entering edge should equal the minimum slack on a leaving edge:

$$\min\{slack(v,w)\colon\ v\notin X,\ w\in X\}$$

$$=\min\{slack(v,w)\colon\ v\in X,\ w\notin X\} \quad \text{for all } X\subseteq V(G). \tag{1}$$

If this were not true, we could improve the solution by changing the arrival time of all latches in a group X violating (1) by some constant.

In fact, the two models are equivalent. If G is strongly connected, (1) is satisfied by a unique $\pi\colon V(G)\to\mathbb{R}$ (up to addition of a constant); see [16]. Hence for this π the vector of slacks in nondecreasing order is lexicographically maximum. This problem is known as the minimum balance problem [16,22].

We shall extend the above observations to a very general model comprising all situations on practical chips. In the general model minimization of the cycle time is not a pure maximum mean cycle problem, but the following more general problem: given a digraph G with edge weights, and a partition of the edge set into red and green edges, find a directed cycle in G maximizing the total weight divided by the number of red edges. This is a special case of the maximum ratio cycle problem (all edge times are 0 or 1), which can be solved in $O(\min\{n^3\log^2 n,\ n^3\log n + mn\log^2 n\log\log n\})$ time [11].

Moreover, the slack balancing problem is not a pure minimum balance problem, but the following (more general) problem: given a digraph G with edge weights, and a partition of the edge set into red and green edges, find a potential $\pi\colon V(G)\to\mathbb{R}$ such that each green edge has nonnegative slack and the vector of slacks of red edges, in nondecreasing order, is lexicographical maximal.

We show that these problems (and extensions discussed in Section 6) can also be solved with a parametric shortest path algorithm in $O(nm + n^2\log n)$ time.

This paper is organized as follows. After introducing the problem of finding an optimum clocking schedule with a very simple model in Section 2, a general formulation is developed in Section 3. In Section 4, it is shown that the problem reduces to a maximum mean weight cycle problem in a digraph. This can be solved by the algorithm of Young et al. [22], of which an outline is given.

Then we develop an algorithm which takes additional objectives into account. In Section 5 we first consider slacks on signal paths. We obtain a solution where as few as necessary signal paths are critical. In Section 6 it is shown how this algorithm can be modified to increase the slacks on the clocktree paths. Moreover, slacks on signal paths and clocktree paths can be maximized simultaneously.

Computational results with recent IBM processor chips in Section 7 demonstrate the power of our method. The cycle time of the chips is improved by between 2.5 and 5.5%. Moreover, the number of critical signal and clocktree paths (with zero or small positive slack) decreases substantially. The running time of the algorithm is reasonable even for latch graphs with several million edges.

2. A simple model

In this section we describe a very simple model for clock scheduling, and show how to optimize the cycle time. This model and the algorithmic solution will be extended in the subsequent sections to reflect very general situations.

A logic chip has a set P of primary (=external) inputs and a set Q of primary outputs. Moreover, there is a set S of storage elements which we assume to store one bit each (e.g. flip-flops or latches). The chip works in time intervals, called cycles, of equal length T. T is called the cycle time of the chip. All the computations of the chip can be described by a boolean function $B : \{0,1\}^{P\cup S} \rightarrow \{0,1\}^{Q\cup S}$. In each cycle B is computed, i.e. the values at the primary inputs and the storage elements are used to compute values at the primary outputs and new values to be stored in S; these will be reused as input of B in the next cycle.

The input signals arrive at the primary inputs with the frequency $1/T$, i.e. at time $\xi_p, \xi_p + T, \xi_p + 2T$, etc. for each $p \in P$. The output signals are expected to arrive at the primary outputs at time $\xi_q, \xi_q + T, \xi_q + 2T$, etc. for each $q \in Q$. The numbers $\xi_p(p \in P)$ and $\xi_q(q \in Q)$ are—together with B—part of the specification of the chip.

The storage elements receive clock signals with the same frequency $1/T$. A special network, the clocktree, distributes one or more clock signals to the storage elements. A simple storage element $s \in S$ works as follows: At certain times $x_s, x_s + T, x_s + 2T$ etc., the result of B for s is expected at the data input of s; at the same time (or slightly later) this value is available at the data output of s and can be used as input for the next computation of B. The time x_s depends on the clock signal arriving at s and is not fixed in advance.

Let G be the graph with $V(G) = P \cup Q \cup S$ whose edges correspond to signal paths: There is an edge (v,w) if the output w of B depends on the input v, i.e. if $B(z)_w \neq B(z')_w$ for some vectors $z, z' \in \{0,1\}^{P\cup S}$ with $z_u = z'_u$ for all $u \in (P \cup S) \setminus v$. (Subscripts denote component vectors.)

To be precise, there might be a path from v to w although w does not depend on v. Such so-called false paths can be ignored if they are detected. However, it is often impossible to detect all false paths (this is *coNP*-hard problem).

For each path from v to w the propagation time of the signal from v to w may vary due to process variations, temperature etc. So we have bounds $t_{vw}^{\min}$ and $t_{vw}^{\max}$ for the minimum and maximum propagation time over all paths from v to w. Since the network without the latches is acyclic, the graph G and the bounds on the propagation times can be computed by simple forward propagation in topological order.

The chip works correctly if every signal arrives in time:

$$x_v + t_{vw}^{\max} \leqslant x_w + T \quad \text{for all } (v,w) \in E(G), \tag{2}$$

and no signal arrives too early (i.e. during the previous cycle):

$$x_v + t_{vw}^{\min} \geqslant x_w \quad \text{for all } (v,w) \in E(G). \tag{3}$$

Here T is the cycle time and x_s ($s \in S$) are variables, while $x_v := \xi_v$ for $v \in P \cup Q$ are given constants. (2) are sometimes called late mode constraints (also known as zero clocking or setup test), (3) are early mode constraints (double clocking or hold test).

In our introductory example (Figs. 1 and 2) we ignored early mode constraints. For example, if the minimum propagation delay from latch D to latch C is 3 units (this is quite realistic if the maximum is 6 units), then the clock signal at D must not arrive more than 3 units earlier than the clock signal at C.

In older designs all x_s ($s \in S$) had approximately the same value which was fixed at an early stage. The main advantage of this so-called zero skew approach is its simplicity. A network distributing the clock signals (the clocktree) can be designed in order to guarantee (almost) zero-skew (i.e. simultaneous clock signals); algorithms for this task have been described by Tsay [20] and Muuss [12]. However, this is also true for prescribed arrival times with nonzero skew.

The zero-skew approach has several disadvantages. The most striking one is that the cycle time can be improved by choosing the x_s individually. We shall show now how to choose the x_s optimally to minimize the cycle time. Later we show how to obtain a solution where only very few paths on the chip are critical.

In future technologies further disadvantages of the zero-skew approach will also become important. If all storage elements switch at the same time, a very large capacitance has to be loaded simultaneously. The effect is that the supply voltage can fluctuate considerably. Moreover, crosstalk on parallel wires, especially in the clock-tree, is a more serious problem with the zero-skew approach. Both effects lead to unpredictable timing behaviour and can cause the chip to fail.

How can we choose the numbers x_s optimally? Let us ignore (3) for a moment. Then minimization of the cycle time T is quite easy: Contract the set $P \cup Q$ in G to a special vertex r. In the resulting graph G' we define edge weights $c((v,w)) := t_{vw}^{\max}$ for $(v,w) \in E(G) \cap (S \times S)$ and

$$c((r,w)) := \max\{t_{pw}^{\max} + \xi_p : p \in P, (p,w) \in E(G)\},$$

$$c((v,r)) := \max\{t_{vq}^{\max} - \xi_q : q \in Q, (v,q) \in E(G)\}.$$

Then we have:

Proposition 2.1 (Vygen [21]). *The maximum mean weight of a directed circuit in G' with respect to c equals the minimum cycle time T such that values x_s ($s \in S$) satisfying (2) exist.*

Proof. Let T and x_s ($s \in S$) be a solution of (2), and let C be any circuit in G', say with vertices $v_0, v_1, \ldots, v_k = v_0$ in this order. Setting $x(r) := 0$ we have

$$\sum_{i=1}^{k} c((v_{i-1}, v_i)) \leqslant \sum_{i=1}^{k} (T + x(v_i) - x(v_{i-1})) = kT,$$

so the mean weight of C is at most T.

On the other hand, if T is the maximum mean weight of a circuit in G', define $c'(e) := T - c(e)$. There is no negative circuit in G' with respect to c', so we can compute a shortest path potential $\pi : V(G') \to \mathbb{R}$ with $\pi(w) \leqslant \pi(v) + c'(v,w)$ for all $(v,w) \in E(G')$. Setting $x(v) := \pi(r) - \pi(v)$ for all latches v we obtain a solution satisfying (2). $\square$

So the problem reduces to a maximum mean cycle problem, which is well-solved [6,22].

3. The general problem

To formulate the general problem we have to take a closer look at the storage elements. The clock input of a storage element $s \in S$ has the value 1 in the time from a_s to b_s, then again from $a_s + T$ to $b_s + T$, from $a_s + 2T$ to $b_s + 2T$ and so on. In the remaining time it has the value 0.

When the clock input has the value 1, the storage element is open, i.e. it stores the value currently seen at the data input. When the clock input has value 0, the stored bit remains unchanged. (Sometimes the roles of 0 and 1 are interchanged, but this does not matter.) At any time, the stored bit is available at the data output for subsequent computations.

To model this situation we need two variables per storage element instead of just one. We may shift the clock input for each $s \in S$ by some value y_s, meaning that the clock input has the value 1 in the time intervals $[a_s + y_s, b_s + y_s], [a_s + y_s + T, b_s + y_s + T], [a_s + y_s + 2T, b_s + y_s + 2T]$ and so on. We assume that the length of the interval $b_s - a_s$ cannot be changed.

Since the shifting times y_s have to be realized by a clocktree, it is reasonable to impose a lower bound l_s and an upper bound u_s on y_s for each storage element s:

$$l_s \leqslant y_s \leqslant u_s \quad (s \in S). \tag{4}$$

Moreover, we have a variable x_s for the time when the data signal is valid at the data input of s. Of course, we must have

$$a_s + y_s \leqslant x_s \leqslant b_s + y_s \quad (s \in S) \tag{5}$$

and the data signal should remain valid within the whole interval $[x_s, b_s + y_s]$. For primary inputs and outputs v we set $y_v = 0$ and $a_v = b_v = x_v$ (this value corresponds to ξ_v in the previous section).

A data signal might encounter more than one storage element per cycle (for example in designs using transparent latches). So for each signal path from v to w it has to be specified whether a signal starting at s within the time interval $[a_v, b_v]$ must arrive before b_w (i.e. in the same cycle) or before $b_w + T$ (i.e. in the next cycle). In the first case we set $\zeta_{vw} := 0$, in the second case we set $\zeta_{vw} := 1$.

Then the late mode constraints read as follows:

$$x_v + t_{vw}^{\max} \leqslant x_w + \zeta_{vw} T \quad \text{for all } (v, w) \in E(G), \tag{6}$$

where G is defined as in the previous section. For the early mode constraints we have to take the whole intervals into account where the storage elements are open:

$$a_v + y_v + t_{vw}^{\min} \geqslant b_w + y_w + (\zeta_{vw} - 1)T \quad \text{for all } (v, w) \in E(G). \tag{7}$$

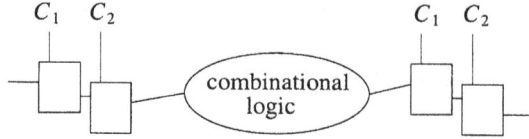

Fig. 3. Two master–slave latches with clock signals C_1 and C_2 and a signal path in between.

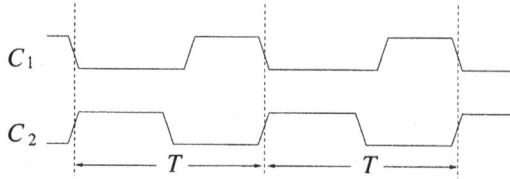

Fig. 4. Clock signals at a master–slave latch: voltages as functions of time (0=low voltage, 1=high voltage).

This describes a quite general model. In the simple model of Section 2 we considered only the case where $a_s = b_s$ for all $s \in S$ and $\zeta_{vw} = 1$ for all $(v, w) \in E(G)$. Although it is technically impossible to generate a clock signal which is 1 at a single point of time only, the simple model has some practical relevance: A commonly used storage element is the so-called master–slave latch. Fig. 3 shows two master–slave latches with a signal path in between. A master–slave latch has two clock inputs, one data input and one data output. It works as if it would consist of two simple storage elements, where the data output of the first one is connected to the data input of the second one. Moreover, the clock input of the second part is roughly the inverse of the clock signal for the first part (see Fig. 4).

As long as the first clock signal C_1 is 1, the value arriving at the data input is stored. If the first clock signal is 0, it does not change anymore. Now if the second clock signal C_2 is 1, the stored bit is visible at the data output, and the new computations begin. If the falling edge of C_1 arrives at the same time as the rising edge of C_2, then such a master–slave latch behaves like a simple latch which is open only at one point of time. However, it is also possible to have two independent clock signals arriving at the latch. This may of course lead to better solutions. Moreover, it is possible to have combinational logic (without storage elements) in between; this common technique is known as cycle stealing [14,19,10,9].

In the above general model a master–slave latch can be represented simply by two simple storage elements s and s'. Then we shall usually have $t_{ss'}^{\min} = t_{ss'}^{\max} = 0$. This may be used to eliminate the variable $x_{s'}$ (Set it to $\max\{x_s, a_{s'} + y_{s'}\}$). However, this increases the number of constraints. It is not a priori clear whether this substitution is computationally favorable or not; indeed this depends on the structure of the design.

In the next section we show how to efficiently solve the above defined linear program, i.e. minimize the cycle time T subject to constraints (4), (5), (6) and (7). After that we shall distribute slacks on signal and clock paths optimally.

Our model is quite flexible. Additional technical constraints can easily be incorporated. For example one can model dynamic circuits by constraints of the same type as
above in a straightforward way. Some other constraints (end-of-cycle test, simultaneous
clock signals for registers) will be mentioned in Section 7.

Another advantage of this model is that there is no need to distinguish between
master–slave latches, simple level-sensitive latches (where $a_s < b_s$; one also speaks
of transparent latches) and edge-triggering flip-flops (where $a_s = b_s$). It includes all
previous models for cycle time optimization, see [4,14,18,17,2,13,21].

4. Computing the optimal cycle time

Observe that each of the inequality constraints (4), (5), (6) and (7) has one or
two x- or y-variables and in addition possibly the special variable T. If a constraint
has two x- or y-variables, they have opposite sign. To have exactly two variables per
inequality we introduce an artificial variable z_0 (corresponding to r in Section 2) which
we assume to have value zero. For technical reasons we substitute $\lambda := -T$ and obtain
a linear program of the following very special type:

$$\max \quad \lambda$$

s.t.

$$z_i + c_{ij} \geqslant z_j \quad \text{for } (i,j) \in E_1, \tag{8}$$

$$z_i + c_{ij} - \lambda \geqslant z_j \quad \text{for } (i,j) \in E_2, \tag{9}$$

where λ and $z_0, z_1, z_2, \ldots, z_n$ are variables, the c_{ij} are constants and $E_1, E_2 \subset \{0, \ldots, n\} \times \{0, \ldots, n\}$. Each constraint (4) corresponds to two elements $(i, 0), (0, i)$ of E_1, each constraint (5) corresponds to one element of E_1, and each of the constraints (6) and (7)
corresponds to an element of E_1 or E_2, depending on the ξ-constant.

Note that assuming $z_0 = 0$ causes no loss of generality since adding a constant to
all variables z_i does not affect feasibility.

Now we translate our linear optimization problem to a network problem. Given the
above linear program we construct a directed graph $G' = (V, E)$ as follows: For each
variable z_i there is a vertex v_i. For a constraint of type (8) we have a directed edge
from vertex v_i to vertex v_j of cost c_{ij}. For a constraint of type (9) we also have an
edge from v_i to v_j, but the cost is $c_{ij} - \lambda$. Such an edge is called parameterized: the
cost of the edge depends on the parameter λ.

As an example, Fig. 5 shows the vertices and edges of G' for two master–slave
latches with a signal path in between (from left to right). Each master–slave latch
consists of two simple latches, and for each simple latch we have an x-variable (on
top) and a y-variable. So there are four variables (vertices) for each master–slave
latch; the artificial variable z_0 is represented by the vertex at the bottom. There are
eight edges for constraints of type (4), eight for type (5), three edges for type (6) and
three for type (7). Three of the edges are parameterized.

We are looking for the maximum value for λ such that values for the variables z_i
exist which fulfill all constraints. It is easy to see that such values exist if and only if

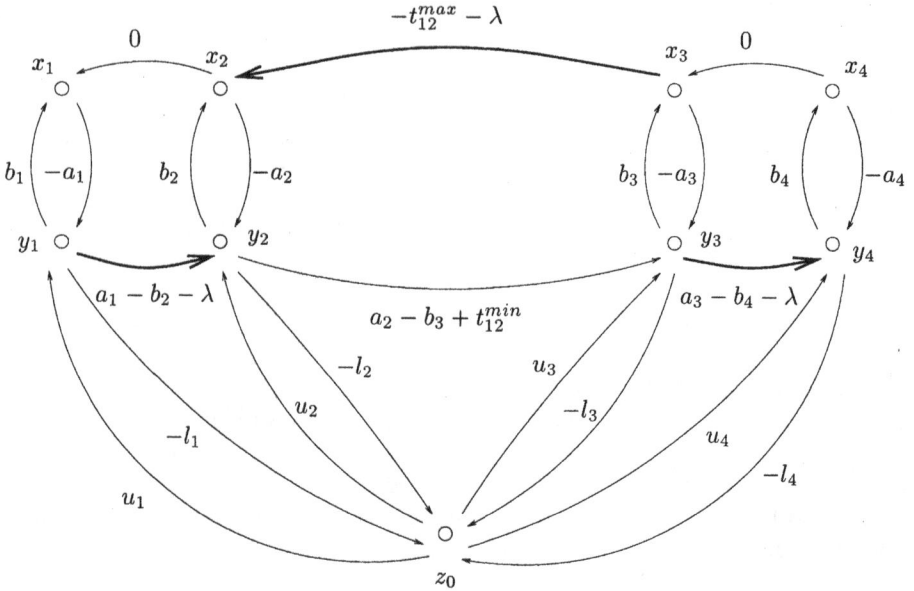

Fig. 5. The vertices and edges in G' for two master–slave latches and a signal path in between as shown in Fig. 3. The bold edges are parameterized.

the digraph G' does not contain a directed circuit of negative cost (negative circuit, for short). This is proved similar to Proposition 2.1; see [18,17,19,2]. In fact, our problem can be formulated as follows: given a digraph G with edge weights, and a partition of the edge set into red and green edges, find a directed circuit in G minimizing the total weight divided by number of red edges.

The above LP with constraints (9) and (8) might be infeasible in some cases (if there is a negative circuit consisting of unparameterized edges only). Our algorithm detects infeasibility and returns the negative circuit(s) causing the problem. Usually one can cope with this by omitting some early mode constraints: these can be met by inserting buffers (increasing the delay of the path). In fact it is usually not good to take all early mode constraints into account because this might increase the optimum value of the LP (hence the cycle time); one usually prefers inserting a buffer. As we describe in Section 7 we incorporate early mode constraints only after having determined the best possible cycle time. In the following we assume that the LP is feasible.

Previous authors solved this LP either by linear programming [4,14] or by binary search with a subroutine testing for a negative circuit [18,17,19,2,13]. However, due to its special structure the problem can be solved more efficiently by a direct combinatorial algorithm of Young et al. [22]. We briefly describe their algorithm (which was originally designed for parametric shortest paths) since we shall extend it in Sections 5 and 6. For a detailed description and an efficient implementation see also Bünnagel [1].

The algorithm computes a sequence $-\infty = \lambda_0 \leqslant \lambda_1 \leqslant \cdots \leqslant \lambda_k$ of values for the parameter λ and a sequence $T_1, \ldots, T_k$ of shortest paths trees in G' from a specified vertex r, the root (in our case we can take the vertex corresponding to the artificial variable z_0: all vertices are reachable from this vertex), such that T_i is a shortest paths tree for all parameters λ with $\lambda_{i-1} \leqslant \lambda \leqslant \lambda_i$ ($i = 1, \ldots, k$). The last value λ_k will be the solution of the linear program.

We start by computing a shortest paths tree for $\lambda = -\sum_{e \in E} |c(e)|$. This value of λ is small enough such that no negative circuit (i.e. directed circuit of negative total cost) exists. The resulting tree is T_1.

Assuming that T_i is already computed we show how to compute λ_i and T_{i+1}. Let P_{rv} be the path form r to v in T_i. We check for each edge $e = (u, v)$ whether the path $P_{ru} + e$ contains more parameterized edges than P_{rv}. If so, $P_{ru} + e$ is a potential pivot path and e is a potential pivot edge. For some value λ_e the path $P_{ru} + e$ will be shorter than P_{rv}.

λ_i is the minimum value λ_e for all potential pivot edges e. One edge $e = (u, v)$ with the minimum value λ_e becomes the pivot edge. We perform a pivot step by deleting the edge with head v from T_i and inserting edge e. The resulting tree is T_{i+1}.

If adding the pivot edge e results in a directed circuit, the algorithm stops. The cost of this directed circuit is zero for λ_e, and $\lambda_e = \lambda_k$ is the maximum value of λ such that G' contains no negative circuit.

The last tree T_k also provides a solution for the variables z_i: one can set z_i to the cost of the path P_{rv_i} in T_k for $\lambda = \lambda_k$. This solution is also called a shortest paths potential.

The worst-case running time of this algorithm (with an efficient implementation) is $O(nm + n^2 \log n)$ where $n = |V(G')|$ and $m = |E(G')|$. However, it is much faster in practice as the experimental results will demonstrate.

We showed how to determine the clocking schedule with the optimum cycle time. However, the solution obtained so far has a serious drawback. Many inequalities of the linear program (in particular all whose corresponding edges belong to the tree T_k) are satisfied with equality. If such a tight inequality corresponds to a signal path, then this path will be critical, i.e. the slack is zero. In the next section we show how the slack can be increased for many critical signal paths. In Section 6 we show how to increase slack on clocktree paths optimally.

We should note that the above method can also be used for static timing analysis with transparent latches, without changing clock arrival times:

5. Balancing slacks on signal paths

Having computed the optimal cycle time T subject to the constraints described in Section 3, we now increase the slack on the signal paths. This is very important: First of all, at the time when the clock schedule is optimized, the propagation delays can only be estimated. Positive slacks on most paths make the chip less sensitive to routing detours, process variations and manufacturing skew. Moreover, if one tries to optimize the cycle time further (e.g. by different logic implementation or placement), only few paths have to be considered.

If $\varepsilon_{vw}^{\max}$ and $\varepsilon_{vw}^{\min}$ are the slacks on the signal path from v to w for late mode and early mode respectively, the constraints (6) and (7) become

$$x_v + t_{vw}^{\max} + \varepsilon_{vw}^{\max} \leqslant x_w + \zeta_{vw} T \tag{6'}$$

and

$$a_v + y_v + t_{vw}^{\min} - \varepsilon_{vw}^{\min} \geqslant b_w + y_w + (\zeta_{vw} - 1)T. \tag{7'}$$

The task is to maximize the slack variables $\varepsilon_{vw}^{\max}$ and $\varepsilon_{vw}^{\min}$ for as many signal paths as possible such that there is a solution for the linear inequalities (4), (5), (6') and (7') for a given cycle time T.

To make this precise we introduce the following partial order relation on the set of all solutions:

Definition 5.1. Let $(\varepsilon_1, \ldots, \varepsilon_k)$ and $(\varepsilon_1', \ldots, \varepsilon_k')$ be two vectors (of slack variables). Let π be a permutation on $\{1, \ldots, k\}$ such that $\varepsilon_{\pi(1)} \leqslant \varepsilon_{\pi(2)} \leqslant \cdots \leqslant \varepsilon_{\pi(k)}$ and π' be a permutation such that $\varepsilon_{\pi'(1)}' \leqslant \varepsilon_{\pi'(2)}' \leqslant \cdots \leqslant \varepsilon_{\pi'(k)}'$. We say that solution $(\varepsilon_1, \ldots, \varepsilon_k)$ is better than solution $(\varepsilon_1', \ldots, \varepsilon_k')$ if $(\varepsilon_{\pi(1)}, \ldots, \varepsilon_{\pi(k)})$ is greater than $(\varepsilon_{\pi'(1)}', \ldots, \varepsilon_{\pi'(k)}')$ in lexicographic order, i.e. if there exists an $l, 1 \leqslant l \leqslant k$, such that $\varepsilon_{\pi(i)} = \varepsilon_{\pi'(i)}'$ for $i < l$ and $\varepsilon_{\pi(l)} > \varepsilon_{\pi'(l)}'$.

The *slack balancing problem* consists of finding a solution of (4), (5), (6') and (7') such that the vector of all slack variables is best possible.

In other words, we look for a solution of (4), (5), (6') and (7') such that the vector of all slack variables in nondecreasing order is lexicographically maximal.

The following example illustrates this definition: Suppose we have four signal paths, one solution with late-mode slacks ($\varepsilon_{vw}^{\max}$) $0, 1, 3, 0$ and early-mode slacks ($\varepsilon_{vw}^{\min}$) $2, 3, 1, 0$ and another solution with late-mode slacks $0, 2, 2, 0$ and early-mode slacks $5, 2, 1, 3$. We sort both solutions with increasing slack regardless of the slack being for late mode or early mode: $(0, 0, 0, 1, 1, 2, 3, 3)$ for the first solution and $(0, 0, 1, 2, 2, 2, 3, 5)$ for the second solution. The second solution is better.

It will be shown that any optimum solution for the slack balancing problem has the same vector of slack variables. We now describe an algorithm which finds this solutions. It proceeds as follows:

The slacks of all signal paths are increased simultaneously until they cannot be increased anymore, i.e. some constraints, which form a directed circuit, are already tight. Then we take the subset of all signal paths on which the slack can still be increased and continue to increase the slack on these signal paths.

The same digraph $G' = (V, E)$ as in Section 4 is constructed, but the costs are different. The optimal cycle time T is already computed and should not change, it becomes part of the cost of the respective edges.

The parameterized edges are now those which correspond to constraints with $\varepsilon_{vw}^{\max}$ or $\varepsilon_{vw}^{\min}$ (i.e. (6') and (7')), all other edges are not parameterized. The parameter λ now represents the slack of all signal paths.

We first compute again the maximum value λ such that G' contains no negative circuit, using the parametric shortest path algorithm described in Section 4. This value is zero (if T was the optimal cycle time), and so is the slack of all signal paths for which the corresponding edges belong to the zero cost directed circuit C found by the algorithm. Increasing the parameter λ of any of the parameterized edges on C is impossible, because it would result in a negative circuit. All edges on C lose their parameter, only the parameter of all other edges is increased.

C is contracted, and the costs of the edges leaving and entering C are adjusted: Let z be the vertex to which C is contracted, and let w be the vertex of C nearest to the root r in the last tree T_k computed. For a vertex v of C denote by $c(P_{wv})$ the cost of the path from w to v in T_k for parameter λ_k.

For an edge $e=(v,u)$ leaving C, i.e. v belongs to C but u does not, the corresponding new edge $e'=(z,u)$ after the contraction gets the cost $c(e')=c(e)+c(P_{wv})$. For an edge $e=(u,v)$ entering C, the corresponding new edge gets the cost $c(e')=c(e)-c(P_{wv})$.

The algorithm continues to increase the parameter λ and to change the tree such that it remains a shortest paths tree until the next directed circuit of zero cost is found. The value of the parameter λ at this state is again the slack of the signal paths for which the corresponding edges belong to the directed circuit.

We can now prove that the solution computed by this algorithm is the optimum solution with respect to Definition 5.1:

Theorem 5.2. *The algorithm described above finds an optimum solution to the slack balancing problem. Moreover, any optimum solution for the slack balancing problem has the same vector of slack variables.*

Proof. Whenever the algorithm finds a directed circuit of zero cost, the parameters of all parameterized edges of the circuit have the same value and so all the slacks of the corresponding signal paths are equal. Increasing the slack of one of these signal paths is only possible by decreasing the slack of another signal path, but this would result in a solution which is worse. □

The algorithm presented here is a modification of the algorithm for the minimum balance problem described by Young et al. [22]. For the minimum balance problem all edges of the directed graph G are parameterized. Here only some edges are parameterized, namely those edges which correspond to constraints of signal paths with variables ε_{vw}^{max} or ε_{vw}^{min}.

This is also the reason why we speak of balancing the slacks: the slacks are increased and distributed "equally" on the signal paths.

With an efficient implementation the worst-case running time of the algorithm described above is $O(nm + n^2 \log n)$.

But note that it is not necessary to run the algorithm to the very end. The algorithm can be stopped at any time, e.g. when the certain value of the parameter λ is reached. Then the slack of the signal paths are only increased up to this value. Since slacks exceeding a certain amount are usually not interesting this option is used in practice; see Section 7.

6. Balancing slacks on clocktree paths

Prescribed arrival times for clock signals at latches cannot be realized since process variations make it impossible to predict the exact arrival times. This is usually taken into account by adding a constant to the cycle time. If one has an interval $[y_s - \varepsilon_s, y_s + \varepsilon_s]$ of valid clock shifts for each latch s (instead of a fixed number y_s) feasible arrival times are much easier to realize and a smaller cycle time can be achieved (see also [13]).

Clocktrees with prescribed skews can be designed by basically the same algorithm as zero-skew clocktrees. Although this can be done quite efficiently, prescribed skews (zero or not) make detours in the clocktree wiring necessary. If one has intervals for the arrival times of the clock signals one can design clocktrees with significantly smaller wirelength; see e.g. [5]. We show how to achieve large intervals for as many latches as possible.

So for each latch s we introduce an additional variable ε_s: the clock signal is 1 in the interval $[a_s + y'_s, b_s + y'_s]$ for some $y'_s \in [y_s - \varepsilon_s, y_s + \varepsilon_s]$. All constraints must be met for all possible values of y'_s within this interval. ε_s can be considered as the slack on the clocktree path ending at latch s.

With these additional variables ε_s for all latches we can reformulate the linear inequalities:

$$a_s + y_s + \varepsilon_s \leqslant x_s \leqslant b_s + y_s - \varepsilon_s, \tag{5''}$$

$$a_v + y_v - \varepsilon_v + t^{\min}_{vw} \geqslant b_w + y_w + \varepsilon_w + (\zeta_{vw} - 1)T. \tag{7''}$$

The problem is solved similarly to the slack balancing problem for signal paths (Section 5), but some modifications are needed since inequality $(7'')$ contains the slack variables ε_v and ε_w of two different latches v and w which are connected by a signal path.

For each constraint $(7'')$ we introduce an additional variable m_{vw} and split the inequality into two:

$$a_v + y_v - \varepsilon_v + t^{\min}_{vw} \geqslant m_{vw},$$
$$m_{vw} \geqslant b_w + y_w + \varepsilon_w + (\zeta_{vw} - 1)T.$$

Now each inequality contains at most one slack variable ε_s, and if it contains the variable ε_s, then it contains also the corresponding variable y_s.

As before we construct a digraph G' with a vertex for each variable and an edge for each inequality. An edge is parameterized if and only if the corresponding inequality. An edge is parameterized if and only if the corresponding inequality contains a slack variable ε_s. But in contrast to the problem of balancing slacks on signal paths (described in Section 5) we now have several inequalities with the same slack variable ε_s.

We use a similar algorithm to that of the previous section. Now it might happen that when a directed zero cost circuit C is found and contracted there is an edge e on C and an edge f not belonging to C with the same slack variable. In this case the value of this slack variable is already determined; the parameter of f must not increase anymore. Edge f (and other edges with the same slack variable) get the cost which they have with the current value of the parameter, they are no longer parameterized.

Observe that all edges which lose their parameter have at least one vertex belonging to C, namely the vertex corresponding to the variable y_s.

The cost of these edges are adjusted with the contraction of the directed circuit, hence the running time of the algorithm does not change.

In Section 5 we have balanced the slacks on the signal paths, in this section those on clocktree paths. We finally show that it is also possible to balance the slacks on signal paths and clocktree paths simultaneously. Constraint (7) is substituted by

$$a_v + y_v - \varepsilon_v + t_{vw}^{\min} - \varepsilon_{vw}^{\min} \geqslant b_w + y_w + \varepsilon_w + (\zeta_{vw} - 1)T. \tag{7'''}$$

We look for a solution of constraints (4), (5''), (6') and (7'''). For each constraint (7''') we introduce two additional variables m'_{vw} and m''_{vw} and replace the constraint by the following three inequalities:

$$a_v + y_v - \varepsilon_v \geqslant m'_{vw},$$

$$m'_{vw} + t_{vw}^{\min} - (\zeta_{vw} - 1)T - \varepsilon_{vw}^{\min} \geqslant m''_{vw},$$

$$m''_{vw} \geqslant b_w + y_w + \varepsilon_w.$$

Then we apply the same algorithm as above. We obtain:

Theorem 6.1. *The slack balancing problem for signal and clocktree paths (constraints (4), (5''), (6') and (7''')) has a unique optimum solution which can be computed in time* $O(nm + n^2 \log n)$ *where* n *is the number of primary inputs, primary outputs and latches and* m *is the number of signal paths.*

Proof. It can be derived in the same way as in the proof of Theorem 5.2 that the optimum solution is unique and that the algorithm finds it.

In order to see that the running time is of the given order, observe that the new vertices added for the constraints have only one incoming edge. Such an edge can never be exchanged during the parametric shortest path algorithm unless it is contracted.

By the results of Young et al. [22] it is sufficient to show that the total number of pivot steps is $O(n^2)$. During the algorithm let X be the set of all original vertices in G' (not those resulting from subdividing edges) plus all vertices which have emerged by contraction. For each vertex $v \in X$ consider $\varphi(v) = 5|X| - \rho(v)$, where $\rho(v)$ is the number of parameterized edges on P_{rv}. $\varphi(v)$ is positive and bounded by $O(n)$ as is $|X|$.

At each pivot step there is at least one vertex $v \in X$ for which $\rho(v)$ increases, hence $\sum_{v \in X} \varphi(v)$ strictly decreases.

Finally, we show that $\sum_{v \in X} \varphi(v)$ does not increase due to contraction. If a directed circuit is contracted, then $\rho(v)$ can decrease by at most two (for incoming and leaving edges of the directed circuit) plus three times the number by which $|X|$ decreases (the number of vertices in X on the directed circuit minus one). This is compensated by the term $5|X|$ in $\varphi(v)$. Moreover, for each new vertex z which enters X by the contraction there is at least one vertex v with $\varphi(z) \leqslant \varphi(v)$ which leaves X.

Hence the expression $\sum_{v \in X} \varphi(v)$ never increases and strictly decreases with each pivot step. Since it is $O(n^2)$ initially and nonnegative throughout, the number of pivot steps is $O(n^2)$. □

7. Computational results

We have implemented the algorithm in C, all runs are on an IBM RISC System/6000 Model 595. Our algorithm has been applied, among others, to the G3 series of IBM S/390 processor chips [3] (L2 and PU) and the latest follow ups (MBA). For details of the design system see Koehl et al. [8] and Kick et al. [7]. Table 1 shows the different chips with target cycle time, number of circuits, nets, pins and primary inputs and outputs (IOs, some of which are bidirectional) and the number of signal paths. See also Fig. 10 for a placement of the MBA chip.

Table 1
Characteristics of three chips

Chip	Target cycle time (ns)	Circuits	Nets	Pins	Primary IOs	Latches	Signal paths
L2	6.5	87 177	103 590	339 351	928	17 032	1 173 132
PU	6.5	164 056	171 666	591 410	744	17 265	2 670 459
MBA	4.46	394 257	402 373	1 441 312	586	40 639	1 475 535

Table 2
Size of the digraph G' constructed for balancing slacks on signal paths

Chip	Nodes	Edges
L2	52 999	2 103 937
PU	68 932	5 433 150
MBA	268 153	3 637 831

Table 3
Total running time in seconds for evaluating the timing of the chip by simple propagation (timing) and creating all constraints (creation)

Chip	Timing	Creation
L2	104.11	419.71
PU	178.67	1377.62
MBA	859.25	1006.31

Table 4
Improvement of the worst slack and the number of critical signal paths for the three different chips

Chip		Worst slack	Number of signal paths with late mode slack $\varepsilon_{vw}^{\max}$ smaller than							
			< -0.2	< -0.1	< 0.0	< 0.1	< 0.2	< 0.3	< 0.4	< 0.5
L2	Before optimization	−0.048	0	0	594	731	740	5781	9541	11938
	After optimization	0.313	0	0	0	0	0	0	0	672
PU	Before optimization	−0.103	0	1	143	1384	11349	51578		
	After optimization	0.060	0	0	0	44	1617	44285		
MBA	Before optimization	−0.224	5	44	400	2633	9901	21780		
	After optimization	−0.051	0	0	28	89	2283	18768		

Table 5
Total running time in seconds for the main optimization routine: for computing the optimal cycle time and for balancing slacks on late and early mode constraints and also clocktree paths up to different values for λ^{late}, λ^{early} and λ^{clock}

Chip	Cycle time	λ^{late}	λ^{early}	λ^{clock}	λ^{late}	λ^{early}	λ^{clock}
		=0.1	=0.1	=0.05	=0.2	=0.2	=0.1
L2	153.21	4.91	6.93	11.56	4.94	31.87	300.54
PU	19.45	22.35	123.33	569.68	113.68	326.17	13004.25
MBA	20.36	37.63	135.48	671.48	81.34	259.43	92490.36

Chip	λ^{late}	λ^{early}	λ^{late}	λ^{early}	λ^{late}	λ^{early}
	=0.3	=0.3	=0.4	=0.4	=0.5	=0.5
L2	5.45	97.48	104.80	634.16	282.21	1608.97
PU	1140.04	1768.60				
MBA	212.60	11011.82				

In addition to the constraints described so far further technical restrictions had to be taken into account. For example, for some master–slave latches it is required that the data signal arrives at the latch before the rising time of the clock signal of the slave latch (end-of-cycle test). In this case one has a constraint of the form

$$x_v + t_{vw}^{\max} \leqslant a_w + y_w + \zeta_{vw} T.$$

Moreover, for certain registers (which can be regarded as sets of latches) simultaneous clock signals are required. In this case only one y-variable for these latches is needed. For implementation reasons we still have separate y-variables and constraints of type $y_v + 0 \leqslant y_w$ and $y_w + 0 \leqslant y_v$. This explains the differences in the number of vertices and edges of the graph G' in Table 2 to the numbers which one would expect by Table 1.

Our program consists of two main parts: first the constraints are generated by simple forward propagation for late mode and early mode constraints. During the propagation we store at each circuit the set of all primary inputs and latches from which this circuit can be reached, along with the maximum propagation delay for late mode, resp. the minimum propagation delay for early mode. Table 3 shows the running times for the generation of all constraints for the three different chips. A detailed description of the timing analysis program can be found in Schietke [15]. The running times for simple propagation (computation of arrival times and slacks only) are shown for comparison.

The second step consists of the main optimization algorithm. Rather than increasing the slacks on late mode and early mode constraints simultaneously, we first increase the slacks on late mode constraints up to a certain value λ^{late} while ignoring all early mode constraints. Then we add all early mode constraints and increase the slack on these constraints up to λ^{early}. For this the late mode constraints are added with unparameterized edges, such that the slack of late mode constraints with slack smaller than λ^{late} does not decrease and the slack of all other late mode constraints remains at least λ^{late}. The reason for treating late and early mode constraints differently is that early

```
------------------------------------------------------------------------------------
      slacks_late
      number of values: 1475535
      minimum value:    -0.223992
      one star represents 300 values.

  0.400 -   0.450 | 13830 | *********************************************
  0.350 -   0.400 | 11409 | **************************************
  0.300 -   0.350 |  8635 | ****************************
  0.250 -   0.300 |  6754 | **********************
  0.200 -   0.250 |  5125 | *****************
  0.150 -   0.200 |  5239 | *****************
  0.100 -   0.150 |  2029 | ******
  0.050 -   0.100 |  1535 | *****
 -0.000 -   0.050 |   698 | **
 -0.050 -  -0.000 |   255 |
 -0.100 -  -0.050 |   101 |
 -0.150 -  -0.100 |    37 |
 -0.200 -  -0.150 |     2 |
 -0.250 -  -0.200 |     5 |
 -0.300 -  -0.250 |     0 |
------------------------------------------------------------------------------------
```

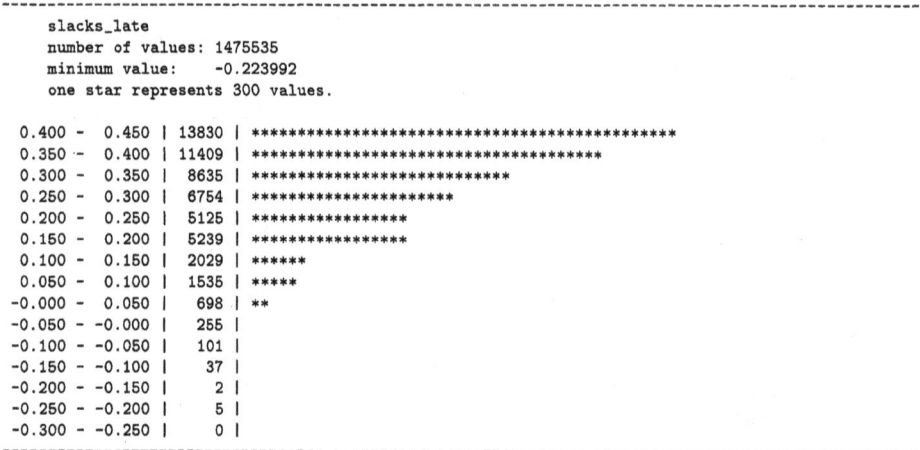

Fig. 6. Slacks of late mode constraints ($\varepsilon_{vw}^{\max}$) without optimization for the MBA.

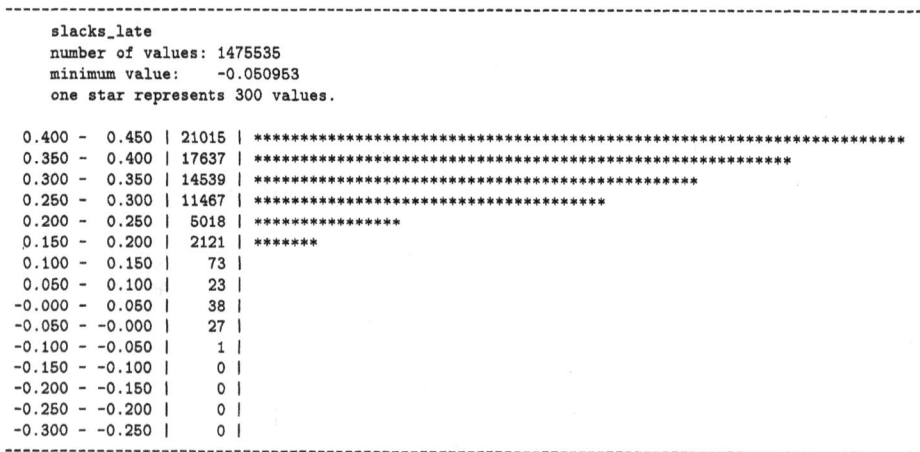

```
------------------------------------------------------------------------------------
      slacks_late
      number of values: 1475535
      minimum value:    -0.050953
      one star represents 300 values.

  0.400 -   0.450 | 21015 | ***********************************************************************
  0.350 -   0.400 | 17637 | **********************************************************
  0.300 -   0.350 | 14539 | *************************************************
  0.250 -   0.300 | 11467 | **************************************
  0.200 -   0.250 |  5018 | ****************
  0.150 -   0.200 |  2121 | *******
  0.100 -   0.150 |    73 |
  0.050 -   0.100 |    23 |
 -0.000 -   0.050 |    38 |
 -0.050 -  -0.000 |    27 |
 -0.100 -  -0.050 |     1 |
 -0.150 -  -0.100 |     0 |
 -0.200 -  -0.150 |     0 |
 -0.250 -  -0.200 |     0 |
 -0.300 -  -0.250 |     0 |
------------------------------------------------------------------------------------
```

Fig. 7. Optimized late mode slacks ($\varepsilon_{vw}^{\max}$) for the MBA.

mode problems can usually be fixed quite efficiently by inserting a buffer ($=2$ inverters). Finally, we balance the slacks on clocktree paths up to a value λ^{clock} as described in Section 6. Again it is assured that the slacks of late and early mode constraints do not decrease below the value to which they were optimized.

Obviously, the result of the optimization depends on the length of the interval given by l_s and u_s specifying by how much the clock signal arrival time can be shifted from its nominal value. For the L2 and PU designers called for $l_s = -0.4$ ns and $u_s = 0.4$ ns for most of the latches. However, for 42.8% of the slave latches and 1.4% of the master latches of the L2 we had a prescribed arrival time of the clock signal, i.e. $l_s = u_s = 0.0$ ns. On the PU 1.4% of all master latches and 25% of all slave latches

```
slacks_early
number of values: 1457513
minimum value:    -0.021301
one star represents 300 values.

  0.400 -   0.450 | 94877 | ***********************************************************************>>>
  0.350 -   0.400 | 79750 | ***********************************************************************>>>
  0.300 -   0.350 | 62978 | ***********************************************************************>>>
  0.250 -   0.300 | 51393 | ***********************************************************************>>>
  0.200 -   0.250 | 55248 | ***********************************************************************>>>
  0.150 -   0.200 | 23346 | ***********************************************************************>>>
  0.100 -   0.150 | 13643 | *******************************************
  0.050 -   0.100 |  9235 | ******************************
 -0.000 -   0.050 |  5743 | *******************
 -0.050 - -0.000  |   140 |
 -0.100 - -0.050  |     0 |
 -0.150 - -0.100  |     0 |
 -0.200 - -0.150  |     0 |
 -0.250 - -0.200  |     0 |
 -0.300 - -0.250  |     0 |
```

Fig. 8. Slacks of early mode constraints ($\varepsilon_{uw}^{\min}$) without optimization for the MBA.

```
slacks_early
number of values: 1457513
minimum value:    -0.125193
one star represents 300 values.

  0.400 -   0.450 | 75226 | ****************************************************************>>>
  0.350 -   0.400 | 64770 | ****************************************************************>>>
  0.300 -   0.350 | 60908 | ****************************************************************>>>
  0.250 -   0.300 | 45244 | ****************************************************************>>>
  0.200 -   0.250 | 20418 | ****************************************************************
  0.150 -   0.200 |   622 | **
  0.100 -   0.150 |   260 |
  0.050 -   0.100 |   344 | *
 -0.000 -   0.050 |   502 | *
 -0.050 - -0.000  |   244 |
 -0.100 - -0.050  |     8 |
 -0.150 - -0.100  |     1 |
 -0.200 - -0.150  |     0 |
 -0.250 - -0.200  |     0 |
 -0.300 - -0.250  |     0 |
```

Fig. 9. Optimized early mode slacks ($\varepsilon_{uv}^{\min}$) for the MBA.

had $l_s = u_s = 0.0$ ns. On the MBA most of the latches had $l_s = -0.4$ ns and $u_s = 0.5$ ns, but 1.99% of the master latches and 9.65% of the slave latches had $l_s = u_s = 0.0$ ns.

Table 4 shows the worst slack of all late mode constraints before and after optimization with respect to the target cycle time mentioned in Table 1. This means that the L2 could run with a cycle time of 6.5 ns + 0.048 ns = 6.548 ns before the optimization and with a cycle time of 6.5 ns − 0.313 ns = 6.187 ns after optimization: the cycle time was improved by 5.5%. Similarly, the cycle time of the PU was improved by 2.5% and the cycle time of the MBA by 3.7%.

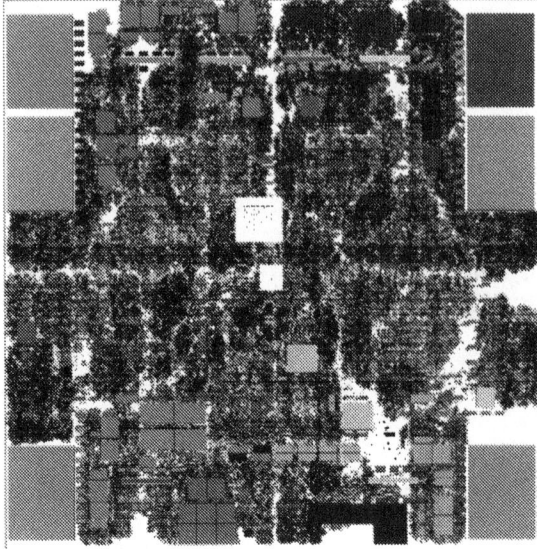

Fig. 10. MBA Placement.

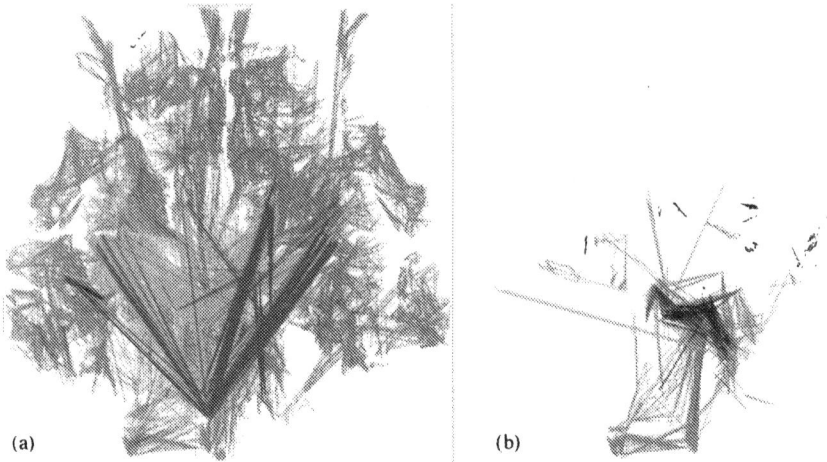

(a)   (b)

Fig. 11. Signal path on MBA with slack < 0.2ns before and after optimization.

Table 4 also shows the number of signal paths with late mode slack smaller than -0.2 ns, -0.1 ns, ... This demonstrates how dramatically the number of critical paths decreases.

In Table 5 the running times for the algorithm for the three chips for computing the optimal cycle time and for balancing slacks for different scenarios with λ^{late}, λ^{early} and

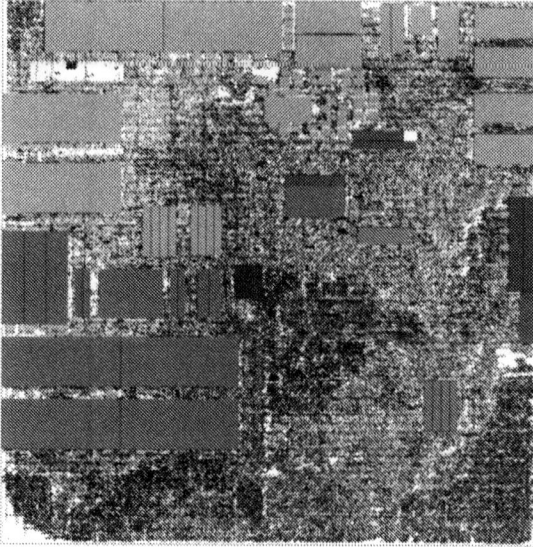

Fig. 12. PU Placement.

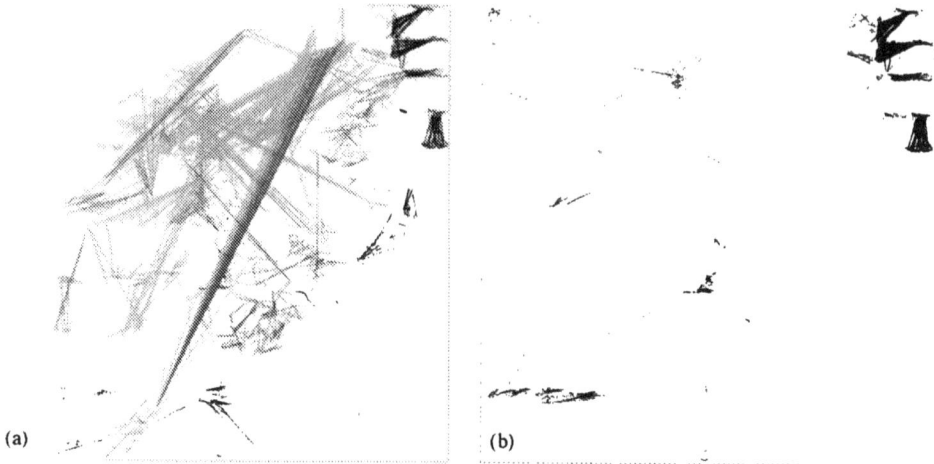

(a)

(b)

Fig. 13. Signal Paths on PU with slack < 0.2ns befor and after optimization.

λ_{clock} are shown. Again, all late mode slacks are taken with respect to the target cycle times.

Fig. 6 shows a frequency distribution of the slacks of all late mode constraints for the MBA for the case that no optimization is possible (i.e. $l_s = u_s = 0.0$ ns for all latches s). The first column shows the different intervals, the second column gives the number of signal paths whose slack is within the interval and the third column

```
------------------------------------------------------------------------------
    y_master_shift
    number of values: 31561
    maximum value:    0.500000
    minimum value:   -0.400000
    one star represents 100 values.

   0.500 -   0.550 |  1196 | **********
   0.450 -   0.500 |    67 |
   0.400 -   0.450 |   214 | **
   0.350 -   0.400 |   381 | ***
   0.300 -   0.350 |   596 | *****
   0.250 -   0.300 |   933 | *********
   0.200 -   0.250 |  1228 | ***********
   0.150 -   0.200 |  1544 | ***************
   0.100 -   0.150 |  1534 | ***************
   0.050 -   0.100 |  2312 | ***********************
   0.000 -   0.050 |  3735 | *************************************
  -0.050 -   0.000 |  3911 | ***************************************
  -0.100 -  -0.050 |  4411 | ********************************************
  -0.150 -  -0.100 |  3290 | *******************************
  -0.200 -  -0.150 |  2332 | ***********************
  -0.250 -  -0.200 |  1437 | **************
  -0.300 -  -0.250 |   804 | ********
  -0.350 -  -0.300 |   451 | ****
  -0.400 -  -0.350 |  1185 | ***********
------------------------------------------------------------------------------
```

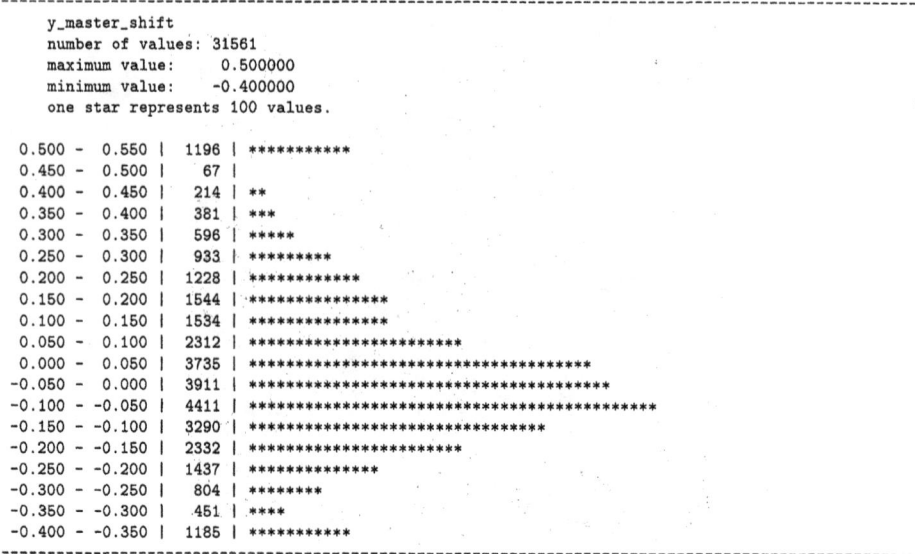

Fig. 14. Shifting time (y) of master latches.

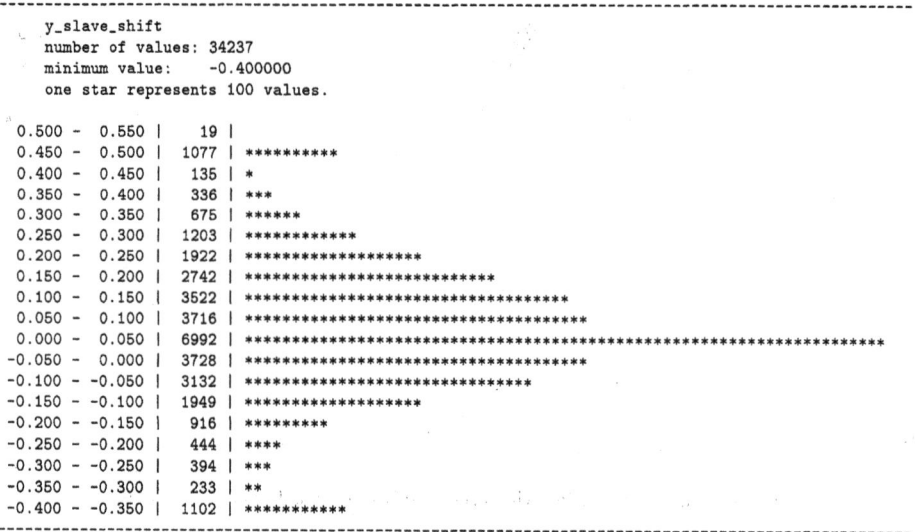

```
------------------------------------------------------------------------------
    y_slave_shift
    number of values: 34237
    minimum value:   -0.400000
    one star represents 100 values.

   0.500 -   0.550 |    19 |
   0.450 -   0.500 |  1077 | *********
   0.400 -   0.450 |   135 | *
   0.350 -   0.400 |   336 | ***
   0.300 -   0.350 |   675 | ******
   0.250 -   0.300 |  1203 | ************
   0.200 -   0.250 |  1922 | *******************
   0.150 -   0.200 |  2742 | ***************************
   0.100 -   0.150 |  3522 | ***********************************
   0.050 -   0.100 |  3716 | *************************************
   0.000 -   0.050 |  6992 | ********************************************************************
  -0.050 -   0.000 |  3728 | *************************************
  -0.100 -  -0.050 |  3132 | *******************************
  -0.150 -  -0.100 |  1949 | *******************
  -0.200 -  -0.150 |   916 | *********
  -0.250 -  -0.200 |   444 | ****
  -0.300 -  -0.250 |   394 | ***
  -0.350 -  -0.300 |   233 | **
  -0.400 -  -0.350 |  1102 | ***********
------------------------------------------------------------------------------
```

Fig. 15. Shifting times (y) of slave latches.

gives a graphical representation of this number by a proportional number of stars. For example, it can be read from Fig. 6 that the MBA without optimization has 101 signal paths with $-0.100 \text{ ns} \leqslant \varepsilon_{vw}^{\max} < -0.050 \text{ ns}$. Fig. 7 shows the corresponding frequency distributions of late mode constraints after optimization for $\lambda = 0.2 \text{ ns}$.

```
-----------------------------------------------------------------------------------------
    e_master
    number of values: 31561
    minimum value:     0.000000
    one star represents 150 values.

  0.100 -  0.110 |  7179 | **********************************************
  0.090 -  0.100 |  1860 | ************
  0.080 -  0.090 |  2110 | **************
  0.070 -  0.080 |  2505 | ****************
  0.060 -  0.070 |  2346 | ***************
  0.050 -  0.060 |  2336 | ***************
  0.040 -  0.050 |  3018 | ********************
  0.030 -  0.040 |  3697 | ************************
  0.020 -  0.030 |  3401 | **********************
  0.010 -  0.020 |  1943 | ************
  0.000 -  0.010 |  1166 | *******
-----------------------------------------------------------------------------------------
```

Fig. 16. Slack on clock tree $_s$ paths of master latches.

```
-----------------------------------------------------------------------------------------
    e_slave
    number of values: 34237
    minimum value:     0.000000
    one star represents 150 values.

  0.100 -  0.110 |  9061 | ***************************************************************
  0.090 -  0.100 |  2717 | ******************
  0.080 -  0.090 |  1960 | *************
  0.070 -  0.080 |  2280 | ***************
  0.060 -  0.070 |  2249 | **************
  0.050 -  0.060 |  2189 | **************
  0.040 -  0.050 |  3169 | *********************
  0.030 -  0.040 |  3867 | *************************
  0.020 -  0.030 |  3143 | ********************
  0.010 -  0.020 |  1871 | ************
  0.000 -  0.010 |  1731 | **********
-----------------------------------------------------------------------------------------
```

Fig. 17. Slack on clock tree $_s$ paths of slave latches.

Figs. 8 and 9 show the same for all early mode constraints. Increasing the slacks on late mode constraints makes the worst slack of all early mode constraints worse, but nevertheless the total number of early mode constraints with negative slack decreases considerably.

Fig. 11a and b shows the effect of slack balancing at a glance. Each line connects the endpoints of a critical path, with respect to the placement shown in Fig. 10. The colours have the following meaning:

• red lines represent signal paths with a negative late mode slack;
• yellow lines represent signal paths with a late mode slack between 0.0 and 0.2 ns;
• blue lines represent signal paths with a negative early mode slack;
• green lines represent signal paths with an early mode slack between 0.0 and 0.2 ns.

The left-hand side is the situation before optimization, the right-hand side shows that after optimization only very few critical areas remain. Fig. 13a and b is the analogous picture for the PU, with respect to the placement shown in Fig. 12. Figs. 14 and 15 show the frequency distribution of the y-variables (shifting times of clock signal arrival times at latches) of all master and slave latches respectively. The values correspond to the achieved slacks in Figs. 7 and 9. One can observe that more than 75% of all y-variables are with in ± 0.2 ns.

Finally, in Figs. 16 and 17 the result of balancing the slacks on clocktree paths is shown. Even though the slacks on late mode and early mode constraints have already been increased up to 0.2 ns, the clock signal for most of the latches still does not have to arrive at exactly the prescribed time.

Acknowledgements

We thank the anonymous referees for comments that helped improving the presentation of this paper.

References

[1] U. Bünnagel, Effiziente Implementierung von Netzwerkalgorithmen, Diploma Thesis, University of Bonn, 1998 (in German).

[2] R.B. Deokar, S. Sapatnekar, A graph-theoretic approach to clock skew optimization, Proceedings of the IEEE International Symposium on Circuits and Systems, 1994, pp. 407–410.

[3] G. Doettling, K.J. Getzlaff, B. Leppla, W. Lipponer, T. Pflueger, T. Schlipf, D. Schmunkamp, U. Wille, S/390 parallel enterprise server generation3: a balanced system and cache structure, IBM J. Res. Dev. 41 (1997) 405–428.

[4] J.P. Fishburn, Clock skew optimization, Trans. Comput. C-39 (1990) 945–951.

[5] J.H. Huang, A.B. Kahng, C.-W.A. Tsao, On bounded-skew routing tree problem, Proceedings of the 32nd Design Automation Conference, 1995, pp. 508–513.

[6] R.M. Karp, A characterization of the minimum mean cycle in a digraph, Discrete Math. 23 (1978) 309–311.

[7] B. Kick, U. Baur, J. Koehl, T. Ludwig, T. Pflueger, Standard-cell-based design methodology for high performance support chips, IBM J. Res. Dev. 41 (1997) 505–514.

[8] J. Koehl, U. Baur, T. Ludwig, B. Kick, T. Pflueger, A flat, timing-driven design system for a high-performance CMOS processor chipset, Proceedings of the Conference "Design, Automation and Test in Europe", 1998, pp. 312–320.

[9] J. Lee, D.T. Tang, C.K. Wang, A timing analysis algorithm for circuits with level-sensitive latches, IEEE Trans. Comput.-Aided Des. Integrated Circuits Systems 15 (1996) 535–543.

[10] I. Lin, J.A. Ludwig, K. Eng, Analyzing cycle stealing on synchronous circuits with level-sensitive latches, Proceedings of the 29th ACM/IEEE Design Automation Conference, 1992, pp. 393–398.

[11] N. Megiddo, Applying parallel computation algorithms in the design of serial algorithms, J. Assoc. Comput. Machinery 30 (1983) 852–865.

[12] K. Muuss, Clockskew Optimierung, Diploma Thesis, University of Bonn, 1994 (in German).

[13] J.L. Neves, E.G. Friedman, Optimal clock skew scheduling tolerant to process, variations, Proceedings of the 33rd Design Automation Conference, 1996, pp. 623–628.

[14] K.A. Sakallah, T.N. Mudge, O.A. Olukotun, $checkT_c$ and $minT_c$: timing verification and optimal clocking of synchronous digital circuits, Proceedings of the IEEE International Conference on Computer-Aided Design, 1990, pp. 552–555.

[15] J. Schietke, Timing-Optimierung beim physikalischen Layout von nicht-hierarchischen Designs hochintegrierter Logikchips, Ph.D. Thesis, University of Bonn, 1999 (in German).

[16] H. Schneider, M.H. Schneider, Towers and cycle covers for max-balanced graph, Math. Oper. Res. 16 (1991) 208–222.

[17] N. Shenoy, R.K. Brayton, Graph algorithms for clock schedule optimization, Proceedings of the IEEE International Conference on Computer-Aided Design, 1992, pp. 132–136.

[18] T.G. Szymanski, Computing optimal clock schedules, Proceedings of the 29th ACM/IEEE Design Automation Conference, 1992, pp. 399–404.

[19] T.G. Szymanski, N. Shenoy, Verifying clock schedules, Proceedings of the IEEE International Conference on Computer-Aided Design, 1992, pp. 124–131.

[20] R.-S. Tsay, Exact zero skew, IEEE Trans. Comput.-Aided Des. 21 (1991) 205–221.

[21] J. Vygen, Plazierung im VLSI-Design und ein zweidimensionales Zerlegungsproblem, Ph.D. Thesis, University of Bonn, 1996 (in German).

[22] N.E. Young, R.E. Tarjan, J.B. Orlin, Faster parametric shortest path and minimum balance algorithms, Networks 21 (1991) 205–221.

ELSEVIER   Discrete Applied Mathematics 123 (2002) 129–154

DISCRETE
APPLIED
MATHEMATICS

Lift-and-project for Mixed 0–1 programming: recent progress [☆]

Egon Balas[*], Michael Perregaard

Graduate School of Industrial Admin., Carnegie Mellon University, Schenley Park, Pittsburgh, PA 15213-3890, USA

Received 28 September 1999; received in revised form 3 May 2000; accepted 15 May 2000

Abstract

This article reviews the disjunctive programming or lift-and-project approach to 0-1 programming, with an emphasis on recent developments. Disjunctive programming is optimization over unions of polyhedra. The first three sections of the paper define basic concepts and introduce the two fundamental results underlying the approach. Thus, section 2 describes the compact higher dimensional representation of the convex hull of a union of polyhedra, and its projection on the original space; whereas section 3 is devoted to the sequential convexifiability of facial disjunctive programs, which include mixed 0-1 programs. While these results originate in Balas' work in the early- to mid-seventies, some new results are also included: it is shown that on the higher dimensional polyhedron representing the convex hull of a union of polyhedra, the maximum edge-distance between any two vertices in 2. Also, it is shown that in the process of sequential convexification of a 0-1 program, fractional intermediate values of the variables can occur only under very special circumstances. The next section relates the above results to the matrix-cone approach of Lovász and Schrijver and of Sherali and Adams. Section 5 introduces the lift-and-project cuts of Balas, Ceria and Cornuéjols from the early nineties, and discusses the cut generating linear program (CGLP), cut lifting and cut strengthening. The next section briefly outlines the branch and cut framework in which the lift-and-project cuts turned out to be computationally useful, while section 7 discusses some crucial aspects of the cut generating procedure: alternative normalizations of (CGLP), complementarity of the solution components, size reduction of (CGLP), and ways of deriving multiple cuts from a disjunction. Finally, section 8 discusses computational results in branch-and-cut mode as well as in cut-and-branch mode. © 2002 Elsevier Science B.V. All rights reserved.

Keywords: Lift-and-project; Disjunctive programming; Mixed 0–1 programming; Cut generation

---

[☆] Research supported by the National Science Foundation through grant DMI-9802773 and by the Office of Naval Research through contract N00014-97-1-0196.
[*] Corresponding author.
E-mail address: eb17@andrew.cmu.edu (E. Balas).

0166-218X/02/$ - see front matter © 2002 Elsevier Science B.V. All rights reserved.
PII: S 0166-218X(01)00340-7

1. Introduction

This is a state-of-the-art survey of the disjunctive programming approach to mixed 0 –1 programming, also known as lift-and-project, with an emphasis on current research on the subject. The foundations of this approach were laid in a July 1974 Technical Report, published 24 years later as an invited paper [1] with a foreword. For additional work on disjunctive programming in the 1970s and 1980s see [2,3,8,9,12,13,17–19,22]. In particular, [2] contains a detailed account of the origins of the disjunctive approach and the relationship of disjunctive cuts to Gomory's mixed integer cut, intersection cuts and others. Disjunctive programming received a new impetus in the early 1990s from the work on matrix cones by Lovász and Schrijver [20], see also Sherali and Adams [21]. The version that led to the computational breakthroughs of the nineties is described in the two papers by Balas et al. [5,6], the first of which discusses the cutting plane theory behind the approach, while the second deals with the branch-and-cut implementation and computational testing. Related recent developments are discussed in [7,4,10,14–16,23,25,26].

In this survey, known results are not proved, but referenced. Results that are proved are believed to be new.

1.1. Disjunctive programming

Disjunctive programming is optimization over unions of polyhedra. While polyhedra are convex sets, their unions of course are not. The name reflects the fact that the objects investigated by this theory can be viewed as the solution sets of systems of linear inequalities joined by the logical operations of conjunction, negation (taking of complement) and disjunction, where the nonconvexity is due to the presence of disjunctions. Pure and mixed integer programs, in particular pure and mixed 0–1 programs can be viewed as disjunctive programs; but the same is true of a host of other problems, like for instance the linear complementarity problem. Our focus will be on pure and mixed 0–1 programs.

The constraint set of a disjunctive program, called a disjunctive set, can be expressed in many different forms, of which the following two extreme ones have special significance. Let

$$P_i := \{x \in \mathbb{R}^n : A^i x \geqslant b^i\}, \quad i \in Q$$

be convex polyhedra, with Q a finite index set and (A^i, b^i) an $m_i \times (n+1)$ matrix, $i \in Q$, and let $P := \{x \in \mathbb{R}^n : Ax \geqslant b\}$ be the polyhedron defined by those inequalities (if any) common to all P_i, $i \in Q$. Then the disjunctive set $\bigcup_{i \in Q} P_i$ over which we wish to optimize some linear function can be expressed as

$$\left\{ x \in \mathbb{R}^n : \bigvee_{i \in Q} (A^i x \geqslant b^i) \right\}, \tag{1}$$

which is its *disjunctive normal form* (a disjunction whose terms do not contain further disjunctions). The same disjunctive set can also be expressed as

$$
\left\{ x \in \mathbb{R}^n : Ax \geqslant b, \bigvee_{h \in Q_j} (d^h x \geqslant d_0^h), \ j = 1, \ldots, t \right\},
\tag{2}
$$

which is its *conjunctive normal form* (a conjunction whose terms do not contain further conjunctions). Here (d^h, d_0^h) is a $(n+1)$-vector for $h \in Q_j$, all j. The connection between (1) and (2) is that each term $A^i x \geqslant b^i$ of the disjunctive normal form (1) contains $Ax \geqslant b$ and exactly one inequality $d^h x \geqslant d_0^h$ of each disjunction of (2) indexed by Q_j for $j = 1, \ldots, t$, and that all distinct systems $A^i x \geqslant b^i$ with this property are present among the terms of (1). See [3] for details on how to go from (1) to (2) and from (2) to (1).

1.2. Two basic ideas

The lift-and-project approach relies mainly on the following two ideas (results), the first of which uses the disjunctive normal form (1), while the second one uses the conjunctive normal form (2):

1. There is a compact representation of the convex hull of a union of polyhedra in a higher dimensional space, which in turn can be projected back into the original space. The first step of this operation may be viewed as lifting, the second step, projection. As a result one obtains the convex hull in the original space.
2. A large class of disjunctive sets, called *facial*, can be convexified sequentially, i.e. their convex hull can be derived by imposing the disjunctions one at a time, generating each time the convex hull of the current set.

2. Compact representation of the convex hull

Theorem 1 (Balas [1]). *Given polyhedra $P_i := \{x \in \mathbb{R}^n : A^i x \geqslant b^i\} \neq \emptyset$, $i \in Q$, the closed convex hull of $\bigcup_{i \in Q} P_i$ is the set of those $x \in \mathbb{R}^n$ for which there exist vectors $(y^i, y_0^i) \in \mathbb{R}^{n+1}$, $i \in Q$, satisfying*

$$
x - \sum (y^i : i \in Q) = 0,
$$

$$
A^i y^i - b^i y_0^i \geqslant 0,
$$

$$
y_0^i \geqslant 0, \quad i \in Q,
$$

$$
\sum (y_0^i : i \in Q) = 1.
\tag{3}
$$

In particular, denoting by $P_Q := \operatorname{conv} \bigcup_{i \in Q} P_i$ the closed convex hull of $\bigcup_{i \in Q} P_i$ and by $\mathscr{P}$ the set of vectors $(x, \{y^i, y_0^i\}_{i \in Q})$ satisfying (3),

(i) if x^* is an extreme point of P_Q, then $(\bar{x}, \{\bar{y}_0^i, \bar{y}_0^i\}_{i \in Q})$ is an extreme point of $\mathscr{P}$, with $\bar{x} = x^*$, $(\bar{y}^k, \bar{y}_0^k) = (x^*, 1)$ for some $k \in Q$, and $(\bar{y}^i, \bar{y}_0^i) = (0, 0)$ for $i \in Q \setminus \{k\}$.

(ii) *if* $(\bar{x}, \{\bar{y}^i, \bar{y}_0^i\}_{i \in Q})$ *is an extreme point of* $\mathscr{P}$, *then* $\bar{y}^k = \bar{x} = x^*$ *and* $\bar{y}_0^k = 1$ *for some* $k \in Q$, $(\bar{y}^i, \bar{y}_0^i) = (0, 0)$, $i \in Q \setminus \{k\}$, *and* x^* *is an extreme point of* P_Q.

Note that in this higher dimensional representation of P_Q, the number of variables and constraints is linear in the number $|Q|$ of polyhedra in the union, and so is the number of facets of $\mathscr{P}$. Note also that in any basic solution of the linear system (3), $y_0^i \in \{0, 1\}$, $i \in Q$, automatically, without imposing this condition explicitly.

Of course, if the set Q is itself exponential in the number of variables, then system (3) becomes unmanageably large. This is the case for instance if we impose simultaneously *all* the integrality conditions of a mixed 0–1 program with p 0–1 variables, in which case we have a disjunction with 2^p terms, one for every p-component 0–1 point. But if we impose only disjunctions that yield a set Q of manageable size, then this representation becomes extremely useful (such an approach is facilitated by the sequential convexifiability of facial disjunctive sets, see below).

In the special case of a disjunction of the form $x_j \in \{0, 1\}$, when $|Q| = 2$ and

$$P_{j0} := \{x \in \mathbb{R}_+^n : Ax \geqslant b, \; x_j = 0\},$$

$$P_{j1} := \{x \in \mathbb{R}_+^n : Ax \geqslant b, \; x_j = 1\},$$

$P_Q := \mathrm{conv}\,(P_{j0} \cup P_{j1})$ is the set of those $x \in \mathbb{R}^n$ for which there exist vectors $(y, y_0), (z, z_0) \in \mathbb{R}_+^{n+1}$ such that

$$x - y - z = 0,$$

$$Ay - by_0 \geqslant 0,$$

$$-y_j = 0,$$

$$Az - bz_0 \geqslant 0,$$

$$z_j - z_0 = 0,$$

$$y_0 + z_0 = 1. \tag{3'}$$

Unlike the general system (3), the system (3'), in which $|Q| = 2$, is of quite manageable size.

2.1. Projection and polarity

In order to generate the convex hull P_Q, and more generally, to obtain valid inequalities (cutting planes) in the space of the original variables, we project $\mathscr{P}$ onto the x-space:

Theorem 2 (Balas [1]). $\mathrm{Proj}_x(\mathscr{P}) = \{x \in \mathbb{R}^n : \alpha x \geqslant \beta \text{ for all } (\alpha, \beta) \in W_0\}$, *where*

$$W_0 := \{(\alpha, \beta) \in \mathbb{R}^{n+1} : \alpha = u^i A^i, \; \beta \leqslant u^i b^i \text{ for some } u^i \geqslant 0, \; i \in Q\}.$$

The polyhedral cone W_0 used to project $\mathscr{P}$ can be shown to be the *reverse polar cone* P_Q^* of P_Q, i.e. the cone of all valid inequalities for P_Q:

Theorem 3 (Balas [1]).

$$P_Q^* := \{(\alpha, \beta) \in \mathbb{R}^{n+1}: \alpha x \geqslant \beta \ for \ all \ x \in P_Q\}$$
$$= \{(\alpha, \beta) \in \mathbb{R}^{n+1}: \alpha = u^i A^i, \ \beta \leqslant u^i b^i \ for \ some \ u^i \geqslant 0, \ i \in Q\}.$$

To turn again to the special case of a disjunction of the form $x_j \in \{0, 1\}$, projecting the system (3′) onto the x-space yields the polyhedron P_Q whose reverse polar cone is

$$P_Q^* = \{(\alpha, \beta) \in \mathbb{R}^{n+1}: \alpha \geqslant uA - u_0 e_j,$$
$$\alpha \geqslant vA + v_0 e_j,$$
$$\beta \leqslant ub,$$
$$\beta \leqslant vb + v_0,$$
$$u, v \geqslant 0\},$$

(where e_j is the jth unit vector.)

One of the main advantages of the higher dimensional representation is that in projecting it back we have an easy criterion to distinguish facets of P_Q from other valid inequalities.

Theorem 4 (Balas [1]). *Assume P_Q is full dimensional. The inequality $\alpha x \geqslant \beta$ defines a facet of P_Q if and only if (α, β) is an extreme ray of the cone P_Q^*.*

2.2. Adjacency on the higher dimensional polyhedron

In the process of generating facets of P_Q, sometimes one would like to list the extreme points of P_Q adjacent to a given extreme point x. The question arises, can one do this by using the adjacency relations on the higher dimensional polyhedron $\mathscr{P}$, for which a linear description is available?

As usual, we call two extreme points, or zero-dimensional faces, of a polyhedron adjacent if they are contained in the same one-dimensional face. In terms of the system of linear inequalities defining the polyhedron, two basic solutions are adjacent (correspond to adjacent extreme points of the polyhedron) if one can associate with them two bases that differ in exactly one column.

From Theorem 2.2 of [1], it follows that every basis for the system (3) defining $\mathscr{P}$ is of the form (modulo row and column permutations)

$$B = \begin{pmatrix} I & -E^1 & \cdots & -E^{q-1} & -I & \\ & B^1 & & & & \\ & & \ddots & & & \\ & & & B^{q-1} & & \\ & & & & B^q & -b^q \\ & \delta^1 & \cdots & \delta^{q-1} & 0 & 1 \end{pmatrix},$$

where B^q is a basis for the system $A^q y^q - s^q = b^q$ (with s^q a vector of surplus variables), $q = |Q|$ (i.e. q plays the role of the index k in statements (i) and (ii) of Theorem 1); further, for $i \in \{1, \ldots, q-1\}$, B^i is a basis for the system $A^i y^i - b^i y_0^i - s^i = 0$; δ^i is a row vector whose entries are all zero if B^i does not contain the column b^i, otherwise the entry corresponding to b^i is 1 and the remaining entries are 0; I is the identity matrix corresponding to the basic components of x (and of y^q), while E^i is the diagonal matrix with 1 in the positions that are basic for both x and y^i, 0 in the remaining positions; and all blanks are zeros. Clearly, the solution corresponding to B is of the form $(y^q, y_0^q) = (x, 1)$ and $(y^i, y_0^i) = 0$ for all $i \in Q \setminus \{q\}$.

Evidently, every extreme point of P_Q is an extreme point of some P_i, $i \in Q$. In the next theorem and its Corollary, we assume that $|Q| \geqslant 2$.

Theorem 5. *Let x^1 and x^2 be arbitrary extreme points of P_Q such that $x^1 \in P_i$, $x^2 \in P_j$, with $i \neq j$. Then there exist vectors (y^{1i}, y_0^{1i}), (y^{2i}, y_0^{2i}), $i \in Q$, such that $(x^1, \{y^{1i}, y_0^{1i}\}_{i \in Q})$ and $(x^2, \{y^{2i}, y_0^{2i}\}_{i \in Q})$ are adjacent extreme points of $\mathcal{P}$.*

Proof. Given x^1, there is exactly one $k \in Q$ such that $(y^{1k}, y_0^{1k}) = (x^1, 1)$ is a feasible solution to $A^k y^{1k} - b^k y_0^{1k} \geqslant 0$. Assigning this value to (y^{1k}, y_0^{1k}) and setting $(y^{1i}, y_0^{1i}) = (0,0)$ for all $i \in Q \setminus \{k\}$ defines an extreme point $(x^1, \{y^{1i}, y_0^{1i}\}_{i \in Q})$ of $\mathcal{P}$. Similarly, given x^2, there is exactly one $\ell \in Q$ such that $(y^{2\ell}, y_0^{2\ell}) = (x^2, 1)$ is a feasible solution to $A^\ell y^{2\ell} - b^\ell y_0^{2\ell} \geqslant 0$; which can be used to define an extreme point $(x^2, \{y^{2i}, y_0^{2i}\}_{i \in Q})$ of $\mathcal{P}$ in which $(y^{2\ell}, y_0^{2\ell}) = (x^2, 1)$ and $(y^{2i}, y_0^{2i}) = 0$ for all $in \in Q \setminus \{\ell\}$. Furthermore, since $x^1 \neq x^2$, $\ell \neq k$.

Now although the two extreme points of $\mathcal{P}$ described above have the values of their components uniquely defined, each can be associated with a multitude of bases: indeed, while the basis associated with the kth subsystem in the case of x^1, and with the ℓth subsystem in the case of x^2, is uniquely determined by the vectors $y^{1k} = x^1$ and $y^{2\ell} = x^2$, respectively, the bases associated with the remaining subsystems, whose variables are all zero, can be chosen freely. Furthermore, moving between such bases entails only degenerate pivots. In view of this, given a basis B associated with $(x^1, \{y^{1i}, y_0^{1i}\}_{i \in Q})$, in which $(y^{1k}, y_0^{1k}) = (x^1, 1)$, one can perform degenerate pivots involving only changes in the sub-basis B^ℓ associated with $(y^{1\ell}, y_0^{1\ell})$ until it is brought to the form required for $(y^{2\ell}, y_0^{2\ell}) = (x^2, 1)$, at which point a nondegenerate pivot can replace the entire basis B associated with $(x^1, \{y^{1i}, y_0^{1i}\}_{i \in Q})$ with a basis B' associated with $(x^2, \{y^{2i}, y_0^{2i}\}_{i \in Q})$. $\square$

Corollary 6. *The maximum edge-distance between any pair of vertices of $\mathcal{P}$ is 2.*

Proof. Let $(x^1, \{y^{1i}, y_0^{1i}\}_{i \in Q})$ and $(x^2, \{y^{2i}, y_0^{2i}\}_{i \in Q})$ be two arbitrary vertices of $\mathcal{P}$, and suppose $x^1 \in P_k$, $x^2 \in P_\ell$. If $k \neq \ell$, then from Theorem 5 there exist vectors $(\bar{y}^{1i}, \bar{y}_0^{1i})$, $(\bar{y}^{2i}, \bar{y}_0^{2i})$, $i \in Q$, such that $(x^1, \{\bar{y}^{1i}, \bar{y}_0^{1i}\}_{i \in Q})$ and $(x^2, \{\bar{y}^{2i}, \bar{y}_0^{2i}\}_{i \in Q})$ are adjacent extreme points of $\mathcal{P}$. Further, from Theorem 1(ii), $(\bar{y}^{1k}, \bar{y}_0^{1k}) = (x^1, 1) = (y^{1k}, y_0^{1k})$ and $\bar{y}^{2\ell}, \bar{y}_0^{2\ell}) = (x^2, 1) = (y^{2\ell}, y_0^{2\ell})$, with $(\bar{y}^{1i}, \bar{y}_0^{1i}) = (y^{1i}, y_0^{1i}) = (0,0)$ for all $i \in Q \setminus \{k\}$, and $(\bar{y}^{2i}, \bar{y}_0^{2i}) = (y^{2i}, y_0^{2i}) = (0,0)$ for all $i \in Q \setminus \{\ell\}$). If $k = \ell$, i.e. x^1 and x^2 belong to the same polyhedron P_k, choose any vertex $(x^3, \{y^{3i}, y_0^{3i}\}_{i \in Q})$ of $\mathcal{P}$ such that $x^3 \notin P_k$.

Then by the argument above, this vertex is adjacent to both $(x^1, \{y^{1i}, y_0^{1i}\}_{i \in Q})$ and $(x^2, \{y^{2i}, y_0^{2i}\}_{i \in Q})$; hence the latter two vertices are at an edge-distance of 2 from each other. $\square$

Thus our answer to the question raised initially is on the whole negative, in the sense that adjacency on $\mathscr{P}$ tells us almost nothing about adjacency on P_Q.

Next we turn to the class of disjunctive programs that are sequentially convexifiable.

3. Sequential convexification

A disjunctive set is called *facial* if every inequality in (2) induces a face of P. Zero–one programs (pure or mixed) are facial disjunctive programs, general integer programs are not. Sequential convexifiability is one of the basic properties that distinguish 0–1 programs from general integer programs.

Theorem 7 (Balas [1]). *Let*

$$D := \left\{ x \in \mathbb{R}^n \colon Ax \geqslant b, \bigvee_{h \in Q_j} (d^h x \geqslant d_0^h), \ j = 1, \ldots, t \right\},$$

where $1 \leqslant t \leqslant n$, $|Q_j| \geqslant 1$ *for* $j = 1, \ldots, t$, *and* D *is facial. Let* $P_D := \mathrm{conv}\,(D)$. *Define*

$$P^0 (=P) := \{ x \in \mathbb{R}^n \colon Ax \geqslant b \}$$

and for $j = 1, \ldots, t$,

$$P^j := \mathrm{conv} \left(P^{j-1} \cap \left\{ x \colon \bigvee_{h \in Q_j} (d^h x \geqslant d_0^h) \right\} \right).$$

Then

$$P^t = P_D.$$

While faciality is a sufficient condition for sequential convexifiability, it is not necessary. A necessary and sufficient condition is given in [9]. The most important class of facial disjunctive programs are mixed 0–1 programs, and for that case Theorem 7 asserts that if we denote

$$P_D := \mathrm{conv}\,\{ x \in \mathbb{R}_+^n \colon Ax \geqslant b, \ x_j \in \{0, 1\}, \ j = 1, \ldots, p \},$$

$$P^0 := \{ x \in \mathbb{R}_+^n \colon Ax \geqslant b \}$$

and define recursively for $j = 1, \ldots, p$

$$P^j := \mathrm{conv}\,(P^{j-1} \cap \{ x \colon x_j \in \{0, 1\} \}),$$

then

$$P^p = P_D.$$

Thus, in principle, a 0–1 program with p 0–1 variables can be solved in p steps. Here each step consists of imposing the 0–1 condition on one new variable and generating all the inequalities that define the convex hull of the set defined in this way.

3.1. Disjunctive rank

Based on this property, one can define the *disjunctive rank* of an inequality $\alpha x \geqslant \beta$ for a mixed 0–1 program as the smallest integer k for which there exists an ordering $\{i,\ldots,i_p\}$ of $\{1,\ldots,p\}$ such that $\alpha x \geqslant \beta$ is valid for P^k. In other words, an inequality is of rank k if it can be obtained by k, but not by fewer than k, applications of the recursive procedure defined above. Clearly, the disjunctive rank of a cutting plane for 0–1 programs is bounded by the number of 0–1 variables. It is known that the number of 0–1 variables is not a valid bound for the Chvatal rank of an inequality.

The above definition of the disjunctive rank is based on using the disjunctions $x_j \in \{0,1\}$, $j = 1,\ldots,p$. Tighter bounds can be derived by using stronger disjunctions. For instance, a 0–1 program whose constraints include the generalized upper bounds $\sum (x_j: j \in Q_i) = 1$, $i = 1,\ldots,t$, with $|Q_i| = |Q_j| = q$, $Q_i \cap Q_j = \emptyset$, $i,j \in \{1,\ldots,t\}$, and $|\bigcup_{i=1}^{t} Q_i| = p$, can be solved as a disjunctive program with the disjunctions

$$\bigvee_{j \in Q_i} (x_j = 1), \quad i = 1,\ldots,t(=p/q)$$

in which case the disjunctive rank of any cut is bounded by the number $t = p/q$ of *GUB* constraints.

Here is a new result concerning the nature of solutions generated during the sequential convexification process. In this process, the following question arises.

3.2. Fractionality of intermediate points

Define

$$P := \{x \in \mathbb{R}^n_+ : Ax \geqslant b\},$$

$$P^1 := \operatorname{conv} \{x \in P : x_1 \in \{0,1\}\}$$

and suppose the system defining P^1 has already been generated, i.e.

$$P^1 := \{x \in P : \alpha^i x \geqslant \beta^i, \ i \in I\}$$

is at hand. Now for $j \in \{2,\ldots,n\}$, consider

$$P^{1j} := \operatorname{conv} \{x \in P^1 : x_j \in \{0,1\}\}.$$

By virtue of the sequential convexifiability of 0–1 programs (Theorem 7),

$$P^{1j} := \operatorname{conv} \{x \in P : x_1 \in \{0,1\}, x_j \in \{0,1\}\};$$

in other words, by imposing the 0–1 condition on x_j we automatically enforce the condition $x_1 \in \{0,1\}$ too. But what about the intermediate solutions generated "on the

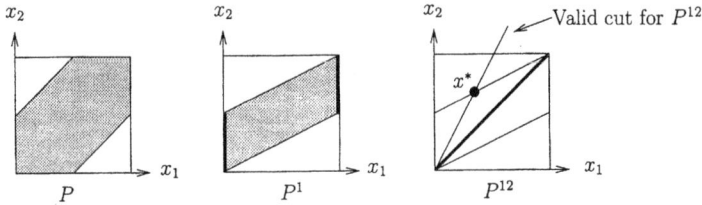

Fig. 1. $0 < x_1^* < 1$, $0 < x_2^* < 1$.

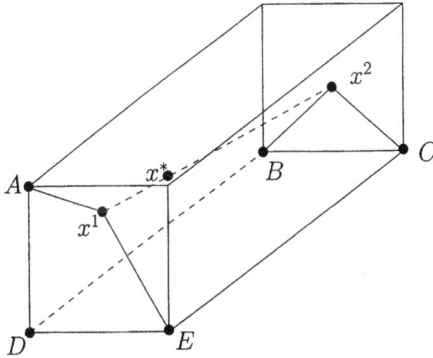

Fig. 2. Hyperplane ABC (not shown) intersects edge $[x^1, x^2]$ in a point x^* with $0 < x_j^* < 1$, $j = 1, 2, 3$.

way" from P^1 to P^{1j} for some j? As we add new cutting planes to our linear program, will the resulting solutions satisfy $x_1 \in \{0, 1\}$? In general, this cannot be guaranteed, as illustrated by the two-dimensional example of Fig. 1, where a valid cut generated in the process of getting from P^1 to P^{12} produces the fractional solution x^*.

The desired property cannot be maintained even if we restrict ourselves to the use of facet defining cutting planes, as illustrated in Fig. 2. Here P^1 is the convex hull of A, D, E, x^1, B, C, x^2, and the cutting plane through the points A, B, C is facet defining for both P^{12} and P^{13}. Yet, this hyperplane intersects the edge $[x^1, x^2]$ in its interior, hence in a point fractional in all three components.

However, a property almost as strong as the one whose absence we have just illustrated, still holds:

Theorem 8. *Let P^1 and P^{1j}, $j \in \{2, \ldots, n\}$, be as above. Let $\alpha x \geqslant \beta$ be a valid inequality for P^{1j}, $j \in \{2, \ldots, n\}$, and let x^* be an extreme point of $P^1 \cap \{x : \alpha x \geqslant \beta\}$. Then $0 < x_1^* < 1$ implies $0 < x_j^* < 1$ for all $j \in \{2, \ldots, n\}$.*

Proof. Suppose $x_j^* \in \{0, 1\}$ for some $j \in \{2, \ldots, n\}$; then $x^* \in P^{1j}$. If x^* is an extreme point of P^{1j}, then from Theorem 7, $x_1^* \in \{0, 1\}$ and we are done. If x^* is not extreme, then $\alpha x^* = \beta$ and there exist points $x^1, x^2 \in P^{1j}$, $x^1 \neq x^2$, such that $x^* = \lambda x^1 + (1 - \lambda)x^2$ for some $0 < \lambda < 1$, and $\alpha x^1 > \beta$, $\alpha x^2 < \beta$. But this contradicts the assumption that $\alpha x \geqslant \beta$ is valid for P^{1j}. □

What is the situation if we use more general cuts, not necessarily valid for all P^{1j}, $j \in \{2, \ldots, n\}$? The result of Theorem 8 can be extended to this case as follows.

Let N be the index set of 0–1 variables, and for any $S \subseteq N$, denote

$$P^S := \text{conv}\{x \in P : x_j \in \{0, 1\}, \ j \in S\}.$$

Suppose we have derived the convex hull of $P \cap \{x : x_j \in \{0, 1\}, \ j \in S_1\}$ for some $S_1 \subset N$, namely

$$P^{S_1} := \{x \in P : \alpha^i x \geqslant \beta^i, \ i \in M\};$$

let $Dx \geqslant d$ be an arbitrary set of inequalities, and denote

$$\tilde{P}^{S_1} := \{x \in P^{S_1} : Dx \geqslant d\}.$$

Theorem 9. *Let $S_2, \ldots, S_q$ be all the distinct minimal subsets of $N \setminus S_1$ such that $Dx \geqslant d$ is valid for $P^{S_1 \cup S_j}$ for $j = 2, \ldots, q$; and let x^* be any extreme point of $\tilde{P}^{S_1}$. Then $0 < x_i^* < 1$ for some $i \in S_1$ implies $0 < x_{k(j)}^* < 1$ for some $k(j) \in S_j$ for each $j \in \{2, \ldots, n\}$.*

Proof. Suppose there exists $j_* \in \{2, \ldots, n\}$ such that $x_k^* \in \{0, 1\}$ for all $k \in S_{j_*}$. Then $x^* \in P^{S_1 \cup S_{j_*}}$. If x^* is an extreme point of $P^{S_1 \cup S_{j_*}}$, then from Theorem 7 $x_i^* \in \{0, 1\}$ for all $i \in S_1$ and we are done. If x^* is not extreme, then there exists an inequality $D_i x \geqslant d_i$ of the system $Dx \geqslant d$ such that $D_i x^* = d_i$, and x^* is the convex combination of two points $x^1, x^2 \in P^{S_1 \cup S_j}$ such that $D_i x^1 > d_i$ and $D_i x^2 < d_i$, a contradiction. □

4. Another derivation of the basic results

The two basic ingredients of our approach, the lifting/projection technique and sequential convexification, can also be derived by the following procedure [5]. Define

$$P := \{x \in \mathbb{R}^n : \tilde{A}x \geqslant \tilde{b}\} \subseteq \mathbb{R}^n$$

and

$$P_D := \text{conv}\{x \in P : x_j \in \{0, 1\}, \ j = 1, \ldots, p\}$$

with the inequalities $x \geqslant 0$ and $x_j \leqslant 1$, $j = 1, \ldots, p$, included in $\tilde{A}x \geqslant \tilde{b}$.

1. Select an index $j \in \{1, \ldots, p\}$. Multiply $\tilde{A}x \geqslant \tilde{b}$ with $1 - x_j$ and x_j to obtain the nonlinear system

$$(1 - x_j)(\tilde{A}x - \tilde{b}) \geqslant 0,$$

$$x_j(\tilde{A}x - \tilde{b}) \geqslant 0. \tag{4}$$

2. Linearize (4) by substituting y_i for $x_i x_j$, $i = 1, \ldots, n$, $i \neq j$, and x_j for x_j^2.
3. Project the resulting polyhedron onto the x-space.

Theorem 10 (Balas et al. [5]). *The outcome of steps* $1, 2, 3$ *is*

$$conv(P \cap \{x: x_j \in \{0, 1\}\}).$$

Corollary 11 (Balas et al. [5]). *Repeating steps* $1, 2, 3$ *for each* $j \in \{1, \ldots, p\}$ *in turn yields* P_D:

The fact that this procedure is isomorphic to the one introduced earlier can be seen by examining the outcome of step 2. In fact, the linearized system resulting from step 2 is precisely $(3')$, the higher dimensional representation of the disjunctive set defined by the constraint $x_j \in \{0, 1\}$ (see [5] for details).

The above 3-step procedure is a streamlined version of the matrix cone procedure of Lovász and Schrijver [20]. The latter involves in step 1 multiplication with $1 - x_j$ and x_j for every $j \in \{1, \ldots, p\}$ rather than just one. While obtaining P_D by this procedure still involves p iterations of steps 1, 2, 3, the added computational cost brings a reward: after each iteration, the coefficient matrix of the linearized system must be positive semidefinite, a condition that can be used in various ways to derive strong bounds or cuts (see [20,7]).

Another similar procedure, due to Sherali and Adams [21], is based on multiplication with every product of the form $(\pi_{j \in J_1} x_j)(\pi_{j \in J_2} (1 - x_j))$, where J_1 and J_2 are disjoint subsets of $\{1, \ldots, p\}$ such that $|J_1 \cup J_2| = t$ for some $1 \leqslant t \leqslant p$. Linearizing the resulting nonlinear system leads to a higher dimensional polyhedron whose strength (tightness) is intermediate between P and P_D, depending on the choice of t: for $t = p$, the polyhedron becomes identical to $\mathscr{P}$ defined by the system (3) for a 0–1 polytope.

5. Generating cuts

The implementation of the disjunctive programming approach into a practical 0–1 programming algorithm had to wait until the early 1990s. It required not only the choice of a specific version of disjunctive cuts, but also a judicious combination of cutting with branching, made possible in turn by the discovery of an efficient procedure for lifting cuts generated in a subspace (for instance, at a node of the search tree) to be valid in the full space (i.e. throughout the search tree).

5.1. Deepest cuts

As mentioned earlier, if P_D is full dimensional, then facets of P_D correspond to extreme rays of the reverse polar cone P_D^*. To generate such extreme rays, for each 0–1 variable x_j that is fractional at the linear programming optimum, we solve a linear program over a normalized version of the cone P_D^* corresponding to the disjunction $x_j = 0 \vee x_j = 1$, with an objective function aimed at cutting off the linear programming optimum $\bar{x}$ by as much as possible. This "cut generating linear program" for the jth variable is of the form

$$
\begin{aligned}
\min \ & \alpha\bar{x} - \beta \\
\text{s.t.} \ & \alpha - uA + u_0 e_j \geqslant 0, \\
& \alpha - vA - v_0 e_j \geqslant 0, \\
& -\beta + ub = 0, \\
& -\beta + vb + v_0 = 0, \\
& u, v \geqslant 0
\end{aligned}
\qquad\qquad (\text{CGLP})_j
$$

and (i) $\beta \in \{1, -1\}$, or (ii) $\sum_j |\alpha_j| \leqslant 1$.

For details, see [5,6].

The normalization constraint (i) or (ii) has the purpose of turning the cone P_D^* into a polyhedron. In case of (i) this is achieved by using $\beta = 1$ or $\beta = -1$, whichever is indicated. In case of (ii), by substituting $\alpha_j^+ - \alpha_j^-$ for α_j, with α_j^+, $\alpha_j^- \geqslant 0$, $\forall j$. A third normalization, proposed later and used in computational experiments subsequent to [6], is

$$
\text{(iii)} \quad \sum_i u_i + u_0 + \sum_i v_i + v_0 = 1.
$$

The merits and demerits of various normalizations will be discussed later.

Solving $(\text{CGLP})_j$ yields a cut $\alpha x \geqslant \beta$, where

$$
\alpha_k = \begin{cases}
\max\{ua_k, va_k\} & k \in N \setminus \{j\} \\
\max\{ua_j - u_0, va_j + v_0\} & k = j,
\end{cases}
$$

with a_k the kth column of A, and

$$
\beta = \min\{ub, vb + v_0\}.
$$

This cut maximizes the amount $\beta - \alpha\bar{x}$ by which $\bar{x}$ is cut off. The experiments of [6] indicated that the most efficient way of generating cuts is to stop short of solving (CGLP) to optimality. This idea was implemented by ignoring those columns of (CGLP) associated with constraints of P not tight at the optimum, except for the lower and upper bounding constraints on the 0–1 variables.

5.2. Cut lifting

In general, a cutting plane derived at a node of the search tree defined by a subset $F_0 \cup F_1$ of the 0–1 variables, where F_0 and F_1 index those variables fixed at 0 and 1, respectively, is only valid at that node and its descendants in the tree (where the variables in $F_0 \cup F_1$ remain fixed at their values). Such a cut can in principle be made valid at other nodes of the search tree, where the variables in $F_0 \cup F_1$ are no longer fixed, by calculating appropriate values for the coefficients of these variables—a procedure called lifting. However, calculating such coefficients is in general a daunting task, which may require the solution of an integer program for every coefficient. One important advantage of the cuts discussed here is that the multipliers u, u_0, v, v_0

obtained along with the cut vector (α, β) by solving (CGLP)$_j$ can be used to calculate by closed form expressions the coefficients α_h of the variables $h \in F_0 \cup F_1$.

While this possibility of calculating efficiently the coefficients of variables absent from a given subproblem (i.e. fixed at certain values) is crucial for making it possible to generate cuts during a branch-and-bound process that are valid throughout the search tree, its significance goes well beyond this aspect. Indeed, most columns of A corresponding to nonbasic components of $\bar{x}$ typically play no role in determining the optimal solution of (CGLP)$_j$ and could therefore be ignored. In other words, the cuts can be generated in a subspace involving only a subset of the variables, and then lifted to the full space. This is the procedure followed in [5,6], where the subspace used is that of the variables indexed by some $R \subset N$ such that R includes all the 0–1 variables that are fractional and all the continuous variables that are positive at the LP optimum. The lifting coefficients for the variables not in the subspace, which are all assumed to be at their lower bound, are then given by

$$\alpha_\ell := \max\{ua_\ell, va_\ell\}, \quad h \in N \setminus R$$

where u and v are the optimal vectors obtained by solving (CGLP)$_j$.

These coefficients always yield a valid lifted inequality. If normalization (i) is used in (CGLP)$_j$, the resulting lifted cut is exactly the same as the one that would have been obtained by applying (CGLP)$_j$ to the problem in the full space. If other normalizations are used, the resulting cut may differ in some coefficients.

5.3. Cut strengthening

The cut $\alpha x \geqslant \beta$ derived from a disjunction of the form $x_j \in \{0, 1\}$ can be strengthened by using the integrality conditions on variables other than x_j, as shown in [8] (see also Section 7 of [2]). Indeed, if x_k is such a variable, the coefficient

$$\alpha_k := \max\{ua_k, va_k\}$$

can be replaced by

$$\alpha_k' := \min\{ua_k + u_0\lceil m_k \rceil, va_k - v_0\lfloor m_k \rfloor\},$$

where

$$m_k := \frac{va_k - ua_k}{u_0 + v_0}.$$

For a proof of this statement, see [2,5] or [6]. The strengthening "works", i.e. produces an actual change in the coefficient, only if

$$|ua_k - va_k| \geqslant u_0 + v_0. \tag{5}$$

Indeed, if (5) does not hold, then either $ua_k > va_k$ and $0 \geqslant m_k > -1$, or $ua_k < va_k$ and $0 \leqslant m_k < 1$; in either case, $\alpha_k' = \alpha_k$. Furthermore, the larger the difference $|ua_k - va_k|$, the more room there is for strengthening the coefficient in question.

This strengthening procedure can also be applied to cuts derived from disjunctions other than $x_j \in \{0, 1\}$, including disjunctions with more than two terms. In the latter case, however, the closed form expression for the value m_k used above has to be replaced by a procedure for calculating those values, whose complexity is linear in the number of terms in the disjunction (see [8] or [2] for details).

5.4. The overall cut generating procedure

Considering what we said about solving the cut generating LP in a subspace and then lifting the resulting cut to the full space and strengthening it, the actual cut generating procedure is not just "lift and project", but rather RLPLS, an acronym for

- RESTRICT the problem to a subspace defined from the LP optimum, and choose a disjunction;
- LIFT the disjunctive set to describe its convex hull in a higher dimensional space;
- PROJECT the polyhedron describing the convex hull onto the original (restricted) space, generating cuts;
- LIFT the cuts into the original full space;
- STRENGTHEN the lifted cuts.

6. Branch-and-cut

No cutting plane approach known at this time can solve large, hard integer programs just by itself. Repeated cut generation tends to produce a flattening of the region of the polyhedron where the cuts are applied, as well as numerical instability which can only partly be mitigated by a tightening of the tolerance requirements. Therefore, the successful use of cutting planes requires their combination with some enumerative scheme. One possibility is to generate cutting planes as long as that seems profitable, thereby creating a tighter linear programming relaxation than the one given originally, and then to solve the resulting problem by branch and bound. Another possibility, known as branch-and-cut, consists of branching and cutting intermittently; i.e., when the cut generating procedure "runs out of steam", move elsewhere in the feasible set by branching. This approach depends crucially on the ability to lift the cuts generated at different nodes of the search tree so as to make them valid everywhere.

The first successful implementation of lift-and-project for mixed 0–1 programming in a branch-and-cut framework was the mixed integer program optimizer (MIPO) code described in [6]. The procedure it implements can be outlined as follows.

Nodes of the search tree (subproblems created by branching) are stored along with their optimal LP bases and associated bounds. Cuts that are generated are stored in a pool. The cut generation itself involves the RLPLS process described earlier. Furthermore, cuts are not generated at every node, but at every kth node, where k is a cutting frequency parameter.

At any given iteration, a subproblem with best (weakest) lower bound is retrieved from storage and its optimal LP solution $\bar{x}$ is recreated. Next, all those cuts in the pool that are tight for $\bar{x}$ or violated by it, are added to the constraints and $\bar{x}$ is updated by reoptimizing the LP.

At this point, a choice is made between generating cuts or skipping that step and going instead to branching. The frequency of cutting is dictated by the parameter k that is problem dependent, and calculated after generating cuts at the root node. Its value is a function of several variables believed to characterize the usefulness of cutting planes for the given problem, primarily the average depth of the cuts obtained. In our experiments, the most frequently used value of k was 8.

If cuts are to be generated, this happens according to the RLPLS scheme described above. First, a subspace is chosen by retaining all the 0–1 variables fractional and all the continuous variables positive at the LP optimum, and removing the remaining variables along with their lower and upper bounds. A lift and project cut is then generated from each disjunction $x_j \in \{0, 1\}$ for j such that $0 < \bar{x}_j < 1$ (this is called a round of cuts). Each cut is lifted and strengthened; and if it differs from earlier cuts sufficiently (the difference between cuts is measured by the angle between their normals), it is added to the pool; otherwise it is thrown away. After generating a round of cuts, the current LP is reoptimized again.

If cuts are not to be generated (or have already been generated), a fractional variable is chosen for branching on a disjunction of the form $x_j \in \{0, 1\}$; i.e., two new subproblems are created, their lower bounds are calculated, and they are stored. The branching variable is chosen as the one with largest (in absolute value) cost coefficient among those whose LP optimal value is closest to 0.5.

7. Variations on the cut generating LP

The solution of the cut generating LP depends on two factors: the objective function and the normalization used. The choice of the former is dictated by the fact that the immediate goal is to cut off the LP optimum by as much as possible. The normalization is a different story.

7.1. Alternative normalizations

It was shown in [1] that if normalization (i) is used, then (CGLP) has a finite minimum if and only if $\bar{x}\lambda \in P_D$ for some $\lambda \in \mathbb{R}_+$. This condition is satisfied for certain classes of problems, for instance set covering (for $\beta = 1$) and set packing (for $\beta = -1$), but not for others, and the absence of a finite minimum leads to complications.

In case of normalizations (ii) or (iii), a different difficulty arises. If P_D is full-dimensional, then the inequality $\alpha x \geqslant \beta$ defines a facet of P_D if and only if (α, β) is an extreme ray of the reverse polar cone P_D^*. If P_D^* is truncated or intersected with a single hyperplane in (α, β)-space, then the extreme points of the resulting polyhedron correspond to extreme rays of P_D^*. But if P_D^* is truncated or intersected by multiple hyperplanes, that will typically result in a polyhedron whose extreme points do not always correspond to extreme rays of P_D^*. This is exactly what happens in the case of normalizations (ii) and (iii). In the case of (ii), the constraint $\sum(|\alpha_j|: j \in N) \leqslant 1$, which requires α to belong to an n-dimensional octahedron, is equivalent to imposing on α the 2^n inequalities $\delta^i \alpha \leqslant 1$, $\delta^i \in \{1, -1\}^n$, $i = 1, \ldots, 2^n$, which define the facets of the octahedron. In the case of (iii), the constraint $\sum_i u_i + u_0 + \sum_i v_i + v_0 = 1$ guarantees that (CGLP)$_j$ will have a finite minimum for every nonnegative objective function. Since the multipliers $u_i, v_i, i = 0, \ldots, m$, are all required to be nonnegative, the normalization (iii) bounds each multipler; and since α is bounded from below by a linear combination of those multipliers, it follows that the objective function of (CGLP)$_j$ is bounded from below for any nonnegative $\bar{x}$.

If (iii) is replaced by

$$\text{(iii')} \quad \sum_i u_i + u_0 + \sum_i v_i + v_0 + \sum_k s_k + \sum_k t_k = 1,$$

where $s_k \geqslant 0$, $t_k \geqslant 0$, $k = 1, \ldots, n$, are surplus variables used to bring (CGLP)$_j$ to equality form, then (iii') bounds P_D^* in every direction. This follows because the surplus variables are now also required to be nonnegative, and thus α is also bounded from above by a linear combination of the multipliers.

Although the higher dimensional cone is truncated by a single hyperplane through either (iii) or (iii'), the outcome in the (α, β)-subspace may correspond to a truncation of P_D^* by multiple hyperplanes, and thus an extreme point of the resulting polyhedron may not correspond to an extreme ray of P_D^*. To avoid this difficulty, we propose another normalization, whose generic form is (iv) $\alpha y = 1$. Let (CGLP)^y denote the problem with this normalization. The advantage of normalization (iv) is that it intersects P_D^* with a single hyperplane in the (α, β)-space and thus has the effect that every extreme point of the resulting polyhedron corresponds to an extreme ray of P_D^*. This does not imply that every extreme point of the higher-dimensional (CGLP)^y corresponds to an extreme ray of P_D^*; but it does imply that if the objective $\min(\alpha\bar{x} - \beta)$ is bounded then there *exists* an optimal extreme point of (CGLP)^y which corresponds to an extreme ray of P_D^*.

Theorem 12. *Let (CGLP)^y be feasible. Then it has a finite minimum if and only if $\bar{x} + y\lambda \in P_D$ for some $\lambda \in \mathbb{R}$.*

Proof. If (CGLP)^y is unbounded in the direction of minimization, then there exists $(\tilde{\alpha}, \tilde{\beta}) \in P_Q^*$ with $\bar{x}^T \tilde{\alpha} < \tilde{\beta}$ and $y^T \tilde{\alpha} = 0$. But then $(\bar{x} + y\lambda)^T \tilde{\alpha} < \tilde{\beta}$ for all $\lambda \in \mathbb{R}$, hence $\bar{x} + y\lambda \notin P_D$.

Conversely, if $\bar{x} + y\lambda \notin P_D$ for all $\lambda \in \mathbb{R}$, there exists $(\hat{\alpha}, \hat{\beta}) \in \mathbb{R}^{n+1}$ such that $\hat{\alpha}x \geqslant \hat{\beta}$ for all $x \in P_D$ and $\hat{\alpha}(\bar{x} + y\lambda) < \hat{\beta}$ for all $\lambda \in \mathbb{R}$, i.e. $\hat{\alpha}y = 0$ and $\hat{\alpha}\bar{x} < \hat{\beta}$. But then $(\hat{\alpha}, \hat{\beta})$ is a direction of unboundedness for (CGLP)^y. □

Theorem 13. *If (CGLP)^y has an optimal solution $(\tilde{\alpha}, \tilde{\beta})$, then*

$$\bar{x}^T \tilde{\alpha} - \tilde{\beta} = \lambda^* := \min\{\lambda : \bar{x} + y\lambda \in P_D\}$$

and

$$(\bar{x} + y\lambda^*)^T \tilde{\alpha} = \tilde{\beta}.$$

Proof. Let $(\tilde{\alpha}, \tilde{\beta})$ be an optimal solution to (CGLP)^y, and define $\lambda^* := \min\{\lambda : \bar{x} + y\lambda \in P_D\}$. Since $\tilde{\alpha}y \neq 0$, there exists $\lambda^0 \in \mathbb{R}$ such that $(\bar{x} + y\lambda^0)^T \tilde{\alpha} = \tilde{\beta}$. Further,

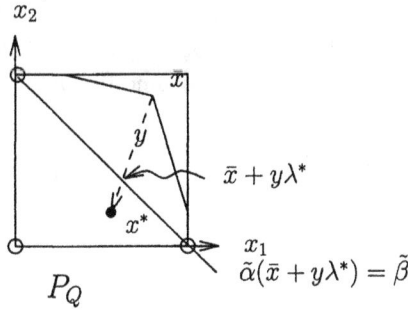

Fig. 3. $y = x^* - \bar{x}$.

$\bar{x}^{\mathrm{T}}\tilde{\alpha} - \tilde{\beta} = y^{\mathrm{T}}\tilde{\alpha}\lambda^0 = \lambda^0$. We claim that $\lambda^0 = \lambda^*$. For suppose $\lambda^0 > \lambda^*$. Then

$$(\bar{x} + y\lambda^*)^{\mathrm{T}}\tilde{\alpha} - \tilde{\beta} = \bar{x}^{\mathrm{T}}\tilde{\alpha} - \tilde{\beta} + y^{\mathrm{T}}\tilde{\alpha}\lambda^*$$

$$= -\lambda^0 + \lambda^* \quad (\text{since } y^{\mathrm{T}}\tilde{\alpha} = 1)$$

$$< 0,$$

i.e. the point $\bar{x} + y\lambda^*$ violates the inequality $\tilde{\alpha}x \geqslant \tilde{\beta}$, contradicting $\bar{x} + y\lambda^* \in P_D$.

Now suppose $\lambda^0 < \lambda^*$. Then there exists a hyperplane $\bar{\alpha}x = \bar{\beta}$ such that $\bar{\alpha}x \geqslant \bar{\beta}$ for all $x \in P_Q$, $\bar{\alpha}(\bar{x} + y\lambda^*) = \bar{\beta}$, and $\bar{\alpha}(\bar{x} + y\lambda^0) < \bar{\beta}$. Further, since $\bar{\alpha}y\lambda^0 < \bar{\alpha}y\lambda^*$ and $\lambda^0 < \lambda^*$, it follows that $\bar{\alpha}y > 0$ and so w.l.o.g. we may assume that $(\bar{\alpha}, \bar{\beta})$ is scaled so as to make $\bar{\alpha}y = 1$. But then

$$\bar{\alpha}\bar{x} - \bar{\beta} < \bar{\alpha}y\lambda^0 = \bar{\alpha}\bar{x} - \bar{\beta},$$

contradicting the optimality of $(\tilde{\alpha}, \tilde{\beta})$ for $(\text{CGLP})^y$. $\square$

Corollary 14. *Let $y := x^* - \bar{x}$ for some $x^* \in P_Q$. Then $(CGLP)_y$ has an optimal solution $(\tilde{\alpha}, \tilde{\beta})$ such that (i) $\tilde{\alpha}\bar{x} < \tilde{\beta}$; and (ii) $\tilde{\alpha}x = \tilde{\beta}$ is the supporting hyperplane of P_Q that intersects the line segment $(\bar{x}, x^*]$ at the point closest to x^*.*

Theorem 13 and Corollary 14 are illustrated in Fig. 3.

7.2. Complementarity of solution components

Consider the linear program $(\text{CGLP})_j$ used to generate a cut from the disjunction $x_j = 0$ or $x_j = 1$. $(\text{CGLP})_j$ has a set of trivial solutions corresponding to the constraint set $Ax \geqslant b$. Assuming that normalization (iii) is used, set $\bar{u}_i = \bar{v}_i = \frac{1}{2}$ for some $i \in M$, $\bar{u}_h = \bar{v}_h = 0$ for all $h \in M \setminus \{i\}$ and $\bar{u}_0 = \bar{v}_0 = 0$. Then $(\tilde{\alpha}, \tilde{\beta}, \bar{u}, \bar{u}_0, \bar{v}, \bar{v}_0)$ is a solution to $(\text{CGLP})_j$ with $\alpha = a^i$ and $\beta = b_i$, i.e., the coefficient vector of the ith constraint of $Ax \geqslant b$. We call a basic solution *nontrivial* if it is not of this type.

Theorem 15. *Any nontrivial basic solution $w := (\alpha, \beta, u, u_0, v, v_0)$ to $(CGLP)_j$ satisfies $u \cdot v = 0$.*

Proof. We assume that (CGLP)$_j$ uses normalization (iii). An analogous reasoning proves the other cases. Let A have rows a_h, $h \in M$, let $\bar{w}$ be a basic solution, and suppose $\bar{u}_i \bar{v}_i > 0$ for some $i \in M$. W.l.o.g., assume $0 < \bar{u}_i \leqslant \bar{v}_i$. Define

$$\hat{u}_h := \begin{cases} 0 & h = i \\ \rho\bar{u}_h & h \in M \setminus \{i\} \end{cases}, \qquad \hat{v}_h := \begin{cases} \rho(\bar{v}_h - \bar{u}_h) & h = i \\ \rho\bar{v}_h & h \in M \setminus \{i\} \end{cases}$$

$$\hat{u}_0 = \rho\bar{u}_0, \quad \hat{v}_0 := \rho\bar{v}_0, \quad \hat{\alpha} := \rho(\bar{\alpha} - \bar{u}_i a_i), \quad \hat{\beta} := \rho(\bar{\beta} - \bar{u}_i b_i),$$

with $\rho := 1/(1 + 2\bar{u}_i)$, and

$$\tilde{u}_h := \begin{cases} 2\sigma\bar{u}_h & h = i \\ \sigma\bar{u}_h & h \in M \setminus \{i\} \end{cases}, \qquad \tilde{v}_h := \begin{cases} \sigma(\bar{v}_h + \bar{u}_h) & h = i \\ \sigma\bar{v}_h & h \in M \setminus \{i\} \end{cases}$$

$$\tilde{u}_0 := \sigma\bar{u}_0, \quad \tilde{v}_0 := \sigma\bar{v}_0, \quad \tilde{\alpha} := \sigma(\bar{\alpha} + \bar{u}_i a_i), \quad \tilde{\beta} := \sigma(\bar{\beta} + \bar{u}_i b_i)$$

with $\sigma := 1/(1 + 2\bar{u}_i)$.

Then $\hat{w}$ and $\tilde{w}$ are both feasible. But since w is nontrivial, $\hat{w} \neq 0 \neq \tilde{w}$ and so it is easily verified that $(1/2\rho)\hat{w} + (1/2\sigma)\tilde{w} = \bar{w}$, which contradicts the assumption that $\bar{w}$ is basic. □

The complementarity property shown in Theorem 15 means that while the two variables u_0, v_0 associated with the inequalities $x_j \leqslant 0$ and $x_j \geqslant 1$ may both be (and typically are) positive at the optimum, the pair (u_i, v_i) associated with the ith inequality of $Ax \geqslant b$ is complementary for every i: at most one of the two variables can be positive. This is a consequence of the intuitively plausible fact, that a given inequality of $Ax \geqslant b$ can be profitably added with a positive multiplier either to one term of the disjunction, or to the other, but not to both. In fact, in addition to the complementarity of the pairs (u_i, v_i), typically both members of many pairs are 0. In other words, some of the inequalities of $Ax \geqslant b$ do not contribute to the improvement of the cut, whichever term of the disjunction they are added to. This has led us to a search for criteria by which to decide for each inequality of $Ax \geqslant b$, whether it should be added to the first term of the disjunction with multiplier u_i, or to the second term with the multiplier v_i, or not included at all in the cut generating LP.

7.3. Reduced-size (CGLP)

After some experimentation with several different criteria, we concluded that the best indicator of the usefulness of the presence of an inequality of $Ax \geqslant b$, $x \geqslant 0$, in one term or the other of the disjunction is to be found in the optimal simplex tableau of the linear program $\max\{cx: x \in P\}$. Namely, suppose we want to build the cut generating LP for the disjunction $x_k \in \{0, 1\}$, where $\bar{x}_k$ is a fractional component of the LP optimum $\bar{x}$. Let the row of the optimal simplex tableau associated with x_k be

$$x_k = \bar{a}_{k0} - \sum_{j \in J} \bar{a}_{kj} x_j, \tag{6}$$

where J is the index set of nonbasic variables and $0 < \bar{a}_{k0} = \bar{x}_k < 1$, and the nonbasic variables are all at their lower bound. Next we restrict the system $Ax \geq b$, $x \geq 0$ to the subspace obtained by removing all nonbasic structural variables and all constraints that are not binding at the LP optimum (hence all basic surplus variables). (Here structural variables are the components of x, whereas surplus variables stand for the components of $s = Ax - b$.) We are then left with only the basic structural variables and the nonbasic surplus variables, and can write the resulting system as

$$Bx_B - s_M = b_M,$$

$$x_B, s_M \geq 0.$$

Note that B has $|M|$ columns and $|M|$ rows and is nonsingular. Multiplying with B^{-1} yields

$$x_B = B^{-1}b_M + B_k^{-1}s_M,$$

a system whose row corresponding to x_k is

$$x_k = B_k^{-1}b_M + B_k^{-1}s_M,$$

where B_k^{-1} is row k of B^{-1}. This is just another way of writing the equation that remains after we remove from (6) the nonbasic structural variables, and replace the notation x_j with s_j for the surplus variables:

$$x_k = \bar{a}_{k0} + \sum_{i \in M}(-\bar{a}_{ki})s_i,$$

where $\bar{a}_{k0} = \bar{x}_k = B_k^{-1}b_M$, and for $i \in M$, $\bar{a}_{ki} = -B_{ki}^{-1}$, with B_{ki}^{-1} the ith component of B_k^{-1}.

The simplest disjunctive cut derived from the condition $x_k \leq 0 \vee x_k \geq 1$, namely the intersection cut from the pair of halfspaces $0 \leq x_k \leq 1$, is known to be (see [2]) $\pi s_M \geq \pi_0$, where

$$\pi_0 = \bar{x}_k(1 - \bar{x}_k)$$

and for $i \in M$,

$$\pi_i := \max\{\pi_i^1, \pi_i^2\}$$

with

$$\pi_i^1 := (\bar{x}_k - 1)B_{ki}^{-1}, \quad \pi_i^2 := \bar{x}_k B_{ki}^{-1}.$$

We wish to construct a basic solution of $(CGLP)_k$, whose (α, β)-component yields the cut $\alpha x \geq \beta$ obtained by expressing $\pi s_M \geq \pi_0$ in terms of x. For this purpose, we write $\alpha = (\alpha_B, \alpha_R)$ where α_B stands for the components associated with the columns of B, and α_R for the components that have been removed. We now define

$$\alpha_B := \pi B, \quad \beta := \pi_0 + \pi b_M,$$

$$u := \pi - \pi^1, \quad v := \pi - \pi^2,$$

$$u_0 := 1 - \bar{x}_k, \quad v_0 := \bar{x}_k. \tag{7}$$

Theorem 16. *The vector* $w:=(\alpha_B, \beta, u, u_0, v, v_0)$ *defined by* (7) *is a basic feasible solution of* $(\text{CGLP})_k$ *with the normalization* $u_0 + v_0 = 1$.

Proof. Since α_R has been removed, the expression $\alpha - uA + u_0 e_k$ reduces to $\alpha_B - uB + u_0 e_k$, with e_k the unit vector in the subspace of α_B. We then have

$$\alpha_B - uB + u_0 e_k = \pi B - (\pi - \pi^1)B + (1 - \bar{x}_k)e_k$$

$$= \pi^1 B + (1 - \bar{x}_k)e_k$$

$$= (\bar{x}_k - 1)B_k^{-1}B + (1 - \bar{x}_k)e_k = 0,$$

since $B_k^{-1}B = e_k$. Next,

$$\alpha_B - vB - v_0 e_k = \pi B - (\pi - \pi^2)B - \bar{x}_k e_k$$

$$= \pi^2 B - \bar{x}_k e_k$$

$$= \bar{x}_k B_k^{-1}B - \bar{x}_k e_k = 0.$$

Further,

$$-\beta + ub_M = -\pi_0 - \pi b_M + (\pi - \pi^1)b_M$$

$$= -\bar{x}_k(1 - \bar{x}_k) - \pi^1 b_M$$

$$= -\bar{x}_k(1 - \bar{x}_k) + (1 - \bar{x}_k)B_k^{-1}b_M = 0,$$

since $B_k^{-1}b_M = \bar{x}_k$. Also,

$$-\beta + vb_M + v_0 = -\pi_0 - \pi b_M + (\pi - \pi^2)b_M + \bar{x}_k$$

$$= -\bar{x}_k(1 - \bar{x}_k) - \pi^2 b_M + \bar{x}_k$$

$$= -\bar{x}_k + (\bar{x}_k)^2 - \bar{x}_k B_k^{-1}b_M + \bar{x}_k = 0.$$

Finally,

$$u_0 + v_0 = (1 - \bar{x}_k) + \bar{x}_k = 1.$$

Furthermore, (7) implies that $u, v \geqslant 0$.

This proves that w is feasible. To see that it is basic, note that there are $2|M| + 2$ constraints satisfied at equality, and the same number of nonnegative variables, whose coefficient vectors are linearly independent. □

Using the basic solution (7), we construct the associated simplex tableau of $(\text{CGLP})_k$, and among the nonbasic variables u_i, v_i, we keep only those with negative reduced cost, while removing the others. Our interpretation that these are the only variables likely to improve the cut (in terms of the chosen objective) is more than born out by our computational experience: as shown in the computational section of this paper, the cuts obtained from this smaller (CGLP) tend to be just as strong as those obtained from the full fledged problem.

Our approach for constructing a starting solution for $(CGLP)_k$ highlights the connection between this lift-and-project cut and the mixed integer Gomory cut, which in this case (since the nonbasic $0–1$ variables have been removed) is identical to the intersection cut from the pair of half-spaces $0 \leqslant x_k \leqslant 1$. Thus the lift-and-project cut can be viewed as a generalization of the mixed integer Gomory cut $\pi s_M \geqslant \pi_0$, where the generalization consists in optimally combining each of the inequalities $\pi^1 s_M \geqslant \pi_0$ and $\pi^2 s_M \geqslant \pi_0$ with some of the constraints of P before taking the component-wise maximum of π^1 and π^2.

7.4. Multiple cuts from a disjunction

There are several ways of deriving more than one cut from a given disjunction. The approach proposed in [4] was to generate several facets of P_Q containing its optimal extreme point x^{opt}. This approach asks for the calculation of x^{opt} (recall, we are talking about a disjunction with two terms, not too expensive to solve), to be used to generate n facets of P_Q containing x^{opt}. The way to accomplish this is to replace the objective function of $(CGLP)_j$ by $\min(x^{opt})^T \alpha - \beta$, which results in a linear program whose optimal solutions $(\alpha, \beta, u, u^0, v, v^0)$ yield all the valid inequalities $\alpha x \geqslant \beta$ (including those that define facets of P_Q) satisfied at equality by x^{opt} (Theorem 1 of [4]). Thus, having obtained one such optimal solution, one may generate all the others by pivoting in columns with zero reduced cost. In theory this is a way of generating all the facets of P_Q that contain x^{opt}. In practice, the massive degeneracy that is typically present in the optimal tableau of this problem makes the procedure of finding alternative optima with the relevant (α, β)-components computationally rather expensive.

An alternative way of generating multiple cuts from the same disjunction is to explore near-optimal solutions to $(CGLP)_j$ by forcing to 0 some component of (u, v) positive in the optimal tableau. This has been explored by Ceria and Pataki [14]; we have also tried it, with results slightly better than those obtained with the previous approach.

Finally, a third way which we found considerably more useful than either of these two, is the following. Having found an optimal solution to $(CGLP)_j$, we go back to the optimal simplex tableau of $\min\{cx : x \in P\}$, and generate all adjacent solutions to $\bar{x}$ obtainable by a single pivot: let these solutions be $x^1, \ldots, x^k$. We then use each one of them in turn to replace $\bar{x}$ in the objective function of $(CGLP)_j$. This yields reasonably good results, to be discussed in the computational section.

8. Computational experience

8.1. Results in branch-and-cut mode

The procedure described in Sections 5 and 6 was implemented in the code MIPO, described in detail in [6]. This implementation of MIPO does not have its own linear programming routine; instead, it calls a simplex code whenever it has to solve or reoptimize an LP. In the experiments of [6] the LP solver used was that of CPLEX

Table 1

OSL		CPLEX		MINTO		MIPO	
First	Second	First	Second	First	Second	First	Second
Ranking by number of search tree nodes							
15	2	0	4	4	10	11	10
Ranking by CPU time							
2	3	6	6	10	3	12	14

2.1. The test bed consisted of 29 test problems from MIPLIB and other sources, ranging in size from about 30 to 9000 0–1 variables, and about 20 to 2000 constraints. The large majority of these problems have a real world origin; they were contributed mostly by people who tried, not always successfully, to solve them. Of the MIPLIB problems, most of those not included into the testbed were omitted as too easily solved by straight branch-and-bound; two problems were excluded because their LP relaxation exceeded our dimensioning. MIPO was compared with MINTO, OSL and CPLEXMIP 2.1 (the most advanced version available at the time of the experiments). The outcome (see [6] for detailed results) is best summarized by showing the number of times a code ranked first, and second, both in terms of search tree nodes and in terms of computing time. This is done in Table 1, whose two parts correspond to Tables 9 and 10 of [6].

In a sense, MIPO turned out to be the most robust among the four codes: it was the only one that managed to solve all 29 test problems, and it ranked first or second in computing time on 26 out of the 29 instances.

Other experiments with lift-and-project in an enumerative framework are reported on in [24], where S. Thienel compares the performance of ABACUS, an object oriented branch-and-cut code, in two different modes of operation, one using lift-and-project cuts and the other using Gomory cuts; with the outcome that the version with lift-and-project cuts is considerably faster on all hard problems, where hard means requiring at least 10 min.

Little experimentation has taken place so far with cuts derived from stronger disjunctions than the 0–1 condition on a single variable. In [7] the MIPO procedure was run on maximum clique problems, where the higher dimensional formulation used to generate cuts was the one obtained by multiplying the constraint set with inequalities of the form $1 - \sum(x_j : j \in S) \geqslant 0$, $x_j \geqslant 0$, $j \in S$, where S is a stable set. This is the same as the higher dimensional formulation derived from the disjunction

$$(x_j = 0, j \in S) \vee (x_{j_1} = 1, x_j = 0, j \in S \setminus \{j_1\}) \vee \ldots \vee$$

$$(x_{j_s} = 1, x_j = 0, j \in S \setminus \{j_s\}),$$

where $s = |S|$. As this disjunction is more powerful than the standard one, the cuts obtained were stronger; but they were also more expensive to generate, and without some specialized code to solve the highly structured cut generating LP's of this formulation, the trade-off between the strength of the cuts and the cost of generating them favored the weaker cuts from the standard disjunction.

8.2. Results in cut-and-branch mode

Bixby et al. [11] report on their computational experience with a parallel branch-and-bound code, run on 51 test problems after generating several types of cuts at the root node. One of the cut types used was disjunctive or lift-and-project cuts, generated essentially as in [6] with normalization (ii), but without restriction to a subspace and without strengthening. Since deriving these cuts in the full space is expensive, the routine generating them was activated only for 4 of the hardest problems. Their addition to the problem constraints reduced the integrality gap by 58.7%, 94.4%, 99.9% and 94.4%, respectively.

In [14], Ceria and Pataki report computational results with a disjunctive cut generator used in tandem with the CPLEX branch and bound code. Namely, the cut generator was used to produce 2 and 5 rounds of cuts from the 0–1 disjunctions for the 50 most promising variables fractional at the LP optimum, after which the resulting problem with the tightened LP relaxation was solved by the CPLEX 5.0 MIP code. This "cut-and-branch" procedure was tested on 18 of the hardest MIPLIB problems and the results were compared to those obtained by using CPLEX 5.0 without the cut generator. The outcome of the comparison can be summarized as follows (see Table 2 of [14] for details).

- The total running time of the cut-and-branch procedure was less than the time without cuts for 14 of the 18 problems; while the opposite happened for the remaining 4 problems.
- Two of the problems solved by cut-and-branch in 8 and 3 min, respectively could not be solved by CPLEX alone in 20 h.
- For six problems the gain in time was more than 4-fold.

Very good results were obtained on two difficult problems outside the above set. One of them, *set1ch*, could not be solved by CPLEX alone, which after exhausting its memory limitations stopped with a solution about 15% away from the optimum. On the other hand, running the cutting plane generator for 10 rounds on this problem produced a lower bound within 1.4% of the integer optimum, and running CPLEX on the resulting tightened formulation solved the problem to optimality in 28 s.

The second difficult problem, *seymour*, was formulated by Paul Seymour in an attempt to find a minimal irreducible configuration in the proof of the four color theorem. It was not solved to optimality until very recently. The value of the LP relaxation is 403.84, and an integer solution of 423.00 was known. The best previous known lower bound of 412.76 was obtained by running a parallel computer with 16 processors for about 60 h, using about 1400 Mbytes of memory. The cut-and-branch procedure applied to this case generated 10 rounds of 50 cuts in about 10.5 h, and produced a lower bound of 413.16, using less than 50 Mbytes of memory. Running CPLEX for another 10 h on this tightened formulation then raised the bound to 414.20. More recently, using the same lift-and-project cuts, but a more potent computing environment, the problem was solved to optimality, cf. Pataki [20a]

Table 2
Computational results with the full size and the reduced (CGLP)

Problem	Optimum of the preprocessed LP	Lower bound after adding cuts		Average number of columns in CGLP		Average number of pivots in CGLP	
		Full CGLP	Reduced CGLP	Full	Reduced	Full	Reduced
P0548	3126	5713	5709	820	305.6	25.62	0.18
P2756	2703	2880	2880	1984	313.4	5.39	0.08
Pp08a	2748	2103	2103	875	490.8	19.26	5.33
Vpm2	10.27	11.05	11.02	873	374.1	37.18	4.50
10-teams	897.0	904.0	904.0	1528	584.2	709.9	258.7
Danoint	62.64	62.65	62.66	1989	637.4	1230.0	619.3
Misc07	1415	1415	1415	763	203.9	154.0	46.2
Pk1	0	0	0	198	48.0	21.67	17.0
Seymour	267.8	271.1	270.8	13022	2338.0	1936.7	158.4
Vpm1	16.43	17.01	17.01	893	388.0	54.64	5.93
Mod010	6532	6533	6543	1315	542.6	258.4	121.6
L152lav	4656	4659	4659	874	366.7	186.4	92.8
Set1ch	30427	35174	35174	2768	1515.8	45.0	6.43

8.3. Results with a reduced-size CGLP

We have extensively tested the reduced-size (CGLP) constructed from the optimal simplex tableau for $\min\{cx: x \in P\}$ by using the complementarity of (u, v), as described in Section 7.

In the experiment summarized in Table 2, we chose 13 of the harder MIPLIB problems, and for each instance we solved, for each 0–1 variable fractional at the LP optimum, a (CGLP) (with normalization (iii)) not restricted to a subspace, in two versions: the full size (CGLP) and the reduced size CGLP. The results are shown in Table 2. Every problem was preprocessed by CPLEX before generating cuts.

It is clear from the table, that the reduced (CGLP), while generating cuts whose strength—as measured by the lower bounds they provide—is fully equal to that of the cuts generated by the full (CGLP), requires a computational effort that is several times smaller.

8.4. Results with multiple cuts from a disjunction

In a computational experiment meant to test the efficiency of generating multiple cuts from the same disjunction, we ran two versions of MIPO on a set of MIPLIB problems. Both versions used the same formula for calculating at the root node the cutting frequency. The first version used the standard approach of generating one cut from each disjunction, whereas the second version generated $q + 1$ cuts from each disjunction, solving each (CGLP)$_j$ for the objective functions $\min\{\alpha x^* - \beta\}$, with $x^* = \bar{x}, x^1, \ldots, x^q$, where $\bar{x}$ is the optimal solution to $\min\{cx: x \in P\}$, and $x^1, \ldots, x^q$ are extreme points of P adjacent to $\bar{x}$, obtainable by one pivot in the simplex tableau

Table 3

	Time (CPU Sec)		Search tree nodes	
Problem	Version 1	Version 2	Version 1	Version 2
C-fat-200-1	413	113	49	31
Pp08CUTS	358	299	2891	1743
San200-0.9-3	2358	1751	495	609
Stein45	1335	1789	20761	24711
Vpm1	295	190	8811	4153
Vpm2	432	399	8901	2877
10 teams	1490	[a]	259	[a]

[a]Time or memory limit exceeded.

associated with $\bar{x}$ (q was limited to at most 0.5 times the number of basic variables). The outcome is shown in Table 3 for the problems that required at least 250 s to solve. As can be seen, the extra work spent on generating more cuts pays off more often than not: the computing time is smaller in version 2 in five out of the seven instances.

References

[1] E. Balas, Disjunctive programming: properties of the convex hull of feasible points, Invited paper, with a foreword by G. Cornuéjols and W. Pulleyblank, Discrete Appl. Math. 89 (1998) 3–44 (originally MSRR# 348, Carnegie Mellon University, July 1974).

[2] E. Balas, Disjunctive programming, Ann. Discrete Math. 5 (1979) 3–51.

[3] E. Balas, Disjunctive programming and a hierarchy of relaxations for discrete optimization problems, SIAM J. Algebraic Discrete Methods 6 (1985) 466–485.

[4] E. Balas, A modified lift-and-project procedure, Math. Programming 79 (1997) 19–31.

[5] E. Balas, S. Ceria, G. Cornuéjols, A lift-and-project cutting plane algorithm for mixed 0–1 programs, Math. Programming 58 (1993) 295–324.

[6] E. Balas, S. Ceria, G. Cornuéjols, Mixed 0–1 programming by lift-and-project in a branch-and-cut framework, Manag. Sci. 42 (1996) 1229–1246.

[7] E. Balas, S. Ceria, G. Cornuéjols, G. Pataki, Polyhedral methods for the maximum clique problem, in: D. Johnson, M. Trick (Eds.), Clique, Coloring and Satisfiability: The Second DIMACS Challenge, The American Mathematical Society, Providence, RI, 1996, pp. 11–27.

[8] E. Balas, R.G. Jeroslow, Strengthening cuts for mixed integer programs, European J. Oper. Res. 4 (1980) 224–234.

[9] E. Balas, J. Tama, J. Tind, Sequential convexification in reverse convex and disjunctive programming, Math. Programming 44 (1989) 337–350.

[10] N. Beaumont, An algorithm for disjunctive programming, European J. Oper. Res. 48 (1990) 362–371.

[11] R. Bixby, W. Cook, A. Cox, E. Lee, Computational experience with parallel mixed integer programming in a distributed environment, Ann. Oper. Res. 90 (1999) 19–45.

[12] C. Blair, Two rules for deducing valid inequalities for 0–1 programs, SIAM J. Appl. Math. 31 (1976) 614–617.

[13] C. Blair, Facial disjunctive programs and sequences of cutting planes, Discrete Appl. Math. 2 (1980) 173–180.

[14] S. Ceria, G. Pataki, Solving integer and disjunctive programs by lift-and-project, in: R.E. Bixby, E.A. Boyd, R.Z. Rios-Mercado (Eds.), IPCO VI, Lecture Notes in Computer Science, Vol. 1412, Springer, Berlin, 1998, pp. 271–283.

[15] S. Ceria, J. Soares, Disjunctive cuts for mixed 0–1 programming: duality and lifting, GSB, Columbia University, 1997.

[16] S. Ceria, J. Soares, Convex programming for disjunctive optimization, GSB, Columbia University, 1997.

[17] R.G. Jeroslow, Cutting plane theory: disjunctive methods, Ann. Discrete Math. 1 (1977) 293–330.

[18] R.G. Jeroslow, Representability in mixed integer programming I: characterization results, Discrete Appl. Math. 17 (1987) 223–243.

[19] R.G. Jeroslow, Logic based decision support: mixed integer model formulation, Ann. Discrete Math. 40 North-Holland, Amsterdam, 1989.

[20] L. Lovász, A. Schrijver, Cones of matrices and set functions and 0–1 optimization, SIAM J. Optim. 1 (1991) 166–190.

[20a] G. Pataki, Private communications, March 2001.

[21] H. Sherali, W. Adams, A hierarchy of relaxations between the continuous and convex hull representations for zero–one programming problems, SIAM J. Discrete Math. 3 (1990) 411–430.

[22] H. Sherali, C. Shetty, Optimization with disjunctive constraints, Lecture Notes in Economics and Mathematical Systems, Vol. 181, Springer, Berlin, 1980.

[23] R.A. Stubbs, S. Mehrotra, A branch-and-cut method for 0–1 mixed convex programming, Department of Industrial Engineering, Northwestern University, 1996.

[24] S. Thienel, ABACUS: A branch-and-cut system, Doctoral Dissertation, Faculty of Mathematics and The Natural Sciences, University of Cologne, 1995.

[25] M. Turkay, I.E. Grossmann, Disjunctive programming techniques for the optimization of process systems with discontinuous investment costs-multiple size regions, Ind. Engng. Chem. Res. 35 (1996) 2611–2623.

[26] H.P. Williams, An alternative explanation of disjunctive formulations, European J. Oper. Res. 72 (1994) 200–203.

DISCRETE
APPLIED
MATHEMATICS

Discrete Applied Mathematics 123 (2002) 155–225

ELSEVIER

Pseudo-Boolean optimization [☆]

Endre Boros*, Peter L. Hammer

RUTCOR, Rutgers University, 640 Bartholomew Road, Piscataway, NJ 08854-8003, USA

Abstract

This survey examines the state of the art of a variety of problems related to pseudo-Boolean optimization, i.e. to the optimization of set functions represented by closed algebraic expressions. The main parts of the survey examine general pseudo-Boolean optimization, the specially important case of quadratic pseudo-Boolean optimization (to which every pseudo-Boolean optimization can be reduced), several other important special classes, and approximation algorithms. © 2002 Elsevier Science B.V. All rights reserved.

1. Introduction

Set functions, i.e. mappings from the family of subsets of a finite ground set to the set of reals, have been present in the mathematical literature for more than a century, with a substantial development of this area starting in the early 1950s. The importance of set functions in game theory and optimization brought them to the full focus of attention of applied mathematicians, especially of those working in operations research. This increased interest in set functions is motivated by their presence in the mathematical models of a wide spectrum of problems occurring in a variety of applications.

Set functions are frequently considered as being defined by an oracle, or more specifically, by an algorithm capable of delivering their values for any subset of the given finite ground set. Any graph parameter (e.g., chromatic number, stability number, etc.) associated to the subgraphs induced by a subset of the vertices of a given graph is an example for such a set function. However, in numerous examples a set function can be defined by a closed algebraic formula, an advantageous

---
[☆] This research was supported in part by the Office of Naval Research (Grant N00014-92-J-1375) and by the National Science Foundation (Grant DMS 98-06389).
* Corresponding author.

E-mail addresses: boros@rutcor.rutgers.edu (E. Boros), hammer@rutcor.rutgers.edu (P.L. Hammer).

special case of an oracle. As a trivial example, we can think of the cardinality of a subset S of the finite ground set $X = \{1, 2, \ldots, n\}$ being given by $|S| = x_1 + x_2 + \cdots + x_n$, where $(x_1, x_2, \ldots, x_n)$ is the characteristic vector of the subset S. Several other examples of set functions given by closed algebraic formulae will be given in Section 3.

The explicit knowledge of closed algebraic representations of set functions makes possible the application of a substantially larger set of mathematical techniques and results for their analysis and optimization. For instance, results of convex and nonconvex analysis, and various other techniques of nonlinear programming can thus be directly applied to problems, which are otherwise discrete in nature. It is important to remark that every set function defined on a finite ground set admits closed form algebraic representations. However, the derivation of such a form can be computationally difficult, or even intractable in some cases.

In this paper we shall focus on those set functions which are defined on a finite ground set and are given by closed algebraic formulae, and shall pay special attention to the case of multi-linear polynomial representations. Because of the analogy with Boolean functions, these functions will be called *pseudo-Boolean*.

The natural connections between pseudo-Boolean functions and nonlinear binary optimization have motivated and strongly influenced some of the first studies in this area (see e.g. [72,87,88,91]). Since then the study of pseudo-Boolean functions grew to a major area of research with hundreds of related publications in the last 30 years (see e.g. [94,95]).

Pseudo-Boolean functions appearing in polynomial (or other algebraic) representation play a major role in optimization models in a variety of areas, including VLSI design (via minimization [17,34]), statistical mechanics (spin glasses [17,50]), reliability theory (fault location [137]), computer science (maximum satisfiability [114]), statistics (clustering [145] and ranking [144]), economics [92], finance [105,120,121], operations research (location [73,141,162]), management science (project selection [165]), discrete mathematics (graphs, hypergraphs and networks [61,74,91,140,156]), manufacturing (production and scheduling [9,48,118]).

Beside optimization problems, pseudo-Boolean functions also appear in many other models of current interest. They constitute for instance the main object of investigation in cooperative game theory, where they are viewed as characteristic functions of games with side-payments [86,132,134,155]. They are used in various models of theoretical physics, where they describe the Hamiltonian energy function of spin glass systems [115,126,167] and of neural networks [6,107,138]. They occur in combinatorial theory, as rank functions of matroids [44,164], or as functions associated with certain graph-parameters, such as stability number, chromatic number, etc. [21,61, 133,156]).

In this paper we present a brief overview of the theory and algorithmic aspects of pseudo-Boolean functions and their optimization. Due to the large volume of related research, our survey is far from complete. We focus on results and techniques specific to this type of representation of set functions, with particular attention being paid to special classes, including quadratic, sub- and supermodular, and hyperbolic pseudo-Boolean functions.

2. Definitions and notations

Let us denote by $\mathbb{R}$ the set of reals, by $\mathbb{Z}$ the set of integers, and let $\mathbb{B} = \{0,1\}$ and $\mathbb{U} = [0,1]$. Further let n denote a positive integer, and let $\mathbf{V} = \{1,2,\ldots,n\}$. For a subset $S \subseteq V$ let us denote by $\mathbf{1}^S \in \mathbb{B}^n$ its characteristic vector, i.e.

$$\mathbf{1}^S_j = \begin{cases} 1 & \text{if } j \in S, \\ 0 & \text{otherwise.} \end{cases}$$

We shall consider functions in n binary variables, denoted by x_1, $x_2,\ldots,x_n$, and shall use $\mathbf{x} = (x_1,\ldots,x_n) \in \mathbb{B}^n$ to denote a binary vector, as well as the vector of these variables. In many situations the binary values 0 and 1 are used to encode unordered bivalent attributes, and play a perfectly symmetric role. It is natural to work both with the variables x_i and with their *complements* $\bar{x}_i \overset{\text{def}}{=} 1 - x_i$, for $i \in V$, called together *literals*. Let $\mathbf{L} = \{x_1, \bar{x}_1,\ldots,x_n, \bar{x}_n\}$ denote the set of literals.

Mappings $f: \mathbb{B}^n \mapsto \mathbb{R}$ are called *pseudo-Boolean functions*. Since there is a one-to-one correspondence between the subsets of $\mathbf{V}$ and the set of binary vectors $\mathbb{B}^n$, these functions are in fact *set functions*, i.e. mappings which associate a real value to every subset of a finite set. In this survey we shall refer to f as a set function, whenever we want to emphasize that the values of f are given in some implicit way (e.g. via an oracle), and we shall call it a pseudo-Boolean function if it is given explicitly by an algebraic expression.

The simplest, and perhaps least efficient, way of representing a pseudo-Boolean function is by a table listing the real values $f(\mathbf{x})$ corresponding to every binary vector $\mathbf{x} \in \mathbb{B}^n$.

As we shall see later, all pseudo-Boolean functions can be uniquely represented as *multi-linear polynomials*, of the form

$$f(x_1,\ldots,x_n) = \sum_{S \subseteq \mathbf{V}} c_S \prod_{j \in S} x_j. \tag{1}$$

By convention, we shall always assume that $\prod_{j \in \emptyset} x_j = 1$.

The size of the largest subset $S \subseteq \mathbf{V}$ for which $c_S \neq 0$ is called the *degree* of f, and is denoted by $\deg(f)$. We shall call a pseudo-Boolean function f *linear* (*quadratic*, *cubic*, etc.) if $\deg(f) \leqslant 1$ ($\leqslant 2, 3$, etc.)

The *size* of an expression (1) is the total number of variable occurrences in it, i.e.

$$\text{size}(f) \overset{\text{def}}{=} \sum_{S: c_S \neq 0} |S|. \tag{2}$$

Let us associate to a pseudo-Boolean function f its ith *derivative*,

$$\Delta_i(\mathbf{x}) \overset{\text{def}}{=} \frac{\partial f}{\partial x_i}(\mathbf{x})$$

$$= f(x_1,\ldots,x_{i-1},1,x_{i+1},\ldots,x_n) - f(x_1,\ldots,x_{i-1},0,x_{i+1},\ldots,x_n) \tag{3}$$

and its ith *residual*

$$\Theta_i(\mathbf{x}) \overset{\text{def}}{=} f(\mathbf{x}) - x_i \Delta_i(\mathbf{x}) \tag{4}$$

for all indices $i \in \mathbf{V}$. Let us notice that the functions Δ_i and Θ_i are themselves pseudo-Boolean functions, which depend on all the variables, but x_i.

Frequently, pseudo-Boolean functions are also represented as *posiforms*, i.e. polynomial expressions in terms of all the literals, having the form

$$\phi(x_1,\ldots,x_n) = \sum_{T \subseteq \mathbf{L}} a_T \prod_{u \in T} u, \tag{5}$$

where $a_T \geqslant 0$ whenever $T \neq \emptyset$. It is also customary to assume that $a_T = 0$ if $\{u, \bar{u}\} \subseteq T$ for some $u \in \mathbf{L}$, since otherwise the product $\prod_{u \in T} u$ is identically zero over $\mathbb{B}^n$.

Similarly to the case of polynomial expressions, let us call the size of the largest subset T of literals for which $a_T \neq 0$ the *degree* of the posiform ϕ, and denote it by $\deg(\phi)$; let us call a posiform ϕ *linear* (*quadratic, cubic*, etc.) if $\deg(\phi) \leqslant 1$ ($\leqslant 2, 3$, etc.) Furthermore, let us measure the *size* of a posiform as the total number of literal occurrences in it, i.e.

$$\text{size}(\phi) \overset{\text{def}}{=} \sum_{T : a_T \neq 0} |T|. \tag{6}$$

For the purpose of analyzing algorithms, the sum of the coefficients

$$A(\phi) \overset{\text{def}}{=} \sum_{T \neq \emptyset} a_T \tag{7}$$

will also be frequently needed.

It is obvious that a posiform (5) determines uniquely a pseudo-Boolean function. However, the reverse is not true: a pseudo-Boolean function can have many different posiforms representing it. Let us also add that while it is computationally easy to generate a posiform from a polynomial expression (1), it might be computationally difficult to generate the unique polynomial expression corresponding to a given posiform.

In the sequel, we shall use the letters x, y, and z to refer to variables, u, v, and w to refer to literals., and bold face letters $\mathbf{x}$, $\mathbf{y}$, $\mathbf{p}$, etc., to denote vectors. The letters f, g and h will usually denote pseudo-Boolean functions as well as their unique multi-linear polynomial expressions, while the greek letters ϕ and ψ will denote posiforms.

In this survey we shall consider minimization and maximization problems; whenever it is not necessary to specify which one is considered, we shall simply speak of optimization (or simply opt) to refer to any one of them.

3. Examples

There are a large number of combinatorial optimization models, which arise naturally or can be formulated easily as pseudo-Boolean optimization problems. In this section we recall some of these (mostly well-known) formulations.

One of the simplest problems of this type is well-known in algorithmic graph theory. Given a graph $G = (V, E)$ with vertex set V and edges set E, a subset $S \subseteq V$ is called *independent*, if no edge of G has both its endpoints in S. In the *maximum independent set problem* we need to find the largest cardinality independent set, the cardinality of which we shall denote by $\alpha(G)$. It is easy to show that

$$\alpha(G) = \max_{\mathbf{x} \in \mathbb{B}^V} \left(\sum_{i \in V} x_i - \sum_{(i,j) \in E} x_i x_j \right),$$

holds for every graph, furthermore that if $\mathbf{x}^* = \mathbf{1}^S$ is a maximizing binary vector of the above quadratic function, then a maximum cardinality independent set S^* of G (in fact with $S^* \subseteq S$) can be obtained in $O(n)$ time.

The complement $V \setminus S$ of an independent set S of G is called a *vertex cover* of the graph. Denoting by $\tau(G) = |V| - \alpha(G)$ the size of a smallest vertex cover of G, it is easy to show that

$$\tau(G) = \min_{\mathbf{x} \in \mathbb{B}^V} \left(\sum_{i \in V} x_i + \sum_{(i,j) \in E} \bar{x}_i \bar{x}_j \right).$$

The above formulations can be extended analogously to the weighted variants of these problems, as well as to hypergraphs. For instance, given a hypergraph $\mathcal{H} \subseteq 2^V$, a subset S of its vertices is called a *vertex cover* of $\mathcal{H}$ (known also as a *hitting set*) if $S \cap H \neq \emptyset$ for all hyperedges $H \in \mathcal{H}$. Denoting by $\tau(\mathcal{H})$ the size of a smallest vertex cover, it can be shown that

$$\tau(\mathcal{H}) = \min_{\mathbf{x} \in \mathbb{B}^V} \left(\sum_{i \in V} x_i + \sum_{H \in \mathcal{H}} \prod_{i \in H} \bar{x}_i \right).$$

This optimization problem can also be viewed as a pseudo-Boolean formulation of the equivalent *set covering* problem (over the transposed hypergraph.)

For a subset S of vertices of a graph $G = (V, E)$, let us denote by $\delta(S)$ the number of edges with exactly one endpoint in S. The *maximum cut* problem is to find a subset S which maximizes $\delta(S)$. Using the characteristic vector $\mathbf{x} = \mathbf{1}^S$ to represent a subset S, we can easily see that

$$\max_{S \subseteq V} \delta(S) = \max_{\mathbf{x} \in \mathbb{B}^V} \left(\sum_{(i,j) \in E} (x_i \bar{x}_j + \bar{x}_i x_j) \right).$$

A graph $G = (V, E)$ is called *signed* if the edge set is partitioned $E = E^+ \cup E^-$ into two (disjoint) subsets, called *positive*, and *negative* edges, respectively. A signed graph is called *balanced* if every cycle of it involves an even number of negative edges. The *signed graph balancing* problem consists in finding a minimum cardinality subset F of edges in a signed graph G, the removal of which makes G

balanced. Denoting by $\sigma(G)$ the size of such a minimum edge set, we can immediately see that

$$\sigma(G) = \min_{\mathbf{x} \in \mathbb{B}^V} \left(\sum_{(i,j) \in E^+} (x_i \bar{x}_j + \bar{x}_i x_j) + \sum_{(i,j) \in E^-} (x_i x_j + \bar{x}_i \bar{x}_j) \right).$$

The *maximum satisfiability* problem, one of the most frequently studied problems in the recent applied mathematics/theoretical computer science literature, has also a natural pseudo-Boolean formulation. In this problem the input consists of a family $\mathscr{C}$ of subsets $C \subseteq \mathbf{L}$ of literals, called *clauses*. We say that a binary assignment $\mathbf{x} \in \mathbb{B}^V$ is *satisfying* such a clause $C \subseteq \mathbf{L}$, if the (Boolean) disjunction of the literals in C takes value 1 (true) for this assignment. The maximum satisfiability problem consists in finding a binary assignment satisfying the maximum number of clauses of $\mathscr{C}$. It is easy to see that $\mathbf{x} \in \mathbb{B}^V$ satisfies a clause C iff $\prod_{u \in C} \bar{u} = 0$. Thus, the problem is equivalent with the pseudo-Boolean maximization problem

$$\max_{\mathbf{x} \in \mathbb{B}^V} \left(\sum_{C \in \mathscr{C}} \left(1 - \prod_{u \in C} \bar{u} \right) \right).$$

In the weighted version of the problem there is also a nonnegative weight a_C associated to every clause $C \in \mathscr{C}$, and the objective is to maximize the total weight of satisfied clauses

$$\max_{\mathbf{x} \in \mathbb{B}^V} \left(\sum_{C \in \mathscr{C}} a_C \left(1 - \prod_{u \in C} \bar{u} \right) \right).$$

4. General pseudo-Boolean functions

4.1. Representations of pseudo-Boolean functions

In this section we review a few basic properties of multi-linear polynomial and posiform representations of a pseudo-Boolean function, starting with the non-uniqueness of posiform representations.

Example 4.1. Let us show first on a small example that pseudo-Boolean functions can have different posiform representations. Consider for this the following two posiforms:

$$\psi_1 = 5x_1 + 4\bar{x}_1 \bar{x}_2 x_3 + 7x_1 x_2 x_4 + 9x_3 \bar{x}_4, \tag{8}$$

and

$$\psi_2 = x_1 + 4x_1 x_2 + 4x_1 \bar{x}_2 \bar{x}_3 + 7x_1 x_2 x_4 + 4\bar{x}_2 x_3 + 9x_3 \bar{x}_4. \tag{9}$$

It is easy to verify that in fact both posiforms define the same pseudo-Boolean function $g : \mathbb{B}^4 \mapsto \mathbb{R}$ given by the table

x				$g(\mathbf{x})$
0	0	0	0	0
0	0	0	1	0
0	0	1	0	13
0	0	1	1	4
0	1	0	0	0
0	1	0	1	0
0	1	1	0	9
0	1	1	1	0
1	0	0	0	5
1	0	0	1	5
1	0	1	0	14
1	0	1	1	5
1	1	0	0	5
1	1	0	1	12
1	1	1	0	14
1	1	1	1	12

Let us observe next that if a pseudo-Boolean function $f : \mathbb{B}^n \mapsto \mathbb{R}$ is represented by a posiform ϕ (5), then

$$a_\emptyset \leqslant \min_{\mathbf{x} \in \mathbb{B}^n} f(\mathbf{x}). \tag{10}$$

This trivial observation will serve us handily, when looking for lower bounds to a minimization problem. Let us show next that in fact

Proposition 1. *Every pseudo-Boolean function $f : \mathbb{B}^n \mapsto \mathbb{R}$ can be represented by a posiform ϕ for which*

$$a_\emptyset = \min_{\mathbf{x} \in \mathbb{B}^n} f(\mathbf{x}).$$

Proof. Let us start by defining $a_\emptyset = \min_{\mathbf{x} \in \mathbb{B}^n} f(\mathbf{x})$, as in the statement. Let us further define for every binary vector $\mathbf{y} \in \mathbb{B}^n$ a corresponding set of literals $L(\mathbf{y}) \subseteq \mathbf{L}$ by setting $L(\mathbf{y}) \overset{\text{def}}{=} \{x_i | y_i = 1, \ i \in \mathbf{V}\} \cup \{\bar{x}_i | y_i = 0, \ i \in \mathbf{V}\}$, and let $a_{L(\mathbf{y})} = f(\mathbf{y}) - a_\emptyset$ for all $\mathbf{y} \in \mathbb{B}^n$. We claim that with this notation

$$\Phi_f(\mathbf{x}) = a_\emptyset + \sum_{\mathbf{y} \in \mathbb{B}^n} a_{L(\mathbf{y})} \prod_{u \in L(\mathbf{y})} u \tag{11}$$

is indeed a posiform representing f. To verify this claim we can notice that $a_{L(\mathbf{y})} \geqslant 0$ for all $\mathbf{y} \in \mathbb{B}^n$ by the above definitions, that the term $(\prod_{u \in L(\mathbf{y})} u)(\mathbf{x}) = 0$ for all binary

vectors $\mathbf{x} \neq \mathbf{y}$, and that this term takes the value 1 if $\mathbf{x} = \mathbf{y}$. Hence, in every $\mathbf{x} \in \mathbb{B}^n$
$$\Phi_f(\mathbf{x}) = a_\emptyset + a_{L(\mathbf{x})} = f(\mathbf{x}). \quad \Box$$

Let us remark that posiform (11) is in fact uniquely defined for every pseudo Boolean function f. We shall call this unique posiform Φ_f the *min-term* representation of f.

Let us prove next that every pseudo-Boolean function has a unique multi-linear polynomial representation.

Proposition 2 (Hammer et al. [87], Hammer and Rudeanu [91]). *Every pseudo-Boolean function $f : \mathbb{B}^n \mapsto \mathbb{R}$ has a unique multi-linear polynomial representation of form* (1).

Proof. We need to show that the values of f over $\mathbb{B}^n$ determine uniquely the coefficients c_S, $S \subseteq V$ in (1). Let us show this by induction on the size of these subsets. Clearly, $f(0, 0, \ldots, 0) = c_\emptyset$, and hence $c_\emptyset$ is uniquely determined by the value $f(\mathbf{1}^0)$.

Let us assume next that the coefficients c_T are already shown to be unique for all subsets T of size less than k, and let $S \subseteq V$ be a subset of size k. By observing that

$$f(\mathbf{1}^S) = \sum_{T \subseteq S} c_T,$$

we obtain

$$c_S = f(\mathbf{1}^S) - \sum_{T \subset S} c_T. \tag{12}$$

Since all terms on the right-hand side have unique values, in view of (12) the same holds for c_S, thus completing the proof. $\quad \Box$

Example 4.2. The function g in Example 4.1 has

$$g(x_1, x_2, x_3, x_4) = 5x_1 + 13x_3 - 4x_1x_3 - 4x_2x_3 - 9x_3x_4 + 4x_1x_2x_3 + 7x_1x_2x_4$$

as its unique multi-linear polynomial form.

It will be useful later to express the derivatives and the residuals of a pseudo-Boolean function in terms of the coefficients of its multi-linear polynomial expression.

Proposition 3. *Given a pseudo-Boolean function f by its multi-linear polynomial expression* (1), *we have the following equalities for its derivatives* (3) *and residuals* (4) *for all indices $i \in V$:*

$$\Delta_i(\mathbf{x}) = \sum_{S \subseteq V \setminus \{i\}} c_{S \cup \{i\}} \prod_{j \in S} x_j \quad \text{and} \quad \Theta_i(\mathbf{x}) = \sum_{S \subseteq V \setminus \{i\}} c_S \prod_{j \in S} x_j. \tag{13}$$

Proof. Immediate from (1) by elementary calculations. $\quad \Box$

Example 4.3. For illustration, if $f(x_1, x_2) = -1 + 3x_1 + 4x_2 - 2x_1x_2$, then $\Delta_1(\mathbf{x}) = 3 - 2x_2$, and $\Theta_1(\mathbf{x}) = -1 + 4x_2$, etc.

4.2. Rounding procedures and derandomization

Let us note that (1) can also be viewed as a real valued expression, which can be evaluated for any real vector $\mathbf{r} \in \mathbb{R}^n$.

Proposition 4 (Boros and Prékopa [37], Rosenberg [148]). *Let us consider a pseudo-Boolean function f given by (1), and let $\mathbf{r} \in \mathbb{U}^n$. Then there exist binary vectors $\mathbf{x}, \mathbf{y} \in \mathbb{B}^n$ for which*

$$f(\mathbf{x}) \leqslant f(\mathbf{r}) \leqslant f(\mathbf{y}),\qquad(14)$$

furthermore, such vectors can be generated in $O(\text{size}(f))$ time.

Proof. We shall give a constructive proof only for the existence of $\mathbf{x}$, since an analogous procedure will work for the existence of $\mathbf{y}$.

ROUNDDOWN $(f, \mathbf{r})$

Initialize: Set $t = 0$ and $\mathbf{q}^0 = \mathbf{r}$.
Loop: While there exists an index $j \in V$ for which $0 < q_j < 1$:
Set $t = t + 1$ and define

$$q_i^t = \begin{cases} q_i^{t-1} & \text{if } i \neq j, \\ 0 & \text{if } i = j \text{ and } \varDelta_j(\mathbf{q}^{t-1}) > 0, \\ 1 & \text{otherwise.} \end{cases}$$

Output: Set $\mathbf{x} = \mathbf{q}^t$.

Let us note first that in every iteration of the above procedure one of the fractional components is changed to an integral one, and hence ROUNDDOWN terminates in at most n iterations by outputting a binary vector $\mathbf{x}$.

Let us also observe that if j is the index chosen in the tth iteration then we have $\varDelta_j(\mathbf{q}^t) = \varDelta_j(\mathbf{q}^{t-1})$ and $\Theta_j(\mathbf{q}^t) = \Theta_j(\mathbf{q}^{t-1})$, since these functions depend only on the variables x_i for $i \neq j$, and since the vectors $\mathbf{q}^t$ and $\mathbf{q}^{t-1}$ differ only in their jth components. Thus the inequality

$$f(\mathbf{q}^{t-1}) - f(\mathbf{q}^t) = (q_j^{t-1} - q_j^t)\varDelta_j(\mathbf{q}^{t-1}) \geqslant 0$$

holds by the update rule in the core of the **Loop**, and hence we have

$$f(\mathbf{x}) = f(\mathbf{q}^t) \leqslant f(\mathbf{q}^{t-1}) \leqslant \cdots \leqslant f(\mathbf{q}^1) \leqslant f(\mathbf{q}^0) = f(\mathbf{r}).$$

To see the claimed complexity, we need to organize these computations carefully. First of all we can preselect the fractional components of $\mathbf{r}$ in $O(n)$ steps, and build a variable-term data structure, and pre-compute the values of the terms of (1) in $O(\text{size}(f))$ steps. After this, each selection in the **Loop** can be executed in constant time, and both the evaluation of $\varDelta_j$ and the update of the values of the terms (of those which depend on the jth component) can be executed in time proportional to the number of occurrences of x_j in (1). Hence the total time of the algorithm after the pre-computations is O(size(f)), thus proving our claim. $\square$

Let us remark that by changing the inequality in the core of the **Loop** of the above procedure, we obtain an increasing sequence of function values, and produce a binary vector **y**. We shall refer to that version of the above procedure as ROUNDUP(f, **r**).

The above simple properties have a number of consequences. To state them, we need a few more definitions.

Let us introduce the notation Argopt$_D(f)$ for the subset consisting of the points of D which are optimal solutions of the optimization problem opt$_{\mathbf{x} \in D} f(\mathbf{x})$. Let us further introduce the generic notations $\mathbf{r}^{\min}$, $\mathbf{r}^{\max}$, $\mathbf{x}^{\min}$ and $\mathbf{x}^{\max}$ to denote an arbitrary optimizing vector for the respective optimization problems, i.e. $\mathbf{r}^{\min} \in \mathrm{Argmin}_{\mathbb{U}^n}(f)$, $\mathbf{r}^{\max} \in \mathrm{Argmax}_{\mathbb{U}^n}(f)$, $\mathbf{x}^{\min} \in \mathrm{Argmin}_{\mathbb{B}^n}(f)$, and $\mathbf{x}^{\max} \in \mathrm{Argmax}_{\mathbb{B}^n}(f)$.

Example 4.4. Returning to the pseudo-Boolean function g in Example 4.1, we can see that

$$\mathrm{Argmin}_{\mathbb{B}^4}(g) = \{(0,0,0,0),(0,1,0,0),(0,0,0,1),(0,1,0,1),(0,1,1,1)\},$$

$$\mathrm{Argmax}_{\mathbb{B}^4}(g) = \{(1,0,1,0),(1,1,1,0)\},$$

$$\mathrm{Argmin}_{\mathbb{U}^4}(g) = \{(0,a,0,b),(0,1,c,1)|0 \leqslant a,b,c \leqslant 1\},$$

and

$$\mathrm{Argmax}_{\mathbb{U}^4}(g) = \{(1,a,1,0)|0 \leqslant a \leqslant 1\}.$$

The first consequence we can draw easily from the above properties is that both the maximization and the minimization of a pseudo-Boolean function over $\mathbb{B}^n$ can in fact be viewed as continuous nonlinear optimization problems over $\mathbb{U}^n$.

Corollary 1. *For any pseudo-Boolean function f*

$$\mathrm{opt}_{\mathbf{x} \in \mathbb{B}^n} f(\mathbf{x}) = \mathrm{opt}_{\mathbf{r} \in \mathbb{U}^n} f(\mathbf{r}).$$

Proof. On the one hand, $\mathbb{U}^n \supset \mathbb{B}^n$ implies that

$$\min_{\mathbf{q} \in \mathbb{U}^n} f(\mathbf{q}) \leqslant \min_{\mathbf{x} \in \mathbb{B}^n} f(\mathbf{x}) \leqslant \max_{\mathbf{x} \in \mathbb{B}^n} f(\mathbf{x}) \leqslant \max_{\mathbf{q} \in \mathbb{U}^n} f(\mathbf{q}). \tag{15}$$

On the other hand, applying ROUNDDOWN(f, $\mathbf{r}^{\min}$), we can obtain by Proposition 4 a binary vector $\mathbf{x} \in \mathbb{B}^n$ for which $f(\mathbf{x}) \leqslant f(\mathbf{r}^{\min})$, and hence $\min_{\mathbf{x} \in \mathbb{B}^n} f(\mathbf{x}) \leqslant f(\mathbf{r}^{\min}) = \min_{\mathbf{q} \in \mathbb{U}^n} f(\mathbf{q})$ follows, implying $\min_{\mathbf{q} \in \mathbb{U}^n} f(\mathbf{q}) \geqslant \min_{\mathbf{x} \in \mathbb{B}^n} f(\mathbf{x})$. This, together with (15) implies that

$$\min_{\mathbf{x} \in \mathbb{B}^n} f(\mathbf{x}) = \min_{\mathbf{q} \in \mathbb{U}^n} f(\mathbf{q}).$$

Similarly, applying RoundUp($f, \mathbf{r}^{\max}$), we can verify that

$$\max_{\mathbf{x} \in \mathbb{B}^n} f(\mathbf{x}) = \max_{\mathbf{q} \in \mathbb{U}^n} f(\mathbf{q}). \qquad \square$$

Another, algorithmic consequence of Proposition 4 is the existence of linear time "rounding" procedures, which can be viewed as efficient ways of "derandomization".

Proposition 5. *Let us assume that the binary variables x_i, $i \in V$ are independent random variables with $\mathrm{Prob}(x_i = 1) = p_i$ and $\mathrm{Prob}(x_i = 0) = q_i \overset{\mathrm{def}}{=} 1 - p_i$ for $i \in V$, where $\mathbf{p} = (p_1, \ldots, p_n) \in \mathbb{U}^n$ is a given vector of the probabilities defining the distribution. Let further $f : \mathbb{B}^n \mapsto \mathbb{R}$ be a pseudo-Boolean function given by (1). Then the expected value of f is*

$$\mathrm{Exp}[f] = f(\mathbf{p}).$$

Proof. Since the sum of the expectations of random variables is the same as the expectation of their sum, it is enough to show that

$$\mathrm{Exp}\left[c_S \prod_{j \in S} x_j \right] = c_S \prod_{j \in S} p_j$$

for an arbitrary term of (1). This latter equality follows readily by the independence of the variables, since that implies

$$\mathrm{Exp}\left[c_S \prod_{j \in S} x_j \right] = c_S \prod_{j \in S} \mathrm{Exp}[x_j] = c_S \prod_{j \in S} p_j. \qquad \square$$

Let us note that only multi-linearity was needed for the above proof, hence the same property will also hold for any posiform.

In view of the above result, we can regard RoundDown and RoundUp as efficient derandomizations, since e.g. for a fractional vector $\mathbf{p} \in \mathbb{U}^n$ a simple probabilistic argument guarantees the existence of a binary vector $\mathbf{x} \in \mathbb{B}^n$ with $f(\mathbf{x}) \leqslant \mathrm{Exp}[f]$, while RoundDown provides an efficient deterministic way of generating such a vector.

This kind of probabilistic arguments are frequently used in combinatorics and in various approximation algorithms for combinatorial optimization problems (see e.g. [5,69,70]), and the above simple rounding procedures are special cases of the "probabilistic rounding" technique based on conditional probabilities, introduced in [37,142,143]. When started from $\mathbf{p} = (\frac{1}{2}, \ldots, \frac{1}{2})$, RoundDown is essentially equivalent with the heuristic procedure proposed in [109].

4.3. Local optima

In this section, following the presentation of [91], we shall recall some necessary conditions of optimality. To avoid unnecessary repetitions, we shall state results only for the case of minimization problems, although similar necessary conditions can be stated for maximization problems, as well.

Let us associate to a binary vector $\mathbf{x} \in \mathbb{B}^n$ its *neighborhood* $N(\mathbf{x})$, defined as

$$N(\mathbf{x}) = \{\mathbf{y} \mid \rho_H(\mathbf{x}, \mathbf{y}) \leqslant 1\}, \tag{16}$$

where $\rho_H(\mathbf{x}, \mathbf{y})$ denotes the so-called *Hamming distance* of the vectors $\mathbf{x}$ and $\mathbf{y}$, defined as the number of components in which these two vectors differ.

Given a pseudo Boolean function $f : \mathbb{B}^n \mapsto \mathbb{R}$, a binary vector $\mathbf{x} \in \mathbb{B}^n$ is called a *local minimum* of f if $f(\mathbf{y}) \geqslant f(\mathbf{x})$ for all neighboring vectors $\mathbf{y} \in N(\mathbf{x})$. Let us add that of course every (global) minimum of f is also a local minimum.

Proposition 6. *Given a pseudo-Boolean function f, a binary vector $\mathbf{x} \in \mathbb{B}^n$ is a local minimum of f if and only if*

$$x_i = \begin{cases} 1 & \text{if } \Delta_i(\mathbf{x}) < 0, \\ 0 & \text{if } \Delta_i(\mathbf{x}) > 0 \end{cases} \tag{17}$$

for all $i \in V$.

Proof. Let us denote by $\mathbf{y}^i$ the binary vector obtained from $\mathbf{x}$ by switching its ith component. Then, by the above definition, $\mathbf{x}$ is a local minimum iff $f(\mathbf{y}^i) \geqslant f(\mathbf{x})$ for all indices $i \in V$. In view of (4) and of the fact that $\mathbf{x}$ and $\mathbf{y}^i$ differ only in their ith components, we have

$$f(\mathbf{y}^i) = y_i^i \Delta_i(\mathbf{y}^i) + \Theta_i(\mathbf{y}^i) = (1 - x_i)\Delta_i(\mathbf{x}) + \Theta_i(\mathbf{x})$$

for $i \in V$. Hence

$$f(\mathbf{y}^i) - f(\mathbf{x}) = (1 - 2x_i)\Delta_i(\mathbf{x})$$

follows for every index $i \in V$. Thus $f(\mathbf{y}^i) \geqslant f(\mathbf{x})$ implies (17), for all $i \in V$. □

Let us note that, given a binary vector $\mathbf{x} \in \mathbb{B}^n$, conditions (17) are very easy to test. Even if the pseudo-Boolean function f is given only by an oracle, the derivatives can be evaluated by $n + 1$ calls to this oracle (by obtaining the values for $f(\mathbf{x})$ and for $f(\mathbf{y}^i)$ for $i \in V$, as in the proof above.)

The above set of conditions inspired a large number of heuristic algorithms, the so called *local search methods*. These algorithms focus on finding a local minimum, in the hope that it turns out to be a global one, as well.

$$\text{LOCALSEARCH}(f, x^0)$$

Input: A pseudo-Boolean function f and a binary vector $\mathbf{x}^0 \in \mathbb{B}^n$. Set $k = 0$. [It is assumed that for every binary vector $\mathbf{x} \in \mathbb{B}^n$ there exists a well defined and computable subset $N(\mathbf{x}) \subseteq \mathbb{B}^n$, called the neighborhood of $\mathbf{x}$.]

Iteration: While there exists $\mathbf{y} \in N(\mathbf{x}^k)$ for which $f(\mathbf{y}) < f(\mathbf{x}^k)$, let $\mathbf{x}^{k+1} = \mathbf{y}$ and set $k = k + 1$.

Output: RETURN the local minimum $\mathbf{x}^k$ of f.

Though, these algorithms tend to work very well and fast in practice, theoretical guarantees for efficiency exist only in some special cases. Finding a local minimum remains, in general, a difficult problem, even in cases when the number of local minima is very large (see e.g. [96,110,157,158]), and even for special classes of pseudo-Boolean functions [82]. A further complication is that the number of local minima can indeed be very large (e.g., exponentially large in the input size of the problem) even for quadratic pseudo-Boolean functions (see e.g. [112]), and obviously, not all of those are equally good solutions. A natural idea to increase the chances that a local minimum is also a global one is to use larger neighborhoods (e.g., considering all points within Hamming distance k, for $k = 2, 3, \ldots$). Unfortunately, not only the computational cost of each iteration increases by this, but there are results indicating that, in a worst-case sense (see e.g. [151,161]) even the use of substantially larger neighborhoods will not necessarily yield better results either. Various techniques have been proposed in the literature for the implicit handling of much larger neighborhoods without a sharp increase in the computational cost of the iterations. Perhaps the most successful and most widely applied such method is the so called *tabu search* algorithm, see e.g. [64,65].

Let us add that continuous local optimization techniques are efficient for the minimization of a convex (or maximization of a concave) function, since any local optimum of those is also a global one. Recognition of convexity or concavity of some continuous extensions of pseudo-Boolean functions is hence important, and considered also in the literature (see e.g. [41,42,124]). Other techniques modify the function to achieve convexity (or concavity) by giving up multi-linearity of the objective function (see e.g. [90,117]).

4.4. Reductions to quadratic optimization

In this section we recall from [149] that the optimization of a pseudo-Boolean function can always be reduced in polynomial time to the optimization of a quadratic pseudo-Boolean function. The basic idea in this reduction is the substitution of the product of two variables by a new one, and the addition of appropriate penalty terms having the role of forcing, at any point of optimum, the new variable to take the value of the product of the two substituted variables.

The following simple observation provides the basic tool for such a substitution.

Observation 1. Assume that $x, y, z \in \mathbb{B}$. Then the following equivalences hold:

$$xy = z \quad \text{iff} \quad xy - 2xz - 2yz + 3z = 0,$$

and

$$xy \neq z \quad \text{iff} \quad xy - 2xz - 2yz + 3z > 0.$$

These equivalences can easily be verified by trying out all 8 possible binary substitutions for the variables x, y, and z. Let us further remark that, of course, the above quadratic expression is not the only one for which such equivalences would hold.

Since a quadratic constraint can be eliminated by inserting the corresponding expression into the objective function as a penalty term (with a large multiplier), the above equivalences suggest the following method for reducing a pseudo-Boolean minimization problem to a quadratic one:

REDUCEMIN(f)

Input: A pseudo-Boolean function f given by its multi-linear polynomial form (1).

Initialize: Set $M \stackrel{\mathrm{def}}{=} 1 + 2\sum_{S \subseteq V} |c_S|$, $m = n$, and $f^n = f$.

Loop: While there exists a subset $S^* \subseteq V$ for which $|S^*| > 2$ and $c_{S^*} \neq 0$ repeat:

    1. Choose two elements i and j from S^* and update
$$c_{\{i,j\}} = c_{\{i,j\}} + M, \text{ set}$$
$$c_{\{i,m+1\}} = c_{\{j,m+1\}} = -2M \text{ and}$$
$$c_{\{m+1\}} = 3M, \text{ and}$$
       for all subsets $S \supseteq \{i,j\}$ with $c_S \neq 0$ define
$$c_{(S \setminus \{i,j\}) \cup \{m+1\}} = c_S \text{ and set } c_S = 0.$$
    2. Define $f^{m+1}(x_1,\dots,x_{m+1}) = \sum_{S \subseteq V} c_S \prod_{k \in S} x_k$, and set $m = m + 1$.

Output: Set $g = f^m$.

Theorem 1 (Rosenberg [149]). REDUCEMIN(f) *terminates in polynomial time in the size of* f, *and produces a pseudo-Boolean function* g *in* m *variables, the size of which is polynomially bounded in* size(f), *and such that*

$$\min_{\mathbf{y} \in \mathbb{B}^m} g(\mathbf{y}) = \min_{\mathbf{x} \in \mathbb{B}^n} f(\mathbf{x}).$$

Proof. In the main loop of the above algorithm we replace each occurrence of $x_i x_j$ by x_{m+1}, and we add the expression $M(x_i x_j - 2x_i x_{m+1} - 2x_j x_{m+1} + 3x_{m+1})$ to the objective function.

Hence, by Observation 1 and by our choice of M we have $f^{m+1}(x_1,\dots,x_{m+1}) = f^m(x_1,\dots,x_m) \leqslant \max_{\mathbf{x} \in \mathbb{B}^m} f^m(\mathbf{x}) < M/2$ whenever $x_{m+1} = x_i x_j$, and $f^{m+1}(x_1,\dots,x_{m+1}) \geqslant M/2$ whenever $x_{m+1} \neq x_i x_j$. Thus,

$$\min_{\mathbf{y} \in \mathbb{B}^{m+1}} f^{m+1}(\mathbf{y}) = \min_{\mathbf{x} \in \mathbb{B}^m} f^m(\mathbf{x}),$$

implying the claimed equality of the minima.

It also follows that the number of those terms in f^{m+1} for which $c_S \neq 0$, $|S| > 2$ is at least one less than in f^m, hence the algorithm must terminate in at most size(f) iterations, proving the claim about complexity. $\square$

Example 4.5. Let us consider the following pseudo-Boolean function:

$$f(x_1, x_2, x_3, x_4, x_5) \stackrel{\mathrm{def}}{=} 5x_1 x_2 - 7x_1 x_2 x_3 x_4 + 2x_1 x_2 x_3 x_5.$$

Applying REDUCEMIN to this function we get initially $M = 1 + 5 + 7 + 2 = 15$, $m = 5$ and $f^5 = f$. We can choose in the **Loop** first $i = 1$ and $j = 2$ and we get

$$c_{1,2} = c_{1,2} + M = 20,$$
$$c_{1,6} = c_{2,6} = -2M = -30,$$

and

$$c_6 = 3M = 45,$$

and for the two terms containing both x_1 and x_2 we get

$$c_{3,4,6} = c_{1,2,3,4} = -7,$$
$$c_{1,2,3,4} = 0,$$
$$c_{3,5,6} = c_{1,2,3,5} = 2,$$

and

$$c_{1,2,3,5} = 0.$$

Thus, we obtain

$$f^6 = 45x_6 + 20x_1x_2 - 30x_1x_6 - 30x_2x_6 - 7x_3x_4x_6 + 2x_3x_5x_6.$$

Since there are still terms of degree higher than 2, we have to repeat the **Loop**, and can choose e.g. $i = 3$ and $j = 6$, yielding

$$c_{3,6} = c_{3,6} + M = 15,$$
$$c_{3,7} = c_{6,7} = -2M = -30,$$

and

$$c_7 = 3M = 45,$$

while for the two terms containing both x_3 and x_6 we get

$$c_{4,7} = c_{3,4,6} = -7,$$
$$c_{3,4,6} = 0,$$
$$c_{5,7} = c_{3,5,6} = 2,$$

and

$$c_{3,5,6} = 0.$$

Therefore, for $m = 7$ we obtain

$$f^7 = 45x_6 + 45x_7 + 20x_1x_2 - 30x_1x_6 - 30x_2x_6 + 15x_3x_6 - 30x_3x_7$$
$$- 7x_4x_7 + 2x_5x_7 - 30x_6x_7.$$

Since there are no non-quadratic terms left, the algorithm terminates outputting $g = f^7$. It is easy to verify that indeed,

$$\min_{\mathbf{x} \in \mathbb{B}^5} f(\mathbf{x}) = \min_{\mathbf{y} \in \mathbb{B}^7} g(\mathbf{y}) = -2$$

and that the relations $x_6 = x_1 x_2$ and $x_7 = x_3 x_6$ provide a one-to-one correspondence between the minima of these functions, i.e. between the sets $\text{Argmin}_{\mathbb{B}^5}(f) = \{(1,1,1,1,0)\}$, and $\text{Argmin}_{\mathbb{B}^7}(f^7) = \{(1,1,1,1,0,1,1)\}$.

One might hope for a more efficient reduction by trying to substitute at a time 3 (or more) variables with a new one. It is easy to see however that the above procedure cannot be generalized in this way:

Observation 2. There is no quadratic pseudo-Boolean function $f(x,y,z,u)$ (in four binary variables) for which $f(x,y,z,u) = 0$ if $u = xyz$, and $f(x,y,z,u) > 0$ whenever $u \neq xyz$.

Let us note also that the number of (new) variables in the final output depends clearly on the selection of the pairs i, j in the **Loop** of the above algorithm. One could try to minimize the number of (new) variables by finding a better selection procedure for these pairs. However, this optimization problem is NP-hard, even for cubic inputs, as shown by the following simple observation.

Observation 3. Let us consider a graph $G = (\mathbf{V}, E)$, and define a pseudo-Boolean function

$$f(x_0, x_1, \ldots, x_n) = \sum_{(i,j) \in E} a_{i,j} x_0 x_i x_j,$$

where the coefficients $a_{i,j}$ are arbitrary reals for all $(i,j) \in E$. Let us denote by m the smallest number for which REDUCEMIN(f) produces a quadratic pseudo-Boolean function g in m binary variables. Then,

$$m = n + \tau(G),$$

where $\tau(G)$ denotes the size of a smallest vertex cover of the graph G.

Clearly, this shows that determining the smallest m is not easier then computing the vertex cover of a graph, which is known to be NP-complete (see e.g. [63]).

4.5. Persistency

Let us return now to posiforms, and review some of their specific properties. Let us first recall the trivial fact that all (nonconstant) terms of (5) have nonnegative coefficients, and hence their sum can never be less than zero. This implies that, in a way, minimizing a posiform is essentially the same as trying to make as many of its terms as possible vanish. This is evident e.g. in the *maximum satisfiability* (or in short MAX-SAT) problem, which is equivalent with the minimization of a given posiform ϕ in which $a_\emptyset = 0$, with the important exception that the objective there is stated as

$$\max_{\mathbf{x} \in \mathbb{B}^n}(A(\phi) - \phi(\mathbf{x})),$$

i.e. as the maximization of the total weight of terms which can be made to vanish simultaneously (and not as the minimization of ϕ.) Recall that $A(\phi)$ denotes the sum of the coefficients of the posiform ϕ, as defined in (7).

This idea also gives us the possibility to recognize whether a partial assignment of binary values to some of the variables is "optimal" in some sense. For this, let us call a binary vector $\mathbf{y} \in \mathbb{B}^S$ corresponding to a subset $S \subseteq \mathbf{V}$ a *partial assignment*. (Note that $\mathbb{B}^n$ is only a simpler and possibly more conventional notation used instead of the more accurate notation $\mathbb{B}^{\mathbf{V}}$.) Furthermore, for a subset $S \subseteq \mathbf{V}$ of indices and a vector $\mathbf{x} \in \mathbb{B}^n$, let us denote by $\mathbf{x}[S] \in \mathbb{B}^S$ the subvector corresponding to the indices in S, i.e. $\mathbf{x}[S] = (x_i | i \in S)$. For a partial assignment $\mathbf{y} \in \mathbb{B}^S$ and for a vector $\mathbf{x} \in \mathbb{B}^n$, let us define the *switch* of $\mathbf{x}$ by $\mathbf{y}$ to be the binary vector $\mathbf{z}$ defined by

$$
z_j = \begin{cases} x_j & \text{if } j \notin S, \\ y_j & \text{if } j \in S \end{cases}
$$

and let us denote it by $\mathbf{z} = \mathbf{x}[S \leftarrow \mathbf{y}]$. For example, if $n = 5$, $S = \{1, 2, 5\}$ and $\mathbf{y}$ is the partial assignment $y_1 = 1$, $y_2 = 0$, and $y_5 = 1$, then the switch of $\mathbf{x} = (1, 1, 1, 0, 0)$ by $\mathbf{y}$ will be the vector $\mathbf{z} = (1, 0, 1, 0, 1)$.

Given a pseudo-Boolean function f, and a partial assignment $\mathbf{y} \in \mathbb{B}^S$ for some subset $S \subseteq \mathbf{V}$, following [79] we shall say that *strong persistency* holds for f at $\mathbf{y}$, if for all $\mathbf{x} \in \text{Argmin}_{\mathbb{B}^n}(f)$ we have $\mathbf{x}[S] = \mathbf{y}$, i.e. if the restriction of all minimizing points of f to S coincide with the partial assignment $\mathbf{y}$. *Weak persistency* is said to hold for f at $\mathbf{y}$ if $\mathbf{x}[S \leftarrow \mathbf{y}] \in \text{Argmin}_{\mathbb{B}^n}(f)$ holds for all $\mathbf{x} \in \text{Argmin}_{\mathbb{B}^n}(f)$, i.e. if a switch by the partial assignment $\mathbf{y}$ in a minimizing point always results in a minimizing point. Clearly, strong persistency implies weak persistency.

There are several examples for persistency observed in the literature, including both weak and strong persistency for quadratic pseudo-Boolean functions [79], weak persistency in an integer programming formulation of vertex covering [129], etc. Since satisfiability problems can also be viewed as testing whether a given posiform has its minimum equal to its constant term, the so-called "autark" assignments introduced in [127] turn out also to be special cases of weak persistency (c.f. [32]).

Example 4.6. Let us return again to the pseudo-Boolean function g in Example 4.1, and its posiform representation ψ_1. If we let $\mathbf{y}^* = (0, 1) \in \mathbb{B}^{\{1,2\}}$ to be a partial assignment to the first two variables, then we can see that for each vector in $\text{Argmin}_{\mathbb{B}^4}(g)$ switching the first two components to $(0, 1)$ yields always a vector in $\text{Argmin}_{\mathbb{B}^4}(g)$. Hence, weak persistency holds for the pseudo-Boolean function g at $\mathbf{y}^*$. Furthermore, at the partial assignment $\mathbf{z}^* = (0) \in \mathbb{B}^{\{1\}}$, both weak and strong persistency hold for g.

Let us add that verifying if weak or strong persistency holds for a pseudo-Boolean function f at a given partial assignment $\mathbf{y} \in \mathbb{B}^S$ is, in general, a difficult task. However, in some special cases posiforms may provide an efficient guarantee for persistency to hold.

Given a posiform (5), and a partial assignment $\mathbf{y} \in \mathbb{B}^S$ for some subset $S \subseteq \mathbf{V}$, we shall say that $\mathbf{y}$ is a *contractor* for ϕ if whenever $a_T > 0$ and there is an index $i \in S$ for which x_i or $\bar{x}_i$ is in T we have $\prod_{u \in T} u(\mathbf{y}) \equiv 0$. In other words, $\mathbf{y}$ is a contractor for ϕ if it makes all those terms of ϕ vanish which involve at least one of the variables x_i, $i \in S$, regardless of the values of the other variables x_j, $j \notin S$.

Let us note that checking if a given partial assignment is a contractor for a posiform ϕ can be done in linear $O(\text{size}(\phi))$ time.

Proposition 7. *Given a posiform ϕ as in* (5), *and a contractor* $\mathbf{y} \in \mathbb{B}^S$ *of it, we have*

$$\phi(\mathbf{x}[S \leftarrow \mathbf{y}]) \leqslant \phi(\mathbf{x})$$

for all binary vectors $\mathbf{x} \in \mathbb{B}^n$.

Proof. Let us denote by $L(S)$ the set of literals corresponding to variables with indices in S, i.e. $L(S) = \{x_i, \bar{x}_i | i \in S\}$. Let us further consider an arbitrary binary vector $\mathbf{x} \in \mathbb{B}^n$, and let $\mathbf{z} = \mathbf{x}[S \leftarrow \mathbf{y}]$ denote its switch by $\mathbf{y}$.

We can see that

$$\left(\sum_{T \subseteq L, T \cap L(S) = \emptyset} a_T \prod_{u \in T} u \right)(\mathbf{x}) = \left(\sum_{T \subseteq L, T \cap L(S) = \emptyset} a_T \prod_{u \in T} u \right)(\mathbf{z}),$$

since these terms involve only variables not belonging to S, and $\mathbf{z}[V \setminus S] = \mathbf{x}[V \setminus S]$ holds by the definition of $\mathbf{z}$. We can also observe that

$$\left(\sum_{T \subseteq L, T \cap L(S) \neq \emptyset} a_T \prod_{u \in T} u \right)(\mathbf{z}) = 0,$$

since $\mathbf{y}$ is assumed to be a contractor for ϕ. Therefore we have

$$\phi(\mathbf{z}) = \left(\sum_{T \subseteq L, T \cap L(S) = \emptyset} a_T \prod_{u \in T} u \right)(\mathbf{z})$$

$$= \left(\sum_{T \subseteq L, T \cap L(S) = \emptyset} a_T \prod_{u \in T} u \right)(\mathbf{x})$$

$$\leqslant \phi(\mathbf{x}),$$

since the coefficients of the nontrivial terms of ϕ are all nonnegative, and the constant term, corresponding to $T = \emptyset$, is included in the summation on the previous line. □

Corollary 2. *If the pseudo-Boolean function f is given as a posiform ϕ, and the partial assignment* $\mathbf{y} \in \mathbb{B}^S$, $S \subseteq \mathbf{V}$ *is a contractor for ϕ, then weak persistency holds for f at* $\mathbf{y}$.

Proof. Immediate by Proposition 7. □

In fact, we shall see later that all of the cases for weak persistency cited in the literature follow from this corollary, and hence can be verified simply by exhibiting an appropriate posiform of the pseudo-Boolean function in question.

Example 4.7. Returning to the pseudo-Boolean function of Example 4.1, we can note first that the vector $\mathbf{y}^*$, as defined in Example 4.6, is a contractor for the posiform ψ_1.

Let us further remark that not all persistencies follow easily by contractors, e.g. $\mathbf{z}^*$, defined in the same example, is not a contractor for ψ_1, though the pseudo-Boolean function g has another posiform ψ_2, given in Example (4.1), for which $\mathbf{z}^*$ is a contractor. Of course, this is not always the case.

As a conclusion of the discussions on persistency and contractors, we can see that posiform representations can provide not only bounds to the optimum values of a pseudo-Boolean function, but also some information about the values of the variables in the optima.

4.6. Basic algorithm

In this section, following [91], we present a general algorithm for finding the optimum of a pseudo-Boolean function, based on the necessary conditions of local optimality presented in Section 4.3. This algorithm, the so called *basic algorithm*, was introduced first in [87,88], and later simplified in [91]. We shall also recall some more recent results from [47], showing that for a special class of problems this algorithm can be implemented to run in polynomial time.

From a theoretical point of view, conditions (17) can be viewed as a characterization of the components of a local minimum in terms of the other components of that vector. This suggests the possibility of finding an expression for a component in terms of the other components, which would hold in all local minima, and thus in all minimum points as well. Such an expression could then be used to eliminate a variable, and substitute the minimization problem with another one having one variable less. Such eliminations schemes are well-known from linear algebra.

Let us first recall briefly how the basic algorithm works. For the sake of notational simplicity, we shall assume that variables are eliminated in the order x_n, $x_{n-1}, \ldots, x_1$.

	BASICALGORITHM (f)
Input:	Let n denote the number of variables. If $n=1$ and $f(1) > f(0)$ then RETURN $x_1^* = 0$, otherwise RETURN $x_1^* = 1$. If $n > 1$ then continue.
Local optimality:	Label the variables, and choose x_n to be eliminated. Determine the pseudo-Boolean function g_n defined by $$g_n(x_1, \ldots, x_{n-1}) = \begin{cases} 1 & \text{if } \Delta_n(x_1, \ldots, x_{n-1}) < 0, \quad \text{and} \\ 0 & \text{otherwise.} \end{cases}$$
Recursion:	Determine $f^{n-1}(x_1, \ldots, x_{n-1}) \stackrel{\text{def}}{=} f(x_1, \ldots, x_{n-1}, g_n(x_1, \ldots, x_{n-1}))$, and obtain the optimal values for $x_1^*, x_2^*, \ldots, x_{n-1}^*$ by calling BASICALGORITHM(f^{n-1}).
Output:	Set $x_n^* = g_n(x_1^*, \ldots, x_{n-1}^*)$, and RETURN the binary vector $\mathbf{x}^* = (x_1^*, \ldots, x_n^*)$.

Of course, the above algorithm needs the specification of what does it mean that a pseudo-Boolean function is "determined", e.g. in the cases of g_n and f^{n-1} appearing in the two main steps of the above procedure. Let us say that we want all pseudo-Boolean functions in the course of this procedure to be represented either by their unique multi-linear polynomial form, or by a posiform (though it might be possible to use some sort of oracle as a description for these functions).

Instead of giving a detailed proof here, for which the reader is referred to [91], we shall illustrate this algorithm on a small example.

Example 4.8. Let us consider the following simple pseudo-Boolean function f in three binary variables, defined by its unique multi-linear polynomial below:

$$f(x_1, x_2, x_3) = 3 + 3x_1 - 2x_1x_2 - 4x_1x_3 + 5x_2x_3.$$

After calling BASICALGORITHM (f), we get $\Delta_3 = -4x_1 + 5x_2$, and thus

$$g_3(x_1, x_2) = \left\{ \begin{array}{ll} 1 & \text{if } x_1 = 1 \quad \text{and} \quad x_2 = 0, \\ 0 & \text{otherwise} \end{array} \right\} = x_1 \bar{x}_2.$$

Thus, we get

$$f^2(x_1, x_2) = f(x_1, x_2, x_1 \bar{x}_2) = 3 + 3x_1 - 2x_1x_2 + x_1\bar{x}_2(5x_2 - 4x_1)$$

$$= 3 - x_1 + 2x_1x_2$$

After calling BASICALGORITHM (f^2), we get $\Delta_2 = 2x_1$, and therefore

$$g_2(x_1) = \left\{ \begin{array}{ll} 1 & \text{never,} \\ 0 & \text{always,} \end{array} \right\} = 0.$$

Hence,

$$f^1(x_1) = f^2(x_1, 0) = 3 - x_1.$$

After the call BBASICALGORITHM (f^1), we have $n=1$, and since $f^1(0) > f^1(1)$, $x_1^* = 1$ is returned.

Then $x_2^* = g_2(x_1^*) = g_2(1) = 0$ is computed and $(x_1^*, x_2^*) = (1, 0)$ is returned.

Finally, $x_3^* = g_3(1, 0) = 1$ is computed, and $\mathbf{x}^* = (1, 0, 1)$ is returned as a minimizer of f.

Computationally, the above procedure can be very expensive, since both determining g_k and computing f^{k-1} can be intractable for inputs of realistic size (in a worst case, the size of both of these functions can be exponential in the input size).

There are however some special cases, when both the size of these functions, and the computing time for these steps can be controlled, and therefore the algorithm can be executed in polynomial time. Such a case was presented and analyzed in [47].

If a pseudo-Boolean function $f : \mathbb{B}^n \mapsto \mathbb{R}$ is given by its unique multi-linear polynomial (1), let us associate to it a graph $G_f = (V, E)$, called its *co-occurrence* graph, in which $(i, j) \in E$ (for $i, j \in V$, $i \neq j$) iff f has a term for which $S \supseteq \{i, j\}$ and $c_S \neq 0$. We shall say that G_f is a *partial k-tree*, if there exists a supergraph $G^* = (V, E^*)$,

$E^* \supseteq E$ and a permutation $\pi \in \mathbb{S}_n$ of the indices in $\mathbf{V}$ such that the set $P^i \stackrel{\text{def}}{=} \{\pi_j | \pi_j < \pi_i,$ $(j,i) \in E^*\} \subseteq \mathbf{V}$ is a clique in G^* and $|P^i| \leqslant k$, for every $i \in \mathbf{V}$.

Theorem 2 (Crama et al. [47]). *If for a pseudo-Boolean function f its co-occurrence graph G_f is a partial k-tree, then* BASICALGORITHM *can be implemented to run in polynomial time in the input size* size(f) *and in* 2^k.

Proof. We shall present only a sketch of the proof here. To simplify notation, let us assume w.l.o.g. that the permutation π in the definition of a partial k-tree is $\pi = (1, 2, 3, \ldots, n)$.

Let us observe first that if $G_f = G_{f^n}$ is a partial k-tree, then Δ_n and thus g_n depend only on at most k of the variables, and hence their multi-linear polynomials have at most 2^k nonzero coefficients, which can be computed recursively by (12), in time polynomial in 2^k. The same applies to the computational needs of f^{n-1}.

Let us observe next, that if the substitution of g_n into $f = f^n$ (when computing f^{n-1}) generates new terms, and changes thus the co-occurrence graph, then for a new edge $(i,j) \in E(G_{f^{n-1}}) \setminus E(G_{f^n})$ we must have $\{i,j\} \subseteq P^n$, and hence the subgraph G^*_{n-1} of G^*, induced by $\mathbf{V} \setminus \{n\}$ is still a supergraph of $G_{f^{n-1}}$, i.e. the input to the next level is also a partial k-tree.

Since there are at most n recursive calls in the algorithm, the claimed complexity follows. $\square$

4.7. Posiform maximization

Several of the results and methods we cited so far concern the minimization of posiforms, or simply the minimization (or maximization) of multi-linear polynomial expressions. While the minimization and the maximization of a polynomial expression are quite obviously equivalent problems (the simple change of sign providing the equivalence), there is no such simple connection between the minimization and maximization of posiforms. There is, of course, a linear transformation between these two problems (via the recursive substitutions $u = 1 - \bar{u}$ for certain literals u), however, this transformation typically changes the structure of the expressions, which is particularly damaging if one tries to translate approximation algorithms.

Let us further remark that posiforms are usually not only more concise representations of pseudo-Boolean functions than multi-linear polynomial expressions, but are also specially suggestive, since both the minimization and the maximization of posiforms have natural and intuitive interpretations.

Let us recall that for a given a graph $G = (V, E)$ its *stability number* $\alpha(G)$ is defined as the maximum size of a maximal independent vertex set (stable set). Furthermore, if there are weights $w : V \mapsto \mathbb{R}_+$ associated to the vertices, then the *weighted stability number* $\alpha_w(G)$ of G is defined as the maximum weight of a maximal independent vertex set, where the weight of a vertex set is the sum of the weights of the vertices in the set.

Given a posiform ϕ as in (5), let us associate to it a (weighted) graph $G_\phi = (\mathcal{T}, E)$, called its *conflict* graph. Vertices of G_ϕ correspond to the non-trivial terms of ϕ, i.e.

$\mathscr{T} = \{T \subseteq \mathbf{L} | T \neq \emptyset, \ a_T > 0\}$. To a vertex $T \in \mathscr{T}$ we shall associate a_T as its weight. We shall say that two such terms $T, T' \in \mathscr{T}$ are in *conflict*, if there is a literal $u \in T$ for which $\bar{u} \in T'$. The edges of G_ϕ correspond to the conflicting pairs of terms, i.e. $E = \{(T, T') | T, T' \in \mathscr{T}, \ T \ \text{and} \ T' \ \text{are in conflict}\}$.

Conflict graphs were introduced in [75] and the following interesting connection was shown:

Theorem 3. *For any posiform ϕ,*

$$\max_{\mathbf{x} \in \mathbb{B}^n} \phi(\mathbf{x}) = a_\emptyset + \alpha_a(G_\phi). \tag{18}$$

Proof. Given a binary vector $\mathbf{x} \in \mathbb{B}^n$, let us observe first that the set $\mathscr{S}_\mathbf{x} \overset{\text{def}}{=} \{T \in \mathscr{T} | T(\mathbf{x}) = 1\}$ consisting of the terms which do not vanish at $\mathbf{x}$ is in fact a stable set of the graph G_ϕ. This is clear, because no two of these terms can have a conflicting literal, since otherwise at least one of them would vanish in the point $\mathbf{x} \in \mathbb{B}^n$. Hence,

$$\phi(\mathbf{x}) = a_\emptyset + \sum_{T \in \mathscr{S}_\mathbf{x}} a_T \leqslant a_\emptyset + \alpha_a(G_\phi)$$

follows for all $\mathbf{x} \in \mathbb{B}^n$, by the definition of the weighted stability number.

Conversely, if $\mathscr{S} \subseteq \mathscr{T}$ is a stable set of G_ϕ, then the terms in $\mathscr{S}$ have no conflicting literals, and thus all literals appearing in these terms can be made simultaneously equal to 1. In other words, for any stable set $\mathscr{S} \subseteq \mathscr{T}$ there exists a binary vector $\mathbf{x}_\mathscr{S} \in \mathbb{B}^n$ such that $T(\mathbf{x}_\mathscr{S}) = 1$ for all $T \in \mathscr{S}$. Applying this observation to a maximum weight stable set $\mathscr{S}^*$, we get

$$a_\emptyset + \alpha_a(G_\phi) = a_\emptyset + \sum_{T \in \mathscr{S}^*} a_T = a_\emptyset + \sum_{T \in \mathscr{S}^*} a_T T(\mathbf{x}_{\mathscr{S}^*}) \leqslant \phi(\mathbf{x}_{\mathscr{S}^*}) \leqslant \max_{\mathbf{x} \in \mathbb{B}^n} \phi(\mathbf{x}). \quad \square$$

Furthermore, it can be shown that, conversely, every weighted graph stability problem corresponds in this way to a posiform maximization.

Theorem 4 (Hammer [75]). *Given a graph $G = (V, E)$ and nonnegative weights $a_i \geqslant 0$ associated to the vertices $i \in V$, there exists a posiform ϕ_G in $n' < |V|$ variables, consisting of $|V|$ terms, and such that*

$$\alpha_a(G) = \max_{\mathbf{x} \in \mathbb{B}^{n'}} \phi_G(\mathbf{x}). \tag{19}$$

Proof. Let us consider a covering of the edges of G by (not necessarily induced) complete bipartite subgraphs of G, i.e. let $\mathscr{B} \overset{\text{def}}{=} \{(A_j, B_j) | j = 1, \ldots, n'\}$ be a family of pairs of subsets of the vertices, such that (i) $A_j \cap B_j = \emptyset$, (ii) $A_j \times B_j \subseteq E$ hold for all $j = 1, \ldots, n'$, and (iii) $E = \bigcup_{j=1}^{n'} A_j \times B_j$. Let us associate to such a covering $\mathscr{B}$ of the edges a posiform

$$\phi_\mathscr{B}(\mathbf{x}) \overset{\text{def}}{=} \sum_{i \in V} a_i \prod_{j : i \in A_j} x_j \prod_{j : i \in B_j} \bar{x}_j. \tag{20}$$

Let us remark that we use the convention of $\prod_{j \in \emptyset} x_1 = 1$, implying that the constant term of $\phi_\mathscr{B}$ is the sum of the weights of the isolated vertices of G. It is easy to verify that for any covering $\mathcal{B}$ of the edges by complete bipartite subgraphs, G is the conflict graph of the posiform $\phi_{\mathcal{B}}$.

Since all edges can be covered by at most $|V| - 1$ stars in any graph, the statement follows from Theorem 3. $\square$

The two theorems above show that there is a very natural and strong equivalence between weighted graph stability and posiform maximization, even extending to approximative solutions and approximation algorithms. This connection shows also that posiform maximization, just like graph stability, is in all likelihood a difficult problem even if one needs only a good approximation (see e.g. [57,136].)

On the other hand, this connection raises the problem of characterizing graph stability problems for which the corresponding posiform maximization problem has special characteristics, e.g., it involves a posiform of bounded degree. For instance the so called *quadratic graphs*, i.e. those for which there exists a family of complete bipartite subgraphs, covering all edges, such that the corresponding posiform (20) is quadratic, were studied in [22,23,43,93]. Clearly, this condition means simply that the edge set of these graphs can be covered with a family $\mathcal{B}$ of complete bipartite subgraphs in such a way that each vertex belongs to at most 2 of the complete bipartite subgraphs in $\mathcal{B}$. In spite of their close formal resemblance with line graphs, no good characterization or recognition algorithm for quadratic graphs is known.

In a similar way in the reverse direction, this connection raises the problem of characterizing those posiforms for which the corresponding conflict graph has some special characteristics, e.g., bipartite, chordless, perfect, etc. For instance, a family of posiforms, having bipartite conflict graphs, was used in [24] to characterize supermodular cubic posiforms.

Let us remark finally that the above connection between graph stability and posiform maximization raises the possibility of applying algebraic manipulations to graph stability problems. Let us demonstrate this idea on a small example.

Example 4.9. Let us consider the graph $G_1 = (V, E)$ given in Fig. 1 (a) and let us cover its edges with the family $\mathcal{B}_1 = \{(A_1, B_1), (A_2, B_2), (A_3, B_3), (A_4, B_4)\}$ of complete bipartite subgraphs consisting of 4 stars; here $A_i = \{i\}$ for $i = 1, \ldots, 4$, while $B_1 = \{2, 3, 7\}$, $B_2 = \{4, 5\}$, $B_3 = \{5, 6, 7\}$, and $B_4 = \{5, 6\}$. The corresponding posiform is

$$\phi_1 = x_1 + \bar{x}_1 x_2 + \bar{x}_1 x_3 + \bar{x}_2 x_4 + \bar{x}_2 \bar{x}_3 \bar{x}_4 + \bar{x}_3 \bar{x}_4 + \bar{x}_1 \bar{x}_3.$$

Since in this example all complete bipartite subgraphs will actually be stars, we indicated these stars in the figures with small arcs along the edges pointing out form the center of the stars.

Applying the identities $\bar{x}_1 x_3 + \bar{x}_1 \bar{x}_3 = \bar{x}_1$, and $x_1 + \bar{x}_1 = 1$ we find that $\phi_1 = 1 + \phi_2$, where

$$\phi_2 = \bar{x}_2 x_4 + \bar{x}_1 x_2 + \bar{x}_2 \bar{x}_3 \bar{x}_4 + \bar{x}_3 \bar{x}_4.$$

The conflict graph $G_2 = G_{\phi_2}$ is shown in Fig. 2, together with another conflict representation of it, corresponding to the following complete bipartite subgraph covering of

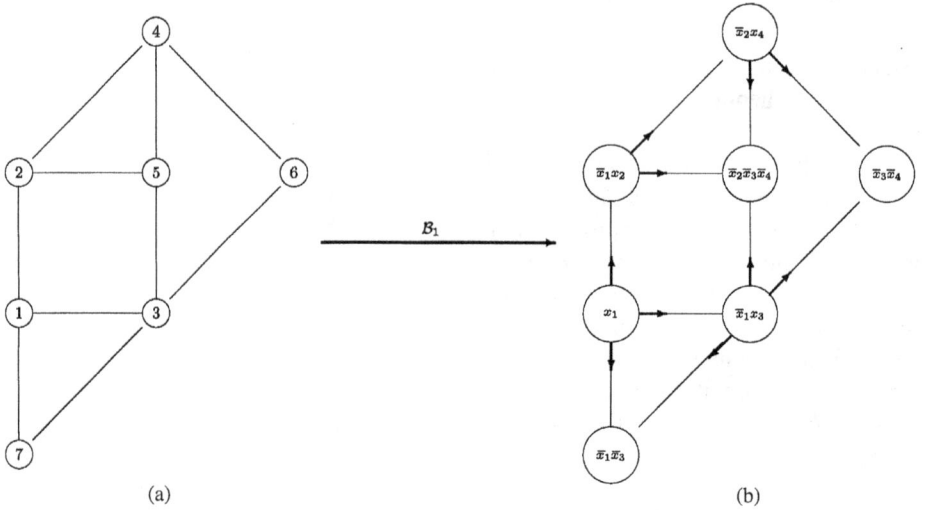

Fig. 1. Graph G_1 of Example 4.9 and its conflict representation using the complete bipartite subgraph cover $\mathscr{B}_1$.

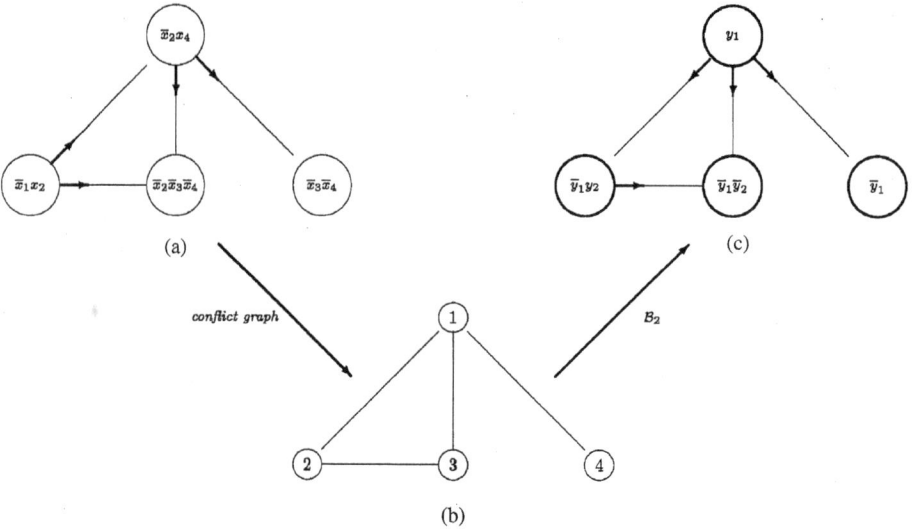

Fig. 2. The conflict graph G_2 of the posiform ϕ_2 in Example 4.9, and its conflict representation using the complete bipartite subgraph cover $\mathscr{B}_2$.

its edge set: $\mathscr{B}_2 = \{(A_1, B_1), (A_2, B_2)\}$, where $A_1 = \{1\}$, $B_1 = \{2, 3, 4\}$, $A_2 = \{2\}$, and $B_2 = \{3\}$. This representation yields the posiform

$$\phi_3 = y_1 + \bar{y}_1 + \bar{y}_1 y_2 + \bar{y}_1 \bar{y}_2.$$

Applying here again the trivial identities $\bar{y}_1 y_2 + \bar{y}_1 \bar{y}_2 = \bar{y}_1$ and $y_1 + \bar{y}_1 = 1$, we obtain

$$\phi_3 = 1 + \bar{y}_1.$$

Summarizing the above, we have

$$\alpha(G_1) = \max_{\mathbf{x} \in \mathbb{B}^4} \phi_1(\mathbf{x})$$

$$= 1 + \max_{\mathbf{x} \in \mathbb{B}^4} \phi_2(\mathbf{x}) = 1 + \alpha(G_2) = 1 + \max_{\mathbf{y} \in \mathbb{B}^2} \phi_3(\mathbf{y})$$

$$= 2 + \max_{y_1 \in \mathbb{B}} \bar{y}_1 = 3.$$

Of course, not all graphs can be handled with the same efficiency. An algorithm, based on a systematic way of applying this type of algebraic manipulations, was proposed in [54], and studied further in e.g. [51,83,84,102–104,106]. Using algebraic manipulations of the type indicated above, the so called *struction algorithm* transforms a graph G into another graph G', such that

$$\alpha(G) = 1 + \alpha(G')$$

holds. Repeating this transformation, one can, in principle, arrive to a trivial graph, and hence can determine the stability number of G.

In reality, the sequence of graphs produced in this way may have an exponential growth in size [106], although several classes of graphs are known [83,84,102–104] where this difficulty does not arise, and the stability number of which can therefore be found in polynomial time by the repeated application of struction. In a recent study [4] it was observed that the application of a single step of the struction algorithm leads to a graph for which the gap between the upper and lower bounds of the stability number (obtained by using some of the usual techniques) is considerably smaller than the corresponding gap for the original graph.

The basic idea of the struction method is to select an arbitrary vertex, say v_0, of a graph G as the "center" of the transformation, impose an arbitrary order, say v_1, $v_2, \ldots, v_k$ among the vertices in its neighborhood $N(v_0) = \{v \in V(G) | (v_0, v) \in E(G)\}$, use a family of stars with centers v_i, $i = 1, \ldots, k$ to cover all the edges which have at least one endpoint in the closed neighborhood $N[v_0] = N(v_0) \cup \{v_0\}$, and cover the remaining edges of the graph by an arbitrary family of other complete bipartite subgraphs. Let ϕ denote the posiform corresponding to this edge covering with complete bipartite subgraphs. It was shown in [54] that this posiform can always be written after some simple algebraic transformations as

$$\phi = 1 + \psi,$$

where ψ is also a posiform, obtainable from ϕ in polynomial time. If G' is the conflict graph of ψ, then we have the equation

$$\alpha(G) = \max_{\mathbf{x} \in \mathbb{B}^n} \phi(\mathbf{x}) = 1 + \max_{\mathbf{x} \in \mathbb{B}^n} \psi(\mathbf{x}) = 1 + \alpha(G').$$

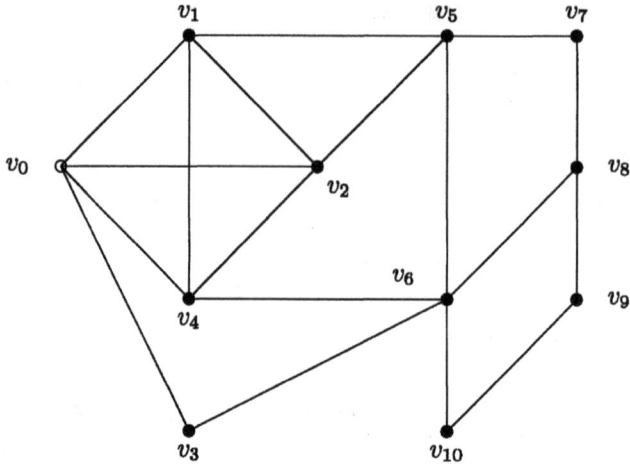

Fig. 3.

Let us illustrate this on a small example $G=(V,E)$ given in Fig. 3, in which all vertex weights are equal to 1.

For example, if we choose vertex v_0 as the center of struction, if the stars used in the transformation are $S_1 = \{\{v_1\}, \{v_0, v_2, v_4, v_5\}\}$, $S_1 = \{\{v_2\}, \{v_0, v_4, v_5\}\}$, $S_3 = \{\{v_3\}, \{v_0, v_6\}\}$, $S_4 = \{\{v_4\}, \{v_0, v_6\}\}$, if we cover the rest of the edges with an arbitrary family $\mathcal{B}$ of bipartite subgraphs, and we denote the variable associated to the star S_i, by x_i, $i = 1, \ldots, 4$, then the posiform corresponding to this edge covering will be

$$\phi(\mathbf{x}) = \bar{x}_1\bar{x}_2\bar{x}_3\bar{x}_4 + x_1 + \bar{x}_1x_2 + x_3 + \bar{x}_1\bar{x}_2x_4 + \psi(\mathbf{x}),$$

where ψ is the posiform corresponding to $\mathcal{B}$. It is easy to see that

$$\bar{x}_1\bar{x}_2\bar{x}_3\bar{x}_4 + x_1 + \bar{x}_1x_2 + x_3 + \bar{x}_1\bar{x}_2x_4 = 1 + x_1x_3 + \bar{x}_1x_2x_3 + \bar{x}_1\bar{x}_2x_3x_4$$

and hence

$$\phi(\mathbf{x}) = 1 + x_1x_3 + \bar{x}_1x_2x_3 + \bar{x}_1\bar{x}_2x_3x_4 + \psi(\mathbf{x}).$$

In fact it was shown in [54] that a similar transformation which allows to write the original posiform as the sum of a positive constant and another posiform can always be carried out. It was also shown that this procedure has a direct and easy graph theoretic interpretation, describing directly the construction of the conflict graph of the new posiform, and by-passing all the algebraic manipulations.

Returning to the example, notice that each of the three new nonconstant terms on the right-hand side above contains exactly two noncomplemented variables, and that a one-to-one correspondence can be established between them and the non-edges of the subgraph induced by $N(v_0)$. The conflict graph of the posiform $\phi(\mathbf{x}) - 1$ on the right-hand side is shown in Fig. 4 (where we have denoted by $v_{i,j}$ the vertex associated to the nonedge (v_i, v_j) in the subgraph induced by $N(v_0)$).

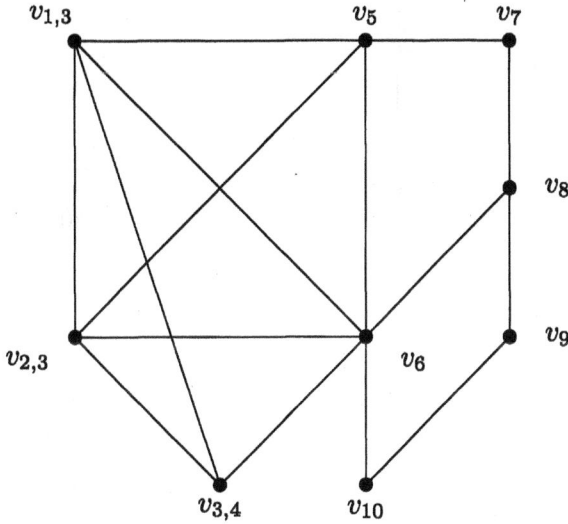

Fig. 4.

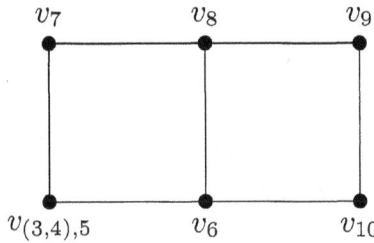

Fig. 5.

Clearly, the stability number of this graph is exactly one unit smaller than that of the original graph. Repeating now the same transformation to the new graph and taking $v_{1,3}$ as the center of the struction, we find the graph in Fig. 5, whose stability number is exactly 2 units smaller than that of the original graph. By executing two additional struction steps, using first v_{10} as center, and then v_8 as center, produces in turn the graphs in Figs. 6(a) and (b), showing that the stability number of the original graph, being 4 units higher than that of the last graph produced, was equal to 5.

4.8. l_2-approximations and applications to game theory

In this section, following [80], we shall consider lower degree l_2-approximations of pseudo-Boolean functions.

Let us denote by $\mathscr{F}_k$ the family of pseudo-Boolean functions of degree at most k $(k = 0, 1, \ldots, n)$. Clearly, $\mathscr{F}_0 \subset \mathscr{F}_1 \subset \cdots \subset \mathscr{F}_n$.

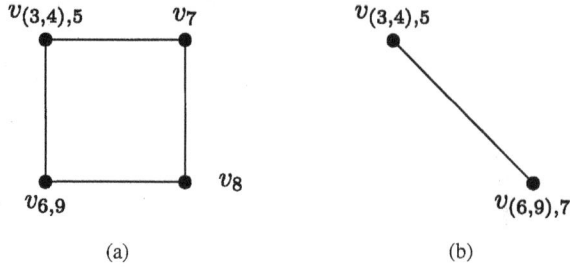

Fig. 6.

Given a pseudo-Boolean function f and a fixed integer $k \geq 0$, let us consider the following approximation problem:

$$\min_{g \in \mathscr{F}_k} \sum_{\mathbf{x} \in \mathbb{B}^n} [g(\mathbf{x}) - f(\mathbf{x})]^2. \tag{21}$$

Proposition 8. *Problem* (21) *has a unique solution.*

Proof. Let us consider pseudo-Boolean functions in n variables represented by their table form, or equivalently, as vectors in $\mathbb{R}^{2^n}$. Then,

$$W_k \overset{\text{def}}{=} \{(g(\mathbf{x})|\mathbf{x} \in \mathbb{B}^n) \,|\, g \in \mathscr{F}_k\} \tag{22}$$

is clearly a linear subspace of $\mathbb{R}^{2^n}$, for $k = 0, 1, \ldots, n$. It is immediate to see then that Problem (21) is equivalent to computing the orthogonal projection of $(f(\mathbf{x})|\mathbf{x} \in \mathbb{B}^n)$ onto W_k, and hence the solution exists and is unique. $\square$

Let us denote in the sequel by $A_k[f] \in \mathscr{F}_k$ the unique solution of Problem (21), which provides the best l_2-approximation of f.

Given a subset $S \subseteq \mathbb{B}^n$, let us denote the average value of the function f over the subset S by $\text{Ave}_S[f]$, i.e.

$$\text{Ave}_S[f] = \frac{1}{|S|} \sum_{\mathbf{x} \in S} f(\mathbf{x}). \tag{23}$$

Given a term $t(\mathbf{x}) = \prod_{u \in T} u$ (where $T \subseteq \mathbf{L}$), let us associate to it the Boolean subcube $\mathbb{B}_T$ defined by

$$\mathbb{B}_T = \{\mathbf{x} \in \mathbb{B}^n | t(\mathbf{x}) = 1\}. \tag{24}$$

Furthermore, let us associate to T the half-integral vector $\mathbf{z}^T \in \{0, \frac{1}{2}, 1\}^n$, defined by

$$(\mathbf{z}^T)_j = \begin{cases} 1 & \text{if } x_j \in T, \\ 0 & \text{if } \bar{x}_j \in T, \text{ and} \\ \frac{1}{2} & \text{otherwise.} \end{cases}$$

Before showing a characterizing property of l_2-approximations, we need to state two simple lemmas.

Lemma 1. *For a pseudo-Boolean function f, and a subset $T \subseteq \mathbf{L}$ we have*

$$\mathrm{Ave}_{\mathbb{B}_T}[f] = f(\mathbf{z}^T).$$

Proof. Follows from Proposition 5. $\square$

Lemma 2. *Given a sequence of m reals, $a_1, a_2, \ldots, a_m$, their average $\gamma = \mathrm{Ave}[a_1, a_2, \ldots, a_m]$ is the unique minimizing point of the expression*

$$\sum_{i=1}^{m} (\gamma - a_i)^2.$$

Proof. This well-known fact follows immediately from the equation

$$0 = \frac{\mathrm{d}}{\mathrm{d}\gamma} \sum_{i=1}^{m} (\gamma - a_i)^2 = 2 \sum_{i=1}^{m} (\gamma - a_i),$$

since $\sum_{i=1}^{m} (\gamma - a_i)^2$ is a strictly convex function of γ. $\square$

We are now ready to state a characterizing property of l_2-approximations.

Theorem 5. *Given a pseudo-Boolean function f in n variables, a function $g \in \mathcal{F}_k$ is equal to $A_k[f]$ if and only if*

$$\mathrm{Ave}_{\mathbb{B}_T}[f] = \mathrm{Ave}_{\mathbb{B}_T}[g] \tag{25}$$

for all subsets of literals $T \subseteq \mathbf{L}$ with $|T| \leqslant k$.

Proof. Let us assume first that $g = A_k[f]$. Choosing an arbitrary subset $T \subseteq \mathbf{L}$ for which $|T| \leqslant k$, let $t(\mathbf{x}) = \prod_{u \in T} u$, and let us consider the optimization problem

$$\min_{\gamma \in \mathbb{R}} \sum_{\mathbf{x} \in \mathbb{B}^n} [g(\mathbf{x}) + \gamma t(\mathbf{x}) - f(\mathbf{x})]^2.$$

On the one hand, we can see immediately that this problem has $\gamma = 0$ as its unique minimum, since $g' = g + \gamma t$ is also a function belonging to $\mathcal{F}_k$ (because of $|T| \leqslant k$), and $g = A_k[f]$ is a unique minimizer of $\sum_{\mathbf{x} \in \mathbb{B}^n} [g'(\mathbf{x}) - f(\mathbf{x})]^2$ in $\mathcal{F}_k$.

On the other hand, we can observe that only those terms can influence the above minimization problem in which $t(\mathbf{x}) \neq 0$, i.e. that the problem can be written equivalently as

$$\min_{\gamma \in \mathbb{R}} \sum_{\mathbf{x} \in \mathbb{B}_T} [g(\mathbf{x}) + \gamma - f(\mathbf{x})]^2 = \min_{\gamma \in \mathbb{R}} \sum_{\mathbf{x} \in \mathbb{B}_T} [\gamma - (f(\mathbf{x}) - g(\mathbf{x}))]^2.$$

Applying Lemma 2 to the sequence $\{(f(\mathbf{x}) - g(\mathbf{x}))| \mathbf{x} \in \mathbb{B}_T\}$, we obtain

$$0 = \mathrm{Ave}_{\mathbb{B}_T}[f - g] = \mathrm{Ave}_{\mathbb{B}_T}[f] - \mathrm{Ave}_{\mathbb{B}_T}[g], \tag{26}$$

thus proving (25).

For the converse direction, let us assume that (25) holds for all subsets $T \subseteq \mathbf{L}$ with $|T| \leq k$. We shall show that these equations determine $g \in \mathscr{F}_k$ uniquely, implying then the statement, since we have just shown above that the same equations hold for $A_k[f] \in \mathscr{F}_k$, too.

For this end, let us rewrite Eqs. (25) with the help of Lemma 1 as

$$g(\mathbf{z}^T) = f(\mathbf{z}^T) \quad \text{for all } T \subseteq \mathbf{L}, \ |T| \leq k \tag{27}$$

and let us assume that both $g \in \mathscr{F}_k$ and $g' \in \mathscr{F}_k$ satisfy all these equalities. Then,

$$g(\mathbf{z}^T) - g'(\mathbf{z}^T) = 0$$

must also hold for all subsets $T \subseteq \mathbf{L}$, $|T| \leq k$. Let us represent $g - g' \in \mathscr{F}_k$ by its unique multilinear polynomial form

$$g(\mathbf{x}) - g'(\mathbf{x}) = \sum_{S \subseteq \mathbf{V}, |S| \leq k} \hat{c}_S \prod_{j \in S} x_j,$$

assume that $g \neq g'$, and denote by S an arbitrary subset for which $\hat{c}_S \neq 0$. Let us further denote by $T_{S'}$ the subset of the literals defined as $T_{S'} = \{x_j | j \in S \setminus S'\} \cup \{\bar{x}_j | j \in S'\}$. Then we have

$$0 = \sum_{S' \subseteq S} (-1)^{|S'|}(g(\mathbf{z}^{T_{S'}}) - g'(\mathbf{z}^{T_{S'}})) = \hat{c}_S,$$

which can be verified by elementary computations. Thus we get $\hat{c}_S = 0$, contradicting the choice of S, and hence our assumption that $g \neq g'$, thus proving the uniqueness of g. $\quad\square$

Let us observe that in fact we have shown somewhat more than the claim of Theorem 5. Namely, if Eqs. (27) hold for two pseudo-Boolean functions, f and g, then g and $A_k[f]$ have the same coefficients in their multilinear polynomials for all terms of degree at most k.

Let us also observe that Eqs. (27), which form a system of linear equations in the coefficients of the multilinear polynomial representing $g = A_k[f]$, provide a simple and direct way of computing the best l_2-approximation $A_k[f]$ of a given pseudo-Boolean function f.

Proposition 9. *Given a pseudo-Boolean function f, the multilinear polynomial expression for $A_k[f]$ can be computed in time polynomial in $N_k^n = \sum_{i=0}^{k} \binom{n}{i}$, by evaluating f in N_k^n points.*

Proof. It can easily be seen that writing the Eqs. (27) e.g. for all sets $T \subseteq \mathbf{L}^+$, which do not involve complemented literals and for which $|T| \leq k$, we obtain a full rank system of linear equations in the N_k^n coefficients of the multilinear polynomial expression of $A_k[f]$. $\quad\square$

Since A_k is an additive operator, perhaps the simplest way to compute the best l_2 approximations is to compute separately the sums of the l_2 approximations of simple terms, e.g.

$$A_k \left[\sum_{S \subseteq \mathbf{V}} c_S \prod_{j \in S} x_j \right] = \sum_{S \subseteq \mathbf{V}} c_S A_k \left[\prod_{j \in S} x_j \right].$$

For instance, the best l_2 approximations were computed in this way in [80] for some low values of k.

Proposition 10 (see Hammer and Holzman [80]). *For $S \subseteq \mathbf{V}$ we have*

$$A_1 \left[\prod_{j \in S} x_j \right] = -\frac{|S| - 1}{2^{|S|}} + \frac{1}{2^{|S|-1}} \sum_{j \in S} x_j,$$

and

$$A_2 \left[\prod_{j \in S} x_j \right] = \frac{(|S| - 1)(|S| - 2)}{2^{|S|+1}} - \frac{|S| - 2}{2^{|S|-1}} \sum_{j \in S} x_j + \frac{1}{2^{|S|-2}} \sum_{\substack{i,j \in S \\ i < j}} x_i x_j. \tag{28}$$

Linear l_2-approximations play an important role in game theory.

Let us consider $\mathbf{V}$ as the set of *players* of a multiperson game, where the subsets $S \subseteq \mathbf{V}$ are called *coalitions*. The game (more precisely, a game with side payments in characteristic function form) is given by a function $v : 2^{\mathbf{V}} \mapsto \mathbb{R}$, for which it is assumed that $v(\emptyset) = 0$. We can view the value $v(S)$ as the "worth" of coalition S. Identifying $2^{\mathbf{V}}$ with $\mathbb{B}^n$ in the usual natural way, we can view all such games as pseudo-Boolean functions v for which $v(0, 0, \ldots, 0) = 0$.

A game is called *simple* if $v(\mathbf{x}) \in \mathbb{B}$ for all $\mathbf{x} \in \mathbb{B}^n$, and v is monotone, i.e. if $v(S) \leq v(S')$ whenever $S \subseteq S'$. In this case, the coalitions $S \subseteq \mathbf{V}$ for which $v(1^S) = 1$ are called *winning*, while the rest are *loosing*. Given a simple game v, one would like to be able to compute real values p_i for $i \in \mathbf{V}$, representing the "influence" of the individual players on the game. Such values are called *power indices*. One of the power indices considered in the literature, and in many applications, computes p_i as the proportion of those loosing coalitions which turn winning if player i joins them. More precisely, the so called *Banzhaf index* (see [15]) is defined as

$$p_i = \text{Ave} \left[\frac{dv}{dx_i} \right]$$

for $i \in \mathbf{V}$. It can be seen easily from our analysis that these indices are in fact the linear coefficients of $A_1[v]$.

Another, similar index was introduced by Shapley (see [154]), and that also can be shown to correspond to the linear coefficients in a best "linear"-approximation of v in a slightly differently weighted norm (see [80]).

5. Quadratic pseudo-Boolean functions

This part is devoted to the study of quadratic pseudo-Boolean functions, a class of functions arising naturally in the formulation of numerous combinatorial optimization problems. The importance of these functions is further explained by the fact—discussed in Section 4.4—that all pseudo-Boolean optimization problems can be reduced to the quadratic case. In this section we shall survey results regarding bounds, structural properties, polyhedral representations, as well as heuristic and exact optimization algorithms for this class of pseudo-Boolean functions.

Let us remark that in the case of quadratic pseudo-Boolean functions, optimization of polynomial representations is computationally equivalent with the optimization of quadratic posiform representations, since transformations between these two forms are polynomial.

For the sake of uniformity, we shall present all results for the case of minimization problems, but obviously, all results mentioned in the sequel can naturally be re-formulated for the case of maximization problems.

Let us start by recalling some results regarding a useful upper bound of quadratic pseudo-Boolean functions, called the *roof dual*, introduced in [79], which was shown to be obtainable through several alternative approaches. Next, we show that these approaches, as well as the bound given by them, can be generalized to a hierarchy of increasingly tighter bounds. Following [135], we shall recall results about polyhedral formulations of these optimization problems, and finally survey some algorithms for quadratic pseudo-Boolean optimization.

Let us denote by $\mathscr{F}_2$ the family of quadratic pseudo-Boolean functions, and specializing the notations introduced earlier, let us assume that they are represented either by their (unique) quadratic polynomial expression

$$f(x_1,\ldots,x_n) = c_0 + \sum_{j=1}^{n} c_j x_j + \sum_{1 \leqslant i < j \leqslant n} c_{ij} x_i x_j \tag{29}$$

or by a quadratic posiform

$$f(x_1,\ldots,x_n) = a_0 + \sum_{u \in \mathbf{L}} a_u u + \sum_{u,v \in \mathbf{L}, u \neq v} a_{uv} uv, \tag{30}$$

where, as before, $\mathbf{L}$ denotes the set of literals, $a_u \geqslant 0$ and $a_{uv} \geqslant 0$ for all $u, v \in \mathbf{L}$.

Let us remark that among the posiforms representing a quadratic pseudo-Boolean function there may also be some having degrees higher than 2.

Example 5.1. Let us consider the following quadratic pseudo-Boolean function:

$$f(x_1, x_2, x_3) = 1 - x_1 - x_2 - x_3 + x_1 x_2 + x_1 x_3 + x_2 x_3$$

along with two different posiform representations of it:

$$f(x_1, x_2, x_3) = -1 + \bar{x}_1 + x_1 x_2 + x_1 x_3 + \bar{x}_2 \bar{x}_3$$

$$= x_1 x_2 x_3 + \bar{x}_1 \bar{x}_2 \bar{x}_3.$$

The equivalence of these expressions can be shown easily by expanding the complementations in the posiforms.

Let us denote by $\mathscr{P}_k$ the family of posiforms of degree at most k, for $k \geqslant 1$, let $\mathscr{P} = \bigcup_{k \geqslant 0} \mathscr{P}_k$, and for $k \geqslant \deg(f)$ let $\mathscr{P}_k(f)$ denote the family of those posiforms of degree at most k, which represent the pseudo-Boolean function f. Let us also denote by $C(\phi)$ the constant term of a posiform $\phi \in \mathscr{P}$, and let us note that $\phi - C(\phi) \geqslant 0$ holds for all posiforms $\phi \in \mathscr{P}$.

5.1. Roof duality

Three different approaches were shown in [79] to yield the same upper bound for a quadratic pseudo-Boolean function $f \in \mathscr{F}_2$. This common upper bound was called the *roof dual* of f. Here we recall these results reformulated for the case of lower bounds, for which the term floor dual would perhaps be more suitable. However, we keep the name *roof dual* in order to emphasize that all the results are perfectly analogous for the case of upper and lower bounds.

Roof duality has been studied in many papers since its introduction in [79], and its strong relation to several other basic techniques was demonstrated in several publications, together with numerous generalizations and algorithmic improvements (see e.g., [1,2,25,28,30,31,36,38,81,94,125,156]).

We recall first the three basic techniques, leading to the same bound, as shown in [179].

5.1.1. Majorization

A term-by-term majorization procedure, introduced in [79], provides a useful upper bound to a pseudo-Boolean function f, by appropriately selecting the linear majorants of the terms of its polynomial representation. Analogously, we can find lower bounds based on linear minorants of the terms.

The method of developing a "good" linear majorant (or minorant) of a pseudo-Boolean function appeared in the literature much earlier (see e.g. [77,89,150]) and was also applied to graph stability [78]. Here we recall the technique from [79], translated to the case of minimization.

A linear function

$$l(\mathbf{x}) = l_0 + l_1 x_1 + \cdots + l_n x_n$$

is called a *linear minorant* of $f \in \mathscr{F}_2$, if $l(\mathbf{x}) \leqslant f(\mathbf{x})$ holds for all $\mathbf{x} \in \mathbb{B}^n$. Let us denote by $\mathscr{F}_1$ the family of linear functions.

A family $\mathscr{S}$ of linear minorants of f is called *complete* if f is the pointwise maximum of these linear functions, i.e. if

$$f(\mathbf{x}) = \max_{l \in \mathscr{S}} l(\mathbf{x})$$

for all $\mathbf{x} \in \mathbb{B}^n$. For a complete family of minorants, we have the equality

$$\min_{\mathbf{x} \in \mathbb{B}^n} f(\mathbf{x}) = \min_{\mathbf{x} \in \mathbb{B}^n} \max_{l \in \mathscr{S}} l(\mathbf{x}).$$

Interchanging the minimization and maximization on the right-hand side above, we obtain a lower bound to the minimum of f:

$$M(f, \mathscr{S}) \stackrel{\text{def}}{=} \max_{l \in \mathscr{S}} \min_{\mathbf{x} \in \mathbb{B}^n} l(\mathbf{x}) \leqslant \min_{\mathbf{x} \in \mathbb{B}^n} f(\mathbf{x}). \tag{31}$$

For an efficiently computable lower bound of this type, we shall consider a complete family of linear minorants, formed by combining "best l_1-norm" linear minorants of the quadratic terms of $f \in \mathscr{F}_2$.

It can easily be seen that all linear minorants $\alpha + \beta x + \gamma y$ of the quadratic term xy, which minimize the sum of the gaps for all four possible binary substitutions, i.e. which are solutions of the problem (in variables α, β and γ)

$$\text{minimize} \quad \sum_{(x,y) \in \mathbb{B}^2} [xy - \alpha - \beta x - \gamma y]$$

$$\text{subject to} \quad xy \geqslant \alpha + \beta x + \gamma y \quad \text{for all } (x, y) \in \mathbb{B}^2$$

are of the form $-\alpha = \beta = \gamma = \lambda$, for any $0 \leqslant \lambda \leqslant 1$. Similarly, it can be shown that the best l_1-norm linear minorants of $-xy$ are of the form $-\lambda x - (1 - \lambda)y$, for any $0 \leqslant \lambda \leqslant 1$.

Given a quadratic pseudo-Boolean function $f \in \mathscr{F}_2$ as in (29), let us define a family of linear minorants of it, by taking the weighted sum of the best l_1-norm linear minorants of its terms, and using as weights the coefficients of the terms, i.e.

$$\mathscr{R}(f) \stackrel{\text{def}}{=} \left\{ \begin{array}{l} c_0 + \displaystyle\sum_{j=1}^{n} c_j x_j \\[2ex] + \displaystyle\sum_{\substack{1 \leqslant i < j \leqslant n \\ c_{ij} > 0}} c_{ij} \lambda_{ij} [x_i + x_j - 1] \\[3ex] + \displaystyle\sum_{\substack{1 \leqslant i < j \leqslant n \\ c_{ij} < 0}} c_{ij} [\lambda_{ij} x_i + (1 - \lambda_{ij}) x_j] \end{array} \middle| \begin{array}{l} 0 \leqslant \lambda_{ij} \leqslant 1 \\[1ex] 1 \leqslant i < j \leqslant n \\[1ex] c_{ij} \neq 0 \end{array} \right\}.$$

In [79] the analogously defined linear majorants are called *roofs*. Let us call here the corresponding lower bound

$$M_2(f) \stackrel{\text{def}}{=} M(f, \mathscr{R}(f)) = \max_{l \in \mathscr{R}(f)} \min_{\mathbf{x} \in \mathbb{B}^n} l(\mathbf{x}) \tag{32}$$

the *roof dual* of f.

It was demonstrated in [79] that the same bound can be computed by solving a linear programming problem. To see this, let us observe first that if $l(\mathbf{x}) = l_0 + l_1 x_1 + \cdots + l_n x_n$, then $\min_{\mathbf{x} \in \mathbb{B}^n} l(\mathbf{x})$ is the maximum value of the linear program

$$l_0 + \sum_{j=1}^{n} z_j \to \max$$

$$l_j \geqslant z_j \quad \text{and} \quad 0 \geqslant z_j \quad \text{for } j = 1, \ldots, n.$$

Replacing then the inner minimization in the right-hand side of (32) with the above maximization, we find that $M_2(f)$ is the maximum value of the following linear program:

$$\text{maximize} \quad c_0 - \sum_{\substack{1 \leqslant i < j \leqslant n \\ c_{ij} > 0}} \lambda_{ij} c_{ij} + \sum_{j=1}^{n} z_j$$

$$\text{subject to} \quad c_j + \sum_{\substack{i \neq j \\ c_{ij} > 0}} \lambda_{ij} c_{ij} + \sum_{\substack{1 \leqslant i < j \\ c_{ij} < 0}} c_{ij}(1 - \lambda_{ij}) + \sum_{\substack{j < i \leqslant n \\ c_{ij} < 0}} c_{ij} \lambda_{ij} \geqslant z_j \qquad (33)$$

$$0 \geqslant z_j \quad \text{for } j = 1, \ldots, n,$$

$$0 \leqslant \lambda_{ij} \leqslant 1 \quad \text{for } 1 \leqslant i < j \leqslant n, \ c_{ij} \neq 0.$$

5.1.2. Linearization

Linearization is a standard, and quite natural technique to reduce nonlinear binary optimization to linear integer programming. The first publication of this nature is perhaps [60], and there were many others to follow, see e.g. [3,12,55,67,68,88,146,163].

The basic idea in all these transformations is to replace a nonlinear term by a new variable, and use linear inequalities to force the new variable to take in all feasible solutions the value of the corresponding term. For instance, to enforce the equality $u = xyz$ for binary variables $x, y, z \in \mathbb{B}$, we can write

$$u \leqslant x, \qquad (34a)$$

$$u \leqslant y, \qquad (34b)$$

$$u \leqslant z, \qquad (34c)$$

$$u \geqslant x + y + z - 2, \qquad (34d)$$

$$u \geqslant 0. \qquad (34e)$$

It is easy to see that for all binary assignments to x, y and z, the only feasible assignment to the variable u is the value xyz. Let us also add that in a minimization problem constraints (34a)–(34c) are redundant if u has a positive objective function coefficient, and constraints (34d)–(34e) are redundant when the objective function coefficient of u is negative (and the other way around for maximization problems.)

Applying this re-formulation to the minimization of $f \in \mathscr{F}_2$ given by (29), and introducing new variables $y_{ij} = x_i x_j$ for all quadratic terms of f, we obtain the following equivalent linear 0–1 programming problem:

$$\text{minimize} \quad c_0 + \sum_{j=1}^{n} c_j x_j + \sum_{1 \leqslant i < j \leqslant n} c_{ij} y_{ij}$$

$$\text{subject to} \quad \left. \begin{array}{l} y_{ij} \geqslant x_i + x_j - 1 \\ y_{ij} \geqslant 0 \end{array} \right\} \quad 1 \leqslant i < j \leqslant n, \ c_{ij} > 0,$$

$$\left. \begin{array}{l} y_{ij} \leqslant x_i \\ y_{ij} \leqslant x_j \end{array} \right\} \quad 1 \leqslant i < j \leqslant n, \ c_{ij} < 0, \qquad (35)$$

$$x_j \in \mathbb{B}, \quad 1 \leqslant j \leqslant n.$$

Replacing in the above formulation the integrality conditions on $\mathbf{x}$ by the constraints $0 \leqslant x_j \leqslant 1$ for $j = 1, \ldots, n$, we obtain a linear programming relaxation, the optimum value of which will be denoted by $L_2(f)$. Clearly, $L_2(f)$ is a lower bound of the minimum of f.

There are several variations of such linearization methods. For instance, we could also consider a quadratic posiform $\phi \in \mathcal{P}_2(f)$, and introduce new variables for the products of literals appearing in ϕ. Assuming that ϕ is represented by the form (30), we obtain the formulation

$$\text{minimize} \quad a_0 + \sum_{u \in \mathbf{L}} a_u u + \sum_{u,v \in \mathbf{L}, u \neq v} a_{uv} y_{uv}$$

$$\text{subject to} \quad \left. \begin{array}{l} y_{uv} \geqslant u + v - 1 \\[2mm] y_{uv} \geqslant 0 \end{array} \right\} \quad u, v \in \mathbf{L}, \ a_{uv} > 0 \tag{36}$$

$$x_j \in \mathbb{B}, \quad 1 \leqslant j \leqslant n.$$

(Here of course we write in the constraints $1 - x_j$ for $u = \bar{x}_j$ and x_j for $u = x_j$.) Let us replace again the integrality constraints $\mathbf{x} \in \mathbb{B}^n$ by $0 \leqslant x_j \leqslant 1$ for $j = 1, \ldots, n$, and let us denote by $L_2(\phi)$ the optimum value of this linear programming relaxation of problem (36). It is obvious again that $L_2(\phi)$ is a lower bound of the minimum of ϕ.

5.1.3. Complementation

The third approach considered in [79] is based on representing a quadratic pseudo-Boolean function $f \in \mathcal{F}_2$ by a quadratic posiform $\phi \in \mathcal{P}_2(f)$, and using the constant term $C(\phi)$ of ϕ as a lower bound of the minimum of f. Then the problem of finding the best lower bound of this type can be formulated as

$$C_2(f) \overset{\text{def}}{=} \max_{\phi \in \mathcal{P}_2(f)} C(\phi). \tag{37}$$

This problem can also be formulated as a linear program. For this, let us consider the coefficients of the posiform ϕ as unknowns, and let us formulate the conditions that this posiform (30) represents the function f given by (29). From these conditions we get the following equalities:

$$c_0 = a_0 + \sum_{j=1}^{n} a_{\bar{x}_j} + \sum_{1 \leqslant i < j \leqslant n} a_{\bar{x}_i \bar{x}_j}, \tag{38a}$$

$$c_j = a_{x_j} - a_{\bar{x}_j} + \sum_{\substack{1 \leqslant i \leqslant n \\ i \neq j}} (a_{\bar{x}_i x_j} - a_{\bar{x}_i \bar{x}_j}) \quad \text{for } j = 1, \ldots, n, \tag{38b}$$

$$c_{ij} = a_{x_j x_j} + a_{\bar{x}_i \bar{x}_j} - a_{x_i \bar{x}_j} - a_{\bar{x}_i x_j} \quad \text{for } 1 \leqslant i < j \leqslant n. \tag{38c}$$

Expressing a_0 from the first equation (38a), we can write $C_2(f)$ equivalently as the maximum value of the linear programming problem

$$\text{maximize} \quad c_0 - \sum_{j=1}^{n} a_{\bar{x}_j} - \sum_{1 \leqslant i < j \leqslant n} a_{\bar{x}_i \bar{x}_j}$$

$$\text{subject to} \quad a_{x_j} - a_{\bar{x}_j} + \sum_{\substack{1 \leqslant i \leqslant n \\ i \neq j}} (a_{\bar{x}_i x_j} - a_{\bar{x}_i \bar{x}_j}) = c_j \quad \text{for } j = 1, \ldots, n, \tag{39}$$

$$a_{x_j x_j} + a_{\bar{x}_i \bar{x}_j} - a_{x_i \bar{x}_j} - a_{\bar{x}_i x_j} = c_{ij} \quad \text{for } 1 \leqslant i < j \leqslant n,$$

$$a_u, a_{uv} \geqslant 0 \quad \text{for } u, v \in \mathbf{L}, \ u \neq v.$$

5.1.4. Equivalence of formulations and persistency

The first major result of [79] is that the three approaches described in the previous three subsections for obtaining a lower bound for the minimization of $f \in \mathscr{F}_2$, are in fact equivalent, in the sense that they all yield the same value:

Theorem 6 (Hammer et al. [79]). *For any quadratic pseudo-Boolean function $f \in \mathscr{F}_2$, we have*

$$M_2(f) = L_2(f) = C_2(f) \leqslant \min_{\mathbf{x} \in \mathbb{B}^n} f(\mathbf{x});$$

moreover, the equality of these lower bounds with the minimum of $f(\mathbf{x})$ can be tested in linear time by solving a 2-SAT problem.

The somewhat technical proof, which the reader can find in [79], is based on showing that formulations (33) and (39) are both equivalent with the dual of (35). Using a similar proof, one can also show that

Theorem 7. *For a quadratic pseudo-Boolean function $f \in \mathscr{F}_2$ and for any posiform representation $\phi \in \mathscr{P}_2(f)$ of it, the equality $C_2(f) = L_2(\phi)$ holds.* $\square$

Some other formulations are also known to be equivalent with roof duality; these include Lagrangean formulations, the so-called "paved duality", etc. (see e.g. [1,2,42, 77,94,125].)

The second major result of [79] provides a strong persistency property for quadratic pseudo-Boolean optimization:

Theorem 8 (Strong Persistency, Hammer et al. [79]). *Given a quadratic pseudo-Boolean function $f \in \mathscr{F}_2$, let $\phi \in \mathscr{P}_2(f)$ be a posiform representing it (given as in (30)), such that $C(\phi) = C_2(\phi)$. Then ϕ has the property that if $a_u > 0$ for some literal $u \in \mathbf{L}$, then $u = 0$ in all binary vectors $\mathbf{x} \in \mathrm{Argmin}(f)$ minimizing f.*

Let us denote by $L(\phi)$ the *linear part* of the posiform (or pseudo-Boolean function) ϕ, e.g. if ϕ is given by (30), then $L(\phi) = a_0 + \sum_{u \in \mathbf{L}} a_u u$. The above theorem implies then that $L(\phi)$ must vanish at any minimum of f, whenever $C(\phi) = C_2(\phi)$. Let us denote by $\mathscr{P}_2^*(f)$ the family of such posiforms, i.e. $\mathscr{P}_2^*(f)=\{\phi\in\mathscr{P}_2(f)|C(\phi)=C_2(f)\}$. The linear parts $L(\phi)$ of such extremal quadratic posiforms $\phi\in\mathscr{P}_2^*(f)$ are the roofs of f. Clearly, $\mathscr{P}_2^*(f)$ is a convex subset of the family of quadratic pseudo-Boolean functions, implying

Proposition 11 (Hammer et al. [79]). *For every $f\in\mathscr{F}_2$, there exists a posiform $\phi^*\in\mathscr{P}_2^*(f)$ such that $a_u>0$ in ϕ^* if the literal u has a nonzero coefficient in some posiform $\phi\in\mathscr{P}_2^*(f)$.*

The linear part $L(\phi^*)$ of such a posiform is called a *master roof* of f.

For a given quadratic pseudo-Boolean function $f\in\mathscr{F}_2$ the convex cone $\mathscr{P}_2^*(f)$ maybe nontrivial, and therefore f may have many roofs. Since the sets of variables whose values in the optimum can be determined by using the strong persistency theorem vary with the particular roofs chosen, it is very important to note that the set of variables which can be fixed using a master roof is the union of all those sets of variables which can be fixed by any particular roof (see e.g. [77,78] for applications of persistency in graph theory).

Even though the roof dual as well as a particular roof of a given quadratic pseudo-Boolean function can be computed in polynomial time by using one of the formulations in the previous subsections, the determination of a master roof may still not be obvious (see e.g. [25]). We shall show in the next subsection that in fact master roofs can be computed with the same effort as needed for finding the roof dual.

Example 5.2. Let us consider the quadratic pseudo-Boolean function

$$f(x_1,x_2,x_3)=6-x_1-4x_2-x_3+3x_1x_2+x_2x_3.$$

It can be shown that the roof dual value of this function is $C_2(f)=2$, and two examples for posiforms belonging to $\mathscr{P}_2^*(f)$ are

$$g_1=2+2x_1+3\bar{x}_1\bar{x}_2+\bar{x}_2\bar{x}_3 \tag{40}$$

and

$$g_2=2+2\bar{x}_2+2x_1x_2+\bar{x}_1\bar{x}_2+\bar{x}_2\bar{x}_3. \tag{41}$$

Clearly, the convex combination of g_1 and g_2

$$g=\tfrac{1}{2}g_1+\tfrac{1}{2}g_2=2+x_1+\bar{x}_2+x_1x_2+2\bar{x}_1\bar{x}_2+\bar{x}_2\bar{x}_3$$

is also a member of $\mathscr{P}_2^*(f)$. As it will be seen in the next section, in fact $2+x_1+\bar{x}_2$ is a master roof of f.

The Strong Persistency Theorem implies then that $x_1=\bar{x}_2=0$, i.e. $x_1=0$ and $x_2=1$ in all minima of f. Hence for this example we have

$$\min_{\mathbf{x}\in\mathbb{B}^3} f(\mathbf{x})=\min_{x_3\in\mathbb{B}} 2=2.$$

The above fortunate example illustrates that the variable fixation implied by persistency can sometimes simplify the minimization problem to a trivial one. Though this cannot be expected in general, the equality $C_2(f) = \min_{\mathbf{x} \in \mathbb{B}^n} f(\mathbf{x})$ can always be tested in polynomial time. To see this, let us compute first a posiform $\phi \in \mathscr{P}_2^*(f)$, then fix those variables which are implied by persistency, remove the constant term, and denote by ϕ' the quadratic posiform obtained in this way. According to the above discussion, the minimization of f is equivalent with the minimization of ϕ', thus $C_2(f) = \min_{\mathbf{x} \in \mathbb{B}^n} f(\mathbf{x})$ if and only if $\min_{\mathbf{x} \in \mathbb{B}^n} \phi'(\mathbf{x}) = 0$. Since ϕ' has only quadratic terms with positive coefficients, the latter is obviously equivalent with a 2-SAT problem, which is solvable in polynomial (in fact in linear) time.

A somewhat different but similar conclusion can also be drawn from the integrality of some of the components of an optimal solution to the linear programming formulation (35):

Theorem 9 (Weak Persistency, Hammer et al. [79] Nemhauser and Trotter [129], Picard and Queyranne [139]) *Let $f \in \mathscr{F}_2$, and let $\tilde{\mathbf{x}}$ be an optimal solution of the linear program* (35), *for which $\tilde{x}_j = 1$ for $j \in O$, and $\tilde{x}_j = 0$ for $j \in Z$ (where O and Z are disjoint subsets of the indices). Then, for any minimizing vector $\mathbf{x}^* \in \operatorname{Argmin}(f)$ switching the components to $x_j^* = 1$ for $j \in O$ and $x_j^* = 0$ for $j \in Z$ will also yield a minimum of f.* $\square$

While strong persistency's main advantage is that it allows a fixation of the values of a subset of variables, which holds in every optimum of the problem, the advantage of weak persistency is the fixation of a (usually) larger set of variables, valid in at least one of the optima of the problem.

5.1.5. Network flow model

According to the above discussion, the computation of the roof dual is a polynomial problem, since it can be carried out by solving a linear program. Because of its usefulness, several even more efficient combinatorial procedures were developed for its solution, see e.g. [30,36,38]. We shall recall here the network flow computation based method of [36], since this is perhaps the most efficient procedure to compute the roof dual, providing at the same time a master roof; moreover this method can also be used to obtain simple proofs for some of the theorems cited in the previous subsection.

Let us assume that the quadratic pseudo-Boolean function $f \in \mathscr{F}_2$ is given as a posiform $\phi \in \mathscr{P}_2(f)$ of form (30). Let us associate to such a quadratic posiform a capacitated directed network $G_\phi = (N,A)$, where the node set is defined as $N = \mathbf{L} \cup \{x_0, \bar{x}_0\}$, where x_0 and $\bar{x}_0$ are two additional symbols representing the constants $x_0 = 1$ and $\bar{x}_0 = 0$, respectively. Let us associate two arcs to every quadratic term of ϕ, namely let us associate to term $c_{uv} uv$ the arcs $(\overrightarrow{u, \bar{v}})$ and $(\overrightarrow{v, \bar{u}})$, and let the capacity of both arcs be $\frac{1}{2} c_{uv}$. Similarly, let us homogenize the linear terms in ϕ by writing $c_u u = c_u u x_0$, and thus associating to this term the arcs $(\overrightarrow{u, \bar{x}_0})$ and $(\overrightarrow{x_0, \bar{u}})$, both with capacities $\frac{1}{2} c_u$. Let us note that the constant term of ϕ was disregarded in this construction.

Conversely, given a directed network $G = (N,A)$ with $N = \mathbf{L} \cup \{x_0, \bar{x}_0\}$, and with nonnegative capacities c_{uv} assigned to the arcs $(\overrightarrow{u, v}) \in A$, we can associate to it a quadratic posiform

$$\phi_G \overset{\text{def}}{=} \sum_{(u,v^{\rightarrow}) \in A} c_{uv} u\bar{v}.$$

Let us remark that in the above definition, all those terms which correspond to arcs entering x_0, or leaving $\bar{x}_0$ must vanish, since $u\bar{x}_0 = \bar{x}_0 v = 0$, due to the assumption $\bar{x}_0 = 0$.

It is easy to see that the above definitions imply.

Proposition 12. *There is a one-to-one correspondence between quadratic posiforms* $\phi \in \mathcal{P}_2$ *for which* $C(\phi) = 0$ *and capacitated directed networks* $G = (N, A)$ *with node set* $N = \mathbf{L} \cup \{x_0, \bar{x}_0\}$. *Furthermore, the involution*

$$G_{\phi_G} = G \quad \text{and} \quad \phi_{G_\phi} = \phi$$

holds for such corresponding pairs.

Let us recall next that a feasible flow in the capacitated network $G = (N, A)$ with source x_0 and sink $\bar{x}_0$ is a mapping $\varphi : A \mapsto \mathbb{R}_+$, satisfying the constraints

$$\varphi(u, v) \leqslant c_{uv} \quad \text{for all arcs } (u,\vec{v}) \in A,$$

and

$$\sum_{(u,v^{\rightarrow}) \in A} \varphi(u, v) = \sum_{(v,w^{\rightarrow}) \in A} \varphi(v, w) \quad \text{for all nodes } v \in \mathbf{L}.$$

Given a capacitated network $G = (N, A)$ and a feasible flow φ in it, let us define the *residual network* $G[\varphi] = (N, A^\varphi)$ by setting the residual capacities

$$c_{uv}^\varphi = \begin{cases} c_{uv} - \varphi(u, v) & \text{for } (u,\vec{v}) \in A, \\ \varphi(u, v) & \text{for } (v,\vec{u}) \in A. \end{cases} \tag{42}$$

Let us observe that, due to the special structure of this network (all arcs come in pairs $(u,\vec{v})$ and $(\bar{v},\overset{\rightarrow}{\bar{u}})$), a feasible flow can always be assumed to be *symmetric*, i.e. such that $\varphi(u, v) = \varphi(\bar{v}, \bar{u})$ holds for every pair u, v.

An *augmenting path of capacity* ε in the residual network is a sequence of nodes $v_0, \ldots, v_k$ such that $v_0 = x_0$, $v_k = \bar{x}_0$, and $c_{v_j, v_{j+1}}^\varphi \geqslant \varepsilon$ for $j = 0, 1, \ldots, k-1$. Let us note that for symmetric feasible flows φ in $G = (N, A)$ the path $\bar{v}_k, \bar{v}_{k-1}, \ldots, \bar{v}_0$ is also augmenting, and these two augmenting paths can actually share arcs! It is well known that all feasible flows can be obtained from the constant zero flow by iteratively increasing the flow along augmenting paths.

If $u_1, u_2, \ldots, u_k$ are literals from $\mathbf{L}$, let us call an expression of the form

$$u_1 + \bar{u}_1 u_2 + \bar{u}_2 u_3 + \cdots + \bar{u}_{k-1} u_k + \bar{u}_k, \tag{43}$$

an *alternating sum*. Let us say that a quadratic posiform ϕ *contains the alternating sum* (43) *with weight* ω, if we have $a_{u_1} \geqslant \omega$, $a_{\bar{u}_j u_{j+1}} \geqslant \omega$ for $j = 1, \ldots, k-1$, and $a_{\bar{u}_k} \geqslant \omega$ for all the corresponding coefficients of ϕ.

Proposition 13. *The following identity holds for alternating sums:*

$$u_1 + \bar{u}_1 u_2 + \bar{u}_2 u_3 + \cdots + \bar{u}_{k-1} u_k + \bar{u}_k = 1 + u_1 \bar{u}_2 + \cdots + u_{k-1} \bar{u}_k. \tag{44}$$

Proof. Immediate by elementary calculation. $\square$

If a quadratic posiform ϕ contains the alternating sum (43) with weight ω, then it can be changed into an equivalent posiform with a larger constant term, as follows: first we re-group the terms of ϕ and write it as

$$\phi = \omega[u_1 + \bar{u}_1 u_2 + \bar{u}_2 u_3 + \cdots + \bar{u}_{k-1} u_k + \bar{u}_k] + \phi',$$

where ϕ' is also a quadratic posiform. Then we apply the identity (44) to get

$$\phi = \omega + \omega[u_1 + \bar{u}_1 u_2 + \bar{u}_2 u_3 + \cdots + \bar{u}_{k-1} u_k + \bar{u}_k] + \phi'.$$

Observing now that there is a one-to-one correspondence between alternating sequences contained in a posiform, and augmenting paths in the corresponding network, we have

Proposition 14. *Let us consider a posiform $\phi \in \mathscr{P}_2$, and a feasible flow φ in the corresponding capacitated network $G = G_\phi$. Then, $x_0, u_1, \ldots, u_k, \bar{x}_0$ is an augmenting path of capacity $\varepsilon > 0$ in $G[\varphi]$ if and only if $u_1 + \bar{u}_1 u_2 + \cdots + \bar{u}_{k-1} u_k + \bar{u}_k$ is an alternating sum of weight ε in the corresponding posiform $\phi_{G[\varphi]}$.*

Proof. Follows easily form the above observations (see [36]). $\square$

Two further consequences of the above are (see [36]):

Proposition 15. *Let $\phi \in \mathscr{P}_2(f)$ for a quadratic pseudo-Boolean function $f \in \mathscr{F}_2$, and let φ be a feasible flow in the corresponding network G_ϕ. Let us denote by $v(\varphi)$ the value of the flow (e.g. the total flow leaving the source), and let $\psi = \phi_{G_\phi[\varphi]}$ denote the posiform corresponding to the residual network. Then we have $C(\phi) + v(\varphi) + \psi \in \mathscr{P}_2(f)$.*

Proposition 16. *If $\phi, \psi \in \mathscr{P}_2(f)$ for a quadratic pseudo-Boolean function f with $C(\phi) < C(\psi)$, then there is an augmenting path in the network G_ϕ.*

Putting all these facts together, we can see that the quadratic posiforms representing a given quadratic pseudo-Boolean function f, and having the largest constant term, are those which correspond to the residual network of a maximum flow in the network associated to any quadratic posiform of f.

Theorem 10 (Boros et al. [36]). *Given a quadratic pseudo-Boolean function f, and a posiform representation $\phi \in \mathscr{P}_2(f)$ of it, let us denote by v^* the maximum flow value in the network G_ϕ. Then*

$$C_2(f) = C(\phi) + v^*.$$

Furthermore, from an easy analysis of the residual network corresponding to a maximum flow, we get the following

Theorem 11. *Let $\phi \in \mathcal{P}_2(f)$ for a quadratic pseudo-Boolean function f, let φ^* denote a maximum flow in $G = G_\phi$, and let $S \subseteq \mathbf{L}$ denote the set of nodes of G which are reachable from x_0 via a path with positive residual capacities. Then, $u(\mathbf{x}^*) = 1$ for all $u \in S$ and for all vectors $\mathbf{x}^* \in \mathrm{Argmin}(f)$ minimizing f.*

Proof. Let us denote by ψ the posiform corresponding to the residual network $G[\varphi^*]$. Then $C(\psi) = 0$ by definition, and every term of ψ with a positive coefficient involving a literal u or $\bar{u}$ for some $u \in S$ involves also a negated literal $\bar{v}$ for some $v \in S$, since there cannot be an arc in $G[\varphi^*]$ with positive residual capacity leaving the set of nodes S. Thus, the partial assignment $\mathbf{y}$ defined by $u(\mathbf{y}) = 1$ for all $u \in S$ is a contractor for ψ, and since $C_2(f) + \psi \in \mathcal{P}_2(f)$, weak persistency holds for $\mathbf{y}$ at the minima of f by Corollary 2. Adding that all nodes in S can be reached via arcs with positive residual capacity in $G[\varphi^*]$, it follows that $u(\mathbf{x}^*) = 1$ must hold for all $\mathbf{x}^* \in \mathrm{Argmin}(f)$, i.e. that strong persistency must also hold for $\mathbf{y}$. □

Example 5.3. Let us illustrate the above discussion on the following example:

$$f(x_1, x_2, x_3, x_4, x_5) = 10 - 4x_1 - 4x_3 - 2x_4 + 4x_1x_2 - 2x_2x_3 + 4x_3x_4 - 2x_4x_5.$$

Substituting the identity $x = 1 - \bar{x}$ for the first variables in each of the quadratic terms having a negative coefficient, as well as in the negative linear terms, we find that

$$\phi = -4 + 4\bar{x}_1 + 6\bar{x}_3 + 2\bar{x}_4 + 2\bar{x}_5 + 4x_1x_2 + 2\bar{x}_2x_3 + 4x_3x_4 + 2\bar{x}_4x_5$$

is a quadratic posiform representing f, i.e. $\phi \in \mathcal{P}_2(f)$. The corresponding network G_ϕ is shown in Fig. 7.

Checking in Fig. 7, we can see that in the network G_ϕ, unit flows can be pushed sequentially through each of the following augmenting paths:

$$x_0 \to x_1 \to \bar{x}_2 \to \bar{x}_3 \to \bar{x}_0 \quad \text{and its twin} \quad x_0 \to x_3 \to x_2 \to \bar{x}_1 \to \bar{x}_0,$$

$$x_0 \to x_3 \to \bar{x}_4 \to \bar{x}_0 \quad \text{and its twin} \quad x_0 \to x_4 \to \bar{x}_3 \to \bar{x}_0,$$

$$x_0 \to x_3 \to \bar{x}_4 \to \bar{x}_5 \to \bar{x}_0 \quad \text{and its twin} \quad x_0 \to x_5 \to x_4 \to \bar{x}_3 \to \bar{x}_0.$$

These augmenting paths correspond, respectively, to the following alternating sequences:

$$\bar{x}_1 + x_1x_2 + \bar{x}_2x_3 + \bar{x}_3 \quad \text{and} \quad \bar{x}_3 + x_3\bar{x}_2 + x_2x_1 + \bar{x}_1,$$

$$\bar{x}_3 + x_3\bar{x}_4 + x_4 \quad \text{and} \quad \bar{x}_4 + x_4x_3 + \bar{x}_3,$$

$$\bar{x}_3 + x_3x_4 + \bar{x}_4x_5 + \bar{x}_5 \quad \text{and} \quad \bar{x}_5 + x_5\bar{x}_4 + x_4x_3 + \bar{x}_3.$$

Since there is no further augmenting path in the residual network shown in Fig. 8, we have arrived to a maximum flow of value $v = 6$. The final network in Fig. 8 corresponds indeed to the quadratic posiform

$$\psi = 2\bar{x}_1 + 2x_1x_2 + 2\bar{x}_1\bar{x}_2 + 2x_2\bar{x}_3 + 4\bar{x}_3\bar{x}_4 + 2x_4\bar{x}_5$$

for which we have $\phi = C(\phi) + v + \psi = 2 + \psi$. Hence $C_2(f) = 2$ and $2 + \psi \in \mathcal{P}_2^*(f)$ follow.

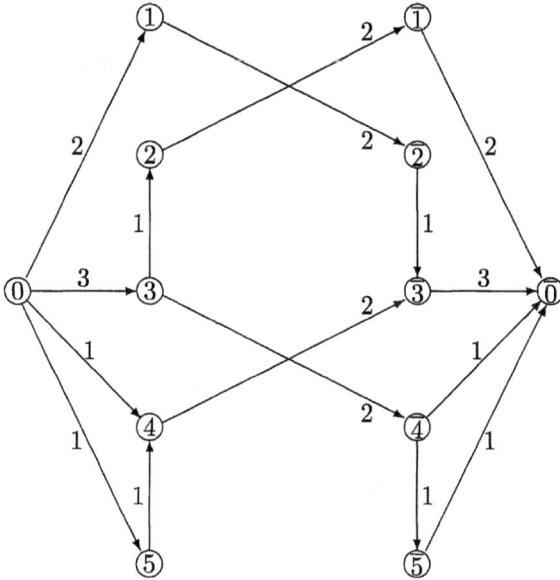

Fig. 7. The capacitated network G_ϕ corresponding to the posiform ϕ of example 5.3.

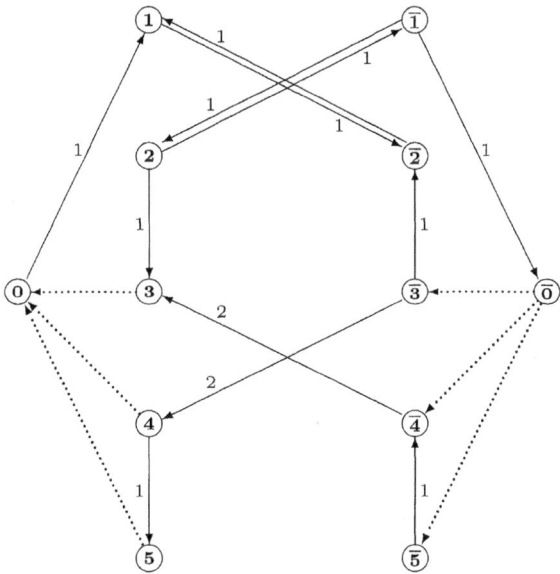

Fig. 8. Residual network of Example 5.3. Only arcs with positive residual capacity are indicated. Dotted arcs, i.e. those which enter the source or leave the sink, have positive capacity but play no role in our analysis.

From the final network, we can see that nodes x_1 and $\bar{x}_2$ can be reached from the source via arcs with positive residual capacity, and hence $x_1 = 1$ and $x_2 = 0$ must hold in all minima of f. Substituting this into the last posiform ψ we can conclude that the minimization of f is equivalent with the minimization of

$$\psi^* = 4\bar{x}_3\bar{x}_4 + 2x_4\bar{x}_5.$$

5.2. Hierarchy of bounds

In this section we study the family of nonnegative pseudo-Boolean functions, along with their generators, and show that the roof dual bound can be extended to a complete hierarchy of bounds.

5.2.1. Cones of positive quadratic pseudo-Boolean functions

We shall view quadratic pseudo-Boolean functions in $|\mathbf{V}| = n$ variables as vectors of the $1 + n + \binom{n}{2}$ coefficients of their unique polynomial form, i.e., as vectors in $\mathbb{R}^{1+n+\binom{n}{2}}$.

For a subset $S \subseteq \mathbf{V}$ of the variables, let us denote by $\mathscr{F}_S$ the family of quadratic pseudo-Boolean functions depending only on variables from S, and let $\mathscr{F}_S^+ \subseteq \mathscr{F}_S$ denote the subfamily of nonnegative ones among them. Clearly, $\mathscr{F}_S$ is a subspace of dimension $1 + |S| + \binom{|S|}{2}$ of $\mathbb{R}^{1+n+\binom{n}{2}}$, and $\mathscr{F}_S^+$ is a convex cone in this subspace. It is also easy to see that $\mathscr{F}_S^+$ is finitely generated, since it is a polyhedral cone, described by the $2^{|S|}$ inequalities requiring the nonnegativity at each of the $2^{|S|}$ binary substitutions to these variables. Let us define next

$$\mathscr{Q}_k \overset{\text{def}}{=} \text{cone}\{\mathscr{F}_S^+ \mid S \subseteq \mathbf{V}, \ |S| \leqslant k\}$$

for $2 \leqslant k \leqslant n$, as the convex cone in $\mathbb{R}^{1+n+\binom{n}{2}}$, generated by the unions of the above cones corresponding to at most k of the variables, chosen in all possible ways. It follows from these definitions that

$$\mathscr{Q}_2 \subseteq \mathscr{Q}_3 \subseteq \cdots \subseteq \mathscr{Q}_n \tag{45}$$

and that all these cones are finitely generated. Let us denote by $\mathscr{B}(\mathscr{Q}_k)$ such a finite set of generators of the cone $\mathscr{Q}_k$. Finally let us note that $\mathscr{Q}_n = \mathscr{F}_V^+$ is the family of all nonnegative quadratic pseudo-Boolean functions.

The characterization of the extremal elements of the above introduced cones is an interesting problem in itself, and as we shall see in the subsequent sections, it has its own importance for several algorithmic problems. The full characterization however seems to be too difficult, and at this time we only have partial solution to it.

Let us consider the following special family of functions, which we recall from [28]:

$$b_{U,\alpha} \overset{\text{def}}{=} \binom{\sum_{u \in U} -\alpha}{2} \tag{46}$$

where $U \subseteq \mathbf{L}$ is a subset of the literals, not containing complemented pairs, and where $\alpha \in \mathbb{Z}$ is an integer. (Here we use $\binom{x}{2} = x(x-1)/2$ for all integers $x \in \mathbb{Z}$.) Clearly, the above defines a pseudo-Boolean function, and by using the identity $u^2 = u$ for the literals $u \in \mathbf{L}$, we can compute the quadratic polynomial representing these functions.

Example 5.4. If $U = \{x, \bar{y}\}$ and $\alpha = 1$, we get

$$
\begin{aligned}
b_{\{x,\bar{y}\},1} &= \frac{(x + \bar{y} - 1)(x + \bar{y} - 2)}{2} \\
&= \frac{x^2 + \bar{y}^2 + 2x\bar{y} - 3x - 3\bar{y} + 2}{2} \\
&= x\bar{y} - x - \bar{y} + 1 \\
&= \bar{x}y.
\end{aligned}
$$

Proposition 17 (Boros et al. [28]). *If $U \subseteq \mathbf{L}$ is a subset of the literals containing no complemented pairs, and α is an integer such that $1 \leqslant \alpha \leqslant |U| - 2$ for $|U| \geqslant 3$, and $\alpha = 1$ for $|U| = 2$, then $b_{U,\alpha} \in \mathcal{B}(\mathcal{Q}_k)$ for $k \geqslant |U|$.*

The above statement implies that in the nested sequence of convex cones (45) there are no equalities:

$$
\mathcal{Q}_2 \subset \mathcal{Q}_3 \subset \cdots \subset \mathcal{Q}_n. \tag{47}
$$

Example 5.5. For instance, if $U = \{u, v, w\} \subseteq \mathbf{L}$ is a subset of 3 distinct literals, and contains no complemented pair, then $b_{U,1} = uvw + \overline{u}\overline{v}\overline{w} = uv + uw + vw - u - v - w + 1 \in \mathcal{Q}_3 \setminus \mathcal{Q}_2$.

For $k \leqslant 3$ the following characterization of the extremal elements of $\mathcal{Q}_k$ is known:

Proposition 18 (Boros et al. [29]). *Let*

$$
\mathcal{B}_2 = \{uv \mid u, v \in \mathbf{L}, \ u \neq v, \ u \neq \bar{v}\}
$$

$$
= \{b_{U,1} \mid U \subseteq \mathbf{L} \text{ containing no complemented literals, } |U| = 2\}, \text{ and}
$$

$$
\mathcal{B}_3 = \{uvw + \overline{u}\overline{v}\overline{w} \mid u, v, w \in \mathbf{L}, \ u \notin \{v, \bar{v}, w, \bar{w}\}, \ v \notin \{w, \bar{w}\}\}
$$

$$
= \{b_{U,1} \mid U \subseteq \mathbf{L} \text{ containing no complemented literals, } |U| = 3\}.
$$

Then, we have $\mathcal{B}(\mathcal{Q}_2) = \mathcal{B}_2$ and $\mathcal{B}(\mathcal{Q}_3) = \mathcal{B}_2 \cup \mathcal{B}_3$.

Not every generator of the cones $\mathcal{Q}_k$ is a function of the form (46). In [33] several families of extremal elements of $\mathcal{Q}_k$, $k \geqslant 10$ are exhibited, which are not in the form of this type.

5.2.2. Complementation, majorization, linearization

Following [28] we shall recall in this section a generalization of the three approaches yielding the roof dual value.

The first approach, the so called complementation, can be extended in a natural way as follows. Let us observe that since every quadratic pseudo-Boolean function f can be represented by a quadratic posiform ϕ (possibly with a negative constant term), there exists a constant C such that $f - C \in \mathcal{Q}_2 \subseteq \mathcal{Q}_k$ for $k \geqslant 2$. Thus, let us define

$$
C_k(f) = \max\{C \in \mathbb{R} \mid f - C \in \mathcal{Q}_k\} \tag{48}
$$

for $k = 2, 3, \ldots, n$. Since $\mathcal{Q}_k$ is a closed convex cone in $\mathbb{R}^{1+n+\binom{n}{2}}$ for $k = 2, 3, \ldots, n$, and since f takes only finitely many different values, in the above definition the maximum exists.

It is clear from the definition of the cones $\mathcal{Q}_k$ that

$$C_2(f) \leqslant C_3(f) \leqslant \cdots \leqslant C_n(f) = \min_{\mathbf{x} \in \mathbb{B}^n} f(\mathbf{x}),$$

implying that the values $C_k(f)$ form a sequence of increasingly better lower bounds of f.

Using a generating set $\mathscr{B}(\mathcal{Q}_k)$ of the cone $\mathcal{Q}_k$, we can also express the lower bound $C_k(f)$ as the optimum of a linear programming problem

$$C_k(f) = \max C$$

$$\text{subject to } f - C = \sum_{b \in \mathscr{B}(\mathcal{Q}_k)} \alpha_b b, \tag{49}$$

$$\alpha_b \geqslant 0 \quad \text{for all } b \in \mathscr{B}(\mathcal{Q}_k),$$

where the equations correspond to the $1 + n + \binom{n}{2}$ coefficients of f.

Using Proposition 18 and the above linear programming problem for $k = 2$, it can be verified that $C_2(f)$ is indeed the roof dual value of f, as defined in the previous sections. The linear programming problem corresponding to $k = 3$ was analyzed in [29], and $C_3(f)$ was introduced there as the *cubic dual* of f. It was also shown that $C_3(f) \geqslant C_2(f)$ with equality holding if and only if $C_2(f) = C_3(f) = \min_{\mathbf{x} \in \mathbb{B}^n} f(\mathbf{x})$, i.e. if and only if these bounds are tight.

The main idea of the second approach, *minorization*, is to consider a family of linear minorants of the quadratic pseudo-Boolean function f, and to find a "best" one among these linear functions. This idea was considered in [12,79], and a generalization, which we recall here appears in [28].

For a *purely* quadratic function h of k variables (i.e., for which $h(0, 0, \ldots, 0) = h(1, 0, \ldots, 0) = \cdots = h(0, \ldots, 0, 1) = 0$) let

$$\mathcal{M}(h) \stackrel{\text{def}}{=} \{l \mid l \text{ linear, } l(\mathbf{x}) \leqslant h(\mathbf{x}) \text{ for all } \mathbf{x} \in \mathbb{B}^k\}$$

denote a family of linear minorants of h over the Boolean vectors.

For a quadratic pseudo-Boolean function f, and for a fixed integer k, let us consider a representation of f of the form

$$f = l + \sum_{j \in J} f_j, \tag{50}$$

where l is linear, and the f_j's are purely quadratic functions of at most k variables. Then, the linear functions of the form

$$p = l + \sum_{j \in J} l_j,$$

with $l_j \in \mathcal{M}(f_j)$ for $j \in J$, are linear minorants of f. Let $\mathcal{M}^k(f)$ denote the set of all linear minorants of f obtained in this way, varying (50) over all possible representations. Then, by definition,

$$\mathcal{M}(f) = \mathcal{M}^n(f) \supseteq \mathcal{M}^{n-1}(f) \supseteq \cdots \supseteq \mathcal{M}^3(f) \supseteq \mathcal{M}^2(f).$$

By defining

$$M_k(f) \overset{\text{def}}{=} \max_{p \in \mathscr{M}^k(f)} \min_{x \in \mathbb{B}^n} p(x), \tag{51}$$

we obtain another series of lower bounds of f

$$\min_{x \in \{0,1\}^n} f(x) = M_n(f) \geqslant M_{n-1}(f) \geqslant \cdots \geqslant M_3(f) \geqslant M_2(f).$$

The elements of $\mathscr{M}^2(f)$ were called *paved upper planes* in [79]. It was also shown there that the bound, obtained by taking the minimum in (51) for $k = 2$ over certain "minimal" elements $p(x)$ of $\mathscr{M}^2(f)$, rather than over all of $\mathscr{M}^2(f)$, is always equal to $C_2(f)$. As a consequence, it follows that $M_2(f) \geqslant C_2(f)$. Later, in [98] it has been proved that $C_2(f) = M_2(f)$ (see also [2,42]).

Using the notation $b = l_b + q_b$ for $b \in \mathscr{B}(\mathcal{Q}_k)$, where l_b denotes the linear part of b, and q_b is purely quadratic, we can obtain a computationally simpler derivation of $M_k(f)$, given by

$$M_k = \max_{\substack{l \text{ is linear, } \alpha_b \geqslant 0 \\ f = l + \sum_{b \in \mathscr{B}(\mathcal{Q}_k)} \alpha_b q_b}} \min_{x \in \mathbb{B}^n} \left[l + \sum_{b \in \mathscr{B}(\mathcal{Q}_k)} \alpha_b l_b \right]. \tag{52}$$

The third approach, *linearization*, is a standard method to represent nonlinear expressions in terms of linear functions and inequalities.

To a quadratic pseudo-Boolean function over $\mathbb{B}^n$

$$f(\mathbf{x}) \overset{\text{def}}{=} q_0 + \sum_{j=1}^{n} q_j x_j + \sum_{1 \leqslant i < j \leqslant n} q_{ij} x_i x_j,$$

we shall associate a linear function over $\mathbb{R}^{n + \binom{n}{2}}$, given by

$$L_f(\mathbf{x}, \mathbf{y}) \overset{\text{def}}{=} q_0 + \sum_{j=1}^{n} q_j x_j + \sum_{1 \leqslant i < j \leqslant n} q_{ij} y_{ij},$$

where $(\mathbf{x}, \mathbf{y})$ denotes the vector $(x_1, \ldots, x_n, y_{12}, \ldots, y_{n-1,n}) \in \mathbb{R}^{n + \binom{n}{2}}$. This association establishes in fact a one-to-one correspondence between quadratic pseudo-Boolean functions in n variables and linear functions in $n + \binom{n}{2}$ variables.

Let us define then a polyhedron for $2 \leqslant k \leqslant n$ by

$$\mathbf{S}^{[k]} \overset{\text{def}}{=} \{(\mathbf{x}, \mathbf{y}) \mid L_b(\mathbf{x}, \mathbf{y}) \geqslant 0 \text{ for all } b \in \mathscr{B}(\mathcal{Q}_k)\}$$

and let

$$L_k(f) \overset{\text{def}}{=} \min L_f(\mathbf{x}, \mathbf{y}) \quad \text{s.t. } (\mathbf{x}, \mathbf{y}) \in \mathbf{S}^{[k]}. \tag{53}$$

It is easy to see by the definition of the cones $\mathcal{Q}_k$, $k = 2, \ldots, n$ that

$$\mathbf{S}^{[2]} \supseteq \mathbf{S}^{[3]} \supseteq \cdots \supseteq \mathbf{S}^{[n]}.$$

It follows by Proposition 18 that $\mathbf{S}^{[2]}$ is the same polyhedron appearing (with slight variations) in many publications (see e.g., [11,31,59,60,67,68,95,122,146]).

It has been shown in [135] that all fractional vertices of $\mathbf{S}^{[2]}$ are cut off by the so called *triangle inequalities*, i.e. by the inequalities $L_b(\mathbf{x}, \mathbf{y}) \geqslant 0$ for $b \in \mathscr{B}(\mathcal{Q}_3)$. A stronger statement was proved in [29]:

Proposition 19 (Boros et al. [29]). *The polyhedron* $\mathbf{S}^{[3]}$ *is the first Chvátal closure of* $\mathbf{S}^{[2]}$.

Using Proposition 18, we can easily see that the roof dual of a quadratic pseudo-Boolean function f is the common value of the three bounds: $C_2(f) = M_2(f) = L_2(f)$. The main result of [28] generalizes this property:

Theorem 12 (Boros et al. [28]). *Given a quadratic pseudo-Boolean function f in n variables, the equalities*

$$C_k(f) = M_k(f) = L_k(f)$$

hold for all $k = 2, 3, \ldots, n$, providing increasingly tighter lower bounds on f, with

$$\min_{\mathbf{x} \in \mathbb{B}^n} f(\mathbf{x}) = C_n(f) = M_n(f) = L_n(f).$$

Definition (53) provides also a simple linear programming formulation to compute these bounds, at least for those cases when $\mathscr{B}(\mathcal{Q}_k)$ is known. For instance, we have

$$\mathbf{S}^{[2]} = \left\{ (\mathbf{x}, \mathbf{y}) \,\middle|\, \begin{array}{rrrl} & & -y_{ij} & \leqslant 0 \\ -x_i & & +y_{ij} & \leqslant 0 \\ & -x_j & +y_{ij} & \leqslant 0 \\ x_i & +x_j & -y_{ij} & \leqslant 1 \\ \multicolumn{4}{l}{\text{for } 1 \leqslant i < j \leqslant n} \end{array} \right\} \tag{54}$$

and

$$\mathbf{S}^{[3]} = \mathbf{S}^{[2]} \cap \left\{ (\mathbf{x}, \mathbf{y}) \,\middle|\, \begin{array}{lll} x_i + x_j + x_k - y_{ij} - y_{ik} - y_{jk} & \leqslant 1 \\ -x_i \qquad\qquad + y_{ij} + y_{ik} - y_{jk} & \leqslant 0 \\ \qquad -x_j \qquad + y_{ij} - y_{ik} + y_{jk} & \leqslant 0 \\ \qquad\qquad -x_k - y_{ij} + y_{ik} + y_{jk} & \leqslant 0 \\ \text{for } 1 \leqslant i < j < k \leqslant n \end{array} \right\}. \tag{55}$$

The sharpness of the roof dual bound, i.e. the validity of $C_2(f) = \min_{\mathbf{x} \in \mathbb{B}^n} f(\mathbf{x})$, can be recognized in $O(|f|)$ time (see [79]), while the sharpness of $C_3(f)$ was shown in [28] to be NP-complete. It was also shown there that $C_2(f) = C_3(f)$ only if these bounds are sharp, while $C_4(f) = C_3(f) < \min_{\mathbf{x} \in \mathbb{B}^n} f(\mathbf{x})$ is possible for some quadratic pseudo-Boolean functions.

5.3. Polyhedra

As for almost all combinatorial optimization problems, polyhedral formulations are quite natural, and are also frequent in the literature of quadratic pseudo-Boolean optimization. In fact, the technique of linearization has already associated some polyhedra with this problem. In this section we survey some of the relevant notions and results, selecting only the most closely related ones, from a very large and active area.

The most natural polytope associated to quadratic pseudo-Boolean optimization is the so called *Boolean quadric* polytope, introduced in [135], and defined by

$$\mathbf{Q} \overset{\text{def}}{=} \{(\mathbf{x}, \mathbf{y}) \mid \mathbf{x} \in \mathbb{B}^n, \ y_{ij} = x_i x_j \text{ for all } 1 \leqslant i < j \leqslant n\}. \tag{56}$$

It was shown in [135] that $\mathbf{S}^{[2]} \supseteq \mathbf{Q}$, that the dimension of both polytopes is $n + \binom{n}{2}$; a number of facet defining inequalites were also introduced in the same paper. These polytopes were also shown to have several other interesting properties.

Let us call the 1-*skeleton* of a polytope P the family of its 1-faces, and let us say that a polytope P has the *Trubin property* with respect to another polytope $Q \subseteq P$ if the 1-skeleton of Q is a subfamily of the 1-skeleton of P (see [13,14,160]).

Proposition 20 (Padberg [135]). *The polytope* $\mathbf{S}^{[2]}$ *has the Trubin property with respect to* $\mathbf{Q}$, *furthermore, all vertices of* $\mathbf{S}^{[2]}$ *are half-integral.*

The relations

$$\mathbf{S}^{[2]} \supseteq \mathbf{S}^{[3]} \supseteq \cdots \supseteq \mathbf{S}^{[n]} = \mathbf{Q}$$

were observed in [28], together with the fact that the facets of $\mathbf{Q}$ correspond in a one-to-one way to the generators of the cone $\mathcal{Q}_n$ of nonnegative quadratic pseudo-Boolean functions.

Proposition 21 (Boros et al. [28]). *A quadratic pseudo-Boolean function f is nonnegative (i.e., $f \in \mathcal{Q}_n$), if and only if the inequality $L_f(\mathbf{x}, \mathbf{y}) \geqslant 0$ is valid for $\mathbf{Q}$. Moreover, $f \in \mathcal{B}(\mathcal{Q}_n)$ if and only if $L_f(\mathbf{x}, \mathbf{y}) \geqslant 0$ is facet defining for $\mathbf{Q}$.*

The fact that $\mathbf{S}^{[2]}$ has the Trubin property with respect to $\mathbf{Q}$ corresponds to the fact that $\mathcal{B}(\mathcal{Q}_2) \subset \mathcal{B}(\mathcal{Q}_n)$. We believe that a more general property must also hold, namely that

$$\mathcal{B}(\mathcal{Q}_2) \subset \mathcal{B}(\mathcal{Q}_3) \subset \cdots \subset \mathcal{B}(\mathcal{Q}_n),$$

implying that, $\mathbf{S}^{[i]}$ has the Trubin property with respect to $\mathbf{S}^{[j]}$ for all $2 \leqslant i < j \leqslant n$.

Another strongly related polytope is the so called *cut polytope*. Let $G = K_{n+1}$ denote the complete graph on vertices $v_0, v_1, \ldots, v_n$, and let $E = E(G)$ denote its edge set. Let us call an edges set $F \subseteq E$ a *cut set*, if the removal of F disconnects G, and let us define $\mathbf{C} \subseteq [0,1]^E$ as the convex hull of the characteristic vectors of cut sets.

It was noted in [49] that the mapping

$$z_{0i} = x_i \quad \text{for } i = 1, \ldots, n,$$

$$z_{ij} = x_i + x_j - 2y_{ij} \quad \text{for } 1 \leqslant i < j \leqslant n \tag{57}$$

is an invertible linear mapping establishing a one-to-one correspondence between $\mathbf{Q}$ and $\mathbf{C}$. Combining this with the $f \leftrightarrow L_f$ association, we get also a one-to-one correspondence between the points of the cut polytope, and the nonnegative quadratic-pseudo-Boolean functions. These mappings help to identify the corresponding results about the facial structure of the cut polytope (see e.g. [18,19,52,53]), the Boolean quadric polytope ([135]) and the conical structure of nonnegative quadratic pseudo-Boolean functions ([28,33]), and are also useful in extending those results.

For instance, the inequalities in (54) and (55) were shown to define facets for **Q** in [135], a fact which also follows easily from the characterization of $\mathscr{B}(\mathscr{Q}_k)$ for $k = 2, 3$ given in the previous section. It can be seen easily by (57) that the inequalities of (55) correspond to the so called *triangle inequalities* for the cut polytope.

Proposition 21 provides also a constructive approach to obtain valid inequalities for **Q** (and hence for **C**), since nonnegative quadratic pseudo-Boolean functions can be constructed by various algebraic techniques.

It is easy to see that if $l : \mathbb{B}^n \to \mathbb{Z}$ is an integer valued linear function, then $l(\mathbf{x})(l(\mathbf{x}) - 1)$ defines a nonnegative quadratic pseudo-Boolean function, and hence $L_{l(l-1)}(\mathbf{x}, \mathbf{y}) \geq 0$ is a valid inequality for **Q**. It was shown in [33] that this family of valid inequalities includes all *hypermetric* inequalities of the cut polytope (see e.g. [52,53]).

A further generalization was introduced in [33].

Proposition 22 (Boros and Hammer [33]). *Let* $1 \leq p \leq n$, $k \geq 0$, *let* $l_j \geq 0$, $j = 1, \ldots, p$ *and* $l_j < 0$ *for* $j = p + 1, \ldots, n$ *be integers, and let* T *be a spanning tree on vertices* $\{1, \ldots, p\}$. *Then the function*

$$f(\mathbf{x}) \overset{\text{def}}{=} \left(\sum_{i=1}^{n} l_i x_i \right) \left(\sum_{i=1}^{n} l_i x_i - (2k + 1) \right) + k(k + 1) \left(\sum_{j=1}^{p} x_j - \sum_{(i,j) \in E(T)} x_i x_j \right) \tag{58}$$

defines a nonnegative quadratic pseudo-Boolean function.

It can be seen that the facets of **C**, defined through the above correspondence, include the *cycle inequalities* (for $k = 1$), and many other families of valid defining inequalities, giving at the same time rise to new families of facet defining inequalities:

Proposition 23 (Boros and Hammer [33]). *Let* $r = k + 1$, $n \geq 3k + 7$, $p = [(n + k + 1)/(k + 2)]$ *for* $k \geq 1$, *and let* $l_1 = \cdots = l_p = r$, $l_{p+1} = \cdots = l_n = -1$. *Let furthermore* T *be an arbitrary spanning tree on the vertices* $\{1, 2, \ldots, p\}$. *Then* f, *defined by* (58), *is an extremal element of* $\mathscr{Q}_n$, *i.e.*, $f \in \mathscr{B}(\mathscr{Q}_n)$.

Let us also remark that the facet class defined in this way for **C** not only includes some facet defining cycle inequalities, but also has polynomial time separation (since a maximum weight spanning tree can be found in polynomial time.)

5.4. Heuristics

There is a large variety of techniques applied in the literature for solving problems that can be modelled by quadratic pseudo-Boolean functions. Many of the published algorithms are either branch-and-bound or branch-and-cut type, exploiting some of the results (bounds, facets, etc.) which were mentioned in the previous subsections, treating essentially those problems as integer programs.

In this brief survey of pseudo-Boolean optimization we shall restrict ourselves to recall a simple but very successful heuristic, which utilizes techniques highly specific to quadratic pseudo-Boolean functions. The DDT algorithm (Devour, Digest and Tidy-up) was introduced in [35] and was later applied successfully for delay-fault testing in VLSI design [39,153], and to a number of other problems transformed to quadratic pseudo-Boolean optimization via a penalty function technique [66].

The basic idea in this heuristic is to represent the problem as the minimization of a quadratic posiform, and then sequentially restrict the Boolean cube to smaller and smaller subsets of it, on which the quadratic terms with the highest coefficients vanish. In this process the restriction is represented as the set of solutions to a quadratic Boolean equation, the consistency of which can be tested efficiently. In each iteration, logical consequences of the current Boolean equation are substituted back into the input formula, which is then simplified, before choosing the next term for elimination.

To make this procedure more precise, let us assume that we would like to minimize the quadratic posiform

$$\phi(\mathbf{x}) = \sum_{T \subset \mathbf{L}} a_T \prod_{u \in T} u, \tag{59}$$

where $a_T \geqslant 0$ for all $T \subseteq \mathbf{L}$, and $a_T > 0$ if and only if $|T| \leqslant 2$.

A straightforward one step implementation of the above idea would be to choose a smallest threshold $\theta \geqslant 0$ for which the quadratic Boolean equation $\bigvee_{T:a_T > \theta} T(\mathbf{x}) = 0$ is still consistent, and then output a solution of this equation. A refined sequential version, the so called *DDT algorithm*, is described below (see [35]).

DDT (DEVOUR–DIGEST–TIDY-UP)

Initialization: Input ϕ given by (59), and set $\mathscr{S} = \emptyset$, and $\tilde{\phi} = \phi$.

Devour: Find a term T of $\tilde{\phi}$ with the largest coefficient, and let
$\mathscr{S} = \mathscr{S} \cup \{T\}$.

Digest: Draw all logical conclusions $\mathscr{C}$ of the Boolean equation
$$\bigvee_{T \in \mathscr{S}} T(\mathbf{x}) = 0 \tag{\$}$$

Tidy-up: Substitute into $\tilde{\phi}$ the consequences $\mathscr{C}$, drawn in the previous step, and simplify the resulting posiform $\tilde{\phi}$. If $\tilde{\phi} \neq const$ then return to **Devour**.

Output: Output a solution $\mathbf{x}$ of the quadratic Boolean equation ($\$$), and STOP.

Let us add that the logical conclusions drawn in step **Digest** are of the form $u = 0$ or $uv = 0$ for some literals u and v, and that all such conclusions can be derived in polynomial time from the quadratic Boolean equation ($\$$). It is important to note that the completeness of this step implies that all terms which have positive coefficients after the substitution of the derived logical consequences in step **Tidy-up** are consistent with the current quadratic equation ($), i.e. any one of those could be added to $\mathscr{S}$ without implying the inconsistency of the quadratic Boolean equation ($).

Instead of giving a formal proof (simple, but technical, see [35]) for the correctness and finiteness of the above procedure, let us illustrate it on a small example.

Example 5.6. Let us consider the minimization of the following quadratic posiform:

$$\phi(x_1, x_2, \ldots, x_5) = 17\bar{x}_1 + 12x_1\bar{x}_2 + 10x_2\bar{x}_4 + 8x_1x_4 + 8x_4\bar{x}_5$$

$$+ 7x_5 + 6x_3\bar{x}_4 + 5\bar{x}_3 + 4x_2x_3 + 4x_4 + \bar{x}_2,$$

in which terms are listed in the decreasing order of their coefficients. The DDT algorithm will take 5 steps before termination, and will output $\mathbf{x} = (1, 1, 0, 0, 0)$, which in this case is indeed a minimum of ϕ:
(1) $T = \bar{x}_1$, $\mathscr{S} = \{\bar{x}_1\}$, $\mathscr{C} = \{x_1 = 1\}$ and $\tilde{\phi} = +13\bar{x}_2 + 12x_4 + 10x_2\bar{x}_4 + 8x_4\bar{x}_5 + 7x_5 + 6x_3\bar{x}_4 + 5\bar{x}_3 + 4x_2x_3$.
(2) $T = \bar{x}_2$, $\mathscr{S} = \{\bar{x}_1, \bar{x}_2\}$, $\mathscr{C} = \{x_1 = x_2 = 1\}$ and $\tilde{\phi} = 14 + 8x_4\bar{x}_5 + 7x_5 + 6x_3\bar{x}_4 + 2x_4 + \bar{x}_3$.
(3) $T = x_4\bar{x}_5$, $\mathscr{S} = \{\bar{x}_1, \bar{x}_2, x_4\bar{x}_5\}$, $\mathscr{C} = \{x_1 = x_2 = 1, x_4\bar{x}_5 = 0\}$ and $\tilde{\phi} = 14 + 7x_5 + 6x_3\bar{x}_4 + 2x_4 + \bar{x}_3$.
(4) $T = x_5$, $\mathscr{S} = \{\bar{x}_1, \bar{x}_2, x_4\bar{x}_5, x_5\}$, $\mathscr{C} = \{x_1 = x_2 = 1, x_4 = x_5 = 0\}$ and $\tilde{\phi} = 15 + 5x_3$.
(5) $T = x_3$, $\mathscr{S} = \{\bar{x}_1, \bar{x}_2, x_4\bar{x}_5, x_5, x_3\}$, $\mathscr{C} = \{x_1 = x_2 = 1, x_3 = x_4 = x_5 = 0\}$ and $\tilde{\phi} = 15$.

A slightly simpler version of the DDT algorithm, in terms of signed graph balancing was also presented in [35], and it was shown that both version can be implemented to run in $O(n|\phi|)$ time.

Let us add that the selection of the next term to be eliminated in the DDT algorithm is based on a very simple "greedy" utility measure, namely the size of the coefficient of that term. One could further modify the above algorithm by making a more careful selection, based on some look-ahead procedure, analyzing several steps in advance before choosing the term to be eliminated, or based on some probabilistic analysis of the expected benefit from eliminating a particular term, etc. Several such variants were recently examined in [66].

6. Special classes

In this section we survey some special classes of pseudo-Boolean functions, limiting our attention to those for which specialized optimization algorithms have been developed.

6.1. Sub- and supermodular functions

Among the most widely studied classes of set functions submodular and supermodular functions play a well-known and special role. A set function $f : 2^V \to \mathbb{R}$ is called *submodular* if

$$f(X) + f(Y) \geqslant f(X \cup Y) + f(X \cap Y) \tag{60}$$

holds for all subsets X, Y of a base set V of n elements. Functions, for which the reverse inequality holds for all subsets $X, Y \subseteq V$ are called *supermodular*, while those for which $f(X) + f(Y) = f(X \cup Y) + f(X \cap Y)$ for all subsets $X, Y \subseteq V$ are called *modular*. Clearly, f is submodular iff $-f$ is supermodular, and vice versa; furthermore, functions which are both sub- and supermodular, are the modular ones.

It is well-known that submodular functions can be minimized in polynomial time (see e.g. [71]), and even strongly polynomial algorithms are available for this task (see e.g. [108,152]). Of course, the same applies to the maximization of supermodular functions.

It is an interesting problem to recognize if a given polynomial expression or a given posiform defines a sub- or supermodular set function.

It is easy to see that a set function is modular if and only if its unique polynomial expression is linear. This is also an easy consequence of the more general characterization of supermodular functions.

Proposition 24 (Fisher et al. [58]). *A pseudo-Boolean function $f : \mathbb{B}^n \to \mathbb{R}$ is supermodular if and only if its second order derivatives*

$$\Delta_{ij}(\mathbf{x}) \overset{\text{def}}{=} \frac{\partial f}{\partial x_i \partial x_j}(\mathbf{x}) \geqslant 0$$

are nonnegative for all $1 \leqslant i < j \leqslant n$ and for all $\mathbf{x} \in \mathbb{B}^n$.

Clearly, f is modular iff both f and $-f$ are supermodular, i.e. if and only if $\Delta_{ij}(\mathbf{x}) \equiv 0$ for all $i \neq j$, and consequently, if and only if it is linear.

For a quadratic pseudo-Boolean function f it follows that f is supermodular iff all quadratic terms of it have nonnegative coefficients, or equivalently, f is submodular iff its quadratic terms are nonpositive. The minimization of such a quadratic submodular function is known to be equivalent with finding a minimum capacity cut in a corresponding network (see e.g. [72,99]). Indeed, let us consider a quadratic posiform of a quadratic submodular function. According to the above, such a posiform ϕ can always be written, possibly after some simple transformations, as

$$\phi(x_1, \ldots, x_n) = \sum_{i \in P} a_i x_i + \sum_{j \in N} a_j \bar{x}_j + \sum_{1 \leqslant i < j \leqslant n} a_{ij} x_i \bar{x}_j,$$

where $P, N \subseteq \mathbf{V}$, and where all the coefficients a_i $(i = P \cup N)$ and a_{ij} $(1 \leqslant i < j \leqslant n)$ are nonnegative. Let us associate then to ϕ a network N_ϕ on the node set $V(N_\phi) = \{s, t\} \cup \mathbf{V}$, with arcs corresponding to the terms of f:

$$A(N_\phi) = \{(s, j) \mid j \in N\} \cup \{(i, t) \mid i \in P\} \cup \{(i, j) \mid 1 \leqslant i < j \leqslant n\},$$

where the capacities of the arcs are defined as $c_{s,j} = a_j$ for $j \in N$, $c_{i,t} = a_i$ for $i \in P$, and $c_{i,j} = a_{ij}$ for $1 \leqslant i < j \leqslant n$. Then the s, t-cuts of this network are in a one-to-one correspondence with the binary vectors: $\mathbf{x} \in \mathbb{B}^n \leftrightarrow S_{\mathbf{x}} \overset{\text{def}}{=} \{s\} \cup \{j \mid x_j = 1\}$. It is easy to check that with these definitions we have

$$\phi(\mathbf{x}) = \sum_{u \in S_{\mathbf{x}}, v \notin S_{\mathbf{x}}} c_{u,v}$$

for all $\mathbf{x} \in \mathbb{B}^n$, and hence the minimum of ϕ will correspond to a minimum capacity cut of N_ϕ.

The above analysis was extended to cubic posiforms, as well.

Proposition 25 (Billionnet and Minoux [24]). *A cubic posiform*

$$\psi(\mathbf{x}) = \sum_{T \subseteq \mathbf{L}} a_T \prod_{u \in T} u,$$

with $a_T \geqslant 0$ for all $T \subseteq \mathbf{L}$, and $a_T = 0$ for all $|T| > 3$, defines a supermodular set function if and only if all terms of it are pure, i.e. if and only if $T \subseteq \{x_1, \ldots, x_n\}$ or $T \subseteq \{\bar{x}_1, \ldots, \bar{x}_n\}$ holds whenever $a_T > 0$.

Based on the above characterization, the maximization of supermodular cubic posiforms was also shown to be equivalent with a network flow computation in [24]. Several other related classes have also been considered in the literature, including almost positive functions, polar functions, and their switch equivalents (see e.g. [41,45,46,100]).

The above results cannot be generalized easily. The recognition of sub(or super) modularity for higher degree posiforms turns out to be a hard problem, unless $P = NP$.

Proposition 26 (Gallo and Simeone [62]). *The recognition of supermodularity of quartic (degree 4) posiforms is co-NP-complete.*

Let us remark finally that even though the minimum of a submodular (or the maximum of a supermodular) set function can be found in polynomial time [71], or even in strongly polynomial time (see [108,152]), the opposite optimization problems, i.e. the maximization of a submodular (or the minimization of a supermodular) set function is NP-hard (see e.g. [58,130]).

Let us further add that a standard greedy procedure for the maximization of a submodular set function provides a $(1 - 1/e)$-approximation of the maximum, as shown in [58,130].

6.2. Half-products

In this section we consider *half-products*, a special subclass of supermodular functions, defined by multilinear polynomial expressions of the following form:

$$f(\mathbf{x}) \stackrel{\text{def}}{=} \sum_{1 \leqslant i < j \leqslant n} a_i b_j x_i x_j - \sum_{i=1}^{n} c_i x_i, \tag{61}$$

where $\mathbf{a} = (a_1, \ldots, a_n)$, $\mathbf{b} = (b_1, \ldots, b_n)$ and $\mathbf{c} = (c_1, \ldots, c_n)$ are nonnegative integer vectors (in fact, the values of a_n and b_1 are not used).

These functions were considered independently in [9,10,118], and have attracted considerable attention, since a number of scheduling problems can be formulated as half-product minimization; examples where this model arises include the scheduling of two machines to minimize total weighted completion time or to minimize the makespan, and the scheduling of a single machine to minimize completion time variance, agreeably weighted completion time variance, or total weighted earliness and tardiness (see e.g. [9,10,40,111,118]). Half-products also occur in physics, e.g. in the infinite range Mattis model of a spin-glass, the energy function is in fact a half-product (see [6,126]).

Half-products clearly form a subfamily of supermodular set functions by Proposition 24, since all quadratic terms of a half-product have a nonnegative coefficient. It can be shown that the minimization of even these special supermodular functions remains NP-hard.

Proposition 27 (Badics and Boros [10]). *The minimization of half-products is NP-hard.*

Still, a fully polynomial time approximation scheme exists for the minimization of half-products, providing another attractive feature of this class.

Proposition 28 (Badics and Boros [10]). *Given a half-product $f(\mathbf{x})$ by (61), let $A = \sum_{i=1}^{n-1} a_i$, and let $\mathbf{x}^*$ denote a minimum of f. Then, for every $\varepsilon > 0$, one can find in $O((n^2 \ln A)/\varepsilon)$ time a binary vector $\mathbf{x}^\varepsilon$, for which*

$$f(\mathbf{x}^\varepsilon) - f(\mathbf{x}^*) \leqslant \varepsilon |f(\mathbf{x}^*)|.$$

The main component of the above approximation scheme is the existence of a pseudo-polynomial dynamic programming algorithm. Introducing

$$g_k(x_1, \ldots, x_k) \overset{\text{def}}{=} \sum_{1 \leqslant i < j \leqslant k} a_i b_j x_i x_j - \sum_{i=1}^{k} c_i x_i,$$

and

$$h_k(x_{k+1}, \ldots, x_n) \overset{\text{def}}{=} \sum_{k < i < j \leqslant n} a_i b_j x_i x_j - \sum_{i=k+1}^{n} c_i x_i$$

for $k = 1, 2, \ldots, n-1$, we can write the half-product f as

$$f(\mathbf{x}) = g_k(x_1, \ldots, x_k) + \left(\sum_{i=1}^{k} a_i x_i \right) \left(\sum_{j=k+1}^{n} b_j x_j \right) + h_k(x_{k+1}, \ldots, x_n).$$

From this it can be seen that if $\mathbf{x}$ and $\mathbf{y}$ are binary vectors for which $x_j = y_j$ for all $j > k$, $g_k(x_1, \ldots, x_k) \leqslant g_k(y_1, \ldots, y_k)$, and $\sum_{i=1}^{k} a_i x_i \leqslant \sum_{i=1}^{k} a_i y_i$, then $f(\mathbf{x}) \leqslant f(\mathbf{y})$. As a consequence, the first k components $x_1^*, \ldots, x_k^*$ of a minimizing vector $\mathbf{x}^*$ of f will correspond to one of the minimal two-dimensional integer vectors of the form $(g_k(x_1, \ldots, x_k), \sum_{i=1}^{k} a_i x_i)$, $\mathbf{x} \in \mathbb{B}^n$, of which we have at most $\sum_{i=1}^{k} a_i \leqslant A$ different ones. Thus, updating recursively for $k = 1, 2, \ldots, n$ these (at most A) different two-dimensional vectors, we can determine the minimum of f in $O(nA)$ steps.

The above dynamic programming idea was further specialized for the case of the so called *ordered symmetric half-products* in [119]. A half-product f is called *symmetric* if it can be written as

$$f(\mathbf{x}) = \sum_{1 \leqslant i < j \leqslant n} a_i b_j (2x_i x_j - x_i - x_j)$$

for some nonnegative integer vectors $\mathbf{a} = (a_1, \ldots, a_n)$ and $\mathbf{b} = (b_1, \ldots, b_n)$. It is called *ordered symmetric* if either $a_1 \leqslant a_2 \leqslant \cdots \leqslant a_n$, or $b_1 \geqslant b_2 \geqslant \cdots \geqslant b_n$.

Proposition 29 (Kubiak [119]). *An ε-approximation of the minimum of an ordered symmetric half-product can be found in $O(n^2/\varepsilon)$ time for any $\varepsilon > 0$.*

6.3. Hyperbolic pseudo-Boolean programming

An interesting special class of pseudo-Boolean optimization is the maximization (or minimization) of the ratio of two linear functions:

$$\max_{\mathbf{x} \in \mathbb{B}^n} f(\mathbf{x}) = \frac{a_0 + \sum_{j=1}^n a_j x_j}{b_0 + \sum_{j=1}^n b_j x_j}. \tag{62}$$

This problem was introduced in [91] as *fractional programming*, and was studied later in [97,128,147]. Its applications include query optimization in data bases and information retrieval (see e.g. [97]).

It is known [91] that (62) has an easy polynomial time solution if

$$b_0 + \sum_{j=1}^n b_j x_j > 0 \quad \text{for all } \mathbf{x} \in \mathbb{B}^n. \tag{63}$$

In order to see the solution of (62), let us observe first that condition (63) implies that strong persistency holds for f at the partial assignment $\mathbf{y}$ defined by

$$y_j = \begin{cases} 1 & \text{if } (a_j > 0 \text{ and } b_j \leqslant 0) \quad \text{or } (a_j = 0 \text{ and } b_j < 0), \\ 0 & \text{if } (a_j < 0 \text{ and } b_j \geqslant 0) \quad \text{or } (a_j = 0 \text{ and } b_j > 0). \end{cases} \tag{64}$$

Let us also note that if $a_j = b_j = 0$ then in fact $f(\mathbf{x})$ does not depend on x_j, and that by substituting $x_j = 1 - x_j'$ whenever $a_j < 0$ and $b_j < 0$ we can obtain an equivalent maximization problem, in which the coefficients of all variables are positive.

Thus, in case (62) holds, we can assume without any loss of generality that

$$b_j > 0 \text{ for } j = 0, 1, \ldots, n, \quad \text{and} \quad a_j > 0 \text{ for } j = 1, 2, \ldots, n. \tag{65}$$

Denoting by $\mathbf{x}^*$ an optimum of (62), let us observe next that $f(\mathbf{x}^*) \geqslant t$ if and only if

$$a_0 - t b_0 + \sum_{j=1}^n (a_j - t b_j) x_j^* \geqslant 0$$

or equivalently, if and only if

$$\max_{\mathbf{x} \in \mathbb{B}^n} \sum_{j=1}^n (a_j - t b_j) x_j \geqslant -a_0 + t b_0.$$

This latter optimization problem is trivial to solve, and for every value of the threshold t its optimal solution is one of the $n+1$ binary vectors of the form

$$x_{j_l} = 1 \quad \text{for } l \leqslant k \quad \text{and} \quad x_{j_l} = 0 \text{ for } l > k \tag{66}$$

for $k = 0, 1, \ldots, n$, where $(j_1, \ldots, j_n)$ is a permutation of the indices such that

$$\frac{a_{j_1}}{b_{j_1}} \geqslant \frac{a_{j_2}}{b_{j_2}} \geqslant \cdots \geqslant \frac{a_{j_n}}{b_{j_n}}.$$

Hence, the optimum of (62) is one of these $n+1$ binary vectors whenever (63) holds. Since these vectors can be generated in $O(n \log n)$ time, and all corresponding values of f can be determined in $O(n)$ time, problem (62) can be solved in this case in $O(n \log n)$ time.

Let us add that an analogous analysis with a similar conclusion can obviously be carried out in case (62) is a minimization problem.

Let us remark next that of course, if $b_0 + b_1 x_1 + \cdots + b_n x_n$ can also take the value zero, then (62) may not have a finite optimum. Furthermore, even if the condition

$$b_0 + \sum_{j=1}^{n} b_j x_j \neq 0 \quad \text{for all } \mathbf{x} \in \mathbb{B}^n \tag{67}$$

holds, but the denominator can take both negative and positive values (e.g. $b_0 < 0$), problem (62) is NP-hard. To see this latter claim, let us consider the problem of deciding if there exists a binary assignment $\mathbf{x} \in \mathbb{B}^n$ for which

$$\sum_{j=1}^{n} s_j x_j = S, \tag{68}$$

where $s_j > 0$, $j = 1, \ldots, n$ and S are given integers. This problem is known as the *subset sum* problem, a well-known NP-complete decision problem. Let us associate to it the hyperbolic pseudo-Boolean optimization problem

$$\max_{\mathbf{x} \in \mathbb{B}^n} \frac{-1}{(-1 - 2S) + \sum_{j=1}^{n} 2 s_j x_j}. \tag{69}$$

It is easy to verify that the maximum of (69) is 1 if and only if (68) has a solution, implying that maximizing (69) cannot be easier than finding a solution to (68).

6.4. Products of linear functions

Another interesting special case of pseudo-Boolean optimization is the maximization (or minimization) of the product of two linear functions over binary variables

$$\max_{\mathbf{x} \in \mathbb{B}^n} f(\mathbf{x}) = l_1(\mathbf{x}) l_2(\mathbf{x}), \tag{70}$$

where

$$l_1(\mathbf{x}) = a_0 + a_1 x_1 + \cdots + a_n x_n, \quad \text{and} \quad l_2(\mathbf{x}) = b_0 + b_1 x_1 + \cdots + b_n x_n.$$

This problem was considered in [76], and was shown to be an NP-hard optimization problem via a reduction from the subset sum problem cited above. The continuous relaxation

$$\max_{\mathbf{q} \in U^n} f(\mathbf{q}) = l_1(\mathbf{q}) l_2(\mathbf{q}), \tag{71}$$

was also considered in [76], and a polynomial $O(n \log n)$ time algorithm was presented for it. Based on this, a branch-and-bound algorithm was developed for (70). Computational experiments were carried out with problems of this type involving up to 10 000 variables.

7. Approximation algorithms

7.1. MAX-SAT and variants

The *maximum satisfiability* problem is one of the central problems of computer science and combinatorial optimization. Given a family of weighted elementary disjunctions (so called *clauses*), the maximum satisfiability problem consists in finding a binary assignment to the Boolean variables which maximizes the total weight of satisfied clauses. Using the equality

$$\bigvee_{u \in C} u = 1 - \prod_{u \in C} \bar{u}$$

for subsets $C \subseteq \mathbf{L}$ of literals, we can reformulate this problem as a pseudo-Boolean optimization problem, as shown in Section 3: Given a family $\mathscr{C}$ of literals, and nonnegative weights $a_C \in \mathbb{R}_+$ associated to the clauses $C \in \mathscr{C}$, the maximum satisfiability problem can be stated as

$$\max_{\mathbf{x} \in \mathbb{B}^n} \sum_{C \in \mathscr{C}} a_C \left(1 - \prod_{u \in C} \bar{u}\right). \tag{72}$$

The maximum satisfiability problem is a well-known NP-hard optimization problem, a common generalization of many other combinatorial optimization problems (e.g., maximum cut in graphs, maximum directed cut in directed graphs, etc.), which is also known to have good approximations, as well as inapproximability results.

For a maximization problem $\max f(\mathbf{x})$ a vector $\tilde{\mathbf{x}}$ is called an *α-approximation*, if

$$\frac{f(\tilde{\mathbf{x}})}{\max f(\mathbf{x})} \geqslant \alpha.$$

It is important to note that this measure of approximability is not invariant under several simple operations, which otherwise do not change the optimization problem. For instance, the same $\tilde{\mathbf{x}}$ vector may not be an α-approximation of the objective function $f(\mathbf{x}) - K$, where K is a nonnegative constant, though the maximization of $f(\mathbf{x}) - K$ is clearly equivalent with the maximization of $f(\mathbf{x})$. It is also important to point out here that while the maximum satisfiability problem is equivalent with the minimization of the corresponding posiform

$$\sum_{C \in \mathscr{C}} a_C \prod_{u \in C} \bar{u}$$

the latter problem cannot be approximated well, unless P = NP. For a more precise treatment of approximation algorithms and related notions of complexity see [136].

Returning to approximations of the maximum satisfiability problem, there are many relevant results to mention here, with a particular increase in research volume in the last decade, due to two important new techniques: on the one hand, semidefinite formulations yielded new efficient approximations to several variants of the maximum satisfiability problem, while, on the other hand, the development of probabilistic proof verification techniques made it possible to prove inapproximability beyond certain rates, assuming P ≠ NP.

Before giving a brief overview of these results, let us recall an extended terminology for the many variants of this problem. MAX SAT refers to the maximum satisfiability problem as stated in (72). If $|C| \leqslant k$ for all clauses $C \in \mathscr{C}$, the problem is called MAX-k-SAT. Particular attention is given in the literature to MAX-2-SAT, which is, as an optimization problem, equivalent with quadratic pseudo-Boolean optimization.

A natural generalization of maximum satisfiability is called *maximum constraint satisfaction problem*, or MAX CSP. Let g be an arbitrary Boolean expression, and let us denote by $g(S)$ the value of this function when applied to the set of literals $S \subseteq \mathbf{L}$. For any fixed Boolean expression g, MAX CSP(g) denotes the following problem: Given a collection $\mathscr{S}$ of subsets of literals, and nonnegative weights $a_S \in \mathbb{R}_+$ associated to these subsets $S \in \mathscr{S}$, find a binary assignment to the variables, which maximizes the function

$$\sum_{S \in \mathscr{S}} a_S g(S). \tag{73}$$

It is easy to see that for most interesting cases in the literature, g has a very short posiform representation, and hence MAX CSP(g) is a special case of pseudo-Boolean optimization in all these cases. For instance, denoting by

$$OR(S) = \bigvee_{u \in S} u = 1 - \prod_{u \in S} \bar{u},$$

we find that MAX CSP(OR) is the maximum satisfiability problem as written in (72), while if

$$AND(S) = \bigwedge_{u \in S} u = \prod_{u \in S} u,$$

then MAX CSP(AND) is the posiform maximization as considered in Section 4.7.

It is customary to indicate in the lower index if the size of the sets S to which these functions are applied are limited in size, e.g. MAX CSP(OR_2) denotes MAX-2-SAT, etc. For instance, if

$$XOR_2(u,v) = u\bar{v} + \bar{u}v$$

then MAX CSP(XOR_2) is in fact a generalization of the maximum cut problem in graphs and signed graphs. Another interesting case is the majority function

$$MAJ_3(u,v,w) = \begin{cases} 1 & \text{if } u+v+w \geqslant 2, \\ 0 & \text{otherwise.} \end{cases}$$

It is easy to see that MAX CSP(MAJ_3) is the problem of finding the maximum number of clauses in a given 3-CNF, which can be switched to a Horn formula, a problem arising in artificial intelligence (see [26]).

A $\frac{1}{2}$-approximation algorithm for MAX-SAT is simply the application of ROUNDUP (see Section 4.1) starting with $(\frac{1}{2}, \frac{1}{2}, \ldots, \frac{1}{2})$, as proposed in [109]. In fact that algorithm provides a $(1 - 1/2^k)$-approximation whenever all terms are of degree k or larger, implying e.g. a $\frac{3}{4}$-approximation for MAX-2-SAT in which there are no linear terms. For the case, when the linear terms are present, but cannot be trivially simplified a $(\sqrt{5} - 1)/2$-approximation (0.618-approximation) was obtained by [123]. In [116] a randomized algorithm is presented yielding a $\frac{2}{3}$-approximation, in expected value. A transformation producing an equivalent problem without linear terms, and yielding a $\frac{3}{4}$-approximation with the help of ROUNDUP starting from $(\frac{1}{2}, \frac{1}{2}, \ldots, \frac{1}{2})$ was presented in [166]. In fact this method for MAX-2-SAT is simply a combination of roof-duality and the ROUNDUP procedure. We shall recall this procedure in more detail in the next subsection.

In [69] a new $\frac{3}{4}$-approximation algorithm was provided by a combination of the rounding method and another $(1 - 1/e)$-approximation obtained via a linear programming formulation. Though this latter method solves the problem by solving a higher dimensional model first, it can be simplified somewhat by using a simple convex majorant of the input posiform, and a robust $\frac{3}{4}$-approximation algorithm can be obtained via convex programming in the original space of the variables [9].

A substantial improvement was presented in [70] based on an $O(n^2)$-dimensional semidefinite formulation of the problem, yielding a 0.87856-approximation for both MAX-2-SAT and MAX CUT (in fact for MAX CSP(XOR_2)). This method also yielded a 0.758-approximation for MAX SAT. This has further been improved to a 0.93109-approximation for MAX-2-SAT and an 0.859-approximation for quadratic posiform maximization (MAX CSP(AND_2)) in [56].

This novel approach of solving a semidefinite relaxation and then rounding the resulted $O(n^2)$-dimensional fractional solution to a binary n-vector has been applied in the last few years to a large variety of combinatorial optimization problems. Among these we mention the $\frac{7}{8}$-approximation for MAX-3-SAT by [113], a $\frac{1}{2}$-approximation for MAX CSP(AND_3), and a $\frac{2}{3}$-approximation for MAX CSP(MAJ_3) by [168]. Let us add that for the latter problem a robust but weaker $\frac{40}{67}$-approximation can be obtained using pseudo-Boolean techniques without solving a large semidefinite relaxation [26].

On the negative side, the development of the theory of probabilistically checkable proofs [7,8] lead to a series of inapproximability results. For instance the results mentioned above for the MAX-3-SAT, MAX CSP(AND_3) and MAX CSP(MAJ_3) problems are all known to be best possible, unless P = NP (see [20,101,159]). It is also known that for MAX-2-SAT it is impossible to obtain in polynomial time an approximation better than $\frac{21}{22}$, unless $P = NP$ (see [159]).

To give a more detailed overview of all these results and techniques is beyond the scope of our survey of pseudo-Boolean optimization. We shall recall only two results, where the applied techniques are specific to the theory of pseudo-Boolean functions. These results also show that a robust and reasonably good approximation can be achieved without solving a high dimensional relaxation.

Let us point out that most approaches which do not employ semidefinite programming used in fact a variant of the ROUNDUP procedure with $(\frac{1}{2}, \ldots, \frac{1}{2})$ as a starting point. We would like to show that by applying algebraic transformation to the input posiform first, and using persistency before applying ROUNDUP (see e.g. [166]), as well as precomputing a better starting point than $(\frac{1}{2}, \ldots, \frac{1}{2})$ (see e.g. [9,27]) can lead to substantial performance improvements.

7.2. $\frac{3}{4}$-approximation of MAX-2-SAT via roof-duality

As a first example, let us describe here a simple proof for the $\frac{3}{4}$-approximation algorithm of [166] for the MAX-2-SAT problem. Let us consider a MAX-2-SAT instance as given by

$$f(\mathbf{x}) = \sum_{u \in L} a_u(1 - \overline{u}) + \sum_{\substack{u,v \in L \\ u \neq v}} a_{uv}(1 - \overline{u}\overline{v}). \tag{74}$$

By introducing

$$\phi(\mathbf{x}) = \sum_{u \in L} a_u \overline{u} + \sum_{\substack{u,v \in L \\ u \neq v}} a_{uv} \overline{u}\overline{v}$$

and recalling that $A(\phi)$ denotes the sum of the coefficients of the posiform ϕ, we can write the above MAX-2-SAT problem as

$$\max_{\mathbf{x} \in \mathbb{B}^n} f(\mathbf{x}) = \max_{\mathbf{x} \in \mathbb{B}^n}(A(\phi) - \phi(\mathbf{x})) = A(\phi) - \min_{\mathbf{x} \in \mathbb{B}^n} \phi(\mathbf{x}). \tag{75}$$

By applying roof-duality (see Section 5.1.3) we can bring the quadratic posiform ϕ to the form

$$\phi(\mathbf{x}) = C_2(\phi) + l(\mathbf{x}) + \psi(\mathbf{x}), \tag{76}$$

where $C_2(\phi)$ is the roof-dual value of ϕ, $l(\mathbf{x})$ is a linear posiform, and $\psi(\mathbf{x})$ is a pure quadratic posiform.

Let us also recall (see Section 5.1.5) that the right-hand side of (76) can be obtained by iteratively applying identities of form (44)

$$u_1 + \overline{u}_1 u_2 + \overline{u}_2 u_3 + \cdots + \overline{u}_{k-1} u_k + \overline{u}_k = 1 + u_1 \overline{u}_2 + \cdots + u_{k-1} \overline{u}_k.$$

Since the sum of coefficients is exactly one more on the left-hand side than on the right-hand side above, the following equation is implied: $A(\phi) = 2C_2(\phi) + A(l) + A(\psi)$. Since both $A(l)$ and $C_2(\phi)$ are nonnegative, we can conclude that

$$A(\psi) \leqslant A(\phi) - C_2(\phi). \tag{77}$$

Let us now denote by $\mathbf{x}^*$ an optimum of the above MAX-2-SAT instance, i.e. a minimum of both ϕ and ψ according to (76). Furthermore, let $\mathbf{x}^r$ denote the binary vector obtained by eliminating first the variables which are fixed by strong persistency (see Theorem 8), and then by applying RoundDown $(\psi(\tfrac{1}{2},\ldots,\tfrac{1}{2}))$ to derive the values of the rest of the variables. Thus, for both vectors

$$l(\mathbf{x}^*) = l(\mathbf{x}^r) = 0 \tag{78}$$

holds by the above definitions, implying by (76) that

$$\min_{\mathbf{x}\in\mathbb{B}^n} \phi(\mathbf{x}) = \phi(\mathbf{x}^*) = C_2(\phi) + \psi(\mathbf{x}^*) \tag{79}$$

and

$$\phi(\mathbf{x}^r) = C_2(\phi) + \psi(\mathbf{x}^r). \tag{80}$$

Furthermore, by Proposition 5 we have

$$\psi(\mathbf{x}^*) \leqslant \psi(\mathbf{x}^r) \leqslant \psi(\tfrac{1}{2},\tfrac{1}{2},\ldots,\tfrac{1}{2}) = \mathrm{Exp}[\psi]. \tag{81}$$

Here the expectation is taken in a probability space, as in Proposition 5, in which the variables are pairwise-independent random variables with probabilities $\mathrm{Prob}(x_j = 1) = \mathrm{Prob}(x_j = 0) = \tfrac{1}{2}$, $j = 1,\ldots,n$. Since here $\mathrm{Prob}(uv = 1) = \tfrac{1}{4}$ for any pair of literals with $u \notin \{v,\bar{v}\}$, and since ψ is a pure quadratic posiform, we obtain from (81) that

$$\psi(\mathbf{x}^r) \leqslant \tfrac{1}{4}A(\psi). \tag{82}$$

Putting all the above together we get

$$\frac{f(\mathbf{x}^r)}{f(\mathbf{x}^*)} = \frac{A(\phi) - C_2(\phi) - \psi(\mathbf{x}^r)}{A(\phi) - C_2(\phi) - \psi(\mathbf{x}^*)} \geqslant \frac{A(\phi) - C_2(\phi) - \tfrac{1}{4}A(\psi)}{A(\phi) - C_2(\phi)} \geqslant \frac{3}{4}.$$

Here the first equality follows by (79) and (80). The first inequality follows by (82) and by $\psi(\mathbf{x}^*) \geqslant 0$, while the last inequality is implied by (77).

Thus, the above inequality shows that $\mathbf{x}^r$ provides a $\tfrac{3}{4}$-approximation of the problem (75). Let us add that $\mathbf{x}^r$ can be derived in $O(n^3)$ time from f, first determining the roof dual (76) by computing the maximum flow in a network of $2n + 2$ nodes (see Section 5.1.5) in $O(n^3)$ time, next fixing some of the variables by strong persistency in $O(|\phi|) = O(n^2)$ time, and finally applying RoundDown $(\psi(\tfrac{1}{2},\ldots,\tfrac{1}{2}))$ in $O(|\phi|) = O(n^2)$ time again.

7.3. $\tfrac{3}{4}$-approximation of MAXSAT via convex majorization

In this section we recall a $\tfrac{3}{4}$-approximation algorithm for the MAX SAT problem from [9]. This algorithm computes a vector $\mathbf{p}^* \in \mathbb{U}^n$ via convex programming without increasing the dimensions of the problem, and then constructs an approximative solution by applying the RoundDown algorithm starting from $\mathbf{p}^*$. The proof of correctness is very analogous to the one given in [69] with the notable difference that here only one starting point is needed for the rounding procedure, and even that point has to be determined only up to a certain fixed precision, allowing thus a faster and more robust computation.

Let us consider a posiform given as

$$\phi(\mathbf{x}) = \sum_{C \in \mathscr{C}} a_C \prod_{u \in C} \bar{u}, \tag{83}$$

where $\mathscr{C}$ is a given family of subsets of literals, and a_C are positive integers for $C \in \mathscr{C}$. Furthermore, let us consider the corresponding MAX SAT instance

$$\max_{\mathbf{x} \in \mathbb{B}^n} \sum_{C \in \mathscr{C}} a_C \left(1 - \prod_{u \in C} \bar{u}\right) = \max_{\mathbf{x} \in \mathbb{B}^n} (A(\phi) - \phi(\mathbf{x})), \tag{84}$$

where $A(\phi) = \sum_{C \in \mathscr{C}} a_C$ denotes again the sum of the coefficients of the posiform ϕ. Let us next consider the following convex majorant of ϕ

$$g(\mathbf{x}) = \sum_{C \in \mathscr{C}} a_C \left(\frac{\sum_{u \in C} \bar{u}}{|C|}\right)^{|C|} \tag{85}$$

formed as the sum of convex termwise majorants. It is easy to verify that

$$g(\mathbf{p}) \geqslant \phi(\mathbf{p}) \quad \text{holds for every } \mathbf{p} \in \mathbb{U}^n. \tag{86}$$

Let us further fix a small positive constant

$$\varepsilon = \frac{3}{16} - \frac{1}{2e} \approx 0.00356. \tag{87}$$

Let $\mathbf{p}^*$ be the minimum of $g(\mathbf{x})$ over $\mathbb{U}^n$ (since g is a smooth convex function, $\mathbf{p}^*$ is unique), and let $\hat{\mathbf{p}} \in \mathbb{U}^n$ be such that

$$g(\hat{\mathbf{p}}) \leqslant g(\mathbf{p}^*) + \varepsilon. \tag{88}$$

It is well-known that such a $\hat{\mathbf{p}}$ can be determined in polynomial time in the size of g, and hence in the size of ϕ, see e.g. [131]. In fact, due to the tolerance ε, most convex optimization algorithms provide such a vector in a very fast and numerically robust way.

Let us denote again by $\mathbf{x}^* \in \mathbb{B}^n$ a minimum of ϕ, and finally let $\mathbf{x}^r \in \mathbb{B}^n$ denote the binary vector obtained by the application of RoundDown ($\psi, \hat{\mathbf{p}}$), unless ϕ is a quadratic posiform and $\phi(\mathbf{x}^*) = 0$, in which case let $\mathbf{x}^r = \mathbf{x}^*$.

Theorem 13 (Badics [9]). *The vector $\mathbf{x}^r$ is a $\frac{3}{4}$-approximative solution of the MAX SAT problem* (84).

Proof. Clearly, if ϕ is a quadratic posiform, the equality $\phi(\mathbf{x}^*) = 0$ can be detected in $O(|\phi|)$ time, by solving the corresponding quadratic Boolean equation. Obviously, $\mathbf{x}^r = \mathbf{x}^*$ is the exact optimum of (84) in this case, and hence the statement holds.

Let us assume in the sequel that either ϕ is not quadratic, or $\phi(\mathbf{x}^*) > 0$, and let us compute $\mathbf{x}^r$ in polynomial time, as described above, by solving a convex minimization problem first, up to a precision of ε, and then running RoundDown ($\psi, \hat{\mathbf{p}}$). We have then the inequalities

$$\phi(\mathbf{x}^r) \leqslant \phi(\hat{\mathbf{p}}) \leqslant g(\hat{\mathbf{p}}) \leqslant g(\mathbf{p}^*) + \varepsilon \leqslant g(\mathbf{x}^*) + \varepsilon \tag{89}$$

and

$$g(\hat{\mathbf{p}}) \leqslant g(\mathbf{p}^*) + \varepsilon \leqslant g(\tfrac{1}{2}, \ldots, \tfrac{1}{2}) + \varepsilon, \qquad (90)$$

implying thus

$$\phi(\mathbf{x}^r) \leqslant \frac{g(\tfrac{1}{2}, \ldots, \tfrac{1}{2}) + g(\mathbf{x}^*)}{2} + \varepsilon. \qquad (91)$$

Let us also introduce the notation

$$l_C = \sum_{u \in C} \bar{u}(\mathbf{x}^*) \qquad (92)$$

for $C \in \mathscr{C}$. With this notation we have $\phi(\mathbf{x}^*) = \sum_{C \in \mathscr{C}: \, l_C = |C|} a_C$, and thus

$$f(\mathbf{x}^*) = A(\phi) - \phi(\mathbf{x}^*) = \sum_{C \in \mathscr{C}: \, l_C < |C|} a_C. \qquad (93)$$

Furthermore, we can write (91) as

$$\begin{aligned} f(\mathbf{x}^r) &= A(\phi) - \phi(\mathbf{x}^r) \\ &\geqslant A(\phi) - \varepsilon - \frac{g(\tfrac{1}{2}, \ldots, \tfrac{1}{2}) + g(\mathbf{x}^*)}{2} \\ &\geqslant -\varepsilon + \sum_{C \in \mathscr{C}} a_C \left(1 - \frac{1}{2} \left(\frac{l_C}{|C|} \right)^{|C|} - \frac{1}{2^{|C|+1}} \right). \end{aligned} \qquad (94)$$

It is easy to verify by elementary calculations that

$$1 - \frac{1}{2} \left(\frac{l_C}{|C|} \right)^{|C|} - \frac{1}{2^{|C|+1}} \geqslant \begin{cases} \frac{1}{4} & \text{if } l_C = |C|, \ |C| > 0, \\ \frac{3}{4} & \text{if } l_C < |C|, \ |C| \leqslant 2, \\ \frac{3}{4} + \varepsilon & \text{if } l_C < |C|, \ |C| \geqslant 3. \end{cases} \qquad (95)$$

Substituting these estimates back into (94) we obtain

$$f(\mathbf{x}^r) \geqslant \frac{3}{4} \sum_{\substack{C \in \mathscr{C} \\ l_C < |C|}} a_C + \frac{1}{4} \sum_{\substack{C \in \mathscr{C} \\ l_C = |C|}} a_C + \varepsilon \left(-1 + \sum_{\substack{C \in \mathscr{C} \\ l_C < |C| \\ |C| \geqslant 3}} a_C \right). \qquad (96)$$

Since the coefficients a_C for $C \in \mathscr{C}$ are assumed to be positive integers, and since $\frac{1}{4} > \varepsilon$, we have

$$\frac{1}{4} \sum_{\substack{C \in \mathscr{C} \\ l_C = |C|}} a_C + \varepsilon \left(-1 + \sum_{\substack{C \in \mathscr{C} \\ l_C < |C| \\ |C| \geqslant 3}} a_C \right) \leqslant 0$$

if and only if

$$\sum_{\substack{C \in \mathscr{C} \\ l_C = |C|}} a_C = 0 \quad \text{and} \quad \sum_{\substack{C \in \mathscr{C} \\ l_C < |C| \\ |C| \geqslant 3}} a_C = 0,$$

i.e, if and only if ϕ is a quadratic posiform and $\phi(\mathbf{x}^*) = 0$. Since this case was treated separately at the beginning of this proof, we can conclude from (96) that in all other cases

$$f(\mathbf{x}^r) \geqslant \frac{3}{4} \sum_{\substack{C \in \mathscr{C} \\ l_C < |C|}} a_C \tag{97}$$

must hold, implying thus by (93) that

$$\frac{f(\mathbf{x}^r)}{f(\mathbf{x}^*)} \geqslant \frac{3}{4}. \quad \square \tag{98}$$

8. Uncited References

[16,85].

References

[1] W.E. Adams, A. Billionnet, A. Sutter, Unconstrained 0–1 optimization and Lagrangean relaxation, First International Colloquium on Pseudo-Boolean Optimization and Related Topics (Chexbres, 1987), Discrete Appl. Math. 29 (2–3) (1990) 131–142.
[2] W.P. Adams, P.M. Dearing, On the equivalence between roof duality and Lagrangian duality for unconstrained 0–1 quadratic programming problems, Discrete Appl. Math. 48 (1994) 1–20.
[3] W.P. Adams, H.D. Sherali, A tight linearization and an algorithm for zero-one quadratic programming problems, Management Sci. 32 (10) (1986) 1274–1290.
[4] G. Alexe, P.L. Hammer, Boolean simplifications of the stability problem in graphs. RUTCOR Research Report, RUTCOR, August, 2001.
[5] N. Alon, J.H. Spencer, The Probabilistic Method, Wiley, New York, 1992.
[6] D.J. Amit, Modeling Brain Function: the World of Attractor Neural Networks, Cambridge University Press, Cambridge, 1989.
[7] S. Arora, C. Lund, R. Motwani, M. Sudan, M. Szegedy, Proof verification and hardness of approximation problems, Proceedings of the 33rd Annual IEEE Symposium on Foundations of Computer Science, Pittsburgh, PA, 1992, pp. 14–23.
[8] S. Arora, M. Safra, Probabilistic checking of proofs: a new characterization of NP, Proceedings of the 33rd Annual IEEE Symposium on Foundations of Computer Science, Pittsburgh, PA, 1992, pp. 2–13.
[9] T. Badics, Approximation of some Nonlinear Binary Optimization Problems, Ph.D. Thesis, RUTCOR, Rutgers University, 1996.
[10] T. Badics, E. Boros, Minimization of Half-Products, Math. Oper. Res. 23 (1998) 649–660.
[11] E. Balas, Extension de l'algorithme additif a la programmation en nombres at a la programmation non lineare, C. R. Acad. Sci. Paris 258 (1967) 5136–5139.
[12] E. Balas, J.B. Mazzola, Nonlinear 0–1 programming: I. Linearization techniques and II. Dominance relations and algorithms, Math. Programming 30 (1984) 1–45.
[13] E. Balas, M. Padberg, On the set covering problem, Oper. Res. 20 (1972) 1152–1161.
[14] E. Balas, M. Padberg, Set partitioning: a survey, SIAM Rev. 18 (1976) 710–760.

[15] J.F. Banzhaf III, Weighted voting does not work: a mathematical analysis, Rutgers Law Rev. 19 (1965) 317–343.

[16] F. Barahona, A solvable case of quadratic 0–1 programming, Discrete Appl. Math. 13 (1) (1986) 23–26.

[17] F. Barahona, M. Grötschel, M. Jünger, G. Reinelt, An application of combinatorial optimization to statistical physics and circuit layout design, Oper. Res. 36 (1988) 493–513.

[18] F. Barahona, M. Grötschel, A.R. Mahjoub, Facets of the bipartite subgraph polytope, Math. Oper. Res. 10 (1985) 340–358.

[19] F. Barahona, A.R. Mahjoub, On the cut polytope, Math. Programming 36 (1986) 157–173.

[20] M. Bellare, O. Goldreaich, M. Sudan, Free bits, PCPsw and non-approximability—towards tight results, Proceedings of the 36th Annual IEEE Symposium on Foundations of Computer Science, Milwaukee, WI, 1995, pp. 422–431.

[21] C. Benzaken, Y. Crama, P.L. Hammer, F. Maffray, More characterizations of triangulated graphs, RUTCOR Research Report 46-87, Rutgers University, 1987.

[22] C. Benzaken, S. Boyd, P.L. Hammer, B. Simeone, Adjoints of pure bidirected graphs, Congr. Numer. 39 (1983) 123–144.

[23] C. Benzaken, P.L. Hammer, B. Simeone, Some remarks on conflict graphs of quadratic pseudo-Boolean functions, Internat. Ser. Numer. Math. 55 (1980) 9–30.

[24] A. Billionnet, M. Minoux, Maximizing a supermodular pseudo-boolean function: a polynomial algorithm for supermodular cubic functions, Discrete Appl. Math. 12 (1) (1985) 1–11.

[25] A. Billionnet, A. Sutter, Persistency in quadratic 0–1 optimization, Math. Programming Ser. A 54 (1) (1992) 115–119.

[26] E. Boros, Maximum renamable Horn sub-CNFs, Discrete Appl. Math. 96–97 (1999) 29–40.

[27] E. Boros, Y. Caro, Z. Füredi, R. Yuster, Covering non-uniform hypergraphs, J. Combin. Theory (B) 82 (2001) 270–284.

[28] E. Boros, Y. Crama, P.L. Hammer, Upper bounds for quadratic 0–1 maximization, Oper. Res. Lett. 9 (1990) 73–79.

[29] E. Boros, Y. Crama, P.L. Hammer, Chvátal cuts and odd cycle inequalities in quadratic 0–1 optimization, SIAM J. Discrete Math. 5 (1992) 163–177.

[30] E. Boros, P.L. Hammer, A max-flow approach to improved roof-duality in quadratic 0–1 minimization, RUTCOR Research Report RRR 15-1989, RUTCOR, March 1989.

[31] E. Boros, P.L. Hammer, The max-cut problem and quadratic 0–1 optimization, polyhedral aspects, relaxations and bounds, Ann. Oper. Res. 33 (1991) 151–180.

[32] E. Boros, P.L. Hammer, A generalization of the pure literal rule for satisfiability problems, RUTCOR Research Report RRR 20-92, RUTCOR, April 1992; DIMACS Technical Report 92-19.

[33] E. Boros, P.L. Hammer, Cut-polytopes, Boolean quadric polytopes and nonnegative quadratic pseudo-Boolean functions, Math. Oper. Res. 18 (1993) 245–253.

[34] E. Boros, P.L. Hammer, M. Minoux, D. Rader, Optimal cell flipping to minimize channel density in VLSI design and pseudo-Boolean optimization, Discrete Appl. Math. 90 (1999) 69–88.

[35] E. Boros, P.L. Hammer, X. Sun, The DDT method for quadratic 0–1 optimization, RUTCOR Research Report RRR 39-1989, RUTCOR, October 1989.

[36] E. Boros, P.L. Hammer, X. Sun, Network flows and minimization of quadratic pseudo-Boolean functions, RUTCOR Research Report RRR 17-1991, RUTCOR, May 1991.

[37] E. Boros, A. Prékopa, Probabilistic bounds and algorithms for the maximum satisfiability problem, Ann. Oper. Res. 21 (1989) 109–126.

[38] J-M. Bourjolly, P.L. Hammer, W.R. Pulleyblank, B. Simeone, Boolean-combinatorial bounding of maximum 2-satisfiability, Computer Science and Operations Research, Williamsburg, VA, Pergamon Press, Oxford, 1992, pp. 23–42.

[39] M.L. Bushnell, I.P. Shaik, U.S. Patent # 5,422,891, Robust delay-fault built-in testing method and its apparatus. June 6, 1995. (Patents pending in EEC, Canada, Japan and South Korea.)

[40] J. Cheng, W. Kubiak, A faster fully polynomial approximation scheme for agreeable weighted completion time variance, Working Paper, Faculty of Business Administration, Memorial University of New-Foundland, 2000.

[41] Y. Crama, Recognition problems for special classes of polynomials in 0–1 variables, Math. Programming 44 (1989) 139–155.

[42] Y. Crama, Concave extensions for nonlinear 0–1 maximization problems, Math. Programming 61 (1993) 53–60.

[43] Y. Crama, P.L. Hammer, Recognition of quadratic graphs and adjoints of bidirected graphs, in: G.S. Bloom, R.L. Graham, J. Malkevitch (Eds.), Combinatorial Mathematics: Proceedings of the Third International Conference, Annals of the New York Academy of Sciences, Vol. 555, 1989, pp. 140–149.

[44] Y. Crama, P.L. Hammer, Bimatroidal independence systems, Z. Oper. Res. 33 (1989) 149–165.

[45] Y. Crama, P.L. Hammer, R. Holzman, A characterization of a cone of pseudo-Boolean functions via supermodularity-type inequalities, in: P. Kall, J. Kohlas, W. Popp, C.A. Zehnder (Eds.), Quantitative Methoden in den Wirtschaftswissenschaften, Springer, Berlin, Heidelberg, 1989, pp. 53–55.

[46] Y. Crama, P.L. Hammer, T. Ibaraki, Strong unimodularity for matrices and hypergraphs, Discrete Appl. Math. 15 (1986) 221–239.

[47] Y. Crama, P. Hansen, B. Jaumard, The basic algorithm for pseudo-Boolean programming revisited, Discrete Appl. Math. 29 (2–3) (1990) 171–185.

[48] Y. Crama, J.B. Mazzola, Valid inequalities and facets for a hypergraph model of the nonlinear knapsack and FMS part-selection problems, Ann. Oper. Res. 58 (1995) 99–128.

[49] C. De Simone, The cut polytope and the Boolean quadric polytope, Discrete Math. 79 (1989/90) 71–75.

[50] C. De Simone, M. Diehl, M. Jünger, P. Mutzel, G. Reinelt, G. Rinaldi, Exact ground states of Ising spin glasses: New experimental results with a branch and cut algorithm, J. Statist. Physics 80 (1995) 487–496.

[51] D. de Werra, On some properties of the struction of a graph, SIAM J. Algebraic Discrete Methods 5 (2) (1984) 239–243.

[52] M. Deza, M. Laurent, Facets for the cut cone I, Math. Programming Ser. A 56 (1992) 121–160.

[53] M. Deza, M. Laurent, Facets for the cut cone II: clique-web inequalities, Math. Programming Ser. A 56 (1992) 161–188.

[54] C. Ebenegger, P.L. Hammer, D. de Werra, Pseudo-Boolean functions and stability of graphs, Ann. Discrete Math. 19 (1984) 83–98.

[55] C.S. Fabian, S. Rudeanu, GH. Weisz, Rezolvarea problemelor de programare pseudobooleana liniara cu ajutorul unui calculator electronic, Studii Si Cercetari 10 (tomul 20) (1968).

[56] U. Feige, M.X. Goemans, Approximating the value of two prover proof systems, with applications to MAX 2SAT and MAX DICUT, Proceedings of the Third Israel Symposium on Theory of Computing and Systems, Tel Aviv, Israel, 1995, pp. 182–189.

[57] U. Feige, S. Goldwasser, L. Lovasz, S. Safra, Interactive proofs and the hardness of approximating cliques, J. ACM 43 (1996) 268–292.

[58] M.L. Fisher, G.L. Nemhauser, L.A. Wolsey, An analysis of approximations for maximizing submodular setfunctions—I, Math. Programming 14 (1978) 265–294.

[59] R. Fortet, L'algebre de Boole en recherche operationelle, Cahiers Études Centre Rech. Oper. 1 (1959) 5–36.

[60] R. Fortet, Applications de l'algebre de Boole en recherche operationelle, Rev. Francaise Rech. Oper. 4 (1960) 17–26.

[61] A.S. Fraenkel, P.L. Hammer, Pseudo-Boolean functions and their graphs, Ann. Discrete Math. 20 (1984) 137–146.

[62] G. Gallo, B. Simeone, On the supermodular knapsack problem, Math. Programming Study 45 (1989) 295–309.

[63] M.R. Garey, D.S. Johnson, Computers and Intractability: an Introduction to the Theory of NP-Completeness, Freeman, San Francisco, 1979.

[64] F. Glover, Future paths for integer programming and links to artificial intelligence, Comput. Oper. Res. 5 (1986) 533–549.

[65] F. Glover, Tabu search—Part I, ORSA J. Comput. 1 (1989) 190–206.

[66] F. Glover, B. Alidaee, C. Rego, G. Kochenberger. One-pass heuristics for large scale unconstrained binary quadratic problems, Technical Report, HCES-09-00, Hearin Center for Enterprise Science, September 2000.

[67] F. Glover, R.E. Woolsey, Further reduction of zero–one polynomial programs to zero–one linear programming problems, Oper. Res. 21 (1973) 156–161.

[68] F. Glover, R.E. Woolsey, Note on converting the 0–1 polynomial programming problems to zero–one linear programming problems, Oper. Res. 22 (1974) 180–181.

[69] M.X. Goemans, D.P. Williamson, New $\frac{3}{4}$-approximation algorithms for the maximum satisfiability problem, SIAM J. Discrete Math. 7 (1994) 656–666.

[70] M.X. Goemans, D.P. Williamson, Improved approximation algorithms for maximum cut and satisfiability problems using semidefinite programming, J. ACM 42 (1995) 1115–1145.

[71] M. Grötschel, L. Lovász, L. Scrijver, The ellipsoid method and its consequences in combinatorial optimization, Combinatorica 1 (1981) 169–197.

[72] P.L. Hammer, Some network flow problems solved with pseudo-Boolean programming, Oper. Res. 13 (1965) 388–399.

[73] P.L. Hammer, Plant location—a pseudo-Boolean approach, Israel J. Technol. 6 (1968) 330–332.

[74] P.L. Hammer, Pseudo-Boolean remarks on balanced graphs, Internat. Ser. Numer. Math. 36 (1977) 69–78.

[75] P.L. Hammer, The conflict graph of a pseudo-Boolean function, Technical Report, Bell Laboratories, August 1978.

[76] P.L. Hammer, P. Hansen, P.M. Pardalos, D.J. Rader Jr., Maximizing the product of two linear functions in 0–1 variables, RUTCOR Research Report RRR 2-97, RUTCOR, February 1997.

[77] P.L. Hammer, P. Hansen, B. Simeone, Upper planes of quadratic 0–1 functions and stability in graphs, in: O. Mangasarian, R.R. Meyer, S.M. Robinson (Eds.), Nonlinear Programming 4, Academic Press, New York, 1981, pp. 395–414.

[78] P.L. Hammer, P. Hansen, B. Simeone, On vertices belonging to all or to no maximum stable sets of a graph, SIAM J. Algebraic Discrete Methods 3 (1982) 511–522.

[79] P.L. Hammer, P. Hansen, B. Simeone, Roof duality, complementation and persistency in quadratic 0–1 optimization, Math. Programming 28 (1984) 121–155.

[80] P.L. Hammer, R. Holzman, Approximations of pseudo-Boolean functions; applications to game theory, ZOR—Methods and Models of Operations Research 36 (1992) 3–21.

[81] P.L. Hammer, B. Kalantari, A bound on the roof duality gap, Combinatorial Optimization, Lecture Notes in Mathematics, Vol. 1403, 1989, pp. 254–257.

[82] P.L. Hammer, T. Liebling, B. Simeone, D. de Werra, From linear separability to unimodality: a hierarchy of pseudo-Boolean functions, ORWP 84-3, Ecole Polytechnique Federale de Lausanne. SIAM J. Discrete Math. 1 (1988) 177–187.

[83] P.L. Hammer, N.V.R. Mahadev, D. de Werra, Stability in CAN-free graphs, J. Combin. Theory (B) 38 (1985) 23–30.

[84] P.L. Hammer, N.V.R. Mahadev, D. de Werra, The struction of a graph: application to CN-free graphs, Combinatorica 5 (1985) 141–147.

[85] P.L. Hammer, U.N. Peled, On the maximization of a pseudo-Boolean function, J. ACM 19 (1972) 265–282.

[86] P.L. Hammer, U.N. Peled, S. Sorensen, Pseudo-Boolean functions and game theory. I. Core elements and Shapley value, Cahiers Centre Etudes Rech. Oper. 19 (1977) 159–176.

[87] P.L. Hammer, I. Rosenberg, S. Rudeanu, On the determination of the minima of pseudo-Boolean functions, Stud. Cerc. Mat. 14 (1963) 359–364 (in Romanian).

[88] P.L. Hammer, I. Rosenberg, S. Rudeanu, Application of discrete linear programming to the minimization of Boolean functions, Rev. Math. Pures Appl. 8 (1963) 459–475.

[89] P.L. Hammer, I. Rosenberg, Linear decomposition of a positive group-Boolean function, in: L. Collatz, W. Wetterling (Eds.), Numerische Methoden bei Optimierung, Vol. II, Birkhauser, Basel, Stuttgart, 1974, pp. 51–62.

[90] P.L. Hammer, A.A. Rubin, Some remarks on quadratic programming with 0–1 variables, Rev. Frances d'Informatique Rech. Oper. 4 (1970) 67–79.

[91] P.L. Hammer, S. Rudeanu, Boolean Methods in Operations Research and Related Areas, Springer, Berlin, Heidelberg, New York, 1968.

[92] P.L. Hammer, E. Shliffer, Applications of pseudo-Boolean methods to economic problems, Theory and Decision 1 (1971) 296–308.

[93] P.L. Hammer, B. Simeone, Quasimonotone boolean functions and bistellar graphs, Ann. Discrete Math. 9 (1980) 107–119.

[94] P.L. Hammer, B. Simeone, Quadratic functions of binary variables. Combinatorial Optimization, Lecture Notes in Mathematics, Vol. 1403, Springer, Berlin, 1989, pp. 1–56.

[95] P. Hansen, Methods of nonlinear 0–1 programming, Ann. Discrete Math. 5 (1979) 53–70.

[96] P. Hansen, The steepest ascent mildest descent heuristic for combinatorial programming, Presented at the Congress on Numerical Methods in Combinatorial Optimization, Capri, March 1986.

[97] P. Hansen, M. Poggi de Aragão, C.C. Ribeiro, Boolean query optimization and the 0–1 hyperbolic sum problem, Ann. Math. Artif. Intell. 1 (1990) 97–109.

[98] P. Hansen, S.H. Lu, B. Simeone, On the equivalence of paved-duality and standard linearization in nonlinear optimization, Discrete Appl. Math. 29 (1990) 187–193.

[99] P. Hansen, B. Simeone, A class of quadratic pseudoboolean functions whose maximization is reducible to a network flow problem, CORR 79-39, Department of Comb. Optim., University of Waterloo.

[100] P. Hansen, B. Simeone, Unimodular functions, Discrete Appl. Math. 14 (1986) 269–281.

[101] J. Hå stad, Some optimal inapproximability results, Proceedings of the 28th Annual ACM Symposium on Theory of Computing, El Paso, TX, 1997, pp. 1–10.

[102] A. Hertz, Quelques utilisations de la struction, Discrete Math. 59 (1–2) (1986) 79–89.

[103] A. Hertz, Polynomially solvable cases for the maximum stable set problem, ARIDAM VI and VII, New Brunswick, NJ, 1991/1992. Discrete Appl. Math. 60 (1995) 195–210.

[104] A. Hertz, On the use of Boolean methods for the computation of the stability number, Discrete Appl. Math. 76 (1–3) (1997) 183–203.

[105] F.S. Hillier, The Evaluation of Risky Interrelated Investments, North-Holland, Amsterdam, 1969.

[106] K.W. Hoke, M.F. Troyon, The struction algorithm for the maximum stable set problem revisited, Discrete Math. 131 (1994) 105–113.

[107] J.J. Hopfield, Neural networks and physical systems with emergent collective computational abilities, Proc. Nat. Acad. Sci. U.S.A. (Biophysics) 79 (1982) 2554–2558.

[108] S. Iwata, L. Fleischer, S. Fujishige, A strongly polynomial-time algorithm for minimizing submodular functions, Proceedings of the 32nd ACM Symposium on Theory of Computing, May, 2000.

[109] D.S. Johnson, Approximation algorithms for combinatorial problems, J. Comput. System Sci. 9 (1974) 256–278.

[110] D.S. Johnson, C.H. Papadimitriou, M. Yannakakis, How easy is local search, Proceedings of the 26th Annual Symposium on the Foundations of Computer Science, 1985, pp. 39–42.

[111] B. Jurisch, W. Kubiak, J. Józefowska, Algorithms for minclique scheduling problems, Discrete Appl. Math. 72 (1997) 115–139.

[112] B. Kalantari, Quadratic functions with exponential number of local maxima, Oper. Res. Lett. 5 (1986) 47–49.

[113] H. Karloff, U. Zwick, A 7/8-approximation algorithm for MAX 3SAT? Proceedings of the 38th Annual IEEE Symposium on Foundations of Computer Science, Miami Beach, FL, 1997, pp. 406–415.

[114] R.M. Karp, Reducibility among combinatorial problems, in: R.G. Miller, J.W. Thatcher (Eds.), Complexity of Computer Computation, Plenum Press, New York, 1972, pp. 85–104.

[115] S. Kirkpatrick, C.D. Gelatt Jr., M.P. Vecchi, Optimization by simulated annealing, Science 220 (1983) 671–680.

[116] R. Kohli, R. Krishnamurti, Average performance of heuristics for satisfiability, SIAM J. Discrete Math. 2 (1989) 508–523.

[117] F. Korner, An effective branch-and-bound algorithm for Boolean quadratic optimization problems, Z. Angew. Math. Mech. 65 (8) (1985) 392–394.

[118] W. Kubiak, New results on the completion time variance minimization, Discrete Appl. Math. 58 (1995) 157–168.

[119] W. Kubiak, Minimization of ordered, symmetric half-products, Research Report, Faculty of Business Administration, Memorial University of Newfoundland, 2000.

[120] D.J. Laughhunn, Quadratic binary programming with applications to capital budgeting problems, Oper. Res. 18 (1970) 454–461.

[121] D.J. Laughhunn, D.E. Peterson, Computational experience with capital expenditure programming models under risk, J. Business Finance 3 (1971) 43–48.

[122] E. Lawler, The quadratic assignment problem, Manage. Sci. 9 (1963) 586–599.

[123] K.J. Lieberherr, E. Specker, Complexity of partial satisfaction, J. ACM 28 (1981) 411–421.

[124] L. Lovász, Submodular functions and convexity, in: A. Bachem, M. Grotschel, B. Korte (Eds.), Mathematical Programming—The State of the Art, Springer, Berlin, 1983, pp. 235–257.

[125] S.H. Lu, A.C. Williams, Roof duality for 0–1 nonlinear optimization, Math. Prog. 37 (1987) 357–360.

[126] D.C. Mattis, Solvable spin systems with random interaction, Phys. Lett. 56 (1976) 412.

[127] B, Monien, E. Speckenmeyer, Solving satisfiability in less than 2^n steps, Discrete Applied Math. 10 (1985) 287–295.

[128] A. Nagih, Sur la résolution des programmes fractionnaires en variables 0–1, Ph.D. Thesis, Université Paris 13, 1996.

[129] G.L. Nemhauser, L.E. Trotter, Vertex packing: structural properties and algorithms, Math. Programming 8 (1978) 232–248.

[130] G.L. Nemhauser, L. Wolsey, Maximizing submodular set functions: formulations and analysis of algorithms, Ann. Discr. Math. 11 (1981) 279–301.

[131] Y. Nesterov, A. Nemirovskii, Interior-Point Polynomial Algorithms in Convex Programming, SIAM Studies in Applied Mathematics, SIAM, Philadelphia, PA, 1994.

[132] J. von Neumann, O. Morgenstern, Theory of Games and Economic Behavior, Princeton University Press, Princeton, NJ, 1944.

[133] J. Nieminen, A linear pseudo-Boolean viewpoint on matching and other central concepts in graph theory, Zastos. Mat. 14 (1974) 365–369.

[134] G. Owen, Game Theory, Academic Press, New York, 1982.

[135] M. Padberg, The Boolean quadric polytope: some characteristics, facets and relatives, Math. Programming 45 (1989) 139–172.

[136] C.H. Papadimitriou, M. Yannakakis, Optimization, approximation, and complexity classes, J. Comput. System Sci. 43 (1991) 425–440.

[137] S.G. Papaioannou, Optimal test generation in combinational networks by pseudo-Boolean programming, IEEE Trans. Comput. 26 (1977) 553–560.

[138] L. Personnaz, I. Guyon, G. Dreyfus, Collective computational properties of neural networks: new learning mechanisms, Phys. Rev. A 34 (1986) 4217–4228.

[139] J.C. Picard, M. Queyranne, On the integer-valued variables in the linear vertex packing problem, Math. Programming 12 (1977) 97–101.

[140] J.C. Picard, H.D. Ratliff, Minimum cuts and related problems, Networks 5 (1975) 357–370.

[141] J.C. Picard, H.D. Ratliff, A cut approach to the rectilinear facility location problem, Oper. Res. 26 (1978) 422–433.

[142] P. Raghavan, Probabilistic construction of deterministic algorithms: approximating packing integer programs, J. Comput. System. Sci. 37 (1988) 130–143.

[143] P. Raghavan, C. Thompson, Randomized rounding: A technique for provably good algorithms and algorithmic proofs, Combinatorica 7 (1987) 365–374.

[144] R.H. Ranyard, An algorithm for maximum likelihood ranking and Slater's i from paired comparisions, British J. Math. Statist. Psychol. 29 (1976) 242–248.

[145] M.R. Rao, Cluster analysis and mathematical programming, J. Amer. Statist. Assoc. 66 (1971) 622–626.

[146] J. Rhys, A selection problem of shared fixed costs and network flows, Manage. Sci. 17 (1970) 200–207.

[147] P. Robillard, (0, 1) hyperbolic programming problems, Naval Res. Logist. Quart. 18 (1971) 47–57.

[148] I.G. Rosenberg, 0–1 optimization and non-linear programming, Rev. Française Automatique, Informatique Rech. Opér. (Série Bleue) 2 (1972) 95–97.

[149] I.G. Rosenberg, Reduction of bivalent maximization to the quadratic case, Cahiers Centre Etudes Rech. Oper. 17 (1975) 71–74.

[150] I.G. Rosenberg, Linear decompositions of positive real function of binary arguments, Utilitas Math. 17 (1980) 17–34.

[151] V. Rödl, C.A. Tovey, Multiple optima in local search, J. Algorithms 8 (2) (1987) 250–259.

[152] A. Schrijver, A combinatorial algorithm minimizing submodular functions in strongly polynomial time. J. Combin. Theory (B) 80 (2000) 346–355.

[153] I.P. Shaik, An optimization approach to robust delay-fault built-in testing, Ph.D. in Electrical Engineering, Rutgers University, 1996.

[154] L.S. Shapley, A value for *n*-person games, in: H.W. Kuhn, A.W. Tucker (Eds.), Contributions to the theory of games II, Princeton University Press, Princeton, NJ, 1953.

[155] I. Shepanik, Use of pseudo-boolean preference functions in metagame analysis, Ph.D. Thesis, Department of System Design, 1973.

[156] B. Simeone, Quadratic 0–1 programming, Boolean functions and graphs, Ph.D. Thesis in Combinatorics and Optimization, Waterloo, 1979.

[157] C.A. Tovey, Hill climbing with multiple local optima, SIAM J. Algebraic Discrete Methods 6 (1985) 384–393.

[158] C.A. Tovey, Low order polynomial bounds on the expected performance of local improvement algorithms, Math. Programming 35 (1986) 193–224.

[159] L. Trevisan, G.B. Sorkin, M. Sudan, D.P. Williamson, Gadgets, approximation, and linear programming (extended abstract). Proceedings of the 37th Annual IEEE Symposium on Foundations of Computer Science, Burlington, VT, 1996, pp. 617–626.

[160] V. Trubin, On a method of solution of integer linear programming problems of special kind, Soviet Math. Dokl. 10 (1969) 1544–1546.

[161] Z. Tuza, Maximum cuts: improvements and local algorithmic analogues of the Edwards–Erdös inequality, Discrete Math. 194 (1999) 39–58.

[162] A. Warszawski, Pseudo-Boolean solutions to multidimensional location problems, Oper. Res. 22 (1974) 1081–1085.

[163] L.G. Watters, Reduction of integer polynomial problems to zero-one linear programming problems, Oper. Res. 15 (1967) 1171–1174.

[164] D.J.A. Welsh, Matroid Theory, Academic Press, London, 1976.

[165] H.M. Weingartner, Capital budgeting of interrelated projects: survey and synthesis, Manage. Sci. 12 (1966) 485–516.

[166] M. Yannakakis, On the approximation of maximum satisfiability, J. Algorithms 17 (1994) 475–502.

[167] A.P. Young, Spin glasses, J. Statist. Phys. 34 (1984) 871–881.

[168] U. Zwick, Approximation algorithms for constraint satisfaction problems involving at most three variables per constraint, Proceedings of the Ninth SODA, 1998, pp. 201–210.

ELSEVIER

Discrete Applied Mathematics 123 (2002) 227–256

DISCRETE
APPLIED
MATHEMATICS

Scheduling and constraint propagation

Peter Brucker*

Department of Mathematics and Computer Science, Universität Osnabrück, D-49069 Osnabrück, Germany

Received 4 October 1999; received in revised form 7 June 2000; accepted 7 August 2000

Abstract

The resource-constrained project-scheduling problem (RCPSP) and some of its generalizations are defined. Furthermore, constraint propagation techniques for these problems and related machine scheduling problems are introduced and possible applications of these techniques in connection with lower bound calculations, branch-and-bound methods, and heuristics are discussed. © 2002 Elsevier Science B.V. All rights reserved.

Keywords: Project scheduling/resource constraints; Machine scheduling; Constraint propagation; Branch-and-bound algorithm; Linear programming

1. Introduction

Scheduling is concerned with the optimal allocation of scarce resources to activities over time. It has been the subject of extensive research since the early 1950s. Much of the early work on scheduling was concerned with the analysis of single-machine systems, parallel-machines systems, and shop problems. Later more complex machine scheduling situations were investigated. General references on sequencing and scheduling are survey papers by Anderson et al. [3], Lawler et al. [30], Lee et al. [32], and the books by Blazewicz et al. [8], Brucker [9], Pinedo [41].

More recently the resource-constrained project scheduling problem (RCPSP) has been investigated. The RCPSP and its generalizations are very general scheduling models which contain almost all complex machine scheduling problems as special cases (cf. [11] for a survey on the RCPSP).

* Tel.: +49-541-9692538; fax: +49-541-9692770.

E-mail address: peter@mathematik.uni-osnabrueck.de (P. Brucker).

Constraint propagation is a powerful tool which is used in connection with various solution methods for scheduling problems. The purpose of this tutorial is to describe applications of constraint propagation techniques to the RCPSP and to machine scheduling problems.

Accordingly, this tutorial has two parts. In the first part we introduce the RCPSP and some of its generalizations. We also describe the relations to machine scheduling problems. The second part describes constraint propagation techniques and shows how they can be used in connection with branch-and-bound algorithms, lower bound calculations, and heuristics.

All scheduling problems considered in this paper are assumed to be deterministic, i.e. the data that define a scheduling problem instance are known with certainty in advance.

In Section 2.1 the RCPSP and its generalizations are defined. Section 2.2 introduces machine scheduling models as special cases of RCPSP-models. In Section 3 constraint propagation techniques are described. They are used to calculate lower bounds for the RCPSP (Section 4.1) and are combined with linear programming to improve these bounds (Sections 4.2–4.4). In Section 5 it is shown how constraint propagation can be used to enhance branch-and-bound algorithms for scheduling problems. A short discussion of some heuristics for the RCPSP which use constraint propagation follows in Section 6. Section 7 contains conclusions and gives an overview on implementations of solution procedures for scheduling problems in which constraint propagation techniques are applied.

2. Scheduling models

The resource-constrained project scheduling problem (RCPSP) is one of the basic scheduling problems. In Section 2.1 we introduce this problem and some of its generalizations. Machine scheduling problems which may be considered as special cases are introduced in Section 2.2.

2.1. The RCPSP and generalizations

The resource-constrained project scheduling problem ($RCPSP$) may be formulated as follows. Given are n *activities* $i = 1, \ldots, n$ are r (*renewable*) *resources* $k = 1, \ldots, r$. A constant amount of R_k units of resource k is available at any time. Activity i must be processed for p_i time units. During this time period a constant amount of r_{ik} units of resource k is occupied. Furthermore, *precedence constraints* are defined between activities. These are given by relations $i \to j$, where $i \to j$ means that activity j cannot start before activity i is completed. The objective is to determine starting times S_i for the activities $i = 1, \ldots, n$ in such a way that

- at each time t the total resource demand is less than or equal to the resource

  availability for each resource type,   (2.1)

- the given precedence constraints are fulfilled, and   (2.2)

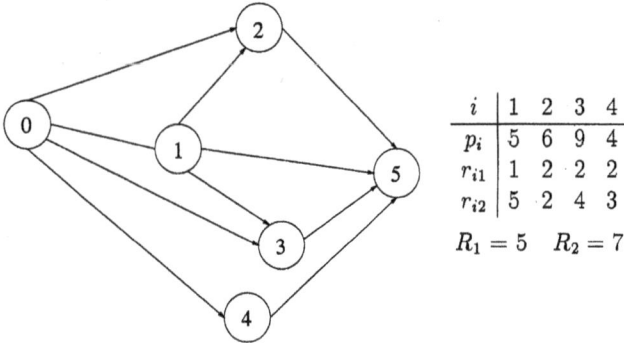

i	1	2	3	4
p_i	5	6	9	4
r_{i1}	1	2	2	2
r_{i2}	5	2	4	3

$$R_1 = 5 \quad R_2 = 7$$

Fig. 1. Project with 4 activities and 2 resources.

- the makespan $C_{\max} = \max_{i=1}^{n} C_i$ is minimized, where $C_i := S_i + p_i$ is assumed to be the completion time of activity i.

The fact that an activity which starts at time S_i finishes at time $S_i + p_i$ implies that activities are not preempted. We may relax this condition by allowing *preemption* (activity splitting). In this case the processing of any activity may be interrupted and resumed at a later date. It will be stated explicitly if we consider models with preemptions.

It is sometimes useful to add a unique *dummy beginning activity* 0 and a unique *dummy termination activity* $n+1$, each with processing time zero. Naturally, we must have $0 \rightarrow i$ and $i \rightarrow n+1$ for all other activities $i = 1, \ldots, n$. The dummy activities require no resources. Furthermore, S_0 is the starting time and S_{n+1} may be interpreted as the makespan of the project.

If preemption is not allowed the vector $S = (S_i)$ defines a *schedule* of the project. S is *feasible* if conditions (2.1) and (2.2) are fulfilled. Schedules may be graphically represented by Gantt charts, as illustrated in Fig. 2.

We may represent the structure of the RCPSP by a so-called *activity-on-node graph* $G = (V, E)$ where V is the set of all activities and $E = \{(i, j) \mid i, j \in V; i \rightarrow j\}$ represents the precedence constraints. For some activity i we define

$$Pred(i) := \{j \mid (j, i) \in E\} \quad \text{and} \quad Succ(i) := \{j \mid (i, j) \in E\}.$$

$Pred(i)$ ($Succ(i)$) is the set of *direct predecessors* (*successors*) of activity i.

Fig. 1 illustrates a small project with four activities (plus the dummy activities 0 and 5) and two resource types. A corresponding feasible schedule is shown in Fig. 2. Next we will discuss some generalizations of the RCPSP-model.

2.1.1. Generalized precedence constraints

Start–start relations of the form

$$S_i + l_{ij} \leqslant S_j, \tag{2.3}$$

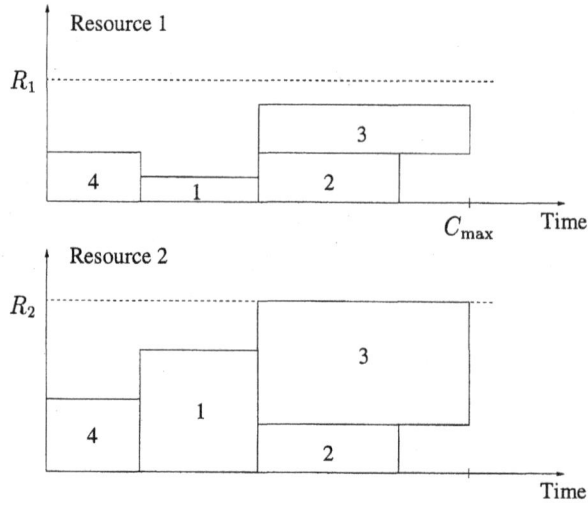

Fig. 2. A feasible schedule.

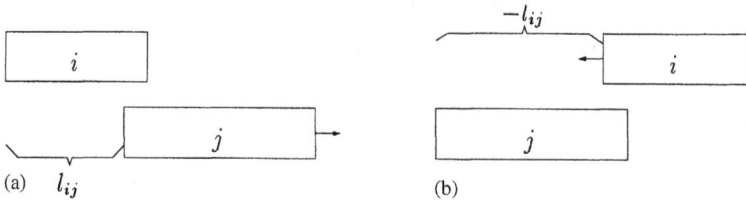

Fig. 3. Positive and negative time-lags.

where l_{ij} is an arbitrary integer number, may be introduced. The interpretation of relation (2.3) depends on the sign of l_{ij}.

If $l_{ij} \geq 0$, then activity j cannot start before l_{ij} time units after the start of activity i. This means that activity j does not start before the starting time of activity i, and l_{ij} is a minimal distance between both starting times (Fig. 3(a)).

If $l_{ij} < 0$, then the earliest start of activity j is $-l_{ij}$ time units before the start of activity i, i.e. activity i cannot start more than $-l_{ij}$ time units later than the starting time of activity j. If $S_j \leq S_i$, this means that $-l_{ij}$ is a maximal distance between both starting times (Fig. 3(b)).

l_{ij} is called *positive* (*negative*) *time-lag* if (2.3) holds and $l_{ij} > 0$ ($l_{ij} < 0$).

Relations (2.3) are very general timing relations between activities. For example, (2.3) with $l_{ij} = p_i$ is equivalent to the precedence relation $i \rightarrow j$. More generally, if there should be a minimal time distance of d_{ij} units between the completion of activity i and the start of activity j then we write $S_i + p_i + d_{ij} \leq S_j$. If for numbers d_{ij}, u_{ij} with $0 \leq d_{ij} \leq u_{ij}$ the inequalities $S_i + p_i + d_{ij} \leq S_j$ and $S_j - u_{ij} - p_i \leq S_i$ hold, then the time between the finishing time of activity i and the starting time of activity j must be at least d_{ij}, but no more than u_{ij}. This includes the special case

$0 \leqslant d_{ij} = u_{ij}$, where activity j must start exactly d_{ij} time units after the completion of activity i.

Also release times r_i and deadlines d_i of activities i can be modeled by the relation (2.3). A *release time* r_i is an earliest starting time, while a *deadline* d_i is a latest finishing time of activity i. To model release times we add the restrictions $S_0 + r_i \leqslant S_i$. To model deadlines we add the restrictions $S_i - (d_i - p_i) \leqslant S_0$. If $r_i \leqslant d_i$, r_i is a release time and d_i is a deadline, then the interval $[r_i, d_i]$ is called a *time window* for activity i. Activity i must be processed completely within its time window.

2.1.2. Nonconstant resource profiles

So far we assumed that each resource k is available in each time period with constant amount R_k. However, we may also consider nonconstant resource profiles $R_k(t)$ where the resource level is time dependent. This allows to model the nonavailability of resources in certain time periods.

2.1.3. Setup times

In a scheduling model with *sequence-dependent setup times* the set of all activities is partitioned into disjoint sets $G_1, \ldots, G_q$, called groups. Associated with each pair (r, t) of group indices is a setup time s_{rt}. If i is directly processed before j for any $i \in G_r$ and $j \in G_t$, then the restriction

$$C_i + s_{rt} \leqslant S_j \qquad (2.4)$$

must be satisfied. Usually, one assumes that $s_{rr} = 0$, i.e. there is no setup if the group G_r does not change, and that the setup times satisfy the triangle inequality

$$s_{rt} + s_{tv} \geqslant s_{rv} \quad \text{for all } r, t, v \in \{1, \ldots, q\}.$$

2.1.4. Other objective functions

Besides the objective of minimizing the makespan $C_{max} := \max_{i=1}^{n} C_i$ one may consider other objective functions $f(C_1, \ldots, C_n)$ depending on the finishing time of the activities. Examples are $\sum_{i=1}^{n} w_i C_i$ (weighted flow time), $\max_{i=1}^{n} L_i$ (maximum lateness), and $\sum_{i=1}^{n} w_i T_i$ (weighted tardiness) where the lateness L_i and tardiness T_i are defined by $L_i := C_i - d_i$ and $T_i = \max\{0, L_i\}$, respectively. Here d_i is a given due date.

However, no constraint propagation results are known for these other objective functions. Thus, we consider only the makespan objective function.

The RCPSP may be used to model various discrete optimization problems like cutting stock problems [21], high school timetabling problems [43] and audit staff scheduling problems [15].

Machine scheduling problems are important special cases of resource-constrained project scheduling problems. These will be discussed in more detail in the next section.

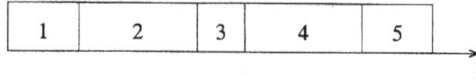

Fig. 4. Single machine schedule.

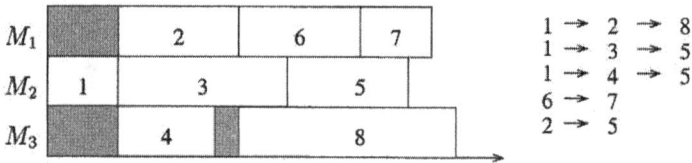

Fig. 5. Identical parallel machines schedule.

2.2. Machine scheduling problems

In this section we will introduce important classes of machine scheduling problems as special cases of the RCPSP. In machine scheduling the activities are usually called jobs or (in the case of shop scheduling problems) operations.

2.2.1. Single-machine scheduling problems

If n jobs $j = 1, \ldots, n$ with processing times p_j and precedences are to be processed on a single machine then we can model this by an RCPSP with one renewable resource, i.e. $r = 1$, where $R_1 = 1$ and $r_{j1} = 1$ for $j = 1, \ldots, n$. A schedule for such a problem with 5 jobs is shown in Fig. 4.

2.2.2. Identical parallel machines

Instead of a single machine there are m machines $M_1, \ldots, M_m$ on which the jobs are to be processed. The processing time p_j of job j does not depend on the machine on which j is processed. A corresponding schedule is shown in Fig. 5. Precedences between jobs are given next to the Gantt chart.

This problem corresponds to an RCPSP with $r = 1$, $R_1 = m$ and $r_{j1} = 1$ for $j = 1, \ldots, n$.

2.2.3. General shop scheduling problems

We have jobs $j = 1, \ldots, n$ and m machines $M_1, \ldots, M_m$. Job j consists of n_j operations $O_{1j}, \ldots, O_{n_j, j}$. Two operations of the same job cannot be processed at the same time. Operation O_{ij} must be processed for p_{ij} time units on a dedicated machine $\mu_{ij} \in \{M_1, \ldots, M_m\}$. Precedence constraints are given between arbitrary operations. Such a general shop scheduling problem can be modeled by an RCPSP with $r = m + n$ renewable resources with $R_k = 1$ for $k = 1, \ldots, m + n$ and $\sum_{j=1}^{n} n_j$ activities O_{ij}. Furthermore, operation O_{ij} uses r_{ijk} units of resource type k, where

$$r_{ijk} = \begin{cases} 1 & \text{if } \mu_{ij} = M_k \text{ or } k = m + j, \\ 0 & \text{otherwise.} \end{cases}$$

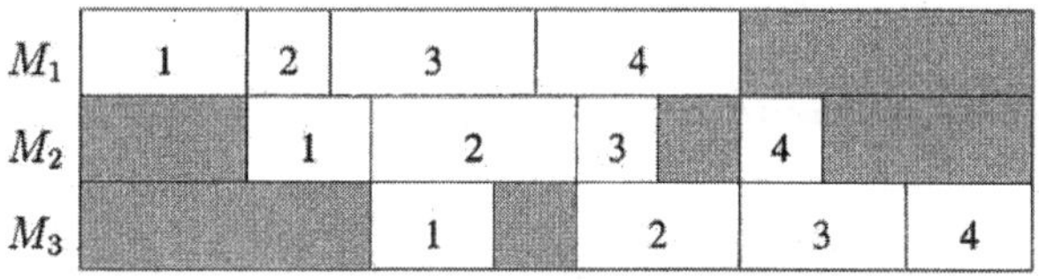

M_1	1	2	3	4		
M_2		1	2	3	4	
M_3			1	2	3	4

Fig. 6. Schedule for a flow shop problem.

The resources $k = 1,\ldots,m$ correspond with the machines while resource $m + j$ ($j = 1,\ldots,n$) is needed to model the fact that different operations of job j cannot be processed at the same time.

Job shop, flow shop, and open shop problems are important special cases of the general shop scheduling problem. These will be discussed next.

Job shop problems: A job shop problem is a general shop scheduling problem with precedences of the form

$$O_{1j} \rightarrow O_{2j} \rightarrow \cdots \rightarrow O_{nj,j}$$

for $j = 1,\ldots,n$, i.e. there are no precedences between operations of different jobs and the precedences between operations of the same job build a chain.

Flow shop problems: A flow shop problem is a special job shop problem with $n_j = m$ for $j = 1,\ldots,n$ and $\mu_{ij} = M_i$ for $i = 1,\ldots,m$, $j = 1,\ldots,n$, i.e. O_{ij} must be processed on M_i. In Fig. 6 a schedule for a flow shop problem with 4 jobs and 3 machines is shown.

Open shop problems: An open shop problem is like a flow shop problem but without precedences between the operations.

2.2.4. Multi-processor task scheduling problems

Again we have n jobs $j = 1,\ldots,n$ and m machines $M_1,\ldots,M_m$. Associated with each job j there is a processing time p_j and a subset of machines $\mu_j \subseteq \{M_1,\ldots,M_m\}$. During the time in which job j is processed it occupies each of the machines in μ_j. Finally, there are precedence constraints.

This problem can be formulated as an RCPSP with $r = m$ renewable resources with $R_k = 1$ for $k = 1,\ldots,r$. Furthermore,

$$r_{jk} = \begin{cases} 1 & \text{if } M_k \in \mu_j, \\ 0 & \text{otherwise.} \end{cases}$$

3. Constraint propagation

With constraint propagation timing restrictions are derived from given ones. These methods are used to tighten the search space in connection with branch-and-bound methods and local search heuristics. Also inconsistency may be derived, which possibly leads to lower bounds for the makespan minimization problem.

Constraint propagation techniques have mainly been developed in connection with machine scheduling problems [7,12,13,18,25]. In Brucker et al. [16] constraint propagation is applied to the RCPSP.

3.1. Basic relations

Given a schedule $S = (S_i)_{i=0}^{n+1}$, for each pair (i,j) of activities $i \neq j$ exactly one of the three relations $i \to j$, $j \to i$, or $i \| j$ holds.

The meaning of $i \to j$ is that activity j does not start before i is finished, i.e. that

$$S_i + p_i \leqslant S_j \tag{3.1}$$

holds. $i \to j$ is called a *conjunction*. A set of conjunctions is denoted by C. The meaning of $i \| j$ is that activities i and j are processed in parallel for at least one time unit. This is the case if and only if neither $i \to j$ nor $j \to i$ holds, i.e.

$$S_i + p_i > S_j \quad \text{and} \quad S_j + p_j > S_i \tag{3.2}$$

hold. $i \| j$ is called a *parallelity relations*. N is a set of parallelity relations.

The negation of $i \| j$ is a relation called *disjunction* and is denoted by $i - j$. $i - j$ means that either $i \to j$ or $j \to i$ holds. D denotes a set of disjunctions.

We are interested in sets C, N which describe a schedule minimizing the makespan. To derive such sets we may start with sets C_0, D_0, N_0 of conjunctions, disjunctions, parallel relations, respectively, which are satisfied by any optimal schedule, and try to derive further relations.

The following relations may be derived immediately from the data of an RCPSP.

An initial set C_0 of conjunctions is given by the set of all precedence relations $i \to j$.

An initial set D_0 of disjunctions may be induced by resource constraints for pairs of activities by setting $i - j \in D_0$ if and only if $r_{ik} + r_{jk} > R_k$ for some renewable resource k $(k = 1, \dots, r)$.

An initial set N_0 of parallelity relations may be derived by a procedure which is more complicated. We assume that $S_0 = 0$, i.e. that no activity can start earlier than time $t = 0$. Furthermore, we assume that an upper bound UB for the $C_{\max}$-value is given, i.e. we have $S_{n+1} \leqslant UB$. For activity i we define a

- *head* r_i by a lower bound for the earliest starting time of i, and a.
- *tail* q_i by a lower bound for the length of the time period between the finishing time of i and the optimal makespan.

Heads and tails may be calculated as follows. Let $G = (V, E)$ be the activity-on-node graph of the RCPSP. The length of a (directed) path p from activity i to j is the sum of processing times of all activities in p excluding the processing time of j. A valid head r_i is the length of a longest path from the dummy activity 0 to i. Symmetrically, a valid tail q_i is the length of a longest path from i to the dummy vertex $n + 1$ minus the processing time p_i of i. Given an upper bound UB one may define $d_i = UB - q_i$, where d_i is the latest finishing time of activity i in any schedule with $C_{\max} \leqslant UB$. Thus, activity i must be processed within its *time window* $[r_i, d_i]$ in any feasible schedule with $C_{\max} \leqslant UB$.

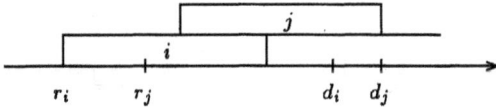

Fig. 7. Overlapping activities.

Let i and j be two activities with time windows $[r_i, d_i]$ and $[r_j, d_j]$. If

$$p_i + p_j > \max\{d_i, d_j\} - \min\{r_i, r_j\} \tag{3.3}$$

then i and j overlap in any feasible schedule with $C_{\max} \leqslant UB$ (cf. Fig. 7).

We define N_0 as the set of all parallelity relations induced in this way, i.e.

$$N_0 = \{i \,\|\, j \mid i \text{ and } j \text{ satisfy } (3.3)\}.$$

3.2. Start–start distance matrix

Let $S = (S_j)_{j=0}^{n+1}$ be a feasible schedule with $S_0 = 0$ and $S_{n+1} \leqslant UB$. Then we have $i \to j$ if and only if (3.1), i.e. $S_j - S_i \geqslant p_i$ holds. Furthermore, we have $i \,\|\, j$ if and only if (3.2) holds, which is equivalent to

$$S_j - S_i \geqslant -(p_j - 1) \quad \text{and} \quad S_i - S_j \geqslant -(p_i - 1) \tag{3.4}$$

because all data are integers. Additionally, for arbitrary activities i, j we have

$$S_i + p_i \leqslant S_{n+1} \leqslant UB \leqslant UB + S_j \quad \text{or} \quad S_j - S_i \geqslant p_i - UB$$

because $S_j \geqslant 0$ and $S_{n+1} \leqslant UB$.

If we define

$$d_{ij} = \begin{cases} 0 & \text{if } i = j, \\ p_i & \text{if } i \to j, \\ -(p_j - 1) & \text{if } i \,\|\, j, \\ p_i - UB & \text{otherwise} \end{cases} \tag{3.5}$$

then the entries of d_{ij} $(i, j = 0, \ldots, n+1)$ of the $(n+2) \times (n+2)$-matrix $D = (d_{ij})$ are lower bounds of the differences $S_j - S_i$, i.e.

$$S_j - S_i \geqslant d_{ij} \quad \text{for all } i, j = 0, \ldots, n+1. \tag{3.6}$$

We call D a start–start distance (SSD-)matrix.

If additionally generalized precedence constraints (2.3) hold which are equivalent to $S_j - S_i \geqslant l_{ij}$ then we may incorporate these constraints by replacing d_{ij} by

$$d_{ij} := \max\{d_{ij}, l_{ij}\}. \tag{3.7}$$

In this case all results derived are also valid for the RCPSP with generalized precedence constraints.

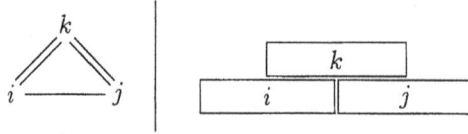

Fig. 8. Symmetric triple.

The relation $S_j - S_i \geq d_{ij}$ has the transitivity property:

$$S_j - S_i \geq d_{ij} \text{ and } S_k - S_j \geq d_{jk} \text{ imply } S_k - S_i \geq d_{ij} + d_{jk} \tag{3.8}$$

(we obtain the third relation by adding the first and second).

Due to (3.8) the matrix $D = (d_{ij})$ may be replaced by its transitive closure $\bar{D} = (\bar{d}_{ij})$, which can be derived from D in time $O(n^3)$ by applying the Floyd–Warshall algorithm (cf. [1]).

The first row of an SSD-matrix represents heads because $S_i = S_i - S_0 \geq d_{0i}$ for each activity. Similarly, the last column of an SSD-matrix contains information on the tails of the activities. More precisely, $d_{i,n+1} - p_i$ is a tail of activity i because we have

$$S_{n+1} - (S_i + p_i) = (S_{n+1} - S_i) - p_i \geq d_{i,n+1} - p_i.$$

From (3.1) and (3.4) we derive:

- $i \to j$ if and only if $d_{ij} \geq p_i$ holds, $\tag{3.9}$

- $i \,\|\, j$ if and only if both $d_{ij} \geq -(p_j - 1)$ and $d_{ji} \geq -(p_i - 1)$ hold. $\tag{3.10}$

If $d_{ii} > 0$ for some activity i then no feasible schedule with $C_{\max} \leq UB$ exists because $d_{ii} > 0$ implies the contradiction $0 = S_i - S_i \geq d_{ii} > 0$.

Calculating the transitive closure of the SSD-matrix is the simplest form of constraint propagation. One may start with an SSD-matrix containing basic information about the RCPSP like the relations in C_0, D_0, N_0, heads, tails, and generalized precedence constraints.

Other methods of constraint propagation will be discussed in the next sections. The results of these propagation techniques are increasing entries of D. In this case a further increase may be possible by applying the Floyd–Warshall algorithm.

3.3. Symmetric triples and extensions

A triple (i, j, k) of activities is called a *symmetric triple* if
- $k \,\|\, i$ and $k \,\|\, j$, and
- i, j, k cannot be processed simultaneously due to resource constraints.

For a symmetric triple (i, j, k) we can add $i - j$ to D (see Fig. 8). This can be seen as follows.

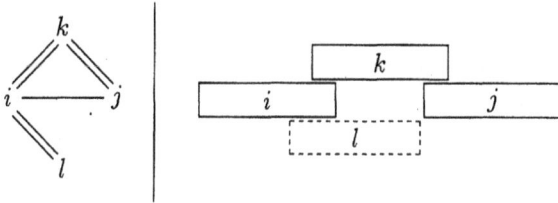

Fig. 9. Extension of a symmetric triple—Condition 1.

Assume that $i \| j$, i.e. i and j are processed jointly at some period t.

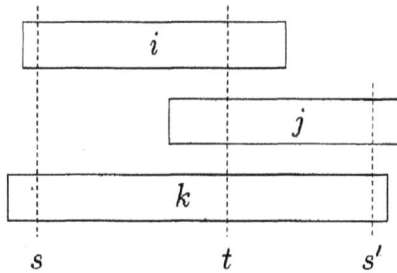

k must be processed jointly with $i(j)$ at some period $s(s')$, where $s < t < s'$ or $s' < t < s$. This implies that at time t all three activities must be processed jointly, which is a contradiction.

All symmetric triples can be found in $O(n|N|r)$ time.

Further relations can be deduced in connection with symmetric triples (i,j,k):

(1) If $l \| i$ and j,k,l cannot be processed simultaneously, then $l - j$ can be added to D (see Fig. 9). If, additionally, $i \to j$ $(j \to i)$ then $l \to j \in C$ $(j \to l \in C)$.

It is not difficult to prove the first part of statement (1): If $l \| j$ then we have

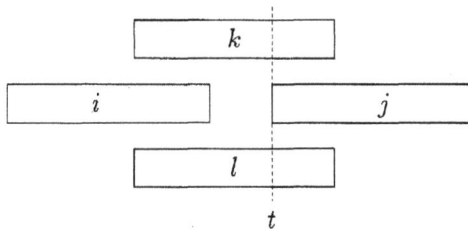

Thus, at time t activities j,k,l must be processed jointly, which is a contradiction. Similarly, the other two claims are true.

(2) Assume that $p_k - 1 \leqslant p_i$, $p_k - 1 \leqslant p_j$, $p_k - 1 \leqslant p_l$, i,k,l cannot be processed jointly, and j,k,l cannot be processed jointly, then $k - l \in D$ (see Fig. 10).

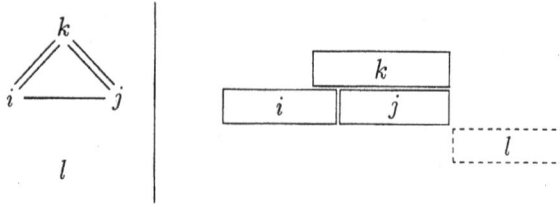

Fig. 10. Extension of a symmetric triple—Condition 2.

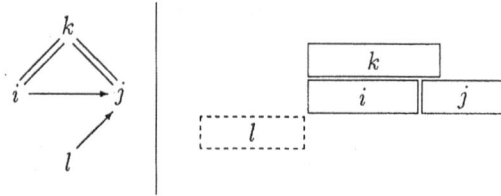

Fig. 11. Extension of a symmetric triple—Condition 3.

That statement (2) is correct can be seen as follows. Because $p_k - 1 \leqslant p_l$, it is not possible that l is processed after $i(j)$ and before $j(i)$. Thus, if $k \| l$ then we have

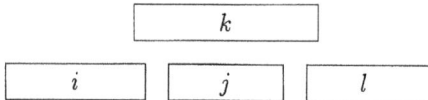

(the role of i and j may be interchanged) which contradicts $p_k - 1 \leqslant p_j$ (i and l occupy at least 2 time units of k).

Other propagation conditions are listed below. The proofs are similar to the proofs for (1) and (2). All the checks can be done in $O(rn^2 |N|)$ time.

(3) Suppose that the conditions $p_k - 1 \leqslant p_i$ and $p_k - 1 \leqslant p_l$ hold. Furthermore, assume that the precedence relations $l \rightarrow j$ and $i \rightarrow j$ are given and i, k and l cannot be processed in parallel. Then the additional conjunction $l \rightarrow k \in C$ can be fixed (see Fig. 11).

(4) Suppose that the conditions $p_k - 1 \leqslant p_j$ and $p_k - 1 \leqslant p_l$ hold. Furthermore, assume that the precedence relations $i \rightarrow j$ and $i \rightarrow l$ are given and j, k and l cannot be processed in parallel. Then the additional conjunction $k \rightarrow l \in C$ can be fixed (see Fig. 12).

(5) Suppose that the conditions $p_k - 1 \leqslant p_i$ and $p_k - 1 \leqslant p_j$ hold. Furthermore, assume that the precedence relations $l \rightarrow i$ and $l \rightarrow j$ ($i \rightarrow l$ and $j \rightarrow l$) are given. Then we may fix the conjunction $l \rightarrow k \in C$ ($k \rightarrow l \in C$) (see Fig. 13).

(6) Let the parallelity relations $l \| i$ and $l \| j$ be given. Then the additional parallelity relation $l \| k$ can be fixed (see Fig. 14).

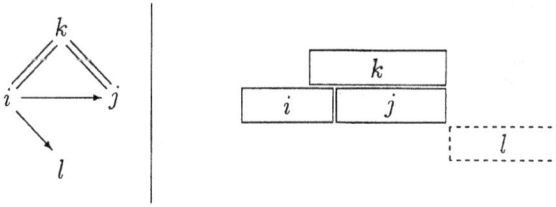

Fig. 12. Extension of a symmetric triple—Condition 4.

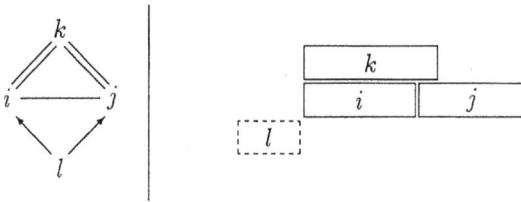

Fig. 13. Extension of a symmetric triple—Condition 5.

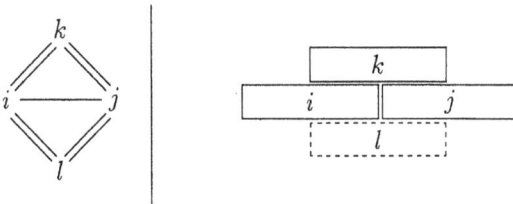

Fig. 14. Extension of a symmetric triple—Condition 6.

3.4. Disjunctive sets

A subset $I \subseteq \{1, \ldots, n\}$ of all (non-dummy) activities is called a *disjunctive set* (or *clique*) if $i - j$ or $i \to j$ or $j \to i$ for all $i, j \in I$, $i \neq j$. Disjunctive sets are cliques in the undirected graph defined by all disjunctions and conjunctions. Examples of disjunctive sets are

- the set of all jobs of a single-machine scheduling problem,
- the set of all operations belonging to the same job of a general shop problem,
- the set of operations to be processed by the same machine of a general shop problem,
- the set of tasks j with $M_i \in \mu_j$ for a fixed machine M_i, i.e., which occupy a machine M_i during processing.

We define $P(I) := \sum_{i \in I} p_i$. Let $[r_i, d_i]$ be a time window for each activity i. Then conjunctions $i \to j$ between jobs of a disjunctive set I may be derived from the following general result.

Theorem 3.1 (Dorndorf et al. [25]). *Let I be a disjunctive set and let $J', J'' \subseteq J \subseteq I$. If*

$$\max_{\substack{i \in J \setminus J' \\ j \in J \setminus J'' \\ i \neq j}} (d_j - r_i) < P(J) \tag{3.11}$$

then an activity in J' must start first or an activity in J'' must end last in J in each feasible schedule.

Proof. If no activity in J' starts first and no activity in J'' ends last in each feasible schedule, then all activities in J must be processed in a time interval of length

$$\max_{\substack{i \in J \setminus J' \\ j \in J \setminus J'' \\ i \neq j}} (d_j - r_i),$$

which is not possible if (3.11) holds. $\square$

By choosing different subsets J' and J'', different tests may be derived. The following tests of this type can be found in the literature.

Input test: If condition (3.11) holds for $J' = \{i\}$ with $i \in J$ and $J'' = \emptyset$, then activity i must start first in J, i.e. we conclude that

$i \to j$ for all $j \in J \setminus \{i\}$

(or short $i \to J \setminus \{i\}$). In this case i is called *input* of J.

Output test: If condition (3.11) holds for $J' = \emptyset$ and $J'' = \{i\}$ with $i \in J$, then activity i must end last in J, i.e. we conclude that

$j \to i$ for all $j \in J \setminus \{i\}$

(or for short $J \setminus \{i\} \to i$). In this case i is called the *output* of J.

Carlier and Pinson [19] developed an $O(|I|^2)$ algorithm to find all inputs and outputs for all sets $J \subseteq I$. The complexity of such an algorithm was improved to $O(|I| \log |I|)$ [12,20].

Input-or-output test: If condition (3.11) holds for $J' = \{i\}$ and $J'' = \{j\}$ with $i, j \in J$, then activity i must start first in J or activity j must end last in J, i.e. i is the input for J or j is output for J.

Dorndorf et al. [25] have designed an $O(|I|^3)$-algorithm for testing the input-or-output conditions.

Input/output negation test: If condition (3.11) holds for $J' = J \setminus \{i\}$ and $J'' = \{i\}$ with $i \in J$, then i must not start first in $J \setminus \{i\}$ (input negation).

If condition (3.11) holds for $J' = \{i\}$ and $J'' = J \setminus \{i\}$ with $i \in J$, then i must not end last in $J \setminus \{i\}$ (output negation).

Baptiste and Le Pape [6] developed an algorithm that tests all interesting J and i with effort $O(|I|^2)$.

The different tests are summarized in Table 1. Notice that for $i \neq j$ the input-or-output test is stronger than the output negation test or the input negation test, and that the input test or the output test is stronger than the input-or-output test.

Table 1
Summary of disjunctive interval consistency tests

Test	$J \setminus J'$	$J \setminus J''$	Conclusion
Input	$J \setminus \{i\}$	J	$i \rightarrow J \setminus \{i\}$
Output	J	$J \setminus \{i\}$	$J \setminus \{i\} \rightarrow i$
Input-or-output	$J \setminus \{i\}$	$J \setminus \{j\}$	$i \rightarrow J \setminus \{i\} \vee J \setminus \{j\} \rightarrow j$
Input negation	$\{i\}$	$J \setminus \{i\}$	$i \nrightarrow J \setminus \{i\}$
Output negation	$J \setminus \{i\}$	$\{i\}$	$J \setminus \{i\} \nrightarrow \{i\}$

The input-or-output test may be combined with the input negation test or with the output negation test leading to the conclusion $J \setminus \{j\} \rightarrow j$ or $i \rightarrow J \setminus \{i\}$. Additional conjunctions $i \rightarrow j$ may be derived by these tests, which generally leads to smaller time windows.

3-set conditions: Another method to deduce conjunctions $i \rightarrow j$ for activities i, j belonging to a disjunctive set I is based on the concept of 3-set conditions.

Consider three activities $i, j, k \in I$. If each of the following conditions:

$$d_k - r_j < p_i + p_j + p_k, \tag{3.12}$$

$$d_i - r_k < p_i + p_j + p_k, \tag{3.13}$$

$$d_i - r_j < p_i + p_j + p_k \tag{3.14}$$

holds, then there exists no feasible schedule in which j is processed before i. Otherwise the possible processing orders of i, j, k are j, i, k or k, j, i of j, k, i. However, none of these processing orders is feasible. Thus, the conjunction $i \rightarrow j$ must hold. (3.13) –(3.14) are called *3-set-conditions*. Brucker et al. [12] developed an $O(n^2)$-algorithm which calculates all conjunctions that can be derived by 3-set conditions for a disjunctive set with n activities. These concepts have been generalized to general shop scheduling problems with sequence-dependent setup times in Brucker and Thiele [17].

3.5. Cumulative resources

If $R_k = 1$ then the set of all activities i which occupy resource k (i.e. with $r_{ik} = 1$) define a disjunctive set. This is not the case if $R_k > 1$. Resources k for which $R_k > 1$ are called *cumulative resources*. For these resources the tests of the previous section are not valid. However, some concepts may be generalized introducing the term "work". For this purpose we consider a fixed resource k. Then $w_i := r_{ik} p_i$ is the *work* needed to process activity i. Let I_k be the set of activities i with $r_{ik} > 0$. For $J \subseteq I_k$ we define $W(J) = \sum_{i \in J} w_i$. $R_k(t_2 - t_1)$ is the work available during the interval $[t_1, t_2]$ $(t_1 < t_2)$.

Similarly to Theorem 3.1 we can prove

Theorem 3.2 (Dorndorf et al. [25]). *Let* $J', J'' \subseteq J \subseteq I_k$. *If*

$$R_k \cdot \max_{\substack{i \in J \setminus J' \\ j \in J \setminus J''}} (d_j - r_i) < W(J) \tag{3.15}$$

then an activity in J' *must start first or an activity in* J'' *must end last.* $\square$

In contrast to (3.11), in (3.15) we can no longer assume that $i \neq j$ because an activity that starts first may now also end last.

Theorem 3.2 can be used to derive tests similar to those listed in Table 1. The meaning of conclusions such as $J \setminus \{i\} \to i$ or $i \to J \setminus \{i\}$ is that i must end after (start before) activities in $J \setminus \{i\}$; in contrast to the disjunctive case this, however, does not imply that it must also start after (end before) $J \setminus \{i\}$.

3.6. Shaving

Let $\triangle_i = [r_i, t_i]$ with $r_i < t_i$ be the (current) domain of activity i, i.e. for any schedule $S = (S_j)_{j=0}^{n+1}$ improving some upper bound *UB* condition $S_i \in [r_i, t_i]$ must be satisfied. If for some integer t with $r_i \leqslant t < t_i$ we can show that no feasible schedule with $S_i \in [r_i, t]$ exists then $\triangle_i$ may be replaced by $[t+1, t_i]$. This process is called *shaving*, which was introduced by Martin and Shmoys [35].

3.7. Constraint propagation procedures

There are different ways to combine the constraint propagation techniques in a constraint propagation procedure. For the RCPSP the following procedure, which is applied to an SSD-matrix $D = (d_{ij})$ and a set of disjunctions, provides good results.

Procedure constraint propagation
1. Calculate transitive distances;
2. Symmetric triples;
3. Fix direct conjunction;
4. Process cliques

Constraint propagation calls the procedures calculate transitive distances, symmetric triples, fix direct conjunctions, and process cliques which may be described as follows.

Calculate transitive distances applies the Floyd–Warshall algorithm to the current SSD-matrix and possibly fixes conjunctions and parallelity relations according to (3.9) and (3.10), respectively. If a diagonal element of the SSD-matrix becomes positive, then we return by indicating that no feasible solution exists.

The procedure *Symmetric triples* derives conjunctions, disjunctions or parallelity relations by systematically applying the checks of Section 3.3. The results are used to update the SSD-matrix.

The procedure *Fix direct conjunctions* is based on the following observation. Assume that for activities i, j, with $i - j \in D$ the inequality $d_{ij} \geqslant -(p_j - 1)$ holds. Then

$$S_j - S_i \geqslant d_{ij} \geqslant -(p_j - 1), \quad \text{i.e.} \quad C_j = S_j + p_j \geqslant S_i + 1 > S_i,$$

which implies $i \to j$. In this case we call $i \to j$ a *direct conjunction*. Fix direct conjunctions fixes all direct conjunctions.

The procedure *process cliques* calculates maximal disjunctive sets (cliques) and applies the input tests and output tests of Section 3.4 to these maximal cliques. The number of all possible maximal cliques may be quite large. Therefore we calculate maximal cliques $I_1, \ldots, I_q$ with $I_1 \cup I_2 \cup \cdots \cup I_q = \{1, \ldots, n\}$ as follows.

To build I_1 we start with some activity i_1 and add an activity i_2 with $i_2 - i_1$. Then we add some activity i_3 with $i_3 - i_1$ and $i_3 - i_2$, etc. I_1 is completed if no activity j exists with $j - i$ for all $i \in I_1$. To build I_2 we start with some activity not belonging to I_1 and apply the same procedure. Note that $i \in I_1$ may be added to I_2. If $I_1, \ldots, I_k$ are built and there is some activity $i \notin I_1 \cup I_2 \cup \cdots \cup I_k$ then we start I_{k+1} with i. The procedure stops if all activities are covered by the constructed cliques. Only I_v with $|I_v| \geq 2$ are used for the tests.

Starting with the set D_0 of disjunctions and with an initial SSD-matrix defined by S_0, N_0 and given heads and tails, we repeat the procedure *constrained propagation* until we detect infeasibility or the SSD-matrix no longer changes.

4. Lower bounds

In this section we will discuss methods for calculating lower bounds for the RCPSP. If LB is a lower bound for some instance of the RCPSP and UB is the solution value given by some heuristic, then $UB - LB$ is an upper bound for the distance between the optimal solution value and UB. Thus, good lower bounds may be used to estimate the quality of heuristic solutions. Lower bounds are also needed for the construction of branch-and-bound algorithms.

4.1. Constructive and destructive lower bounds

A *constructive lower bound* for the RCPSP may be derived by solving a relaxation of the RCPSP, which is less complex than the original problem. In a relaxation of an optimization problem certain restrictions which make the problem hard are eliminated. For example, if we eliminate the resource constraints of the RCPSP, then the optimal makespan of the resulting problem is equal to the length of a longest path in the activity-on-node network, which can be calculated efficiently. Stronger constructive lower bounds based on a linear programming formulation have been developed by Mingozzi et al. [36] and Möhring et al. [38].

Destructive lower bounds are based on a different idea. Assume that UB is a fictitious upper bound, i.e. a guess value, for the RCPSP. If we can prove that no feasible schedule with $C_{\max} \leq UB$ exists, then $UB + 1$ is a valid lower bound for the RCPSP. We may apply constraint propagation to prove infeasibility. If we do not succeed we may try with a smaller UB value. If we succeed, we may repeat this process with a larger UB value. Such a repetition can be organized by *incremental steps* or by *binary search*. An incremental step increases UB by an appropriate value $\Delta \geq 1$. When applying binary search we start with a valid upper bound UB and a valid lower bound

LB. If for the fictitious upper bound $UB' := \lceil (UB + LB)/2 \rceil$ we can prove infeasibility, then we replace LB by $UB' + 1$. Otherwise UB is replaced by UB'. Destructive lower bounds have been considered by Klein and Scholl [28].

In the next sections we discuss methods for calculating constructive and non-constructive lower bounds which use linear programming and column generation [5,14].

4.2. An LP-based constructive lower bound

A set X of activities is called *feasible* if all $i \in X$ can be processed simultaneously, i.e.

- there are no conjunctions and disjunctions between pairs $i, j \in X$, and
- $\sum_{i \in X} r_{ik} \leq R_k$ for $k = 1, \ldots, r$.

A feasible set is *non-dominated* if it is not a proper subset of some feasible set.

Let $X_1, X_2, \ldots, X_q$ be the one-element sets $\{i\}$ ($i = 1, \ldots, n$) and all nondominated feasible sets for an RCPSP and $a_j \in \{0,1\}^n$ the incidence vectors of X_j ($j = 1, \ldots, q$), i.e.

$$a_{ij} = \begin{cases} 1 & \text{if } i \in X_j, \\ 0 & \text{otherwise.} \end{cases}$$

We assume that $X_i = \{i\}$ for $i = 1, \ldots, n$.

Denote by x_j the number of time units in which all activities in X_j are processed jointly. Then the following linear program provides a lower bound for the RCPSP. It relaxes conjunctions by treating them as disjunctions and allows preemption.

$$\min \; \sum_{j=1}^{q} x_j \tag{4.1}$$

$$\text{s.t.} \; \sum_{j=1}^{q} a_{ij} x_j \geq p_i \quad (i = 1, \ldots, n) \tag{4.2}$$

$$x_j \geq 0 \quad (j = 1, \ldots, q). \tag{4.3}$$

Clearly, (4.1)–(4.3) provides a lower bound because an optimal solution for the RCPSP with objective value $C_{\max}^*$ provides a solution x^* for (4.1)–(4.3) such that $\sum_{j=1}^{q} x_j^* = C_{\max}^*$.

Unfortunately, the number q of all nondominated feasible sets grows exponentially with the number n of activities. For $n = 60$ ($n = 90$) we have approximately $q = 300\,000$ ($q = 8\,000\,000$).

Mingozzi et al. [36] considered a restriction of the dual of (4.1)–(4.3) which they solved heuristically. We solve (4.1)–(4.3) directly by column generation techniques.

4.3. Column generation procedure for the RCPSP

Column generation techniques are based on the fact that when applying the revised simplex method it is sufficient to store the current basis B and to have a procedure at hand which calculates an entering column a, or which proves that such a column does not exist. Such a procedure is called a *pricing procedure*. Usually, the pricing procedure calculates a set of columns which may enter the basis during the next iterations. These columns are added to a so-called *working set* of columns. Some other nonbasic columns may also be deleted from the working set.

A generic column generation algorithm may be formulated as follows.

Algorithm Column Generation

1. INITIALIZE;
2. WHILE CALCULATE_COLUMNS produces new columns DO
     BEGIN
3.   INSERT_DELETE_COLUMNS;
4.   OPTIMIZE
     END

INITIALIZE: provides a basic solution and working set to start with.
CALCULATE_COLUMNS: is the pricing procedure.
INSERT_DELETE_COLUMNS: organizes insertion and deletion of columns.
OPTIMIZE: solves the linear program restricted to the working set of columns.

Several INSERT_DELETE_COLUMNS procedures may be provided, OPTIMIZE can be done by an LP-solver like CPLEX.

Thus, only INITIALIZE and CALCULATE_COLUMNS must be implemented problem-dependently.

Next we apply column generation to (4.1)–(4.3). In this case we have the following procedures.

INITIALIZE

As a starting basic for the first phase we use $e_i \in \mathbb{R}^n$ $(i = 1, \ldots, n)$ and set $x_i := p_i$. These variables correspond to the one-element sets $\{i\}$. We add the columns corresponding with the slack variables for the constraints in (4.2) to the working set.

INSERT_DELETE_COLUMNS

We use a DELETE-strategy which never deletes the slack variables from the working set, i.e. $-e_i$ is always in B or in the working set.

CALCULATE_COLUMNS

After applying OPTIMIZE for the values of the dual variables we always have $y_i \geqslant 0$ $(i = 1, \ldots, n)$ because

- $-y_i = -ye_i = c_i = 0$ if $-e_i$ is in the basis, or
- $-y_i = -ye_i \leqslant c_i = 0$, i.e. $y_i \geqslant 0$ if $-e_i$ is in the working set.

We sort the activities such that

$$y_1 \geqslant y_2 \geqslant \cdots \geqslant y_{n_0} > 0, \quad y_{n_0+1} = \cdots = y_n = 0.$$

Thus, it is sufficient to generate nondominated vectors $(a_1, \ldots, a_{n_0})$. These can be extended to a nondominated vector $(a_1, \ldots, a_n)$ by adding $a_i = 1$ for all $i = n_0 + 1, \ldots, n$ which maintain feasibility.

We calculate incidence vectors of relevant nondominated sets by the following enumeration.

Nodes of the enumeration tree are *partial strings* $(a_1, \ldots, a_l)$ with $1 \leqslant l \leqslant n_0$. $(a_1, \ldots, a_l)$ has at most two sons. $(a_1, \ldots, a_l, 0)$ is the *right son*, the *left son* $(a_1, \ldots, a_l, 1)$ exists if and only if $(a_1, \ldots, a_l, 1, 0, \ldots, 0)$ corresponds to a feasible set.

Sets are generated in lexicographic order by depth first search. They must satisfy

$$\sum_{i=1}^{n} y_i \cdot a_i > 1 \tag{4.4}$$

$$\sum_{i=1}^{n} r_{ik} \cdot a_i \leqslant R_k \quad (k = 1, \ldots, r) \tag{4.5}$$

$$a_i \cdot a_j = 0 \quad (i - j \in D, i \to j \in C \text{ or } j \to i \in C) \tag{4.6}$$

$$a_i \in \{0, 1\} \quad (i = 1, \ldots, n). \tag{4.7}$$

Each time a partial column $a^* = (a_1^*, \ldots, a_{n_0}^*)$ satisfying (4.4)–(4.7) has been found, we search only for further columns a with

$$\sum_{i=1}^{n_0} y_i a_i > \sum_{i=1}^{n_0} y_i a_i^*. \tag{4.8}$$

This ensures that only nondominated columns $a \in \{0, 1\}^{n_0}$ are calculated.

Proof. Assume that a column a satisfying (4.8) is dominated. Then at least one component a_k of a can be changed from 0 to 1 without violating feasibility. But then a dominating column a' with $a_k' = 1$ has already been considered due to the lexicographic scan.

Let a^* be the best column found so far when we consider a. Since a' has been considered before, we have

$$\sum_{i=1}^{n_0} y_i a_i^* \geqslant \sum_{i=1}^{n_0} y_i a_i' \geqslant \sum_{i=1}^{n_0} y_i a_i + y_k > \sum_{i=1}^{n_0} y_i a_i,$$

which contradicts (4.8). $\square$

Note that if we search the enumeration tree completely, then a nondominated column a which maximizes $\sum_{i=1}^{n} y_i a_i$ is found.

We may apply the following dominance rule: If we set $a_k = 0$ and

$$\sum_{v=1}^{k-1} y_v a_v + \sum_{v=k+1}^{n_0} y_v \leqslant \sum_{i=1}^{n_0} y_i a_i^* \quad \text{holds,}$$

then we can backtrack.

We stop the search process if one of the following conditions holds:

- we have found a given number of columns, or
- we have found at least one column and the ratio between the number l of generated leaves and the number l^u of leaves satisfying (4.8) exceeds a certain limit s (i.e. $l \geqslant l^u \cdot s$).

The generated columns are added to the working set. If the number of columns in the working set exceeds a given size, some arbitrary nonbasic columns are deleted.

4.4. An LP-based destructive method

A method which provides very good lower bounds for the RCPSP is a destructive method which combines constraint propagation with linear programming techniques. In most cases these lower bounds are close to a corresponding upper bound achieved by a heuristic (see [14] for computational results).

Given a fictitious upper bound *UB*, one first tries to prove infeasibility by constraint propagation. The constraint propagation procedure also provides us with time windows $[r_i, d_i]$ for the activities $i = 1, \ldots, n$. If we cannot prove infeasibility, we use the time windows in an LP-formulation of the feasibility problem and then try to prove infeasibility by solving the linear program.

For the LP-formulation let $z_0 < z_1 < \cdots < z_\tau$ be the ordered sequence of all different r_i- and d_i-values. We define

$$I_t := [z_{t-1}, z_t] \quad \text{for } t = 1, \ldots, \tau$$

and let F_t be the set of all activities i which can be scheduled in interval I_t (i.e. satisfying $r_i \leqslant z_{t-1} < z_t \leqslant d_i$). Finally, denote the feasible subsets of F_t by X_{jt} ($j = 1, \ldots, f_t$), and the incidence vector corresponding with X_{jt} by a^{jt}.

We consider a preemptive relaxation of the feasibility problem where x_{jt} denotes the number of time units in which all activities in X_{jt} are processed simultaneously.

$$\sum_{t=1}^{\tau} \sum_{j=1}^{f_t} a_i^{jt} x_{jt} \geqslant p_i \quad (i = 1, \ldots, n) \tag{4.9}$$

$$\sum_{j=1}^{f_t} x_{jt} \leqslant z_t - z_{t-1} \quad (t = 1, \ldots, \tau) \tag{4.10}$$

$$x_{jt} \geqslant 0 \quad (t = 1, \ldots, \tau; \ j = 1, \ldots, f_t). \tag{4.11}$$

If (4.9)–(4.11) has no feasible solution, then $UB+1$ is a lower bound for the RCPSP. For the infeasibility check we introduce artificial variables u_t $(t=1,\ldots,\tau)$ and solve

$$\min \ \sum_{t=1}^{\tau} u_t \tag{4.12}$$

$$\text{s.t.} \ \sum_{t=1}^{\tau}\sum_{j=1}^{f_t} a_i^{jt} x_{jt} \geqslant p_i \quad (i=1,\ldots,n) \tag{4.13}$$

$$-\sum_{j=1}^{f_t} x_{jt} + u_t \geqslant -z_t + z_{t-1} \quad (t=1,\ldots,\tau) \tag{4.14}$$

$$x_{jt} \geqslant 0 \quad (t=1,\ldots,\tau;\ j=1,\ldots,f_t) \tag{4.15}$$

$$u_t \geqslant 0 \quad (t=1,\ldots,\tau). \tag{4.16}$$

System (4.9)–(4.11) is infeasible if and only if the optimal solution value of (4.12)–(4.16) is positive.

(4.12)–(4.16) will be solved by column generation techniques.

4.4.1. Initialization

For each activity $i \in \{1,\ldots,n\}$ let $I_{t(i)}$ be the first interval in which i can be processed. Then the starting working set consists of the columns corresponding with the one-element sets $\{i\}$ $(i=1,\ldots,n)$ in the interval $I_{t(i)}$, the artificial variables u_t $(t=1,\ldots,\tau)$ and the slack variables according to (4.14).

4.4.2. Calculate columns

The dual constraints for (4.12)–(4.16) are of the form

$$\sum_{i=1}^{n} a_i^{jt} y_i - w_t \leqslant 0 \quad (t=1,\ldots,\tau;\ j=1,\ldots,f_t) \tag{4.17}$$

$$w_t \leqslant 1 \quad (t=1,\ldots,\tau) \tag{4.18}$$

$$y_i \geqslant 0 \quad (i=1,\ldots,n) \tag{4.19}$$

$$w_t \geqslant 0 \quad (t=1,\ldots,\tau). \tag{4.20}$$

Thus, we have to find a column $a^{jt} \in \{0,1\}^n$ violating (4.17), i.e. satisfying

$$\sum_{i=1}^{n} a_i^{jt} y_i - w_t > 0$$

for an index $t \in \{1,\ldots,\tau\}$. This can be done by an enumeration procedure similarly to that in Section 4.3.

5. A branch-and-bound procedure

In this section we present a branch-and-bound procedure based on the constraint propagation techniques introduced in Section 3. The nodes of the corresponding enumeration tree represent sets of feasible solutions which are defined by conjunctions, disjunctions, and parallelity relations. We branch at such a node by splitting the set into disjoint subsets. This is accomplished by adding either the parallelity relations $i \| j$ or the disjunction $i - j$ for some activities i, j.

First we will introduce schedule schemes which are used to represent the vertices of the enumeration tree mathematically.

Let C, D, $N \subseteq V \times V$ be disjoint relations, where C is a set of conjunctions, D is a set of disjunctions, and N is a set of parallelity relations. D and N are symmetric, i.e. with $i - j \in D$ ($i \| j \in N$) also $j - i \in D$ ($j \| i \in N$), while C is antisymmetric, i.e. if $i \to j \in C$ then $j \to i \notin C$. If $i - j \in D$ then it is not specified whether $i \to j$ or $j \to i$. Based on C, D, N we define

$$F = \{(i,j) \mid i,j \in V; \ i \neq j; \ i \to j \notin C, \ j \to i \notin C, \ i - j \notin D, \ i \| j \notin N\}.$$

The relations in F are called *flexibility relations*. These relations are denoted by $i \sim j$. Note that F is symmetric. The tuple (C, D, N, F) with the property that the sets C, D, N, F are disjoint and for any $i, j \in V$ with $i \neq j$ exactly one of the relations $i \to j$, $j \to i$, $i - j$, $i \| j$, or $i \sim j$ holds is called a scheduling scheme. A scheduling scheme (C, D, N, F) defines a (possible empty) set $\mathcal{S}(C, D, N, F)$ of schedules. More precisely, $\mathcal{S}(C, D, N, F)$ is the set of all schedules with the following properties:

- i is finished when j starts if $i \to j \in C$,
- i and j are processed in parallel for at least one time unit if $i \| j \in N$, and
- i and j are not processed in parallel if $i - j \in D$.

$\mathcal{S}_f(C, D, N, F)$ denotes the corresponding set of feasible schedules, i.e. all schedules in $\mathcal{S}(C, D, N, F)$ which additionally satisfy the resource constraints.

If C_0, D_0, N_0 are defined as in Section 3.1 and F_0 are the corresponding flexibility relations, then $\mathcal{S}_f(C_0, D_0, N_0, F_0)$ is the set of all feasible schedules.

Now we consider a scheduling scheme $(C, D, N, \emptyset)$ with no flexibility relations. Then the scheduling scheme $(C', \emptyset, N, \emptyset)$ is called a *transitive orientation* of $(C, D, N, \emptyset)$ if

- $(C', \emptyset, N, \emptyset)$ is derived from $(C, D, N, \emptyset)$ by changing each disjunction $i - j \in D$ into either $i \to j$ or $j \to i$,
- C' is transitive, i.e. $i \to j$ and $j \to k$ imply $i \to k$.

If $(C', \emptyset, N, \emptyset)$ is a transitive orientation then the directed graph (V, C') is acyclic.

Let $(C', \emptyset, N, \emptyset)$ be a transitive orientation of a scheduling scheme. Then we may calculate a corresponding (not necessarily feasible) schedule as follows.

Consider the acyclic directed graph (V, C'). For each activity $i \in V$ we calculate the length r_i of a longest path from 0 to i. The length of such a path p from 0 to i is the sum of all processing times of activities on p except i. The schedule in which each activity i starts at time r_i is called the earliest start schedule (*ES*-schedule). We denote this schedule, which depends on C' only, by $\mathcal{S}_{ES}(C')$. $\mathcal{S}_{ES}(C')$ need not be feasible for the RCPSP because the capacity constraints may be violated. Möhring [37] has shown that all ES-schedules $\mathscr{S}_{ES}(C')$ for arbitrary transitive orientations $(C', \emptyset, N, \emptyset)$ of a schedule scheme $(C, D, N, \emptyset)$ have the same C_{max}-value. Furthermore, the following theorem holds.

Theorem 5.1. Let $(C', \emptyset, N, \emptyset)$ be a transitive orientation of a scheduling scheme $(C, D, N, \emptyset)$ and let $\mathscr{S}_{ES}(C')$ be the corresponding ES-schedule. If $\mathscr{S}_{ES}(C')$ is feasible then $\mathscr{S}_{ES}(C')$ dominates all schedules in $\mathscr{S}_f(C, D, N, \emptyset)$. Otherwise $\mathscr{S}_f(C, D, N, \emptyset)$ is empty.

Proof. Assume that $S_{ES}(C')$ is infeasible, i.e. there is a time period t such that activities of a set H are processed in parallel during t and for some resource type k we have $\sum_{i \in H} r_{ik} > R_k$. If a feasible schedule $S \in \mathscr{S}_f(C, D, N, \emptyset)$ exists then for at least two activities $i, j \in H$ one of the relations $i \to j \in C$ or $j \to i \in C$ or $i - j \in D$, i.e. $i \| j \notin N$ holds. This implies $i \to j \in C'$ or $j \to i \in C'$, which contradicts the definition of H.

Now consider an arbitrary feasible schedule $S \in \mathscr{S}_f(C, D, N, \emptyset)$. S defines a transitive orientation $(C'', \emptyset, N, \emptyset)$ of $(C, D, N, \emptyset)$. The corresponding ES-schedule $S_{ES}(C'')$ dominates S. Due to the result of Möhring $S_{ES}(C')$ and $S_{ES}(C'')$ have the same C_{max}-value. Thus, $S_{ES}(C')$ dominates S as well. □

The following branch-and-bound algorithm for solving the RCPSP calls a recursive procedure Branch-and-Bound (C, D, N, F) which is formulated afterwards.

Branch-and-bound RCPSP

1. $UB := \sum_{i=1}^{n} p_i$;
2. Branch-and-bound (C_0, D_0, N_0, F_0).

Branch-and-bound (C,D,N,F)

1.   Constraint_Propagation (C, D, N, F);
2.   IF $F = \emptyset$ THEN
      BEGIN
3.   Calculate a transitive orientation $(C', \emptyset, N, \emptyset)$ corresponding with $(C, D, N, \emptyset)$;
4.   Calculate the earliest start schedule $S_{ES}(C')$;
5.   IF $S_{ES}(C')$ is feasible and the makespan $C_{max}(C')$ of
6. $S_{ES}(C')$ satisfies $C_{max}(C') < UB$ THEN $UB := C_{max}(C')$
7.   ELSE RETURN
      END
    ELSE
      BEGIN
8.   Calculate a lower bound $LB(C, D, N, F)$ for $\mathscr{S}_f(C, D, N, F)$;
9.   IF $LB(C, D, N, F) < UB$ THEN
      BEGIN
10.   Choose $i \sim j \in F$;

11.   Branch-and-bound $(C, D \cup \{i - j\}, N, F \setminus \{i \sim j\})$;
12.   Branch-and-bound $(C, D, N \cup \{i \| j\}, F \setminus \{i \sim j\})$
           END
        END

The procedure Constraint_Propagation (C, D, N, F) applies the constraint propagation algorithm described in Section 3.7 to C, D, N and returns the new schedule scheme.

In a different branch-and-bound approach a feasible schedule is constructed from left to right by adding an activity in each step. More precisely, each vertex of the enumeration tree corresponds with a partial schedule and the branching process consists of extending the partial schedule in different ways by adding an eligible activity. In general, depth-first-search is used in order to keep the memory requirements low. Different methods using different branching schemes and pruning methods have been developed: Patterson et al. [40] and Sprecher [44] use the concept of a precedence tree, the methods of Christofides et al. [22] and Demeulemeester and Herroelen [24] are based on delay alternatives, while Stinson et al. [45] use extension alternatives. Mingozzi et al. [36] use a slightly different approach.

Another branching scheme for the RCPSP is based on domains. Each node of the enumeration tree is defined by the domains $\triangle_i$ of all activities i. For branching, an activity i is chosen and its domain $\triangle_i = [r_i, t_i]$ is replaced either by $\triangle_i^l = [r_i, t]$ (left branch) or by $\triangle_i^r = [t + 1, t_i]$ (right branch) where $r_i \leqslant t < t_i$. In a branch-and-bound algorithm which solves the RCPSP with generalized precedence constraints, Dorndorf et al. [26] use this branching scheme with $t = r_i$. Thus, by branching to the left, the starting time of activity i will be fixed. By the application of constraint propagation techniques they reduced the search space considerably.

Shop scheduling problems and multi-processor task scheduling problems can be presented by mixed graphs $G = (V, C, D)$ where C is a set of directed arcs (defining the precedence relations) and D is a set of (undirected) edges (which correspond to all disjunctions). The vertices $i \in V$ correspond to the operations or tasks. A feasible schedule corresponds to an orientation of all edges such that the resulting directed graph has no cycles. The corresponding makespan of such an schedule is the length of the longest path in this graph. The length of a path p is the sum of processing times of all activities in p. A common branching scheme for this type of problems consists of choosing an undecided disjunction $i - j$ and replacing it by either the conjunction $i \to j$ or $j \to i$. Furthermore, the constraint propagation methods for disjunctive sets apply.

6. Heuristics for the RCPSP

To solve an RCPSP of medium to large size in reasonable time one has to apply heuristic methods. The following types of heuristics have been applied: priority based heuristics [2,23,29,31,46], local search heuristics [5,33,34,42,47,48], and genetic algorithms [39]. The most important heuristics are based on a list scheduling procedure which constructs a feasible schedule for a given permutation or list of all activities.

6.1. A list-scheduling algorithm

Let L be a list of all activities $0, 1, \ldots, n, n+1$. For two activities i, j we write $i \prec j$ if in L activity i is listed before j. All lists considered are assumed to be *compatible* with the precedences $i \to j \in C$, i.e. we always have $i \prec j$ if $i \to j \in C$.

The following Algorithm List-Schedule calculates a feasible schedule $S = H(L)$ using the list L. In this algorithm PS is the set of finishing times of scheduled activities.

Algorithm List-Schedule (L)

1. $PS := \{0\}$;
2. WHILE $L \neq \emptyset$ DO
    BEGIN
3.   Eliminate the next activity j from L;
4. $r_j := \max\{S_i + p_i \mid i \to j \in C\}$;
5.   Calculate the earliest starting time $t \in PS$, $t \geqslant r_j$ such that activity j can be scheduled without violating the resource constraints;
6.   Schedule j at time $S_j := t$;
7. $PS := PS \cup \{S_j + p_j\}$
    END.

During the algorithm for all $t \in PS$ we keep track of the amounts $r_k(t)$ of resource R_k used by all activities being processed or starting at time t. Using these $r_k(t)$-values it can be checked whether a new activity can be started at time $t \in PS$ without violating the resource constraints. The complexity of this algorithm is $O(n^2 r)$.

It is not difficult to show that a list L^* always exists such that $S^* = H(L^*)$ is optimal.

6.2. List-scheduling heuristics

The simplest list-scheduling heuristics are based on priority rules. For our cases a list is constructed by some rule.

A local search for the RCPSP may be organized on the set of all compatible lists. For constructing a neighbor of some list L we may apply operators depending on $S = H(L)$, which provide the next list in the search. Several methods exist to define such operators. For example, the following operators have been used in Baar et al. [5] in connection with a tabu-search heuristic. They generalize operators defined for the job shop problem, which are based on critical paths.

Let L be a list and S be the corresponding list-schedule. Consider the directed graph $G_S = (V, A_S)$, where V is the set of all activities and

$$(i, j) \in A_S \quad \text{if and only if} \quad S_i + p_i = S_j.$$

A *critical path* with respect to S is a path in G_S from 0 to $n+1$. At least one critical path always exists. An arc (i, j) belonging to a critical path is called a *critical arc* if

(i, j) does not belong to C. To improve the schedule S at least one critical arc must become noncritical. Therefore, we construct neighbor lists of the list L in such a way that critical arcs in S may become noncritical. Next, we describe how to construct such a neighborhood.

First, we calculate a critical path $CP(S)$ with a minimal number of critical arcs.

Then the neighborhood is defined by three types of operators. Each depends on the critical path $CP(S)$. The idea of these operators is to cancel a critical arc (i, j).

1. shift$_{ij}$ is a *shift-operator* which is defined for a critical arc $(i, j) \in CP(S)$, with $i \prec j$ in L. It moves activity i together with all activities t satisfying

   $$i \rightarrow t \in C \quad \text{and} \quad i \prec t \prec j$$

   immediately after j. This can be illustrated by

   $$L = \ldots i \ldots t \ldots j \ldots \Rightarrow L' = \ldots jit \ldots$$

2. bshift$_{ij}$ is a *backshift-operator* which is defined for a critical arc $(i, j) \in CP(S)$ with $j \prec i$ in L. It moves the first activity u which satisfies the conditions

   $$i \prec u \quad \text{and} \quad (i, u) \notin C.$$

   immediately before i. Such a backshift is illustrated by

   $$L = \ldots j \ldots i \ldots l \ldots u \ldots \Rightarrow L' = \ldots j \ldots ui \ldots l \ldots$$

   Note that there is no l with $i \prec l$ and $l \rightarrow u$.
3. The *frontshift-operator* fshift$_{ij}$ is defined symmetrically. It can be illustrated by

   $$L = \ldots u \ldots l \ldots j \ldots i \ldots \Rightarrow L' = \ldots l \ldots ju \ldots i \ldots$$

Constraint propagation techniques may be used to reduce the set of compatible lists and the set of neighbors. Another type of heuristics which apply constraint propagation techniques are truncated branch-and-bound methods. A truncated enumeration tree is constructed to produce many feasible solutions from which the best is chosen. During the search process constraint propagation is applied using the best actual solution value to further truncate the enumeration tree.

7. Concluding remarks

We have discussed the use of constraint propagation techniques in combination with solution methods for the resource constrained project scheduling problem and some of the most important machine scheduling problems. Rather than presenting algorithmic details and computational results we tried to survey some of the main concepts. Readers interested in the specific algorithms are referred to Tables 2–4 in which algorithmic details and computational results are documented. Tables 2–4 contain besides references a description of the problems and the specific constraint propagation techniques used in these papers.

Table 2
Branch-and-bound algorithms

Job shop problem	
Carlier and Pinson [19]	Input–output tests,
Applegate and Cook [4]	Branching by fixing disjunctions
Brucker et al. [13]	
Martin and Shmoys [35]	Input–output tests, shaving, time-oriented branching
Open shop problem	
Dorndorf et al. [27]	Input–output tests, input–output negation tests, shaving, branching by fixing disjunctions
RCPSP	
Brucker et al. [16]	Transitive closure, symmetric triples, input–output tests, branching by adding disjunctions or parallelity relations
Dorndorf et al. [26]	Input–output tests, shaving time-oriented branching
General shop problems with sequence-dependent setup times	
Brucker and Thiele [17]	Input–output tests, 3-set conditions, branching by fixing disjunctions

Table 3
Lower bound calculations

Job shop problem	
Martin and Shmoys [35]	Shaving
RCPSP	
Klein and Scholl [28]	Simple constraint propagation methods
Brucker and Knust [14]	Transitive closure, symmetric triples, input–output tests

Table 4
Heuristics

Job shop problem	
Brucker and Brinkkoetter [10]	Input–output tests, shaving, truncated branch-and-bound

Acknowledgements

The author is grateful to the anonymous referees for their helpful comments on an earlier draft of the paper.

References

[1] R.K. Ahuja, T.L. Magnanti, J.B. Orlin, Network Flows, Prentice-Hall, Englewood Cliffs, NJ, 1993.
[2] R. Alvarez-Valdés, J.M. Tamarit, Heuristic algorithms for resource-constrained project scheduling: a review and an empirical analysis, in: R. Slowinski, J. Weglarz (Eds.), Advances in Project Scheduling, Elsevier, Amsterdam, 1989, pp. 113–134.

[3] E.J. Anderson, C.A. Glass, C.N. Pott, Machine scheduling, in: A. Aarts, J.K. Lenstra (Eds.), Local Search in Combinatorial Optimization, Wiley, Chichester, 1997, pp. 361–414.

[4] D. Applegate, W. Cook, A computational study of the job-shop scheduling problem, ORSA J. Comput. 3 (1991) 149–156.

[5] T. Baar, P. Brucker, S. Knust, Tabu search algorithms and lower bounds for the resource-constrained project scheduling problem, in: S. Voss, S. Martello, I.H. Osman, C. Roucairol (Eds.), Meta-Heuristics, Kluwer Academic Publishers, Dordrecht, 1999.

[6] P. Baptiste, C. Le Pape, Edge-finding constraint propagation algorithms for disjunctive and cumulative scheduling, Proceedings of the 15th Workshop of the U.K. Planning Special Interest Group, Liverpool, UK, 1996.

[7] P. Baptiste, A theoretical and experimental study of resource constrained propagation, Ph.D. Thesis, Université de Technologie de Compiègne, 1998.

[8] J. Blazewicz, K. Ecker, E. Pesch, G. Schmidt, J. Weglarz, Scheduling Computer and Manufacturing Processes, Springer, Berlin, 1996.

[9] P. Brucker, Scheduling Algorithms, Springer, Berlin, 2001.

[10] P. Brucker, W. Brinkkötter, Solving open benchmark problems for the job-shop problem, J. Scheduling 4 (2001) 53–64.

[11] P. Brucker, A. Drexl, R. Möhring, K. Neumann, E. Pesch, Resource-constrained project scheduling: notation, classification, model, and methods, Euro. J. Oper. Res. 112 (1999) 3–41.

[12] P. Brucker, B. Jurisch, A. Krämer, The job-shop problem and immediate selection, Ann. Oper. Res. 50 (1994a) 73–114.

[13] P. Brucker, B. Jurisch, B. Sievers, A branch & bound algorithm for the job-shop problem, Discrete Appl. Math. 49 (1994b) 107–127.

[14] P. Brucker, S. Knust, A linear programming and constraint propagation based lower bound for the RCPSP, Euro. J. Oper. Res. 127 (2000) 355–362.

[15] P. Brucker, D. Schumacher, A new tabu search procedure for an audit-scheduling problem, J. Scheduling 2 (1999) 157–173.

[16] P. Brucker, S. Knust, A. Schoo, O. Thiele, A branch and bound algorithm for the resource-constrained project scheduling problem, Euro. J. Oper. Res. 107 (1998) 272–288.

[17] P. Brucker, O. Thiele, A branch & bound method for the general-shop problem with sequence dependent setup times, Oper. Res. Spektrum 18 (1996) 145–161.

[18] J. Carlier, E. Pinson, An algorithm for solving the job-shop problem, Manage. Sci. 35 (1989) 164–176.

[19] J. Carlier, E. Pinson, A practical use of Jackson's preemptive schedule for solving the job shop problem, Ann. Oper. Res. 26 (1990) 269–287.

[20] J. Carlier, E. Pinson, Adjustment of heads and tails for the job-shop problem, Euro. J. Oper. Res. 78 (1994) 146–161.

[21] V. Chvatal, Linear Programming, W.H. Freeman and Company, New York, 1983.

[22] N. Christofides, R. Alvarez-Valdés, J.M. Tamarit, Project scheduling with resource constraints: a branch and bound approach, Euro. J. Oper. Res. 29 (1987) 262–273.

[23] E.W. Davis, J.H. Patterson, A comparison of heuristic and optimum solutions in resource-constrained project scheduling, Manage. Sci. 21 (1975) 944–955.

[24] E. Demeulemeester, W. Herroelen, A branch-and-bound procedure for the multiple resource-constrained project scheduling problem, Manage. Sci. 38 (1992) 1803–1818.

[25] U. Dorndorf, T.P. Huy, E. Pesch, A survey of interval capacity consistency tests for time- and resource-constrained scheduling, in: J. Weglarz (Ed.), Project Scheduling, Kluwer Academic Publishers, Dordrecht, 1998.

[26] U. Dorndorf, E. Pesch, T. Phan-Huy, A time-oriented branch-and-bound algorithm for resource constrained project scheduling with generalized precedence constraints, Management Science 46 (2000) 1365–1384.

[27] U. Dorndorf, E. Pesch, T. Phan-Huy, Solving open shop problem, J. Scheduling 4 (2001) 157–174.

[28] R. Klein, A. Scholl, Computing lower bounds by destructive improvement—an application to resource-constrained project scheduling, Euro. J. Oper. Res. 112 (1999) 322–346.

[29] R. Kolisch, A. Drexl, Adaptive search for solving hard project scheduling problems, Naval Res. Logist. 43 (1996) 23–40.

[30] E.L. Lawler, J.K. Lenstra, A.H.G. Rinnooy Kan, D.B. Shmoys, Sequencing and scheduling: algorithms and complexity, in: S.C. Graves, A.H.G. Rinnooy Kan, P. Zipkin (Eds.), Handbooks in Operations Research and Management Science, Vol. 4: Logistic of Production and Inventory, North-Holland, Amsterdam, 1993, pp. 445–522.

[31] S.R. Lawrence, Resource constrained project scheduling: an experimental investigation of heuristic scheduling techniques, Working Paper, Graduate School of Industrial Administration, Carnegie-Mellon University, Pittsburgh, 1984.

[32] C.-Y. Lee, L. Lei, M. Pinedo, Current trends in deterministic scheduling, Ann. Oper. Res. 70 (1997) 1–41.

[33] V.J. Leon, B. Ramamoorthy, Strength and adaptability of problem-space based neighborhoods for resource-constrained scheduling, OR Spektrum 17 (1995) 173–182.

[34] K. Nonobe, T. Ibaraki, Formulation and tabu search algorithm for the resource constrained project scheduling problem (RCPSP), Working Paper, Department of Applied Mathematics and Physics, Kyoto University, 1999.

[35] P. Martin, D.B. Shmoys, A new approach to computing optimal schedules for the job-shop scheduling problem, Proceedings of the Fifth International IPCO Conference, 1996.

[36] A. Mingozzi, V. Maniezzo, S. Ricciardelli, An exact algorithm for the resource-constrained project scheduling based on a new mathematical formulation, Manage. Sci. 44 (1998) 714–729.

[37] R.H. Möhring, Algorithm aspects of comparability graphs and interval graphs, in: I. Rival (Ed.), Graphs and Order: the Role of Graphs in the Theory of Ordered Sets and its Applications, NATO Advanced Science Institute Series, Math. Phys. Sci. Ser., C, 1985, pp. 41–101.

[38] R.H. Möhring, A.S. Schulz, F. Stork, M. Uetz, Resource constrained project scheduling: computing lower bounds by solving minimum cut problems, in Algorithms – ESA'99, ed. Nesetril, J. Lecture Notes in Computer Science 1643, Proc. 7th Annual European Symp. on Algorithms, Springer, Berlin, 1999, pp. 139–150.

[39] S. Mori, C.C. Tseng, A genetic algorithm for multi-mode resource constrained project scheduling problem, Euro. J. Oper. Res. 100 (1997) 134–141.

[40] J.H. Patterson, R. Słowiński, F.B. Talbot, J. Węglarz, An algorithm for a general class of precedence and resource constrained scheduling problems, in: R. Słowiński, J. Węglarz (Eds.), Advances in Project Scheduling, Elsevier Science, Amsterdam, 1989, pp. 3–28.

[41] M. Pinedo, Scheduling: Theory, Algorithms, and Systems, Prentice-Hall, Englewood Cliffs, NJ, 1995.

[42] S.E. Sampson, E.N. Weiss, Local search techniques for the generalized RCPSP, Naval Res. Logist. Quart. 40 (1993) 665–675.

[43] A. Schaerf, Tabu search techniques for large high-school timetabling problems, in: Proceedings of the 13th National Conference on Artificial Intelligence (AAAI-96), AAAI Press/MIT Press, Portland, USA, 1996, pp. 363–368.

[44] A. Sprecher, Scheduling resource-constrained projects competitively at modest memory requirements. *Management Science*, 46 (2000) 710–723.

[45] J.P. Stinson, E.W. Davis, B.M. Khumawala, Multiple resource-constrained scheduling using branch and bound, AIIE Trans. 10 (1978) 252–259.

[46] A. Thesen, Heuristic scheduling of activities under resource and precedence restrictions, Manage. Sci. 23 (1976) 412–422.

[47] P.R. Thomas, S. Salhi, A tabu search approach for the resource constrained project scheduling problem, J. Heuristics 4 (1998) 123–139.

[48] M.G.A. Verhoeven, Tabu search for resource-constrained scheduling, Euro. J. Oper. Res. 106 (1998) 266–276.

ELSEVIER

Discrete Applied Mathematics 123 (2002) 257–302

DISCRETE
APPLIED
MATHEMATICS

Selected topics on assignment problems [☆]

Rainer E. Burkard

Technische Universität Graz, Institut für Mathematik, Steyrergasse 30, A-8010 Graz, Austria

Received 2 December 1999; received in revised form 4 April 2000; accepted 10 April 2000

Abstract

We survey recent developments in the fields of bipartite matchings, linear sum assignment and bottleneck assignment problems and applications, multidimensional assignment problems, quadratic assignment problems, in particular lower bounds, special cases and asymptotic results, biquadratic and communication assignment problems. © 2002 Elsevier Science B.V. All rights reserved.

MSC: 90C27; 90B80; 68Q25; 90C05

Keywords: Assignment problem; Bottleneck assignment problem; Multidimensional assignment problem; Quadratic assignment problem; Special cases; Asymptotic theory; Quadratic bottleneck assignment problem; Biquadratic assignment problem; Communication assignment problem

1. Introduction

Assignment problems deal with the question how to assign n items (jobs, students) to n other items (machines, tasks). Their underlying combinatorial structure is an *assignment,* which is nothing else than a bijective mapping φ between two finite sets of n elements. In the optimization problem we are looking for the *best* assignment, i.e., we have to optimize some objective function which depends on the assignment φ. Assignments can be represented in different ways. The bijective mapping between two finite sets V and W can be represented in a straight forward way by a *perfect matching* in a bipartite graph $G = (V, W; E)$, where the vertex sets V and W have n vertices. Edge $(i, j) \in E$ is an edge of the perfect matching iff $j = \varphi(i)$, cf. Fig. 1

E-mail address: burkard@opt.math.tu-graz.ac.at (R.E. Burkard).

[☆] This research has been supported by Spezialforschungsbereich F 003 "Optimierung und Kontrolle", Projektbereich Diskrete Optimierung.

$$\varphi = \begin{pmatrix} 1 & 2 & 3 & 4 \\ 2 & 4 & 3 & 1 \end{pmatrix}$$

$$X_\varphi = \begin{pmatrix} 0 & 1 & 0 & 0 \\ 0 & 0 & 0 & 1 \\ 0 & 0 & 1 & 0 \\ 1 & 0 & 0 & 0 \end{pmatrix}$$

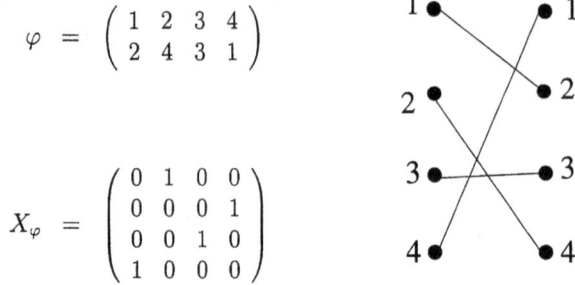

Fig. 1. Different representations of assignments.

By identifying the sets V and W we get the representation of an assignment by a *permutation*. Every permutation φ of the set $N = \{1,\ldots,n\}$ corresponds in a unique way to a *permutation matrix* $X_\varphi = (x_{ij})$ with $x_{ij} = 1$ for $j = \varphi(i)$ and $x_{ij} = 0$ for $j \neq \varphi(i)$. This matrix X_φ can be viewed as adjacency matrix of the bipartite graph G representing the perfect matching, see Fig. 1.

The set of all assignments (permutations) of n items will be denoted by $\mathscr{S}_n$ and has $n!$ elements. We can describe this set by the following constraints called *assignment constraints*.

$$\sum_{i=1}^{n} x_{ij} = 1 \quad \text{for all } j = 1,\ldots,n,$$

$$\sum_{j=1}^{n} x_{ij} = 1 \quad \text{for } i = 1,\ldots,n,$$

$$x_{ij} \in \{0,1\} \quad \text{for all } i,j = 1,\ldots,n. \tag{1}$$

The set of all matrices $X = (x_{ij})$ fulfilling the assignment constraints will be denoted by $\mathbf{X}_n$.

When we replace the conditions $x_{ij} \in \{0,1\}$ in (1) by $x_{ij} \geqslant 0$, we get a *doubly stochastic matrix*. The set of all doubly stochastic matrices forms the *assignment polytope* P_A. Birkhoff [15] showed that the assignments correspond uniquely to the vertices of P_A. Thus every doubly stochastic matrix can be written as convex combination of permutation matrices.

Theorem 1.1 (Birkhoff [15]). *The vertices of the assignment polytope correspond uniquely to permutation matrices.*

Flows in networks offer another model to describe assignments. Let $G = (V, W; E)$ be a bipartite graph with $|V| = |W| = n$. We embed G in the network $\mathscr{N} = (N, A, c)$ with *node set* N, *arc set* A and *arc capacities* c. The node set N consists of a *source* s, a *sink* t and the vertices of $V \cup W$. The source is connected to every node in V by a directed arc of capacity 1, every node in W is connected to the sink by a directed

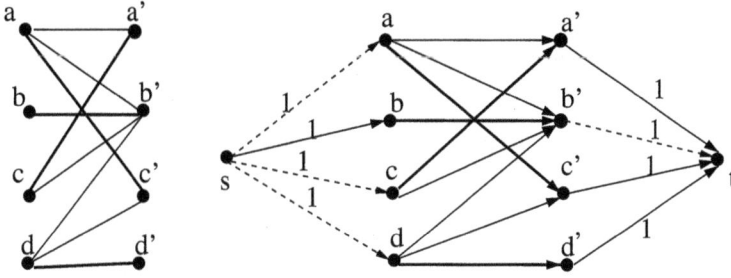

Fig. 2. Perfect matching in a bipartite graph and corresponding network flow model. A minimum cut is given by $\{s, b, b'\}$. The dashed arcs lie in the cut. Thus the cut has value 4.

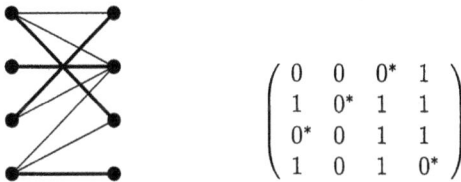

$$\begin{pmatrix} 0 & 0 & 0^* & 1 \\ 1 & 0^* & 1 & 1 \\ 0^* & 0 & 1 & 1 \\ 1 & 0 & 1 & 0^* \end{pmatrix}$$

Fig. 3. 0–1 matrix model of a bipartite graph and the corresponding minimum vertex cover which corresponds to the cut in Fig. 2. An assignment is given by the entries marked by $*$.

arc of capacity 1, and every arc in E is directed from V to W and supplied with an infinite capacity. The *maximum network flow problem* asks for a flow with maximum value $z(f)$. Obviously, a maximum integral flow in the special network constructed above corresponds to a matching with maximum cardinality, see Fig. 2. A *cut* in the network $\mathcal{N}$ is a subset C of the node set N with $s \in C$ and $t \notin C$. The value $u(C)$ of cut C is defined as (Fig. 3)

$$u(C) := \sum_{\substack{x \in C, \, y \notin C \\ (x, y) \in A}} c(x, y),$$

where $c(u, v)$ is the capacity of the arc (x, y).

Ford and Fulkerson's famous *Max Flow-Min Cut Theorem* [69] states that the value of a maximum flow equals the minimum value of a cut. This max flow-min cut theorem can directly be translated in König's Matching Theorem [99]. Given a bipartite graph G, a *vertex cover* (cut) in G is a subset of its vertices such that every edge is incident with at least one vertex in this set.

Theorem 1.2 (König's Matching Theorem [99]). *In a bipartite graph the minimum number of vertices in a vertex cover equals the maximum cardinality of a matching.*

Let us now formulate this theorem in the language of 0–1 matrices. Given a bipartite graph $G = (V, W; E)$ with $|V| = |W| = n$, we define the zero-adjacency matrix B of G

as $(n \times n)$ matrix $B = (b_{ij})$ where

$$b_{ij} := \begin{cases} 0 & \text{if } (i,j) \in E, \\ 1 & \text{if } (i,j) \notin E. \end{cases}$$

A *zero-cover* is a subset of the rows and columns of matrix B which contains all 0 elements. A row (column) which is an element of a zero-cover is called a *covered row* (*covered column*). Now we get

Theorem 1.3. *There exists an assignment φ with $b_{i\varphi(i)} = 0$ for all $i = 1,\ldots,n$, if and only if the minimum zero cover has n elements.*

Since a maximum matching corresponds uniquely to a maximum flow in the corresponding network $\mathcal{N}$, we can construct a zero-cover in the zero-adjacency matrix B by means of a minimum cut C in this network: if node $i \in V$ of the network does not belong to the cut C, then row i is an element of the zero-cover. Analogously, if node $j \in W$ belongs to the cut C, then column j is an element of the zero-cover.

2. Perfect matchings

In this section we deal with the question, whether there exists an assignment (i.e. a perfect matching) in a given bipartite graph or not. A basic answer to this question is provided by Hall's Marriage Theorem [83]. For a vertex $v \in V$ let $N(v)$ be the set of its neighbors, i.e., the set of all vertices $w \in W$ which are connected with v by an edge in E. Thus $N(v)$ contains the "friends" of v. Moreover, for any subset V' of V let $N(V') = \bigcup_{v \in V'} N(v)$.

Theorem 2.1 (Marriage Theorem [83]). *Let $G = (V, W; E)$ be a bipartite graph with $|V| = |W|$. G contains an assignment (perfect matching, marriage) if and only if for all subsets V' of V:*

$$|V'| \leqslant |N(V')| \quad \text{(Hall condition).}$$

When we want to apply this theorem for a special graph we have to check exponentially many subsets V' of V. Hopcroft and Karp [88] gave a polynomial-time algorithm to decide this question. They construct a perfect matching in $O(|E|\sqrt{|V|})$ steps, if it exists. This is done by a careful analysis of an algorithm for finding a maximum flow in a network with arc capacities 1.

Alt et al. [5] improve the complexity for dense graphs by a fast matrix scanning technique and obtain an $O(|V|^{1.5}\sqrt{|E|/\log|V|})$ implementation for the Hopcroft–Karp algorithm. They save the factor $\log|V|$ by storing the occurring 0–1 matrices (e.g. the adjacency matrix of the graph) in blocks of length $\log|V|$ as RAM-words which can be processed in constant time.

A randomized algorithm can decide even faster, whether G contains an assignment or not. This algorithm is based on the following theorem of Tutte [148]. The *Tutte matrix* $A(x) = (x_{ij})$ of an (undirected) graph $G = (V, E)$ is a skew symmetric matrix

$$\begin{pmatrix} 0 & 0 & 0 & x_{14} & x_{15} & x_{16} \\ 0 & 0 & 0 & x_{24} & 0 & 0 \\ 0 & 0 & 0 & x_{34} & 0 & 0 \\ -x_{14} & -x_{24} & -x_{34} & 0 & 0 & 0 \\ -x_{15} & 0 & 0 & 0 & 0 & 0 \\ -x_{16} & 0 & 0 & 0 & 0 & 0 \end{pmatrix}$$

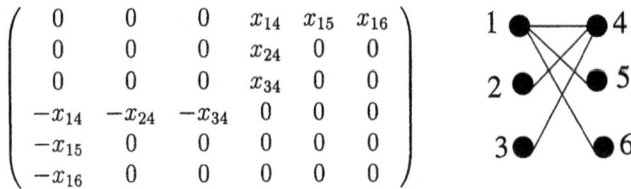

Fig. 4. The Tutte matrix of a bipartite graph.

with indeterminate entries x_{ij}, $x_{ij} = -x_{ji}$, where $x_{ij} = x_{ji} \equiv 0$, iff (i, j) is not an edge of G, see Fig. 4.

Theorem 2.2 (Tutte [148]). *Let $A(G)$ be the Tutte matrix of graph $G = (V, E)$. There exists a perfect matching in G, if and only if the determinant of $A(G)$ is not identically equal to 0.*

This theorem can now be used in the following algorithm.

Algorithm.

1. Generate randomly the values of x_{ij}, $1 \leqslant i$, $j \leqslant n$ from the set $\{1, 2, \ldots, |E|^2\}$.
2. Compute $\det A(G)$.
3. If $\det A(G) \neq 0$, stop. Graph G contains an assignment. Otherwise goto Step 1, unless a prespecified number r of repetitions is already reached.

Due to (Coppersmith and Vinograd [53]) the determinant in Step 2 can be computed in $O(n^{2.376})$ steps. The algorithm errs, if it yields the value 0 for the determinant, but $\det A(G)$ is not identically equal to 0. This happens if the random numbers generated in Step 1 hit by chance a root of the polynomial $\det A(G)$. According to a result of Schwarz [142] this happens with a probability equal to $1/|E|$. Thus the algorithm errs after r repetitions with a probability $(1/|E|)^r$.

This algorithm is faster than the best known deterministic algorithm. Note, however, that this procedure does not provide an assignment explicitly. A similar parallel algorithm which also provides an assignment, is due to Mulmuley et al. [115]. Their algorithm requires $O(|V|^{3.5}|E|)$ processors to find a perfect matching in $O(\log^2 |V|)$ time.

3. The time slot assignment problem

As an application of perfect matchings we consider the following problem from telecommunication. For remitting data from one sending earth station via a satellite to another receiving earth station, the so-called *time division multiple access* (*TDMA*) *technique* may be used. This technique buffers first the data to be remitted in the ground stations. Then they are sent in very short data bursts to the satellite. There they are received by transponders and again transmitted to the earth, namely to the receiving earth stations. A *transponder* connects one sending station with a receiving station. For a fixed time interval of variable length λ_k the n sending stations are connected with the receiving stations via the n transponders onboard the satellite, i.e., a certain *switch mode* is applied. Mathematically, a switch mode P_k corresponds to a permutation matrix, whose 1-entries show the current connections. After a short while the connections onboard the satellite are simultaneously changed and in the next time interval new pairs of sending and receiving stations are connected via transponders, i.e., a new switch mode is applied.

The *time slot assignment problem* tackles the questions as to which switch modes should be applied and how long each of them lasts such that a given amount of data can be remitted in the shortest possible time. Given an $(n \times n)$ traffic matrix $T = (t_{ij})$, where t_{ij} describes the amount of information to be remitted from the ith sending station to the jth receiving station, we have to determine the switch modes P_k, $k = 1, 2, \ldots$, and the nonnegative lengths λ_k of the corresponding time slots during which the switch modes P_k are applied, such that all data are remitted in the shortest possible time. This leads to the following mathematical model:

$$\min \quad \sum_k \lambda_k$$

s.t.

$$\sum_k \lambda_k p_{ij}^{(k)} \geq t_{ij}, \quad \text{for } 1 \leq i, \ j \leq n,$$

$$\lambda_k \geq 0, \quad \text{for all } k. \tag{2}$$

This problem can be solved optimally by the following algorithm of complexity $O(n^4)$, see e.g. [19]. First we assume that the traffic matrix T has constant row and column sums. Otherwise, let t^* be the maximum value of the row and column sums. Now we can fill up the matrix in a straightforward way by increasing some elements such that all row and column sums are equal to t^*. Since due to the Theorem of Birkhoff a doubly stochastic matrix is the convex combination of permutation matrices, we can write the traffic matrix T as a weighted sum of switch modes. The sum of weights equals t^*, i.e., this decomposition is optimal, since no two elements of the same row or column can be remitted at the same time.

Algorithm (Matrix decomposition).

1. Let $k := 1$.
2. Construct a bipartite graph with $|V| = |W| = n$ and the following edges:
   $$(i, j) \in E \quad \text{iff} \quad t_{ij} > 0.$$

3. Find a perfect matching φ_k corresponding to a switch mode P_k in this graph.
4. Let $\lambda_k := \min\{t_{i\varphi(i)}\}$.
5. Form $T := T - \lambda_k P_k$.
6. If $T \neq \mathbf{0}$, set $k := k + 1$ and goto 2; otherwise stop.

Note that in every iteration we have a matrix T with constant row and column sums. Therefore there exists a perfect matching in Step 2 and it can be found by Hopcroft and Karp's procedure. We also know from the dimension $(n - 1)^2$ of the assignment polytope that T is decomposed in at most $n^2 - 2n + 2$ different switch modes. Now we can question, whether this mathematically optimal procedure leads to a technically feasible solution. Onboard the satellite are about 40 transponders. A time interval which has to be split up into different time slots has a length of about 2 ms. But it is technically infeasible to switch about 1600 times within 2 ms from one mode to another. Therefore one has to restrict the number of switch modes. But then, unfortunately, the time slot assignment problem with the additional restriction that the number of switch modes is $O(n)$, becomes $\mathcal{NP}$-hard, see [137]. Thus one has to solve the constraint time slot assignment problem by heuristics, see e.g. the comments on balanced linear assignment problems at the end of Section 4.3.

Another possibility consists in considering a different model. According to a proposal of Lewandowski et al. [106] a system of $2n$ switch modes $P_1, \ldots, P_n$, $Q_1, \ldots, Q_n$ could be fixed onboard the satellite, where

$$\sum_k P_k = \sum_l Q_l = \mathbf{1}$$

and $\mathbf{1}$ is the matrix with 1-entries only. For such a setting the time slot assignment problem can be transformed in an ordinary linear sum assignment problem. For details see [19].

4. Linear assignment problems

4.1. A general solution method for algebraic linear assignment problems

Linear assignment problems can be solved by only *adding, subtracting* and *comparing* the cost coefficients. A careful reasoning shows moreover that a subtraction $b - a$ occurs only in the case that the cost element a is not greater than b. Thus, when we want to determine $b - a$, we ask for an element c such that $a + c = b$. These considerations lead to an algebraic model, originally introduced in Burkard et al. [36], which allows to formulate and solve linear assignment problems within a general framework. The underlying algebraic structure is a so-called *d-monoid*, i.e., a totally ordered commutative semigroup $(H, *, \leqslant)$ with composition $*$ and linear order relation $\leqslant$, which fulfills in addition the following axiom

$$\text{If } a \leqslant b, \quad \text{then there exists an element } c \in H \text{ such that } a * c = b. \tag{3}$$

Given n^2 cost coefficients $c_{ij} \in H$, the *algebraic linear assignment problem* can be formulated as

$$\min_{\varphi} (c_{1\varphi(1)} * c_{2\varphi(2)} * \cdots * c_{n\varphi(n)}). \tag{4}$$

It is possible to solve algebraic assignment problems in $O(n^4)$ steps. But if we additionally assume that the *weak cancellation rule*

$$\text{If } a * c = b * c, \quad \text{then either } a = b \text{ or } a * c = b \tag{5}$$

holds in $(H, *, \leqslant)$, we can solve these problems in $O(n^3)$ steps, see [43]. For further results in this direction, consult the survey on algebraic optimization by Burkard and Zimmermann [44].

Special examples for d-monoids which obey the weak cancellation rule are:
- $H = \mathbb{R}$ with addition as composition and the usual order relation. This model leads to linear sum assignment problems (LSAP):

$$\min_{\varphi} (c_{1\varphi(1)} + c_{2\varphi(2)} + \cdots + c_{n\varphi(n)}).$$

- H is the set of extended real numbers $\bar{\mathbb{R}}$ (including $-\infty$) with the usual order relation. The composition is defined by $a * b := \max(a, b)$. This model leads to linear bottleneck assignment problems:

$$\min_{\varphi} \max\{c_{1\varphi(1)}, c_{2\varphi(2)}, \ldots, c_{n\varphi(n)}\}.$$

- $H = \mathbb{R}^n$, the composition is the vector addition and the order relation is the lexicographical order. This leads to lexicographical sum assignment problems.
- $H = \bar{\mathbb{R}} \times \mathbb{R}$, $\leqslant$ is the lexicographical order and the composition is defined by

$$(a, b) * (c, d) := \begin{cases} (a, b) & \text{if } a \geqslant c, \\ (a, b + d) & \text{if } a = c. \end{cases}$$

This leads to the so-called *time–cost assignment problem*. Let a denote a time and let b denote the corresponding cost. In a time cost assignment problem we want to find an assignment which first minimizes the maximum occurring time. Secondly, under all solutions which yield this time, a solution with minimum cost is to be found. Such problems occur if n customers have to be served as fast a possible, e.g. under emergency aspects, and then a cost minimal optimal solution should be found.

Linear (algebraic) assignment problems can be solved by transforming the $n \times n$ cost matrix $C = (c_{ij})$ until we find a zero cover which has n elements (cf. Theorem 1.3). Thus we have to define *zero elements* in H, we have to describe the transformations of the cost matrix C which we shall call *admissible transformations*, and we have to explain in which way these admissible transformations should be applied in order to get finally a minimum zero cover.

Let us start with the definition of admissible transformations. We abbreviate the objective function value of permutation φ with respect to the cost matrix C by

$$z[C, \varphi] := c_{1\varphi(1)} * c_{2\varphi(2)} * \cdots * c_{n\varphi(n)}.$$

Definition 4.1 (Admissible transformation). A transformation T of the $n \times n$ matrix $C = (c_{ij})$ to the matrix $\bar{C} = (\bar{c}_{ij})$ is called *admissible* with *index* $z(T)$, if

$$z[C, \varphi] = z(T) * z[\bar{C}, \varphi]$$

for all $\varphi \in \mathscr{S}_n$.

If we perform an admissible transformations T after an admissible transformation S, we get again an admissible transformation. If S and T have the indices $z(S)$ and $z(T)$, respectively, their composition has index $z(S) * z(T)$. The definition above states a property of admissible transformations, but does not explain, how we can get it. This is provided by the following theorem:

Theorem 4.2 (Admissible transformations for assignment problems [36]). *Let* $I, J \subseteq \{1, 2, \ldots, n\}$, $m := |I| + |J| - n \geq 1$, *and* $c := \min\{c_{ij} : i \in I, j \in J\}$. *Then the transformation* $C \mapsto \bar{C}$ *defined by*

$$\bar{c}_{ij} * c = c_{ij}, \quad for \ i \in I, \ j \in J$$

$$\bar{c}_{ij} = c_{ij} * c, \quad for \ i \notin I, \ j \notin J$$

$$\bar{c}_{ij} = c_{ij}, \quad otherwise$$

is admissible with $z(T) = c * c * \cdots * c$, *where the expression on the right-hand side contains* m *factors.*

Note that we make use of (3) in the first line of the definition of $\bar{c}_{ij}$!

In the semigroup H the role of 0-elements is replaced by so-called *dominated elements*. An element $a \in H$ is dominated by an element $z \in H$, if $a * z = z$. Thus in $(\mathbb{R}, \leqslant, +)$ the 0 is dominated by any other number. Now we can formulate the following optimality criterion:

Theorem 4.3. *Let* $T : C \to \bar{C}$ *be an admissible transformation such that there exists a permutation* $\hat{\varphi}$ *with the following properties:*
1. $z(T) * \bar{c}_{ij} \geqslant z(T)$,
2. $z[\bar{C}, \hat{\varphi}] * z(T) = z(T)$.
Then $\hat{\varphi}$ *is an optimal assignment with value* $z(T)$.

The first property in Theorem 4.3 says that all cost coefficients $\bar{c}_{ij}$ are "non-negative" (with respect to $z(T)$). The second property of Theorem 4.3 says that the current objective function value is already dominated by $z(T)$, i.e., has value "0".

Proof. Let φ be an arbitrary permutation. According to Definition 4.1 and properties (1) and (2) of the proposition above we get:

$$z[C, \varphi] = z(T) * z[\bar{C}, \varphi] \geqslant z(T) = z(T) * z[\bar{C}, \hat{\varphi}] = z[C, \hat{\varphi}].$$

Therefore $\hat{\varphi}$ is optimal. $\square$

Now we have to specify in which way the admissible transformations should be applied. This is stated in the following algorithm:

4.1.1. Algorithm for solving linear algebraic assignment problems

1. Perform *row reductions* in matrix C, i.e., perform admissible transformations with $I = \{k\}$, $J = \{1, 2, \ldots, n\}$. Start with $k = 1$ and let $z := z(T)$ be the corresponding

index. Continue with $k = 2, \ldots, n$ and update $z := z * z(T)$. Afterwards all elements in the transformed matrix are "nonnegative" with respect to z, namely $\bar{c}_{ij} * z \geqslant z$.

2. Perform *column reductions*, i.e., perform admissible transformations with $I = \{1, 2, \ldots, n\}$, $J = \{k\}$, for $k = 1, 2, \ldots, n$. Afterwards every row and column in the transformed cost matrix contains at least one element which is dominated by z. All other elements remain non-negative with respect to z.

3. Determine a maximum matching in the following bipartite graph $G = (V, W; E)$, where V contains the row indices of the transformed cost matrix, W the column indices and $(i, j) \in E$, iff $\bar{c}_{ij} * z = z$.

4. If the maximum matching is perfect, then stop: the optimal solution is given by this matching and z is the optimal value of the objective function. Otherwise, go to Step 5.

5. Determine a minimum cover of the transformed cost coefficients dominated by z. This cover yields the new index sets I and J. I contains the indices of the uncovered rows, J contains the indices of uncovered columns.

6. Perform an admissible transformation determined by the new index sets I and J as in Theorem 4.2, update $z := z * z(T)$, and go to Step 3.

It is rather straightforward to show that this algorithm yields an optimal solution of the algebraic assignment problem after at most $n^2 - 2n + 3$ admissible transformations. If the composition $*$ is specialized to "+", the algorithm described above becomes a variant of the Hungarian method. If the composition is specialized to the *max* operation, then we obtain the bottleneck assignment problem and the above algorithm is a variant of the threshold method.

Finally we address a case where the solution of an algebraic assignment problem can be stated explicitly. We say a cost matrix $C = (c_{ij})$ fulfills the *algebraic Monge property*, if it fulfills the following conditions:

$$c_{ij} * c_{kl} \leqslant c_{il} * c_{kj}, \quad \text{for } 1 \leqslant i < k \leqslant n, \; 1 \leqslant j < l \leqslant n. \tag{6}$$

We can show that the following theorem holds, see [87,38].

Theorem 4.4. *If the cost matrix C of an algebraic linear assignment problem fulfills the algebraic Monge property* (6), *then this assignment problem is solved by the identical permutation id, defined by $id(i) = (i)$ for all $i = 1, 2, \ldots, n$.*

The Monge property depends on the proper numbering of the rows and columns of the matrix. A matrix C is called a *permuted Monge matrix*, if there exists a pair of permutations (φ, ψ) such that the matrix $C^{(\varphi, \psi)} = (c_{\varphi(i)\psi(j)})$ obtained from C by permuting its rows according to φ and its columns according to ψ, is a Monge matrix, see [38]. The problem of recognizing permuted algebraic Monge matrices is rather subtle in the general case. It can be shown that this problem is NP-hard, if $n \geqslant 3$ and the ordered semigroup fulfills no additional property. It becomes, however, polynomially solvable, if for instance a weak cancellation rule (4.1) is fulfilled. For details see [38].

4.2. Linear sum assignment problems

Linear sum assignment problems (LSAP) belong to the classical problems of mathematical programming. They occur mainly as subproblems in more complex situations like the travelling salesman problem, vehicle routing problems, personnel assignments and similar problems from practice. An interesting application in railway systems is described by Neng [119], who considers the problem of assigning engines to trains due to traffic constraints and formulates this problem as a linear assignment problem.

A large number of algorithms, sequential and parallel, has been developed for the LSAP, e.g. primal-dual algorithms, simplex-like methods, cost operation algorithms, forest algorithms and relaxation approaches. For a survey on these methods and available computer programs see the recent article of Burkard and Çela [24] or the annotated bibliography of Dell'Amico and Martello [56]. It should be pointed out that nowadays it is possible to solve large scale dense LSAPs (with $n \approx 10^6$) within a couple of minutes, see [105].

Whereas $O(n^3)$ is the best worst case complexity for sequential linear sum assignment algorithms, an algorithm with expected running time of $O(n^2 \log n)$ was developed by Karp [94] in the case of independent and uniformly distributed cost coefficients c_{ij} in $[0,1]$. This algorithm is a special implementation of the classical shortest augmenting path algorithm. It uses priority queues to compute shortest augmenting paths in $O(n^2 \log n)$ time which yields a worst case time complexity of $O(n^3 \log n)$.

For very large problems there is a need for good and fast heuristics. Karp et al. [96] developed a fast heuristic which runs in $O(n \log n)$ time in the worst case and $O(n)$ expected time. In the case of uniformly distributed cost coefficients in $[0,1]$ it provides a solution whose value is, smaller than $3 + O(n^{-a})$, for some $a > 0$, with probability $1 - O(n^{-a})$. The basic idea is to construct a "cheap" sparse subgraph of the given graph. Then it is shown that this sparse subgraph contains a perfect matching with high probability. If the subgraph does not contain a perfect matching a solution for the original LSAP instance is determined in a greedy way.

It has already been pointed out at the end of Subsection 4.1 that the identical permutation is an optimal solution of an LSAP if its cost matrix fulfills a Monge condition. This remains even true, when the Monge condition is relaxed to the so-called *weak Monge property* (cf. [58])

$$c_{ii} + c_{kl} \leqslant c_{il} + c_{ki}, \quad \text{for } 1 \leqslant i < k \leqslant n, \ 1 \leqslant i < l \leqslant n. \tag{7}$$

Analogously, it can be shown that the permutation φ defined by $\varphi(i) = n - i + 1$ for all i, is an optimal solution of an LSAP with an *Anti-Monge cost matrix* C, i.e. a cost matrix $C = (c_{ij})$ fulfilling

$$c_{ij} + c_{kl} \geqslant c_{il} + c_{kj} \quad \text{for } 1 \leqslant i < k \leqslant n, \ 1 \leqslant j < l \leqslant n. \tag{8}$$

As was pointed out in the last subsection, Monge properties depend on the proper numbering of the rows and columns of the considered matrix. In the case of sum problems, Deǐneko and Filonenko [54] designed an $O(n^2)$ algorithm which decides, whether an $n \times n$ matrix C is a permuted Monge matrix. Moreover, if C is a permuted Monge matrix, the algorithm constructs the appropriate permutations φ, ψ for the rows and the columns within this time bound. As a consequence, the LSAP with a permuted Monge cost matrix can be solved in $O(n^2)$ time. The reader is referred to Burkard et al. [38] for a detailed discussion of Monge properties, and a description of the algorithm of Deĭneko and Filonenko.

An important special case of an LSAP with a permuted Monge cost matrix arises if the cost coefficients have the form

$$c_{ij} = u_i v_j \quad \text{for all } i, j$$

with non-negative numbers u_i and v_j. Such an LSAP can simply be solved in $O(n \log n)$ time by ordering the elements u_i and v_j, see the following theorem on minimum and maximum scalar products:

Theorem 4.5 (Hardy et al. [85]). *Let $0 \leqslant u_1 \leqslant \cdots \leqslant u_n$ and $0 \leqslant v_1 \leqslant \cdots \leqslant v_n$. Then for any permutation φ*

$$\sum_{i=1}^{n} u_i v_{n+1-i} \leqslant \sum_{i=1}^{n} u_i v_{\varphi(i)} \leqslant \sum_{i=1}^{n} u_i v_i.$$

4.3. Linear bottleneck assignment problems

Linear bottleneck assignment problems (LBAP) have the form

$$\min_{\varphi} \max_{1 \leqslant i \leqslant n} c_{i\varphi(i)}. \tag{9}$$

They were introduced by Fulkerson et al. [74] and occur e.g. in connection with assigning jobs to parallel machines so as to minimize the latest completion time. Another application occurs in locating objects in space. Let us consider n objects which are detected by two sensors at geographically different sites. Each sensor measures the angle under which the object can be seen, i.e., it provides n lines, on which the objects lie. The location of every object is found by intersecting the appropriate lines. The pairing of the lines is modeled as follows: let c_{ij} be the smallest distance between the ith line from sensor 1 and the jth line from sensor 2. Due to small errors during the measurements, c_{ij} might even be greater than 0 if some object is determined by lines i and j. Solving an LBAP with cost matrix $C = (c_{ij})$ leads to very good results in practice (cf. [16] who used, however, linear sum assignment problems instead of the error-minimizing bottleneck problems). A similar technique can be used for tracking missiles in space. If their locations at two different times t_1 and t_2 are known, we compute the (squared) Euclidean distances between any pair of old and new locations and solve the corresponding linear bottleneck assignment problem in order to match the points in the right way.

Considering bottleneck assignment problems, Gross [78] proved the following min–max theorem which was a starting point of the theory on blocking systems, see [63].

Theorem 4.6 (Gross [78]). *Let $N = \{1, 2, \ldots, n\}$ and let $\mathscr{S}_n$ be the set of all permutations φ of N. Then the following min–max equality holds for an arbitrary $n \times n$*

matrix $C = (c_{ij})$ *with elements* c_{ij} *drawn from a totally ordered set:*

$$\min_{\varphi \in \mathscr{S}_n} \max_{i \in N} c_{i\varphi(i)} = \max_{\substack{I,J \subseteq N \\ |I|+|J|=n+1}} \min_{i \in I, j \in J} c_{ij}. \tag{10}$$

Note the relationship to Theorem 4.2, where we perform a transformation with $\min_{i \in I, j \in J} c_{ij}$. Indeed, the algorithm of Section 4.1 leads to so-called *threshold algorithms* for solving bottleneck assignment problems: A threshold algorithm alternates between two phases. In the first phase a cost element c_{ij}^*—the *threshold value*—is chosen and a matrix $\bar{C}$ is defined by

$$\bar{c}_{ij} := \begin{cases} 1 \text{ if } c_{ij} > c_{ij}^*, \\ 0 \text{ if } c_{ij} \leqslant c_{ij}^*. \end{cases}$$

In the second phase it is checked, whether the bipartite graph with zero-adjacency matrix $\bar{C}$ contains a perfect matching or not. The smallest value c_{ij}^* for which the corresponding bipartite graph contains a perfect matching is the optimum value of the LBAP (9).

There are several ways to implement such a threshold algorithm. One possibility is to order the cost elements increasingly and to apply a binary search in the first phase. This leads to an $O(T(n) \log n)$ algorithm, where $T(n)$ is the time complexity for checking the existence of a perfect matching.

Another possibility is to mimic the Hungarian method. We start with

$$c^* := \max_{i,j} \left(\min_i c_{ij}, \min_j c_{ij} \right) \tag{11}$$

and grow bottleneck augmenting paths as long as the matrix $\bar{C}$ does not contain an assignment with objective function value equal to 0, see [122,76,59,45,57]. FORTRAN codes for this method can be found in the book by Burkard and Derigs [31] and in Carpaneto and Toth [45]. The implementations differ in the determination of a starting solution and in the applied data structures. One of the most efficient implementations is described in Derigs [57]. A thorough recent investigation on computational issues concerning LBAPs can be found in Pferschy [126]. Among others Pferschy proposes an implementation using sparse subgraphs.

An algorithm with the currently best (theoretical) time complexity is obtained by combining the threshold approach with augmenting paths. This idea goes back to Gabow and Tarjan [75], who designed an algorithm with worst time complexity $O(m\sqrt{n} \log n)$ for the LBAP whose underlying bipartite graph G has $2n$ vertices and m edges. For dense graphs this bound has been improved further by Punnen and Nair [133]. According to these authors we first solve

$$\min_{M \in F^*} \max_{(i,j) \in M} c_{ij}, \tag{12}$$

where F^* is the set of all matchings in G which differ from a maximum matching by at most $n\sqrt{n/m}$ edges. This problem can be solved in $O(n^{1.5}\sqrt{m})$ time by combining the maximum matching algorithm of Alt et al. [5] (see Section 1) with binary search. Then this solution is extended to an optimal solution of the bottleneck assignment problem by growing at most $n\sqrt{n/m}$ augmenting paths. Every augmenting path can be completed in $O(m)$ time, see e.g. [147]. Thus the overall complexity of the algorithm becomes $O(n\sqrt{nm})$.

Pferschy [125] describes an algorithm with expected running time $O(n^2)$. Thus this algorithm is linear in m in the case that the graph G is dense. Pferschy's algorithm uses again the idea of thinning out the original problem by considering only the $2n \log n$ cheapest edges. This can be done in $O(n^2)$ time by using a linear selection algorithm for finding the $2n \log n$-smallest edge. In the second step the LBAP on the sparse subgraph is solved by using the method of Gabow and Tarjan. This yields $O((n \log n)^{3/2})$ additional elementary operations. Finally the solution is completed to a perfect matching in the full graph by applying again the algorithm of Gabow and Tarjan. But this completion step is only necessary with a low probability, namely with a probability less than $O(1/\sqrt{n \log n})$. Thus, the expected time needed by the completion is $O(n^2)$, and we get an overall expected running time of $O(n^2)$. This algorithm provides not only a good bound in terms of complexity, but is also very efficient and simple to use.

The so-called *balanced assignment problems* which were introduced by Martello et al. [110] are related to bottleneck assignment problems. Given a real $n \times n$ matrix $C = (c_{ij})$, the balanced assignment problem can be formulated as

$$\min_{\varphi} \left[\max_i c_{i\varphi(i)} - \min_i c_{i\varphi(i)} \right].$$

For solving this problem the authors sort the entries c_{ij} non-decreasingly and propose the following $O(n^4)$ procedure:

4.3.1. Algorithm for balanced assignment problems

1. Solve the corresponding bottleneck assignment problem. Let φ^* be its optimal solution.
2. Define

$$l := \min_i c_{i\varphi^*(i)}, \quad u := \max_i c_{i\varphi^*(i)}.$$

   If $l = u$, stop. A balanced solution has been found. Otherwise go to Step 3.
3. Delete in C all elements $\leq l$ and $> u$ and grow augmenting paths.
   If there exists a perfect matching in the corresponding graph, set φ^* equal to that solution and go to Step 2.
   If no solution exists, then go to Step 4.
4. If u is already the maximum element of matrix C, stop. The present solution is optimal. Otherwise increase u to the next larger element and return to Step 3.

These balanced linear assignment problems can be used in a heuristic for decomposing traffic matrices arising from TDMA-systems (see Section 3) in at most n switch modes (cf. [8], who used LBAPs instead of balanced linear assignment problems in this context). Given the traffic matrix T, let φ^* be an optimal solution of the balanced assignment problem with coefficient matrix T. We set $\lambda_1 := \max_{1 \leq i \leq n} t_{i\varphi^*(i)}$ and forbid the elements $t_{i\varphi^*(i)}$, $i = 1, 2, \ldots, n$. With this new matrix we solve the next balanced assignment problem and determine λ_2. We continue in this way until all elements of T are forbidden. The rational behind this approach is that during the application of a fixed switch mode all involved stations should have about the same workload.

5. Multidimensional assignment problems

5.1. General multidimensional assignment problems

Multi-dimensional (sometimes called *multi-index*) assignment problems (MAP) have been introduced by Pierskalla [127] as natural extensions of linear assignment problems. The most prominent representatives of this class are axial and planar 3-dimensional assignment problems, which are treated in the subsections below. An annotated bibliography on this subject can be found in [22].

The axial MAP can be written as

$$\min \quad \sum_{i_1=1}^{n} \cdots \sum_{i_d=1}^{n} c_{i_1 \cdots i_d} x_{i_1 \cdots i_d}$$

$$\text{s.t.} \quad \sum_{i_2=1}^{n} \cdots \sum_{i_d=1}^{n} x_{i_1 \cdots i_d} = 1, \quad i_1 = 1, \ldots, n,$$

$$\sum_{i_1=1}^{n} \cdots \sum_{i_{k-1}=1}^{n} \sum_{i_{k+1}=1}^{n} \cdots \sum_{i_d=1}^{n} x_{i_1 \cdots i_d} = 1,$$

$$\text{for } k = 2, \ldots, d-1, \quad \text{and} \quad i_k = 1, 2, \ldots, n,$$

$$\sum_{i_1=1}^{n} \cdots \sum_{i_{d-1}=1}^{n} x_{i_1 \cdots i_d} = 1, \quad i_d = 1, \ldots, n,$$

$$x_{i_1 \cdots i_d} \in \{0, 1\} \quad \text{for } 1 \leqslant i_1, i_2, \ldots, i_d \leqslant n, \tag{13}$$

with n^d cost coefficients $c_{i_1 \cdots i_d}$.

More simply, we just asks for $d-1$ permutations $\varphi_1, \varphi_2, \ldots, \varphi_{d-1}$ which minimize the objective function

$$\sum_{i=1}^{n} c_{i \varphi_1(i) \varphi_2(i) \ldots \varphi_{d-1}(i)}.$$

The multidimensional assignment problem is NP-hard in general, but in the case that the array of the cost coefficients is a Monge array (see [38]), it is solved by the identical permutations $\varphi_i = id$, for $i = 1, 2, \ldots, d-1$. Burkard et al. [42] have shown, however, that the MAP remains NP-hard for $d \geqslant 3$, if the cost array fulfills an Anti Monge condition. (Remember that for $n = 2$ an LSAP with an Anti-Monge cost matrix is solved by the permutation φ with $\varphi(i) = n - i + 1$.) This implies that *maximizing* the objective function with cost elements drawn from a Monge array is NP-hard.

General multidimensional assignment problems have recently been considered to model data association problems in connection with multi-target tracking and multi-sensor surveillance, see Poore [128] and Poore et al. [131] for more details. These authors solve the occurring MAPs by Lagrangean relaxation methods, see [130,132]. A numerical study of data association problems arising in multi-target and multi-sensor tracking is given in Poore [129]. Greedy randomized adaptive search (GRASP) heuristics for multidimensional assignment problems arising in multitarget tracking and data association have been proposed by Murphey et al. [116,117].

Pusztaszeri et al. [134] describe another interesting MAP which arises in the context of tracking elementary particles. By solving a five-dimensional assignment problem, they reconstruct tracks of charged elementary particles generated by the Large Electron-Positron Collider at CERN in Geneva.

5.2. Axial three-dimensional assignment problems

Consider n^3 cost coefficients c_{ijk}. The three-dimensional axial assignment problem (3-DAP) can be described with the help of two permutations φ and ψ as

$$\min_{\varphi,\psi\in\mathscr{S}_n} \sum_{i=1}^n c_{i\varphi(i)\psi(i)}. \tag{14}$$

Its name stems from the formulation

$$\min \sum_{i=1}^n \sum_{j=1}^n \sum_{k=1}^n c_{ijk}x_{ijk}$$

$$\text{s.t.} \sum_{j=1}^n \sum_{k=1}^n x_{ijk} = 1, \quad i = 1,2,\ldots,n,$$

$$\sum_{i=1}^n \sum_{k=1}^n x_{ijk} = 1, \quad j = 1,2,\ldots,n,$$

$$\sum_{i=1}^n \sum_{j=1}^n x_{ijk} = 1, \quad k = 1,2,\ldots,n,$$

$$x_{ijk} \in \{0,1\} \quad \text{for all } 1 \leqslant i,j,k \leqslant n, \tag{15}$$

where the 1-s on the right-hand side are assigned to positions at the axes of a 3-dimensional array. The sum over the corresponding "flat" in the array must equal the amount assigned to the position on the axis, see Fig. 5.

According to (14) 3-DAP has $(n!)^2$ feasible solutions. Karp [93] showed that the 3-DAP is $\mathcal{NP}$-hard.

Euler [65] started the investigation of the axial 3-index assignment polytope, i.e., the convex hull of feasible solutions to problem (15). He considers the role of odd cycles for a class of facets of this polytope. Independently, Balas and Saltzman [10] investigate in detail the polyhedral structure of the three-index assignment polytope. They show that this polytope has dimension $n^3 - 3n + 2$ and they describe an $O(n^4)$ separation algorithm for facets induced by certain cliques. Balas and Qi [9] and Qi et al. [135] continue the above work.

Several branch and bound algorithms were proposed for solving 3-DAPs. Most of these algorithms split the current problem into two subproblems by fixing one variable x_{ijk} to 1 and to 0, respectively. Balas and Saltzman [11] introduced a branching strategy which exploits the structure of the problem and allows to fix several variables at each branching node. Hansen and Kaufman [84] describe a primal-dual method similar to

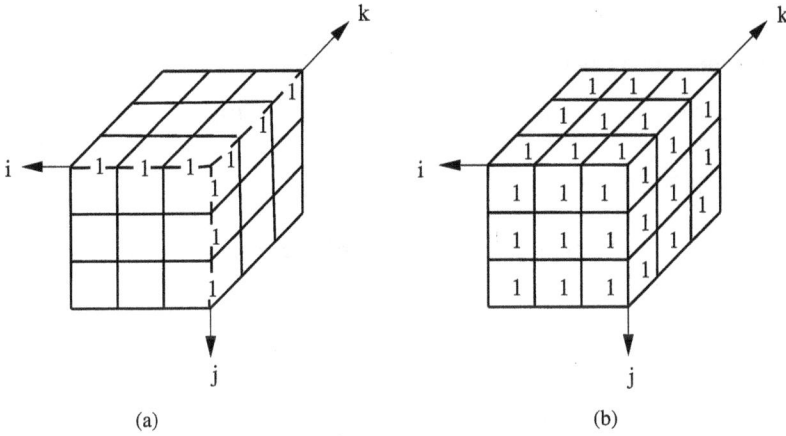

Fig. 5. (a) A geometric representation of the axial 3-dimensional assignment problem for $n = 3$. (b) A geometric representation of the planar 3-dimensional assignment problem for $n = 3$.

the Hungarian method for linear assignment problems. First as many 0-elements as possible are generated among the cost coefficients by generalized row reductions and column reductions. Then a covering problem is solved for these 0-elements. If the covering number is smaller than n, further 0-elements are generated by means of an admissible transformation similar to that in Section 4.1. If the covering number equals n, the corresponding stability problem in a hypergraph is solved. This stability problem replaces the determination of a maximum matching in the 2-dimensional case. The minimum number of (hyper-)edges in a cover is in general strictly larger than the cardinality of a stable set. If the stability number equals n, an optimal solution has been found, otherwise a branching is performed. Note the similarity of this method with the approach in Section 4.1 for solving algebraic assignment problems. Due to this similarity the same algorithm solves the "*algebraic*" version of the 3-DAP with cost coefficients drawn from a d-monoid. The subproblems of finding a minimum cover and a maximum stable set are in the 3-dimensional case, however, $\mathcal{NP}$-hard.

Due to Burkard and Fröhlich [35] admissible transformations for 3-DAPs have the following form. Let us first introduce some notation. Let $N := \{1, 2, \ldots, n\}$, $I, J, K \subseteq N$ and denote $\bar{I} := N \setminus I$, $\bar{J} = N \setminus J$, $\bar{K} = N \setminus K$. With these definitions we get

Theorem 5.1 (Admissible transformations for the 3-DAP [35]). *Let a 3-DAP with cost coefficients c_{ijk}, $i, j, k = 1, 2 \ldots, n$, be given. For $I, J, K \subseteq N$ with $m := n - (|I| + |J| + |K|) \geq 1$ and*

$$c := \min\{c_{ijk} : (i, j, k) \in \bar{I} \times \bar{J} \times \bar{K}\}$$

we define

$$\bar{c}_{ijk} := c_{ijk} - c, \quad (i, j, k) \in \bar{I} \times \bar{J} \times \bar{K}$$

$$\bar{c}_{ijk} := c_{ijk} + c, \quad (i, j, k) \in (\bar{I} \times J \times K) \cup (I \times \bar{J} \times K) \cup (I \times J \times \bar{K})$$

$$\bar{c}_{ijk} := c_{ijk} + 2c, \quad (i,j,k) \in I \times J \times K$$

$$\bar{c}_{ijk} := c_{ijk}, \quad otherwise.$$

Then, for any feasible solution φ, ψ of the 3-DAP we have

$$\sum_{i=1}^{n} c_{i\varphi(i)\psi(i)} = \sum_{i=1}^{n} \bar{c}_{i\varphi(i)\psi(i)} + mc.$$

Row and column reductions are special cases of the admissible transformations described by the above theorem.

Strong lower bounds are essential for a branch and bound procedure. In the case of 3-DAP lower bounds can be computed by the following Lagrangean relaxation approach. Let us take two blocks of the constraints in (15) into the objective function via Lagrangean multipliers (cf. [35]):

$$L(\pi, \varepsilon) := \min \left\{ \sum_{i=1}^{n} \sum_{j=1}^{n} \sum_{k=1}^{n} (c_{ijk} + \pi_j + \varepsilon_i) x_{ijk} - \sum_{j=1}^{n} \pi_j - \sum_{i=1}^{n} \varepsilon_i \right\}$$

such that

$$\sum_{i=1}^{n} \sum_{j=1}^{n} x_{ijk} = 1, \quad k = 1, 2, \ldots, n$$

$$x_{ijk} \in \{0, 1\}, \quad 1 \leqslant i, j, k \leqslant n$$

$$\pi \in \mathbb{R}^n, \ \varepsilon \in \mathbb{R}^n.$$

Since $L(\pi, \varepsilon)$ is a concave function, we can use a subgradient method for finding its maximum.

5.2.1. Algorithm for maximizing $L(\pi, \varepsilon)$

1. Start with $r = 0$, $\pi^r := \varepsilon^r := 0$.
2. Use a greedy algorithm to minimize $L(\pi^r, \varepsilon^r)$. Let x_{ijk}^r be the corresponding optimal solution.
3. Define

$$v_{i_0}^r := |\{x_{i_0,j,k}^r : x_{i_0,j,k} = 1\}| - 1 \quad \text{for } i_0 = 1, 2, \ldots, n$$

and

$$w_{j_0}^r := |\{x_{i,j_0,k}^r : x_{i,j_0,k} = 1\}| - 1 \quad \text{for } j_0 = 1, 2, \ldots, n.$$

4. If $v^r = w^r = (0, 0, \ldots, 0)$, then the maximum is reached. Terminate.
5. If a prespecified number of iterations is not yet reached, update π and ε by setting

$$\pi^{r+1} := \pi^r + \lambda_r w^r,$$

$$\varepsilon^{r+1} := \varepsilon^r + \lambda_r v^r,$$

where λ_r is a suitable step length. Go to Step 2.
Otherwise terminate.

Another subgradient procedure for solving a Lagrangean relaxation of the 3-DAP together with computational considerations has been described by Frieze and Yadegar [72]. Burkard and Rudolf [41] report on satisfactory computational results obtained by an algorithm which uses the classical branching rule combined with a reduction step in every node of the search tree. The lower bound computation is done by applying the above described subgradient optimization procedure.

There exists a number of polynomially solvable special cases of the 3-DAP. As mentioned in Section 5.1, the 3-DAP becomes polynomially solvable, if the cost coefficients are taken from a 3-dimensional Monge array (see [38]). Burkard et al. [42] investigate 3-DAPs with decomposable cost coefficients, where $c_{ijk} = u_i v_j w_k$ and u_i, v_j, and w_k are non-negative. They show that the maximization version of this problem is polynomially solvable, whereas the minimization is in general $\mathcal{NP}$-hard. Moreover, several polynomially solvable special cases of the minimization problem are identified.

5.3. Planar three-dimensional assignment problems

Planar 3-dimensional assignment problems (3-PAP) have the form:

$$\min \ \sum_{i=1}^{n}\sum_{j=1}^{n}\sum_{k=1}^{n} c_{ijk} x_{ijk}$$

s.t.

$$\sum_{i=1}^{n} x_{ijk} = 1, \quad j,k = 1,2,\ldots,n,$$

$$\sum_{j=1}^{n} x_{ijk} = 1, \quad i,k = 1,2,\ldots,n,$$

$$\sum_{k=1}^{n} x_{ijk} = 1, \quad i,j = 1,2,\ldots,n,$$

$$x_{ijk} \in \{0,1\}, \quad i,j,k = 1,2,\ldots,n. \tag{16}$$

3-PAPs play a crucial role in the context of time tabling problems. For a geometric interpretation of planar 3-dimensional assignment problems see Fig. 5. Every "flat" in the three-dimensional array x_{ijk} must contain a (2-dimensional) assignment. Thus the feasible solutions of the 3-PAP correspond to *Latin squares*. Fig. 6 shows a feasible solution for a 3-PAP with $n = 3$: number 1 represents the assignment in the lowest horizontal flat, number 2 shows the assignment in the medium flat, and 3 represents the assignment in the upper flat. Due to this interpretation, the number of feasible solutions of a 3-PAP of size n equals the number of Latin squares of order n, and hence increases very fast. Due to Bammel and Rothstein [12] the number of feasible solutions for a 3-PAP with $n = 9$ is $9! \cdot 8! \cdot 377,597,570,964,258,816$.

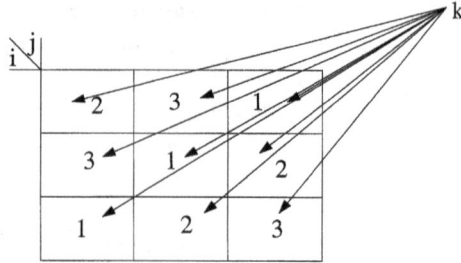

Fig. 6. A latin square representing a feasible solution of the planar 3-dimensional assignment problem of size $n = 3$.

Frieze [71] showed that the 3-PAP is $\mathcal{NP}$-hard. A partial description of the polyhedral structure of the 3-PAP polytope can be found in Euler et al. [66]. See also the related study on time tabling polyhedra by Euler and Verge [67].

There are not many algorithms known for the 3-PAP. The first branch and bound algorithm goes back to Vlach [150]. Vlach computes lower bounds by applying (generalized) row and column reductions similar to those used in the case of 3-DAPs. Another branch and bound procedure for the planar 3-index assignment problem has been described by Magos and Miliotis [109] who also report computational results. Later, Magos [108] used similar ideas for implementing a tabu search algorithm for 3-PAPs. A move in the neighborhood of some Latin square is completely determined by changing the contents of a certain cell (i, j). This affects at least 4 and at most $2n$ other cells which have to be adapted accordingly. Since there are n^2 cells in total and each cell may take $n - 1$ new values, the neighborhood size lies between $n(n - 1)/2$ and $n^2(n - 1)/4$. This neighbourhood structure has two nice properties: first, the change in the objective function value after each move can be computed in linear time. Secondly, not all moves have to be evaluated in each iteration: all moves which put a certain element in a certain cell imply the same change in the objective function, independently from the solution to which they are applied. The numerical results with this algorithm show a good trade-off between computation time and solution quality for 3-PAP instances of size up to $n = 14$.

6. Quadratic assignment problems

6.1. Problem statement, applications and complexity

The quadratic assignment problem (QAP) was introduced by Koopmans and Beckmann [100] in 1957 as a mathematical model for the location of indivisible economical activities. For a comprehensive survey of this field see [28], for an annotated bibliography contact [22]. First we give a description of a QAP as a locational problem. Let us assign n facilities to n locations with the cost being proportional to the flow between the facilities multiplied with their distances plus costs associated with a facility being placed at a certain location. The objective is to allocate each facility at a location such that the total cost is minimized. Thus we are given three $n \times n$ matrices, the flow matrix $A = (a_{ij})$, the distance matrix $B = (b_{kl})$ and matrix $C = (c_{ik})$, where c_{ik} is the cost of placing facility i at location k. The QAP in Koopmans–Beckmann form can now be written as

$$\min_{\varphi \in \mathcal{S}_n} \sum_{i=1}^{n} \sum_{j=1}^{n} a_{ij} b_{\varphi(i)\varphi(j)} + \sum_{i=1}^{n} c_{i\varphi(i)}. \tag{17}$$

Each individual product $a_{ij} b_{\varphi(i)\varphi(j)}$ is the cost caused by assigning facility i to location $\varphi(i)$ and facility j to location $\varphi(j)$. An instance of the QAP with input matrices A, B and C will be denoted by $QAP(A, B, C)$.

A more general version of the QAP was introduced by Lawler [103]. Lawler considers a four-dimensional cost array $C = (c_{ijkl})$ instead of the two matrices A and B, where the linear term C can be added to the elements c_{ijij}. Thus we get the general form of a QAP as

$$\min_{\varphi \in \mathcal{S}_n} \sum_{i=1}^{n} \sum_{j=1}^{n} c_{ij\varphi(i)\varphi(j)}. \tag{18}$$

It is astonishing how many real life applications can be modeled as QAPs. A natural application in location theory was used by Dickey and Hopkins [60] in a *campus planning* model. The problem consists of planning the sites of n buildings on a campus, where b_{kl} is the distance from site k to site l, and a_{ij} is the traffic intensity between building i and building j. The objective is to minimize the total weekly walking distance between the buildings.

In addition to facility location QAPs appear in applications such as layout problems, backboard wiring, computer manufacturing, scheduling, process communications and turbine balancing (see Section 6.5). In the field of ergonomics Burkard and Offermann [39] showed that QAPs can be applied to *typewriter keyboard design*. The problem is to arrange the keys on a keyboard such as to minimize the time needed to write some text. Let the set of integers $N = \{1, 2, \ldots, n\}$ denote the set of symbols to be arranged. Then a_{ij} denotes the frequency of the appearance of the pair of symbols i and j. The entries of the distance matrix b_{kl} are the times needed to press the key in position l after pressing the key in position k. A permutation $\varphi \in \mathcal{S}_n$ describes an assignment of symbols to keys. An optimal solution φ^* for the QAP minimizes the average time for writing a text. A similar application related to an ergonomic design is the development of control boards in order to minimize eye fatigue by McCormick [112]. Further applications concern the ranking of archeological data [101], the ranking of a team in a relay race [86], scheduling parallel production lines [77], and analyzing chemical reactions for organic compounds [149].

By replacing the sums in the objective function of a QAP by the maximum operation, we get the so-called *quadratic bottleneck assignment problem* (BQAP). A BQAP (in Koopmans–Beckmann form) can be formulated as

$$\min_{\varphi} \max_{1 \leq i, j \leq n} a_{ij} b_{\varphi(i)\varphi(j)}.$$

The first occurrence of the BQAP is due to Steinberg [145] and arises as an application in backboard wiring while trying to minimize the maximum length of the involved wires. Another important application of the BQAP, the *bandwidth minimization problem* stems from numerical analysis. In the bandwidth problem we want to find a permutation of the rows and columns of a given matrix such that after permuting the rows and columns the new matrix has minimum bandwidth. It is easy to see that this problem can be modeled as a special BQAP with a 0–1 flow matrix which has an 1-entry, iff the given matrix has at this position a non-zero entry. Defining the distance matrix $B = (b_{kl})$ by $b_{kl} := |k - l|$ and solving the corresponding BQAP leads to an optimal solution of the bandwidth minimization problem.

Besides these applications basically all QAP applications give rise to a BQAP model as well, because it often makes sense to minimize the largest cost instead of the overall cost incurred by some decision.

In contrast to linear assignment problems, quadratic assignment problems remain among the hardest combinatorial optimization problems. Marzetta and Brüngger [111] report recently on the solution of a Koopmans Beckmann problem of size $n = 25$ of the QAP-library (see Section 6.4 and Burkard et al. [37]). The inherent difficulty for solving QAPs is also reflected by their computational complexity. Sahni and Gonzalez [141] showed that the QAP is $\mathcal{NP}$-hard and that even finding an approximate solution within some constant factor from the optimum value cannot be done in polynomial time unless $\mathcal{P} = \mathcal{NP}$. These results hold even for Koopmans–Beckmann QAPs with coefficient matrices fulfilling the triangle inequality, see Queyranne [136]. The *linear dense arrangement problem*, however, which is a special Koopmans–Beckmann QAP, admits a *polynomial time approximation scheme* (*PTAS*), see Arora et al. [6]. In the linear dense arrangement problem matrix A is the distance matrix of n points which are regularly spaced on a line, i.e., points with abscissae given by $x_p = p$, $p = 1, \ldots, n$ and B is a dense 0–1 matrix, i.e., the number of 1-entries in B is in $\Omega(n^2)$.

Recently it has been shown that even local search is hard in the case of the QAP. It can be shown (see [124] for details) that the QAP with respect to the pairwise exchange neighborhood structure as well as with respect to a Lin–Kernighan-like neighbourhood structure [118] is $\mathcal{PLS}$-complete. This implies that the time complexity of a local search method for the QAP using either of the two mentioned neighbourhood structures is exponential in the worst case. Moreover, from results of Papadimitriou and Wolfe [123] follows that deciding whether a given local optimal solution of the QAP is also globally optimum, is $\mathcal{NP}$-complete.

The quadratic bottleneck assignment problem is $\mathcal{NP}$-hard as well, since it contains the bottleneck travelling salesman problem as special case and therefore the problem to decide whether a given graph contains a Hamiltonian cycle or not. To see this, consider a $QAP(A, B)$ where A is the adjacency matrix of the given graph and B is the permutation matrix of a cyclic permutation.

6.2. Different problem formulations and linearizations

There exist different, but equivalent mathematical formulations for QAPs which stress different structural characteristics of the problem and lead to different solution approaches. It is immediate that we can write (17) as an integer quadratic program of the form

$$\min \quad \sum_{i=1}^{n}\sum_{j=1}^{n}\sum_{k=1}^{n}\sum_{l=1}^{n} a_{ij}b_{kl}x_{ik}x_{jl} + \sum_{i,j=1}^{n} c_{ij}x_{ij} \tag{19}$$

$$\text{s.t.} \quad \sum_{i=1}^{n} x_{ij} = 1, \quad j = 1,2,\ldots,n, \tag{20}$$

$$\sum_{j=1}^{n} x_{ij} = 1, \quad i = 1,2,\ldots,n, \tag{21}$$

$$x_{ij} \in \{0,1\}, \quad i,j = 1,2,\ldots,n. \tag{22}$$

We can formulate this Koopmans–Beckmann QAP in a more compact way by defining an *inner product* between matrices. Let the inner product of two real $n \times n$ matrices A, B be defined by

$$\langle A, B \rangle := \sum_{i=1}^{n}\sum_{j=1}^{n} a_{ij}b_{ij}.$$

Given some $n \times n$ matrix B and a permutation $\varphi \in \mathscr{S}_n$ with the associated permutation matrix $X_\varphi \in \mathbf{X}_n$, we have

$$X_\varphi B X_\varphi^{\mathrm{T}} = (b_{\varphi(i)\varphi(j)}). \tag{23}$$

Thus a Koopmans–Beckmann QAP can be written as

$$\min \quad \langle A, XBX^{\mathrm{T}} \rangle + \langle C, X \rangle$$

$$\text{s.t.} \quad X \in \mathbf{X}_n. \tag{24}$$

Formulation (17) together with (23) lead immediately to the so-called *trace formulation* of a Koopmans–Beckmann problem. Recall that the trace on an $n \times n$ matrix is defined as sum of its diagonal elements. Therefore

$$\sum_{i=1}^{n}\sum_{j=1}^{n} a_{ij}b_{\varphi(i)\varphi(j)} = tr(A\bar{B})$$

with $\bar{B} = XB^{\mathrm{T}}X^{\mathrm{T}}$. Since $tr(CX^{\mathrm{T}}) = \sum_{i=1}^{n} c_{i\varphi(i)}$, the QAP in (24) can be formulated as

$$\min \quad tr(AXB^{\mathrm{T}} + C)X^{\mathrm{T}}$$

$$\text{s.t.} \quad X \in \mathbf{X}_n. \tag{25}$$

The trace formulation of the QAP appeared first in Edwards [64], and was used by Finke et al. [68] to introduce eigenvalue bounds for QAPs (see Section 6.3.2).

In the general case, let the coefficients c_{ijkl} be the entries of an $n^2 \times n^2$ matrix S such that c_{ijkl} lies in row $(i-1)n+k$ and column $(j-1)n+l$. Since $x^{\mathrm{T}}[1/2(S+S^{\mathrm{T}})]x = x^{\mathrm{T}}Sx$, we can assume that S is symmetric. The addition of a constant to the entries of the main diagonal of S does not change the optimal solutions of the corresponding QAP, it simply adds a constant to the objective function value. Thus we can assume that S is positive definite or we can also assume that S is negative definite. Let $x = (x_{11}, x_{12}, \ldots, x_{1n}, x_{21}, \ldots, x_{nn})^{\mathrm{T}} = (x_1, \ldots, x_{nn})^{\mathrm{T}}$. Then we can write a QAP as quadratic convex program (quadratic concave program) in the form

$$\min \quad x^{\mathrm{T}} S x$$

$$\text{s.t.} \quad \sum_{i=1}^{n} x_{ij} = 1, \quad j = 1, 2, \ldots, n,$$

$$\sum_{j=1}^{n} x_{ij} = 1, \quad i = 1, 2, \ldots, n,$$

$$x_{ij} \geqslant 0, \quad i, j = 1, 2, \ldots, n. \tag{26}$$

where S is symmetric and positive (negative) definite.

Many authors have proposed methods for linearizing the quadratic form in the objective function (19) by introducing additional variables. Lawler [103] replaces the quadratic terms $x_{ij} x_{kl}$ in the objective function of (18) by n^4 variables

$$y_{ijkl} := x_{ij} x_{kl}, \quad i, j, k, l = 1, 2, \ldots, n, \tag{27}$$

and obtains in this way a 0–1 linear program with $n^4 + n^2$ binary variables and $n^4 + 2n + 1$ constraints. The QAP (19)–(22) can be written as a 0–1 linear program in the following form (see [103,20])

$$\min \quad \sum_{i,j=1}^{n} \sum_{k,l=1}^{n} c_{ijkl} y_{ijkl}$$

$$\text{s.t.} \quad \sum_{i,j=1}^{n} \sum_{k,l=1}^{n} y_{ijkl} = n^2, \tag{28}$$

$$x_{ij} + x_{kl} - 2y_{ijkl} \geqslant 0, \quad i, j, k, l = 1, 2, \ldots, n,$$

$$y_{ijkl} \in \{0, 1\}, \quad i, j, k, l = 1, 2, \ldots, n,$$

$$(x_{ij}) \in \mathbf{X}_n. \tag{29}$$

Kaufman and Broeckx [97] introduced a linearization which yields the smallest number of additionally introduced variables and constraints. Their model employs n^2 real variables, n^2 binary variables and $n^2 + 2n$ constraints. Frieze and Yadegar [73] get a mixed integer linear programming formulation for the QAP with n^4 real variables, n^2 binary variables and $n^4 + 4n^3 + n^2 + 2n$ constraints.

Based on a linearization technique for general 0–1 polynomial programs due to Adams and Sherali [2,3], Adams and Johnson [1] present an 0–1 linear integer programming formulation for the QAP, which resembles to a certain extent the linearization of Frieze and Yadegar. They show that a QAP can be written as the following mixed 0–1 linear program

$$\min \quad \sum_{i,j=1}^{n} \sum_{k,l=1}^{n} c_{ijkl} y_{ijkl}$$

s.t.

$$\sum_{i=1}^{n} y_{ijkl} = x_{kl}, \quad j,k,l = 1,2,\dots,n,$$

$$\sum_{j=1}^{n} y_{ijkl} = x_{kl}, \quad i,k,l = 1,2,\dots,n,$$

$$y_{ijkl} = y_{klij}, \quad i,j,k,l = 1,2,\dots,n,$$

$$y_{ijkl} \geqslant 0, \quad i,j,k,l = 1,2,\dots,n,$$

$$(x_{ij}) \in \mathbf{X}_n, \tag{30}$$

where each y_{ijkl} represents the product $x_{ij}x_{kl}$. The above formulation contains n^2 binary variables x_{ij}, n^4 continuous variables y_{ijkl} and $n^4 + 2n^3 + 2n$ constraints excluding the non-negativity constraints on the continuous variables. Although a significant smaller formulation in terms of the number of variables and constraints could be obtained, the structure of the above formulation is favorable for solving QAPs approximately by means of the Lagrangean dual.

Closely related to linearizations are polyhedral studies of the QAP which have recently be performed by Barvinok [13], Jünger and Kaibel [89,90], Kaibel [91], and Padberg and Rijal [121]. The *QAP polytope* can be defined by using the following graph theoretical model:

For each $n \in \mathbb{N}$ consider a graph $G_n = (V_n, E_n)$ with vertex set $V_n = \{(i,j): 1 \leqslant i,j \leqslant n\}$ and edge set $E_n = \{((i,j),(k,l)): i \neq k, j \neq l\}$. The maximum cliques in G_n have cardinality n and correspond to permutations. Given an instance of the QAP with coefficients c_{ijkl} we introduce the coefficients c_{ijij} as vertex weights and c_{ijkl} with $i \neq k$, $j \neq l$ as weight of the edge $((i,j),(k,l))$. Solving the QAP is equivalent to finding a maximal clique with minimum total vertex weight and edge weight. For each maximum clique C in G_n we denote its incidence vector by (x^C, y^C), where

$$x_{ij}^C = \begin{cases} 1 & \text{if } (i,j) \text{ is a vertex of the clique } C, \\ 0 & \text{otherwise} \end{cases}$$

and

$$y_{ijkl}^C = \begin{cases} 1 & \text{if } (i,j),(k,l) \text{ is an edge of clique } C, \\ 0 & \text{otherwise.} \end{cases}$$

The QAP polytope QAP_n is then defined as convex hull of all vectors (x^C, y^C), where C is a maximum clique in G_n. It turns out that *the traveling salesman polytope* and *the linear ordering polytope* are projections of QAP_n, and that QAP_n is a face of the *Boolean quadric polytope*, see [91].

Barvinok [13], Padberg and Rijal [121], as well as Jünger and Kaibel [89] compute independently the dimension of QAP_n and show that the inequalities $y_{ijkl} \geq 0$, $i \neq k$, $j \neq l$, are facet defining. These facets are usually called the *trivial facets* of QAP_n. Moreover, Padberg and Rijal [121], and Jünger and Kaibel [89] show that the affine hull of QAP_n is described by the following equations which are linearly independent:

$$\sum_{i=1}^{n} x_{ij} = 1, \quad 1 \leqslant j \leqslant n-1, \tag{31}$$

$$\sum_{j=1}^{n} x_{ij} = 1, \quad 1 \leqslant i \leqslant n, \tag{32}$$

$$-x_{kl} + \sum_{i=1}^{k-1} y_{ijkl} + \sum_{i=k+1}^{n} y_{klij} = 0 \quad \begin{array}{l} 1 \leqslant j \neq l \leqslant n, 1 \leqslant k \leqslant n-1, \\ \text{or} \quad 1 \leqslant l < j \leqslant n,\ k = n, \end{array} \tag{33}$$

$$-x_{ij} + \sum_{l=1}^{j-1} y_{ijkl} + \sum_{l=j+1}^{n} y_{ijkl} = 0 \quad \begin{array}{l} 1 \leqslant j \leqslant n, 1 \leqslant i \leqslant n-3, \\ i < k \leqslant n-1 \quad \text{or} \\ 1 \leqslant j \leqslant n-1,\ i = n-2, \\ k = n-1, \end{array} \tag{34}$$

$$-x_{kj} + \sum_{l=1}^{j-1} y_{ilkj} + \sum_{l=j+1}^{n} y_{ilkj} = 0 \quad \begin{array}{l} 1 \leqslant j \leqslant n-1, 1 \leqslant i \leqslant n-3, \\ i < k \leqslant n-1. \end{array} \tag{35}$$

Summarizing we get the following theorem:

Theorem 6.1.

(i) *The affine hull of the QAP polytope QAP_n is given by the linear equations (31)–(35). These equations are linearly independent. The rank of the system is $2n(n-1)^2 - (n-1)(n-2)$, for $n \geq 3$.*

(ii) *For $n \geq 3$ the dimension of QAP_n is equal to $1+(n-1)^2+n(n-1)(n-2)(n-3)/2$.*

(iii) *The inequalities $y_{ijkl} \geq 0$, $i < k$, $j \neq l$, define facets of QAP_n.*

Padberg and Rijal [121] identify additionally two classes of valid inequalities for QAP_n, the *clique* inequalities and the *cut* inequalities, where the terminology is related to the graph G_n. The authors specify some conditions under which the cut inequalities are not facet defining. It is an open problem, however, to identify facet defining inequalities within these classes. Further valid inequalities, the so-called *box inequalities* have been described by Kaibel [91]. These inequalities are obtained by exploiting the relationship between the Boolean quadric polytope and the QAP polytope. For box inequalities it can be decided in polynomial time whether they are facet defining or not, and in the latter case some dominating facet defining inequality can be derived.

Similar results have been obtained for the *symmetric QAP polytope $SQAP_n$* arising in the case that at least one of the coefficient matrices A or B in a Koopmans–Beckmann problem is symmetric. $SQAP_n$ is defined by means of a hypergraph $H_n = (V_n, F_n)$, where V_n is the same set of vertices as in graph G_n and F_n is the set of hyperedges $\{(i,j),(k,l),(i,l),(k,j)\}$ for all $i \neq k$, $j \neq l$. A set $C \subset V_n$ is called a clique in H_n if it is a clique in G_n. Again, the incidence vector (x^C, y^C) of a clique C is introduced by,

$$x_{ij} = \begin{cases} 1, & \text{if } (i,j) \text{ is a vertex of clique } C \\ 0, & \text{otherwise} \end{cases}$$

and

$$y_{ijkl} = \begin{cases} 1, & \text{if } i < k, \ l \neq j, \ \{(i,j),(k,l),(i,l),(k,j)\} \text{ is a hyperedge of clique } C \\ 0, & \text{otherwise.} \end{cases}$$

The convex hull of all incidence vectors (x^C, y^C) is called the symmetric QAP polytope $SQAP_n$.

Padberg and Rijal [121] and Jünger and Kaibel [90] give the following minimal description for the affine hull of $SQAP_n$:

$$\sum_{j=1}^{n} x_{ij} = 1, \quad 1 \leqslant i \leqslant n, \tag{36}$$

$$\sum_{i=1}^{n} x_{ij} = 1, \quad 1 \leqslant j \leqslant n-1, \tag{37}$$

$$-x_{ij} - x_{kj} + \sum_{l=1}^{j-1} y_{ilkj} + \sum_{l=j+1}^{n} y_{ijkl} = 0 \quad \begin{matrix} 1 \leqslant i < k \leqslant n, \\ 1 \leqslant j \leqslant n, \end{matrix} \tag{38}$$

$$-x_{kj} - x_{kl} + \sum_{i=1}^{k-1} y_{ijkl} + \sum_{i=k+1}^{n} y_{kjil} = 0 \quad \begin{matrix} 1 \leqslant k \leqslant n, \\ 1 \leqslant j \leqslant n-3, \\ 1 \leqslant j < l \leqslant n-1. \end{matrix} \tag{39}$$

The results concerning $SQAP_n$ can be summarized in the following theorem:

Theorem 6.2.
  (i) *The affine hull of the symmetric QAP polytope $SQAP_n$ is described by the linear equations (36)–(39). These equations are linearly independent, their rank is $n^2(n-2) + 2n - 1$.*
 (ii) *The dimension of $SQAP_n$ is equal to $(n-1)^2 + n^2(n-3)^2/4$.*
(iii) *The inequalities $y_{ijkl} \geqslant 0$ for $i < k$, $j < l$, and $x_{ij} \geqslant 0$ for $1 \leqslant i, j \leqslant n$, define facets of $SQAP_n$.*
 (iv) *For each $i < k$ and for all $J \subseteq \{1, 2, \ldots, n\}$ the* row curtain *inequalities*

$$-\sum_{j \in J} x_{ij} + \sum_{\substack{j, l \in J \\ j < l}} y_{ijkl} \leqslant 0$$

are valid for SQAP$_n$. *For each* $j < l$ *and for all* $I \subseteq \{1, 2, \ldots, n\}$ *the* column curtain inequalities

$$-\sum_{i \in I} x_{ij} + \sum_{\substack{i,k \in I \\ i < k}} y_{ijkl} \leqslant 0$$

are valid for SQAP$_n$.

All curtain inequalities with $3 \leqslant |I|, |J| \leqslant n - 3$ *define facets of* SQAP$_n$. *The other curtain inequalities define faces which are contained in trivial facets of* SQAP$_n$.

(v) *The separation problem for curtain inequalities is* $\mathcal{NP}$-*hard.*

6.3. Lower bounds

Since QAPs are $\mathcal{NP}$-hard, good lower bounds are of eminent importance for solving these problems by implicit enumeration procedures like branch and bound. We require for a good bound that it is not too hard to compute, that it can easily be evaluated for subsets of the problem which occur after some branching and, finally, that it is tight. There are many different proposals for deriving bounds. In the following we survey briefly bounds based on linearizations of the QAP and eigenvalue bounds which are related to the trace formulation of the QAP.

6.3.1. Bounds based on linearizations

Let us consider a Koopmans–Beckmann problem QAP(A,B,C). W.l.o.g. we can assume that all entries in the matrices A and B are non-negative. For each row index i let $\hat{a}_{(i,.)}$ be the $(n-1)$-dimensional vector obtained from the ith row of A by deleting the element a_{ii}. Similarly define $\hat{b}_{(k,.)}$ for every row k of matrix B. According to Theorem 4.5 we get the minimum scalar product $\langle a, b \rangle^- := \min_\varphi \sum_{i=1}^n a_i b_{\varphi(i)}$ of two non-negative vectors $a, b \in \mathbb{R}^n$ by sorting the elements of a non-decreasingly and the elements of b non-increasingly. Thus $\langle \hat{a}_{(i,.)}, \hat{b}_{(k,.)} \rangle^-$ is the minimum cost which occurs if index i is mapped to index k. In order to find a lower bound on the value of a QAP we first compute the n^2 minimum scalar products $\langle \hat{a}_{(i,.)}, \hat{b}_{(k,.)} \rangle^-$ and define a new cost matrix $L = (l_{ik})$ by

$$l_{ik} = a_{ii} b_{kk} + c_{ik} + \langle \hat{a}_{(i,.)}, \hat{b}_{(k,.)} \rangle^-. \tag{40}$$

We obtain the Gilmore–Lawler lower bound GLB for the Koopmans–Beckmann QAP by solving the linear assignment problem with cost matrix L. The appropriate sorting of the rows and columns of A and B can be done in $O(n^2 \log n)$ time. The computation of all l_{ik} takes $O(n^3)$ time and the same amount of time is needed to solve the last LAP. Thus the Gilmore–Lawler bound for Koopmans–Beckmann problems can be computed in $O(n^3)$ time. Thus it is easy to compute, but it deteriorates fast as the size of the problems increases. A very similar procedure is possible for general QAPs, see e.g. the handbook article of Burkard and Çela [24].

The Gilmore–Lawler bound can be strengthened by splitting the coefficients a_{ij} and b_{kl} and thus transfering some amount from the quadratic part of the objective function to the linear part of the objective function. This can be done by defining new coefficients $\bar{a}_{ij}, \bar{b}_{kl}, \lambda_i$ and μ_k by the formulas

$$a_{ij} = \bar{a}_{ij} + \lambda_i,$$

$$b_{kl} = \bar{b}_{kl} + \mu_k,$$

where the amounts λ_i and μ_k are suitably chosen, e.g. as row minima. Such a *reduction* was first used by Conrad [52] and later independently investigated by many researchers (see [17,140,64,73]). Similar procedures can also be applied to the bottleneck QAP, see [18].

There are several bounding strategies which are closely related to the Gilmore–Lawler bound and reductions. One of them is the bounding strategy of Hahn and Grant [82]. This procedure combines GLB ideas with reduction steps in a dual framework. Other possibilities are exploited by reformulation methods in which the coefficients of the problem are changed such that the new problem has the same objective function value as the original problem for any permutation matrix $X = (x_{ij}) \in \mathbf{X}_n$, but a stronger bound can be derived. Reformulation rules stem from Carraresi and Malucelli [46] and from Assad and Xu [7].

A further way to strengthen GLB stems from Frieze and Yadegar [73]. These authors start from their linearization and include some of the constraints via Lagrangean multipliers in the objective function. The corresponding Lagrangean problem can be solved by subgradient methods and yields sharper bounds than GLB. A similarly approach is used by Adams and Johnson [1]. They add the so-called *complementary constraints*

$$y_{ijkl} = y_{klij}$$

to the objective function via Lagrangean multipliers α_{ijkl} and obtain a Lagrangean relaxation $AJ(\alpha)$ of the following form:

$$\min \sum_{\substack{i=1}}^{n} \sum_{\substack{j=1 \\ j>i}}^{n} \sum_{k=1}^{n} \sum_{\substack{l=1 \\ l \neq k}}^{n} (c_{ijkl} - \alpha_{ikjl}) y_{ikjl}$$

$$- \sum_{\substack{i=1}}^{n} \sum_{\substack{j=1 \\ j<i}}^{n} \sum_{k=1}^{n} \sum_{\substack{l=1 \\ l \neq k}}^{n} (c_{ijkl} - \alpha_{jlik}) y_{ikjl} + \sum_{i=1}^{n} \sum_{k=1}^{n} a_{ik} b_{ik} x_{ik}$$

s.t.

$$\sum_{j=1}^{n} y_{ijkl} = x_{ik}, \quad 1 \leqslant i, k, l \leqslant n,$$

$$\sum_{l=1}^{n} y_{ijkl} = x_{ik}, \quad 1 \leqslant i, j, k \leqslant n,$$

$$0 \leqslant y_{ijkl} \leqslant 1, \quad 1 \leqslant i, j, k, l \leqslant n,$$

$$x_{ik} \in \mathbf{X}_n.$$

Let $\theta(\alpha)$ denote the optimal value of $AJ(\alpha)$. Then $\max_\alpha \theta(\alpha)$ equals the optimal value of the continuous relaxation of the Adams–Johnson linearization, see (30). Adams and Johnson show that for each fixed set of the multipliers α the problem $AJ(\alpha)$ can be solved efficiently by solving $n^2 + 1$ LAPs. Moreover they develop an iterative dual ascent procedure to solve approximately the above maximization problem which leads to the Adams–Johnson bound (AJB). This bound generalizes and unifies all previously mentioned bounds like GLB, the reduction bounds as well as the bound of Assad and Xu which can be obtained for special settings of the Lagrangean multipliers α_{ijkl}. It does not comprise, however, the bounds of Carraresi and Malucelli and the Hahn–Grant bound (HGB). Karisch et al. [92] showed recently that both, AJB and HGB can be obtained from the dual of the continuous relaxation of the MILP formulation (30) proposed by Adams and Johnson. They propose an iterative algorithm to solve this dual approximately and show that AJB, HGB, and all other Gilmore–Lawler-like bounds including the Carraresi–Malucelli bound can be obtained by applying this algorithm with specific settings for the control parameters. The same authors identify a setting for the parameters which seems to provide a bounding algorithm with a better time/quality trade-off than all the other mentioned bounding procedures.

6.3.2. Eigenvalue bounds

The trace formulation of a Koopmans–Beckmann problem can be used to derive a new class of bounds, the so-called *eigenvalue bounds* which were introduced by Finke et al. [68]. When implemented carefully, these techniques produce bounds of good quality in comparison to Gilmore–Lawler-like bounds. The eigenvalue bounds are, however, expensive in terms of computation time and deteriorate quickly when lower levels of a branch and bound tree are searched (see [50]).

Let us start from a Koopmans–Beckmann QAP with symmetric matrices A and B. In this case all eigenvalues $\lambda_1, \lambda_2, \ldots, \lambda_n$ of matrix A and $\mu_1, \mu_2, \ldots, \mu_n$ of matrix B are real. We collect these eigenvalues in vectors λ and μ, respectively. The matrices A and B possess diagonalizations of the form $A = PAP^{\mathrm{T}}$ and $B = QMQ^{\mathrm{T}}$ with orthogonal matrices P and Q and diagonal matrices $\Lambda = \mathrm{diag}(\lambda)$, $M = \mathrm{diag}(\mu)$. Since the columns $p_1, p_2, \ldots, p_n$ of P and $q_1, q_2, \ldots, q_n$ of Q form orthonormal bases, matrix S defined by $s_{ij} := \langle p_i, q_j \rangle^2$ is a doubly stochastic matrix which, due to Birkhoff's Theorem, can be written as convex combination of permutation matrices:

$$S = \sum_{\varphi \in \mathscr{S}_n} \alpha_\varphi X_\varphi.$$

It is easy to see that

$$\mathrm{tr}\, AB = \lambda^{\mathrm{T}} S \mu = \sum_{\varphi \in \mathscr{S}_n} \alpha_\varphi \langle \lambda, X_\varphi \mu \rangle$$

which implies

$$\langle \lambda, \mu \rangle^- \leqslant \mathrm{tr}\, AB \leqslant \langle \lambda, \mu \rangle^+.$$

Since XBX^{T} has the same eigenvalues as B, we get

Theorem 6.3 (Eigenvalue bounds for the QAP).

$$\langle \lambda, \mu \rangle^- \leqslant tr\, AXBX^{\mathrm{T}} \leqslant \langle \lambda, \mu \rangle^+.$$

By applying reduction techniques to the quadratic part of the objective function significant improvements can be achieved. Hadley et al. [81] consider also the case of non-symmetric QAPs and develop for them eigenvalue bounds by means of Hermitian matrices.

A more general approach to eigenvalue based lower bounding techniques was employed by Hadley et al. [79]. Consider the following sets of $n \times n$ matrices, where I is the $n \times n$ identity matrix and $\mathbf{1} := (1, \ldots, 1)^{\mathrm{T}}$ is the n-dimensional vector of all ones:

$$\mathcal{O}_n := \{X \colon X^{\mathrm{T}} X = I\} \qquad \text{set of } orthogonal\ n \times n\ matrices,$$

$$\mathcal{E}_n := \{X \colon X\mathbf{1} = X^{\mathrm{T}}\mathbf{1} = \mathbf{1}\}, \text{ set of } n \times n\ matrices\ with\ row \tag{41}$$
$$\qquad\qquad\qquad\qquad\qquad\qquad\ and\ column\ sums\ equal\ to\ one,$$

$$\mathcal{N}_n := \{X \colon X \geqslant 0\}, \qquad \text{set of } non\text{-}negative\ n \times n\ matrices.$$

It is a well known result that $\mathbf{X}_n = \mathcal{O}_n \cap \mathcal{E}_n \cap \mathcal{N}_n$. The above characterization of $\mathbf{X}_n$ implies that we get a relaxation of the QAP, if we delete one or two of the matrix sets $\mathcal{O}_n, \mathcal{E}_n$ and $\mathcal{N}_n$ in the intersection $\mathbf{X}_n = \mathcal{O}_n \cap \mathcal{E}_n \cap \mathcal{N}_n$. With respect to Theorem 6.3, Rendl and Wolkowicz [138] show that

$$\min_{X \in \mathcal{O}_n} tr(AXBX^{\mathrm{T}}) = \langle \lambda, \mu \rangle^-,$$

$$\max_{X \in \mathcal{O}_n} tr(AXBX^{\mathrm{T}}) = \langle \lambda, \mu \rangle^+.$$

In other words, the lower bound on the quadratic part of the QAP as obtained by the eigenvalue bound is derived by relaxing the feasible set to the set of orthogonal matrices.

A tighter relaxation was proposed in [80], where the set of permutation matrices was relaxed to $\mathcal{O}_n \cap \mathcal{E}_n$. The authors incorporate $\mathcal{E}_n$ in the objective function by exploiting the fact that for any $X \in \mathbf{X}_n$ the vector of ones is both a left and right eigenvector with eigenvalue 1.

The above considerations lead directly to bounds based on *semidefinite relaxations*. Let $\mathcal{Z}_n$ be the set of $n \times n$ 0–1 matrices. Then

$$\mathbf{X}_n = \mathcal{Z}_n \cap \mathcal{E}_n = \mathcal{Z}_n \cap \mathcal{O}_n.$$

Thus we can write a QAP in trace form as

$$\min\ tr(AXBX^{\mathrm{T}} - CX^{\mathrm{T}})$$

s.t.

$$XX^{\mathrm{T}} = X^{\mathrm{T}}X = I,$$
$$X\mathbf{1} = X^{\mathrm{T}}\mathbf{1} = \mathbf{1},$$
$$x_{ij}^2 - x_{ij} = 0.$$

From this formulation semidefinite programming relaxations of the QAP can be obtained, see [154]. These relaxations are solved by interior point methods or cutting plane methods. The quality of the bounds obtained in this way is competitive with the best existing lower bounds for the QAP. However, due to prohibitively high computation time requirements, the use of such approaches as basic bounding procedures within branch and bound algorithms is up to now not feasible.

6.4. Solution methods for QAPs

Since QAPs are notoriously hard to solve to optimality, there is a special need for good heuristics. It was a great surprise that even straightforward implementations of metaheuristics like *simulated annealing (SA)* or *tabu search (TS)* perform very well. An explanation for this behaviour will be given in Section 7.3.

The first simulated annealing algorithm for QAPs was published in Burkard and Rendl [40], soon after the first author heard a lecture of Černý [48] in Prague on the possibility to apply this thermodynamically motivated simulation procedure to the travelling salesman problem. (Černý recognized this possibility independently about at the same time as Kirkpatrick et al. [98].) Improved simulated annealing (SA) algorithms for the QAP have been proposed by several other authors, e.g. by Wilhelm and Ward [153] and Connolly [51].

Tabu search methods for the QAP have been proposed among others by Skorin-Kapov [143] and Taillard [146]. Taillard uses a so-called *robust tabu search* where the size of the tabu list is randomly chosen between a maximum and a minimum value. Battiti and Tecchiolli [14] developed the *reactive tabu search* which involves a mechanism for adopting the size of the tabu list. Reactive tabu search aims at improving the robustness of the algorithm. The algorithm notices when a cycle occurs, i.e., when a certain solution is revisited, and increases the tabu list size according to the length of the detected cycle. Computational results show that generally the reactive tabu search outperforms other tabu search algorithms for the QAP. More recently, parallel implementations of tabu search have been proposed, see e.g. [49]. Tabu search algorithms allow a natural parallel implementation by dividing the burden of the search in the neighborhood among several processors.

Among *genetic algorithms* for the QAP the approach due to Ahuja et al. [4] seems to outperform the others. It is a hybrid algorithm which combines features from greedy algorithms with ideas from genetic algorithms.

There is a number of codes available for solving QAPs. The reader is referred to QAPLIB, a library on quadratic assignment problems which is maintained by Burkard et al. [37]. QAPLIB contains programs, test instances with best known results and references with respect to QAPs and can be found at http://www.opt.math.tu-graz.ac.at/~karisch/qaplib. In particular FORTRAN codes for the GLB (up to the size $n=256$) and a branch and bound algorithm due to Burkard and Derigs [31] can be downloaded from the QAPLIB web page. Further one can find there the source file of a FORTRAN implementation of the simulated annealing algorithm of Burkard and Rendl [40]. Recently, Espersen, Karisch, Çela, and Clausen developed QAPpack which is a JAVA package containing a branch and bound algorithm to solve the QAP. QAPpack contains several different bounds and can be found at http://www.imm.dtu.dk/~te/QAPpack.

The source file of a C++ implementation of the simulated annealing algorithm of Connolly [51], due to Taillard, can be downloaded from Taillard's web page at . Also the source file of a PASCAL implementation of Taillard's robust tabu search algorithm can be found there.

Finally, the source file of a FORTRAN implementation of Li and Pardalos' generator for QAP instances with known optimal solution [107] can be obtained by sending an email to coap@math.ufl.edu with subject line send 92006.

6.5. Polynomially solvable special cases

Since QAPs are $\mathcal{NP}$-hard, the question arises, in which cases they can be solved in polynomial time. This means for Koopmans–Beckmann problems that the matrices A and B must have a special structure which enables a polynomial-time algorithm.

In contrast to the traveling salesman problem it turns out that the QAP with both coefficient matrices being Monge or Anti-Monge is $\mathcal{NP}$-hard, whereas the complexity of a QAP with one coefficient matrix being Monge and the other one being Anti-Monge is still open, see [26,47]. The case where A is a Monge matrix, B is a chess-board matrix and the size n of the problem is even, $n = 2m$, is solved by the permutation

$$\varphi(k) := \begin{cases} i, & \text{if} \quad k = 2i - 1, \ 1 \leqslant i \leqslant m, \\ m + i, & \text{if} \quad k = 2i, \ 1 \leqslant i \leqslant m. \end{cases}$$

The computational complexity in the case where n is odd, is open. Here a matrix $B = (b_{ij})$ is called a *chess-board matrix*, if its entries are given by $b_{ij} = (-1)^{i+j}$. A few other versions of the QAP involving Monge and Anti-Monge matrices with additional structural properties can be solved by dynamic programming.

Other special cases of the QAP involve matrices with a specific diagonal structure e.g. *circulant* and *Toeplitz matrices*. An $n \times n$ matrix $A = (a_{ij})$ is called a *Toeplitz matrix* if there exist numbers $c_{-n+1}, \ldots, c_{-1}, c_0, c_1, \ldots, c_{n-1}$ such that $a_{ij} = c_{j-i}$, for all i, j. A matrix A is called a *circulant matrix* if it is a Toeplitz matrix and the generating numbers c_i fulfill the conditions $c_i = c_{i-n}$, for $0 \leqslant i \leqslant n - 1$. In other words, a Toeplitz matrix has constant entries along lines parallel to the diagonal, whereas a circulant is given by its first row and the entries of the ith row resembles the first row shifted circularly by $i - 1$ places to the right.

QAPs with one Anti-Monge (Monge) matrix and one Toeplitz (circulant) matrix remain $\mathcal{NP}$-hard unless additional conditions are imposed on the coefficient matrices. A well studied problem is the so called Anti-Monge–Toeplitz QAP where the rows and columns of the Anti-Monge matrix are non-decreasing, see [29]. It has been shown that this problem is $\mathcal{NP}$-hard and contains as a special case the so called *turbine runner problem* introduced by Mosewich [114] and formulated as a QAP by Laporte and Mercure [102]. In the turbine runner problem we are given n blades to be welded in regular spacing around the cylinder of the turbine. Due to inaccuracies in the manufacturing process the masses m_i of the blades differ slightly and consequently the gravity center of the system does not lie on the rotation axis of the cylinder, leading to instabilities. In an effort to make the system as stable as possible, it is desirable to locate the blades so as to minimize the distance between the center of gravity and the rotation axis. Mathematically, this problem can be formulated as QAP of the form

$$\min_{\varphi} \sum_{i=1}^{n} \sum_{j=1}^{n} \cos\left(\frac{2(i-j)\pi}{n}\right) m_{\varphi(i)} m_{\varphi(j)}.$$

Note that matrix A with $a_{ij} = \cos(2(i-j)\pi/n)$ is a periodic Toeplitz matrix, whereas matrix $B = (b_{kl})$ with $b_{kl} := m_k m_l$ is an Anti-Monge matrix, if the masses m_k are sorted decreasingly. It turns out that the *maximization* version of this problem is polynomially solvable, whereas the minimization of the objective function (6.5) is $\mathcal{NP}$-hard.

Further polynomially solvable special cases of the Anti-Monge–Toeplitz QAP arise if additional constraints e.g. *benevolence* or *k-benevolence* are imposed on the Toeplitz matrix. These conditions are expressed in terms of properties of the generating function of these matrices, see [29].

All polynomially solvable QAPs described above, where A is an Anti-Monge (Monge) matrix and B is a Toeplitz (circulant) matrix are *constant permutation QAPs*. This means that an optimal solution can be specified explicitly. The technique used to prove this fact and to identify the optimal permutation is called *reduction to extremal rays*. This technique exploits two facts: first, the involved matrix classes form cones, and secondly, the objective function of the QAP is linear with respect to each of the coefficient matrices. These two facts allow us to restrict the investigations to instances of the QAP with 0–1 coefficient matrices which are extremal rays of the above mentioned cones.

The identification of polynomially solvable special cases of the QAP which are not constant permutation QAPs and can be solved algorithmically remains a challenging open question.

A subclass of Monge matrices are the so-called *Kalmanson matrices*. A matrix $A = (a_{ij})$ is a *Kalmanson matrix*, if it is symmetric and its entries satisfy the following inequalities for all indices i, j, k, l, $i < j < k < l$:

$$a_{ij} + a_{kl} \leqslant a_{ik} + a_{jl}, \quad a_{il} + a_{jk} \leqslant a_{ik} + a_{jl}.$$

(For more information on Monge, Anti-Monge and Kalmanson matrices, and their properties the reader is referred to the survey article of Burkard et al. [38].) The QAP(A,B) with a Kalmanson matrix A and a Toeplitz matrix B has been investigated by Deĭneko and Woeginger [55]. The computational complexity of this problem is an open question, but analogously as in the case of the Anti-Monge–Toeplitz QAP, polynomially solvable versions of the problem are obtained by imposing additional constraints to the Toeplitz matrix.

7. Asymptotic results for assignment problems

Assignment problems show an interesting behaviour when their size tends to infinity. Whereas for linear assignment problems the gap between best and worst solution tends to infinity as the problem size increases, the best and the worst solution of quadratic assignment problems tend almost surely to the same value, when the size of the problems increases. Let us describe first what is known about the asymptotic behaviour of linear sum assignment problems. Throughout this section we assume that the cost coefficients of the problems are independent random variables with a common prespecified distribution.

7.1. Asymptotic results for linear sum assignment problems

Closely related with the question about the expected behaviour of the optimal value of a linear sum assignment problem is the question whether a random bipartite graph admits a perfect matching or not. The existence of a perfect matching in a bipartite graph G is intuitively connected with the number of edges of G and the fact that G does not contain isolated vertices. Walkup [152] considers the class $\mathcal{G}(n,d)$ of *directed* bipartite graphs $G = (V, W; E)$ with $|V| = |W| = n$, where each vertex has out-degree d. Let $P(n, d)$ be the probability that a graph chosen randomly from $\mathcal{G}(n,d)$ contains a perfect matching. Walkup shows that $P(n, 1)$ tends to 0, but for all $d \geqslant 2$, $P(n, d)$ tends to 1 as n approaches infinity. (Notice that due to the Marriage Theorem 2.1 the existence of a perfect matching in an undirected bipartite graph, regular of degree d, is trivial.) Using the above result Walkup [151] shows in a following paper that 3 is an upper bound on the expected optimal value of the LSAP in the case that the cost coefficients c_{ij} are independent random variables uniformly distributed on $[0, 1]$.

Four years later Karp [95] improved the upper bound on the optimum objective function value of an LSAP to 2. Both, Walkup's and Karp's proofs are non-constructive and cannot be exploited in heuristics for producing assignments with expected optimal value within the given bounds.

Independent and uniformly distributed cost elements c_{ij} on $[0, 1]$ lead immediately to independent and uniformly distributed cost elements $\bar{c}_{ij} := 1 - c_{ij}$ on $[0, 1]$. Therefore we can derive from the following equality

$$\max_{\varphi} \sum_{i=1}^{n} \bar{c}_{i\varphi(i)} = n - \min_{\varphi} \sum_{i=1}^{n} c_{i\varphi(i)}$$

that the maximum objective function value of a linear assignment problem tends to infinity as the problem size increases. Thus the gap between minimum and maximum objective function values of a LSAP becomes arbitrarily large, when the problem size increases.

Lower bounds for the expected optimal value of the LSAP with independent and uniformly distributed costs c_{ij} on $[0, 1]$ are given by Lazarus [104]. The author exploits weak duality and evaluates the expected value of the dual objective function $\sum_i u_i + \sum_j v_j$ achieved after row and column reductions, see the algorithm in Section 4.1. By computations involving first order statistics it can be shown that the expected value of $\sum_i u_i + \sum_j v_j$—which is a lower bound for the expected optimal value of the LSAP—is of order $1 + 1/e + \log n/n$. This yields a bound of 1.368. Moreover, Lazarus evaluates the maximum number of 0-entries in the cost matrix after row and column reductions. It turns out that the probability of finding an optimal assignment only after row and column reductions tends to 0 as n tends to infinity.

The lower bound on the expected optimal value of an LSAP was improved by Olin [120] to 1.51, which is currently the best value known. Olin considers first the solution of the dual obtained by row and column reductions and improves this solution of the dual by adding the second smallest element of each row to all elements in that row. Then the largest among the terms added to the elements of a column is subtracted from this column. This transformation leads to a new dual feasible solution which yields an expected value of the objective function equal to 1.47. By applying an analogous transformation starting from the columns of the reduced cost matrix, the bound is increased to 1.51. Mézard and Parisi [113] conjectured that the expected optimal value of an LSAP is $\pi^2/6 = 1.645$, if the cost coefficients are independent and uniformly distributed random variables in $[0, 1]$. Indeed, Donath [61] observed in his computational experiments a value close to 1.6. For further asymptotic results concerning different distribution functions of the cost coefficients c_{ij} see the recent handbook article of Burkard and Çela [24].

7.2. Asymptotic results for linear bottleneck assignment problems

Pferschy [125] investigates the asymptotic behavior of linear bottleneck assignment problems. He shows that the expected value of an optimal solution of LBAP tends towards the lower end of the range of cost coefficients for any bounded distribution function when the size n of the problem increases. In particular he shows:

Theorem 7.1. If $\sup\{x|F(x) < 1\} < \infty$, then the optimal solution Z_n of a random LBAP with cost coefficients distributed according to distribution function F satisfies

$$\lim_{n \to \infty} E[Z_n] = \inf\{x|F(x) > 0\}.$$

In the case of uniformly distributed cost coefficients in $[0, 1]$ Pferschy derives the following lower and upper bounds for $E(Z_n)$.

Theorem 7.2. Let $B(x, y)$ be the Beta function. Then we get for $n > 78$:

$$E[Z_n] < 1 - \left[\frac{2}{n(n+2)}\right]^{2/n} \frac{n}{n+2} + \frac{123}{610n}$$

and

$$E[Z_n] \geq 1 - nB\left(n, 1 + \frac{1}{n}\right) = \frac{\ln n + 0.5749}{n} + O\left(\frac{\ln^2 n}{n^2}\right).$$

7.3. Asymptotic results for quadratic sum assignment problems

In contrast to linear assignment problems it can be shown that under mild probabilistic assumptions the ratio between "best" and "worst" values of the objective function of a QAP approaches 1 as the size of the problem tends to infinity. This is a very strange and interesting asymptotic behaviour which was at first proved for QAPs by Burkard and Fincke [33] in 1983. Later, in 1985, a whole class of combinatorial optimization problems showing this behaviour was found, see Burkard and Fincke [34]. As a consequence of this behaviour we can expect that *every heuristic finds an almost optimal solution when applied to QAP instances which are large enough.* On the other hand this behaviour shows that the landscape of objective function values for QAPs is very flat. This implies that it will be difficult for branch and bound procedures to detect the true optimal solution. Indeed, Dyer et al. [62] showed the following result for QAPs whose coefficient matrices have independently distributed random entries with a common distribution.

Theorem 7.3 (Dyer et al. [62]). *Consider any branch and bound algorithm for solving a QAP with randomly generated coefficients which have finite expected values, variances and third moments. Assume that the branch and bound algorithm assigns one index in each step and employs a Gilmore–Lawler bound. Then the number of branched nodes explored is at least $n^{(1-o(1))n/4}$ with a probability tending to 1 as the size n of the QAP tends to infinity.*

Burkard and Fincke [33] investigate the relative difference between the worst and the best value of the objective function for Koopmans–Beckmann QAPs. First the Euclidean case is considered, where A is the distance matrix of independently and uniformly distributed points in the unit square. Then they consider the general case where the entries of matrices A and B are independent random variables taken from a uniform distribution on $[0,1]$. In both cases it is shown that the relative difference between the best and worst solution values approaches 1 with a probability tending to 1 as the size of the problem tends to infinity. These results were strengthened by Frenk et al. [70] as well as by Rhee [139]. Their results can be summarized in the following theorem:

Theorem 7.4 (Frenk et al. [70] and Rhee [139]). *Consider a sequence of QAPs($A^{(n)}$, $B^{(n)}$) whose coefficients are independently distributed random variables in $[0,M]$ with expected values $E(A)$ and $E(B)$, resp., all entries of A (and B, resp.) having the same distribution. Denote by $Z(A^{(n)}, B^{(n)}, \varphi)$ the value of the QAP with respect to permutation φ. Then there exists a constant K_1 (which does not depend on n), such that the following inequality holds almost surely:*

$$\limsup_{n \to \infty} \frac{\sqrt{n}}{\sqrt{\log n}} \left| \frac{Z(A^{(n)}, B^{(n)}, \varphi)}{n^2 E(A)E(B)} - 1 \right| \leq K_1.$$

Moreover, let Y be a random variable defined by

$$Y = Z(A^{(n)}, B^{(n)}, \varphi_{\text{opt}}^{(n)}) - n^2 E(A)E(B),$$

where $\varphi_{\text{opt}}^{(n)}$ is an optimal solution of $QAP(A^{(n)}, B^{(n)})$. Then there exists another constant K_2, also independent of the size of the problem, such that

$$\frac{1}{K_2} n^{3/2} (\log n)^{1/2} \leqslant E(Y) \leqslant K_2 n^{3/2} (\log n)^{1/2},$$

$$P\{|Y - E(Y)| \geqslant t\} \leqslant 2 \exp\left(\frac{-t^2}{4n^2 \|A\|_\infty^2 \|B\|_\infty^2}\right)$$

for each $t \geqslant 0$, where $E(Y)$ denotes the expected value of variable Y and $\|A\|_\infty$ ($\|B\|_\infty$) is the so-called row sum norm of matrix A (B) defined by $\|A\|_\infty = \max_{1 \leqslant i \leqslant n} \sum_{j=1}^n |a_{ij}|$.

Later it was shown by Burkard and Fincke [34] that this strange asymptotic behaviour is related to the ratio between the number of (non-zero) cost coefficients and the logarithm of the number of feasible solutions of a combinatorial optimization problem. Consider a sequence P_n of combinatorial optimization (minimization) problems defined on finite ground sets E_n. A feasible solution of the problem can be represented by a subset F of E_n. Let $\mathcal{F}_n$ denote the class of all feasible solutions of problem P_n. Further let $c_n : E_n \to \mathbb{R}^+$ be the cost coefficients and $z_n(F) := \sum_{e \in F} c_n(e)$ be the objective function value of a feasible solution F. Then the following theorem can be shown:

Theorem 7.5 (Burkard and Fincke [34], Szpankowski [144] and Burkard and Çela [23]). *Consider a sequence of combinatorial optimization problems P_n with the following properties*:

- *All feasible solutions of problem P_n have the same cardinality $|F_n|$.*
- *Every element $e \in E_n$ occurs in the same number of feasible solutions F_n.*
- *The cost elements $c_n(e)$, $e \in E_n$ are i.i.d. random variables in $[0, M]$.*
- $\lim_{n \to \infty} \frac{\log |\mathcal{F}_n|}{|F_n|} = 0.$

Then all *solution values converge almost surely to n times the expected value of the cost coefficients $c_n(e)$, i.e., all solutions have asymptotically almost surely the same value.*

This theorem was first proven in 1985 by Burkard and Fincke [34] in a weaker form, namely assuring convergence in probability. The authors gave, however, explicit bounds for the asymptotic behaviour. Szpankowski [144] showed 1995 the convergence almost surely. Recently, Burkard and Çela [23] derived this theorem from thermodynamical considerations using the Boltzmann distribution. This is of particular interest, since simulated annealing is a simulation tool, which yields excellent results for these problems and stems from the same background. Note that the condition

$$\lim_{n \to \infty} \frac{\log |\mathcal{F}_n|}{|F_n|} = 0$$

is fulfilled for QAPs, but not for LAPs. Both problems have $|\mathcal{F}_n| = n!$ feasible solutions, but a feasible solution of a LAP has n coefficients, whereas a feasible solution of a QAP has n^2 coefficients in the objective function.

7.4. Asymptotic results for quadratic bottleneck assignment problems

Similar results as in the sum case hold also for quadratic bottleneck assignment problems. Let $z_{opt}(n)$ be the minimum objective function value of a BQAP of size n and let z_{wor} be the maximum objective function value. In Burkard and Fincke [32] the following theorem has been proven:

Theorem 7.6. *Let a_{ij} and b_{kl} be independent and uniform $[0, 1]$ random variables for $1 \leqslant i, j, k, l \leqslant n$. Then*

$$\lim_{n \to \infty} \mathbf{P} \left\{ \frac{z_{wor} - z_{opt}}{z_{opt}} \leqslant \left(\left(\frac{n}{2 \log n} \right)^{1/2} - 1 \right)^{-1} \right\} = 1.$$

In the case of general combinatorial optimization problems it has been shown by Burkard and Fincke [34] that this asymptotic behaviour relies again on the ratio of the logarithm of the number of feasible solutions to the number of non-zero cost coefficients in a feasible solution. If this ratio tends to 0 for increasing problem sizes then all solutions have almost the same value in probability.

8. Further assignment models

8.1. Biquadratic assignment problems

In connection with VLSI synthesis programmable logic arrays (PLA) have to be implemented. We want to find an encoding of states such that the actual implementation by flip-flops is of minimum size. In the case of data flip-flops this leads to a quadratic assignment problem. If, however, toggle flip-flops are used, a new problem arises which is called *biquadratic assignment problem (BiQAP)*. Mathematically we can formulate a BiQAP as follows. Let two arrays $A = (a_{ijkl})$ and $B = (b_{mpst})$ with $1 \leqslant i, j, k, l, m, p, s, t \leqslant n$ be given. The BiQAP can then be stated as:

$$\min_{\varphi \in \mathscr{S}_n} \sum_{i=1}^{n} \sum_{j=1}^{n} \sum_{k=1}^{n} \sum_{l=1}^{n} a_{ijkl} b_{\varphi(i)\varphi(j)\varphi(k)\varphi(l)}.$$

Burkard et al. [27] compute lower bounds for the BiQAP derived from lower bounds of the QAP. The computational results show that these bounds are weak and deteriorate as the size of the problem increases. This observation suggests that branch and bound methods will only be effective on very small instances. For larger instances efficient heuristics that find good-quality approximate solutions are needed. Several heuristics for the BiQAP have been developed by Burkard and Çela [21], in particular deterministic improvement methods and variants of simulated annealing and tabu search algorithms. Computational experiments on test problems of size up to $n = 32$, for which the optimum objective function value is known beforehand, suggest that a specific simulated annealing algorithm is best suited for this kind of problems.

BiQAPs show the same asymptotic behaviour as QAPs as has been shown and analyzed in detail in Burkard et al. [27].

8.2. Communication assignment problems

Let us consider n communication centers $C_1, C_2, \ldots, C_n$ which have to be assigned to the vertices of an undirected, connected graph. Each communication center C_i transmits messages to every other center C_j at a rate of t_{ij} messages per time unit. In case that there is no direct connection between the centers C_i and C_j the messages are routed via intermediate centers. Once the communication centers are assigned to the vertices of the graph, the messages have to be routed. We distinguish between two different routing patterns: In the *single path model* for every pair of communication centers a single path in the graph is selected and the whole traffic is routed along this path. In the *fractional model* the traffic is split into several parts which are routed along different paths from the origin to the destination. For any fixed assignment φ and any fixed routing pattern ρ we denote the overall amount of traffic passing through center C_i as $N(C_i; \varphi, \rho)$ and call this the *noise* at center C_i.

The communication assignment problem asks for an assignment of the communication centers to the vertices of the underlying graph and for a routing pattern such that the maximum noise becomes as small as possible:

$$\min_{\varphi, \rho} \max_{1 \leqslant i \leqslant n} N(C_i; \varphi, \rho).$$

In Burkard et al. [30] it has been shown that the communication assignment problem is $\mathcal{NP}$-hard even for such simple graphs as paths, cycles and star graphs with branch length 3. By using a perfect matching algorithm and binary search the communication assignment problem can be solved in polynomial time in star graphs of branch length $\leqslant 2$. Further it has been shown that in so-called double star graphs the single path version is $\mathcal{NP}$-hard, whereas the fractional model can be solved in polynomial time by a sequence of $O(n^2)$ linear programs with $n - 2$ variables and $n^2 + n$ constraints each. Moreover, the authors develop a branch and bound algorithm for solving this problem in trees. Later, Burkard et al. [25] designed various heuristics like simulated annealing and tabu search for this problem. The good performance of these heuristics seems again be related to the asymptotic behaviour of communication assignment problems. These problems show again a similar asymptotic behaviour as QAPs as can be seen in the following theorem:

Theorem 8.1. *Consider a sequence of communication assignment problem instances whose entries t_{ij} are i.i.d. random variables on a compact interval $[0, M]$. Then the ratio between maximum noise and minimum noise approaches 1 with a probability tending to 1 as the size of the problems tends to infinity.*

References

[1] W.P. Adams, T.A. Johnson, Improved linear programming-based lower bounds for the quadratic assignment problem, in: P.M. Pardalos, H. Wolkowicz (Eds.), Quadratic Assignment and Related

Problems, DIMACS Series on Discrete Mathematics and Theoretical Computer Science, Vol. 16, AMS, Providence, RI, 1994, pp. 43–75.

[2] W.P. Adams, H.D. Sherali, A tight linearization and an algorithm for zero-one quadratic programming problems, Management Sci. 32 (1986) 1274–1290.

[3] W.P. Adams, H.D. Sherali, Linearization strategies for a class of zero-one mixed integer programming problems, Oper. Res. 38 (1990) 217–226.

[4] R.K. Ahuja, J.B. Orlin, A. Tivari, A descent genetic algorithm for the quadratic assignment problem, Computers and Operations Research 27 (2000) 917–934.

[5] H. Alt, N. Blum, K. Mehlhorn, M. Paul, Computing maximum cardinality matching in time $O(n^{1.5}\sqrt{m/\log n})$, Inform. Process. Lett. 37 (1991) 237–240.

[6] S. Arora, A. Frieze, H. Kaplan, A new rounding procedure for the assignment problem with applications to dense graph arrangement problems, Proceedings of the 37th Annual IEEE Symposium on Foundations of Computer Science (FOCS), 1996, pp. 21–30.

[7] A.A. Assad, W. Xu, On lower bounds for a class of quadratic 0–1 programs, Oper. Res. Lett. 4 (1985) 175–180.

[8] E. Balas, P.R. Landweer, Traffic assignment in communications satellites, Oper. Res. Lett. 2 (1983) 141–147.

[9] E. Balas, L. Qi, Linear-time separation algorithms for the three-index assignment polytope, Discrete Appl. Math. 43 (1993) 1–12.

[10] E. Balas, M.J. Saltzman, Facets of the three-index assignment polytope, Discrete Appl. Math. 23 (1989) 201–229.

[11] E. Balas, M.J. Saltzman, An algorithm for the three-index assignment problem, Oper. Res. 39 (1991) 150–161.

[12] S.E. Bammel, J. Rothstein, The number of 9×9 Latin squares, Discrete Math. 11 (1975) 93–95.

[13] A.I. Barvinok, Computational complexity of orbits in representations of symmetric groups, Adv. Soviet Math. 9 (1992) 161–182.

[14] R. Battiti, G. Tecchiolli, The reactive tabu search, ORSA J. Comput. 6 (1994) 126–140.

[15] G. Birkhoff, Tres observaciones sobre el algebra lineal, Rev. univ. nac. Tucumán (A) 5 (1946) 147–151.

[16] W.L. Brogan, Algorithm for ranked assignments with applications to multiobject tracking, J. Guidance 12 (1989) 357–364.

[17] R.E. Burkard, Die Störungsmethode zur Lösung quadratischer Zuordnungsprobleme, Oper. Res. Verfahren 16 (1973) 84–108.

[18] R.E. Burkard, Quadratische Bottleneckprobleme, Oper. Res. Verfahren 18 (1974) 26–41.

[19] R.E. Burkard, Time-slot assignment for TDMA-systems, Computing 35 (1985) 99–112.

[20] R.E. Burkard, Locations with spatial interactions: the quadratic assignment problem, in: P.B. Mirchandani, R.L. Francis (Eds.), Discrete Location Theory, Wiley, New York, 1991, pp. 387–437.

[21] R.E. Burkard, E. Çela, Heuristics for biquadratic assignment problems and their computational comparison, European J. Oper. Res. 83 (1995) 283–300.

[22] R.E. Burkard, E. Çela, Quadratic and three-dimensional assignments, in: M. Dell'Amico, F. Maffioli, S. Martello (Eds.), Annotated Bibliographies in Combinatorial Optimization, Wiley, Chichester, 1997, pp. 373–391.

[23] R.E. Burkard, E. Çela, An asymptotical study of combinatorial optimization problems by means of statistical mechanics. SFB Report 133, Institute of Mathematics, Technical University Graz, Austria, 1998.

[24] R.E. Burkard, E. Çela, Linear assignment problems and extensions, in: D.-Z. Du, P.M. Pardalos (Eds.), Handbook of Combinatorial Optimization, Vol. 4, Kluwer Academic Publishers, Dordrecht, 1999.

[25] R.E. Burkard, E. Çela, T. Dudás, A communication assignment problem on trees: heuristics and asymptotic behaviour, in: P.M. Pardalos, D.W. Hearn, W.W. Hager (Eds.), Network Optimization, Lecture Notes in Economics and Mathematical Systems, Vol. 450, Springer, Berlin, 1997, pp. 127–155.

[26] R.E. Burkard, E. Çela, V.M. Demidenko, N.N. Metelski, G.J. Woeginger, Perspectives of easy and hard cases of the quadratic assignment problems, SFB Report 104, Institute of Mathematics, Technical University Graz, Austria, 1997.

[27] R.E. Burkard, E. Çela, B. Klinz, On the biquadratic assignment problem, in: P.M. Pardalos, H. Wolkowicz (Eds.), Quadratic Assignment and Related Problems, DIMACS Series on Discrete Mathematics and Theoretical Computer Science, Vol. 16, AMS, Providence, RI, 1994, pp. 117–146.

[28] R.E. Burkard, E. Çela, P. Pardalos, L.S. Pitsoulis, The quadratic assignment problem, in: D.-Z. Du, P.M. Pardalos (Eds.), Handbook of Combinatorial Optimization, Vol. 3, Kluwer Academic Publishers, Dordrecht, 1998, pp. 241–339.

[29] R.E. Burkard, E. Çela, G. Rote, G.J. Woeginger, The quadratic assignment problem with an Anti-Monge and a Toeplitz matrix: easy and hard cases, Math. Programming 82 (1998) 125–158.

[30] R.E. Burkard, E. Çela, G.J. Woeginger, A minimax assignment problem in treelike communication networks, European J. Oper. Res. 87 (1995) 670–684.

[31] R.E. Burkard, U. Derigs, Assignment and Matching Problems: Solution Methods with FORTRAN Programs, Springer, Berlin, 1980.

[32] R.E. Burkard, U. Fincke, On random quadratic bottleneck assignment problems, Math. Programming 23 (1982) 227–232.

[33] R.E. Burkard, U. Fincke, The asymptotic probabilistic behavior of the quadratic sum assignment problem, Z. Oper. Res. 27 (1983) 73–81.

[34] R.E. Burkard, U. Fincke, Probabilistic asymptotic properties of some combinatorial optimization problems, Discrete Appl. Math. 12 (1985) 21–29.

[35] R.E. Burkard, K. Fröhlich, Some remarks on 3-dimensional assignment problems, Methods of Oper. Res. 36 (1980) 31–36.

[36] R.E. Burkard, W. Hahn, U. Zimmermann, An algebraic approach to assignment problems, Math. Programming 12 (1977) 318–327.

[37] R.E. Burkard, S.E. Karisch, F. Rendl, QAPLIB—a quadratic assignment problem library, J. Global Optim. 10 (1997) 391–403. An on-line version is available via World Wide Web at the following URL: http://www.opt.math.tu-graz.ac.at/~karisch/qaplib/

[38] R.E. Burkard, B. Klinz, R. Rudolf, Perspectives of Monge properties in optimization, Discrete Appl. Math. 70 (1996) 95–161.

[39] R.E. Burkard, J. Offermann, Entwurf von Schreibmaschinentastaturen mittels quadratischer Zuordnungsprobleme, Z. Oper. Res. 21 (1977) B121–B132 (in German).

[40] R.E. Burkard, F. Rendl, A thermodynamically motivated simulation procedure for combinatorial optimization problems, European J. Oper. Res. 17 (1984) 169–174.

[41] R.E. Burkard, R. Rudolf, Computational investigations on 3-dimensional axial assignment problems, Belgian J. Oper. Res. 32 (1993) 85–98.

[42] R.E. Burkard, R. Rudolf, G.J. Woeginger, Three dimensional axial assignment problems with decomposable cost coefficients, Discrete Appl. Math. 65 (1996) 123–169.

[43] R.E. Burkard, U. Zimmermann, Weakly admissible transformations for solving algebraic assignment and transportation problems, Math. Programming Study 12 (1980) 1–18.

[44] R.E. Burkard, U. Zimmermann, Combinatorial optimization in linearly ordered semimodules: a survey, in: B. Korte (Ed.), Modern Applied Mathematics, North-Holland, Amsterdam, 1982, pp. 392–436.

[45] G. Carpaneto, P. Toth, Algorithm for the solution of the bottleneck assignment problem, Computing 27 (1981) 179–187.

[46] P. Carraresi, F. Malucelli, A reformulation scheme and new lower bounds for the QAP, in: P. Pardalos, H. Wolkowicz (Eds.), Quadratic Assignment and Related Problems, DIMACS Series in Discrete Mathematics and Theoretical Computer Science, Vol. 16, AMS, Providence, RI, 1994, pp. 147–160.

[47] E. Çela, The Quadratic Assignment Problem: Theory and Algorithms, Kluwer Academic Publishers, Dordrecht, The Netherlands, 1998.

[48] V. Černý , Thermodynamical approach to the traveling salesman problem: an efficient simulation algorithm, J. Optim. Theory Appl. 45 (1985) 41–51.

[49] J. Chakrapani, J. Skorin-Kapov, Massively parallel tabu search for the quadratic assignment problem, Ann. Oper. Res. 41 (1993) 327–342.

[50] J. Clausen, S.E. Karisch, M. Perregaard, F. Rendl, On the applicability of lower bounds for solving rectilinear quadratic assignment problems in parallel, Comput. Optim. Appl. 10 (1998) 127–147.

[51] D.T. Connolly, An improved annealing scheme for the QAP, European J. Oper. Res. 46 (1990) 93–100.

[52] K. Conrad, Das Quadratische Zuweisungsproblem und zwei seiner Spezialfälle, Mohr-Siebeck, Tübingen, 1971.

[53] D. Coppersmith, S. Vinograd, Matrix multiplication via arithmetic progressions, J. Symbolic Comput. 9 (1990) 251–280.

[54] V.G. Deĭneko, V.L. Filonenko, On the reconstruction of specially structured matrices, Aktualnyje Problemy EVM i Programmirovanije, Dnjepropetrovsk, DGU, 1979 (1990) 251–280 (in Russian).

[55] V.G. Deĭneko, G.J. Woeginger, A solvable case of the quadratic assignment problem, Oper. Res. Lett. 22 (1998) 13–17.

[56] M. Dell'Amico, S. Martello, Linear Assignment, in: M. Dell'Amico, F. Maffioli, S. Martello (Eds.), Annotated Bibliographies in Combinatorial Optimization, Wiley, Chichester, 1997, pp. 355–371.

[57] U. Derigs, Alternate strategies for solving bottleneck assignment problems—analysis and computational results, Computing 33 (1984) 95–106.

[58] U. Derigs, O. Goecke, R. Schrader, Monge sequences and a simple assignment algorithm, Discrete Appl. Math. 15 (1986) 241–248.

[59] U. Derigs, U. Zimmermann, An augmenting path method for solving linear bottleneck assignment problems, Computing 19 (1978) 285–295.

[60] J.W. Dickey, J.W. Hopkins, Campus building arrangement using TOPAZ, Transportation Res. 6 (1972) 59–68.

[61] W.E. Donath, Algorithms and average-value bounds for assignment problems, IBM J. Res. Develop. 13 (1969) 380–386.

[62] M.E. Dyer, A.M. Frieze, C.J.H. McDiarmid, On linear programs with random costs, Math. Programming 35 (1986) 3–16.

[63] J. Edmonds, D.R. Fulkerson, Bottleneck extrema, J. Combin. Theory 8 (1970) 299–306.

[64] C.S. Edwards, A branch and bound algorithm for the Koopmans–Beckmann quadratic assignment problem, Math. Programming Stud. 13 (1980) 35–52.

[65] R. Euler, Odd cycles and a class of facets of the axial 3-index assignment polytope, Applicationes Math. (Zastosowania Matematyki) 19 (1987) 375–386.

[66] R. Euler, R.E. Burkard, R. Grommes, On Latin squares and the facial structure of related polytopes, Discrete Math. 62 (1986) 155–181.

[67] R. Euler, H. Le Verge, Time-tables, polyhedra and the greedy algorithm, Discrete Appl. Math. 65 (1996) 207–221.

[68] G. Finke, R.E. Burkard, F. Rendl, Quadratic assignment problems, Ann. Discrete Math. 31 (1987) 61–82.

[69] L.R. Ford, D.R. Fulkerson, Maximal flow through a network, Canad. J. Math. 8 (1956) 399–404.

[70] J.B.G. Frenk, M. van Houweninge, A.H.G. Rinnooy Kan, Asymptotic properties of the quadratic assignment problem, Math. Oper. Res. 10 (1985) 100–116.

[71] A.M. Frieze, Complexity of a 3-dimensional assignment problem, European J. Oper. Res. 13 (1983) 161–164.

[72] A.M. Frieze, L. Yadegar, An algorithm for solving 3-dimensional assignment problems with application to scheduling in a teaching practice, J. Oper. Res. Soc. 32 (1981) 989–995.

[73] A.M. Frieze, J. Yadegar, On the quadratic assignment problem, Discrete Appl. Math. 5 (1983) 89–98.

[74] R. Fulkerson, I. Glicksberg, O. Gross, A production line assignment problem, Technical Report RM-1102, The Rand Corporation, Sta. Monica, CA, 1953.

[75] H.N. Gabow, R.E. Tarjan, Algorithms for two bottleneck optimization problems, J. Algorithms 9 (1988) 411–417.

[76] R. Garfinkel, An improved algorithm for the bottleneck assignment problem, Oper. Res. 19 (1971) 1747–1751.

[77] A.M. Geoffrion, G.W. Graves, Scheduling parallel production lines with changeover costs: Practical applications of a quadratic assignment/LP approach, Oper. Res. 24 (1976) 595–610.

[78] O. Gross, The bottleneck assignment problem, Technical Report P-1630, The Rand Corporation, Sta. Monica, CA, 1959.

[79] S.W. Hadley, F. Rendl, H. Wolkowicz, Bounds for the quadratic assignment problem using continuous optimization techniques, Proceedings of the First Integer Programming and Combinatorial Optimization Conference (IPCO), University of Waterloo Press, 1990, pp. 237–248.

[80] S.W. Hadley, F. Rendl, H. Wolkowicz, A new lower bound via projection for the quadratic assignment problem, Math. Oper. Res. 17 (1992) 727–739.

[81] S.W. Hadley, F. Rendl, H. Wolkowicz, Nonsymmetric quadratic assignment problems and the Hoffman-Wielandt inequality, Linear Algebra Appl. 58 (1992) 109–124.

[82] P. Hahn, T. Grant, Lower bounds for the quadratic assignment problem based upon a dual formulation, Oper. Res. 46 (1998) 912–922.

[83] Ph. Hall, On representatives of subsets, J. London Math. Soc. 10 (1935) 26–30.

[84] P. Hansen, L. Kaufman, A primal-dual algorithm for the three-dimensional assignment problem, Cahiers du CERO 15 (1973) 327–336.

[85] G.H. Hardy, J.E. Littlewood, G. Pólya, Inequalities, Cambridge University Press, London, 1952.

[86] D.R. Heffley, Assigning runners to a relay team, in: S.P. Ladany, R.E. Machol (Eds.), Optimal Strategies in Sports, North-Holland, Amsterdam, 1977, pp. 169–171.

[87] A.J. Hoffman, On simple linear programming problems, in: V. Klee (Ed.), Convexity, Proceedings of Symposia in Pure Mathematics, Vol. 7, AMS, Providence, RI, 1963, pp. 317–327.

[88] J.E. Hopcroft, R.M. Karp, An $n^{5/2}$ algorithm for maximum matchings in bipartite graphs, SIAM J. Comput. 2 (1973) 225–231.

[89] M. Jünger, V. Kaibel, The QAP polytope, and the star transformation, Discr. Appl. Math. 111 (2001) 283–306.

[90] M. Jünger, V. Kaibel, On the SQAP polytope, SIAM J. Optim. 11 (2000) 444–463.

[91] V. Kaibel, Polyhedral Combinatorics of the Quadratic Assignment Problem, Ph.D. Thesis, Universität zu Köln, Germany, 1997.

[92] S.E. Karisch, E. Çela, J. Clausen, T. Espersen, A dual framework for lower bounds of the quadratic assignment problem based on linearization, Computing 63 (1999) 351–403.

[93] R.M. Karp, Reducibility among combinatorial problems, in: R.E. Miller, J.W. Thatcher (Eds.), Complexity of Computer Computations, Plenum Press, New York, 1972, pp. 85–103.

[94] R.M. Karp, An algorithm to solve the $m \times n$ assignment problem in expected time $O(mn \log n)$, Networks 10 (1980) 143–152.

[95] R.M. Karp, An upper bound on the expected cost of an optimal assignment, in: D.S. Johnson, T. Nishizeki, A. Nozaki, H.S. Wilf (Eds.), Perspectives in Computing, Vol. 15, Academic Press, Boston, 1987, pp. 1–4.

[96] R.M. Karp, A.H.G. Rinnooy Kan, R.V. Vohra, Average case analysis of a heuristic for the assignment problem, Math. Oper. Res. 19 (1994) 513–522.

[97] L. Kaufman, F. Broeckx, An algorithm for the quadratic assignment problem using Benders' decomposition, European J. Oper. Res. 2 (1978) 204–211.

[98] S. Kirkpatrick, C.D. Gelatt, M.P. Vecchi, Optimization by simulated annealing, Science 220 (1983) 671–680.

[99] D. König, Graphok és matrixok, Mat. Fiz. Lapok 38 (1931) 116–119.

[100] T.C. Koopmans, M.J. Beckmann, Assignment problems and the location of economic activities, Econometrica 25 (1957) 53–76.

[101] J. Krarup, P.M. Pruzan, Computer-aided layout design, Math. Programming Study 9 (1978) 75–94.

[102] G. Laporte, H. Mercure, Balancing hydraulic turbine runners: a quadratic assignment problem, European J. Oper. Res. 35 (1988) 378–382.

[103] E.L. Lawler, The quadratic assignment problem, Management Sci. 9 (1963) 586–599.

[104] A.J. Lazarus, Certain expected values in the random assignment problem, Oper. Res. Lett. 14 (1993) 207–214.

[105] Y. Lee, J.B. Orlin, On very large scale assignment problems, in: W.W. Hager, D.W. Hearn, P.M. Pardalos (Eds.), Large Scale Optimization: State of the Art, Kluwer Academic Publishers, Dordrecht, The Netherlands, 1994, pp. 206–244.

[106] J.L. Lewandowski, J.W.S. Liu, C.L. Liu, SS/TDMA time slot assignment with restricted switching modes, IEEE Trans. Comm. COM-31 (1983) 149–154.

[107] Y. Li, P.M. Pardalos, Generating quadratic assignment test problems with known optimal permutations, Comput. Optim. Appl. 1 (1992) 163–184.

[108] D. Magos, Tabu search for the planar three-index assignment problem, J. Global Optim. 8 (1996) 35–48.

[109] D. Magos, P. Miliotis, An algorithm for the planar three-index assignment problem, European J. Oper. Res. 77 (1994) 141–153.

[110] S. Martello, W.R. Pulleyblank, P. Toth, D. de Werra, Balanced optimization problems, Oper. Res. Lett. 3 (1984) 275–278.

[111] A. Marzetta, A. Brüngger, A dynamic programming bound for the quadratic assignment problem, in: Computing and Combinatorics: Fifth Annual International Conference COCOON'99, Lecture Notes in Computer Science, Vol. 1627, Springer, Heidelberg, 1999, pp. 339–348.

[112] E.J. McCormick, Human Factors Engineering, McGraw-Hill, New York, 1970.

[113] M. Mézard, G. Parisi, On the solution of the random link matching problems, J. Phys. 48 (1987) 1451–1459.

[114] J. Mosevich, Balancing hydraulic turbine runners—a discrete combinatorial optimization problem, European J. Oper. Res. 26 (1986) 202–204.

[115] K. Mulmuley, U.V. Vazirani, V.V. Vazirani, Matching is as easy as matrix inversion, Combinatorica 7 (1987) 105–113.

[116] R. Murphey, P.M. Pardalos, L.S. Pitsoulis, A GRASP for the multitarget multisensor tracking problem, in: P.M. Pardalos, D.-Z. Du (Eds.), Network Design: Connectivity and Facilities Location, DIMACS Series on Discrete Mathematics and Theoretical Computer Science, Vol. 40, AMS, Providence, RI, 1998, pp. 277–302.

[117] R. Murphey, P.M. Pardalos, L.S. Pitsoulis, A Parallel GRASP for the Data Association Multidimensional Assignment Problem, in: Parallel Processing of Discrete Problems, The IMA Volumes in Mathematics and its Applications, Vol. 106, Springer, Berlin, 1998, pp. 159–180.

[118] K.A. Murthy, P. Pardalos, Y. Li, A local search algorithm for the quadratic assignment problem, Informatica 3 (1992) 524–538.

[119] B. Neng, Zur Erstellung von optimalen Triebfahrzeugplänen, Z. Oper. Res. 25 (1981) B159–B185.

[120] B. Olin, Asymptotic Properties of Random Assignment Problems, Ph.D. Thesis, Division of Optimization and Systems Theory, Department of Mathematics, Royal Institute of Technology, Stockholm, 1992.

[121] M.W. Padberg, M.P. Rijal, Location, Scheduling, Design and Integer Programming, Kluwer Academic Publishers, Boston, 1996.

[122] E.S. Page, A note on assignment problems, Comput. J. 6 (1963) 241–243.

[123] C.H. Papadimitriou, D. Wolfe, The complexity of facets resolved, Proceedings of the 25th Annual IEEE Symposium on the Foundations of Computer Science (FOCS), 1985, pp. 74–78.

[124] P.M. Pardalos, F. Rendl, H. Wolkowicz, The quadratic assignment problem: a survey and recent developments, in: P.M. Pardalos, H. Wolkowicz (Eds.), Quadratic Assignment and Related Problems, DIMACS Series on Discrete Mathematics and Theoretical Computer Science, Vol. 16, AMS, Providence, RI, 1994, pp. 1–42.

[125] U. Pferschy, The random linear bottleneck assignment problem, RAIRO Oper. Res. 30 (1996) 127–142.

[126] U. Pferschy, Solution methods and computational investigations for the linear bottleneck assignment problem, Computing 59 (1997) 237–258.

[127] W.P. Pierskalla, The multidimensional assignment problem, Oper. Res. 16 (1968) 422–431.

[128] A.B. Poore, Multidimensional assignment formulation of data association problems arising from multitarget and multisensor tracking, Comput. Optim. Appl. 3 (1994) 27–54.

[129] A.B. Poore, A numerical study of some data association problems arising in multitarget tracking, in: W.W. Hager, D.W. Hearn, P.M. Pardalos (Eds.), Large Scale Optimization: State of the Art, Kluwer Academic Publishers, Dordrecht, The Netherlands, 1994, pp. 339–361.

[130] A.B. Poore, N. Rijavec, Partitioning multiple data sets: multidimensional assignments and Lagrangian relaxation, in: P.M. Pardalos, H. Wolkowicz (Eds.), Quadratic Assignment and Related Problems, DIMACS Series in Discrete Mathematics and Theoretical Computer Science, Vol. 16, AMS, Providence, RI, 1994, pp. 317–342.

[131] A.B. Poore, N. Rijavec, M. Liggins, V. Vannicola, Data association problems posed as multidimensional assignment problems: problem formulation, in: O.E. Drummond (Ed.), Signal and Data Processing of Small Targets, SPIE, Bellingham, WA, 1993, pp. 552–561.

[132] A.B. Poore, A.J. Robertson III, A new Lagrangean relaxation based algorithm for a class of multidimensional assignment problems, Comput. Optim. Appl. 8 (1997) 129–150.

[133] A.P. Punnen, K.P.K. Nair, Improved complexity bound for the maximum cardinality bottleneck bipartite matching problem, Discrete Appl. Math. 55 (1994) 91–93.

[134] J. Pusztaszeri, P.E. Rensing, T.M. Liebling, Tracking elementary particles near their primary vertex: a combinatorial approach, J. Global Optim. 16 (1995) 422–431.

[135] L. Qi, E. Balas, G. Gwan, A new facet class and a polyhedral method for the three-index assignment problem, in: D.-Z. Du (Ed.), Advances in Optimization, Kluwer Academic Publishers, Dordrecht, 1994, pp. 256–274.

[136] M. Queyranne, Performance ratio of heuristics for triangle inequality quadratic assignment problems, Oper. Res. Lett. 4 (1986) 231–234.

[137] F. Rendl, On the complexity of decomposing matrices arising in satellite communication, Oper. Res. Lett. 4 (1985) 5–8.

[138] F. Rendl, H. Wolkowicz, Applications of parametric programming and eigenvalue maximization to the quadratic assignment problem, Math. Programming 53 (1992) 63–78.

[139] W.T. Rhee, Stochastic analysis of the quadratic assignment problem, Math. Oper. Res. 16 (1991) 223–239.

[140] C. Roucairol, A reduction method for quadratic assignment problems, Oper. Res. Verfahren 32 (1979) 183–187.

[141] S. Sahni, T. Gonzalez, P-complete approximation problems, J. Assoc. Comput. Mach. 23 (1976) 555–565.

[142] J.T. Schwarz, Fast probabilistic algorithms for verification of polynomial identities, J. ACM 27 (1980) 701–717.

[143] J. Skorin-Kapov, Tabu search applied to the quadratic assignment problem, ORSA J. Comput. 2 (1990) 33–45.

[144] W. Szpankowski, Combinatorial optimization problems for which almost every algorithm is asymptotically optimal!, Optimization 33 (1995) 359–367.

[145] L. Steinberg, The backboard wiring problem: a placement algorithm, SIAM Rev. 3 (1961) 37–50.

[146] E. Taillard, Robust tabu search for the quadratic assignment problem, Parallel Comput. 17 (1991) 443–455.

[147] R.E. Tarjan, Data Structures and Network Algorithms, SIAM, Philadelphia, PA, 1983.

[148] W.T. Tutte, The factorization of linear graphs, J. London Math. Soc. 22 (1947) 107–111.

[149] I. Ugi, J. Bauer, J. Friedrich, J. Gasteiger, C. Jochum, W. Schubert, Neue Anwendungsgebiete für Computer in der Chemie, Angew. Chem. 91 (1979) 99–111.

[150] M. Vlach, Branch and bound method for the three-index assignment problem, Ekonom.-Mat. Obzor 12 (1967) 181–191.

[151] D.W. Walkup, On the expected value of a random assignment problem, SIAM J. Comput. 8 (1979) 440–442.

[152] D.W. Walkup, Matching in random regular bipartite digraphs, Discrete Math. 31 (1980) 59–64.

[153] M.R. Wilhelm, T.L. Ward, Solving quadratic assignment problems by simulated annealing, IEEE Trans. 19 (1987) 107–119.

[154] Q. Zhao, S.E. Karisch, F. Rendl, H. Wolkowicz, Semidefinite relaxations for the quadratic assignment problem, J. Combin. Optim. 2 (1998) 71–109.

Discrete Applied Mathematics 123 (2002) 303–338

DISCRETE
APPLIED
MATHEMATICS

ELSEVIER

Ideal clutters [☆]

Gérard Cornuéjols[a],*, Bertrand Guenin[b]

[a]*GSIA, Carnegie Mellon University, Schenley Park, Pittsburgh, PA 15213-3890, USA*
[b]*Department of Combinatorics and Optimization, Faculty of Mathematics, University of Waterloo, Waterloo, Ontario, Canada N2L 3G1*

Received 4 October 1999; accepted 13 March 2000

Abstract

The Operations Research model known as the Set Covering Problem has a wide range of applications. See for example the survey by Ceria, Nobili and Sassano and edited by Dell'Amico, Maffioli and Martello (Annotated Bibliographies in Combinatorial Optimization, Wiley, New York, 1997). Sometimes, due to the special structure of the constraint matrix, the natural linear programming relaxation yields an optimal solution that is integer, thus solving the problem. Under which conditions do such integrality properties hold? This question is of both theoretical and practical interest. On the theoretical side, polyhedral combinatorics and graph theory come together in this rich area of discrete mathematics. In this tutorial, we present the state of the art and open problems on this question. © 2002 Elsevier Science B.V. All rights reserved.

Keywords: Ideal clutter; Ideal matrix; Set covering; Integer polyhedron; Width–length inequality; Max Flow Min Cut property

1. Introduction

A *clutter* $\mathscr{C}$ is a family $E(\mathscr{C})$ of subsets of a finite ground set $V(\mathscr{C})$ with the property that $A_1 \nsubseteq A_2$ for all distinct $A_1, A_2 \in E(\mathscr{C})$. $V(\mathscr{C})$ denotes the set of *vertices* and $E(\mathscr{C})$ the set of *edges* of $\mathscr{C}$. A clutter is *ideal* if $\{x \geqslant \mathbf{0} : x(A) \geqslant 1 \text{ for all } A \in E(\mathscr{C})\}$ is an *integral polyhedron*, i.e. all its extreme points have $0, 1$ coordinates. Here $x(A)$ denotes $\sum_{i \in A} x_i$. This concept is also known under the name of *width–length property*, *weak Max Flow Min Cut property* or $\mathcal{2}_+$-*MFMC property*. We prefer the term "ideal" because it stresses the parallel with "perfection".

[☆] This work was supported in part by NSF Grants DMI-0098427, DMI-9802773, DMS-9509581, DMS 96-32032 and ONR Grant N00014-97-1-0196.
* Corresponding author.
E-mail address: gc0v@andrew.cmu.edu (G. Cornuéjols).

A clutter is *trivial* if it has no edge or if it has the empty set as unique edge. Given a nontrivial clutter $\mathscr{C}$, we define $M(\mathscr{C})$ to be a $0,1$ matrix whose columns are indexed by $V(\mathscr{C})$, whose rows are indexed by $E(\mathscr{C})$ and where $M_{ij} = 1$ if and only if the vertex corresponding to column j belongs to the edge corresponding to row i. In other words, the rows of $M(\mathscr{C})$ are the characteristic vectors of the sets in $E(\mathscr{C})$. Note that the definition of $M(\mathscr{C})$ is unique up to permutation of rows and permutation of columns. $M(\mathscr{C})$ contains no dominating row, since $\mathscr{C}$ is a clutter (A vector $r \in F$ is said to be *dominating* if there exists $v \in F$ distinct from r such that $r \geqslant v$). A $0,1$ matrix M containing no dominating rows is called a *clutter matrix*. Given any $0,1$ clutter matrix M, we denote by $\mathscr{C}(M)$ the unique clutter for which $M(\mathscr{C}(M)) = M$. The $0,1$ matrix M is *ideal* if the clutter $\mathscr{C}(M)$ is ideal. Clearly, $\mathscr{C}(M)$ is ideal if and only if $\{x \geqslant 0 : Mx \geqslant 1\}$ is an integral polyhedron. In this tutorial we present the state of the art and open problems on ideal clutters and matrices. Parts of the tutorial overlap with [10].

1.1. Blockers

A *transversal* of a clutter $\mathscr{C}$ is a set of vertices that intersects all the edges. The *blocker* $b(\mathscr{C})$ of a clutter $\mathscr{C}$ is the clutter with $V(\mathscr{C})$ as vertex set and the minimal transversals of $\mathscr{C}$ as edge set. That is, $E(b(\mathscr{C}))$ consists of the minimal members of $\{B \subseteq V(\mathscr{C}): |B \cap A| \geqslant 1 \text{ for all } A \in E(\mathscr{C})\}$. In other words, the rows of $M(b(\mathscr{C}))$ are the minimal $0,1$ vectors x^{T} such that x belongs to the polyhedron $P(\mathscr{C}) = \{x \geqslant 0 : M(\mathscr{C})x \geqslant 1\}$.

Example 1.1. Let G be a graph and s, t be distinct nodes of G. If $\mathscr{C}$ is the clutter of st-paths, then $b(\mathscr{C})$ is the clutter of minimal st-cuts.

Edmonds and Fulkerson [17] observed that $b(b(\mathscr{C})) = \mathscr{C}$. Before proving this property, we make the following remark.

Remark 1.2. Let $\mathscr{H}$ and $\mathscr{K}$ be two clutters defined on the same vertex set. If
(i) every edge of $\mathscr{H}$ contains an edge of $\mathscr{K}$ and
(ii) every edge of $\mathscr{K}$ contains an edge of $\mathscr{H}$,
then $\mathscr{H} = \mathscr{K}$.

Theorem 1.3. *If $\mathscr{C}$ is a clutter, then $b(b(\mathscr{C})) = \mathscr{C}$.*

Proof. Let A be an edge of $\mathscr{C}$. The definition of $b(\mathscr{C})$ implies that $|A \cap B| \geqslant 1$, for every edge B of $b(\mathscr{C})$. So A is a transversal of $b(\mathscr{C})$, i.e. A contains an edge of $b(b(\mathscr{C}))$.

Now let A be an edge of $b(b(\mathscr{C}))$. We claim that A contains an edge of $\mathscr{C}$. Suppose otherwise. Then $V(\mathscr{C}) - A$ is a transversal of $\mathscr{C}$ and therefore it contains an edge B of $b(\mathscr{C})$. But then $A \cap B = \emptyset$ contradicts the fact that A is an edge of $b(b(\mathscr{C}))$. So the claim holds.

Now the theorem follows from Remark 1.2. □

Two $0, 1$ matrices of the form $M(\mathscr{C})$ and $M(b(\mathscr{C}))$ are said to form a *blocking pair*. The next theorem is an important result due to Lehman [32]. It states that, for a blocking pair A, B of $0, 1$ matrices, the polyhedron P defined by

$$Ax \geqslant \mathbf{1}, \tag{1}$$

$$x \geqslant \mathbf{0} \tag{2}$$

is integral if and only if the polyhedron Q defined by

$$Bx \geqslant \mathbf{1}, \tag{3}$$

$$x \geqslant \mathbf{0} \tag{4}$$

is integral. The proof of this result uses the following remark.

Remark 1.4. (i) The rows of B are exactly the $0, 1$ extreme points of P.
   (ii) If an extreme point x of P satisfies $x^{\mathrm{T}} \geqslant \lambda^{\mathrm{T}} B$ where $\lambda_i \geqslant 0$ and $\sum \lambda_i = 1$, then x is a $0, 1$ extreme point of P.

Proof. (i) follows from the fact that the rows of B are the minimal $0, 1$ vectors in P.
   To prove (ii), note that x is an extreme point of $P_I = \{\chi: \chi^{\mathrm{T}} \geqslant \lambda^{\mathrm{T}} B \text{ where } \lambda_i \geqslant 0 \text{ and } \sum \lambda_i = 1\}$ for otherwise x would be a convex combination of distinct $x^1, x^2 \in P_I$ and, since $P_I \subseteq P$, this would contradict the assumption that x is an extreme point of P. Now (ii) follows by observing that the extreme points of P_I are exactly the rows of B. $\square$

Theorem 1.5 (Lehman [32]). *A clutter is ideal if and only if its blocker is.*

Proof. By Theorem 1.3, it suffices to show that if P defined by (1)–(2) is integral, then Q defined by (3)–(4) is also integral.
   Let a be an arbitrary extreme point of Q. By (3), $Ba \geqslant \mathbf{1}$, i.e. $a^{\mathrm{T}}x \geqslant 1$ is satisfied by every x such that x^{T} is a row of B. Since P is an integral polyhedron, it follows from Remark 1.4(i) that $a^{\mathrm{T}}x \geqslant 1$ is satisfied by all the extreme points of P. By (4), $a \geqslant 0$. Therefore $a^{\mathrm{T}}x \geqslant 1$ is satisfied by all points in P. Furthermore, $a^{\mathrm{T}}x = 1$ for some $x \in P$. Now, by linear programming duality, we have

$$1 = \min\{a^{\mathrm{T}}x: x \in P\} = \max\{\lambda^{\mathrm{T}}\mathbf{1}: \lambda^{\mathrm{T}}A \leqslant a^{\mathrm{T}}, \lambda \geqslant 0\}.$$

Therefore, by Remark 1.4(ii), a is a $0, 1$ extreme point of Q. $\square$

1.2. Related concepts

Let $M \neq 0$ be a $0, 1$ clutter matrix and consider the following pair of dual linear programs.

$$\min\{wx: x \geqslant \mathbf{0}, Mx \geqslant \mathbf{1}\} \tag{5}$$

$$= \max\{y\mathbf{1}: y \geqslant \mathbf{0}, yM \leqslant w\}. \tag{6}$$

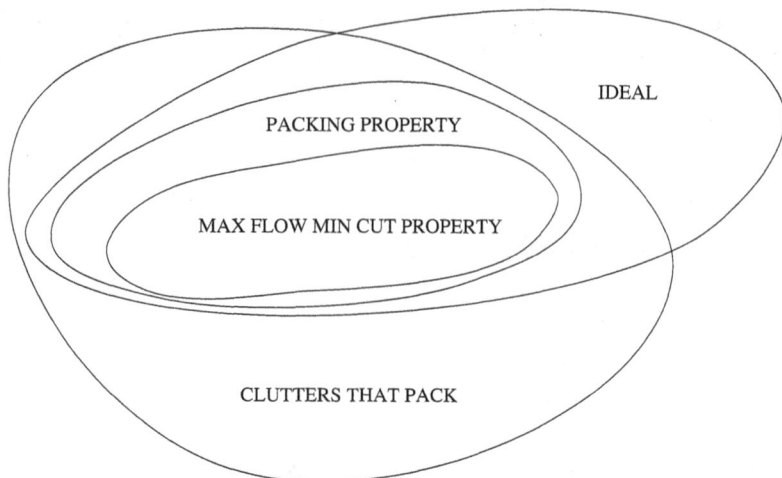

Fig. 1. Classes of clutters.

The clutter $\mathscr{C}(M)$ is ideal if (5) has an optimal solution vector x that is integral for all $w \geqslant 0$. Next, we consider concepts that involve integrality in both the primal and the dual problems.

Definition 1.6. The clutter $\mathscr{C}(M)$ *packs* if both (5) and (6) have optimal solution vectors x and y that are integral when $w = \mathbf{1}$.

Definition 1.7. The clutter $\mathscr{C}(M)$ has the packing property if both (5) and (6) have optimal solution vectors x and y that are integral for all vectors w with components equal to $0, 1$ or $+\infty$.

Definition 1.8. The clutter $\mathscr{C}(M)$ has the *Max Flow Min Cut* property (or MFMC property) if both (5) and (6) have optimal solution vectors x and y that are integral for all nonnegative integral vectors w.

Clearly, the MFMC property for a clutter implies the packing property which itself implies that the clutter packs (see Fig. 1). Conforti and Cornuéjols [6] conjectured that, in fact, the MFMC property and the packing property are identical. This conjecture is still open.

Conjecture 1.9. *A clutter has the MFMC property if and only if it has the packing property.*

Clearly, the MFMC property implies idealness. In fact, the packing property implies idealness.

Theorem 1.10. *If a clutter has the packing property, then it is ideal.*

This follows from a result of Lehman [33] that we will prove in Section 4.

A linear system $Ax \geq b$ is *total dual integral* (TDI) if the linear program $\min wx$ subject to $Ax \geq b$ has an integral optimal dual solution y for every integral w for which the linear program has a finite optimum. Edmonds and Giles [18] proved that, if $Ax \geq b$ is TDI and b is integral, then $P = \{x: Ax \geq b\}$ is an integral polyhedron. The proof of the Edmonds–Giles theorem can be found in Schrijver [49, pp. 310–311], or Nemhauser and Wolsey [37, pp. 536–537]. It follows that $\mathscr{C}(M)$ has the MFMC property if and only if (6) has an optimal integral solution y for all nonnegative integral vectors w.

Definition 1.11. Let k be a positive integer. The clutter $\mathscr{C}(M)$ has the *1/k-MFMC property* if it is ideal and, for all nonnegative integral vectors w, the linear program (6) has an optimal solution vector y such that ky is integral.

When $k = 1$, this definition reduces to that of the MFMC property. If $\mathscr{C}(M)$ has the $1/k$-MFMC property, then it also has the $1/q$-MFMC property for every integer q that is a multiple of k.

Example 1.12. Let $V(\mathscr{C})$ be the set of edges of K_4 and let $E(\mathscr{C})$ be the set of triangles of K_4. The reader can verify that $\mathscr{C}$ is ideal, does not have the MFMC property and, in fact, does not pack. Whereas $b(\mathscr{C})$ is ideal, packs and, in fact, has the MFMC property.

1.3. Deletion, contraction and minor

Let $\mathscr{C}$ be a clutter. For $j \in V(\mathscr{C})$, the *contraction* $\mathscr{C}/j$ and *deletion* $\mathscr{C} \setminus j$ are clutters defined as follows: both have $V(\mathscr{C}) - \{j\}$ as vertex set, $E(\mathscr{C}/j)$ is the set of minimal members in $\{S - \{j\}: S \in E(\mathscr{C})\}$ and $E(\mathscr{C} \setminus j) = \{S: j \notin S \in E(\mathscr{C})\}$.

Contractions and deletions of distinct vertices can be performed sequentially, and it is easy to show that the result does not depend on the order.

Proposition 1.13. *For a clutter $\mathscr{C}$ and distinct vertices j_1, j_2,*
  (i) $(\mathscr{C} \setminus j_1) \setminus j_2 = (\mathscr{C} \setminus j_2) \setminus j_1$,
  (ii) $(\mathscr{C}/j_1)/j_2 = (\mathscr{C}/j_2)/j_1$,
  (iii) $(\mathscr{C} \setminus j_1)/j_2 = (\mathscr{C}/j_2) \setminus j_1$.

Proof. Use the definitions of contraction and deletion! □

Definition 1.14. A clutter $\mathscr{D}$ obtained from $\mathscr{C}$ by a sequence of deletions and contractions is a *minor* of $\mathscr{C}$.

If V_1 and V_2 are disjoint subsets of $V(\mathscr{C})$, we let $\mathscr{C}/V_1 \setminus V_2$ be the minor obtained from $\mathscr{C}$ by contracting all vertices of V_1 and deleting all vertices of V_2. If $V_1 \neq \emptyset$ or $V_2 \neq \emptyset$, the minor is *proper*.

Proposition 1.15. *For a clutter $\mathscr{C}$ and $U \subset V(\mathscr{C})$,*
  (i) $b(\mathscr{C} \setminus U) = b(\mathscr{C})/U$,
  (ii) $b(\mathscr{C}/U) = b(\mathscr{C}) \setminus U$.

Proof. Use the definitions of contraction, deletion and blocker! □

We leave it as an exercise to prove the following result.

Proposition 1.16. *If a clutter is ideal, then so are all its minors.*

Contracting $j \in V(\mathscr{C})$ corresponds to setting $x_j = 0$ in the set covering constraints $Mx \geq 1$ of (5) since column j is removed from M as well as the resulting dominating rows. Deleting j corresponds to setting $x_j = 1$ since column j is removed from M as well as all the rows with a 1 in column j.

Corollary 1.17. *Let M be a $0,1$ matrix. The following are equivalent.*
- *The polyhedron $\{x \geq 0, \ Mx \geq 1\}$ is integral.*
- *The polytope $\{0 \leq x \leq 1, \ Mx \geq 1\}$ is integral.*

2. st-cuts and st-paths

Consider a digraph (N, A) with $s, t \in N$. Let $\mathscr{C}$ be the clutter where $V(\mathscr{C}) = A$ and where $E(\mathscr{C})$ is the family of st-paths.

Theorem 2.1 (Ford and Fulkerson [20]). *The clutter $\mathscr{C}$ has the MFMC property.*

This theorem is a restatement of the famous Max Flow Min Cut theorem of Ford–Fulkerson: for any nonnegative integral arc capacities w, the minimum capacity of an st-cut equals the maximum number of st-paths such that every arc $a \in A$ belongs to at most w_a of the paths. Indeed, the Ford–Fulkerson theorem states that both (5) and (6) have optimal solutions that are integral.
    Theorem 2.1 implies that $\mathscr{C}$ is ideal and therefore the polyhedron

$$\{x \in \mathfrak{R}_+^A : x(P) \geq 1 \text{ for all } st\text{-paths } P\}$$

is integral. Its extreme points are the minimal st-cuts. In the remainder, it will be convenient to refer to minimal st-cuts simply as st-cuts.
    As a consequence of Lehman's theorem (Theorem 1.5), the clutter of st-cuts is also ideal. So the polyhedron

$$\{x \in \mathfrak{R}_+^A : x(C) \geq 1 \text{ for all } st\text{-cuts } C\}$$

is integral. In fact, it is easy to show that the clutter of st-cuts has the MFMC property.

2.1. The width–length inequality

In a network, the product of the minimum number of edges in an st-path by the minimum number of edges in an st-cut is at most equal to the total number of edges in the network. This width–length inequality can be generalized to any nonnegative edge lengths ℓ_e and widths w_e: the minimum length of an st-path times the minimum width of an st-cut is at most equal to the scalar product $\ell^{\mathrm{T}}w$. This width–length inequality was observed by Moore and Shannon [36] and Duffin [16]. A length and a width can be defined for any clutter and its blocker. Interestingly, Lehman [32] showed that the width–length inequality can be used as a characterization of idealness.

Theorem 2.2 (Width–length inequality, Lehman [32]). *For a clutter $\mathscr{C}$ and its blocker $b(\mathscr{C})$, the following statements are equivalent.*
- *$\mathscr{C}$ and $b(\mathscr{C})$ are ideal;*
- $\min\{w(C)\colon C \in E(\mathscr{C})\} \times \min\{\ell(D)\colon D \in E(b(\mathscr{C}))\} \leqslant w^{\mathrm{T}}\ell$ *for all $\ell, w \in \mathfrak{R}_+^n$.*

Proof. Let $A = M(\mathscr{C})$ and $B = M(b(\mathscr{C}))$ be the blocking pair of $0, 1$ matrices associated with $\mathscr{C}$ and $b(\mathscr{C})$, respectively.

First we show that if $\mathscr{C}$ and $b(\mathscr{C})$ are ideal then, for all $\ell, w \in R_+^n$, $\alpha\beta \leqslant w^{\mathrm{T}}\ell$ where $\alpha := \min\{w(C)\colon C \in E(\mathscr{C})\}$ and $\beta := \min\{\ell(D)\colon D \in E(b(\mathscr{C}))\}$.

If $\alpha = 0$ or $\beta = 0$, then this clearly holds.

If $\alpha > 0$ and $\beta > 0$, we can assume w.l.o.g. that $\alpha = \beta = 1$ by scaling ℓ and w. So $Aw \geqslant \mathbf{1}$, i.e. w belongs to the polyhedron $P := \{x \geqslant \mathbf{0},\ Ax \geqslant \mathbf{1}\}$. Therefore w is greater than or equal to a convex combination of the extreme points of P, which are the rows of B by Remark 1.4(i) since P is an integral polyhedron. It follows that $w^{\mathrm{T}} \geqslant \lambda^{\mathrm{T}}B$ where $\lambda \geqslant \mathbf{0}$ and $\sum_i \lambda_i = 1$. Similarly, one shows that $\ell^{\mathrm{T}} \geqslant \mu^{\mathrm{T}}A$ where $\mu \geqslant \mathbf{0}$ and $\sum_i \mu_i = 1$. Since $BA^{\mathrm{T}} \geqslant J$, where J denotes the matrix of all 1's, it follows that

$$w^{\mathrm{T}}\ell \geqslant \lambda^{\mathrm{T}}BA^{\mathrm{T}}\mu \geqslant \lambda^{\mathrm{T}}J\mu = 1 = \alpha\beta.$$

Now we prove the converse. Let $\mathscr{C}$ be a nontrivial clutter and let w be any extreme point of $P := \{x \geqslant \mathbf{0}\colon Ax \geqslant \mathbf{1}\}$. Since $Aw \geqslant \mathbf{1}$, it follows that $\min\{w(C)\colon C \in E(\mathscr{C})\} \geqslant 1$. For any point z in $Q := \{z \geqslant \mathbf{0}\colon Bz \geqslant \mathbf{1}\}$, we also have $\min\{z(D)\colon D \in E(b(\mathscr{C}))\} \geqslant 1$. Using the hypothesis, it follows that $w^{\mathrm{T}}z \geqslant 1$ is satisfied by all points z in Q. Furthermore, equality holds for at least one $z \in Q$. Now, by linear programming duality,

$$1 = \min\{w^{\mathrm{T}}z\colon z \in Q\} = \max\{\mu^{\mathrm{T}}\mathbf{1}\colon \mu^{\mathrm{T}}B \leqslant w^{\mathrm{T}}, \mu \geqslant \mathbf{0}\}.$$

It follows from Remark 1.4(ii) that w is a $0, 1$ extreme point of P. Therefore, $\mathscr{C}$ is ideal. By Theorem 1.5, $b(\mathscr{C})$ is also ideal. $\square$

2.2. Two-commodity flows

Let G be an undirected graph and let $\{s_1, t_1\}$ and $\{s_2, t_2\}$ be two pairs of nodes of G. A *two-commodity cut* is a set of edges separating each of the pairs $\{s_1, t_1\}$ and $\{s_2, t_2\}$. A *two-commodity path* is an $s_1 t_1$-path or an $s_2 t_2$-path.

For any edge capacities $w \in \mathfrak{R}_+^{E(G)}$, Hu [30] showed that a minimum capacity two-commodity cut can be obtained by solving the linear program (5) where M is the incidence matrix of two-commodity paths versus edges.

Theorem 2.3 (Hu [30]). *The clutter of two-commodity paths is ideal.*

Hence, the polyhedron

$$x(P) \geqslant 1 \quad \text{for all two-commodity paths } P,$$

$$x_e \geqslant 0 \quad \text{for all } e \in E(G)$$

is integral.

Using Lehman's theorem (Theorem 1.5), the polyhedron

$$x(C) \geqslant 1 \quad \text{for all two-commodity cuts } C,$$

$$x_e \geqslant 0 \quad \text{for all } e \in E(G)$$

is integral.

The clutters of 2-commodity paths and of 2-commodity cuts do not pack, but both have the 1/2-MFMC property (Hu [35] and Seymour [54], respectively).

The clutter of multicommodity paths is not always ideal for more than two commodities, but conditions on the graph G and the source–sink pairs $\{s_1, t_1\}, \ldots, \{s_k, t_k\}$ have been obtained under which it is ideal. See [45,42,34,21] for examples.

3. *T*-cuts and *T*-joins

Consider a connected graph G with nonnegative edge weights w_e, for $e \in E(G)$. The Chinese Postman Problem consists in finding a minimum weight closed walk going through each edge at least once (the edges of the graph represent streets where mail must be delivered and w_e is the length of the street). Equivalently, the postman must find a minimum weight set of edges $J \subseteq E(G)$ such that $J \cup E(G)$ induces an Eulerian graph, i.e. J induces a graph the odd degree nodes of which coincide with the odd degree nodes of G. Since $w \geqslant \mathbf{0}$, we can assume w.l.o.g. that J is acyclic. Such an edge set J is called a *postman set*.

The problem is generalized as follows. Let G be a graph and T a node set of G of even cardinality. An edge set J of G is called a *T-join* if it induces an acyclic graph the odd degree nodes of which coincide with T. For disjoint node sets S_1, S_2, let (S_1, S_2) denote the set of edges with one endnode in S_1 and the other in S_2. A *T-cut* is a minimal edge set of the form $(S, V(G) - S)$ where S is a set of nodes with $|T \cap S|$ odd. Clearly every *T*-cut intersects every *T*-join.

Edmonds and Johnson [19] considered the problem of finding a minimum weight *T*-join. One way to solve this problem is to reduce it to the perfect matching problem in a complete graph K_p, where $p = |T|$. Namely, compute the lengths of shortest paths in G between all pairs of nodes in T, use these values as edge weights in K_p and find a minimum weight perfect matching in K_p. The union of the corresponding paths in G is a minimum weight T-join. Edmonds and Johnson developed a direct primal-dual algorithm for the minimum weight T-join problem and, as a by-product, obtained that the clutter of T-cuts is ideal.

Theorem 3.1 (Edmonds and Johnson [19]). *The polyhedron*

$$x(C) \geqslant 1 \quad \text{for all } T\text{-cuts } C, \tag{7}$$

$$x_e \geqslant 0 \quad \text{for all } e \in E(G) \tag{8}$$

is integral.

In the next section, we give a non-algorithmic proof of this theorem suggested by Pulleyblank [46].

The Edmonds–Johnson theorem together with the fact that the blocker of an ideal clutter is ideal (Theorem 1.3 of Lehman) implies that the clutter of T-joins is also ideal. That is the polyhedron

$$x(J) \geqslant 1 \quad \text{for all } T\text{-joins } J,$$

$$x_e \geqslant 0 \quad \text{for all } e \in E(G)$$

is integral.

The clutter of T-cuts does not pack, but it has the 1/2-MFMC property (Seymour [57]). The clutter of T-joins does not have the 1/2-MFMC property (there is an example requiring multiplication by 4 to get an integer dual), but it may have the 1/4-MFMC property (open problem). Another intriguing conjecture is the following. Recall that, in a graph G, a *postman set* is a T-join where T coincides with the nodes of G having odd degree.

Conjecture 3.2 (Conforti and Johnson [9]). *The clutter of postman sets packs in graphs noncontractible to the Petersen graph.*

If true, this implies the four color theorem! Indeed, the special case where G is cubic is Tutte's conjecture, recently proved by Robertson et al. [48].

3.1. Proof of the Edmonds–Johnson theorem

First, we prove the following lemma. For $v \in V(G)$, let $\delta(v)$ denote the set of edges incident with v. A *star* is a tree where one node is adjacent to all the other nodes.

Lemma 3.3. *Let $\tilde{x}$ be an extreme point of the polyhedron*

$$x(\delta(v)) \geqslant 1 \quad \text{for all } v \in T, \tag{9}$$

$$x_e \geqslant 0 \quad \text{for all } e \in E(G). \tag{10}$$

The connected components of the graph $\tilde{G}$ induced by the edges such that $\tilde{x}_e > 0$ are either

(i) *odd cycles with nodes in T and edges $\tilde{x}_e = 1/2$, or*
(ii) *stars with nodes in T, except possibly the center, and edges $\tilde{x}_e = 1$.*

Proof. Every connected component C of $\tilde{G}$ is either a tree or contains a unique cycle, since the number of edges in C is at most the number of inequalities (9) that hold with equality.

Assume first that C contains a unique cycle. Then (9) holds with equality for all nodes of C, which are therefore in T. Now C is a cycle since, otherwise, C has a pendant edge e with $\tilde{x}_e = 1$ and therefore C is disconnected, a contradiction. If C is an even cycle, then by alternately increasing and decreasing $\tilde{x}$ around the cycle by a small ε ($-\varepsilon$ respectively), $\tilde{x}$ can be written as a convex combination of two points satisfying (9) and (10). So (i) must hold.

Assume now that C is a tree. Then (9) holds with equality for at least $|V(C)| - 1$ nodes of C. In particular, it holds with equality for at least one node of degree one. Since C is connected, this implies that C is a star and (ii) holds.   □

Proof of Theorem 3.1. In order to prove the theorem, it suffices to show that every extreme point $\tilde{x}$ of the polyhedron (7)–(8) is the incidence vector of a T-join. We proceed by induction on the number of nodes of G.

Suppose first that $\tilde{x}$ is an extreme point of the polyhedron (9)–(10). Consider a connected component of the graph $\tilde{G}$ induced by the edges such that $\tilde{x}_e > 0$ and let S be its node set. Since $\tilde{x}(S, V(G) - S) = 0$, it follows from (7) that S contains an even number of nodes of T. By Lemma 3.3, $\tilde{G}$ contains no odd cycle, showing that $\tilde{x}$ is an integral vector. Furthermore, $\tilde{x}$ is the incidence vector of a T-join since, by Lemma 3.3 again, the component of $\tilde{G}$ induced by S is a star and $|S \cap T|$ even implies that the center is in T if and only if the star has an odd number of edges.

Assume now that $\tilde{x}$ is not an extreme point of the polyhedron (9)–(10). Then there is some T-cut $C = (V_1, V_2)$ with $|V_1| \geq 2$ and $|V_2| \geq 2$ such that

$$\tilde{x}(C) = 1.$$

Let $G_1 = (V_1 \cup \{v_2\}, E_1)$ be the graph obtained from G by contracting V_2 to a single node v_2. Similarly, $G_2 = (V_2 \cup \{v_1\}, E_2)$ is the graph obtained from G by contracting V_1 to a single node v_1. The new nodes v_1, v_2 belong to T. For $i = 1, 2$, let $\tilde{x}^i$ be the restriction of $\tilde{x}$ to E_i. Since every T-cut of G_i is also a T-cut of G, it follows by induction that $\tilde{x}^i$ is greater than or equal to a convex combination of incidence vectors of T-joins of G_i. Let $\mathscr{T}_i$ be this set of T-joins. Each T-join in $\mathscr{T}_i$ has exactly one edge incident with v_i. Since $\tilde{x}^1$ and $\tilde{x}^2$ coincide on the edges of C, it follows that the T-joins of $\mathscr{T}_1$ can be combined with those of $\mathscr{T}_2$ to form T-joins of G and that $\tilde{x}$ is greater than or equal to a convex combination of incidence vectors of T-joins of G. Since $\tilde{x}$ is an extreme point, it is the incidence vector of a T-join.   □

We have just proved that the clutter of T-cuts is ideal. It does not have the MFMC property in general graphs. However, Seymour proved that it does in bipartite graphs.

Seymour also showed that, in a general graph, if the edge weights w_e are integral and their sum is even in every cycle, then the dual variables can be chosen to be integral in an optimum solution.

3.2. st-T-cuts

Goemans and Ramakrishnan [24] introduced a generalization of st-cuts, T-cuts and two-commodity cuts as follows. In a graph G, let s, t be two distinct nodes and let T be a node set of even cardinality. An st-T-cut is a T-cut $\delta(U) := \{uv \in E : u \in U, v \notin U\}$ where U contains exactly one of s or t. The st-cut clutter is obtained when $T = \{s, t\}$, the T-cut clutter is obtained when t is an isolated node and the two-commodity cut clutter is obtained when $T = \{s', t'\}$.

Recently, Guenin [27] characterized exactly when the clutter of st-T-cuts is ideal. This generalizes theorems of Hu (Theorem 2.3) and Edmonds–Johnson (Theorem 3.1).

4. Minimally nonideal matrices

Lehman (Theorem 1.5) showed that ideal $0, 1$ matrices always come in pairs (if M is ideal, so is its blocker $b(M)$) and that the width–length inequality is in fact a characterization of idealness (recall Theorem 2.2). Another important result of Lehman about ideal $0, 1$ matrices is the following.

Theorem 4.1 (Lehman [33]). *For a $0, 1$ matrix A, the following statements are equivalent*:

(i) *the matrix A is ideal*,

(ii) $\min\{cx : Ax \geqslant \mathbf{1}, x \geqslant \mathbf{0}\}$ *has an integral optimal solution x for all $c \in \{0, 1, +\infty\}^n$.*

The fact that (i) implies (ii) is an immediate consequence of the definition of idealness. The difficult part of Lehman's theorem is that (ii) implies (i). The main purpose of this section is to prove this result. This is done by studying properties of minimally nonideal matrices.

4.1. Lehman's characterization

A $0, 1$ matrix A is *minimally nonideal* (*mni*) if

(i) A contains no dominating row,

(ii) $Q(A) := \{x \geqslant \mathbf{0} : Ax \geqslant \mathbf{1}\}$ is not an integral polyhedron,

(iii) For every $i = 1, \ldots, n$, both $Q(A) \cap \{x : x_i = 0\}$ and $Q(A) \cap \{x : x_i = 1\}$ are integral polyhedra.

If A is *mni*, the clutter $\mathscr{C}(A)$ is also called *mni*. Equivalently, a clutter $\mathscr{C}$ is *mni* if it is not ideal but all its proper minors are ideal.

For $t \geqslant 2$ integer, let $\mathscr{J}_t$ denote the clutter with $t + 1$ vertices and edges corresponding, respectively, to the points and lines of the finite degenerate projective plane. Namely, $V(\mathscr{J}_t) := \{0, \ldots, t\}$, and $E(\mathscr{J}_t) := \{\{1, \ldots, t\}, \{0, 1\}, \{0, 2\}, \ldots, \{0, t\}\}$.

A matrix A is *isomorphic* to a matrix B if B can be obtained from A by a permutation of rows and a permutation of columns.

Let J denote a square matrix all of whose entries are 1's, and let I be the identity matrix. Given a *mni* matrix A, let $\bar{x}$ be an extreme point of the polyhedron $Q(A):=\{x \geqslant \mathbf{0}: Ax \geqslant \mathbf{1}\}$ with fractional components. The maximum row submatrix $\bar{A}$ of A such that $\bar{A}\bar{x}=\mathbf{1}$ is called a *core* of A. So A has a core for each fractional extreme point of $Q(A)$.

Theorem 4.2 (Lehman [33]). *Let A be a mni matrix and $B = b(A)$. Then*
 (i) *A has a unique core $\bar{A}$ and B has a unique core $\bar{B}$;*
 (ii) *$\bar{A}$ and $\bar{B}$ are square matrices;*
 (iii) *Either A is isomorphic to $M(\mathscr{J}_t)$, $t \geqslant 2$, or the rows of $\bar{A}$ and $\bar{B}$ can be permuted so that*

$$\bar{A}\bar{B}^{\mathrm{T}} = J + dI$$

 for some positive integer d.

Lehman's proof of this theorem is rather terse. Seymour [58], Padberg [44] and Gasparyan et al. [22] give more accessible presentations of Lehman's proof. In the next section, we present a proof of Lehman's theorem following Padberg's polyhedral point of view.

Bridges and Ryser [2] studied square matrices Y, Z that satisfy the matrix equation $YZ = J + dI$.

Theorem 4.3 (Bridges and Ryser [2]). *Let Y and Z be $n \times n$ $0,1$ matrices such that $YZ = J + dI$ for some positive integer d. Then*
 (i) *each row and column of Y has the same number r of ones, each row and column of Z has the same number s of ones with $rs = n + d$,*
 (ii) *$YZ = ZY$.*

Proof. It is straightforward to check that $(J + dI)^{-1} = (1/d)I - (1/d(n+d))J$. Hence

$$YZ = J + dI \Rightarrow YZ\left(\frac{1}{d}I - \frac{1}{d(n+d)}J\right) = I \Rightarrow Z\left(\frac{1}{d}I - \frac{1}{d(n+d)}J\right)Y = I$$

i.e. $\quad ZY = \dfrac{1}{n+d}ZJY + dI = \dfrac{1}{n+d}\mathbf{s}\mathbf{r}^{\mathrm{T}} + dI$

where $\mathbf{s}:=Z\mathbf{1}$ and $\mathbf{r}:=Y^{\mathrm{T}}\mathbf{1}$.

It follows that, for each i and j, $n+d$ divides $r_i s_j$. On the other hand, the trace of the matrix ZY is equal to the trace of YZ, which is $n(d + 1)$. This implies $(1/(n + d))(\sum_1^n s_i r_i) = n$ and, since $s_i > 0$ and $r_i > 0$, we have $r_i s_i = n + d$. Now consider distinct i, j. Since $r_i s_i = r_j s_j = n+d$ and $n+d$ divides $r_i s_j$ and $r_j s_i$, it follows that $r_i = r_j$ and $s_i = s_j$. Therefore, all columns of Z have the same sum s and all rows of Y have the same sum r. Furthermore, $ZY = J + dI$ and, by symmetry, all columns of Y have the same sum and all rows of Z have the same sum. $\square$

Theorems 4.2 and 4.3 have the following consequence.

Corollary 4.4. *Let A be a mni matrix nonisomorphic to $M(\mathscr{J}_t)$. Then it has a nonsingular row submatrix $\bar{A}$ with exactly r ones in every row and column. Moreover, rows of A not in $\bar{A}$ have at least $r + 1$ ones.*

This implies the next result, which is a restatement of Theorem 4.1.

Corollary 4.5. *Let A be a $0,1$ matrix. The polyhedron $Q(A) = \{x \in R_+^n : Ax \geqslant 1\}$ is integral if and only if $\min\{wx : x \in Q(A)\}$ has an integral optimal solution for all $w \in \{0, 1, \infty\}^n$.*

Note that Theorem 1.10 mentioned in the introduction follows from Corollary 4.5.

Let A be a *mni* matrix nonisomorphic to $M(\mathscr{J}_t)$ and let B be its blocker. Let $\bar{A}$ be the unique core of A and $\bar{B}$ be the unique core of B. Define, $\mathscr{A} := \mathscr{C}(A)$, $\mathscr{B} := \mathscr{C}(B)$, $core(\mathscr{A}) := \mathscr{C}(\bar{A})$, $core(\mathscr{B}) := \mathscr{C}(\bar{B})$. Corollary 4.4 implies that $core(\mathscr{A})$ (resp. $core(\mathscr{B})$) is the set of edges of $\mathscr{A}$ (resp. $\mathscr{B}$) of minimum cardinality. Let L be the edge of $core(\mathscr{A})$ which corresponds to the ith row of $\bar{A}$ and let U be the edge of $core(\mathscr{B})$ which corresponds to the ith row of $\bar{B}$. Theorem 4.2 states that $\bar{A}\bar{B}^T = J + dI$. It follows that L intersects every edge of $core(\mathscr{B})$ exactly once except for U which is intersected $d + 1$ times. We say that L and U are *mates*. It follows from Theorem 4.3(ii) that $\bar{A}\bar{B}^T = \bar{B}^T\bar{A} = J + dI$. In particular for every column j of $\bar{B}$, $col(\bar{B}, j)^T\bar{A} = 1 + de_j$. We can restate this as follows.

Corollary 4.6. *Let A and B be mni matrices which are not isomorphic to $M(\mathscr{J}_t)$. Suppose $\bar{A}$ has r ones per row and $\bar{B}$ has s ones per row. Let j be the index of a column of $\bar{B}$. Let $L_1, \ldots, L_s$ be the edges of $core(\mathscr{A})$ corresponding to the rows of $\bar{A}$ whose indices are given by the characteristic set of column j of $\bar{B}$. Then $L_1 - \{j\}, \ldots, L_s - \{j\}$ are pairwise disjoint, and exactly $d + 1$ of these edges contain j.*

The previous corollary implies immediately.

Remark 4.7. *Let $\mathscr{A}$ be a mni clutter distinct from $\mathscr{J}_t$. Let C_1, C_2 be edges of $core(\mathscr{A})$ and let U_1, U_2 be their mates. If $e \in U_1 \cap U_2$ then $L_1 \cap L_2 \subseteq \{e\}$ and if $e \in C_1 \cap C_2$ then $U_1 \cap U_2 \subseteq \{e\}$.*

4.1.1. Proof of Lehman's theorem

Let A be an $m \times n$ *mni* matrix, $\bar{x}$ a fractional extreme point of $Q(A) := \{x \in R_+^n : Ax \geqslant 1\}$ and $\bar{A}$ a core of A. That is, $\bar{A}$ is the maximal row submatrix of A such that $\bar{A}\bar{x} = 1$. For simplicity of notation, assume that $\bar{A}$ corresponds to the first p rows of A, i.e. the entries of $\bar{A}$ are a_{ij} for $i = 1, \ldots, p$ and $j = 1, \ldots, n$. Since A is *mni*, every component of $\bar{x}$ is nonzero. Therefore $p \geqslant n$ and $\bar{A}$ has no row or column containing only 0's or only 1's.

The following easy result will be applied to the bipartite representation G of the $0, 1$ matrix $J - \bar{A}$ where J denotes the $p \times n$ matrix of all 1's, namely ij is an edge of G if and only if $a_{ij} = 0$, for $1 \leqslant i \leqslant p$ and $1 \leqslant j \leqslant n$. Let $d(u)$ denote the degree of node u.

Lemma 4.8 (de Bruijn and Erdös [14]). *Let $(I \cup J, E)$ be a bipartite graph with no isolated node. If $|I| \geqslant |J|$ and $d(i) \geqslant d(j)$ for all $i \in I$, $j \in J$ such that $ij \in E$, then $|I| = |J|$ and $d(i) = d(j)$ for all $i \in I$, $j \in J$ such that $ij \in E$.*

Proof. $|I| = \sum_{i \in I} (\sum_{j \in N(i)} 1/(d(i))) \leqslant \sum_{i \in I} \sum_{j \in N(i)} 1/d(j) = \sum_{j \in J} \sum_{i \in N(j)} 1/d(j) = |J|$. Now the hypothesis $|I| \geqslant |J|$ implies that equality holds throughout. So $|I| = |J|$ and $d(i) = d(j)$ for all $i \in I$, $j \in J$ such that $ij \in E$. □

The key to proving Lehman's theorem is the following lemma.

Lemma 4.9. *$p = n$ and, if $a_{ij} = 0$ for $1 \leqslant i, j \leqslant n$, then row i and column j of $\bar{A}$ have the same number of ones.*

Proof. Let x^j be defined by

$$x_k^j = \begin{cases} \bar{x}_k & \text{if } k \neq j, \\ 1 & \text{if } k = j, \end{cases}$$

and let F_j be the face of $Q(A) \cap \{x_j = 1\}$ of smallest dimension that contains x^j. Since A is *mni*, F_j is an integral polyhedron. The proof of the lemma will follow unexpectedly from computing the dimension of F_j.

The point x^j lies at the intersection of the hyperplanes in $\bar{A}x = \mathbf{1}$ such that $a_{kj} = 0$ (at least $n - \sum_{k=1}^{p} a_{kj}$ such hyperplanes are independent since $\bar{A}$ has rank n) and of the hyperplane $x_j = 1$ (independent of the previous hyperplanes). It follows that

$$dim(F_j) \leqslant n - \left(n - \sum_{k=1}^{p} a_{kj} + 1 \right) = \sum_{k=1}^{p} a_{kj} - 1.$$

Choose a row a^i of $\bar{A}$ such that $a_{ij} = 0$. Since $x^j \in F_j$, it is greater than or equal to a convex combination of extreme points b^ℓ of F_j, say $x^j \geqslant \sum_{\ell=1}^{t} \gamma_\ell b^\ell$, where $\gamma > 0$ and $\sum \gamma_\ell = 1$.

$$1 = a^i x^j \geqslant \sum_{\ell=1}^{t} \gamma_\ell a^i b^\ell \geqslant 1. \tag{11}$$

Therefore, equality must hold throughout. In particular $a^i b^\ell = 1$ for $\ell = 1, \ldots, t$. Since b^ℓ is a $0, 1$ vector, it has exactly one nonzero entry in the set of columns k where $a_{ik} = 1$. Another consequence of the fact that equality holds in (11) is that $x_k^j = \sum_{\ell=1}^{t} \gamma_\ell b_k^\ell$ for every k where $a_{ik} = 1$. Now, since $x_k^j > 0$ for all k, it follows that F_j contains at least $\sum_{k=1}^{n} a_{ik}$ linearly independent points b^ℓ, i.e.

$$dim(F_j) \geqslant \sum_{k=1}^{n} a_{ik} - 1.$$

Therefore, $\sum_{k=1}^{n} a_{ik} \leqslant \sum_{k=1}^{p} a_{kj}$ for all i, j such that $a_{ij} = 0$.

Now Lemma 4.8 applied to the bipartite representation of $J - \bar{A}$ implies that $p = n$ and

$$\sum_{k=1}^{n} a_{ik} = \sum_{k=1}^{n} a_{kj} \quad \text{for all } i, j \text{ such that } a_{ij} = 0. \quad \square$$

Lemma 4.10. *$\bar{x}$ has exactly n adjacent extreme points in $Q(A)$, all with $0, 1$ coordinates.*

Proof. By Lemma 4.9, exactly n inequalities of $A\bar{x} \geqslant 1$ are tight, namely $\bar{A}\bar{x} = 1$. In the polyhedron $Q(A)$, an edge adjacent to $\bar{x}$ is defined by $n - 1$ of the n equalities in $\bar{A}x = 1$. Moving along such an edge from $\bar{x}$, at least one of the coordinates decreases. Since $Q(A) \in R_+^n$, this implies that $\bar{x}$ has exactly n adjacent extreme points on $Q(A)$. Suppose $\bar{x}$ has a fractional adjacent extreme point $\bar{x}'$. Since A is *mni*, $0 < \bar{x}'_j < 1$ for all j. Let $\bar{A}'$ be the $n \times n$ nonsingular submatrix of A such that $\bar{A}'\bar{x}' = 1$. Since $\bar{x}$ and $\bar{x}'$ are adjacent on $Q(A)$, $\bar{A}$ and $\bar{A}'$ differ in only one row. W.l.o.g. assume that $\bar{A}'$ corresponds to rows 2 to $n + 1$. Since A contains no dominating row, there exists j such that $a_{1j} = 0$ and $a_{n+1,j} = 1$. Since $\bar{A}'$ cannot contain a column with only 1's, $a_{ij} = 0$ for some $2 \leqslant i \leqslant n$. But now, Lemma 4.8 is contradicted with row i and column j in either $\bar{A}$ or $\bar{A}'$. $\square$

Lemma 4.10 has the following implication. Let $\bar{B}$ denote the $n \times n$ $0, 1$ matrix whose rows are the extreme points of $Q(A)$ adjacent to $\bar{x}$. By Remark 1.4(i), $\bar{B}$ is a submatrix of B. By Lemma 4.10, $\bar{B}$ satisfies the matrix equation

$$\bar{A}\bar{B}^{\mathrm{T}} = J + D,$$

where J is the matrix of all 1's and D is a diagonal matrix with positive diagonal entries $d_1, \ldots, d_n$.

Lemma 4.11. *Either*
(i) *$\bar{A} = \bar{B}$ are isomorphic to $M(\mathscr{J}_t)$, for $t \geqslant 2$, or*
(ii) *$D = dI$, where d is a positive integer.*

Proof. Consider the bipartite representation G of the $0, 1$ matrix $J - \bar{A}$.
Case 1: G is connected.
Then it follows from Lemma 4.9 that

$$\sum_{k} a_{ik} = \sum_{k} a_{kj} \quad \text{for all } i, j. \tag{12}$$

Let α denote this common row and column sum.

$$(n + d_1, \ldots, n + d_n) = 1^{\mathrm{T}}(J + D) = 1^{\mathrm{T}}\bar{A}\bar{B}^{\mathrm{T}} = (1^{\mathrm{T}}\bar{A})\bar{B}^{\mathrm{T}} = \alpha 1^{\mathrm{T}}\bar{B}^{\mathrm{T}}.$$

Since there is at most one d, $1 \leqslant d < \alpha$, such that $n + d$ is a multiple of α, all d_i must be equal to d, i.e. $D = dI$.
Case 2: G is disconnected.

Let $q \geqslant 2$ denote the number of connected components in G and let

$$\bar{A} = \begin{pmatrix} K_1 & & 1 \\ & \cdots & \\ 1 & & K_q \end{pmatrix},$$

where K_t are $0, 1$ matrices, for $t = 1, \ldots, q$. It follows from Lemma 4.9 that the matrices K_t are square and $\sum_k a_{ik} = \sum_k a_{kj} = \alpha_t$ in each K_t.

Suppose first that $\bar{A}$ has no row with $n-1$ ones. Then every K_t has at least two rows and columns. We claim that, for every j, k, there exist i, l such that $a_{ij} = a_{ik} = a_{lj} = a_{lk} = 1$. The claim is true if $q \geqslant 3$ or if $q = 2$ and j, k are in the same component (simply take two rows i, l from a different component). So suppose $q = 2$, column j is in K_1 and column k is in K_2. Since no two rows are identical, we must have $\alpha_1 \geqslant 1$, i.e. $a_{ij} = 1$ for some row i of K_1. Similarly, $a_{lk} = 1$ for some row l of K_2. The claim follows.

For each row b of $\bar{B}$, the vector $\bar{A}b^{\mathrm{T}}$ has an entry greater than or equal to 2, so there exist two columns j, k such that $b_j = b_k = 1$. By the claim, there exist rows a_i and a_l of $\bar{A}$ such that $a_i b^{\mathrm{T}} \geqslant 2$ and $a_l b^{\mathrm{T}} \geqslant 2$, contradicting the fact that $\bar{A}b^{\mathrm{T}}$ has exactly one entry greater than 1.

Therefore $\bar{A}$ has a row with $n-1$ ones. Now it is routine to check that $\bar{A}$ is isomorphic to $M(\mathcal{J}_t)$, for $t \geqslant 2$. $\square$

To complete the proof of Theorem 4.2, it only remains to show that the core $\bar{A}$ is unique and that $\bar{B}$ is a core of B and is unique.

If $\bar{A} = M(\mathcal{J}_t)$ for some $t \geqslant 2$, then the fact that A has no dominated rows implies that $A = \bar{A}$. Thus $B = \bar{B} = M(\mathcal{J}_t)$. So, the theorem holds in this case.

If $\bar{A}\bar{B}^{\mathrm{T}} = J + dI$ for some positive integer d, then, by Theorem 4.3, all rows of $\bar{A}$ contain r ones. Therefore, $\bar{x}_j = 1/r$, for $j = 1, \ldots, n$. The feasibility of $\bar{x}$ implies that all rows of A have at least r ones, and Lemma 4.9 implies that exactly n rows of A have r ones. Now $Q(A)$ cannot have a fractional extreme point $\bar{x}'$ distinct from $\bar{x}$, since the above argument applies to $\bar{x}'$ as well. Therefore A has a unique core $\bar{A}$. Since $\bar{x}$ has exactly n neighbors in $Q(A)$ and they all have s components equal to one, the inequality $\sum_1^n x_i \geqslant s$ is valid for the $0, 1$ points in $Q(A)$. This shows that every row of B has at least s ones and exactly n rows of B have s ones. Since B is mni, $\bar{B}$ is the unique core of B. $\square$

4.2. Examples of mni clutters

Let $Z_n = \{0, \ldots, n-1\}$. We define addition of elements in Z_n to be addition modulo n. Let $k \leqslant n-1$ be a positive integer. For each $i \in Z_n$, let C_i denote the subset $\{i, i+1, \ldots, i+k-1\}$ of Z_n. Define the *circulant* clutter $\mathscr{C}_n^k$ by $V(\mathscr{C}_n^k) := Z_n$ and $E(\mathscr{C}_n^k) := \{C_0, \ldots, C_{n-1}\}$.

Lehman [32] gave three infinite classes of minimally nonideal clutters: $\mathscr{C}_n^2$, $n \geqslant 3$ odd, their blockers, and the degenerate projective planes $\mathcal{J}_n$, $n \geqslant 2$.

Conjecture 4.12 (Cornuéjols and Novick [13]). *There exists n_0 such that, for $n \geq n_0$, all mni matrices have a core isomorphic to $\mathcal{C}_n^2$, $\mathcal{C}_n^{(n+1)/2}$ for $n \geq 3$ odd, or $\mathcal{J}_n$, for $n \geq 2$.*

However, there exist several known "small" *mni* matrices that do not belong to any of the above classes. For example, Lehman [32] noted that $\mathcal{F}_7$ is *mni*. $\mathcal{F}_7$ is the clutter with 7 vertices and 7 edges corresponding to points and lines of the Fano plane (finite projective geometry on 7 points):

$$M(\mathcal{F}_7) = \begin{pmatrix} 1 & 1 & 0 & 1 & 0 & 0 & 0 \\ 0 & 1 & 1 & 0 & 1 & 0 & 0 \\ 0 & 0 & 1 & 1 & 0 & 1 & 0 \\ 0 & 0 & 0 & 1 & 1 & 0 & 1 \\ 1 & 0 & 0 & 0 & 1 & 1 & 0 \\ 0 & 1 & 0 & 0 & 0 & 1 & 1 \\ 1 & 0 & 1 & 0 & 0 & 0 & 1 \end{pmatrix}.$$

Let K_5 denote the complete graph on five nodes and let $\mathcal{O}_{K_5}$ denote the clutter whose vertices are the edges of K_5 and whose edges are the odd cycles of K_5 (the triangles and the pentagons). Seymour [53] noted that $\mathcal{O}_{K_5}$, $b(\mathcal{O}_{K_5})$, and $\mathcal{C}_9^2$ with the extra edge $\{3, 6, 9\}$ are *mni*.

Ding [15] found the following *mni* clutter: $V(\mathcal{D}_8) := \{1, \ldots, 8\}$ and

$$E(\mathcal{D}_8) := \{\{1, 2, 6\}, \{2, 3, 5\}, \{3, 4, 8\}, \{4, 5, 7\}, \{2, 5, 6\}, \{1, 6, 7\},$$

$$\{4, 7, 8\}, \{1, 3, 8\}\}.$$

Cornuéjols and Novick [13] characterized the *mni* circulant clutters $\mathcal{C}_n^k$. They showed that the following ten clutters are the only *mni* $\mathcal{C}_n^k$ for $k \geq 3$:

$$\mathcal{C}_5^3, \ \mathcal{C}_8^3, \ \mathcal{C}_{11}^3, \ \mathcal{C}_{14}^3, \ \mathcal{C}_{17}^3, \ \mathcal{C}_7^4, \ \mathcal{C}_{11}^4, \ \mathcal{C}_9^5, \ \mathcal{C}_{11}^6, \ \mathcal{C}_{13}^7.$$

Independently, Qi [47] discovered $\mathcal{C}_9^5$ and $\mathcal{C}_{11}^6$ and Ding [15] discovered $\mathcal{C}_8^3$.

Let $\mathcal{T}_{K_5}$ denote the clutter whose vertices are the edges of K_5 and whose edges are the triangles of K_5 (interestingly, $M(\mathcal{T}_{K_5})$ is also the node–node adjacency matrix of the Petersen graph). It can be shown that $\mathcal{T}_{K_5}$, $core(b(\mathcal{T}_{K_5}))$ and their blockers are *mni*. Often, when a *mni* clutter $\mathcal{H}$ has the property that $core(\mathcal{H})$ and $core(b(\mathcal{H}))$ are also *mni*, many more *mni* clutters can be constructed from $\mathcal{H}$ and from $b(\mathcal{H})$, see [13]. For example, Cornuéjols and Novick [13] have constructed more than one thousand mni clutters from $\mathcal{T}_{K_5}$. More results can be found in [39].

Lütolf and Margot [35] designed a computer program that enumerates possible cores of minimally nonideal matrices. It first enumerates the square $0, 1$ matrices Y, Z that satisfy the matrix equation $YZ = J + dI$, and then checks that the covering polyhedron has a unique fractional extreme point. Lütolf and Margot [35] enumerated all square *mni* matrices of dimension at most 12×12 and found 20 such matrices (previously, only 15 were known); they found 13 new square *mni* matrices of dimensions 14×14 and

17×17; and they found 38 new nonsquare *mni* matrices with 11, 14 and 17 columns with nonisomorphic cores. The overwhelming majority of these examples have $d = 1$: Only three cores with $d = 2$ are known (namely $\mathscr{F}_7, \mathscr{T}_{K_5}$ and the core of its blocker) and none with $d \geqslant 3$.

A clutter $\mathscr{C}$ is *minimally nonpacking* if it does not pack, but all its proper minors do. If $\mathscr{C}$ is minimally nonpacking, then $M(\mathscr{C})$ is also said to be minimally nonpacking.

Theorem 4.13 (Cornuéjols et al. [12]). *Let A be a mni matrix nonisomorphic to $M(\mathscr{J}_t), t \geqslant 2$. If A is minimally nonpacking, then $d = 1$.*

Conjecture 4.14 (Cornuéjols et al. [12]). *Let A be a mni matrix nonisomorphic to $M(\mathscr{J}_t)$, $t \geqslant 2$. Then A is minimally nonpacking if and only if $d = 1$.*

Using a computer program, this conjecture was verified for all known minimally nonideal matrices with $n \leqslant 14$.

Proof of Theorem 4.13. We show that, if $\mathscr{C} \neq \mathscr{J}_t$ is a mni clutter with $d > 1$ then $\mathscr{C}$ is not minimally nonpacking. Let L be an edge of $core(\mathscr{C})$ and let U be its mate. Let $r := |L|$ and $s := |U|$. Let i be any vertex in $L \cap U$ and let $I := (L - U) \cup \{i\}$.

Claim 1. *Every transversal of $\mathscr{C} \setminus I$ has cardinality at least $s - 1$.*

Proof. It suffices to show that every transversal of $core(\mathscr{C}) \setminus I$ has cardinality at least $s - 1$. Suppose there exists a transversal T of $core(\mathscr{C}) \setminus I$ with $|T| \leqslant s - 2$. Let j be any vertex in $U - \{i\}$. By Corollary 4.6, L is among the s edges of $core(\mathscr{C})$ that pairwise intersect at most in $\{j\}$. Since $I \subseteq L - \{j\}$, there are $s - 1$ edges of $core(\mathscr{C}) \setminus I$ that pairwise intersect at most in $\{j\}$. Therefore, $|T| \leqslant s - 2$ implies $j \in T$. By symmetry among the vertices of $U - \{i\}$, it follows that $U - \{i\} \subseteq T$. So in particular $|T| \geqslant s - 1$, a contradiction. □

Suppose $\mathscr{C} \setminus I$ packs. Then it follows from Claim 1 that $\mathscr{C} \setminus I$ contains $s - 1$ disjoint edges $L_1, \ldots, L_{s-1}$.

Claim 2. *None of $L_1, \ldots, L_{s-1}$ are edges of $core(\mathscr{C})$.*

Proof. Suppose that L_1 is an edge of $core(\mathscr{C})$ and let U_1 be its mate. Then $U_1 - (I \cup L_1)$ contains an edge T in $b(\mathscr{C})/(I \cup L_1)$. By assumption $|L_1 \cap U_1| = d + 1 \geqslant 3$. Thus

$$|T| \leqslant |U_1 - L_1| = |U_1| - (d + 1) = s - (d + 1) \leqslant s - 3.$$

By Proposition 1.15, T is a transversal of $\mathscr{C} \setminus (I \cup L_1)$. But $L_2, \ldots, L_{s-1}$ are disjoint edges of $\mathscr{C} \setminus (I \cup L_1)$, which implies that every transversal of $\mathscr{C} \setminus (I \cup L_1)$ has cardinality at least $s - 2$, a contradiction. □

By Corollary 4.4, the edges $L_1, \ldots, L_{s-1}$ have cardinality at least $r + 1$. Moreover they do not intersect I. Therefore we must have:

$$(r + 1)(s - 1) \leqslant n - |I| = (rs - d) - (r - d) = rs - r.$$

Thus $r \leqslant 1$, a contradiction.   □

4.3. A conjecture

As a parallel to Theorem 4.1, we can restate Conjecture 1.9 as follows.

Conjecture 4.15 (Conforti and Cornuéjols [6]). *For a $0, 1$ matrix A, the following statements are equivalent*:
  (i) *the matrix A has the MFMC property*,
  (ii) $\min\{cx: Ax \geqslant 1, \ x \geqslant 0\}$ *has an integral optimal dual solution y for all $c \in \{0, 1, +\infty\}^n$.*

4.4. Ideal minimally nonpacking clutters

Minimally nonpacking clutters are either ideal or minimally nonideal. This follows from Theorem 1.10. Theorem 4.13 above discussed the minimally nonideal case. In this section, we discuss the ideal case. The clutter of triangles of K_4 is such an example: this clutter has 6 vertices (the 6 edges of K_4) and 4 edges (the 4 triangles of K_4 viewed as edge sets) and it is denoted by Q_6.

A clutter is *binary* if its edges have an odd intersection with its minimal transversals. Seymour [53] showed that Q_6 is the only ideal minimally nonpacking binary clutter. However, there are ideal minimally nonpacking clutters that are not binary, such as

$$\begin{pmatrix} 1 & 1 & 0 & 1 & 0 & 1 & 0 \\ 1 & 1 & 0 & 0 & 1 & 0 & 1 \\ 0 & 0 & 1 & 0 & 1 & 1 & 0 \\ 0 & 0 & 1 & 1 & 0 & 0 & 1 \\ 1 & 0 & 1 & 1 & 0 & 1 & 0 \\ 0 & 1 & 1 & 0 & 1 & 0 & 1 \end{pmatrix}.$$

Note that, for this clutter, the minimum size of a transversal is 2. Other examples can be found in [12] but none is known with a minimum transversal of size greater than 2. Interestingly, all ideal minimally nonpacking clutters with a transversal of size 2 share strong structural properties with Q_6. A clutter $\mathscr{C}$ has the Q_6-*property* if $M(\mathscr{C})$ has 4 rows such that every column restricted to this set of rows contains two 0's and two 1's and each such 6 possible $0, 1$ vectors occurs at least once.

Theorem 4.16 (Cornuéjols et al. [12]). *Every ideal minimally nonpacking clutter with a transversal of size 2 has the Q_6-property.*

Conjecture 4.17 (Cornuéjols et al. [12]). *Every ideal minimally nonpacking clutter has a transversal of size 2.*

It is proved in [12] that this conjecture would imply Conjecture 1.9 or, equivalently, Conjecture 4.15.

5. Odd cycles in graphs

In this section, we consider the clutter $\mathscr{H}$ of odd cycles in a graph. Seymour [53] characterized exactly the graphs for which $\mathscr{H}$ has the MFMC property and Guenin [26] characterized exactly when $\mathscr{H}$ is ideal.

For edge weights $w \in \mathfrak{R}_+^{E(G)}$, consider the minimization problem (5). Recall that an integral solution to (5) is the incidence vector of a transversal T of $\mathscr{H}$. Since T intersects all odd cycles, $E(G) - T$ induces a bipartite graph. Therefore, a minimal transversal T of $\mathscr{H}$ is the complement of a cut $\delta(U)$. In particular, when $\mathscr{H}$ is ideal, (5) finds a cut of maximum weight in G, i.e. (5) solves the famous *max cut* problem.

5.1. Planar graphs

Orlova and Dorfman [42] showed that the clutter $\mathscr{H}$ of odd cycles is ideal when G is planar.

Theorem 5.1 (Orlova and Dorfman [42]). *In a planar graph, the clutter of odd cycles is ideal.*

Proof. Let G be a planar graph and D its dual. The bounded faces of G form a cycle basis. Thus any odd cycle of G is a symmetric difference of faces, an odd number of which are odd faces. Faces of G correspond to nodes of D. Let T be the set of odd degree nodes of D. An odd cycle of G corresponds to an edge set of D of the form $\delta(U)$ where $|U \cap T|$ has odd cardinality, i.e. a T-cut of D. The clutter of T-cuts in D is ideal by the Edmonds–Johnson theorem (Theorem 3.1) and therefore so is the clutter of odd cycles in G. $\square$

When $G = K_5$, the complete graph on 5 nodes, the clutter $\mathscr{H}$ of odd cycles is not ideal since $x_j = \frac{1}{3}$ for $j = 1, \dots, 10$ is a fractional extreme point of the polyhedron $\{x \in R_+^{10}: M(\mathscr{H})x \geq 1\}$.

Barahona [1] observed that Theorem 5.1 has the following generalization.

Theorem 5.2 (Barahona [1]). *In a graph not contractible to K_5, the clutter of odd cycles is ideal.*

This follows from a famous theorem of Wagner [62] stating that any edge-maximal graph not contractible to K_5 can be constructed recursively by pasting plane triangulations and copies of V_8 along K_3's and K_2's, where V_8 is the cycle $v_1, v_2, \dots, v_8, v_1$ with chords $v_i v_{i+4}$ for $i = 1, 2, 3, 4$.

Is there a converse to Barahona's theorem? In particular, is it true that, if the clutter of odd cycles is ideal in a graph G, then G is not contractible to K_5? The answer to the second question is no. For example, insert a node of degree 2 on every edge of K_5. The graph is now bipartite and the clutter of odd cycles has become the trivial clutter, which is ideal! The problem is that contraction of an edge changes odd cycles into even cycles and vice versa. To get a converse to Barahona's theorem, one needs to redefine contraction appropriately. It is convenient to work in the more general context of signed graphs.

5.2. Signed graphs

Consider a graph G and a subset S of its edges. The pair (G, S) is called a *signed graph*. A subset X of edges of G is *odd* (resp. *even*) if $|X \cap S|$ is odd (resp. even). A set $S' \subseteq E(G)$ is a *signature* of (G, S) if (G, S') has the same odd cycles as (G, S).

Consider a signed graph (G, S) and let $\delta(U)$ be a cut of G. Since $\delta(U)$ intersects every cycle with even parity, $S \triangle \delta(U)$ is a signature of (G, S). We call the operation which consists of replacing S by $S \triangle \delta(U)$ a *signature-exchange*. In a signed graph (G, S), *deleting* an edge means removing it from the graph. *Contracting* an edge e means first (if necessary) doing a signature-exchange so that the edge e is even (i.e. not in the signature) and then removing the edge and identifying its endnodes.

Let E' and E'' be disjoint edge sets. One can readily verify that all the signed graphs obtained by deleting the edges in E' and contracting the edges in E'' are identical (up to signature-exchanges), no matter in which order the contractions and deletions are performed. A signed graph obtained from (G, S) by a sequence of contractions and deletions and signature-exchanges is called a *minor* of (G, S).

Let $\mathcal{H}$ denote the clutter of odd cycles of a signed graph (G, S). It is easy to check that every minor of $\mathcal{H}$ is the clutter of odd cycles of a signed graph (G', S') obtained as a minor of (G, S). A signed complete graph K_r on r nodes is called an *odd-K_r* if all its edges are odd. Guenin proved the following theorem.

Theorem 5.3 (Guenin [26]). *The clutter of odd cycles of a signed graph (G, S) is ideal if and only if (G, S) has no odd-K_5 minor.*

A clutter is *binary* (see Section 6) if its edges and its minimal transversals intersect in an odd number of vertices. The clutter of odd cycles in a signed graph is a binary clutter. Theorem 5.3 is a special case of a famous conjecture of Seymour [53,56] (Conjecture 6.9) on ideal binary clutters. In [53], Seymour characterized the binary clutters that have the MFMC property. Specialized to the clutter of odd cycles, this theorem is the following.

Theorem 5.4 (Seymour [53]). *The clutter of odd cycles of a signed graph (G, S) has the MFMC property if and only if (G, S) has no odd-K_4 minor.*

5.3. Schrijver's proof of Guenin's theorem

One direction of Guenin's theorem is easy: If the clutter of odd cycles is ideal for a signed graph (G, S), then (G, S) has no odd-K_5 minor. Thus the essence of Theorem 5.3 is the converse. Schrijver [50] obtained a shorter proof for this result, which curtails the technical and case-checking part of Guenin's proof.

Schrijver's proof which we give next (albeit with a different presentation along the lines of the proof of Theorem 5.8 see [23]) relies on the following two lemmas on *mni* binary clutters. These lemmas were also used in Guenin's original proof. Observe at the outset that $\mathscr{J}_t$ is not binary.

Lemma 5.5. *Let $\mathscr{H}$ be a mni binary clutter and C_1, C_2 be edges in core($\mathscr{H}$). If $C \subseteq C_1 \cup C_2$ and C is an edge of $\mathscr{H}$ then $C = C_1$ or $C = C_2$.*

Proof. Let C be an edge of $\mathscr{H}$ contained in $C_1 \cup C_2$. Then (Proposition 6.1) $C_1 \triangle C_2 \triangle C$ contains an edge of $\mathscr{H}$, say C'. This implies that $C \cup C' \subseteq C_1 \cup C_2$ and $C \cap C' \subseteq C_1 \cap C_2$ (for if $e \in C \cap C'$ then $e \notin C_1 \triangle C_2$). Hence $|C| + |C'| \leq |C_1| + |C_2|$. So C, C' are also of minimum cardinality, and C, C' are edges of core($\mathscr{H}$). Let B be the mate of C. Since $\mathscr{H}$ is binary, $|C \cap B|$ is odd, hence at least 3. It follows that either, $|C_1 \cap B| \geq 2$ or $|C_2 \cap B| \geq 2$. This implies that C_1 or C_2 is the mate of B, i.e. $C = C_1$ or $C = C_2$. $\square$

Lemma 5.6. *Let $\mathscr{H}$ be a mni binary clutter. For any $e \in V(\mathscr{H})$ there exist edges C_1, C_2, C_3 of core($\mathscr{H}$) and edges B_1, B_2, B_3 of core($b(\mathscr{H})$) such that*
  (i) $C_1 \cap C_2 = C_1 \cap C_3 = C_2 \cap C_3 = \{e\}$
  (ii) $B_1 \cap B_2 = B_1 \cap B_3 = B_2 \cap B_3 = \{e\}$
  (iii) *For distinct $i, j \in \{1, 2, 3\}$ we have $C_i \cap B_j = \{e\}$. For $i \in \{1, 2, 3\}$ we have $|C_i \cap B_i| = d + 1$ where $d + 1$ is odd and $d + 1 \geq 3$.*

Proof. Corollary 4.6 states that there exist s edges $C_1, \dots, C_s$ of core($\mathscr{H}$) such that $C_1 - \{e\}, \dots, C_s - \{e\}$ are pairwise disjoint. Moreover, exactly $d + 1 \geq 2$ of these edges, say $C_1, \dots, C_d$, contain vertex e. As $\mathscr{H}$ is binary, $d + 1$ is odd (since $d + 1 = |C \cap B|$ for any pair of mates C, B). Thus $d + 1 \geq 3$ and (i) follows. Let B_1, B_2, B_3 be the mates of C_1, C_2, C_3. For $i \in \{1, 2, 3\}$ we have: $|C_i \cap B_i| = d + 1 > 1$; $C_1 - \{e\}, \dots, C_s - \{e\}$ disjoint; and $|B_i| = s$. Then $e \in B_i$ as B_i intersects each $C_1, \dots, C_s$. Since $e \in C_1 \cap C_2 \cap C_3$, it follows from Remark 4.7 that $B_i \cap B_j \subseteq \{e\}$ for all distinct $i, j \in \{1, 2, 3\}$. Hence, (ii) holds. Finally (iii) holds since B_1, B_2, B_3 are the mates of C_1, C_2, C_3. $\square$

A key ingredient in Schrijver's proof is the following lemma. The particular version presented here was given in [23].

Lemma 5.7. *Let $G = (V, E)$ be a graph, let e be an edge of G with endnodes x and y, let (Y_0, Y_1, Y_2, Y_3) be disjoint subsets of V, and let P_1, P_2, and P_3 be internally node disjoint xy-paths in $G \setminus e$. Moreover, suppose that*
  (1) $x, y \in Y_0$ *and, for $i \in \{0, 1, 2, 3\}$, Y_i is a stable set of $G \setminus e$,*

(2) *for $i \in \{1,2,3\}$, $V(P_i) \subseteq Y_0 \cup Y_i$, and*
(3) *for distinct $i,j \in \{1,2,3\}$, there exists a path from $V(P_i)$ to $V(P_j)$ in $G[Y_i \cup Y_j]$.*
Then $(G, E(G))$ has a minor isomorphic to odd-K_5.

Proof. Suppose otherwise, and let G be a counterexample minimizing $|V(G)| + |E(G)|$. For distinct $i,j \in \{1,2,3\}$, let P_{ij} be a path from $V(P_i)$ to $V(P_j)$ in $G[Y_i \cup Y_j]$. (We assume that $P_{ij} = P_{ji}$.) By the minimality of G, we have $E(G) := \{e\} \cup P_1 \cup P_2 \cup P_3 \cup P_{12} \cup P_{23} \cup P_{13}$, and $V(G) := V(P_1) \cup V(P_2) \cup V(P_3) \cup V(P_{12}) \cup V(P_{23}) \cup V(P_{13})$.

Suppose that G has a node v of degree 2, and define $G' := G/\delta_G(v)$. Note that, $(G, E(G))/\delta_G(v) = (G', E(G'))$, and that G' satisfies the conditions of the lemma. However, this contradicts the minimality of G, and, hence, G has no nodes of degree 2. Thus, we see that $Y_0 = \{x, y\}$, and, for each $i \in \{1,2,3\}$, P_i has exactly one internal node, say v_i. Now, the neighbors of x are v_1, v_2, v_3, and y, and the neighbors of y are v_1, v_2, v_3, and x. Moreover, since G has no nodes of degree 2, we also conclude that $Y_1 = V(P_{12}) \cap V(P_{13})$, $Y_2 = V(P_{12}) \cap V(P_{23})$, and $Y_3 = V(P_{13}) \cap V(P_{23})$. Therefore, $|Y_1| = |Y_2| = |Y_3|$.

If $|Y_1| = 1$, then $(G, E(G))$ is isomorphic to odd-K_5, so we may assume that $|Y_1| > 1$. For distinct $i,j \in \{1,2,3\}$, let e_{ij} be the edge on P_{ij} that is incident with v_i. Let $G' := G \setminus \{e_{13}, e_{32}, e_{21}\}/\{e_{12}, e_{23}, e_{31}\}$, and, for distinct $i,j \in \{1,2,3\}$, let $P'_{ij} := P_{ij} - \{e_{ij}, e_{ji}\}$. Now let $Y'_1 := V(P'_{12}) \cap V(P'_{13})$, let $Y'_2 := V(P'_{12}) \cap V(P'_{23})$, let $Y'_3 := V(P'_{13}) \cap V(P'_{23})$, and let $Y'_0 := \{x, y\}$. Note that, $(G', E(G'))$ is a minor of $(G, E(G))$ and that G' satisfies the conditions of the lemma. However, this contradicts the minimality of G. □

Given a graph G and $U \subseteq V(G)$, the subgraph of G induced by U is denoted $G[U]$.

Proof of Theorem 5.3. Let $\mathscr{H}$ be a *mni* clutter of odd cycles of a signed graph (G, S). We will show that (G, S) contains an odd-K_5 minor. Fix an edge $e \in E(G)$, with endnodes say x and y. Let C_1, C_2, C_3 be the sets of $core(\mathscr{H})$ and let B_1, B_2, B_3 be the sets of $core(b(\mathscr{H}))$ given in Lemma 5.6.

Claim 1. *For distinct $i,j \in \{1,2,3\}$ the odd cycles C_i and C_j have no common node other than x, y.*

Proof. Otherwise $(C_i \cup C_j) - \{e\}$ contains a path P from x to y different from $C_i - \{e\}$ and $C_j - \{e\}$. By Lemma 5.5, $(C_i \cup C_j) - \{e\}$ contains no odd cycle. Hence, P and $C_i - \{e\}$ have the same parity and so $P \cup \{e\}$ is an odd cycle in $C_i \cup C_j$, contradicting Lemma 5.5. □

Since $\mathscr{H}$ is binary, B_i $(i = 1, 2, 3)$ is a signature. It follows that for distinct $i,j \in \{1,2,3\} B_i \triangle B_j$ intersects all cycles with even parity; i.e. $B_i \triangle B_j$ is a cut of G. Moreover, $e \notin B_i \triangle B_j$. Therefore, for distinct $i,j \in \{1,2,3\}$, there exists $U_{ij} \subseteq V(G)$ such that $\delta(U_{ij}) = B_i \triangle B_j$ and $x, y \notin U_{ij}$. Note that

$$\delta(U_{12} \triangle U_{13} \triangle U_{23}) = \delta(U_{12}) \triangle \delta(U_{13}) \triangle \delta(U_{23}) = \emptyset.$$

Moreover, $x, y \notin U_{12} \triangle U_{13} \triangle U_{23}$ and G is connected. Therefore, $U_{12} \triangle U_{13} \triangle U_{23} = \emptyset$. Let $Y_1 := U_{12} \cap U_{13}$, $Y_2 := U_{12} \cap U_{23}$, $Y_3 := U_{13} \cap U_{23}$, and let $Y_0 = V(G) - (Y_1 \cup Y_2 \cup Y_3)$.

Claim 2. *For distinct $i, j, k \in \{1, 2, 3\}$, the edge set $B_i - \{e\}$ consists of all edges with one endnode in Y_0 and the other in Y_i and all edges with one endnode in Y_j and the other in Y_k.*

Proof. We may assume $i = 1$. Since $B_1 - \{e\}$, $B_2 - \{e\}$, $B_3 - \{e\}$ are pairwise disjoint, $B_1 - \{e\} = (B_1 \triangle B_2) \cap (B_1 \triangle B_3)$. But $\delta(U_{12}) = B_1 \triangle B_2$ and $\delta(U_{13}) = B_1 \triangle B_3$. Thus the edges of $B_1 - \{e\}$ are exactly the edges in both $\delta(U_{12})$ and $\delta(U_{13})$. $\square$

For each $i \in \{1, 2, 3\}$ let $P_i := C_i - \{e\}$, thus P_i is an xy-path. Recall that for distinct $i, j, k \in \{1, 2, 3\}$, $C_i \cap (B_j \cup B_k) = \{e\}$. It follows together with Claim 2 that for each $i \in \{1, 2, 3\}$, $V(P_i) \subseteq Y_0 \cup Y_i$. Moreover, since $|C_i \cap B_i| > 1$, $P_i \cap V(Y_i) \neq \emptyset$.

Claim 3. *For distinct $i, j \in \{1, 2, 3\}$, there exists a path P_{ij} from $V(P_i)$ to $V(P_j)$ in $G[Y_i \cup Y_j]$.*

Proof. Recall, $U_{ij} = Y_i \cup Y_j$. It suffices to prove that $G[U_{ij}]$ is connected. If not, there is an $X \subseteq U_{ij}$ such that $\delta(X)$ is a non-empty proper subset of $\delta(U_{ij})$. Then $B_i \triangle \delta(X)$ is contained in $B_i \cup B_j$ but is distinct from B_i and B_j. Since $B_i \triangle \delta(X)$ is a signature, it contains an element of $b(\mathcal{H})$, a contradiction with Lemma 5.5. $\square$

Let $B := B_1 \triangle B_2 \triangle B_3$. Then B is a signature for (G, S). Let $T := \{e\} \cup P_1 \cup P_2 \cup P_3 \cup P_{12} \cup P_{13} \cup P_{23}$. Each edge in $T - \{e\}$ is in at most one of the sets B_1, B_2, B_3. Therefore, the odd edges of $(G, B)[T]$ are e and any edge whose endnodes are in different parts of (Y_0, Y_2, Y_2, Y_3). Let (G', S') be the signed graph obtained from $(G, B)[T]$ by contracting the edges in $T - B$; thus $S' = E(G')$. For $i \in \{1, 2, 3\}$, let $P'_i = P_i \cap B$; for distinct $i, j \in \{1, 2, 3\}$, let $P'_{ij} = P_{ij} \cap B$; and for $l \in \{0, 1, 2, 3\}$ let Y'_l be the set of nodes of G' corresponding to Y_i. Now by Lemma 5.7, we see that (G', S') contains an odd-K_5 minor, as required. $\square$

5.3.1. Cycling

Let (G, S) be a signed graph. Weights $w \in Z_+^{E(G)}$ are called *Eulerian* if $w(\delta(v))$ is even for every $v \in V(G)$. We say that the clutter of odd cycles of (G, S) is *cycling* [56] if (6) and (5) have both optimum integer solutions for all Eulerian edge-weights. Note that the clutter of odd cycles of odd-K_5 is not cycling. However, it is the only obstruction to the property.

Theorem 5.8 (Geelen and Guenin [23]). *The clutter of odd cycles of a signed graph (G, S) is cycling if and only if (G, S) has no minor isomorphic to odd-K_5.*

Let $\mathcal{H}$ be the clutter of odd cycles of a signed graph (G, S). Suppose that $\mathcal{H}$ is cycling and let $w \in Z_+^{E(G)}$. Now, $2w$ is Eulerian, so there exists an integral optimal solution x to (5) with respect to the weights $2w$. Clearly, x is also optimal with respect to w. Hence, if $\mathcal{H}$ is cycling it is also ideal (see Corollary 4.5). Thus Theorem 5.8 implies Theorem 5.3 and the fact that for clutters of odd cycles, the property of being ideal is the same as cycling. Using the same trick as above, of doubling the edge-capacities, we also obtain the following result.

Corollary 5.9. *If the clutter of odd cycles of a signed graph is ideal then it has the 1/2-MFMC property.*

5.3.2. Odd st-walks

Guenin [27] considers the following generalization of the odd cycle clutter. Let (G, S) be a signed graph and let s, t be two nodes of G. A subset of edges of G is an *odd st-walk* if it is an odd *st*-path or the union of an even *st*-path P and an odd cycle C where P and C share at most one node. The odd cycle clutter is obtained when $s = t$.

Guenin characterized exactly when this clutter is ideal. This generalizes Theorem 5.3.

6. Binary clutters

A clutter is *binary* if its edges and its minimal transversals intersect in an odd number of vertices. It follows from the definition that a clutter is binary if and only if its blocker is binary. An equivalent formulation is given by Lehman.

Proposition 6.1 (Lehman [31], see also Seymour [51]). *A clutter $\mathscr{C}$ is binary if and only if, for any three edges S_1, S_2, S_3 of $\mathscr{C}$, the set $S_1 \triangle S_2 \triangle S_3$ contains an edge of $\mathscr{C}$.*

Proof. Let $\mathscr{C}$ be a binary clutter and $S = S_1 \triangle S_2 \triangle S_3$ where $S_1, S_2, S_3 \in E(\mathscr{C})$. Since every minimal transversal T has an odd intersection with S_1, S_2 and S_3, we have $S \cap T \neq \emptyset$. Therefore S contains an edge of $\mathscr{C}$.

Conversely, assume that for any three edges S_1, S_2, S_3 of $\mathscr{C}$, the set $S_1 \triangle S_2 \triangle S_3$ contains an edge of $\mathscr{C}$. We leave it as an exercise to show that, for any odd number of edges $S_1, \ldots, S_k$ of $\mathscr{C}$, the set $S_1 \triangle \cdots \triangle S_k$ contains an edge of $\mathscr{C}$. Now consider any $S \in E(\mathscr{C})$, $T \in E(b(\mathscr{C}))$ and let $S \cap T = \{x_1, \ldots, x_k\}$. Since $T - x_i$ is not a transversal of $\mathscr{C}$, there exists an edge S_i of $\mathscr{C}$ such that $T \cap S_i = \{x_i\}$. It follows that $T \cap (S \triangle S_1 \triangle \cdots \triangle S_k) = \emptyset$. Therefore $S \triangle S_1 \triangle \cdots \triangle S_k$ does not contain an edge of $\mathscr{C}$. It follows that k is odd. $\square$

Let $\mathscr{P}_4$ be the clutter with four vertices and the following three edges:

$$E(\mathscr{P}_4) = \{\{1, 2\}, \{2, 3\}, \{3, 4\}\}.$$

One can easily show that neither $\mathscr{P}_4$ nor $\mathscr{J}_t$ is a binary clutter, for $t \geqslant 2$. Seymour proved the following.

Theorem 6.2 (Seymour [51]). $\mathscr{C}$ *is a binary clutter if and only if* $\mathscr{C}$ *has no minor* $\mathscr{P}_4$ *or* $\mathscr{J}_t$, *for* $t \geqslant 2$.

The following clutters (and their blockers!) are examples of binary clutters.

Example 6.3. The clutter of st-cuts in a graph.

Example 6.4. The clutter of two-commodity cuts in a graph.

Example 6.5. The clutter of T-joins in a graft (G, T).

Example 6.6. The clutter of odd cycles in a signed graph.

Example 6.7. The clutter of st-T-cuts.

Example 6.8. The clutter of odd st-walks.

6.1. Seymour's conjecture

Recall (Section 4.2) that $\mathscr{F}_7$ denotes the clutter with 7 vertices and 7 edges corresponding to points and lines of the Fano plane (finite projective geometry on 7 points). It is easy to verify that $\mathscr{F}_7$ is binary, *mni* and that $b(\mathscr{F}_7) = \mathscr{F}_7$.

Let K_5 denote the complete graph on five vertices. We let $\mathcal{O}_{K_5}$ denote the binary clutter whose vertices are the edges of K_5 and whose edges are the odd cycles of K_5. So $\mathcal{O}_{K_5}$ has 10 edges of cardinality three and 12 edges of cardinality five. $\mathcal{O}_{K_5}$ is binary and *mni*. It follows that $b(\mathcal{O}_{K_5})$ is binary and *mni*.

Conjecture 6.9 (Seymour [53]). *A binary clutter is ideal if and only if it contains no* $\mathscr{F}_7$, $\mathcal{O}_{K_5}$ *or* $b(\mathcal{O}_{K_5})$ *minor*.

6.2. Binary matroids

In the remainder of this section, we present results of Novick–Sebö [40] and Cornuéjols–Guenin [11] on ideal binary clutters. We adopt a matroidal point of view. See Oxley's excellent textbook [43] on matroid theory for background material.

A matroid is *binary* if it can be represented over $GF(2)$.

Example 6.10. The Fano matroid F_7 has the following binary representation:

$$\begin{pmatrix} 1 & 0 & 0 & 1 & 1 & 0 & 1 \\ 0 & 1 & 0 & 1 & 0 & 1 & 1 \\ 0 & 0 & 1 & 0 & 1 & 1 & 1 \end{pmatrix}.$$

Given a matroid M, the dual matroid is denoted by M^*. A binary matroid is *regular* if it has no F_7 or F_7^* minor [61].

Let M be a matroid with element set U and let k be a positive integer. A *k-separation* of M is a partition (U_1, U_2) of U such that $|U_1| \geq k$, $|U_2| \geq k$ and $r(U_1)+r(U_2) \leq r(U)$ $+k-1$. The matroid M is *k-connected* if it has no $(k-1)$-separation. The k-separation is *strict* if $|U_1| > k$, $|U_2| > k$. A matroid is *internally k-connected* if it has no strict $(k-1)$-separation.

Theorem 6.11 (Seymour [55]). *Every 3-connected, internally 4-connected regular matroid is graphic, cographic or a 10-element matroid R_{10}.*

Theorem 6.12 (Seymour [56]). *Let M be a 3-connected binary matroid with no F_7 minor. Then M is regular or $M = F_7^*$.*

6.3. Signed matroid

Let M be a binary matroid and $S \subseteq V(M)$ a subset of its elements. The pair (M, S) is called a *signed matroid*, and S is called the *signature* of M. We say that a circuit C of M is *odd* (resp. *even*) if $|C \cap S|$ is odd (resp. even).

Proposition 6.13. *The odd circuits of a signed matroid form a binary clutter.*

Proof. Consider a signed matroid (M, S) and let C_1, C_2, C_3 be three odd circuits. Since S intersects each of C_1, C_2, C_3 with odd parity, so does $L = C_1 \triangle C_2 \triangle C_3$. Since M is binary, L is a disjoint union of circuits (see for example [46, Theorem 9.1.2]). One of these circuits must be odd since $|L \cap S|$ is odd. The result now follows from Proposition 6.1. $\square$

Let M be a binary matroid. Any nontrivial binary clutter obtained as the odd circuit clutter of the signed matroid (M, S), for some S, is called a *source* of M. Any nontrivial binary clutter $\mathcal{H}$ such that every circuit of M is of the form $T_1 \triangle T_2$, for $T_1, T_2 \in E(\mathcal{H})$, is called a *lift* of M. One can show that a lift of M is the blocker of a source of M^*.

In a binary matroid, any circuit C and cocircuit D have an even intersection (see for example [46, Theorem 9.1.2]). So, if D is a cocircuit, then (M, S) and $(M, S\triangle D)$ have exactly the same odd circuits.

Remark 6.14. Let (M, S) be a signed matroid and $\mathcal{H}$ the clutter of its odd circuits.
- $\mathcal{H} \setminus e$ is the clutter of odd circuits of the signed matroid $(M \setminus e, S - \{e\})$.
- If $e \notin S$, then $\mathcal{H}/e$ is the clutter of odd circuits of the signed matroid $(M/e, S)$.
- If $e \in S$ is not a loop of M, then $\mathcal{H}/e$ is the clutter of odd circuits of the signed matroid $(M/e, S\triangle D)$ where D is any cocircuit containing e.
- If $e \in S$ is a loop of M, then $\mathcal{H}/e$ is a trivial clutter.

Given a nontrivial binary clutter $\mathcal{H}$, the minimal sets in $E(\mathcal{H}) \cup \{T_1 \triangle T_2: T_1, T_2 \in E(\mathcal{H})\}$ form the circuits of a binary matroid $u(\mathcal{H})$. This binary matroid is called the *up matroid* of $\mathcal{H}$. Since $\mathcal{H}$ is binary, the minimal transversals of $\mathcal{H}$

intersect with odd parity exactly the circuits of $u(\mathcal{H})$ that are edges of $\mathcal{H}$. It follows that $\mathcal{H}$ is the clutter of odd circuits of the signed matroid $(u(\mathcal{H}), S)$ where S is any minimal transversal of $\mathcal{H}$. Moreover, this representation is essentially unique (see for example [11]):

Proposition 6.15. *Let (M, S) and (M', S') be signed matroids that have the same clutter of odd circuits $\mathcal{H}$. If M is a 2-connected matroid and $\mathcal{H}$ is a nontrivial clutter, then $M = M' = u(\mathcal{H})$.*

To prove this, we use the following result of Lehman [31] (see [46, Theorem 4.3.2 or Exercise 9 of Section 9.3]).

Theorem 6.16 (Lehman [31]). *Let t be an element of a 2-connected binary matroid M. The circuits of M not containing t are of the form $C_1 \triangle C_2$ where C_1 and C_2 are circuits of M containing t.*

Proof of Proposition 6.15. Let N be the binary matroid with elements $V(M) \cup \{t\}$ and circuits $\Gamma = C$ when C is an even circuit of (M, S) and $\Gamma = C \cup \{t\}$ when C is an odd circuit of (M, S). Define N' similarly from (M', S'). Since $\mathcal{H}$ is nontrivial, at least one circuit of N contains t and some $x \neq t$. Since M is 2-connected, for every pair of elements in $V(M)$, there is a circuit of M containing both. So for x and any $v \in V(N)$, there is a circuit of N containing both. It follows that, for any pair of elements in $V(N)$, there is a circuit containing both. So N is 2-connected. Furthermore, every $v \in V(\mathcal{H})$ belongs to an edge of $\mathcal{H}$. So N' is 2-connected as well. By Theorem 6.16, a 2-connected matroid is uniquely determined by the set of circuits containing any fixed element. In particular, N and N' are uniquely determined by the circuits containing t. This implies $N = N'$. Since $M = N/t$ and $M' = N'/t$, it follows that $M = M' = u(\mathcal{H})$. $\square$

Proposition 6.17 (Novick and Sebö [40]). *A binary clutter $\mathcal{H}$ is the odd cycle clutter of a signed graph if and only if $u(\mathcal{H})$ is a graphic matroid.*

A binary clutter $\mathcal{H}$ is the T-cut clutter of a graft if and only if $u(\mathcal{H})$ is a cographic matroid.

The next result relates the minors of the matroid $u(\mathcal{H})$ to the minors of the clutter $\mathcal{H}$. For a clutter $\mathcal{H}$ and $v \notin V(\mathcal{H})$, the clutter $\mathcal{H}^+$ has vertex set $V(\mathcal{H}) \cup \{v\}$ and edge set $\{A \cup \{v\}: A \in E(\mathcal{H})\}$.

Theorem 6.18 (Cornuéjols and Guenin [11]). *Let $\mathcal{H}$ be a nontrivial binary clutter such that its up matroid $u(\mathcal{H})$ is 2-connected, and let N be a 2-connected binary matroid. Then $u(\mathcal{H})$ has N as a minor if and only if $\mathcal{H}$ has $\mathcal{H}_1$ or $\mathcal{H}_2^+$ as a minor, where $\mathcal{H}_1$ is a source of N and $\mathcal{H}_2$ is a lift of N.*

To prove this, we use the following result of Brylawski [3] and Seymour [52] (see [46, Proposition 4.3.6]).

Theorem 6.19 (Brylawski [3]; Seymour [52]). *Let M be a 2-connected matroid and N a 2-connected minor of M. For any $i \in V(M) - V(N)$, at least one of $M \setminus i$ or M/i is 2-connected and has N as a minor.*

Proof of Theorem 6.18. $\mathcal{H}$ is the clutter of odd circuits of the signed matroid (M, S) where $M = u(\mathcal{H})$ and S is a minimal transversal of $\mathcal{H}$.

Suppose first that $\mathcal{H}$ has a minor $\mathcal{H}_1$ that is a source of N. Then $\mathcal{H}_1$ is nontrivial and it follows from Remark 6.14 that $\mathcal{H}_1$ is the clutter of odd circuits of a signed matroid (N', S') where N' is a minor of M. Since $\mathcal{H}_1$ is nontrivial and N is 2-connected, $N = N' = u(\mathcal{H}_1)$ by Proposition 6.15. So N is a minor of M.

Suppose now that $\mathcal{H}$ has a minor $\mathcal{H}_2^+$ where $\mathcal{H}_2$ is a lift of N. Let t be the vertex of $V(\mathcal{H}_2^+) - V(\mathcal{H}_2)$. Since $\mathcal{H}_2^+$ is a nontrivial minor of $\mathcal{H}$, it is the clutter of odd circuits of a signed matroid (N', S') where N' is a minor of M. Since $u(\mathcal{H}_2^+)$ is 2-connected, $N' = u(\mathcal{H}_2^+)$ by Proposition 6.15. So N' is 2-connected. Therefore, by Theorem 6.16 and the definition of lift, $N = N' \setminus t$. So N is a minor of M.

Now we prove the converse. Suppose that M has N as minor and does not satisfy the theorem. Let $\mathcal{H}$ be such a counterexample with smallest number of vertices. Clearly, N is a proper minor of M as otherwise $u(\mathcal{H}) = N$, i.e. $\mathcal{H}$ is a source of N. By Theorem 6.19, for every $i \in V(M) - V(N)$, one of $M \setminus i$ and M/i is 2-connected and has N as a minor. Suppose first that M/i is 2-connected and has an N minor. Since M is 2-connected, i is not a loop of M and therefore $\mathcal{H}/i$ is nontrivial by Remark 6.14, a contradiction to the choice of $\mathcal{H}$ with smallest number of vertices. Thus, for every $i \in V(M) - V(N)$, $M \setminus i$ is 2-connected and has an N minor. By minimality, $\mathcal{H} \setminus i$ must be trivial. It follows from Remark 6.14 that all odd circuits of (M, S) use i. As $M = u(\mathcal{H})$, even circuits of M do not use i.

We claim that $V(M) - V(N) = \{i\}$. Suppose not and let $j \neq i$ be an element of $V(M) - V(N)$. The set of circuits of (M, S) using j is exactly the set of odd circuits. It follows that the elements i, j must be in series in M. But then $M \setminus i$ is not connected, a contradiction.

Therefore $V(M) - V(N) = \{i\}$ and $M \setminus i = N$. As the circuits of (M, S) using i are exactly the odd circuits of (M, S), it follows that column i of $\mathcal{H}$ consists of all 1's, i.e. $\mathcal{H} = \mathcal{H}_2^+$. By Theorem 6.16 applied to i and M, every circuit of N is of the form $T_1 \triangle T_2$ where $T_1, T_2 \in E(\mathcal{H}_2)$. So $\mathcal{H}_2$ is a lift of N. □

6.4. k-Connectedness of binary clutters

A binary clutter $\mathcal{H}$ has a *k-separation* if $u(\mathcal{H})$ has a k-separation, i.e. there exists a partition (U_1, U_2) of $V(\mathcal{H})$ such that $|U_1| \geqslant k$, $|U_2| \geqslant k$ and $r(U_1) + r(U_2) \leqslant r(V(\mathcal{H})) + k - 1$. The k-separation is *strict* if $|U_1| > k$, $|U_2| > k$. The binary clutter $\mathcal{H}$ is *k-connected* if it has no $(k-1)$-separation. It is *internally k-connected* if it has no strict $(k-1)$-separation.

Theorem 6.20 (Cornuéjols and Guenin [11]). *Minimally nonideal binary clutters are 3-connected.*

The minimally nonideal binary clutter F_7 has a 3-separation. So minimally nonideal clutters are not 4-connected in general. However they are internally 4-connected.

Theorem 6.21 (Cornuéjols and Guenin [11]). *Minimally nonideal binary clutters are internally 4-connected.*

Conjecture 6.22. *Minimally nonideal binary clutters are internally 5-connected.*

Let Q_6 be the clutter where $V(Q_6)$ is the set of edges of K_4 and $E(Q_6)$ the set of triangles of K_4. The next result proves Seymour's conjecture (Conjecture 6.9) for the class of clutters that do not have $\mathcal{Q}_6^+$ or $b(\mathcal{Q}_6)^+$ as a minor. Recall that the definition of $\mathcal{H}^+$ is given in Section 6.3.

Theorem 6.23 (Cornuéjols and Guenin [11]). *A binary clutter is ideal if it does not have $\mathcal{F}_7$, $\mathcal{O}_{K_5}$, $b(\mathcal{O}_{K_5})$, $\mathcal{Q}_6^+$, or $b(\mathcal{Q}_6)^+$ as a minor.*

Proof. It suffices to show that every *mni* clutter $\mathcal{H}$ contains one of the minors in the statement of the theorem. □

Claim 1. *The result holds if $u(\mathcal{H})$ has no F_7^* minor.*

Proof. When $u(\mathcal{H}) = R_{10}$, then $\mathcal{H}$ is one of the sources of R_{10}. We leave it as an exercise to show that R_{10} has 6 sources. One such source is $b(\mathcal{O}_{K_5})$ and the other five are ideal.

When $u(\mathcal{H})$ is graphic, then $\mathcal{H}$ is ideal if and only if $\mathcal{H}$ has no $\mathcal{O}_{K_5}$ minor, by Proposition 6.17 of Novick–Sebö and Guenin's theorem (Theorem 5.3).

When $u(\mathcal{H})$ is cographic, then $\mathcal{H}$ is ideal, by Proposition 6.17 of Novick–Sebö and the Edmonds–Johnson theorem (Theorem 3.1).

By the connectivity results (Theorems 6.20 and 6.21), $u(\mathcal{H})$ is 3-connected and internally 4-connected. So, by Seymour's theorem (Theorem 6.11), the result holds when $u(\mathcal{H})$ is a regular matroid.

Now consider the case when $u(\mathcal{H})$ is not regular. Another theorem of Seymour (Theorem 6.12) shows that $u(\mathcal{H}) = F_7$. So $\mathcal{H}$ is a source of F_7. It is easy to verify that F_7 has three sources. Two of these sources are ideal and the third is the clutter $\mathcal{F}_7$. So the result holds. □

Claim 2. *The result holds if $u(\mathcal{H})$ has an F_7^* minor.*

Proof. By Theorem 6.18, $u(\mathcal{H})$ has an F_7^* minor if and only if $\mathcal{H}$ has $\mathcal{H}_1$ or $\mathcal{H}_2^+$ as a minor, where $\mathcal{H}_1$ is a source of F_7^* and $\mathcal{H}_2$ is a lift of F_7^*. One can easily verify that F_7^* has one source and three lifts. The source is Q_6^+, which is one of the excluded minors in the statement of the theorem. For the three lifts $\mathcal{H}_2$ of F_7^*, one can check that $\mathcal{H}_2^+$ contains $\mathcal{F}_7$, $\mathcal{Q}_6^+$ and $b(\mathcal{Q}_6)^+$ as minors, respectively, which are excluded minors in the statement of the theorem. □

The class of clutters of T-cuts is closed under minor taking. Moreover, it is not hard to check that none of the five excluded minors of Theorem 6.23 are clutters of T-cuts. Thus Theorem 6.23 implies that clutters of T-cuts are ideal, and thus that their blocker, the clutters of T-joins are ideal. Hence Theorem 6.23 implies the Edmonds–Johnson theorem (Theorem 3.1). Similarly, the class of clutters of odd circuits is closed under minor taking. Moreover, it can be shown that $\mathcal{O}_{K_5}$ is the only clutter of odd circuits among the five excluded minors. It follows that Theorem 6.23 also implies Guenin's theorem (Theorem 5.3). Note, however, that the proof of Theorem 6.23 uses these two results.

7. Ideal 0, ±1 matrices

The concept of ideal $0,1$ matrix can be extended to a $0,\pm1$ matrix. Given a $0,\pm1$ matrix A, denote by $n(A)$ the column vector whose ith component is the number of -1's in the ith row of matrix A. The $0,\pm1$ matrix A is *ideal* if its fractional generalized set covering polytope $Q(A) = \{x\colon Ax \geqslant 1 - n(A), 0 \leqslant x \leqslant 1\}$ only has integral extreme points.

7.1. Propositional logic

In propositional logic, *atomic propositions* $x_1, \ldots, x_j, \ldots, x_n$ can be either *true* or *false*. A *truth assignment* is an assignment of "true" or "false" to every atomic proposition. A *literal* is an atomic proposition x_j or its negation $\neg x_j$. A *clause* is a disjunction of literals and is *satisfied* by a given truth assignment if at least one of its literals is true.

A survey of the connections between propositional logic and integer programming can be found in Hooker [28], Truemper [60] or Chandru and Hooker [5].

A truth assignment satisfies the set S of clauses

$$\bigvee_{j \in P_i} x_j \vee \left(\bigvee_{j \in N_i} \neg x_j \right) \quad \text{for all } i \in S$$

if and only if the corresponding $0,1$ vector satisfies the system of inequalities

$$\sum_{j \in P_i} x_j - \sum_{j \in N_i} x_j \geqslant 1 - |N_i| \quad \text{for all } i \in S.$$

The above system of inequalities is of the form

$$Ax \geqslant 1 - n(A). \tag{13}$$

Given a set S of clauses, the *satisfiability problem* (SAT) consists in finding a truth assignment that satisfies all the clauses in S or show that none exists. Equivalently, SAT consists in finding a $0,1$ solution x to (13) or show that none exists.

Given a set S of clauses (the premises) and a clause C (the conclusion), *logical inference* in propositional logic consists of deciding whether every truth assignment that satisfies all the clauses in S also satisfies the conclusion C.

To the clause C, using transformation (13), we associate an inequality

$$cx \geqslant 1 - n(c),$$

where c is a $0, +1, -1$ vector. Therefore C cannot be deduced from S if and only if the integer program

$$\min\{cx\colon Ax \geqslant \mathbf{1} - n(A), x \in \{0,1\}^n\} \tag{14}$$

has a solution with value $-n(c)$.

The above problems are NP-hard in general but can be solved efficiently for Horn clauses, clauses with at most two literals and several related classes [4,59]. A set S of clauses is *ideal* if the corresponding $0, \pm 1$ matrix A defined in (13) is ideal. If S is ideal, it follows from the definition that the satisfiability and logical inference problems can be solved by linear programming.

Remark 7.1. *Let S be an ideal set of clauses. If every clause of S contains more than one literal then, for every atomic proposition x_j, there exist at least two truth assignments satisfying S, one in which x_j is true and one in which x_j is false.*

Proof. Since the point $x_j = 1/2$, $j = 1, \ldots, n$ belongs to the polytope $Q(A) = \{x\colon Ax \geqslant \mathbf{1} - n(A), 0 \leqslant x \leqslant \mathbf{1}\}$ and $Q(A)$ is an integral polytope, then the above point can be expressed as a convex combination of $0, 1$ vectors in $Q(A)$. Clearly, for every index j, there exists in the convex combination a $0, 1$ vector with $x_j = 0$ and another with $x_j = 1$. $\qquad \square$

Let S be an ideal set of clauses. A consequence of Remark 7.1 is that the satisfiability problem can be solved more efficiently than by general linear programming.

Theorem 7.2 (Conforti and Cornuéjols [7]). *Let S be an ideal set of clauses. Then S is satisfiable if and only if a recursive application of the following procedure stops with an empty set of clauses.*

Recursive step.

If $S = \emptyset$, then S is satisfiable.

If S contains a clause C with a single literal (unit clause), *set the corresponding atomic proposition x_j so that C is satisfied. Eliminate from S all clauses that become satisfied and remove x_j from all the other clauses. If a clause becomes empty, then S is not satisfiable* (unit resolution).

If every clause in S contains at least two literals, choose any atomic proposition x_j appearing in a clause of S and add to S an arbitrary clause x_j or $\neg x_j$.

It is easy to modify the above algorithm in order to solve the logical inference problem when S is an ideal set of clauses.

7.2. Relating ideal $0, \pm 1$ matrices to ideal $0, 1$ matrices

This section follows [8]. Hooker [29] was the first to relate idealness of a $0, \pm 1$ matrix to that of a family of $0, 1$ matrices. These results were strengthened by Guenin [25] and by Nobili, Sassano [38].

A *prime implication* of $Q(A)$ is a generalized set covering inequality $ax \geqslant 1 - n(a)$ that is satisfied by all the $0, 1$ vectors in $Q(A)$ but is not dominated by any other such generalized set covering inequality. A *row monotonization* of A is any $0, 1$ matrix obtained from a row submatrix of A by multiplying some of its columns by -1. A row monotonization of A is *maximal* if it is not a proper submatrix of any row monotonization of A.

Theorem 7.3 (Hooker [29]). *If A is a $0, \pm 1$ matrix such that $Q(A)$ contains all of its prime implications, then A is ideal if and only if all the maximal row monotonizations of A are ideal $0, 1$ matrices.*

In [25], the idealness of a $0, \pm 1$ matrix A is linked to the idealness of a single $0, 1$ matrix as follows. Given a $0, \pm 1$ matrix A, let P and R be $0, 1$ matrices of the same dimension as A, such that $P_{ij} = 1$ if and only if $A_{ij} = 1$, and $R_{ij} = 1$ if and only if $A_{ij} = -1$. The matrix

$$D_A = \left[\begin{array}{c|c} P & R \\ \hline I & I \end{array} \right]$$

is the $0, 1$ *extension* of A. Note that the transformation $x^+ = x$ and $x^- = 1 - x$ maps every vector x in $Q(A)$ into a vector in $\{(x^+, x^-) \geqslant 0 \colon Px^+ + Rx^- \geqslant 1, x^+ + x^- = 1\}$. So $Q(A)$ corresponds to the face of $Q(D_A)$, obtained by setting the inequalites $x^+ + x^- \geqslant 1$ at equality.

Theorem 7.4 (Guenin [25]). *Let A be a $0, \pm 1$ matrix such that $Q(A)$ contains all of its prime implications. Then A is ideal if and only if the $0, 1$ matrix D_A is ideal.*

Furthermore A is ideal if and only if $\min\{cx \colon x \in Q(A)\}$ has an integer optimum for every vector $c \in \{0, \pm 1, \pm \infty\}^n$.

In [38], a condition for a $0, \pm 1$ matrix A to be ideal, without assuming that $Q(A)$ contains all of its prime implications is given as follows. Given a $0, \pm 1$ matrix A, let a^1 and a^2 be two rows of A, such that there is one index k such that $a_k^1 a_k^2 = -1$ and, for all $j \neq k$, $a_j^1 a_j^2 = 0$. A *disjoint implication* of A is the $0, \pm 1$ vector $a^1 + a^2$. The matrix A^+ obtained by recursively adding all disjoint implications and removing all dominated rows is called the *disjoint completion* of A.

Theorem 7.5 (Nobili and Sassano [38]). *Let A be a $0, \pm 1$ matrix. Then A is ideal if and only if D_{A^+} is an ideal $0, 1$ matrix, where A^+ is the disjoint completion of A.*

Let J be a subset of columns of a $0, \pm 1$ matrix A. The *deletion* of J consists of removing all columns in J, all rows with at least one 1 in a column of J and rows that become dominated. The *contraction* of J consists of removing all columns in J, all rows with at least one -1 in a column of J and rows that become dominated. The *semi-deletion* of J consists of removing all rows with a 1 in at least one column of J and then all zero columns. The *semi-contraction* of J consists of removing all rows with at least one -1 in a column of J and then all zero columns.

Nobili and Sassano define a *weak minor* of a $0, \pm 1$ matrix A to be any submatrix that can be obtained from A by a sequence of deletions, contractions, semi-deletions and semi-contractions. They define A to be *minimally nonideal* if A is not ideal but every weak minor of A is ideal. The usefulness of this concept comes from the fact that a $0, \pm 1$ matrix A is minimally nonideal if and only D_A is a minimally nonideal $0, 1$ matrix.

For $n \geqslant 3$, the following $n \times n$ $0, \pm 1$ matrix, denoted $\tilde{J}_n$, is minimally nonideal:

$$\tilde{J}_n = \begin{pmatrix} -1 & 1 & 1 & 1 & 1 & 1 \\ 1 & 1 & 0 & 0 & 0 & 0 \\ 1 & 0 & 1 & 0 & 0 & 0 \\ 1 & 0 & 0 & 1 & 0 & 0 \\ 1 & 0 & 0 & 0 & 1 & 0 \\ 1 & 0 & 0 & 0 & 0 & 1 \end{pmatrix}.$$

Nobili and Sassano [38] give the following characterization of minimally nonideal $0, \pm 1$ matrices.

Theorem 7.6 (Nobili and Sassano [38]). *Let A be a $0, \pm 1$ matrix with n columns. Then A is minimally nonideal if and only if A is a switching of $\tilde{J}_n$, after permutation of rows and columns, or A is a switching of a minimally nonideal $0, 1$ matrix or A contains an $n \times n$ submatrix B with two nonzeroes per row and per column and $\det(B) = \pm 2$ and all rows in A but not in B have at least three nonzeroes.*

References

[1] F. Barahona, The Max Cut problem in graphs not contractible to K_5, Oper. Res. Lett. 2 (1983) 107–111.
[2] W.G. Bridges, H.J. Ryser, Combinatorial designs and related systems, J. Algebra 13 (1969) 432–446.
[3] T.H. Brylawski, A decomposition for combinatorial geometries, Trans. Amer. Math. Soc. 171 (1972) 235–282.
[4] V. Chandru, J.N. Hooker, Extended Horn sets in propositional logic, J. ACM 38 (1991) 205–221.
[5] V. Chandru, J.N. Hooker, Optimization Methods for Logical Inference, Wiley, New York, 1999.
[6] M. Conforti, G. Cornuéjols, Clutters that pack and the Max Flow Min Cut property: a conjecture, in: W.R. Pulleyblank, F.B. Shepherd (Eds.), The Fourth Bellairs Workshop on Combinatorial Optimization, 1993.
[7] M. Conforti, G. Cornuéjols, A class of logic problems solvable by linear programming, J. ACM 42 (1995) 1107–1113.
[8] M. Conforti, G. Cornuéjols, A. Kapoor, K. Vušković, Perfect, ideal and balanced matrices, in: M. Dell'Amico, F. Maffioli, S. Martello (Eds.), Annotated Bibliographies in Combinatorial Optimization, Wiley, New York, 1997, 81–94.
[9] M. Conforti, E.L. Johnson, Two min-max theorems for graphs noncontractible to a four wheel, preprint, IBM Thomas J. Watson Research Center, 1987.
[10] G. Cornuéjols, Combinatorial optimization: packing and covering, CBMS-NSF Regional Conference Series in Applied Mathematics, 74, SIAM, Philadelphia, PA, 2001.
[11] G. Cornuéjols, B. Guenin, Ideal binary clutters, connectivity and a conjecture of Seymour, preprint, Fields Institute, Toronto, 2000, SIAM J. Discrete Math., to appear.
[12] G. Cornuéjols, B. Guenin, F. Margot, The packing property, Math. Program. A 89 (2000) 113–126.

[13] G. Cornuéjols, B. Novick, Ideal 0,1 matrices, J. Combin. Theory B 60 (1994) 145–157.

[14] N.G. de Bruijn. P. Erdös, On a combinatorial problem, Proceedings of Kon. Ned. Akad. v. Wetensch, Vol. 51, pp. 1277–1279.

[15] G. Ding, personal communication, 1989.

[16] R.J. Duffin, The extremal length of a network, J. Math. Anal. Appl. 5 (1962) 200–215.

[17] J. Edmonds, D.R. Fulkerson, Bottleneck Extrema, J. Combin. Theory B 8 (1970) 299–306.

[18] J. Edmonds, R. Giles, A min-max relation for submodular functions on graphs, Ann. Discrete Math. 1 (1977) 185–204.

[19] J. Edmonds, E.L. Johnson, Matchings, Euler tours and the Chinese postman problem, Math. Program. 5 (1973) 88–124.

[20] L.R. Ford Jr., D.R. Fulkerson, Flows in Networks, Princeton University Press, Princeton, NJ, 1962.

[21] A. Frank, Edge-disjoint paths in planar graphs, J. Combin. Theory B 39 (1985) 164–178.

[22] G.S. Gasparyan, M. Preissmann, A. Sebö, Imperfect and nonideal clutters: a common approach, preprint, 1999.

[23] J. Geelen, B. Guenin, Packing odd-circuits, preprint, 2001, J. Combin. Theory B, to appear.

[24] M.X. Goemans, V.S. Ramakrishnan, Minimizing submodular functions over families of sets, Combinatorica 15 (1995) 499–513.

[25] B. Guenin, Perfect and ideal 0,±1 matrices, Math. Oper. Res. 23 (1998) 322–338.

[26] B. Guenin, A characterization of weakly bipartite graphs, in: R.E. Bixby, E.A. Boyd, R.Z. Rios-Mercado (Eds.), Integer Programming and Combinatorial Optimization, Lecture Notes in Computer Science, Vol. 1412, Springer, Berlin, 1998, pp. 9–22, J. Combin. Theory B, to appear.

[27] B. Guenin, Integral polyhedra related to even cycle and even cut matroids, preprint, Fields Institute, Toronto, 2000.

[28] J.N. Hooker, A quantitative approach to logical inference, Decision Support Systems 4 (1988) 45–69.

[29] J. Hooker, Resolution and the Integrality of Satisfiability Polytopes, Math. Program. 74 (1996) 1–10.

[30] T.C. Hu, Multicommodity network flows, Oper. Res. 11 (1963) 344–360.

[31] A. Lehman, A solution of the Shannon switching game, J. SIAM 12 (4) (1964) 687–725.

[32] A. Lehman, On the width–length inequality, Mimeographic Notes, 1965, published: Math. Program. 17 (1979) 403–417.

[33] A. Lehman, On the width–length inequality and degenerate projective planes, unpublished manuscript, 1981, published: W. Cook, P.D. Seymour (Eds.), Polyhedral Combinatorics, DIMACS Series in Discrete Mathematics and Theoretical Computer Science 1, American Mathematical Society, Providence, RI, 1990, pp. 101–105.

[34] M.V. Lomonosov, Combinatorial approaches to multiflow problems, Discrete Appl. Math. 11 (1985) 1–94.

[35] C. Lütolf, F. Margot, A catalog of minimally nonideal matrices, Math. Methods Oper. Res. 47 (1998) 221–241.

[36] E.F. Moore, C.E. Shannon, Reliable circuits using less reliable relays, J. Franklin Instit. 262 (1956) 204–205.

[37] G.L. Nemhauser, L.A. Wolsey, Integer and Combinatorial Optimization, Wiley, New York, 1988.

[38] P. Nobili, A. Sassano, (0,±1) ideal matrices, Math. Program. 80 (1998) 265–281.

[39] B. Novick, Ideal 0,1 matrices, Ph.D. Dissertation, Carnegie Mellon University, 1990.

[40] B. Novick, A. Sebö, On combinatorial properties of binary spaces, in: E. Balas, J. Clausen (Eds.), Integer Programming and Combinatorial Optimization, Lecture Notes in Computer Science, Vol. 920, Springer, Berlin, 1995, 212–227.

[41] H. Okamura, P.D. Seymour, Multicommodity flows in planar graphs, J. Combin. Theory B 31 (1981) 75–81.

[42] G.I. Orlova, Y.G. Dorfman, Finding the maximum cut in a graph, Izvestija Akademii Nauk SSSR, Tehničeskaja Kibernetika 3 (1972) 155–159 (in Russian), English translation: Eng. Cybernet. 10 (1972) 502–506.

[43] J.G. Oxley, Matroid Theory, Oxford University Press, Oxford, 1992.

[44] M.W. Padberg, Lehman's forbidden minor characterization of ideal 0,1 matrices, Discrete Math. 111 (1993) 409–420.

[45] B.A. Papernov, Feasibility of multicommodity flows, in: A.A. Friedman (Ed.), Studies in Discrete Optimization, Nauka, Moscow, 1976, pp. 230–261 (in Russian).

[46] W.R. Pulleyblank, personal communication, 1992.

[47] L. Qi, On the set covering polytope and the MFMC-clutter, preprint, School of Mathematics, University of New South Wales, 1989.

[48] N. Robertson, D.P. Sanders, P.D. Seymour, R. Thomas, in preparation.

[49] A. Schrijver, Theory of Linear and Integer Programming, Wiley, New York, 1986.

[50] A. Schrijver, A short proof of Guenin's characterization of weakly bipartite graphs, preprint, 2001, J. Combin. Theory B, to appear.

[51] P. Seymour, The forbidden minors of binary clutters, J. Combin. Theory B 22 (1976) 356–360.

[52] P.D. Seymour, A note on the production of matroid minors, J. Combin. Theory B 22 (1977) 289–295.

[53] P. Seymour, The matroids with the Max-Flow Min-Cut property, J. Combin. Theory B 23 (1977) 189–222.

[54] P. Seymour, A two-commodity cut theorem, Discrete Math. 23 (1978) 177–181.

[55] P. Seymour, Decomposition of regular matroids, J. Combin. Theory B 28 (1980) 305–359.

[56] P. Seymour, Matroids and multicommodity flows, European J. Combin. 2 (1981) 257–290.

[57] P. Seymour, On odd cuts and plane multicommodity flows, Proc. London Math. Soc. 42 (1981) 178–192.

[58] P.D. Seymour, On Lehman's width–length characterization, in: W. Cook, P.D. Seymour (Eds.), Polyhedral Combinatorics, DIMACS Series in Discrete Mathematics and Theoretical Computer Science 1, American Mathematical Society, Providence, RI, 1990, pp. 107–117.

[59] K. Truemper, Polynomial theorem proving I. Central matrices, Technical Report UTDCS 34-90, 1990.

[60] K. Truemper, Effective Logic Computation, Wiley, New York, 1998.

[61] W.T. Tutte, A homotopy theorem for matroids I, II, Trans. Amer. Math. Soc. 88 (1958) 144–160, 161–174.

[62] K. Wagner, Über eine Eigenschaft der ebenen Komplexe, Math. Ann. 114 (1937) 570–590.

For further reading

G.B. Dantzig, Linear Programming and Extensions, Princeton University Press, Princeton, 1963.

M. Dell'Amico, F. Maffioli, S. Martello (Eds.), Annotated Bibliographies in Combinatorial Optimization, Wiley, New York, 1997.

B. Guenin, On packing and covering polyhedra, Ph.D. Dissertation, Carnegie Mellon University, 1998.

P.D. Seymour, A forbidden minor characterization of matroid ports, Quart. J. Math. Oxford (2) 27 (1976) 407–413.

D.B. West, Introduction to Graph Theory, Prentice-Hall, Englewood Cliffs, NJ, 1996.

ELSEVIER

Discrete Applied Mathematics 123 (2002) 339–361

DISCRETE
APPLIED
MATHEMATICS

Production planning problems in printed circuit board assembly

Yves Crama[a,*], Joris van de Klundert[b], Frits C.R. Spieksma[b]

[a] *Ecole d'Administration des Affaires, University of Liège, Boulevard du Rectorat 7 (B31),
4000 Liège, Belgium*
[b] *Department of Applied Economics, Katholieke Universiteit Leuven, Naamsestraat 69, B-3000 Leuven,
Belgium*

Received 5 January 2000; accepted 16 October 2000

Abstract

This survey describes some of the main optimization problems arising in the context of production planning for the assembly of printed circuit boards. The discussion is structured around
a hierarchical decomposition of the planning process into distinct optimization subproblems, ad-
dressing issues such as the assignment of board types to machine groups, the allocation of
component feeders to individual machines, the determination of optimal production sequences,
etc. The paper reviews the literature on this topic with an emphasis on the most recent developments, on the fundamental structure of the mathematical models and on the relation between
these models and some 'environmental' variables such as the layout of the shop or the product
mix. © 2002 Elsevier Science B.V. All rights reserved.

Keywords: Production planning; Scheduling; Sequencing; Printed circuit boards; Integer programming
models

1. Introduction, facts and figures

The assembly of printed circuit boards (PCBs) has generated a huge amount of
industrial activity over the last 20 years. PCBs are consumed as inputs by three major
industrial sectors: computers, telecommunications and consumer electronics represented
72.5% of the total consumption in 1998 [47]. Although it seems difficult to gather
precise figures, Nakahara [47] indicates that world PCB production grew by 5% in

---

* Corresponding author. Tel.: +32-4-366-3077; fax: +32-4-366-2821.
E-mail addresses: y.crama@ulg.ac.be (Y. Crama), j.vandeklundert@math.unimaas.nl (J. van de Klundert),
frits.spieksma@econ.kuleuven.ac.be (F.C.R. Spieksma).

0166-218X/02/$ - see front matter © 2002 Elsevier Science B.V. All rights reserved.
PII: S0166-218X(01)00345-6

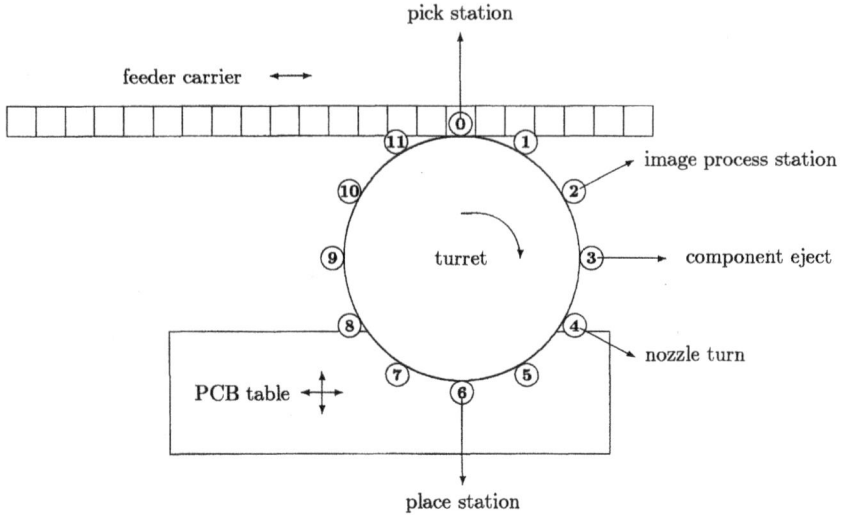

Fig. 1. A machine of the Fuji CP family.

1998, to a total value of roughly $35 billion. The top 15 countries accounted for 92% of this worth, with Japan and the USA producing more than 50% of the total output.

Over the years, PCB production has evolved from a labor-intensive activity to a highly automated one, characterized by steady innovations at the level of design and manufacturing processes. Nowadays, programmed automation has gained the upper-hand in assembly operations. In their description of benchmark PCB assembly factories, Mody et al. [43] estimate that, in industrialized countries, a typical shop features 25–30 machines, for a total equipment value exceeding $1.5 million.

These sophisticated machines perform a large number of high speed, high precision assembly operations requiring various tools and components. Some operating features of the Fuji CPII placement machines are mentioned for instance by Bard et al. [12] (see Section 2 and Fig. 1 for the terminology): turret rotation speed of one station per 0.15 s; table movement speed of 20 mm per 0.15 s; feeder carrier speed of one slot per 0.15 s; duration of picking or placement actions: 0.10 s; placement rate of over 12,000 components/h; error rate of less than 1 in 10,000.

The competition faced by PCB manufacturers creates a need for production efficiency which is achieved—depending on the specific market—by assembling either a few product types in large volumes or a large variety of products in small volumes. Jain et al. [32] compared three of Hewlett–Packard's production sites and describe production characteristics ranging from low mix (less than 20 board types) high volume operation (batches of more than 100 units) to high mix (150 board types) low volume operation (batches of 10 to 25 boards). A detailed discussion of manufacturing flexibility in PCB assembly is provided by Suarez et al. [53], who mention a plant producing only two board models and another one producing more than 2000 different models! The plant studied by Feo et al. [24] assembles 20,000–80,000 boards/month, but Mody et al. [43] consider an output of 40,000 boards/year to be more typical.

All the above features interfere with numerous constraints and conflicting managerial objectives to pose challenging production planning problems. In fact, in the conclusions of their study, Mody et al. [43] point out that PCB manufacturers, both in less developed countries and in newly industrialized countries, will need (among other factors) to increase process efficiency and to master production planning and control in order to improve their competitive situation.

In order to cope effectively with such requirements, decision support systems based on specialized planning and scheduling models may prove a major asset for PCB producers. Many researchers have investigated such models for PCB assembly and have published numerous papers on this topic in the operations research, industrial engineering and production management literature. We are going to review some of this literature, with an emphasis on the most recent developments, on the fundamental structure of the mathematical models and on the relation between these models and some 'environmental' variables such as the layout of the shop or the product mix, with the hope and ambition to provide useful guidance to the reader. For complementary viewpoints or additional information, we refer the reader to excellent previous surveys by McGinnis et al. [43] or Ahmadi [1]. Extensive bibliographic references can also be found on several Internet sites: http://www.econ.kuleuven.ac.be/tew/academic/kwantmet/ members/frits/Bibliography/bibliogr.htm (Crama, van de Klundert and Spieksma), http:// www.Fabtime.com/library.htm (Robinson), http://www.eas.asu.edu/~masml (Fowler and Runger), and http://www.cs.utu.fi/scheduling/Default.htm (Nevalainen et al.).

2. Generic assembly process

Before discussing the fundamental issues involved in the PCB production planning process, it is necessary to give a description of the generic steps involved in the assembly of a printed circuit board.

For our purpose, PCB assembly consists in placing (inserting, mounting) a number of electronic components of prespecified types at prespecified locations on a bare board. Several hundred components of a few distinct types (resistors, capacitors, transistors, integrated circuits, etc.) may be placed on each board.

An automated PCB shop involves several computerized machines (or workstations), possibly with different characteristics, which take care of the assembly operations (see e.g. [24] for a pictorial representation of such a shop). The stations may be linked by a material handling system which allows for some flexibility in routing the boards through the shop. In this case, we will say that the shop is a *flexible* or *decoupled cell*. Most often, however, the machines are laid out into distinct *assembly lines*, or *coupled systems*, and a conveyor connects the machines within each line.

As already mentioned, the placement machines may be of various types. From the point of view of the operations researcher, this is somewhat unfortunate, since the technological characteristics of the equipment influences the nature of some of the planning problems to be solved and the formulation of the associated models. We will have opportunities to return to this point. For the time being, let us settle for a generic description of the placement machines (see e.g. [42] or [23] for more details).

Each machine essentially consists of a *worktable*, a *feeder carrier* (or magazine, or rack) and a *pick-and-place device* (see Fig. 1 for an example). The worktable holds the PCB during the placement operations. Depending on the machine, the table can either be stationary or mobile in the X–Y plane. The components to be placed on the PCB are released by *component feeders* which have to be loaded into the slots of the carrier prior to production. Usually, the carrier can move by translation along an X-axis. Finally, the pick-and-place device allows to retrieve each component from the appropriate feeder and to place it on the board. Very different designs and operating modes exist for the pick-and-place device. Sometimes, it can only move in the Y–Z plane (see e.g. [38]). In other cases, it features 12 workheads arranged circularly on a turret: in each pick-and-place operation, head 0 picks a component while head 6 places another one; thereafter, the device rotates by 30° and a similar operation is repeated (see Fig. 1 from [12,18]). Yet other types of designs are described by Ball and Magazine [10], Ahmadi et al. [3], Leipälä and Nevalainen [37], Crama et al. [19], van Laarhoven and Zijm [35], Francis et al. [26], etc.

McGinnis et al. [42] use the term *machine cycle* to designate a series of consecutive operations beginning with a component retrieval, ending with a component placement and consisting of only one retrieval and one placement. This allows to classify placement machines into two major categories: *sequential* machines are those for which each machine cycle involves exactly one component (the same component is gripped and immediately placed) while *concurrent* machines are those for which each cycle involves the retrieval of one component and the placement of a previously retrieved component (concurrent machines may perform several operations simultaneously). The Fuji CP machine illustrated in Fig. 1 is a concurrent machine.

3. Planning hierarchy

Production planning decisions are frequently formulated in a hierarchical framework where they decompose into long term (strategic), medium term (tactical) and short term (operational) issues. There remains quite a lot of freedom, however, as to the 'best' decomposition to be used in a given situation. The answer to this question depends, among others, on

- characteristics of the product mix (diversity of PCB types, batch sizes, etc.),
- characteristics of the equipment (layout, number of machines, details of the operating mode, etc.),
- managerial policy regarding for instance the frequency of setups or the willingness to redesign the lines on a regular basis.

See e.g. [24] for a global vantage point on the planning process. It should be noted that very similar issues come up in the management of flexible manufacturing systems; see e.g. [50] and a comparison of PCB and FMS environment in Ammons et al. [6].

In this paper, we consider the long-term decisions to be given and we concentrate on tactical and operational decisions. In particular, we assume the *demand mix* and the *shop layout* to be fixed exogenously. Under these conditions, the production planning process must (at least) address the following list of subproblems SP1 to SP8. It must determine:

SP1. an *assignment* of PCB types to product families and to machine groups (cells or lines);

SP2. an *allocation* of component feeders to machines;

SP3. for each PCB type, a *partition* of the set of component locations on this board type, indicating which components are going to be placed by each machine;

SP4. for each machine group, a *sequence* of the PCB types, indicating in which order the board types will be produced on these machines;

SP5. for each machine, the *location of feeders* on the carrier;

SP6. for each pair consisting of a machine and a PCB type, a *component placement sequence*, that is a sequence of the placement operations to be performed by the machine on this board type;

SP7. for each pair consisting of a machine and a PCB type, a *component retrieval plan*, that is, for each component on the board, a rule indicating from which feeder this component should be retrieved;

SP8. for each pair consisting of a machine and a PCB type, a *motion control specification*, that is, for each component, a specification of where the pick-and-place device should be located when it picks or places the component.

(Alternative hierarchical decomposition schemes have been proposed by various authors; see e.g. [1,43,56], etc.)

Observe that problem SP1 is posed at the level of the whole assembly shop and involves all products to be assembled, SP2–SP4 usually arise for each product family at the level of assembly lines or cells, and SP5–SP8 deal with individual machines.

Decisions SP1–SP8 must be made in such a way as to optimize some criterion of production performance. The criterion which is most commonly considered in the literature is *makespan minimization* or, in the context of repetitive assembly, *cycle time minimization*. Other criteria may also be of importance, but are less frequently tackled; for instance, van Zante-de Fokkert and de Kok [56] formulate a variant of SP1 with the objective to minimize the sum of assembly, setup and inventory holding costs.

The above list of decisions covers a wide variety of situations. In any specific one, however, some of the subproblems may become vacuous. For instance, it is quite common to assume that only one feeder is available for each type of component (due to the inventory costs of components). In such a case, subproblems SP3 and SP7 vanish altogether: indeed, subproblem SP3 only arises when a same feeder type is loaded on several machines and subproblem SP7 only arises when a same feeder type is loaded in several slots of a machine.

On the other hand, a host of operational details may encumber the description of the fundamental planning decisions and are frequently omitted in the literature. Some of these details could easily be taken into account, as they only affect the value of certain parameters of the models (for instance, the speed of the pick and place device may depend on the type of the components that it carries). Others, however, may have a significant impact on the formulation and on the complexity of the optimization models (for instance, long translations of the feeder carrier are to be avoided, as they are responsible for additional shocks and wear of the carrier; some feeders may occupy more than one slot on the carrier; etc.).

All in all, however, the major difficulty with the list SP1–SP8 is that all its subproblems are tightly intertwined. This fact has been underlined by virtually all researchers in the field (see e.g. [1]). Not only does the formulation of any subproblem heavily depend on the solution computed for problems of *higher* level, but it also depends, in a very significant way, on the solution of problems of *lower level*. This is true, of course, of any hierarchical decomposition scheme, but appears to be especially troublesome in the present case. As a consequence, several authors have adopted solution procedures which iterate between subproblems, rather than one-pass procedures through the list of decisions.

In this survey, for the ease of exposition, we are going to tackle problems SP1–SP8 in reverse order, starting from detailed scheduling questions to finish with the more encompassing (and arguably, more crucial) tactical questions. Thus, we are successively going to consider single machine single product problems (Section 4), then single machine multi-product problems (Section 5), before we turn to the more realistic multi-machine, multi-product environment and a discussion of issues surrounding setup decisions (Section 6).

4. Single machine, single board type problems

Let us first consider the case where a single PCB type must be repeatedly assembled on a single machine, with the objective of makespan (or cycle time) minimization. In this case, the only subproblems to be solved are:

SP5. feeder location;
SP6. placement sequencing;
SP7. component retrieval;
SP8. motion control.

Van Laarhoven and Zijm [35] emphasize the fact that the latter decisions (as opposed to other planning and scheduling decisions) are directly relevant to the *production preparation* function, which leads to the specification of the numerical control programs guiding the assembly operations for each particular PCB.

Let us now discuss each of these problems in turn, starting at the 'bottom' of the hierarchy.

4.1. Motion control (SP8)

Suppose that feeder locations have been determined, that a component placement sequence is given and that it is known for each location where the component to be placed must be retrieved from (that is, a component retrieval plan is known). In this situation, there may still remain one decision left to make: for placement machines that feature a pick-and place-device that can move in the X–Y plane, as well as a rack and a table that can move in the X-direction, one must determine where the device meets the rack (resp. the board) to pick (resp. to place) the appropriate components. Greedy approaches that avoid waiting times for the pick-and-place device are suggested in Su et al. [52] and Wang et al. [59]. These studies also demonstrate the potential makespan gain when allowing non-static pick and place points versus static ones and try to compute placement sequences and feeder locations (see Subsections 4.3) that minimize makespan.

4.2. Component retrieval (SP7)

Assume now that feeder locations and a component placement sequence have been determined. If several component feeders of a same type have been assigned to more than one carrier slot, it becomes necessary to decide from which feeder each component should be retrieved. Of course, different decisions for a specific component may result in different assembly makespans for the board. This issue is raised by Bard et al. [12] for the Fuji CPII machine (see Fig. 1) and is further investigated by Crama et al. [17]. It is also briefly mentioned by Ahmadi et al. [2].

The complexity of the component retrieval problem depends very much on the modus operandi of the placement machines. For most sequential machines, it can be modeled and solved as a shortest path problem. The same holds true for the Fuji CPII machine if the start of a pick activity coincides with the start of a place activity. However, the problem becomes much less trivial when we lift this (restrictive) assumption. Crama et al. [17] show that the problem can still be solved in polynomial time by dynamic programming, but that a slight generalization is already NP-hard.

4.3. Feeder location and placement sequencing (SP5 and SP6)

Starting with [22], numerous researchers have investigated the joint problem of feeder location and placement sequencing. Let us sketch a formulation of this problem for a sequential machine. We let n denote the number of components to be placed, $f(i)$ denote the feeder delivering component i ($i = 1, \ldots, n$) and C denote the number of slots available in the rack. The 0–1 decision variables are

$x_{ij} = 1$ iff component j is placed directly after component i ($i, j = 1, \ldots, n$),

$y_{f(i),s} = 1$ iff a feeder for component i is stored in slot s ($i = 1, \ldots, n$,

$s = 1, \ldots, C$).

Using these variables we can write down the following model:

$$\text{minimize} \sum_{i=1}^{n} \sum_{j=1}^{n} \sum_{s=1}^{C} c_{ijs} x_{ij} y_{f(j),s} \tag{1}$$

s.t. x describes a Hamiltonian path, $\tag{2}$

y describes a feasible assignment, $\tag{3}$

where c_{ijs} denotes the time elapsed between placing component i and placing component j when the feeder $f(j)$ is stored in slot s. For any fixed assignment of feeders to carrier slots, the placement sequencing problem is (essentially) a traveling salesman problem or shortest Hamiltonian path problem (this is true for sequential as well as for concurrent machines). It is easy to understand, however, that the 'distance' or travel time between successive placements is influenced by the location of the feeders, since a 'pick' operation takes place between successive insertions. Conversely, given any sequence of placement operations, the feeder location problem displays the structure of a linear (or, for some types of machines, quadratic) assignment problem, where the 'cost' of assigning a feeder to a particular slot depends on the movements to be performed to and from this slot. Alternatively, the feeder location problem can also be modeled as a facility location problem.

These observations motivate a popular algorithmic approach which consists in tackling both problems simultaneously by iterating between (heuristic) solutions of the feeder location problem and the placement sequencing problem. This approach was initiated (in another manufacturing framework) by Walas and Askin [58] and was also used by Leipälä and Nevalainen [37] or by Broad et al. [13] for PANASERT machines, by Crama et al. [19] for CSM-60 placement machines, by Egbelu et al. [23], Foulds and Hamacher [25], Leon and Peters [38], Moyer and Gupta [45], etc. Recently Altinkemer et al. [5] have proposed an integrated model and an algorithm which reduces the solution of (SP5)–(SP6) to a number of vehicle routing problems. If the vehicle routing subproblems are solved within an ε-error guarantee, then the same guarantee holds for the integrated model.

In order to conduct a finer analysis of the theoretical properties of the models, some authors have rather elected to focus on one of the two subproblems: they explicitly assume to have a solution of one of the two problems and investigate the properties of the second one. Ahmadi et al. [2], for instance, consider the feeder location problem for the DYNAPERT placement machine, *given* a component placement sequence (the placement sequence could arise in the course of the iterative procedures mentioned above, or could be obtained by simple traveling salesman heuristics like those described by Gaboune et al. [27]). They show that, in their setting, the feeder location problem is NP-hard and they provide an approximation algorithm with worst-case ratio $\frac{3}{2}$. Bard et al. [12] address a similar problem for the Fuji CPII. They propose a quadratic integer programming formulation which they attack by Lagrangian relaxation techniques. Moyer and Gupta [44] or Dikos et al. [21] also treat the component placement sequence as an input.

Conversely Drezner and Nof [22], Ball and Magazine [10] or van Laarhoven and Zijm [35] assume that the feeder location problem has been computed first (by solving a linear assignment model in which the total placement time of all the components retrieved from a given feeder is roughly approximated). For known feeder locations, the placement sequence problem can then be tackled in a second phase.

Notice that, even for fixed feeder locations, modeling the elapsed time between two successive placements may not be entirely straightforward. Independently of the physical distance between such successive placements, the elapsed time is clearly limited from below by the time required to carry out a series of unavoidable operations (e.g., for Fuji CP machines: pick a component, rotate the turret by $30°$, move the feeder carrier, and so on). This gives rise to so-called 'free' movements, whose execution time is 'masked' by the execution time of unavoidable operations. For concurrent machines, in particular, this results in complex 'distance metrics' in the formulations of the placement sequencing problem, but also raises opportunities for improved sequencing. These aspects are discussed by Ahmadi et al. [2], Ahmadi et al. [3], Bard et al. [12], Crama et al. [18], Egbelu et al. [23], Grotzinger [29], etc.

In simpler cases, the special structure of the distance metrics can sometimes be exploited to derive tailor-made heuristics (see [10,26], etc.). Viczián [57] shows that the algorithm proposed in [26] has worst-case ratio equal to $\frac{3}{2}$. Van Laarhoven and Zijm [35] use a simulated annealing heuristic to compute a near-optimal placement sequence.

Finally, observe that, if several feeders of a same type have been assigned to the machine, then the formulation of the placement sequencing problem becomes somewhat tricky. Indeed, the 'distance' between successive placements is now influenced by the solution of the component retrieval subproblem... which we solved (in Section 4.2) under the assumption that the component placement sequence was known! To get around this difficulty, Crama et al. [18] solve the placement sequencing subproblem by an exchange heuristic in which the component retrieval plan is kept fixed over a number of successive iterations and reoptimized once in a while.

5. Single machine, multiple board type problems

As we will see below, a placement machine may frequently be setup for a family of boards (*family setup*, see Section 6), rather than for a unique board type. In such a case, the feeder location problem must be solved simultaneously for all boards in the family, as opposed to placement sequencing which can be solved anew, and independently, for each board type. Thus, there arises an obvious asymmetry between the two subproblems and some of the approaches mentioned in the previous section may become less manageable.

In this multiple-board setting, the feeder location problem can be viewed as follows: we want to

minimize $makespan(\varphi)$

s.t. φ is a feasible feeder assignment,

where $makespan(\varphi)$ is a very complex function of the assignment φ, since it depends on the solution of the placement sequencing problem for *all* boards in the family. The literature on this problem is extremely scarce. As in the single-board version, it is possible to use iterative heuristics which alternate between the computation of tentative feeder assignments and of placement sequences for all board types. This approach is described in [42,38]. Notice, however, that it may involve the solution of a large number of traveling salesman problems. For instance, with three machines and nine board types (as in [18]), 27 traveling salesman problems must be solved for each feeder assignment. If local search is used in order to improve the location of feeders, then the number of TSP instances may grow very large.

In order to reduce the computational burden of the procedure, Crama et al. [18] suggest to rely on a very fast approximation of the objective function *makespan*(φ), which can be used for optimizing feeder locations by local search. In their experiments, the approximation accelerates the search and proves quite accurate.

Dikos et al. [21] develop a genetic algorithm for the feeder location problem with multiple board types, under the assumption that placement sequences are known in advance.

There does not seem to be much more work on the multi-board version of the feeder location and placement sequencing problems: in view of the practical relevance of these problems, there is here ample opportunity for further research.

6. Multiple machines: setup policies

When more than one board type is to be produced over the planning horizon, a policy has to be adopted regarding the conditions under which new feeder setups can be performed. A feeder setup may affect the allocation of component feeders to the machines as well as the location of feeders on the carriers (cf. problems SP2 and SP5 in Section 3). Observe however that, because of interdependencies between the various subproblems, setup policy actually encompasses a broader set of issues, partially reflected in problems SP1–SP5. The practical importance of setup policies cannot be overestimated: Jain et al. [32] mention for instance that, at some Hewlett–Packard shops, over 50% of the production time is spent in setups.

Several types of setup policies have been identified in the PCB literature (see e.g. [6,9,32,42,38] etc. Notice that similar distinctions have also been established in the literature on tool management for flexible manufacturing systems; see e.g. [16,28,49,51]). For a given family of board types to be produced over the planning horizon, a possible typology of setup policies goes as follows:

(a) *tear-down* setups [32] (also called *single unique setup* [42] or *complete setup* [9]): between the assembly of successive board types, all feeders are removed and a new setup is performed;

(b) *partial* setups [9,38]: the removal and replacement of feeders is allowed between successive board types; there are several variants of this idea, to be discussed in Section 6.3;

(c) *family setups* [42]: no feeder setup is allowed between successive boards in the family; thus, the assembly line (or cell) must have sufficient carrier capacity to accommodate all the feeders required by the family.

Ammons et al. [6] provide a nice review of setup policies in connection with machine grouping, product grouping and component allocation issues. We would like to emphasize here that the setup policy adopted by a plant is, to a large extent, influenced by its product mix (which we assumed earlier to be exogenously given). In the sequel, we will refine the formulation of problems SP1–SP8 under different setup hypotheses.

For the ease of exposition, we start the discussion with the most clear-cut situations, i.e. tear-down policy (Section 6.1) and family setups (Section 6.2), and we finish with the more complex case of partial setups (Section 6.3).

6.1. Tear-down policy

Consider first the tear-down policy. This policy appears to be most adequate when the product mix displays a small variety of PCB types, assembled in relatively large batches. In this case, the high setup times incurred under the tear-down policy can be offset by the productivity gains resulting from customized feeder allocation and location decisions.

Under the tear-down policy, most of the planning hierarchy collapses to a collection of simpler questions bearing on a single board type. Essentially, the tear-down policy reduces the planning problem to a single board multiple machine situation. For instance, the issue of PCB sequencing (SP4) vanishes and the feeder location problem (SP5) is solved anew for each PCB type.

The major remaining decisions concern the allocation of feeders and of placement operations to machines, i.e. SP2 and SP3. For an assembly line, the most appropriate model formulation requires to allocate the feeders and the operations so as to minimize (an estimate of) the workload of the bottleneck machine. Such models have been used, for instance, by Crama et al. [19] or van Laarhoven and Zijm [35] for a single PCB type, i.e. in a tear-down policy framework. We will come back to such models in Section 6.2, for multiple board types. Once these problems have been solved, the remaining problems (feeder location, placement sequencing and component retrieval and motion control specification, viz. subproblems SP5–SP8) are single machine problems that have already been discussed in Section 4.

6.2. Family setups

Family setups appear adequate when there is a high (to medium) variety of PCB types, assembled in small (to medium) batches. Indeed, in such situation, the assembly time to be gained from improved feeder allocation/location for each individual board type may not compensate for additional setup time. Some plant managers also prefer to avoid frequent setups which may easily lead to human errors, and thus, to quality and/or productivity losses.

In practice, family setups may actually arise in (at least) two different frameworks. In both cases, we may assume that, prior to the start of the planning horizon, the PCBs to be produced over the given horizon have been partitioned into families (possibly, a unique family). Then,

- either each family is assigned to a distinct group of machines (assembly line or workcell) and each group is setup once for the assembly of the whole family;
- or the families are successively produced on the same line (or in the same workcell) and a new setup is performed before the production of each family.

According to the typology presented above, the second situation should be classified in the category of 'partial setups', but it shares in fact all the characteristics of family

setups. In particular, the question that naturally arises in both cases is (cf. SP1): how
to assign PCB types to product families and—in the first case—to machine groups?

6.2.1. Assignment of PCB types to product families and to machine groups (SP1)

Assigning PCB types to product families is a decision very much akin to those
considered in the group technology (GT) literature on 'cell formation' or in the FMS
literature on 'job grouping' (see e.g. [14,16,50,55]).

In the GT framework, products are grouped by a clustering algorithm based on
component commonality between boards. The 'capacity' of the feeder carriers is not
directly taken into account by classical clustering procedures, which must therefore be
adapted in an ad hoc fashion; see e.g. [46] for an illustration.

The FMS job grouping model on the other hand, explicitly takes the carrier capacity
into account. In its best known version, the objective function of this problem attempts
to minimize the number of families to be formed. This model has been extensively
studied, both from a computational and from a theoretical point of view (see [55,20]
and the survey in [16]). It provides a reasonable proxy of the makespan minimization
problem when all the families have to be produced on a single line of machines and
when the setup time strongly dominates the assembly time.

By contrast, in the multi-line (or multi-cell) setting, the number of machine groups
is fixed a priori. Hence, a more adequate formulation of SP1 concentrates on the
allocation of product types to machine groups so as to minimize the workload of the
most heavily loaded machine group (here, a product family is defined as the collection
of PCB types assigned to a same machine group). The resulting model is akin to
bin packing or parallel machine scheduling models. In order to formulate SP1 as an
integer programming problem, let $i = 1,\ldots,I$ denote the available machine groups, let
$k = 1,\ldots,K$ denote PCB types, let $j = 1,\ldots,J$ denote the feeders to be used, let a_{ik} be
the estimated assembly time for all boards of type k on machine group i, let N_i be the
total (aggregated) capacity of all feeder carriers of the machines in group i and let δ_{jk}
be a 0–1 parameter which takes value 1 if PCB type k requires feeder j and value 0
otherwise. The 0–1 decision variables are

$y_{ik} = 1$ if board type k is assigned to machine group i,
$z_{ij} = 1$ if feeder j is set up on machine group i

and the model can be written as

$$\text{minimize} \quad \max_{i=1,\ldots,I} \sum_{k=1}^{K} a_{ik} y_{ik} \tag{4}$$

$$\text{s.t.} \quad \sum_{i=1}^{I} y_{ik} = 1 \quad \text{for all } k, \tag{5}$$

$$\sum_{j=1}^{J} z_{ij} \leqslant N_i \quad \text{for all } i, \tag{6}$$

$$\delta_{jk} y_{ik} \leqslant z_{ij} \quad \text{for all } i,j,k, \tag{7}$$

$$y_{ik} \in \{0,1\} \quad \text{for all } i,k, \tag{8}$$

$$z_{ij} \in \{0,1\} \quad \text{for all } i,j. \tag{9}$$

A distinguishing feature of the above model is that the machine groups are viewed as completely decoupled (each product type is processed by exactly one group—see constraint (5)), in agreement with the layout and the organization of many assembly shops. Moreover, the model differs from feeder allocation (SP2) or feeder location (SP5) models since it assigns feeders to groups of machines, rather than to individual machines or individual slots, and since it treats feeder capacity at an aggregated level only (constraint (6)).

This type of integer programming model has not been widely studied in the literature. Hillier and Brandeau [31] propose a model (BIP4) which is very similar to (4)–(9), except that its objective is to minimize total assembly cost (or time) rather than to balance the workload. They develop an exact algorithm and a heuristic based on Lagrangian relaxation. In a more general model (where partial setups are allowed), Balakrishnan and Vanderbeck [9] propose to minimize the setup cost, but add an upper-bound on the allowed workload per machine group (so, when restricted to family setups, their model is essentially equivalent to (4)–(9); see Section 6.3 for more details). They attack this model by column generation techniques. Finally, it should be noted that model (4)–(9) shares very obvious similarities with some of the integer programming models proposed for the job grouping problem in the FMS literature (see e.g. [20]).

A difficulty with the above model is that the total assembly time (a_{ik}) is very difficult to estimate, since it depends in a complex way on the set of PCBs which are allocated to each machine group and thus, on the solution of remaining subproblems in the list SP1–SP8.

To proceed, let us now assume that there is a unique family of boards to be produced by an assembly line or cell (i.e., let us assume that the family formation problem has been solved) and let us turn to the remaining subproblems.

6.2.2. Feeder allocation for assembly lines (SP2 and SP3)

Consider a single assembly line which is to be set up (once) for the production of a family of PCB types, say types $1,\ldots,K$. In this setting, it is usually assumed that production takes place in *batch mode*, where batch k consists of d_k boards of type $k = 1,\ldots,K$. Provided all batch sizes are moderately large, this implies that the issue of PCB sequencing (SP4) can be disregarded altogether, as it will not affect performance in a significant way. The remaining issues to be addressed concern the feeder allocation problem (SP2) and, if relevant, the auxiliary problem SP3 (recall that SP3 only arises if feeders containing a same component type have been assigned to several machines). Then, once SP2 and SP3 have been solved, the planning problem is reduced to a collection of single machine single board subproblems (one for each machine in the line), as in Section 4.

McGinnis et al. [42] suggest that, for SP2–SP3, the most appropriate objective function consists in minimizing the sum over all board types of the makespans of these board types on their bottleneck machines. Of course, different types of PCBs, and therefore different batches, may have different bottleneck machines. For simplicity, let us restrict our attention to the feeder allocation problem (SP2) by assuming that each component feeder can only be used once. Let $t_{km}(x)$ denote the assembly time of a board of type k on machine m induced by some feeder allocation x ($k = 1, \ldots, K$ and $m = 1, \ldots, M$). With X denoting the set of feasible feeder allocations, the objective function may be specified as follows (compare with (4)):

$$\min_{x \in X} \sum_{k=1}^{K} d_k \max_{m=1,\ldots,M} t_{km}(x). \tag{10}$$

Observe that setup times do not appear in (10) under the assumption of family setups.

In order to write a more complete formulation, let (similarly to the previous section) $j = 1, \ldots, J$ denote the feeders to be used, let p_{jkm} be the estimated placement time by machine m of all components of type j on a board of type k, and let C_m be the carrier capacity on machine m. The 0–1 decision variables are

$$x_{jm} = 1 \quad \text{if feeder } j \text{ is set up on machine } m$$

for $j = 1, \ldots, J$, $m = 1, \ldots, M$, and a model for SP2 can be written as

$$\text{minimize} \quad \sum_{k=1}^{K} d_k \max_{m=1,\ldots,M} \sum_{j=1}^{J} p_{jkm} x_{jm}, \tag{11}$$

$$\text{s.t.} \quad \sum_{m=1}^{M} x_{jm} = 1 \quad \text{for all } j, \tag{12}$$

$$\sum_{j=1}^{J} x_{jm} \leqslant C_m \quad \text{for all } m, \tag{13}$$

$$x_{jm} \in \{0, 1\} \quad \text{for all } j, m. \tag{14}$$

This model can be linearized by substituting new variables t_k for the max-operators in the objective function (11). This leads to

$$\text{minimize} \quad \sum_{k=1}^{K} d_k t_k \tag{15}$$

$$\text{s.t.} \quad \sum_{m=1}^{M} x_{jm} = 1 \quad \text{for all } j, \tag{16}$$

$$\sum_{j=1}^{J} x_{jm} \leqslant C_m \quad \text{for all } m, \tag{17}$$

$$\sum_{j=1}^{J} p_{jkm} x_{jm} \leqslant t_k \quad \text{for all } k, m, \tag{18}$$

$$x_{jm} \in \{0, 1\} \quad \text{for all } j, m. \tag{19}$$

The assembly times p_{jkm} must be roughly estimated, since feeder allocation, feeder location and placement sequencing decisions will eventually interfere with each other to determine the exact assembly time of each board.

Ammons et al. [6] consider a slightly more general feeder allocation model than (15)–(19) by allowing for multiple copies of each feeder type and for partial setups. They solve this mixed integer programming model by branch-and-bound. They mention, however, that (15) provides a poor approximation of the actual makespan when multiple board types are involved.

Crama et al. [18] handle the same objective function and simultaneously solve the feeder allocation and location problems (SP2 and SP5) by local search. Using some of the ideas mentioned in Section 5, they can anticipate on the solution of the placement sequencing problem and are able to obtain close estimates of the actual makespan.

Lapierre et al. [36] consider an integer programming model similar to (15)–(19), but which explicitly incorporates feeder location decisions. They use Lagrangian relaxation techniques to solve it.

Lin and Tardif [39] consider the objective function (15) in a stochastic environment characterized by uncertain demand and machine breakdowns. They propose and solve a stochastic mixed-integer programming formulation of the problem.

6.2.3. Feeder allocation and production sequencing for flexible cells (SP2–SP4)

Consider now a flexible workcell which is to be set up for the assembly of a family of board types $1, \ldots, K$. Contrary to the case of assembly lines, production can be assumed here to take place in *mixed mode*, with several types of PCBs circulating simultaneously in the cell. The PCB sequencing subproblem SP4 gains therefore more importance and must be taken into account in the formulation of the feeder allocation problem SP2 (here again, we assume for simplicity that each feeder type can be allocated to one machine only and that SP3 vanishes accordingly).

Integer programming models for SP2 have been proposed by several authors. In one of the earliest papers in this vein, Ammons et al. [7] describe a bicriterion model which simultaneously attempts to achieve workload balance and to minimize the number of visits of each board to the machines. The second objective can be viewed as a proxy for material handling utilization and work-in-process, but also aims at reducing the complexity of the subsequent sequencing problem (SP4). Klincewicz and Rajan [33] (see also [48]) formulate a very similar model in which workload balance is incorporated into the constraints rather than in the objective function. In order to state their model, denote the 0–1 decision variables by

$x_{jm} = 1$ if feeder type j is set up on machine m,

$y_{km} = 1$ if board type k must visit machine m.

Let d_k denote the number of boards of type k, let p_{jm} be the estimated placement time of all components of type j by machine m, let $\delta_{jk} = 1$ (resp. 0) if PCB type k requires (resp. does not require) feeder j and let T_- (resp. T_+) be a lower bound (resp. upper bound) on the total workload of each machine (i.e., on the makespan of the cell), for $j = 1, \ldots, J$, $k = 1, \ldots, K$, $m = 1, \ldots, M$. The model in [33] is

$$\text{minimize} \quad \sum_{k=1}^{K} d_k \sum_{m=1}^{M} y_{km}, \tag{20}$$

$$\text{s.t.} \quad \sum_{m=1}^{M} x_{jm} = 1 \quad \text{for all } j, \tag{21}$$

$$\sum_{j=1}^{J} x_{jm} \leqslant C_m \quad \text{for all } m, \tag{22}$$

$$\delta_{jk} x_{jm} \leqslant y_{km} \quad \text{for all } j, k, m, \tag{23}$$

$$\sum_{j=1}^{J} p_{jm} x_{jm} \geqslant T_- \quad \text{for all } m, \tag{24}$$

$$\sum_{j=1}^{J} p_{jm} x_{jm} \leqslant T_+ \quad \text{for all } m, \tag{25}$$

$$x_{jm} \in \{0, 1\} \quad \text{for all } j, m, \tag{26}$$

$$y_{km} \in \{0, 1\} \quad \text{for all } k, m. \tag{27}$$

Klincewicz and Rajan [33] solve this model by a GRASP heuristic. Ammons et al. [7] handle their bicriterion formulation by several heuristic procedures (of the bin packing type for workload balance and of the clustering type for the number of visits) which allow them to put more or less emphasis on each criterion. Another variant of SP2 is proposed by Askin et al. [8]: their objective is to allocate feeders so as to minimize the maximum workload across machines and, simultaneously, to form 'homogeneous' groups of PCBs so as to equalize the assembly time of each PCB within a group. They propose ad hoc heuristics based on similarity measures for the solution of this problem.

Let us now turn to the sequencing subproblem (SP4). This question seems to have been addressed by very few authors. Askin et al. [8] note that, in the framework of flexible cells, problem SP4 resembles the classical open shop scheduling model: given the allocation of feeders to machines (SP2), the assembly of each PCB of type k requires a list of operations $(O_{k1}, \ldots, O_{kJ_k})$, where O_{ki} denotes the placement of component i by machine m_i (where m_i is the machine holding component i). The problem consists in defining the start time of each operation so as to minimize the assembly makespan. Notice that the processing time of each operation O_{ki} is not completely determined as long as the remaining subproblems SP5–SP8 have not been solved, but it can usually be reasonably approximated.

After having solved the feeder allocation problem as indicated above, Askin et al. [8] construct a production schedule by applying specialized heuristics from the open shop literature. These heuristics make explicit use of the 'homogeneous' groups of PCBs formed in the first phase.

Lofgren et al. [40] assume that the allocation of feeders to machines (SP2) is given. They focus on a single board type but consider a situation where precedence constraints exist between the assembly operations to be performed on the boards. They attempt to determine a routing of the boards through the shop so as to minimize the number of visits to machines. They reformulate this problem as a linear ordering problem on a directed graph and they analyze the complexity and worst-case performance of approximation algorithms for this problem. They conjecture that, unless $P = NP$, there does not exist a polynomial time algorithm with finite worst case ratio for their model.

Ahmadi and Wurgaft [4] also assume that the allocation of feeders to machines is given and allow for precedence relations among operations, but they explicitly consider multiple PCB types. In order to synchronize the flow of products in the assembly cell, they are interested in finding large subsets of PCBs for which the precedence relations form an acyclic digraph. Alternatively, they propose to determine the smallest number of operations to be replicated so as to remove all cycles from the precedence graph (replicating an operation is roughly equivalent to using multiple copies of a same feeder type; thus, this question is related, in its spirit, to subproblems SP2–SP3).

6.3. Partial setups

Let us turn, finally, to partial setup policies. As mentioned earlier, there exist numerous variants of these strategies, among which:

(b1) *decompose and sequence* [42]: for each PCB type, the feeders loaded on the machines are exactly those required by the bill-of-materials of this board type; between each pair of successive board types, only those changes are performed which are strictly needed;

(b2) some feeders remain permanently on the machines, the other ones are changed as required by the next PCB type to be produced; the decision as to which feeders are permanent or temporary is explicitly incorporated in the optimization process [6,9];

(b3) some feeders are permanently assigned to the machines for reasons which are exogenous to the optimization models [34];

(b4) *partition and repeat* [15,42]: a new feeder setup is performed after all board types have been *partially* processed by the machines; incomplete boards accumulate as work-in-process.

Being intermediate between tear-down and family setups, partial setups clearly provide the most flexibility and allow, in principle, for optimal reduction of the production makespan. The efficiency tradeoff between family setups and (various types of) partial setups has been discussed, for instance, by Ammons et al. [6], Günther et al. [30], Jain et al. [32], Leon and Peters [38], Maimon et al. [41]. More research is needed on this topic (as already mentioned by McGinnis et al. [42]).

When partial setup is used, all subproblems SP1–SP8 become tightly interconnected. In particular, the sequence in which the different types of boards are produced determines the feeders to be loaded and unloaded when a new setup is performed and thus, largely determines the setup time. So, it becomes even more difficult to decouple product grouping, feeder allocation and board sequencing than in the case of family setups: an 'optimal' assignment of products to machine groups is one for which there exists a sequence of board types entailing few feeder changeovers.

These remarks explain that, under the assumption of partial setups, several researchers have linked problems SP1–SP4 to tool switching models investigated in the FMS literature. For a single machine, a well-known tool switching model can be stated as follows: given a family of boards $k = 1, \ldots, K$, their respective bills-of-materials (described by the parameters δ_{jk}, as in Section 6.2.1) and the feeder carrier capacity C, determine the sequence of boards and the corresponding allocation of feeder types to be loaded on the machines so as to minimize the total number of feeder changeovers. This model, which has close links to the *decompose and sequence* policy, was introduced in a seminal paper by Tang and Denardo [54] (see [16] for a review of the literature on this model). Its connection with PCB assembly was observed by Bard [11]. Jain et al. [32] relied explicitly on this model for a case study on setup optimization at Hewlett–Packard.

When several machines are available for assembly, however, the overall objective of makespan minimization, including setup time *and* assembly time, must be taken into account (since assembly time is influenced by feeder allocation decisions). This objective is not adequately reflected by tool switching models. Therefore, there arises a need for more general models. Such models are proposed by Balakrishnan and Vanderbeck [9] or Ammons et al. [6].

Balakrishnan and Vanderbeck [9] describe a model for product assignment SP1. They postulate that component types are to be partitioned into two classes: *permanent* and *temporary*. Temporary feeders are loaded on the machines as needed and unloaded whenever a batch is completed. With the same generic notations as in Section 6.2.1, let

$y_{ik} = 1$ if board type k is assigned to machine group i,

$z_{ij} = 1$ if feeder j is set up permanently on machine group i,

$v_{ijk} = 1$ if feeder j is set up temporarily on machine group i
          to assemble board type k,

let b_k be the number of batches of type k to be produced over the planning horizon and let T be an upper-bound on the workload of each machine group ($i = 1, \ldots, I$, $k = 1, \ldots, K$, $j = 1, \ldots, J$). The model is

$$\text{minimize} \quad \sum_{k=1}^{K} b_k \sum_{i=1}^{I} \sum_{j=1}^{J} \delta_{jk} v_{ijk}, \tag{28}$$

$$\text{s.t.} \quad \sum_{i=1}^{I} y_{ik} = 1 \quad \text{for all } k, \tag{29}$$

$$\sum_{j=1}^{J} z_{ij} + \sum_{j=1}^{J} \delta_{jk} v_{ijk} \leqslant N_i \quad \text{for all } i, k, \tag{30}$$

$$\delta_{jk} y_{ik} \leqslant z_{ij} + v_{ijk} \quad \text{for all } i, j, k, \tag{31}$$

$$\sum_{k=1}^{K} a_{ik} y_{ik} \leqslant T \quad \text{for all } i, \tag{32}$$

$$y_{ik} \in \{0, 1\} \quad \text{for all } i, k, \tag{33}$$

$$z_{ij} \in \{0, 1\} \quad \text{for all } i, j, \tag{34}$$

$$v_{ijk} \in \{0, 1\} \quad \text{for all } i, j, k. \tag{35}$$

Balakrishnan and Vanderbeck [9] use a column generation approach to solve this model.

The assumption underlying objective function (28) is that all temporary feeders are removed when assembly of the corresponding batch is completed, even if the same feeder is required by the next board type. This is in contrast with the FMS tool switching model mentioned above, where feeders are assumed to remain on the machine if they are common to successive board types. Removing all feeders, however, allows to reoptimize their location between the production of successive batches (subproblem SP5; FMS tool switching models do not take the location of feeders into account.)

Model (28)–(35) can be viewed as a generalization of (4)–(9). Indeed, ruling out partial setups amounts to setting all variables v_{ijk} to zero. Then, searching for the minimum feasible workload T in (29)–(35) is equivalent to solving the family setup model (4)–(9). On the other hand, when T is very large, model (28)–(35) places the emphasis on setup minimization.

A related model is proposed by Ammons et al. [6] for the allocation of feeders to machines on an assembly line (SP2). These authors develop fast heuristics or use an LP-based branch-and-bound code (MINTO) for the solution of their model.

The above-mentioned models use a rough approximation of the assembly time per board, denoted a_{ik}. Leon and Peters [38] use instead an iterative procedure to obtain more accurate estimates.

7. Conclusions

In this paper, we have reviewed some of the literature on process planning for the optimization of PCB assembly. In our view, some of the most noticeable recent trends in this field have been:

- the consideration of multiple board types in the solution of feeder location and placement sequencing models;
- the development of integer programming models and algorithms for product grouping and feeder allocation subproblems.

More research along these two lines is still needed. In particular, there seems to be a lack of techniques to determine the *global* quality of various solution methods. Indeed, in most practical situations, one needs to resort to heuristic (as opposed to exact) methods to deal with the size and complexity of the optimization problems that arise as part of the planning hierarchy SP1–SP8 described in Section 3. However, few methods are able to give an indication of the global quality of the heuristic solutions produced, of the adequacy of different models, or, for that matter, of the adequacy of the hierarchical decomposition itself.

Finally, one of the aims of this survey has been to facilitate the classification of problems and models found in the literature on PCB assembly. Unfortunately, access to this literature is oftentimes obscured by the fact that the description of the production environment involved and of the problems tackled is insufficiently clear. In order to help readers find their way in forthcoming research publications, we would like to advocate that all authors mention (at least) the following typology elements in their papers:

- shop layout (decoupled workcells, one assembly line, several assembly lines, etc.);
- characteristics of the product mix (high volume—low variety, low volume—high variety, etc.);
- setup policy (see Section 6);
- relevant characteristics of the placement machines (sequential, concurrent, etc.);
- decisions to be taken, according to the list SP1–SP8.

We believe that providing such information would improve communication between research teams and would foster new developments in the field.

Acknowledgements

The authors are grateful to Endre Boros and Peter Hammer for giving them the opportunity and motivation to assemble this survey. They acknowledge the partial financial support of ONR (Grant N00014-92-J-1375), of NATO (Grant CRG931531) and of the European Network DONET (contract number ERB TMRX-CT98-0202).

References

[1] R.H. Ahmadi, A hierarchical approach to design, planning, and control problems in electronic circuit card manufacturing, in: R.K. Sarin (Ed.), Perspectives in Operations Management, Kluwer Academic Publishers, Dordrecht, 1993, pp. 409–429.

[2] J. Ahmadi, R. Ahmadi, H. Matsuo, D. Tirupati, Component fixture partitioning/sequencing for printed circuit board assembly with concurrent operations, Oper. Res. 43 (1995) 444–457.

[3] J. Ahmadi, S. Grotzinger, D. Johnson, Component allocation and partitioning for a dual delivery placement machine, Oper. Res. 36 (1988) 176–191.

[4] R.H. Ahmadi, H. Wurgaft, Design for synchronized flow manufacturing, Manage. Sci. 40 (1994) 1469–1483.

[5] K. Altinkemer, B. Kazaz, M. Köksalan, H. Moskowitz, Optimization of printed circuit board manufacturing: integrated modeling and algorithms, European J. Oper. Res. 124 (2000) 409–421.

[6] J.C. Ammons, M. Carlyle, L. Cranmer, G.W. DePuy, K.P. Ellis, L.F. McGinnis, C.A. Tovey, H. Xu, Component allocation to balance workload in printed circuit card assembly systems, IIE Trans. 29 (1997) 265–275.

[7] J.C. Ammons, C.B. Lofgren, L.F. McGinnis, A large scale machine loading problem in flexible assembly, Ann. Oper. Res. 3 (1985) 319–332.

[8] R.G. Askin, M. Dror, A.J. Vakharia, Printed circuit board family grouping and component allocation for a multimachine, open shop assembly cell, Nav. Res. Logist. 41 (1994) 587–608.

[9] A. Balakrishnan, F. Vanderbeck, A tactical planning model for mixed-model electronics assembly operations, Oper. Res. 47 (3) (1999) 395–409.

[10] M.O. Ball, M.J. Magazine, Sequencing of insertions in printed circuit board assembly, Oper. Res. 36 (1988) 192–201.

[11] J.F. Bard, A heuristic for minimizing the number of tool switches on a flexible machine, IIE Trans. 20 (1988) 382–391.

[12] J.F. Bard, R.W. Clayton, T.A. Feo, Machine setup and component placement in printed circuit board assembly, Int. J. Flexible Manuf. Systems 6 (1994) 5–31.

[13] K. Broad, A. Mason, M. Rönnqvist, M. Frater, Optimal robotic component placement, J. Oper. Res. Soc. 47 (1996) 1343–1354.

[14] J.L. Burbidge, The Introduction of Group Technology, Wiley, New York, 1975.

[15] T.F. Carmon, O.Z. Maimon, E.M. Dar-El, Group set-up for printed circuit board assembly, Int. J. Production Res. 27 (1989) 1795–1810.

[16] Y. Crama, Combinatorial optimization models for production scheduling in automated manufacturing systems, European J. Oper. Res. 99 (1997) 136–153.

[17] Y. Crama, O.E. Flippo, J.J. van de Klundert, F.C.R. Spieksma, The component retrieval problem in printed circuit board assembly, Int. J. Flexible Manuf. Systems 8 (1996) 287–312.

[18] Y. Crama, O.E. Flippo, J.J. van de Klundert, F.C.R. Spieksma, The assembly of printed circuit boards: a case with multiple machines and multiple board types, European J. Oper. Res. 98 (1997) 457–472.

[19] Y. Crama, A.W.J. Kolen, A.G. Oerlemans, F.C.R. Spieksma, Throughput rate optimization in the automated assembly of printed circuit boards, Ann. Oper. Res. 26 (1990) 455–480.

[20] Y. Crama, A.G. Oerlemans, F.C.R. Spieksma, Production Planning in Automated Manufacturing, Springer, Berlin, 1996.

[21] A. Dikos, P.C. Nelson, T.M. Tirpak, W. Wang, Optimization of high-mix printed circuit card assembly using genetic algorithms, Ann. Oper. Res. 75 (1997) 303–324.

[22] Z. Drezner, S. Nof, On optimizing bin picking and insertion plans for assembly robots, IIE Trans. 16 (1984) 262–270.

[23] P.J. Egbelu, C.-T. Wu, R. Pilgaonkar, Robotic assembly of printed circuit boards with component feeder location considerations, Production Planning Control 7 (2) (1996) 162–175.

[24] T.A. Feo, J.F. Bard, S.D. Holland, Facility-wide planning and scheduling of printed wiring board assembly, Oper. Res. 43 (1995) 219–230.

[25] L.R. Foulds, H.W. Hamacher, Optimal bin location and sequencing in printed circuit board assembly, European J. Oper. Res. 66 (1993) 279–290.

[26] R.L. Francis, H.W. Hamacher, C.-Y. Lee, S. Yeralan, Finding placement sequences and bin locations for cartesian robots, IIE Trans. 26 (1994) 47–59.

[27] B. Gaboune, G. Laporte, F. Soumis, Optimal strip sequencing strategies for flexible manufacturing operations in two and three dimensions, Int. J. Flexible Manuf. Systems 6 (1994) 123–135.

[28] A.E. Gray, A. Seidmann, K.E. Stecke, A synthesis of decision models for tool management in automated manufacturing, Manage. Sci. 39 (1993) 549–567.

[29] S. Grotzinger, Feeder assignment models for concurrent placement machines, IIE Trans. 24 (1992) 31–46.

[30] H.O. Günther, M. Gronalt, R. Zeller, Job sequencing and component set-up on a surface mount placement machine, Production Planning Control 9 (1998) 201–211.

[31] M.S. Hillier, M.L. Brandeau, Optimal component assignment and board grouping in printed circuit board assembly, Oper. Res. 46 (1998) 675–689.

[32] S. Jain, M.E. Johnson, F. Safai, Implementing setup optimization on the shop floor, Oper. Res. 44 (1996) 843–851.

[33] J.G. Klincewicz, A. Rajan, Using GRASP to solve the component grouping problem, Naval Res. Logist. 41 (1994) 893–912.

[34] C. Klomp, J. van de Klundert, F.C.R. Spieksma, S. Voogt, The feeder rack assignment problem in PCB assembly: a case-study, Int. J. Prod. Econom. 64 (2000) 399–407.

[35] P.J.M. van Laarhoven, W.H.M. Zijm, Production preparation and numerical control in PCB assembly, Int. J. Flexible Manuf. Systems 5 (1993) 187–207.

[36] S.D. Lapierre, L. De Bargis, F. Soumis, Balancing printed circuit board assembly line systems, Int. J. Production Res. 38 (2000) 3899–3911.

[37] T. Leipälä, O. Nevalainen, Optimization of the movements of a component placement machine, European J. Oper. Res. 38 (1989) 167–177.

[38] V.J. Leon, B.A. Peters, Replanning and analysis of partial setup strategies in printed circuit board assembly systems, Int. J. Flexible Manuf. Systems 8 (1996) 389–412.

[39] W.-L. Lin, V. Tardif, Component partitioning under demand and capacity uncertainty in printed circuit board assembly, Int. J. Flexible Manuf. Systems 11 (1999) 159–176.

[40] C.B. Lofgren, L.F. McGinnis, C.A. Tovey, Routing printed circuit cards through an assembly cell, Oper. Res. 39 (1991) 992–1004.

[41] O.Z. Maimon, E.M. Dar-El, T.F. Carmon, Set-up saving schemes for printed circuit boards assembly, European J. Oper. Res. 70 (1993) 177–190.

[42] L.F. McGinnis, J.C. Ammons, M. Carlyle, L. Cranmer, G.W. DePuy, K.P. Ellis, C.A. Tovey, H. Xu, Automated process planning for printed circuit card assembly, IIE Trans. 24 (1992) 18–30.

[43] A. Mody, R. Suri, M. Tatikonda, Keeping pace with change: international competition in printed circuit board assembly, Ind. Corporate Change 4 (1995) 583–613.

[44] L.K. Moyer, S.M. Gupta, SMT feeder slot assignment for predetermined component placement paths, J. Electron. Manuf. 6 (1996) 173–192.

[45] L.K. Moyer, S.M. Gupta, Simultaneous component sequencing and feeder assignment for high speed chip shooter machines, J. Electron. Manuf. 6 (1996) 271–305.

[46] O.Z. Maimon, A. Shtub, Role of similarly measures in PCB grouping procedures, Int. J. Production Res. 30 (1992) 973–983.

[47] H. Nakahara, PCB output 1998, Printed Circuit Fabrication, June 1999.

[48] A. Rajan, M. Segal, Assigning components to robotic workcells for electronic assembly, AT & T Technical J. 68 (1989) 93–102.

[49] M.S. Sodhi, A. Agnetis, R.G. Askin, Tool addition strategies for flexible manufacturing systems, Int. J. Flexible Manuf. Systems 6 (1994) 287–310.

[50] K.E. Stecke, Formulation and solution of nonlinear integer production planning problems for flexible manufacturing systems, Manage. Sci. 29 (1983) 273–288.

[51] K.E. Stecke, I. Kim, A study of part type selection approaches for short-term production planning, Int. J. Flexible Manuf. Systems 1 (1988) 7–29.

[52] Y.S. Su, C. Wang, P.J. Egbelu, D.J. Cannon, A dynamic points specification approach to sequencing robot moves for PCB assembly, Int. J. Comput. Integr. Manuf. 8 (1995) 448–456.

[53] F.F. Suarez, M.A. Cusumano, C.H. Fine, An empirical study of manufacturing flexibility in printed circuit board assembly, Oper. Res. 44 (1996) 223–240.

[54] C.S. Tang, E.V. Denardo, Models arising from a flexible manufacturing machine, Part I: Minimization of the number of tool switches, Oper. Res. 36 (1988) 767–777.

[55] C.S. Tang, E.V. Denardo, Models arising from a flexible manufacturing machine, Part II: Minimization of the number of switching instants, Oper. Res. 36 (1988) 778–784.

[56] J.L. van Zante-de Fokkert, T.G. de Kok, The simultaneous determination of the assignment of items to resources, the cycle times, and the reorder intervals in repetitive PCB assembly, Ann. Oper. Res. 92 (1999) 381–401.

[57] I. Viczián, Finding placement sequences and bin locations for cartesian robots, A working paper of the University of Würzburg, 1993.

[58] R.A. Walas, R.G. Askin, An algorithm for NC turret punch press tool location and hit sequencing, IIE Trans. 16 (1984) 280–287.

[59] C. Wang, L. Ho, D.J. Cannon, Heuristics for assembly sequencing and relative magazine assignment for robotic assembly, Comput. Ind. Eng. 34 (1998) 423–431.

For further reading

A. Agnetis, R.G. Askin, M.S. Sodhi, Tool addition strategies for flexible manufacturing systems, Int. J. Flexible Manuf. Systems 6 (1994) 287–310.

R.H. Ahmadi, P. Kouvelis, Staging problem of a dual delivery pick-and-place machine in printed circuit card assembly, Oper. Res. 42 (1994) 81–91.

R. Tyler, View from Europe, Printed Circuit Fabrication, February 1999.

S. Voogt, Short term scheduling in PCB assembly, Philips Report CTR 597-93-0106, 1993.

T.A. Younis, T.M. Cavalier, On locating part bins in a constrained layout area for an automated assembly process, Comput. Ind. Eng. 18 (1990) 111–118.

![N·H ELSEVIER logo]

Discrete Applied Mathematics 123 (2002) 363–378

DISCRETE
APPLIED
MATHEMATICS

Discrete location problems with push–pull objectives

Jakob Krarup[a,*], David Pisinger[a], Frank Plastria[b]

[a] *DIKU, Department of Computer Science, University of Copenhagen, Universitetsparken 1, DK-2100 Copenhagen, Denmark*
[b] *BEIF, Vrije Universiteit Brussel, Pleinlaan 2, B-1050 Brussels, Belgium*

Received 12 November 1999; received in revised form 15 December 2000; accepted 26 January 2001

Abstract

The models within operational research concerned with locational decisions mostly either consider only the positive effects, pulling the facilities towards demand, or only negative effects, pushing the facilities away from the places affected by the facilities' nearness. In real-world situations both of these opposing forces are at work. We give an overview of a number of push–pull models, yielding alternative ways to incorporate both types of effects simultaneously. The discussion is restricted to models of combinatorial optimisation and includes indications of reduction to standard models and/or algorithmic approaches where possible. © 2002 Elsevier Science B.V. All rights reserved.

1. Introduction

Within the broad interface between operations research (OR) and computer science, a major application area concerns *locational decisions*. The facilities to be located are normally regarded as "*friendly*" since closeness, in general, is viewed as an attractive property whereas the opposite applies for their *obnoxious* counterparts.

Any modern society, however, exhibits facilities which at the same time provide some kind of service to the community but also have certain negative effects to their environment. Typical examples are production plants which provide goods but may be polluting, or sport facilities which should be well accessible but may generate quite annoying effects like noise, congestion or even vandalism, etc. This dual nature

---

* Corresponding author. Tel.: +45 35 321450; fax: +45 35 321401.
E-mail addresses: krarup@diku.dk (J. Krarup), pisinger@diku.dk (D. Pisinger), frank.plastria@vub.ac.be (F. Plastria).

0166-218X/02/$ - see front matter © 2002 Elsevier Science B.V. All rights reserved.
PII: S0166-218X(01)00346-8

characterising many realistic location problems have largely been neglected in location modelling up to the mid 1970s. Studies of models recognising both effects simultaneously, however, have by now become an established research area still gaining further momentum.

The highly suggestive term *push–pull* replacing the less idiomatic *repulsion–attraction* and other variants appears to have been first proposed in the excellent treatise on objectives in locational decisions by Eiselt and Laporte [12]. Pull-objectives like minimising transport costs or worst-case travel time, or maximising reachable population or market share characterise the vast majority of papers which have appeared since the foundations of (modern) discrete location theory were laid in the mid 1960s. Among the myriads of models considered, however, only four of these—at times referred to as *prototype location problems*—have played and do still play a particularly dominant role: the *p-median problem* (*p*-MP), the *p-centre problem* (*p*-CP), the *uncapacitated facility location problem* (UFLP), and the *quadratic assignment problem* (QAP). Despite the seeming simplicity of their underlying assumptions, these models have provided important, quantitative bases for the investigation of numerous practical locational decision problems. They have been used both as optimisation models in their own right or have been employed as subroutines in more integrated models. Finally, due to the large number of extensions available, each of these four prototype problems can be viewed as the foremost member of a *family* of discrete location problems.

A common feature for the first three of these models is that they all involve a finite set of potential sites for the facilities to be located and a finite set of *users* or markets having a prespecified demand for a common good which, in principle, can be provided by any facility. Whereas the locations of the facilities to be established is to be determined, the locations of the users are assumed known and invariant. For QAP, however, the "users" are the facilities themselves with "demands" expressed in terms of interaction or *communication* between each pair of these.

In these models, the objective, be it *minisum* for *p*-MP, UFLP, QAP or *minimax* for *p*-CP, is a *pull*-objective, tending in some way to favour nearness of facilities with users or between facilities. We then talk about location of *attracting* or *desirable* facilities. Most applications of combinatorial optimisation in location have fallen in this category. Note in this context that the three 0–1 programming problems known as *set covering*, *set partitioning*, and *set packing*, claimed to be among the most applicable models in combinatorial optimisation (see e.g. the recent survey [41]), all are special cases of UFLP.

A much smaller, but growing part of the literature studies models aiming at maximising distance to closest facility or minimising total effects on the surrounding population. This leads to *push*-objectives like maxmin distance, dispersion and minimal covering since nearness between facilities and users or among the facilities themselves is discouraged.

Models with push-objectives, in general, are surveyed in [15,35]. We distinguish two types here.

- *Undesirable* or *(ob)noxious* facility location problems, in which a set of *subjects* are specified which are (potentially) (negatively) affected by the facilities, the objective aiming at minimising these detrimental effects, which may be dangerous to health (noxious) or to lifestyle (obnoxious).

- *Dispersion* models, in which there are only facilities to be located in such a way as to affect each other the least possible.

One may note that for the location of a single facility, only the former type may appear, and that a good part of the literature concerned makes use of models in a continuous environment (see [35]). There are two good reasons for this: firstly, undesirable effects such as heat, noise, radiation and pollution typically spread continuously over space, and an adequate description of them must involve the spatial relative position between emittor and receptor; secondly, continuous single-facility models lead often to challenging problems within fields like nonlinear optimisation and computational geometry. Multiple obnoxious facilities location models and dispersion models are typically amenable for (or are reduced to) discrete models, so that combinatorial optimisation applies.

In this paper, we are interested in models that include both desirable and undesirable aspects, in particular those which fall within the general scope of combinatorial optimisation.

Several authors attribute the seminal papers in this area to Goldman and Dearing [18] and Church and Garfinkel [10] which appeared in 1975 and 1978, respectively. An even earlier source, however, is the 1941-volume "What is Mathematics?" by Courant and Robbins [11]. Courant and Robbins consider a famous problem, allegedly first formulated in the early 1600s by Fermat, together with kind of a companion problem termed *complementary problem* (ComP). For a given triangle, Fermat asks for a fourth point such that the sum of its euclidean distances, each weighted by $+1$, to the three given points is minimised. ComP differs from Fermat in that the weight associated with one of these points in -1 instead of $+1$.

To the best of our knowledge, ComP is the first example of a location problem dealing explicity with both positive and negative weights. Unfortunately, the solution proposed in [11]—with the proof left as an exercise for the reader—is incorrect. The correct solution for all triangles is provided in Krarup's 1998-paper [26].

Several situations may be distinguished:

- When only inter-facility interactions are considered, some being *pull*, and some *push*, we will call this a *semi-dispersion* model. Such models are discussed in Section 4.
- Otherwise, the description of the desirable part will involve a set I of users which *pull* the facilities towards them, whereas the undesirable part will be concerned with a set K of *subjects* which *push* the facilities away. It is easier to start by viewing the problem as *bi-objective* with the attractions and the repulsions as two quite opposing objectives, each of which might still be multi-objective (cf. with [8]). Several strategies then exist:
  1. Treat the bi-objective problem as really bi-objective and derive (approximations to) the Pareto-set, not an easy task for multifacility location problems, but see [16]. In continuous single-facility setting this has been attempted by e.g. Ndiaye [32] and Carrizosa et al. [7]. Some proposals of models falling in this category are given in Section 3.

2. Fix a bound on the obnoxious effects as a (set of) constraint(s) and optimise the desirable objective. This lead to standard pull objectives with additional *restrictions*, either around a set of subjects (e.g. [21,33,1]), or between facilities ($K = \emptyset$) (see [29,31]).

3. Fix a bound for the desirable objective part and optimise the undesirable objective. We do not know any work along this approach. In principle, this would lead to standard push objectives but with an additional budget constraint on the attraction cost.

4. Combine the push and pull objectives into a single objective.

5. When users are simultaneously subjects, and vice versa, i.e. $I = K$, we have models for which Eiselt and Laporte [12] use the term *balancing* objectives. The main examples are *equity* objectives like those discussed by Erkut [14] and Marsh and Schilling [28]. Best-known examples are minimisation of variance of user-distances to facilities (see e.g. [30,20,6]).

6. In the general situation, however, the set of users I and the set of subjects K are unrelated (but possibly overlapping). This type of problems, which we suggest to call *general push–pull* models, has not as yet received the attention they deserve in view of the large application potential. Some new models of this type will be explored in the sequel in Section 2.

In this paper, we give an overview of some of the classical location models of combinatorial optimisation type and explore a few of the different ways discussed above to take the semi-obnoxious character of the facilities into account. We will restrict our discussion to some general push–pull models, two bi-criterion models, and a few semi-dispersion models. All the models considered are NP-hard to solve and thus the solution techniques must rely on some kind of enumerative methods. Heuristic approaches may also be applied for the solution of large sized instances.

Since semi-obnoxious facilities may have an attraction as well as a repulsion to the users (I) or subjects (K), we will introduce the term *individual* to be either of these (or both). Since we consider combinatorial problems, we have a finite set $I \cup K$ of individuals, while users and subjects may coincide since I and K are not necessarily disjoint.

In the mathematical formulation of the combinatorial models, we will let the binary variable x_j take on the value 1 whenever a facility $j \in J$ is established. For all models involving users to be serviced by facilities, we shall assume these facilities of unlimited capacity, that is, any established facility can in principle service all users. This assumption implies the existence of an optimal solution in which no user is serviced by more than one facility. In turn, this so-called *single-assignment property* implies that the decision variables allocating users to established facilities are 0–1 variables as well as are the variables linking affected subjects to facilities. Another consequence of the *single-assignment property* to be further elaborated upon in Section 2.1 is that all users can be assumed to have unity demand for the service provided by the facilities.

The binary variable y_{ij} takes on the value 1 when user $i \in I$ is serviced by facility $j \in J$ and the variable z_{kj} has the value 1 whenever subject $k \in K$ is affected by facility $j \in J$. Although all decision variables ultimately must take the values 0 or 1 only, we

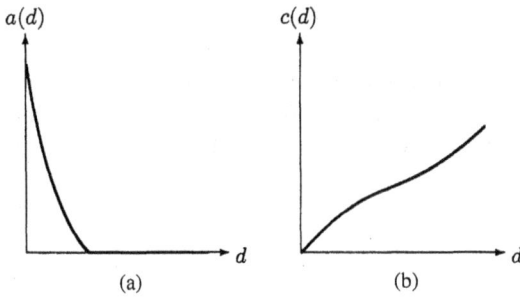

Fig. 1. (a) "Cost" $a(d)$ of a obnoxious facility as function of the distance d. (b) Transportation cost $c(d)$ of a friendly facility as function of the distance d.

may frequently relax this integer requirement on y_{ij} and z_{kj} to $y_{ij}, z_{kj} \in [0, 1]$ since the models automatically will return integer values of y_{ij} and z_{kj} whenever x_j is 0 or 1.

The cost of establishing facility j is given by f_j, and facility j may then satisfy all of user i's demand, $i \in I$, at cost c_{ij}. The repulsive part of an open facility j affects the subjects $k \in K$ with an amount of a_{kj}.

Since most objective functions involve some kind of distance measure, it is important to notice that for location problems with semi-obnoxious facilities, one may often use two different metrics for the attractive and the repulsive part of the objective, since attractiveness is measured by reachability through a network of typically streets while repulsion typically is spread through the air. Thus, it is often natural to use shortest path distance along a network or any extraneously given transport cost-matrix for the pull part, while euclidean-like distances (possibly modified through winds or terrain inclination) or any extraneously given pollution load cost-matrix is used for the push part.

If the obnoxious effect of a facility is expressed in terms of the "transportation costs", one will typically have a decreasing function of the distance measure (see Fig. 1a) until a given distance is reached from where on the noxious effect is negligible. The attractive part of a facility will naturally be expressed by ordinary transportation costs, nondecreasing with distance (see Fig. 1b).

2. General push–pull models

These models consider facility location problems in which a set of *individuals* are specified, which are (potentially) (negatively) affected by the facilities. The objective is to minimise the (semi)noxious effect while maintaining ordinary supply constraints.

2.1. Uncapacitated facility location problems with additive noxious effects

The constituents of the UFLP are: a finite set I of users $i \in I$ and a finite set J of sites for potential facilities indexed by $j \in J$. Furthermore, for each user-facility pair (i, j), let c_{ij} be the total *variable cost* of serving all of user i's demand from facility

j and let f_j be the *fixed cost* of establishing facility j, for all j. "Fixed" means that f_j is to be paid only if facility j actually is established and f_j is then independent of the number of users ($\geqslant 1$) served by that facility.

In a push–pull version of UFLP, we assume that each individual is affected (perhaps negligibly so) by each facility, and that the effects can be expressed as costs. In this first model, we also assume that the costs associated to the effects of the different facilities on a subject are additive.

The "cost" c_{ij} may include measures of the distance from user i to facility j as well as of the time or cost of serving user i from facility j. For example, c_{ij} may be interpreted as $c_{ij} = w_i(h_j + t_{ij})$, where w_i is the number of units demanded by user i, h_j is the per unit cost of operating facility j (including variable production and administrative costs, etc.), and t_{ij} is the transportation cost of shipping one unit to user i from facility j. With c_{ij} so defined, we can without loss of generality assume all users to have unit demand. Finally, no capacity constraints are imposed on the number of users that each potential facility can serve.

In its original form, UFLP is to open a subset of facilities and assign each user to exactly one of them such that the sum of the fixed and the variable costs becomes minimised. Note that the number of facilities to be established, as opposed to e.g. p-MP and p-CP, is not prespecified but results from an optimal solution.

For our purpose, however, we extend UFLP by a set of K subjects, which may be affected by the facilities in a negative way. For all pairs (k, j), $k \in K$, $j \in J$, let a_{kj} express the noxious effect of facility j on individual k. The value a_{kj} is a nonnegative number assumed to equal zero if the noxious effect is below a certain threshold and hence negligible. Whichever unit is used to measuring "noxious effect", we must here convert it to its monetary equivalent such that "costs" and "noxious effect" become additive.

Like in UFLP we introduce the 0–1 decision variables, $x_j = 1$ if facility j is established, and $y_{ij} = 1$ if user i is served by facility j. The inclusion of the obnoxious effects then just adds one term to the objective function:

$$\text{minimise} \quad \left(\sum_{i \in I} \sum_{j \in J} c_{ij} y_{ij} + \sum_{j \in J} f_j x_j \right) + \sum_{k \in K} \sum_{j \in J} a_{kj} x_j$$

$$\text{subject to} \quad \sum_{j \in J} y_{ij} = 1, \quad i \in I,$$

$$y_{ij} \leqslant x_j, \quad i \in I, \ j \in J,$$

$$x_j \in \{0, 1\}, \quad y_{ij} \in [0, 1], \quad i \in I, \ j \in J. \tag{1}$$

If we include the obnoxious effect of facility j in the fixed cost of the facility, we get the new costs $\tilde{f}_j = f_j + \sum_{k \in K} a_{kj}$. The extended problem reduces thereby to an ordinary UFLP with objective function

$$\text{minimise} \quad \sum_{i \in I} \sum_{j \in J} c_{ij} y_{ij} + \sum_{j \in J} \tilde{f}_j x_j. \tag{2}$$

2.2. Uncapacitated facility location problems with minimal covering as noxious effects

In several situations, however, it is not reasonable to assume that the obnoxious effects are additive. If e.g. the facilities poison the subsoil water, every affected subject in some predefined radius will have to connect to the public water supply instead of using his own pump. This cost, however, is constant and does not depend on the number of facilities located close to the subject, but rather on the fact whether some facility is sufficiently close to affect it or not. Since this is an obnoxious effect of minimal covering type we obtain a different model which may be formulated as follows:

$$\text{minimise} \left(\sum_{i \in I} \sum_{j \in J} c_{ij} y_{ij} + \sum_{j \in J} f_j x_j \right) + \sum_{k \in K} a_k z_k$$

$$\text{subject to } \sum_{j \in J} y_{ij} = 1, \quad i \in I,$$

$$y_{ij} \leqslant x_j, \quad i \in I, \; j \in J,$$

$$x_j \leqslant z_k, \quad j \in C_k, \; k \in K,$$

$$x_j \in \{0,1\}, \; y_{ij}, z_k \in [0,1], \quad i \in I, \; j \in J, \; k \in K. \tag{3}$$

The variable z_k here is used to indicate whether subject k is affected by *any* facility, and a_k is the cost of affecting the subject. The set of possible facilities located close enough to affect subject k is denoted by C_k. In this way, the last inequality of model (3) has the effect, that if any of the facilities $j \in C_k$ are established, then z_k is set to 1 and thus the cost a_k is included in the objective. If at the other hand subject k is unaffected by the facilities ($x_j = 0$ for all $j \in C_k$), these constraints are redundant and z_k remains unconstrained. Since by assumption $a_k \geqslant 0$, the optimisation will result in setting $z_k = 0$. Note that this behaviour remains true with continuously relaxed z_k.

If $a_k < 0$, additional constraints of the form $z_k \leqslant \sum_{j \in C_k} x_j$ can be added to the model to oblige z_k to become 0 as soon as all $x_j = 0$ for $j \in C_k$.

3. Two bi-objective models

As mentioned in the introduction, it may be more appropriate to handle the push and pull objectives separately, thus obtaining bi-objective problems, two proposals along the lines of previous section are formulated below.

3.1. Minsum–minsum

For UFLP (3), it is not evident at all that the push and pull parts of the objective can be readily combined into a single objective which in addition is linear. Therefore, it might be better to just keep them apart as separate objectives, and use a bi-objective approach. The formulation is then easily adapted as follows:

$$\text{minimise} \left(\sum_{i \in I} \sum_{j \in J} c_{ij} y_{ij} + \sum_{j \in J} f_j x_j \right)$$

$$\text{minimise} \sum_{k \in K} a_k z_k$$

$$\text{subject to} \sum_{j \in J} y_{ij} = 1, \quad i \in I,$$

$$y_{ij} \leqslant x_j, \quad i \in I, \ j \in J,$$

$$x_j \leqslant z_k, \quad j \in C_k, \ k \in K,$$

$$x_j \in \{0,1\}, \ y_{ij}, z_k \in [0,1], \quad i \in I, \ j \in J, \ k \in K. \tag{4}$$

Because of the presence of the binary variables and in spite of linearity in both the two objectives and the constraints, it cannot be guaranteed that every nondominated solution will be discovered by a weighting approach, the subproblems being of type (3). Therefore, this bi-objective minsum–minsum model must be considered as a different one.

The Pareto set of this problem leads to full trade-off information between the two objectives, invaluable information to any decision maker to finally single out the best solution, which might be a nonextreme one.

3.2. Minsum–maxmin

Another more stringent push objective is the worst-case like maxmin type, where one wants to minimise the maximal effect any subject feels from any of the facilities. When this objective is considered next to the minsum pull objective of the UFLP, yet another bi-objective model arises.

As before we use a_{kj} to expresses the noxious effect of a facility at site j on subject k. Introducing $\tilde{a}_j = \max_{k \in K} a_{kj}$ and an auxiliary variable z for the second objective, we arrive at the linear, bi-objective MIP formulation:

$$\text{minimise} \left(\sum_{i \in I} \sum_{j \in J} c_{ij} y_{ij} + \sum_{j \in J} f_j x_j \right)$$

$$\text{minimise} \ z$$

$$\text{subject to} \sum_{j \in J} y_{ij} = 1, \quad i \in I,$$

$$y_{ij} \leqslant x_j, \quad i \in I, \ j \in J,$$

$$\tilde{a}_j x_j \leqslant z, \quad j \in J,$$

$$z \geqslant 0,$$

$$x_j \in \{0,1\}, \ y_{ij} \in [0,1], \quad i \in I, \ j \in J. \tag{5}$$

4. Semi-dispersion models

In the dispersion models, no subjects are present, but the facilities should be located in such a way as to affect each other the least/most possible. Among the dispersion models to be considered here are the quadratic assignment problem, quadratic knapsack problem, *p*-dispersion problem and the *p*-defence-sum Problem.

4.1. The quadratic assignment problem

Even before the appearance of the seminal papers [10,18], push–pull objectives in discrete optimisation were encountered in 1972 [25] by Krarup within the framework of *layout design.*

Among the projects undertaken at that time by his consulting company was the design of a university hospital, "Klinikum Regensburg" to be built in Regensburg, Germany. An invitation to submit tenders was issued to a number of architects. As was explicitly stated in the announcement of the competition, among the criteria to be considered in the evaluation of proposals was the usual *pull*-objective known from the Koopmans–Beckmann variant of the more general *quadratic assignment problem* (QAP): find a layout minimising the sum of communication times distance (CDIST) taken over all pairs of departments to be located. If we use the binary variables x_{fj} to take on the value 1 if and only if facility $f \in F$ is located at site $j \in J$, we get the following integer programming model:

$$\text{minimise} \quad \sum_{f \in F} \sum_{g \in F} \sum_{j \in J} \sum_{m \in J} a_{fg} b_{jm} x_{fj} x_{gm}$$

$$\text{subject to} \quad \sum_{f \in F} x_{fj} = 1, \quad j \in J,$$

$$\sum_{j \in J} x_{fj} = 1, \quad f \in F,$$

$$x_{fj} \in \{0,1\}, \quad f \in F, \ j \in J. \tag{6}$$

In this model, a_{fg} denotes the communication between facilities f and g, while b_{jm} denotes the distance between sites j and m. Moreover, we assume that $|J| = |F|$.

To assess the architects' proposals regarding this criterion only, the task was to find "a best lower bound" on (6) or, if at all possible, provably optimal solutions to a series of instances of QAP including those two which later on were included in QAPLIB as *Krarup 30a/b* and eventually solved to optimality in 1999 by Hahn. An account of Hahn's approach and the underlying case study is provided in the joint paper [19].

An extension of this model [27] also takes qualitative and "aesthetic" criteria into account. Within a hospital context, typical *push*-objectives may arise from unpleasant odours or noise, exposure to daylight, a view to the crematorium, etc. Since such push–pull objectives cannot readily or meaningfully be translated into a problem formulation aiming at minimising some CDIST only, the use of an *interactive mode of operation* was instead advocated.

An instance of QAP consists, in general, of n, the number of objects to be located, and two square matrices, both of size $n \times n$, and with nonnegative elements representing "communication" and "distance", respectively. For e.g. a hospital, we can indeed imagine the desirability of keeping two departments far apart and assume that this is reflected by a negative entry in the communication matrix. By means of a scaling technique, however, negative entries appearing in one of two matrices defining a QAP can always be replaced by nonnegative numbers as shown by Pardalos and Rosen [34]. A comprehensive survey on theory and algorithms for the QAP is found in [9].

4.2. The quadratic knapsack problem

A different semi-dispersion model is the *quadratic knapsack problem* (QKP). Assume that a number of facilities may be established at some predefined locations. Each facility $j \in J$ has an establishing cost f_j, and the sum of the costs should be kept below a given budget B. The establishment of facility j has an associated gain a_{jj} in the objective function, and if both facilities h and j are established then a (positive or negative) gain a_{hj} is obtained for the interaction between k and j. With decision variables $x_j \in \{0, 1\}$ indicating whether facility j is established or not, we get the following formulation:

$$\text{maximise} \sum_{h \in J} \sum_{j \in J} a_{hj} x_h x_j$$

$$\text{subject to} \sum_{j \in J} f_j x_j \leqslant B,$$

$$x_j \in \{0, 1\}, \quad j \in J. \tag{7}$$

Solution techniques for the QKP with positive coefficients a_{hj} have been considered in [17,23,5]. These are all based on branch-and-bound techniques, thus derivation of strong upper bounds play a central role. Gallo et al. [17] presented upper bounds based on *upper planes*. Since the objective function can be written $\sum_{h \in J} (\sum_{j \in J} a_{hj} x_j) x_h$, we may derive an upper bound on the expression inside the parenthesis for each $h = 1, \ldots, |J|$ as

$$\bar{a}_h = \max \left\{ \sum_{j \in J} a_{hj} x_j \colon \sum_{j \in J} f_j x_j \leqslant B; \ x_j \in \{0, 1\}, j \in J \right\}. \tag{8}$$

An upper bound u on (7) may then be found as a solution to the problem

$$u = \max \left\{ \sum_{h \in J} \bar{a}_h x_k : \sum_{h \in J} f_h x_h \leqslant B; \ x_h \in \{0, 1\}, h \in J \right\}. \tag{9}$$

Both problems (8) and (9) are recognised as ordinary 0–1 knapsack problems. If the integer requirements in the two subproblems are relaxed, then each continuous knapsack problem can be solved in linear time and hence the relaxed bound is derived in $O(|J|^2)$ time.

Caprara et al. [5] improved this bound by observing that the profit sum $a_{hj} + a_{jh}$, which is obtained if both facilities h and j are established, may be split in an arbitrary way among a_{hj} and a_{jh}. In this way tighter bounds of the form (9) can be derived. The optimal splitting of the profits is found by subgradient optimisation techniques.

Both of the above solution techniques may be extended to handle instances with negative profits. Subproblems (8) and (9) should simply be transformed to equivalent problems with positive coefficients using the techniques described in [38]. In this way, valid upper bounds are derived, and the same framework can be used to solve the main problem. Computational experiments with semiobnoxious QKP are reported in [37]. The conclusion of these experiments is that the hardness of a problem increases with the amount of negative profits a_{hj} in the instance.

4.3. The p-defence-sum problem

In the p-defence-sum problem (PDSP), exactly p out of $|J|$ facilities should be opened. There is no user or demand components, but each facility has a degree of noxiousness to the other facilities. Such problems appear e.g. when locating radio transmitters: transmitters using the same frequency should be located far away from each other in order to minimise interference, while it may be desirable to locate transmitters using different frequencies close to each other to minimise establishing costs. Let a_{hj} be a nonnegative square matrix defined for $h, j \in J$ defining some kind of distance between locations. It is common to assume that $a_{jj} = 0$.

The PDSP wishes to maximise the distance sum between all pairs of established facilities, thus having the following quadratic formulation:

$$\text{maximise} \ \sum_{h \in J} \sum_{j \in J} a_{hj} x_h x_j$$

$$\text{subject to} \ \sum_{j \in J} x_j = p,$$

$$x_j \in \{0, 1\}, \quad j \in J. \tag{10}$$

The PDSP remains NP-hard even if the distances satisfy the triangle inequality [22]. No approximation algorithm with fixed ratio-bound ρ have been found for PDSP in the general case, but Kortsarz and Peleg [24] gave an approximation algorithm with variable approximation ratio of $O(|J|^{0.3885})$. A different approach is to consider the case where $p = c|J|$ for a constant $c < 1$. In this case, Srivastav and Wolf [40] presented an approximation algorithm with ratio-bound $\rho(c) = 2.073$, for $c = \frac{1}{2}$ and $\rho(c) = 4.189$ for $c = \frac{1}{4}$. If the triangle inequality is satisfied, an approximation algorithm with ratio $\rho = 4$ has been presented by Ravi et al. [39].

Only a few exact algorithms for PDSP have been presented. In [36] it is described how the upper plane technique presented by Gallo et al. [17] for QKP can be generalised to this problem. The upper planes are now defined as

$$\bar{a}_h = \max \left\{ \sum_{j \in J} a_{hj} x_j \colon \sum_{j \in J} x_j = p; \ x_j \in \{0, 1\}, \ j \in J \right\}. \tag{11}$$

An upper bound u on (10) is then found as

$$u = \max \left\{ \sum_{h \in J} \bar{a}_h x_h \colon \sum_{h \in J} x_h = p; \ x_h \in \{0, 1\}, \ h \in J \right\}. \tag{12}$$

Both problems (11) and (12) ask for the p largest values among a set of n different values, which easily can be found in $O(n)$ time through a median search algorithm. As for the QKP, it is possible to tighten the bounds by splitting the distance sum $a_{hj} + a_{jh}$, among a_{hj} and a_{jh} in an appropriate way.

4.4. The p-dispersion problem

A different variant of PDSP is the p-dispersion problem (PDP), where the objective is to maximise the minimum distance between any pair of facilities. Assuming that $a_{hj} \geqslant 0$ this problem may be formulated as

maximise r

subject to $r \leqslant a_{hj} x_h x_j$,

$$\sum_{j \in J} x_j = p,$$

$$x_j \in \{0, 1\}, \quad j \in J. \tag{13}$$

Even if the distances satisfy the triangle inequality, the problem remains Np-hard [13]. In the general form Ravi et al. [39] showed that the p-dispersion problem cannot be approximated by a fixed ratio ρ unless Np=p. If the triangle inequality is satisfied, an approximation ratio of $\rho = 2$ can be obtained, and this is also a lower bound. Erkut [13] described some simple upper bounds for PDP based on splitting the variables in two sets A and B, where A is the set of already fixed variables and B is the set of free variables. These bounds are applied in a branch-and-bound algorithm which is capable of solving instances with up to 40 facilities.

In [36] it is shown that the PDP (13) may be solved by means of the PDSP (10): for a fixed value of r we define $\tilde{a}_{hj}$ as 1 if $a_{hj} \geqslant r$ and 0 otherwise. Then the problem may be written as

$$\text{maximise} \sum_{h \in J} \sum_{j \in J} \tilde{a}_{hj} x_h x_j$$

$$\text{subject to} \sum_{j \in J} x_j = p,$$

$$x_j \in \{0, 1\}, \quad j \in J. \tag{14}$$

If a solution of value $n(n-1)$ is found, then (13) is feasible for the present value of r. Since r must take on one of the a_{hj} values, at most $O(n^2)$ different values of r need to be considered, and thus binary search can be used to determine the optimal value of r. Notice that (14) is the problem of finding a clique of size p in a graph, for which several approaches have been presented in the literature. Computational results in [36] indicate that the decomposition into a number of clique problems leads to a faster algorithm than that proposed by Erkut.

Both the p-defence-sum and the p-dispersion model may be transformed to problems with only positive coefficients a_{hj} by adding a sufficiently large constant M to all entries a_{hj}. In PDP this will have the effect that the optimal solution value is increased by M, while in PDSP, the solution value is increased by Mp^2, as we choose exactly p facilities.

5. Polynomially solvable cases

For optimisation problems defined on a network, polynomially solvable *special cases* can at times be identified if the underlying network exhibits a special structure. The network could be a simple path, a star, a tree, or a cycle, or it could be identified in terms of more complex forbidden subgraphs. This is the line of approach taken in [4] for the 1-*median problem in a network* (or 1-MP for short) with both positive and negative weights: find a vertex x minimising the sum $f(x)$ of the weighted shortest path distances from itself to all other vertices.

It is shown in [4] that 1-MP is solvable in linear time in the number of vertices of the network N if N is a *cactus*, that is, N is connected and no two cycles have more than one vertex in common. A cactus can be decomposed into a number of *blocks* where each block is either a simple cycle or a tree, here referred to as a *graft*, spanning a subset of vertices.

The algorithm is based on the observation that the *difference* $f(y) - f(x)$ between the objective function values for a pair (x, y) of adjacent vertices can be calculated in *constant* time. This allows the (locally) *optimising vertex* for a graft to be determined in linear time since each of its vertices is visited exactly once.

The difficulty with a *cycle* as opposed to a graft is that any pair of vertices is connected by two edge-disjoint paths. We need, therefore, in a well-defined way to talk of shortest paths as being either *clockwise* or *counterclockwise*. It is, nevertheless, possible also for a cycle to devise a data structure such that the linearity of the resulting algorithm is preserved.

If a single cycle and a single graft can be processed in linear time such that all information so to speak is collected in a single vertex, then the same procedure will work for cacti with two or more cycles since no two cycles by definition have more than one vertex in common. This completes the correct proof of the linear algorithm.

The direction of research initiated by Burkard and Krarup [4] is further pursued in [2] who consider the 2-median problem in a tree with positive and negative weights associated with its vertices. For location problems defined on a network, the points representing facilities can, in principle, be the vertices of the network or anywhere along its edges. The PMP with *positive* weights, however, possesses the so-called *vertex optimality property* which asserts the existence of an optimal solution with the facilities located at a subset of p vertices. Therefore, the p-median problem with positive weights reduces to finding a subset of p vertices such that the overall sum of the weighted distance between each vertex and the *closest* facility becomes minimised. Thus, whereas the objective function for related locational decision problems with *positive* weights depending on the context more or less suggests itself, it is not obvious what actually should be optimised in cases dealing with both positive and negative weights and with more than a single facility to be located.

Two objective functions are proposed in [2]:

(A) minimise the sum of the minimal weight times the distance,

(B) minimise the sum of the weight times the shortest path distance.

It is seen that (A) and (B) are identical if all weights are positive but indeed may be different when some weights are negative. Let 2-MP(A) and 2-MP(B) be the corresponding 2-*median problems*. The results obtained can be summarised as follows:

2-MP(A): If the network has a *cycle*, the *vertex optimality property* does not apply. For a *tree*, however, this property shown here is to hold. This appealing feature is utilised to derive an $O(n^2)$-algorithm for finding a 2-median on a tree with n vertices. The basic idea is roughly via a sequence of *edge deletions* to decompose the tree into two parts and then to apply the linear algorithm proposed in [4] for finding a 1-median for each of the two subtrees. For specially structured trees, the computation of the two 1-medians in each step can be speeded up such that the time complexity reduces to $O(n \log n)$ for an extended star and $O(n)$ for a path. Furthermore, for the general p-median problem on a path, an $O(pn^2)$-algorithm is devised.

2-MP(B): For the 2-median problem on a tree with n vertices with pos/neg weights and with objective function (B), however, the vertex optimality property does not apply. Thus, if the set of potential sites for the two facilities to be located no longer is restricted to be the vertices of the tree but in addition can be any point along any edge, 2-MP(B) becomes far more intricate than 2-MP(A). 2-MP(B) is, therefore, considered under the assumption that the two points representing a 2-median do not coincide. It is furthermore necessary to assume the existence of an optimal solution satisfying this requirement. Finally, the edge deletion procedure employed for 2-MP(A) does no longer work.

For 2-MP(B) it is shown [2] that: (1) there exists an optimal solution such that at least one of the two medians is a vertex, and (2) if no vertex pair represents an optimal solution, then there exists an optimal solution in terms of a pair m_1, m_2 of points such that both m_1 and the *midpoint* between m_1 and m_2 are vertices of the tree. Whereas the point set of feasible locations is infinite, these two observations assert the existence of a *finite* set of points of cardinality $O(n^3)$ which includes a pair of points representing an optimal 2-median. This result allows for an $O(n^3)$-algorithm for 2-MP(B) or a $O(n^2)$-algorithm if the tree is a path. If only vertices are considered as potential locations for the 2-median, it is finally shown that (B) is solvable in $O(n^2)$ time.

In closing it should be mentioned that similar models of robust location models have been investigated in [3].

Acknowledgements

The third author thanks E. Carrizosa for some fruitful modelling discussions. Furthermore, the useful suggestions made by two anonymous referees are gratefully acknowledged.

References

[1] J. Brimberg, H. Juel, A minisum model with forbidden regions for locating a semi-desirable facility in the plane, Location Sci. 6 (1998) 109–120.
[2] R.E. Burkard, E. Cela, H. Dollani, 2-medians in trees with pos/neg weights, Discrete Appl. Math. 105 (2000) 51–71.
[3] R.E. Burkard, H. Dollani, Robust location problems with pos/neg weights on a tree, Report SFB Optimisation and Control No. 148, Institute of Mathematics, Technical University of Graz, April 1999.
[4] R.E. Burkard, J. Krarup, A linear algorithm for the pos/neg-weighted 1-median problem on a cactus, Computing 60 (1998) 193–215.
[5] A. Caprara, D. Pisinger, P. Toth, Exact solution of the quadratic knapsack problem, INFORMS J. Comput. 11 (1999) 125–137.
[6] E. Carrizosa, Minimizing the variance of euclidean distances, Stud. Locational Anal. 12 (1999) 101–118.
[7] E. Carrizosa, E. Conde, M.D. Romero-Morales, Location of a semiobnoxious facility, a biobjective approach, in: R. Caballero, F. Ruiz, R.E. Steuer (Eds.), Advances in Multiple Objective and Goal Programming, Springer, Berlin, 1997.
[8] E. Carrizosa, F. Plastria, Location of semi-obnoxious facilities, Stud. Locational Anal. 12 (1999) 1–27 (previous version appeared also in Semi-plenary papers book, EURO XV Conference, Barcelona, 1997).
[9] E. Çela, The Quadratic Assignment Problem: Theory and Algorithms, Kluwer Academic Publishers, Dordrecht, 1998.
[10] R.L. Church, R.S. Garfinkel, Locating an obnoxious facility on a network, Trans. Sci. 12 (1978) 107–118.
[11] R. Courant, H. Robbins, What is Mathematics?, Oxford University Press, Oxford, 1941.
[12] H.A. Eiselt, G. Laporte, Objectives in location problems, in: Z. Drezner (Ed.), Facility Location. A Survey of Application and Methods, Springer, New York, 1995, pp. 151–180.
[13] E. Erkut, The discrete p-dispersion problem, European J. Oper. Res. 46 (1990) 48–60.
[14] E. Erkut, Inequality measures for location problems, Location Sci. 1 (1993) 199–217.
[15] E. Erkut, S. Neuman, Analytical models for locating undesirable facilities, European J. Oper. Res. 40 (1989) 275–291.
[16] E. Erkut, S. Neuman, A multiobjective model for locating undesirable facilities, Ann. Oper. Res. 40 (1992) 209–227.

[17] G. Gallo, P.L. Hammer, B. Simeone, Quadratic knapsack problems, Math. Programming 12 (1980) 132–149.
[18] A.J. Goldman, P.M. Dearing, Concepts of optimal location for partially noxious facilities, ORSA Bull. 23 (1) (1975) B31.
[19] P. Hahn, J. Krarup, A hospital facility layout problem finally solved, J. of Intelligent Manufacturing 12, Nos. 5/6 (2001).
[20] J. Halpern, O. Maimon, Accord and conflict among several objectives in locational decisions on tree networks, in: J.F. Thisse, H.G. Zoller (Eds.), Locational Analysis of Public Facilities, North-Holland, Amsterdam, 1983, pp. 301–314.
[21] H.W. Hamacher, S. Nickel, Restricted planar location problems and applications, Nav. Res. Logistics 42 (1995) 967–992.
[22] P. Hansen, I.D. Moon, Dispersing facilities on a network, Presentation at the TIMS/ORSA Joint National Meeting, Washington D.C, 1988.
[23] C. Helmberg, F. Rendl, R. Weismantel, Quadratic knapsack relaxations using cutting planes and semidefinite programming, in: W.H. Cunningham, S.T. McCormick, M. Queyranne (Eds.), Proceedings of the Fifth IPCO Conference, Lecture Notes in Computer Science, Vol. 1084, Springer, Berlin, 1996, pp. 175–189.
[24] G. Kortsarz, D. Peleg, On choosing a dense subgraph, in: Proceedings of the 34th Annual IEEE Symposium on Foundation of Computer Science, 1993, pp. 692–701.
[25] J. Krarup, Quadratic assignment, Data 3/72 (1972) 12–15.
[26] J. Krarup, On a "Complementary Problem" of Courant and Robbins, Location Sci. 6 (1998) 337–354.
[27] J. Krarup, P.M. Pruzan, Computer-aided layout design, Math. Programming Stud. 9 (1978) 75–94.
[28] M.T. Marsh, D.A. Schilling, Equity measures in facility location analysis: a review and a framework, European J. Oper. Res. 74 (1994) 1–17.
[29] I.D. Moon, S.S. Chaudhry, An analysis of network location problems with distance constraints, Management Sci. 30 (1984) 290–307.
[30] R.L. Morrill, J. Symons, Efficiency and inequality aspects of optimum location, Geogr. Anal. 9 (1977) 215–225.
[31] A.T. Murray, R.L. Church, Using proximity restrictions for locating undesirable facilities, Stud. Locational Anal. 12 (1999) 81–99.
[32] M. Ndiaye, Efficience sous contraintes et théorie de la localisation, Ph.D., Université de Bourgogne, Dijon, France, 1996, unpublished.
[33] S. Nickel, Bicriteria and restricted 2-facility Weber problems, Math. Methods Oper. Res. 45 (1997) 167–195.
[34] P. Pardalos, J.B. Rosen, Constrained Global Optimization, Springer, Berllin, 1987.
[35] F. Plastria, Optimal location of undesirable facilities: a selective overview, JORBEL 36 (1996) 109–127.
[36] D. Pisinger, Exact solution of p-dispersion problems, Technical Report 99/14, DIKU, University of Copenhagen, Denmark, 1999.
[37] D. Pisinger, On quadratic knapsack problems with push–pull objectives, Technical Report 00/12, DIKU, University of Copenhagen, Denmark, 2000.
[38] D. Pisinger, P. Toth, Knapsack problems, in: D-Z. Du, P. Pardalos (Eds.), Handbook of combinatorial Optimization, Vol. 1, Kluwer Academic Publishers, Dordrecht, 1998, pp. 299–428.
[39] S.S. Ravi, D.J. Rosenkrantz, G.K. Tayi, Heuristics and special case algorithms for dispersion problems, Oper. Res. 42 (1994) 299–310.
[40] A. Srivastav, K. Wolf, Finding dense subgraphs with semidefinite programming, in: K. Jansen, J. Rolim (Eds.), Approximation Algorithms for Combinatorial Optimization, Lecture Notes in Computer Science, Vol. 1444, Springer, Berlin, 1998, pp. 181–191.
[41] R.R. Vemuganti, Applications of set covering, set packing and set partitioning models: a survey, in: D-Z. Du, P. Pardalos (Eds.), Handbook of Combinatorial Optimization, Vol. 1, Kluwer Academic Publishers, Dordrecht, 1998, pp. 573–746.

N·H

ELSEVIER

Discrete Applied Mathematics 123 (2002) 379–396

DISCRETE
APPLIED
MATHEMATICS

Recent advances on two-dimensional bin packing problems

Andrea Lodi, Silvano Martello*, Daniele Vigo

Dipartimento di Elettronica, Informatica e Sistemistica, University of Bologna, Viale Risorgimento 2, 40136 - Bologna, Italy

Received 4 October 1999; received in revised form 1 June 2001; accepted 25 June 2001

Abstract

We survey recent advances obtained for the two-dimensional bin packing problem, with special emphasis on exact algorithms and effective heuristic and metaheuristic approaches. © 2002 Elsevier Science B.V. All rights reserved.

1. Introduction

In the *two-dimensional bin packing problem* (2BP) we are given a set of n rectangular *items* $j \in J = \{1, \ldots, n\}$, each having *width* w_j and *height* h_j, and an unlimited number of finite identical rectangular *bins*, having width W and height H. The problem is to allocate, without overlapping, all the items to the minimum number of bins, with their edges parallel to those of the bins. It is assumed that the items have fixed orientation, i.e., they cannot be rotated.

Problem 2BP has many industrial applications, especially in cutting (wood and glass industries) and packing (transportation and warehousing). Certain applications may require additional constraints and/or assumptions, some of which are discussed in the final section of this paper.

The special case where $w_j = W$ $(j = 1, \ldots, n)$ is the famous *one-dimensional bin packing problem* (1BP): partition n elements, each having an associated size h_j, into the minimum number of subsets so that the sum of the sizes in each subset does not exceed a given capacity H. Since 1BP is known to be strongly NP-hard, the same holds for 2BP.

---

* Corresponding author. Tel.: +39 051 209 3022; fax: +39 051 209 3073.

E-mail addresses: alodi@deis.unibo.it (A. Lodi), smartello@deis.unibo.it (S. Martello), dvigo@deis.unibo.it (D. Vigo).

Gilmore and Gomory [26] proposed the first model for two-dimensional packing problems, by extending their column generation approach for 1BP (see [24,25]). Beasley [5] considered a variant of 2BP (the *cutting stock problem*), and gave an Integer Linear Programming formulation based on the use of discrete coordinates at which items may be allocated. A similar model has been introduced by Hadjiconstantinou and Christofides [30]. Very recently, Fekete and Schepers [19] proposed a new model based on a graph-theoretical characterization of the problem, while Lodi et al. [40] presented, for the special case where the items have to be packed "by levels" (see below) ILP models involving a polynomial number of variables and constraints.

In this paper we survey recent advances obtained for the two-dimensional bin packing problem, with special emphasis on exact algorithms and effective heuristic and metaheuristic approaches. Concerning heuristics, we will only consider *off-line* algorithms, for which it is assumed that the algorithm has full knowledge of the whole input. (The reader is referred to [15], for a survey on *on-line* algorithms, which pack each item as soon as it is encountered, without knowledge of the next items.)

Up to the mid-Nineties, almost all results in the literature concerned heuristic algorithms. In the next section, we first review some basic algorithms having relevant implications on the topic of the present survey, and discuss more recent results. The following sections are devoted to other results obtained in the last few years: lower bounds (Section 3), exact algorithms (Section 4) and metaheuristic approaches (Section 5). We conclude by discussing some variants of the problem (Section 6) and future directions of research (Section 7). For many of the above techniques we summarize the results of computational experiments. For some upper and lower bounds, worst-case results are also discussed.

Without loss of generality, we will assume throughout the paper that all input data are positive integers, and that $w_j \leqslant W$ and $h_j \leqslant H$ ($j = 1, \ldots, n$).

2. Upper bounds

Most of the off-line algorithms from the literature are of greedy type, and can be classified in two families:
- *one-phase algorithms* directly pack the items into the finite bins;
- *two-phase algorithms* start by packing the items into a single *strip*, i.e., a bin having width W and infinite height. In the second phase, the strip solution is used to construct a packing into finite bins.

In addition, most of the approaches are *level algorithms*, i.e., the bin/strip packing is obtained by placing the items, from left to right, in rows forming *levels*. The first level is the bottom of the bin/strip, and subsequent levels are produced by the horizontal line coinciding with the top of the tallest item packed on the level below. Three classical strategies for the level packing have been derived from famous algorithms for the one-dimensional case. In each case, the items are initially sorted by non-decreasing height and packed in the corresponding sequence. Let j denote the current item, and s

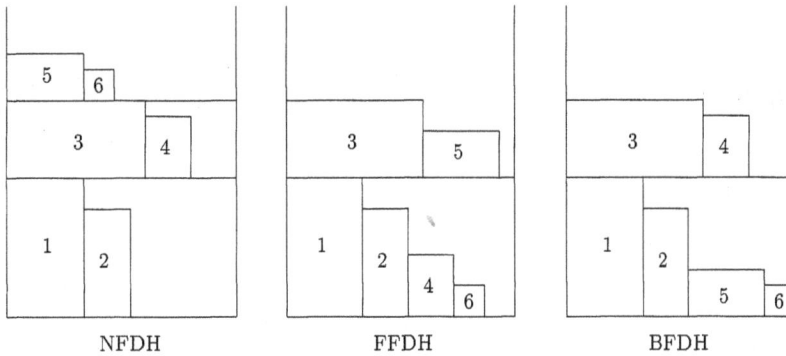

Fig. 1. Level packing strategies.

the last created level:

- *Next-Fit Decreasing Height* (NFDH) strategy: item j is packed left justified on level s, if it fits. Otherwise, a new level $(s:=s+1)$ is created, and j is packed left justified into it;
- *First-Fit Decreasing Height* (FFDH) strategy: item j is packed left justified on the first level where it fits, if any. If no level can accommodate j, a new level is initialized as in NFDH;
- *Best-Fit Decreasing Height* (BFDH) strategy: item j is packed left justified on that level, among those where it fits, for which the unused horizontal space is a minimum. If no level can accommodate j, a new level is initialized as in NFDH.

The above strategies are illustrated through an example in Fig. 1.

In what follows we assume, unless otherwise specified, that the items are initially sorted by non-increasing height.

2.1. Strip packing

Coffman et al. [12] analyzed NFDH and FFDH for the solution of the *two-dimensional strip packing problem*, in which one is required to pack all the items into a strip of minimum height, and determined their asymptotic worst-case behavior. Given a minimization problem P and an approximation algorithm A, let $A(I)$ and $OPT(I)$ denote the value produced by A and the optimal solution value, respectively, for an instance I of P. Coffman et al. [12] proved that, if the heights are normalized so that $\max_j\{h_j\} = 1$, then

$$NFDH(I) \leqslant 2 \cdot OPT(I) + 1 \tag{1}$$

and

$$FFDH(I) \leqslant \tfrac{17}{10} \cdot OPT(I) + 1 \tag{2}$$

Both bounds are *tight* (meaning that the multiplicative constants are as small as possible) and, if the h_j's are not normalized, only the additive term is affected. Observe the similarity of (1) and (2) with famous results on the one-dimensional counterparts of NFDH and FFDH (algorithms *Next-Fit* and *First-Fit*, respectively, see [34]).

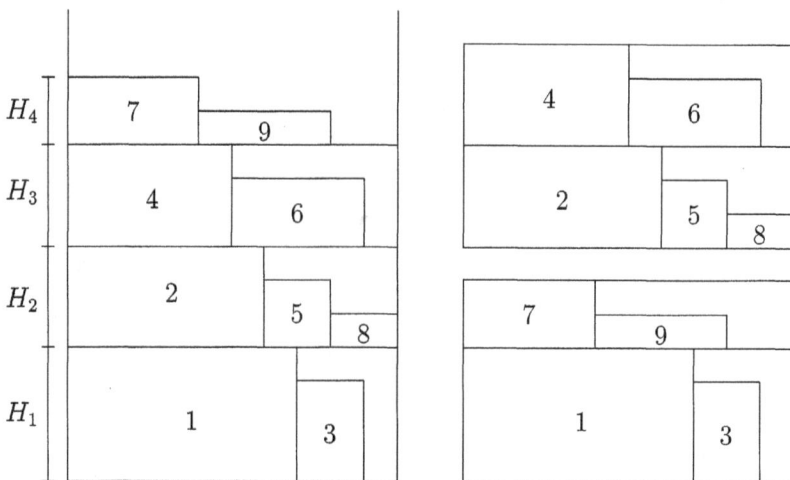

Fig. 2. First and second phase of algorithm HFF.

Any algorithm requiring item sorting is obviously $\Omega(n \log n)$. Both NFDH and FFDH can be implemented so as to require $O(n \log n)$ time, by using the appropriate data structures adopted for the one-dimensional case (see [33]).

Several other papers on the strip packing problem can be found in the literature: see, e.g., [3,45,8,28,2,4,32,46]. The algorithm of Baker et al. [3] is considered in Section 2.4, while the other results, which have not been directly used for the finite bin case, are beyond the scope of this survey, and will not be discussed here.

2.2. Bin packing: two-phase algorithms

A two-phase algorithm for the finite bin packing problem, called *Hybrid First-Fit* (HFF), was proposed by Chung et al. [11]. In the first phase, a strip packing is obtained through the FFDH strategy. Let $H_1, H_2, \ldots$ be the heights of the resulting levels, and observe that $H_1 \geqslant H_2 \geqslant \ldots$. A finite bin packing solution is then obtained by heuristically solving a one-dimensional bin packing problem (with item sizes H_i and bin capacity H) through the *First-Fit Decreasing* algorithm: initialize bin 1 to pack level 1, and, for increasing $i = 2, \ldots$, pack the current level i into the lowest indexed bin where it fits, if any; if no bin can accommodate i, initialize a new bin. An example is shown in Fig. 2. Chung et al. [11] proved that, if the heights are normalized to one, then

$$HFF(I) \leqslant \tfrac{17}{8} \cdot OPT(I) + 5. \tag{3}$$

The bound is not proved to be tight: the worst example gives $HFF(I) = \tfrac{91}{45} \cdot (OPT(I) - 1)$. Both phases can be implemented so as to require $O(n \log n)$ time.

Berkey and Wang [7] proposed and experimentally evaluated a two-phase algorithm, called *Finite Best-Strip* (FBS), which is a variation of HFF. The first phase is performed by using the BFDH strategy. In the second phase, the one-dimensional bin

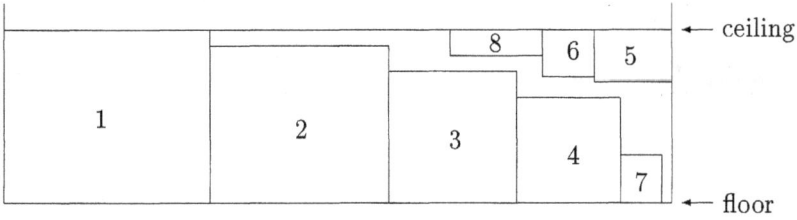

Fig. 3. Algorithm FC.

packing problem is solved through the *Best-Fit Decreasing* algorithm: pack the current level in that bin, among those where it fits (if any), for which the unused vertical space is a minimum, or by initializing a new bin. (For the sake of uniformity, *Hybrid Best-Fit* would be a more appropriate name for this algorithm.)

Let us consider now another variation of HFF, in which the NFDH strategy is adopted in the first phase, and the one-dimensional bin packing problem is solved through the *Next-Fit Decreasing* algorithm: pack the current level in the current bin if it fits, or initialize a new (current) bin otherwise. Due to the next-fit policy, this algorithm is equivalent to a one-phase algorithm in which the current item is packed on the current level of the current bin, if possible; otherwise, a new (current) level is initialized either in the current bin (if enough vertical space is available), or in a new (current) bin. Frenk and Galambos [23] analyzed the resulting algorithm, *Hybrid Next-Fit* (HNF), by characterizing its asymptotic worst-case performance as a function of $\max_j\{w_j\}$ and $\max_j\{h_j\}$. By assuming that the heights and widths are normalized to one, the worst performance occurs for $\max_j\{w_j\} > \frac{1}{2}$ and $\max_j\{h_j\} \geqslant \frac{1}{2}$, and gives:

$$HNF(I) \leqslant 3.382\ldots \cdot OPT(I) + 9 \tag{4}$$

where $3.382\ldots$ is an approximation for a tight but irrational bound. The three algorithms above can be implemented so as to require $O(n \log n)$ time. The next two algorithms have higher worst-case time complexities, although they are, in practice, very fast and effective.

Lodi et al. [37,39] presented an approach (*Floor-Ceiling*, FC) which extends the way items are packed on the levels. Denote the horizontal line defined by the top (resp. bottom) edge of the tallest item packed on a level as the *ceiling* (resp. *floor*) of the level. The previous algorithms pack the items, from left to right, with their bottom edge on the level floor. Algorithm FC may, in addition, pack them, from right to left, with their top edge on the level ceiling. The first item packed on a ceiling can only be one which cannot be packed on the floor below. A possible floor-ceiling packing is shown in Fig. 3. In the first phase, the current item is packed, in order of preference: (i) on a ceiling (provided that the requirement above is satisfied), according to a best-fit strategy; (ii) on a floor, according to a best-fit strategy; (iii) on the floor of a new level. In the second phase, the levels are packed into finite bins, either through the Best-Fit Decreasing algorithm or by using an exact algorithm for the one-dimensional

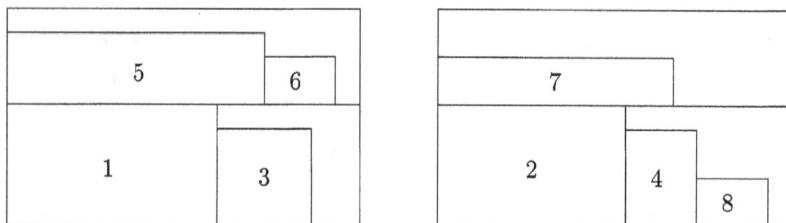

Fig. 4. Algorithm FFF.

bin packing problem, halted after a prefixed number of iterations. The implementation of the first phase given in [37] requires $O(n^3)$ time, while the complexity of the second one obviously depends on the selected algorithm.

Another level packing strategy based on the exact solution of induced subproblems is adopted in the *Knapsack Packing* (KP) algorithm proposed by Lodi et al. [39]. In the (binary) *knapsack problem* one has to select a subset of n elements, each having an associated profit and weight, so that the total weight does not exceed a given capacity and the total profit is a maximum. The first phase of algorithm KP packs one level at a time as follows. The first (tallest) unpacked item, say j^*, initializes the level, which is then completed by solving an associated knapsack problem instance over all the un-packed items, where: (i) the knapsack capacity is $W - w_{j^*}$; (ii) the weight of an item j is w_j; (iii) the profit of an item j is its area $w_j h_j$. Finite bins are finally obtained as in algorithm FC. Algorithm KP (as well as algorithm FC above) may require the solution of NP-hard subproblems, producing a non-polynomial time complexity. In practice, however, the execution of the codes for the NP-hard problems is always halted after a prefixed (small) number of iterations, and in almost all cases, the optimal solution is obtained before the limit is reached (see the computational experiments in [39] and in Section 2.5).

2.3. Bin packing: one-phase algorithms

Two one-phase algorithms were presented and experimentally evaluated by Berkey and Wang [7].

Algorithm *Finite Next-Fit* (FNF) directly packs the items into finite bins exactly in the way algorithm HNF of the previous section does. (Papers [7,23] appeared in the same year.)

Algorithm *Finite First-Fit* (FFF) adopts instead the FFDH strategy. The current item is packed on the lowest level of the first bin where it fits; if no level can accommodate it, a new level is created either in the first suitable bin, or by initializing a new bin (if no bin has enough vertical space available). An example of application of FFF is given in Fig. 4.

Both algorithms can be implemented so as to require $O(n \log n)$ time.

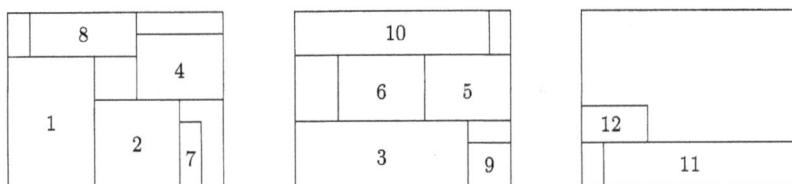

Fig. 5. Algorithm AD.

2.4. Bin packing: non-level algorithms

We finally consider algorithms which do not pack the items by levels. All the algorithms discussed in the following are one-phase.

The main non-level strategy is known as *Bottom-Left* (BL), and consists in packing the current item in the lowest possible position, left justified. Baker et al. [3] analyzed the worst-case performance of the resulting algorithm for the strip packing problem, and proved that: (i) if no item ordering is used, BL may be arbitrarily bad; (ii) if the items are ordered by non-increasing width then $BL(I) \leqslant 3 \cdot OPT(I)$, and the bound is tight.

Berkey and Wang [7] proposed the BL approach for the finite bin case. Their *Finite Bottom-Left* (FBL) algorithm initially sorts the items by non-increasing width. The current item is then packed in the lowest position of any initialized bin, left justified; if no bin can allocate it, a new one is initialized. The computer implementation of algorithm BL was studied by Chazelle [9], who gave a method for producing a packing in $O(n^2)$ time. The same approach was adopted by Berkey and Wang [7].

Lodi et al. [39] proposed a different non-level approach, called *alternate directions* (AD). The method is illustrated in Fig. 5. The algorithm initializes L bins (L being a lower bound on the optimal solution value, see Section 3) by packing on their bottoms a subset of the items, following a best-fit decreasing policy (items 1, 2, 3, 7 and 9 in Fig. 5, where it is assumed that $L = 2$). The remaining items are packed, one bin at a time, into *bands*, alternatively from left to right and from right to left. As soon as no item can be packed in either direction in the current bin, the next initialized bin or a new empty bin (the third one in Fig. 5, when item 11 is considered) becomes the current one. The algorithm has $O(n^3)$ time complexity.

2.5. Computational experiments

Probabilistic analysis and experimental tests are two classical methods for evaluating the expected behavior of approximation algorithms. The former technique is fully illustrated in the book by Coffman and Lueker [13], where specific results on two-dimensional bin packing and strip packing algorithms can be found in Chapter 7. More recent results are in [14]. In this section we summarize the outcome of a series of computational experiments aimed at analyzing the typical behavior of the

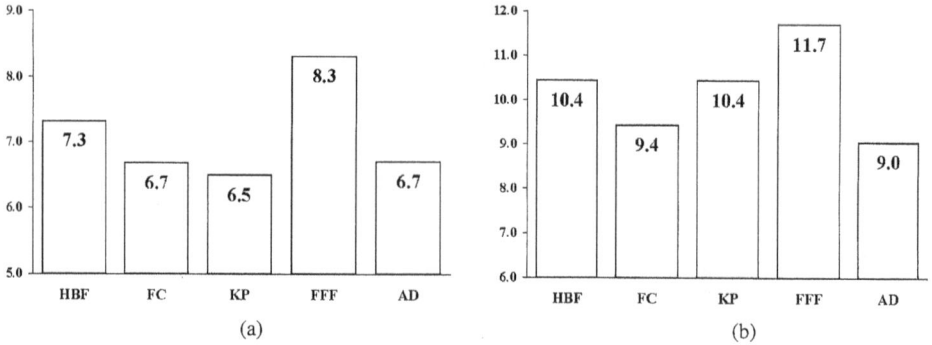

Fig. 6. Average percentage deviations from lower bound: (a) Martello and Vigo instance, Classes 1–4; (b) Berkey and Wang instances, Classes 5–10.

main heuristic algorithms on some classes of instances. The benchmark consists of 500 random instances, with $n \in \{20, 40, 60, 80, 100\}$. Ten different classes of instances were used.

The first four classes were proposed by Martello and Vigo [43], and are based on the generation of items of four different types:

Type 1: w_j uniformly random in $[\frac{2}{3}W, W]$, h_j uniformly random in $[1, \frac{1}{2}H]$.

Type 2: w_j uniformly random in $[1, \frac{1}{2}W]$, h_j uniformly random in $[\frac{2}{3}H, H]$.

Type 3: w_j uniformly random in $[\frac{1}{2}W, W]$, h_j uniformly random in $[\frac{1}{2}H, H]$.

Type 4: w_j uniformly random in $[1, \frac{1}{2}W]$, h_j uniformly random in $[1, \frac{1}{2}H]$.

Class k ($k \in \{1, 2, 3, 4\}$) is then obtained by generating an item of type k with probability 70%, and of the remaining types with probability 10% each. The bin size is always $W = H = 100$.

The next six classes have been proposed by Berkey and Wang [7]:

Class 5: $W = H = 10$, w_j and h_j uniformly random in $[1, 10]$.

Class 6: $W = H = 30$, w_j and h_j uniformly random in $[1, 10]$.

Class 7: $W = H = 40$, w_j and h_j uniformly random in $[1, 35]$.

Class 8: $W = H = 100$, w_j and h_j uniformly random in $[1, 35]$.

Class 9: $W = H = 100$, w_j and h_j uniformly random in $[1, 100]$.

Class 10: $W = H = 300$, w_j and h_j uniformly random in $[1, 100]$.

For each class and value of n, ten instances have been generated. The 500 instances, as well as the generator code, are available on the internet at http://www.or.deis.unibo.it/ORinstances/2BP/.

Fig. 6 summarizes the results, by giving, for each algorithm, the average percentage deviation of the heuristic solution value from a lower bound value, computed as $\max\{L_2, L_3\}$ (see Section 3), with respect to the 200 instances of Classes 1–4 (Fig. 6 (a)) and to the 300 instances of Classes 5–10 (Fig. 6 (b)). The results show that algorithms FC, KP and AD have the best behavior, clearly superior to the classical approaches.

3. Lower bounds

Good lower bounds on the optimal solution value are important both in the implementation of exact enumerative approaches and in the empirical evaluation of approximate solutions. The simplest bound for 2BP is the *Continuous Lower Bound*

$$L_0 = \left\lceil \frac{\sum_{j=1}^{n} w_j h_j}{WH} \right\rceil$$

computable in linear time. Martello and Vigo [43] determined the absolute worst-case behavior of L_0:

$$L_0(I) \geqslant \tfrac{1}{4} \cdot OPT(I)$$

where $L_0(I)$ and $OPT(I)$ denote the value produced by L_0 and the optimal solution value, respectively, for an instance I of problem P. The bound is tight, as shown by the example in Fig. 7. The result holds even if rotation of the items (by any angle) is allowed.

In many cases, the value provided by L_0 can be inadequate (too small) for an effective use within an exact algorithm. A better (tighter) bound was proposed by Martello and Vigo [43]. Given any integer value q, $1 \leqslant q \leqslant \tfrac{1}{2}W$, let

$$K_1 = \{j \in J : w_j > W - q\}, \tag{5}$$

$$K_2 = \{j \in J : W - q \geqslant w_j > \tfrac{1}{2}W\}, \tag{6}$$

$$K_3 = \{j \in J : \tfrac{1}{2}W \geqslant w_j \geqslant q\}. \tag{7}$$

and observe that no two items of $K_1 \cup K_2$ may be packed side by side into a bin. Hence, a lower bound L_1^W for the sub-instance given by the items in $K_1 \cup K_2$ can be obtained by using any lower bound for the 1BP instance defined by element sizes h_j ($j \in K_1 \cup K_2$) and capacity H (see [42,17]). A lower bound for the complete instance is then obtained by taking into account the items in K_3, since none of them may be packed besides an item of K_1:

$$L_2^W(q) = L_1^W + \max\left\{0, \left\lceil \frac{\sum_{j \in K_2 \cup K_3} w_j h_j - (HL_1^W - \sum_{j \in K_1} h_j)W}{WH} \right\rceil\right\}. \tag{8}$$

A symmetric bound $L_2^H(q)$ is clearly obtained by interchanging widths and heights. By observing that both bounds are valid for any q, we have an overall lower bound:

$$L_2 = \max\left(\max_{1 \leqslant q \leqslant \frac{1}{2}W} \{L_2^W(q)\}, \max_{1 \leqslant q \leqslant \frac{1}{2}H} \{L_2^H(q)\} \right) \tag{9}$$

It is shown in [43] that, for any instance of 2BP, the value produced by L_2 is no less than that produced by L_0, and that L_2 can be computed in $O(n^2)$ time.

Martello and Vigo [43] also proposed a computationally more expensive lower bound, which in some cases improves on L_2. Given any pair of integers (p, q), with

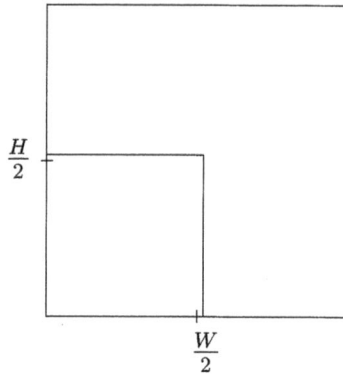

Fig. 7. Worst-case of the continuous lower bound.

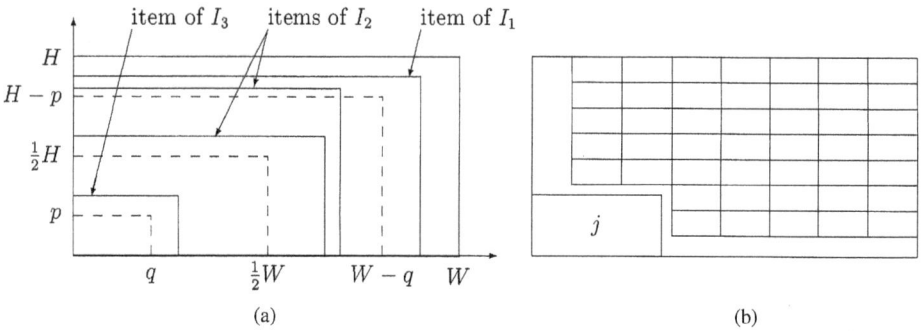

Fig. 8. (a) Items in I_1, I_2 and I_3; (b) relaxed instance with reduced items.

$1 \leqslant p \leqslant \frac{1}{2}H$ and $1 \leqslant q \leqslant \frac{1}{2}W$, define:

$$I_1 = \{j \in J : h_j > H - p \text{ and } w_j > W - q\}, \tag{10}$$

$$I_2 = \{j \in J \setminus I_1 : h_j > \tfrac{1}{2}H \text{ and } w_j > \tfrac{1}{2}W\}, \tag{11}$$

$$I_3 = \{j \in J : \tfrac{1}{2}H \geqslant h_j \geqslant p \text{ and } \tfrac{1}{2}W \geqslant w_j \geqslant q\} \tag{12}$$

(see Fig. 8(a)), and observe that: (i) $I_1 \cup I_2$ is independent of (p, q); (ii) no two items of $I_1 \cup I_2$ may be packed into the same bin; (iii) no item of I_3 fits into a bin containing an item of I_1. A valid lower bound can thus be computed by adding to $|I_1 \cup I_2|$ the minimum number of bins needed for those items of I_3 that cannot be packed into the bins used for the items of I_2. Such a bound can be determined by considering a relaxed instance where each item $i \in I_3$ has the minimum size, i.e., $h_i = p$ and $w_i = q$. Given a bin containing an item j, the maximum number of $p \times q$ items that can be packed into the bin is (see Fig. 8(b)):

$$m(j, p, q) = \left\lfloor \frac{H}{p} \right\rfloor \left\lfloor \frac{W - w_j}{q} \right\rfloor + \left\lfloor \frac{W}{q} \right\rfloor \left\lfloor \frac{H - h_j}{p} \right\rfloor - \left\lfloor \frac{H - h_j}{p} \right\rfloor \left\lfloor \frac{W - w_j}{q} \right\rfloor \quad (13)$$

Hence, for any pair (p, q) a valid lower bound is

$$L_3(p, q) = |I_1 \cup I_2| + \max \left\{ 0, \left\lceil \frac{|I_3| - \sum_{j \in I_2} m(j, p, q)}{\lfloor \frac{H}{p} \rfloor \lfloor \frac{W}{q} \rfloor} \right\rceil \right\} \quad (14)$$

so an overall bound is

$$L_3 = \max_{1 \leq p \leq \frac{1}{2} H, \ 1 \leq q \leq \frac{1}{2} W} \{ L_3(p, q) \}. \quad (15)$$

Lower bound L_3 can be computed in $O(n^3)$ time. No dominance relation exists between L_2 and L_3.

A general bounding technique for bin and strip packing problems in one or more dimensions, based on the use of dual feasible functions, was recently proposed by Fekete and Schepers [22,20].

4. Exact algorithms

An enumerative approach for the exact solution of 2BP was presented by Martello and Vigo [43]. The items are initially sorted in non-increasing order of their area. A reduction procedure tries to determine the optimal packing of some bins, thus reducing the size of the instance. A first incumbent solution, of value z^*, is then heuristically obtained.

The algorithm is based on a two-level branching scheme:

- *outer branch-decision tree*: at each decision node, an item is assigned to a bin without specifying its actual position;
- *inner branch-decision tree*: a feasible packing (if any) for the items currently assigned to a bin is determined, possibly through enumeration of all the possible patterns.

The outer branch-decision tree is searched in a depth-first way, making use of the lower bounds described in the previous section. Whenever it is possible to establish that no more unassigned items can be assigned to a given initialized bin, such a bin is *closed*: an initialized and not closed bin is called *active*. At level k $(k = 1, \ldots, n)$, item k is assigned, in turn, to all the active bins and, possibly, to a new one (if the total number of active and closed bins is less than $z^* - 1$).

The feasibility of the assignment of an item to a bin is first heuristically checked. A lower bound $L(I)$ is computed for the instance I defined by the items currently assigned to the bin: if $L(I) > 1$, a backtracking follows. Otherwise, heuristic algorithms are applied to I: if a feasible single-bin packing is found, the outer enumeration is resumed. If not, the inner branching scheme enumerates all the possible ways to pack I into a bin through the *left-most downward* strategy (see [29]): at each level, the next item is placed, in turn, into all positions where it has its left edge adjacent either to the

Table 1
Number of instances, out of ten, solved to proved optimality

	Class										
n	1	2	3	4	5	6	7	8	9	10	Total
20	10	10	10	10	10	10	9	10	10	10	99
40	8	7	10	9	10	10	10	10	10	5	89
60	8	7	10	2	7	4	7	7	8	10	70
80	7	3	10	—	3	10	—	10	—	10	53
100	7	6	8	—	1	10	—	10	1	2	45
Total	40	33	48	21	31	44	26	47	29	37	356

right edge of another item or to the left edge of the bin, and its bottom edge adjacent either to the top edge of another item or to the bottom edge of the bin. As soon as a feasible packing is found for all the items of I, the outer enumeration is resumed. If no such packing exists, an outer backtracking is performed.

Whenever the current assignment is feasible, the possibility of closing the bin is checked through lower bound computations.

Table 1 gives the results of computational experiments performed, with a Fortran 77 implementation, on the 500 instances described in Section 2.5. The entries give, for each class and value of n, the number of instances (out of ten) solved to proved optimality within a time limit of 300 CPU seconds on a PC Pentium 200 MHz.

Fekete and Schepers [21] recently derived from their graph-theoretical model [19] an alternative enumerative approach.

5. Metaheuristics

In recent years, metaheuristic techniques have become a popular tool for the approximate solution of hard combinatorial optimization problems. (See [1,27] for general introductions to the field.) Lodi et al. [37–39] developed effective tabu search algorithms for 2BP and for some of the variants discussed in the next section. We briefly describe here the unified tabu search framework given in [39], whose main characteristic is the adoption of a search scheme and a neighborhood which are independent of the specific packing problem to be solved. The framework can thus be used for virtually any variant of 2BP, by simply changing the specific deterministic algorithm used for evaluating the moves within the neighborhood search.

Given a current solution, the moves modify it by changing the packing of a subset S of items, trying to empty a specified *target bin*. Let S_i be the set of items currently packed into bin i: the target bin t is the one minimizing, over all bins i, the function

$$\varphi(S_i) = \alpha \frac{\sum_{j \in S_i} w_j h_j}{WH} - \frac{|S_i|}{n} \tag{16}$$

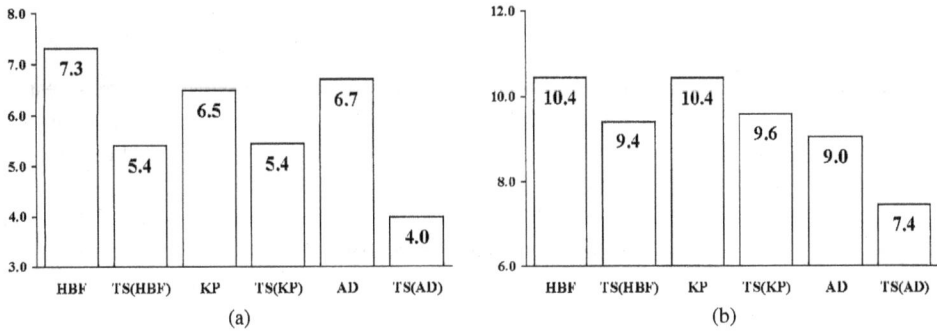

Fig. 9. Average percentage deviations from lower bound: (a) Martello and Vigo instances, Classes 1–4; (b) Berkey and Wang instances, Classes 5–10.

(α is a pre-specified positive weight), which gives a measure of the easiness of emptying the bin. It favors, indeed, target bins packing a small area and a relatively large number of items.

Once the target bin has been selected, subset S is defined so as to include one item, j, from the target bin and the current contents of k other bins. The new packing for S is obtained by executing an appropriate heuristic algorithm A on S. The value of parameter k, which defines the size and the structure of the current neighborhood, is automatically updated during the search.

If the move packs the items of S into k (or less) bins, i.e., item j has been removed from the target bin, a new item is selected, a new set S is defined accordingly, and a new move is performed. Otherwise S is changed by selecting a different set of k bins, or a different item j from the target bin (if all possible configurations of k bins have been attempted for the current j).

If the algorithm gets stuck, i.e., the target bin is not emptied, the neighborhood is enlarged by increasing the value of k, up to a prefixed upper limit. There are a tabu list and a tabu tenure for each value of k.

An initial incumbent solution is obtained by executing algorithm A on the complete instance, while the initial tabu search solution consists of packing one item per bin. In special situations, a move is followed by a diversification action. The execution is halted as soon as a proven optimal solution is found, or a time limit is reached.

Fig. 9 shows the impact of tabu search for three of the heuristics described in Section 2: notation TS(A) indicates that algorithm A is used within the tabu search. The figure gives the average percentage deviations of the heuristic solution value (without and with tabu search) from the best known lower bound value, with respect to the 200 instances of Classes 1–4 (Fig. 9(a)) and to the 300 instances of Classes 5–10 (Fig. 9(b)), as described in Section 2.5. The tabu search was performed with a time limit of 60 s on a Silicon Graphics INDY R10000sc (195 MHz). The results show that the use of tabu search considerably improves the performance of all algorithms.

6. Variants

Two-dimensional bin packing problems occur in several real-world contexts, especially in cutting and packing industries. As a consequence, a number of variants arises, according to specific applications. In most cases the additional requirements concern *orientation* and/or *guillotine cutting*.

In the bin and strip packing problems considered so far we have assumed that the items have a fixed orientation (i.e., they cannot be rotated), and that no restriction is imposed on the cutting patterns. In certain real-world contexts, item rotation (usually by 90°) may be allowed in order to produce better packings. In addition, many practical cutting contexts may impose that the items are obtained through a sequence of guillotine cuts, i.e., edge-to-edge cuts parallel to the edges of the bin. For example, rotation is not allowed when the items are articles to be paged in newspapers or are pieces to be cut from decorated or corrugated stock units, whereas it is allowed in the cutting of plain materials and in most packing contexts. The guillotine constraint is usually imposed by technological characteristics of the automated cutting machines, whereas it is generally not present in packing applications.

Lodi et al. [39] proposed the following typology for the four possible cases produced by the above two characterizations:

2BP|O|G: the items are oriented (O) and guillotine cutting (G) is required;
2BP|R|G: the items may be rotated by 90° (R) and guillotine cutting is required;
2BP|O|F: the items are oriented and cutting is free (F);
2BP|R|F: the items may be rotated by 90° and cutting is free.

(The problem considered so far is thus 2BP|O|F.) The following references are examples of industrial applications involving the above variants. A problem of trim-loss minimization in a crepe-rubber mill, studied by Schneider [44], induces subproblems of 2BP|O|G type; fuzzy two-dimensional cutting stock problems arising in the steel industry, discussed by Vasko et al. [47], are related to 2BP|R|G; the problem of optimally placing articles and advertisements in newspapers and yellow pages, studied by Lagus et al. [36], falls into the 2BP|O|F case; finally, several applications of 2BP|R|F are considered by Bengtsson [6].

An algorithm for one of the variants may obviously guarantee solutions which are feasible for others. The complete set of compatibilities between algorithms and problems is shown in Fig. 10, where A_{XY} is an algorithm for 2BP$|X|Y$ and an edge $(A_{XY}, 2BP|Q|T)$ indicates that A_{XY} produces solutions feasible for 2BP$|Q|T$.

It is easily seen that all the level algorithms described in Section 2 directly produce guillotine packings, the only exception being FC. Adaptations of FC which modify the way items are packed on the ceilings so as to preserve the guillotine constraint (with or without rotation) were presented by Lodi et al. [37,39].

Most of the other heuristics of Section 2 can be modified so as to handle rotation and/or guillotine cutting. Lodi et al. [39] proposed the following new algorithms for the considered variants.

For 2BP|O|G, algorithm KP of Section 2.2 (denoted as KP$_{OG}$ in [39]) directly produces guillotine packings.

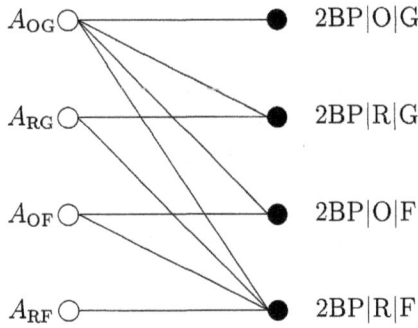

Fig. 10. Compatibilities between algorithms and problems.

For 2BP|R|G, algorithm KP$_{RG}$ modifies KP as follows. The items are initially sorted according to non-increasing value of their shortest edge, and *horizontally oriented* (i.e., with their longest edge as the base): this orientation is always used for the level initializations. For each level, say of height h^*, the knapsack instance includes each unpacked item, either in vertical orientation, if its size does not exceed h^*, or in horizontal orientation, otherwise. Once a feasible finite bin solution has been obtained from the resulting levels, an alternative solution is derived by considering the set of the items currently packed in each level as a pseudo-item, rotating it whenever possible, and applying algorithm KP to the resulting instance.

For 2BP|R|F, algorithm TP$_{RF}$ sorts the items according to non-increasing area, horizontally orients them and reserves L empty bins (L a lower bound on the optimal solution value, see [16]). The algorithm packs one item at a time (either in an existing bin, or by initializing a new one) in a so-called *normal position* (see Christofides and Whitlock [10]), i.e., with its bottom edge touching either the bottom of the bin or the top edge of another item, and with its left edge touching either the left edge of the bin or the right edge of another item. The choice of the bin and of the packing position is done by evaluating a *score*, defined as the percentage of the item perimeter which touches either the bin or other already packed items.

We finally mention that other variants of the two-dimensional bin packing problem can be found in the literature. For example, in guillotine cutting, an upper bound (usually two or three) may be imposed on the number of *stages* (rounds of cuts having the same direction) that are needed to obtain all the items: see, e.g., [31,40,41]. Note that all the level algorithms of Section 2 but FC produce two-stage packings (with trimming). In certain practical applications a secondary objective can also be of interest, namely the maximization of the unused area in one bin, so as to produce a possibly large trim to be used later: see, e.g., [6,18] for 2BP|R|F.

Note that the metaheuristic approach of Section 5 solves all of the above variants, by appropriately changing the deterministic algorithm used for evaluating the moves within the neighborhood search.

7. Conclusions and directions of research

We surveyed recent advances on exact algorithms and effective off-line heuristic and metaheuristic approaches to the two-dimensional bin packing problem, and to some variants typically arising in practical applications. The computational experiments show that many instances of small to moderate size can be solved to optimality, while larger instances can only be handled by approximation algorithms.

In order to increase the effectiveness of the exact approaches, especially those based on implicit enumeration, tighter lower bounds should be devised, capable of fathoming a larger number of branch-decision nodes. In addition, other exact methods, such as, for example, column generation techniques, may lead to interesting results. As far as approximate solutions are concerned, an important open problem is to find polynomial-time approximation schemes. A fully polynomial-time approximation scheme was recently developed, for the two-dimensional strip packing problem, by Kenyon and Rémila [35]. For 2BP, instead, to our knowledge, no polynomial-time approximation scheme is known.

Acknowledgements

We thank the Ministero dell'Università e della Ricerca Scientifica e Tecnologica (MURST) and the Consiglio Nazionale delle Ricerche (CNR), Italy, for the support given to this project. The computational experiments have been executed at the Laboratory of Operations Research of the University of Bologna (Lab.O.R.). We are indebted to two anonymous referees for useful comments.

References

[1] E. Aarts, J.K. Lenstra (Eds.), Local Search in Combinatorial Optimization, John Wiley & Sons, Chichester, 1997.

[2] B.S. Baker, D.J. Brown, H.P. Katseff, A 5/4 algorithm for two-dimensional packing, J. Algorithms 2 (1981) 348–368.

[3] B.S. Baker, E.G. Coffman Jr., R.L. Rivest, Orthogonal packing in two dimensions, SIAM J. Comput. 9 (1980) 846–855.

[4] B.S. Baker, J.S. Schwarz, Shelf algorithms for two-dimensional packing problems, SIAM J. Comput. 12 (1983) 508–525.

[5] J.E. Beasley, An exact two-dimensional non-guillotine cutting tree search procedure, Oper. Res. 33 (1985) 49–64.

[6] B.E. Bengtsson, Packing rectangular pieces—a heuristic approach, Comput. J. 25 (1982) 353–357.

[7] J.O. Berkey, P.Y. Wang, Two dimensional finite bin packing algorithms, J. Oper. Res. Soc. 38 (1987) 423–429.

[8] D.J. Brown, An improved BL lower bound, Inform. Process. Lett. 11 (1980) 37–39.

[9] B. Chazelle, The bottom-left bin packing heuristic: An efficient implementation, IEEE Trans. on Comput. 32 (1983) 697–707.

[10] N. Christofides, C. Whitlock, An algorithm for two-dimensional cutting problems, Oper. Res. 25 (1977) 30–44.

[11] F.K.R. Chung, M.R. Garey, D.S. Johnson, On packing two-dimensional bins, SIAM J. Algebraic Discrete Meth. 3 (1982) 66–76.

[12] E.G. Coffman Jr., M.R. Garey, D.S. Johnson, R.E. Tarjan, Performance bounds for level-oriented two-dimensional packing algorithms, SIAM J. Comput. 9 (1980) 801–826.

[13] E.G. Coffman Jr., G.S. Lueker, Probabilistic Analysis of Packing and Partitioning Algorithms, John Wiley & Sons, Chichester, 1992.

[14] E.G. Coffman Jr., P.W. Shor, Packings in two dimensions: Asymptotic average-case analysis of algorithms, Algorithmica 9 (1993) 253–277.

[15] J. Csirik, G. Woeginger, On-line packing and covering problems, in: Online algorithms, Lecture Notes in Computer Science, eds. F. Amos and G. Woeginger, Vol. 144, Springer, Berlin, 1998, pp. 147–177.

[16] M. Dell'Amico, S. Martello, D. Vigo, A lower bound for the non-oriented two-dimensional bin packing problem, Discrete Appl. Math. 2002, to appear.

[17] M. Dell'Amico, S. Martello, Optimal scheduling of tasks on identical parallel processors, ORSA J. Comput. 7 (1995) 191–200.

[18] A. El-Bouri, N. Popplewell, S. Balakrishnan, A. Alfa, A search based heuristic for the two-dimensional bin-packing problem, INFOR 32 (1994) 265–274.

[19] S.P. Fekete, J. Schepers, On more-dimensional packing I: Modeling, Technical Report ZPR97-288, Mathematisches Institut, Universität zu Köln, 1997.

[20] S.P. Fekete, J. Schepers, On more-dimensional packing II: Bounds, Technical Report ZPR97-289, Mathematisches Institut, Universität zu Köln, 1997.

[21] S.P. Fekete, J. Schepers, On more-dimensional packing III: Exact algorithms, Technical Report ZPR97-290, Mathematisches Institut, Universität zu Köln, 1997.

[22] S.P. Fekete, J. Schepers, New classes of lower bounds for bin packing problems, in: Integer Programming and Combinatorial Optimization (IPCO 98), Lecture Notes in Computer Science, eds. R.E. Bixby, E.A. Boyd and R.Z. Ríos-Mercado, Vol. 1412, Springer, Berlin, 1998, pp. 257–270.

[23] J.B. Frenk, G.G. Galambos, Hybrid next-fit algorithm for the two-dimensional rectangle bin-packing problem, Computing 39 (1987) 201–217.

[24] P.C. Gilmore, R.E. Gomory, A linear programming approach to the cutting stock problem, Oper. Res. 9 (1961) 849–859.

[25] P.C. Gilmore, R.E. Gomory, A linear programming approach to the cutting stock problem—part II, Oper. Res. 11 (1963) 863–888.

[26] P.C. Gilmore, R.E. Gomory, Multistage cutting problems of two and more dimensions, Oper. Res. 13 (1965) 94–119.

[27] F. Glover, M. Laguna, Tabu Search, Kluwer Academic Publishers, Boston, 1997.

[28] I. Golan, Performance bounds for orthogonal oriented two-dimensional packing algorithms, SIAM J. Comput. 10 (1981) 571–582.

[29] E. Hadjiconstantinou, N. Christofides, An exact algorithm for general, orthogonal, two-dimensional knapsack problems, Eur. J. Oper. Res. 83 (1995) 39–56.

[30] E. Hadjiconstantinou, N. Christofides, An exact algorithm for the orthogonal, 2-D cutting problems using guillotine cuts, Eur. J. Oper. Res. 83 (1995) 21–38.

[31] M. Hifi, Exact algorithms for large-scale unconstrained two and three staged cutting problems, In Contribution à la Résolution de Quelques Problèmes Difficiles de l'Optimisation Combinatoire, Thèse d'Habilitation à Diriger des Recherches en Informatique, Université de Versailles–Saint Quentin en Yvelines, 1998.

[32] S. Høyland, Bin-packing in 1.5 dimension, In Proceedings of the Scandinavian Workshop on Algorithm Theory, Lecture Notes in Computer Science, Vol. 318, Springer, Berlin, 1988, pp. 129–137.

[33] D.S. Johnson, Near-optimal bin packing algorithms, PhD Thesis, MIT, Cambridge, MA, 1973.

[34] D.S. Johnson, A. Demers, J.D. Ullman, M.R. Garey, R.L. Graham, Worst-case performance bounds for simple one-dimensional packing algorithms, SIAM J. Comput. 3 (1974) 299–325.

[35] C. Kenyon, E. Rémila, A near-optimal solution to a two-dimensional cutting stock problem, Math. Oper. Res. 25 (2000) 645–656.

[36] K. Lagus, I. Karanta, J. Ylä-Jääski, Paginating the generalized newspaper: A comparison of simulated annealing and a heuristic method. Proceedings of the Fifth International Conference on Parallel Problem Solving from Nature, Berlin, 1996, pp. 549–603.

[37] A. Lodi, S. Martello, D. Vigo, Neighborhood search algorithm for the guillotine non-oriented two-dimensional bin packing problem, in: S. Voss, S. Martello, I.H. Osman, C. Roucairol (Eds.),

Meta-Heuristics: Advances and Trends in Local Search Paradigms for Optimization, Kluwer Academic Publishers, Boston, 1998, pp. 125–139.

[38] A. Lodi, S. Martello, D. Vigo, Approximation algorithms for the oriented two-dimensional bin packing problem, Eur. J. Oper. Res. 112 (1999) 158–166.

[39] A. Lodi, S. Martello, D. Vigo, Heuristic and metaheuristic approaches for a class of two-dimensional bin packing problems, INFORMS J. Comput. 11 (1999) 345–357.

[40] A. Lodi, S. Martello, D. Vigo, Heuristic algorithms for the three-dimensional bin packing problem, Eur. J. Oper. Res. 2002, to appear.

[41] A. Lodi, M. Monaci, Integer linear programming models for two-staged cutting stock problems, Technical Report OR/00/12, DEIS-Università di Bologna, 2000, to appear in Mathematical Programming.

[42] S. Martello, P. Toth, Knapsack Problems: Algorithms and Computer Implementations, John Wiley & Sons, Chichester, 1990.

[43] S. Martello, D. Vigo, Exact solution of the two-dimensional finite bin packing problem, Manage. Sci. 44 (1998) 388–399.

[44] W. Schneider, Trim-loss minimization in a crepe-rubber mill; optimal solution versus heuristic in the 2 (3)-dimensional case, Eur. J. Oper. Res. 34 (1988) 273–281.

[45] D. Sleator, A 2.5 times optimal algorithm for packing in two dimensions, Inform. Process. Lett. 10 (1980) 37–40.

[46] A. Steinberg, A strip-packing algorithm with absolute performance bound 2, SIAM J. Comput. 26 (1980) 401–409.

[47] F.J. Vasko, F.E. Wolf, K.L. Stott, A practical solution to a fuzzy two-dimensional cutting stock problem, Fuzzy Sets and Systems 29 (1989) 259–275.

![NH logo] ELSEVIER

Discrete Applied Mathematics 123 (2002) 397–446

DISCRETE
APPLIED
MATHEMATICS

Cutting planes in integer and mixed integer programming ☆

Hugues Marchand[a,1], Alexander Martin[b], Robert Weismantel[c,2], Laurence Wolsey[d,*]

[a]*Electrabel, Boulevard du Regent 8, 1000 Brussels, Belgium*
[b]*Fachbereich Mathematik, Technische Universität Darmstadt, Schlossgartenstr, 7, 64289 Darmstadt, Germany*
[c]*Fakultät für Mathematik, IMO Otto-von-Guericke Universität Magdeburg Universitätsplatz 2 D-39106 Magdeburg, Germany*
[d]*CORE and INMA, Université Catholique de Louvain, 1348 Louvain-la-Neuve, Belgium*

Received 9 September 1999; received in revised form 30 March 2001; accepted 2 April 2001

Abstract

This survey presents cutting planes that are useful or potentially useful in solving mixed integer programs. Valid inequalities for (i) general integer programs, (ii) problems with local structure such as knapsack constraints, and (iii) problems with 0–1 coefficient matrices, such as set packing, are examined in turn. Finally, the use of valid inequalities for classes of problems with structure, such as network design, is explored. © 2002 Elsevier Science B.V. All rights reserved.

MSC: 90C10; 90C11

Keywords: Mixed integer programming; Cutting planes

☆ Research carried out with financial support of the project TMR-DONET nr. ERB FMRX-CT98-0202 of the European Community.
* Corresponding author. Tel.: +32-10-47-4307; fax: +32-10-47-4301.
E-mail address: wolsey@core.ucl.ac.be (L. Wolsey).
[1] Work carried out at the Department of Operational Research London School of Economics, Houghton Street, London WC2A 2AE, United Kingdom.
[2] Supported by a "Gerhard-Hess-Forschungsförderpreis" (WE 1462/2-1) of the German Science Foundation (DFG).

0. Introduction

This survey is devoted to cutting planes that are useful or potentially useful in solving mixed integer programs. This topic is important because (a) improving formulations with cutting planes is of interest independently of the algorithm used to solve the problem, and (b) linear programming based branch-and-bound with cuts added, known as *branch-and-cut*, is now one of the most widespread and successful tools for solving mixed integer programs.

The paper is divided into four sections. First, we discuss ways of generating cuts for general integer programs (IPs) $\max\{c^{\mathrm{T}}x\colon Ax = b,\ x \in \mathbb{Z}_+^n\}$ and mixed integer programs (MIPs) $\max\{c^{\mathrm{T}}x + h^{\mathrm{T}}y\colon Ax + Gy = b, x \in \mathbb{Z}_+^n, y \in \mathbb{R}_+^p\}$ independently of any problem structure. It was shown theoretically in the 70s and 80s that Gomory's mixed integer cuts, simple disjunctive cuts and mixed integer rounding cuts are based on the same disjunctive argument. In the 90s it has been shown, starting with lift-and-project (disjunctive) cuts, how all three types of cuts can be successfully used computationally.

In Section 2.1 we look at IPs and MIPs with some local structure, starting with knapsack sets. Whereas cover inequalities for 0–1 knapsack sets were studied and used in the 70s and 80s, attention has switched to various generalizations of 0–1 knapsack sets, in particular several mixed knapsack sets containing one or more continuous variables. These have a richer polyhedral structure then the pure knapsack sets, and arise very naturally in mixed integer programs. We introduce lifting, an important technique for strengthening valid inequalities and obtaining facet-defining inequalities. Still in the context of knapsack sets, we also introduce a new way to derive valid inequalities, starting from feasible solutions. Knapsack constraints arise when studying IPs whose constraint matrices have general integer coefficients. We also consider problems with 0–1 coefficient matrices. A natural starting point is the set packing problem. The basic inequalities for the set packing polytope based on cliques and cycles are derived, as well as the separation problem for such inequalities. We then briefly examine the generalizations of these inequalities to general independence systems.

A major challenge is to produce stronger inequalities in a way that is easily characterized, and potentially useful for computation. In Section 3 some steps in this direction are examined. A procedure to mix mixed integer rounding inequalities is presented, and also a way to extend formulations of certain combinatorial optimization problems to include set packing relaxations. Though their computational significance is still to be demonstrated, we discuss polynomial algorithms relating to the Lovász-Schrijver lift-and-project procedure, semi-definite optimization and clique separation.

In Section 4 we look at four important problem classes, ranging from network design to electricity generation, and try to indicate the state-of-the-art in terms of known strong cutting planes, and their use in computation.

We assume that readers are familiar with elementary terminology of valid inequalities and polyhedra, see for instance [78,95,107] for an in-depth treatment. See also [62] for a recent discussion of computational issues.

1. General cutting planes

In this section, we discuss methods of generating cutting planes for general mixed integer programs without exploiting any problem structure. As we will see, in certain cases these methods provide a complete linear description of the polyhedron under consideration. As a warm-up we start with the pure integer case and describe the well known Chvátal–Gomory cutting planes. We will see that this approach (based on a rounding argument) fails if continuous variables are involved. Methods that apply to the general mixed integer case are based on a disjunctive argument, and we will discuss three of them.

1.1. Pure integer programs

Consider a pure integer program $\min\{c^{\mathrm{T}}x: x \in X\}$ where $X = \{x \in \mathbb{Z}_{+}^{n}: Ax = b\}$ and A, b are integer. Gomory and later Chvátal found distinct but closely related ways of finding a linear description of $\mathrm{conv}(X)$. We begin with

1.1.1. Chvátal's geometric view
By definition a polyhedron P is integer if every face contains an integer point. By the integer Farkas lemma (see, for instance, [95] Corollary 4.1a) this in turn is equivalent to the fact that every supporting hyperplane contains an integer vector. The idea is now to look at every supporting hyperplane of $P = \{x \in \mathbb{R}_{+}^{n}: Ax = b\}$ and shift it closer to $P_I = \mathrm{conv}(X)$ until it contains an integer point.

Let $\{x \in \mathbb{R}^{n}: h^{\mathrm{T}}x = \vartheta\}$ be a supporting hyperplane of P with $P \subseteq \{x \in \mathbb{R}^{n}: h^{\mathrm{T}}x \leqslant \vartheta\}$ and h integer. Let

$$Q^{1} := \bigcap_{(h,\vartheta) \in \theta} \{x \in \mathbb{R}^{n}: h^{\mathrm{T}}x \leqslant \lfloor \vartheta \rfloor\}, \tag{1}$$

where θ denotes the set of all supporting hyperplanes of P with integer left-hand side. Obviously, $P_I \subseteq Q^{1}$. At first sight it is not obvious that Q^{1} is again a polyhedron, because there are infinitely many supporting hyperplanes. However, it turns out that Q^{1} is again a polyhedron. This allows us to continue the process and apply the same procedure to Q^{1}. With

$$Q^{0} := P \quad \text{and} \quad Q^{t+1} := (Q^{t})^{1}$$

we have

$$P = Q^{0} \supseteq Q^{1} \supseteq \cdots \supseteq P_I.$$

Chvátal shows that P_I is obtained this way after a finite number of iterations when P is a polytope, and Schrijver shows the result when P is an arbitrary polyhedron.

Theorem 1.1 (Chvátal [30], Schrijver [94]). *Let P be a rational polyhedron. Then*
(i) Q^{1} *is a polyhedron.*
(ii) $Q^{t} = P_I$ *for some finite t.*

The question remains how to generate hyperplanes on demand, i.e., how to find $(h, \vartheta) \in \theta$ that cuts off the current (fractional) solution of the LP relaxation $\min\{c^{\mathrm{T}}x \colon x \in P\}$. Gomory [46,48] gives an answer to this question.

1.1.2. Gomory's algorithmic view

Let x^* be an optimal solution of the LP relaxation $\min\{c^{\mathrm{T}}x \colon x \in P\}$, $P \subseteq R^n_+$ and $B \subseteq \{1, \ldots, n\}$ be a basis of A with $x^*_B = A_B^{-1}b - A_B^{-1}A_N x_N$ and $x^*_N = 0$, where $N = \{1, \ldots, n\} \setminus B$.

If x^* is integral, we terminate with an optimal solution for $\min\{c^{\mathrm{T}}x \colon x \in X\}$. Otherwise, one of the values x^*_B must be fractional. Let $i \in B$ be some index with $x^*_i \notin \mathbb{Z}$. Since every feasible integral solution $x \in X$ satisfies $x_B = A_B^{-1}b - A_B^{-1}A_N x_N$,

$$A_{i\cdot}^{-1}b - \sum_{j \in N} A_{i\cdot}^{-1}A_{\cdot j}x_j \in \mathbb{Z}, \tag{2}$$

where $D_{i\cdot}$ and $D_{\cdot j}$ denotes the ith row and jth column of some matrix D, respectively. The term on the left remains integral when adding integer multiples of x_j, $j \in N$, or an integer to $A_{i\cdot}^{-1}b$. We obtain

$$f(A_{i\cdot}^{-1}b) - \sum_{j \in N} f(A_{i\cdot}^{-1}A_{\cdot j})x_j \in \mathbb{Z}, \tag{3}$$

where $f(\alpha) = \alpha - \lfloor \alpha \rfloor$, for $\alpha \in \mathbb{R}$. Since $0 \leqslant f(\cdot) < 1$ and $x \geqslant 0$, we conclude that

$$f(A_{i\cdot}^{-1}b) - \sum_{j \in N} f(A_{i\cdot}^{-1}A_{\cdot j})x_j \leqslant 0,$$

or equivalently,

$$\sum_{j \in N} f(A_{i\cdot}^{-1}A_{\cdot j})x_j \geqslant f(A_{i\cdot}^{-1}b) \tag{4}$$

is valid for P_I. Moreover, it is violated by the current linear programming solution x^*, since $x^*_N = 0$ and $f(A_{i\cdot}^{-1}b) = f(x^*_i) > 0$. After subtracting $x_i + \sum_{j \in N} A_{i\cdot}^{-1}A_{\cdot j}x_j = A_{i\cdot}^{-1}b$ from (4) we obtain

$$x_i + \sum_{j \in N} \lfloor A_{i\cdot}^{-1}A_{\cdot j} \rfloor x_j \leqslant \lfloor A_{i\cdot}^{-1}b \rfloor, \tag{5}$$

which is, when the right-hand side is not rounded, a supporting hyperplane with integer left-hand side, and thus a member of θ. Moreover, adding this inequality to the system $Ax = b$ preserves the property that all data are integral. Thus, the slack variable that is to be introduced for the new inequality can be required to be integer as well and the whole procedure can be iterated. In fact, Gomory [49] proves that with a particular choice of the generating row such cuts lead to a finite algorithm, i.e., after adding a finite number of inequalities, an integer optimal solution is found. Thus, it provides an alternative proof for Theorem 1.1.

Given P and a general point $x^* \in P$, the *separation problem* for Chvátal–Gomory inequalities is to determine whether $x^* \in P^1$, and if not to find an inequality $h^{\mathrm{T}}x \leqslant \lfloor \vartheta \rfloor$ cutting off x^*. An efficient procedure has been proposed when $h = uA$ with u restricted to be a $\{0, \frac{1}{2}\}$ vector [28], but the general problem has been shown to be NP-hard [108].

1.2. Mixed integer programs

The two approaches discussed so far fail when both integer and continuous variables are present. Chvátal's approach fails because the right-hand side cannot be rounded down in (1). Gomory's approach fails since it is no longer possible to add integer multiples to continuous variables to derive (3) from (2). For instance, $\frac{1}{3}+\frac{1}{3}x_1-2x_2 \in \mathbb{Z}$ with $x_1 \in \mathbb{Z}_+$, $x_2 \in \mathbb{R}_+$ has a larger solution set than $\frac{1}{3}+\frac{1}{3}x_1 \in \mathbb{Z}$. As a consequence, we cannot guarantee that the coefficients of the continuous variables are non-negative and therefore show the validity of (4). Nevertheless, it is possible to derive valid inequalities using the following *disjunctive argument*.

Observation 1.2. *Let $(a^k)^\mathrm{T}x \leqslant \alpha^k$ be a valid inequality for a polyhedron $P^k \subseteq R^{n+}$ for $k = 1, 2$. Then,*

$$\sum_{i=1}^{n} \min(a_i^1, a_i^2)x_i \leqslant \max(\alpha^1, \alpha^2)$$

is valid for both $P^1 \cup P^2$ and $\mathrm{conv}(P^1 \cup P^2)$.

This observation applied in different ways yields valid inequalities for the mixed integer case. We present three methods that are all more or less based on Observation 1.2.

1.2.1. Gomory's mixed integer cuts

Consider again the situation in (2), where x_i, $i \in B$, is required to be integer. We use the following abbreviations $\bar{a}_j = A_{i\cdot}^{-1}A_{\cdot j}$, $\bar{b} = A_{i\cdot}^{-1}b$, $f_j = f(\bar{a}_j)$, $f_0 = f(\bar{b})$, and $N^+ = \{j \in N : \bar{a}_j \geqslant 0\}$ and $N^- = N \setminus N^+$. Expression (2) is equivalent to $\sum_{j \in N} \bar{a}_j x_j = f_0 + k$ for some $k \in \mathbb{Z}$. We distinguish two cases, $\sum_{j \in N} \bar{a}_j x_j \geqslant 0$ and $\sum_{j \in N} \bar{a}_j x_j \leqslant 0$. In the first case,

$$\sum_{j \in N^+} \bar{a}_j x_j \geqslant f_0$$

must hold. In the second case, we have $\sum_{j \in N^-} \bar{a}_j x_j \leqslant f_0 - 1$, which is equivalent to

$$-\frac{f_0}{1-f_0} \sum_{j \in N^-} \bar{a}_j x_j \geqslant f_0.$$

Now we apply Observation 1.2 to the disjunction $P^1 = P \cap \{x : \sum_{j \in N} \bar{a}_j x_j \geqslant 0\}$ and $P^2 = P \cap \{x : \sum_{j \in N} \bar{a}_j x_j \leqslant 0\}$ and obtain the valid inequality

$$\sum_{j \in N^+} \bar{a}_j x_j - \frac{f_0}{1-f_0} \sum_{j \in N^-} \bar{a}_j x_j \geqslant f_0. \tag{6}$$

This inequality may be strengthened in the following way. Observe that the derivation of (6) remains unaffected when adding integer multiples to integer variables. By doing this we may put each integer variable either in the set N^+ or N^-. If a variable is in N^+, the final coefficient in (6) is $\bar{a}_j$ and thus the best possible coefficient after adding integer multiples is $f_j = f(\bar{a}_j)$. In N^- the final coefficient in (6) is $(f_0/(1 - f_0))\bar{a}_j$ and thus $f_0(1 - f_j)/(1 - f_0)$ is the best choice. Overall, we obtain the best possible coefficient by using $\min(f_j, f_0(1 - f_j)/(1 - f_0))$. This yields Gomory's mixed integer cut [47]

$$
\sum_{\substack{j: \, f_j \leqslant f_0 \\ j \text{ integer}}} f_j x_j + \sum_{\substack{j: \, f_j > f_0 \\ j \text{ integer}}} \frac{f_0(1 - f_j)}{1 - f_0} x_j
$$

$$
+ \sum_{\substack{j \in N^+ \\ j \text{ non-integer}}} \bar{a}_j x_j - \sum_{\substack{j \in N^- \\ j \text{ non-integer}}} \frac{f_0}{1 - f_0} \bar{a}_j x_j \geqslant f_0. \tag{7}
$$

Gomory [47] shows that an algorithm based on iteratively adding these inequalities solves $\min\{c^\mathsf{T}x : x \in X\}$ with $X = \{x \in \mathbb{Z}_+^p \times \mathbb{R}_+^{n-p} : Ax = b\}$ in a finite number of steps provided $c^\mathsf{T}x \in \mathbb{Z}$ for all $x \in X$.

1.2.2. Mixed-integer-rounding cuts

Consider the following elementary mixed integer set $X = \{(x, y) \in \mathbb{Z} \times \mathbb{R}_+ : x - y \leqslant b\}$ with $b \in \mathbb{R}$ and the inequality

$$
x - \frac{1}{1 - f(b)} y \leqslant \lfloor b \rfloor. \tag{8}
$$

Proposition 1.3 (Nemhauser and Wolsey [78,79]). *Inequality (8) is valid for* $\mathrm{conv}(X)$.

Proof. Consider the disjunction $P^1 = X \cap \{(x, y) : x \leqslant \lfloor b \rfloor\}$ and $P^2 = X \cap \{(x, y) : x \geqslant \lfloor b \rfloor + 1\}$. For P^1 we immediately see that

$$
(x - \lfloor b \rfloor)(1 - f(b)) \leqslant y
$$

is valid by adding the inequalities $x - \lfloor b \rfloor \leqslant 0$ and $0 \leqslant y$ scaled with weights $1 - f(b)$ and 1. For P^2 we combine $-(x - \lfloor b \rfloor) \leqslant -1$ and $x - y \leqslant b$ with weights $f(b)$ and 1 to obtain

$$
(x - \lfloor b \rfloor)(1 - f(b)) \leqslant y.
$$

Thus, Observation 1.2 implies that $(x - \lfloor b \rfloor)(1 - f(b)) \leqslant y$ is valid for $\mathrm{conv}(P^1 \cup P^2) = \mathrm{conv}(X)$. $\square$

The basic observation expressed in Proposition 1.3 can now be extended to more general situations. Consider the following mixed integer set:

$$
X = \{(x, y) \in \mathbb{Z}_+^n \times \mathbb{R}_+ : a^\mathsf{T}x - y \leqslant b\},
$$

with $a \in \mathbb{R}^n, b \in \mathbb{R}$. We take $f_i = f(a_i)$ and $f_0 = f(b)$ in the sequel.

Proposition 1.4 (Nemhauser and Wolsey [78,79]). *The inequality*

$$\sum_{i=1}^{n} \left(\lfloor a_i \rfloor + \frac{(f_i - f_0)^+}{1 - f_0} \right) x_i - \frac{1}{1 - f_0} y \leqslant \lfloor b \rfloor \tag{9}$$

is valid for conv(X), *where* $v^+ = \max(0, v)$ *for* $v \in \mathbb{R}$. *Inequality* (9) *is called a* mixed integer rounding (MIR) inequality.

Proof. Relax $a^{\mathsf{T}}x - y \leqslant b$ to $\sum_{i \in N^1} \lfloor a_i \rfloor x_i + \sum_{i \in N^2} a_i x_i - y \leqslant b$, where $N^1 = \{i \in \{1, \ldots, n\}: f_i \leqslant f_0\}$ and $N^2 = \{1, \ldots, n\} \setminus N^1$. Applying Proposition 1.3 to $w - z \leqslant b$ with $w = \sum_{i \in N^1} \lfloor a_i \rfloor x_i + \sum_{i \in N^2} \lceil a_i \rceil x_i \in \mathbb{Z}$ and $z = y + \sum_{i \in N^2}(1 - f_i)x_i \geqslant 0$ yields

$$w - \frac{z}{1 - f_0} \leqslant \lfloor b \rfloor. \tag{10}$$

Substituting w and z in (10) gives (9). □

MIR inequalities imply Gomory's mixed integer cuts (7) when applied to the mixed integer set $X = \{(x, y^-, y^+) \in \mathbb{Z}_+^n \times \mathbb{R}_+^2: a^{\mathsf{T}}x + y^+ - y^- = b\}$. To see this consider the relaxation $a^{\mathsf{T}}x - y^- \leqslant b$ of X. Proposition 1.4 gives

$$\sum_{i=1}^{n} \left(\lfloor a_i \rfloor + \frac{(f_i - f_0)^+}{1 - f_0} \right) x_i - \frac{1}{1 - f_0} y^- \leqslant \lfloor b \rfloor.$$

Subtracting the original inequality $a^{\mathsf{T}}x + y^+ - y^- = b$ gives Gomory's mixed integer cut (7).

Nemhauser and Wolsey [79] discuss MIR inequalities in a more general setting. They prove that MIR inequalities provide a complete description for any mixed 0–1 polyhedron. Marchand and Wolsey [69,70] show that certain strong cutting planes for structured mixed integer programs can be derived as MIR inequalities. They also show their computational effectiveness in solving general mixed integer programs.

1.2.3. Lift-and-project cuts

The idea of "lift and project" is to consider the integer programming problem, not in the original space, but in some space of higher dimension (lifting). Then inequalities found in this higher dimensional space are projected back to the original space resulting in tighter integer programming formulations. Versions of this approach differ in how the lifting and the projection are performed, see [10,66,96]. All approaches only apply to 0–1 mixed integer programming problems. We explain the ideas in [10] in more detail and show the connections and differences to [66,96].

The validity of the procedure is based on a trivial observation.

Observation 1.5. *If* $c_0 + c^{\mathsf{T}}x \geqslant 0$ *and* $d_0 + d^{\mathsf{T}}x \geqslant 0$ *are valid inequalities for* X, *then* $(c_0 + c^{\mathsf{T}}x)^{\mathsf{T}}(d_0 + d^{\mathsf{T}}x) \geqslant 0$ *is valid for* X.

Consider a 0–1 program $\min\{c^{\mathsf{T}}x: x \in X\}$ with $X = \{x \in \{0, 1\}^p \times \mathbb{R}^{n-p}: Ax \leqslant b\}$, in which the system $Ax \leqslant b$ already contains the trivial inequalities $0 \leqslant x_i \leqslant 1$ for $i = 1, \ldots, p$. Let $P = \{x \in \mathbb{R}^n: Ax \leqslant b\}$ and $P_I = $ conv(X). Consider the following procedure.

Algorithm 1.6 (Lift-and-Project).

1. *Select an index $j \in \{1, \ldots, p\}$.*
2. *Multiply $Ax \leqslant b$ by x_j and $1 - x_j$ giving*

$$(Ax)x_j \leqslant bx_j,$$

$$(Ax)(1 - x_j) \leqslant b(1 - x_j) \tag{11}$$

and substitute $y_i := x_i x_j$ for $i = 1, \ldots, n$, $i \neq j$ and $x_j := x_j^2$ (lifting). Call the resulting polyhedron $L_j(P)$.
3. *Project $L_j(P)$ back to the original space by eliminating variables y_i. Call the resulting polyhedron P_j.*

The following theorem shows that the jth component of each vertex of P_j is either zero or one.

Theorem 1.7 (Balas et al. [11]). $P_j = \mathrm{conv}(P \cap \{x \in \mathbb{R}^n : x_j \in \{0, 1\}\})$.

For any sequence of indices $i_1, \ldots, i_t \in \{1, \ldots, p\}, t \geqslant 1$ let

$$P_{i_1, i_2, \ldots, i_t} := (\ldots (P_{i_1})_{i_2} \ldots)_{i_t}.$$

A repeated application of Algorithm 1.6 yields P_I.

Theorem 1.8 (Balas et al. [10]). $P_{i_1, \ldots, i_t} = \mathrm{conv}(P \cap \{x \in \mathbb{R}^n : x_{i_h} \in \{0, 1\}, \ h = 1, \ldots, t\})$.

Theorem 1.8 shows that the result does not depend on the order in which one applies Algorithm 1.6 to the selected variable. Thus, we may write $P_{\{i_1, \ldots, i_t\}}$ instead of $P_{i_1, \ldots, i_t}$ and $P_{\{1, \ldots, p\}} = P_I$.

The problem that remains in order to implement Algorithm 1.6 is to carry out Step 3. Let $L_j(P) = \{(x, y) : Dx + By \leqslant d\}$. Then the projection of $L_j(P)$ onto the x-space can be described by

$$P_j = \{x : (u^T D)x \leqslant u^T d \text{ for all } u \in C\},$$

where $C = \{u : u^T B = 0, \ u \geqslant 0\}$. Thus, the problem of finding a valid inequality in Step 3 of Algorithm 1.6 that cuts off a current (fractional) solution x^* can be solved by the linear program

$$\begin{aligned} \max \ &u^T(Dx^* - d), \\ &u \in C. \end{aligned} \tag{12}$$

This linear program is unbounded, if there is a violated inequality, since C is a polyhedral cone. For algorithmic convenience C is often truncated by some "normalizing set", see [10]. If an integer variable x_j that attains a fractional value in a basic feasible solution is used to determined the index j in Algorithm 1.6, then an optimal solution to (11) indeed cuts off x^*.

The computational merits of lift-and-project cuts to solve real-world problems are discussed in [10,11].

There is a close connection between the lift-and-project method and disjunctive programming. In fact, Theorem 1.7 states that $P_j = \text{conv}(P^0 \cup P^1)$ where $P^0 := P \cap \{x \in \mathbb{R}^n: x_j = 0\}$ and $P^1 := P \cap \{x \in \mathbb{R}^n: x_j = 1\}$. The inequalities obtained by projecting $L_j(P)$ onto the x-space may be viewed as inequalities obtained from the disjunction of P into P^0 and P^1. Thus, lift-and-project is a specialization of disjunctive programming, see, for instance, [8,60] for further details on this issue.

Observation 1.5 can be applied to a more general setting. For the ease of exposition we assume that our mixed integer program is indeed a pure integer program, i.e., $p = n$. Sherali and Adams [96] suggest lifting the problem to a higher dimensional space by multiplying $Ax \leqslant b$ by every product $(\prod_{j \in J_1} x_j)(\prod_{j \in J_2}(1 - x_j))$ such that $J_1, J_2 \subseteq \{1, \ldots, n\}$ are disjoint and $|J_1 \cup J_2| = d$ for some fixed value $d \in \{1, \ldots, n\}$. They linearize the problem by setting $x_i = x_i^k$, $2 \leqslant k \leqslant d + 1$, and by replacing every product $\prod_{j \in J} x_j$ by a single variable y_J for $J \subseteq \{1, \ldots, n\}$. Thereafter, the high-dimensional problem is projected to the space of x-variables. If $d = n$ is chosen, then this procedure directly yields a linear description of P_I.

Setting $d = 1$, the first step of the above procedure leads to the system

$$(Ax)x_j \leqslant bx_j \quad \text{for } j = 1, \ldots, n,$$

$$(Ax)(1 - x_j) \leqslant b(1 - x_j) \quad \text{for } j = 1, \ldots, n.$$

Setting $y_{ij} = x_i x_j$ for $1 \leqslant i < j \leqslant n$, and then projecting back to the original space leads to a polyhedron $N(P) \subseteq \bigcap_{j=1}^n P_j$. It is clear from Theorem 1.7 that this tighter procedure must be repeated at most n times to terminate with P_I. Lovász and Schrijver [66] studied this projection in more detail. They note that if $x_0 = 1$, then the product of two valid inequalities

$$(c_0 + c^{\mathrm{T}}x)^{\mathrm{T}}(d_0 + d^{\mathrm{T}}x) = c^{\mathrm{T}} \begin{pmatrix} x_0 \\ x \end{pmatrix} (x_0, x^{\mathrm{T}})d = c^{\mathrm{T}}Xd \geqslant 0,$$

where $X = \binom{x_0}{x}(x_0, x^{\mathrm{T}})$ is a symmetric and positive semidefinite matrix. This is pursued in Section 2.3.

We want to emphasize here that in contrast to the pure integer case none of the cutting plane procedures presented yields a finite algorithm for general mixed integer programs. Gomory needs an integer restricted objective function, and the other two provide finiteness only for 0–1 mixed integer programs. Cook, Kannan, and Schrijver [33] present the so-called split cuts. These cuts are again based on Observation 1.2 and may be viewed as special disjunctive cuts. They turn out to be equivalent to MIR inequalities [79]. However, Cook et al., show that the split cuts in combination with a certain rounding technique, which is based on the idea of discretizing the continuous variables, suffice to generate the mixed integer hull of a polyhedron. See also [83].

2. Simple structures

Above we have looked at valid inequalities for IPs and MIPs. If we restrict our attention to a single constraint, or a small subset of constraints, even a general problem may exhibit some "local" structure. For example, all variables appearing in a constraint may be 0–1 variables, or a small part of the MIP may be a network flow problem. Here, we look at ways to obtain stronger inequalities by using such local structure.

2.1. Knapsacks and cover inequalities

The concept of a cover has been used extensively in the literature to derive valid inequalities for (mixed) integer sets. In this section, we first show how to use this concept to derive cover inequalities for the 0–1 knapsack set. We then discuss how to extend these inequalities to more complex mixed integer sets.

Consider the 0–1 knapsack set

$$K = \left\{ x \in \{0,1\}^N : \sum_{j \in N} a_j x_j \leqslant b \right\}$$

with non-negative coefficients, i.e., $a_j \geqslant 0$ for $j \in N$ and $b \geqslant 0$. The set $C \subseteq N$ is a *cover* if

$$\lambda = \sum_{j \in C} a_j - b > 0. \tag{13}$$

In addition, the cover C is said to be *minimal* if $a_j \geqslant \lambda$ for all $j \in C$. To each cover C, we can associate a simple valid inequality which states that "not all variables x_j for $j \in C$ can be set to one simultaneously".

Proposition 2.1 (Balas [7], Hammer et al. [57], Padberg [85], Wolsey [103]). *Let* $C \subseteq N$ *be a cover. The cover inequality*

$$\sum_{j \in C} x_j \leqslant |C| - 1 \tag{14}$$

is valid for K. *Moreover, if* C *is minimal, then the inequality* (14) *defines a facet of* $\mathrm{conv}(K_C)$ *where* $K_C = K \cap \{x : x_j = 0, \ j \in N \setminus C\}$.

Example 2.2. Consider the 0–1 knapsack set

$$K = \{x \in \{0,1\}^6 : 5x_1 + 5x_2 + 5x_3 + 5x_4 + 3x_5 + 8x_6 \leqslant 17\}.$$

$C = \{1,2,3,4\}$ is a minimal cover for K and the corresponding cover inequality

$$x_1 + x_2 + x_3 + x_4 \leqslant 3$$

defines a facet of $\mathrm{conv}(\{x \in \{0,1\}^4 : 5x_1 + 5x_2 + 5x_3 + 5x_4 \leqslant 17\})$.

If a cover C is not minimal, then it is easily seen that the corresponding cover inequality is redundant, i.e., it is the sum of a minimal cover inequality and some upper bound constraints.

As described in the next subsection, lifting can be used to strengthen cover inequalities and to obtain a large class of facet-defining inequalities for $\mathrm{conv}(K)$ called *lifted cover inequalities*. Generalizations of cover inequalities can be found in [43,99,106]

where the polyhedral structures of, respectively, the 0–1 knapsack set with generalized upper bounds constraints, the 0–1 knapsack with precedence constraints and the
multiple 0–1 knapsack set are studied. Lifted cover inequalities have been used successfully in general purpose branch-and-cut algorithms to tighten the formulation of
0–1 integer programs [36,54]. In [13], it is shown how minimal covers, lifting and
complementation (replacing the binary variable x_j by its complement $1 - \bar{x}_j$) can be
used to obtain all the non-trivial facets of the 0–1 integer programming polytope with
positive coefficients.

The concept of cover is also useful in the study of the polyhedral structure of
problems containing both 0–1, integer and continuous variables. Consider the mixed
0–1 knapsack set

$$S = \left\{ (x,s) \in \{0,1\}^N \times \mathbb{R}_+ : \sum_{j \in N} a_j x_j \leqslant b + s \right\}$$

with non-negative coefficients, i.e., $a_j \geqslant 0$ for $j \in N$ and $b \geqslant 0$.

Proposition 2.3 (Marchand and Wolsey [71]). *Let $C \subseteq N$ be a cover, i.e., C is a
subset of N satisfying* (13). *The inequality*

$$\sum_{j \in C} \min(a_j, \lambda) x_j \leqslant \sum_{j \in C} \min(a_j, \lambda) - \lambda + s \tag{15}$$

is valid for S. Moreover, the inequality (15) *defines a facet of* $\mathrm{conv}(S_C)$ *where $S_C =
S \cap \{x: x_j = 0, j \in N \setminus C\}$.*

Note here that each cover C gives rise to a cover inequality that defines a facet
of $\mathrm{conv}(S_C)$. This is in contrast to the pure integer case where only minimal covers
induce facets.

Example 2.4. Consider the mixed 0–1 knapsack set

$$S = \{(x,s) \in \{0,1\}^6 \times \mathbb{R}_+ : 5x_1 + 5x_2 + 5x_3 + 5x_4 + 3x_5 + 8x_6 \leqslant 17 + s\}.$$

Taking $C' = \{1,2,3,6\}$ a (non-minimal) cover for S, the associated cover inequality

$$5x_1 + 5x_2 + 5x_3 + 6x_6 \leqslant 15 + s$$

defines a facet of $\mathrm{conv}(\{(x,s) \in \{0,1\}^4 \times \mathbb{R}_+ : 5x_1 + 5x_2 + 5x_3 + 8x_6 \leqslant 17 + s\})$.

Cover inequalities of the form (15) can be used to derive valid inequalities for more
complex mixed integer sets. We illustrate this observation by showing how to derive
valid inequalities for an elementary flow model consisting of inflow arcs with capacities
and fixed costs, and a constraint on the total inflow.

Consider the (flow) set

$$X = \left\{ (x, y) \in \{0, 1\}^N \times \mathbb{R}_+^N : \sum_{j \in N} y_j \leqslant b, y_j \leqslant a_j x_j, j \in N \right\}$$

and let $C \subseteq N$ be a (flow) cover, i.e., C is a subset of N satisfying (13). In $\sum_{j \in N} y_j \leqslant b$, ignore y_j for $j \in N \setminus C$ and replace y_j by $a_j x_j - s_j$ for $j \in C$ where $s_j \geqslant 0$ is a slack variable. We obtain

$$\sum_{j \in C} a_j x_j \leqslant b + \sum_{j \in C} s_j.$$

Using Proposition 2.3, we have that the following inequality is valid for X

$$\sum_{j \in C} \min(a_j, \lambda) x_j \leqslant \sum_{j \in C} \min(a_j, \lambda) - \lambda + \sum_{j \in C} s_j,$$

or equivalently, substituting $a_j x_j - y_j$ for s_j,

$$\sum_{j \in C} [y_j + (a_j - \lambda)^+ (1 - x_j)] \leqslant b.$$

Proposition 2.5 (Padberg et al. [86]). *Let $C \subseteq N$ be a flow cover (C is a subset of N satisfying (13)) with $\max_{j \in C} a_j > \lambda$. The flow cover inequality*

$$\sum_{j \in C} [y_j + (a_j - \lambda)^+ (1 - x_j)] \leqslant b \tag{16}$$

is a facet-defining inequality for conv(X).

Flow models have been extensively studied in the literature. Various generalizations of the flow cover inequality (16) have been derived for more complex flow models. In [100], a family of flow cover inequalities is described for a general single node flow model containing variable lower and upper bounds. Generalizations of flow cover inequalities to lot-sizing and capacitated facility location problems can also be found, respectively, in [2,87]. Flow cover inequalities have been used successfully in general purpose branch-and-cut algorithms to tighten formulations of mixed integer sets [52,53,101]. See Example 2.8 and Section 3.

Cover inequalities appear also in other contexts. In [29] cover inequalities are derived for the knapsack set with general integer variables. Unfortunately, in this case, the resulting inequalities do not define facets of the convex hull of the knapsack set restricted to the variables defining the cover. More recently, the notion of cover has been used to define families of valid inequalities for the complementarity knapsack set [39].

2.2. Lifting

The lifting technique is a general approach that has been used in a wide variety of contexts to strengthen valid inequalities. For simplicity of exposition, we first illustrate the main concepts related to this technique by lifting binary variables in a 0–1 knapsack set.

Consider the 0–1 knapsack set

$$K = \left\{ x \in \{0,1\}^N : \sum_{j \in N} a_j x_j \leqslant b \right\}$$

and let M be a subset of N. Suppose that we have an inequality,

$$\sum_{j \in M} \pi_j x_j \leqslant \pi_0, \tag{17}$$

which is valid for $K_M = K \cap \{x : x_j = 0, j \in N \setminus M\}$. The *lifting problem* is to find the lifting coefficients $\{\pi_j\}_{j \in N \setminus M}$ so that

$$\sum_{j \in N} \pi_j x_j \leqslant \pi_0 \tag{18}$$

is valid for K. Ideally, we would like inequality (18) to be "strong" (i.e., if inequality (17) defines a face of high dimension of $\mathrm{conv}(K_M)$, we would like the inequality (18) to define a face of high dimension of $\mathrm{conv}(K)$).

2.2.1. Sequential lifting

One way of obtaining coefficients $\{\pi_j\}_{j \in N \setminus M}$ is to apply *sequential lifting*: lifting coefficients π_j are evaluated one after another. More specifically, the coefficient π_k is computed for a given $k \in N \setminus M$ so that

$$\pi_k x_k + \sum_{j \in M} \pi_j x_j \leqslant \pi_0 \tag{19}$$

is valid for $K_{M \cup \{k\}}$. This can be done by considering the *lifting function*

$$\Phi_M(u) = \min \left\{ \pi_0 - \sum_{j \in M} \pi_j x_j : \sum_{j \in M} a_j x_j \leqslant b - u, x \in \{0,1\}^{|M|} \right\}. \tag{20}$$

Proposition 2.6 (Sequential lifting [85]). *Suppose $K_{M \cup \{k\}} \cap \{x : x_k = 1\} \neq \emptyset$. Inequality (19) is valid for $K_{M \cup \{k\}}$ if $\pi_k \leqslant \Phi_M(a_k)$. Moreover, if $\pi_k = \Phi_M(a_k)$ and (17) defines a face of dimension t of $\mathrm{conv}(K_M)$, then (19) defines a face of $\mathrm{conv}(K_{M \cup \{k\}})$ of at least dimension $t + 1$.*

If one now intends to lift a second variable, then it becomes necessary to update the function Φ_M. Specifically, if $k \in N \setminus M$ was introduced first with a lifting coefficient π_k, then the lifting function becomes

$$\Phi_{M \cup \{k\}}(u) = \min \left\{ \pi_0 - \sum_{j \in M \cup \{k\}} \pi_j x_j : \sum_{j \in M \cup \{k\}} a_j x_j \leqslant b - u, x \in \{0,1\}^{|M|+1} \right\},$$

so in general, function Φ_M can decrease as more variables are lifted in. As a consequence, lifting coefficients depend on the order in which variables are lifted and therefore different lifting sequences often lead to different valid inequalities.

Example 2.7. Consider the 0–1 knapsack set

$$K = \{x \in \{0,1\}^6: 5x_1 + 5x_2 + 5x_3 + 5x_4 + 3x_5 + 8x_6 \leqslant 17\}$$

and let $M = \{1,2,3,4\}$. The inequality

$$x_1 + x_2 + x_3 + x_4 \leqslant 3$$

is valid for $K_{\{1,2,3,4\}}$. Lifting variable x_5 and then variable x_6 leads to

$$x_1 + x_2 + x_3 + x_4 + x_5 + x_6 \leqslant 3.$$

However, lifting variable x_6 and then variable x_5 leads to

$$x_1 + x_2 + x_3 + x_4 + 2x_6 \leqslant 3.$$

It can be checked that both inequalities define facets of $\mathrm{conv}(K)$.

One of the key questions to be dealt with when implementing such a lifting approach is how to compute lifting coefficients π_j. To perform "exact" sequential lifting (i.e., to compute at each step the lifting coefficient given by the lifting function), we have to solve a sequence of integer programs. In the case of the lifting of variables for the 0–1 knapsack set this can be done efficiently using a dynamic programming approach related to the following recursion formula:

$$\Phi_{M \cup \{k\}}(u) = \min[\Phi_M(u), \Phi_M(u + a_k) - \Phi_M(a_k)].$$

Using such a lifting approach, facet-defining inequalities for the 0–1 knapsack set have been derived [14,7,57,103,85] and embedded in a branch-and-bound framework to solve to optimality particular types of 0–1 integer programs [36].

We now indicate how the lifting ideas can be extended to treat variables fixed to values other than zero, and to handle more than one variable at a time.

Lifting a binary variable fixed to one
Consider the binary knapsack set

$$K_{M \cup \{k\}} = \left\{ x \in \{0,1\}^{|M|+1}: \sum_{M \cup \{k\}} a_j x_j \leqslant b + a_k \right\}.$$

Note that with $x_k = 1$, this reduces to the set K_M for which a (facet-defining) inequality $\sum_{j \in M} \pi_j x_j \leqslant \pi_0$ of $\mathrm{conv}(K_M)$ is given. So here we ask for what values of π_k, the inequality

$$\sum_{j \in M} \pi_j x_j + \pi_k(1 - x_k) \leqslant \pi_0$$

is valid for $K_{M \cup \{k\}}$.

The inequality is valid by construction when $x_k = 1$, and when $x_k = 0$, it is valid if and only if $\pi_k \leqslant \Phi_M(-a_k)$. It follows that

$$\sum_{j \in M} \pi_j x_j - \Phi_M(-a_k)x_k \leqslant \pi_0 - \Phi_M(-a_k)$$

is facet-defining for conv($K_{M\cup\{k\}}$). Note that an alternative way to derive this inequality is to work with the complemented variable $\bar{x}_k = 1 - x_k$, which is fixed to zero and then lifted.

Lifting a variable upper bound pair fixed to zero
Consider the set

$$X_{M\cup\{k\}} = \left\{(x,y)\in\{0,1\}^{|M|+1}\times\mathbb{R}_+^{|M|+1}: \sum_{M\cup\{k\}} y_j \leqslant b, y_j \leqslant a_j x_j, j\in M\cup\{k\}\right\}.$$

Note that with $(x_k,y_k)=(0,0)$, this reduces to the flow set over

$$X_M = \left\{(x,y)\in\{0,1\}^M\times\mathbb{R}_+^M: \sum_{j\in M} y_j \leqslant b, y_j \leqslant a_j x_j, j\in M\right\}.$$

Now suppose that the inequality

$$\sum_{j\in M}\pi_j x_j + \sum_{j\in M}\mu_j y_j \leqslant \pi_0$$

is valid and facet-defining for conv(X_M).
As before, let

$$\Psi_M(u) = \min\left\{\pi_0 - \sum_{j\in M}\pi_j x_j - \sum_{j\in M}\mu_j y_j: \sum_{j\in M} y_j \leqslant b - u,\right.$$

$$\left. y_j \leqslant a_j x_j, j\in M, (x,y)\in\{0,1\}^{|M|}\times\mathbb{R}_+^{|M|}\right\}.$$

Now the inequality

$$\sum_{j\in M}\pi_j x_j - \sum_{j\in M}\mu_j y_j + \pi_k x_k + \mu_k y_k \leqslant \pi_0$$

is valid if and only if $\pi_k + \mu_k u \leqslant \Psi_M(u)$ for all $0 \leqslant u \leqslant a_k$, ensuring that all the feasible points with $(x_k,y_k)=(1,u)$ satisfy the inequality.

So the inequality defines a facet if the affine function $\pi_k + \mu_k u$ lies below the function $\Psi_M(u)$ in the interval $[0,a_k]$ and touches it in two points different from $(0,0)$, thereby increasing the number of affinely independent tight points by the number of new variables. In [53] it is also shown how to lift the pair (x_k,y_k) when y_k has been fixed to a_k and x_k to 1.

Example 2.8. (i) Consider the 0–1 knapsack set

$$K = \{x\in\{0,1\}^5: 5x_1 + 5x_2 + 5x_3 + 5x_4 + 7x_5 \leqslant 22\}.$$

Fixing $x_5=1$, we obtain as before that $\sum_{j=1}^4 x_j \leqslant 3$ is facet-defining for conv($K_{\{1,2,3,4\}}$). As $\Phi(-7)=3-4=-1$,

$$x_1 + x_2 + x_3 + x_4 + x_5 \leqslant 4$$

is facet-defining for conv(K).

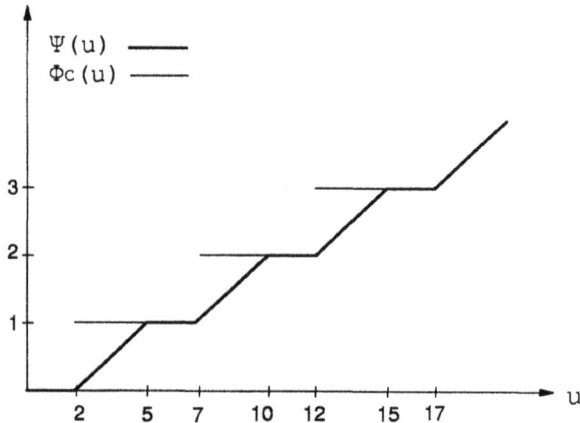

Fig. 1. Functions $\Phi_C(u)$ and $\Psi(u)$.

(ii) Consider the flow set

$$X' = \{(x,y) \in \{0,1\}^5 \times \mathbb{R}_+^5 : \sum_{j=1}^{5} y_j \leqslant 17, y_j \leqslant 5x_j, j = 1,\ldots,4, y_5 \leqslant 8x_5\}.$$

Fixing $(x_5, y_5) = (0,0)$, the flow cover inequality (16)

$$y_1 + y_2 + y_3 + y_4 - 2x_1 - 2x_2 - 2x_3 - 2x_4 \leqslant 9$$

is facet-defining for the resulting set $\mathrm{conv}(X_M)$.

The function $\Psi_M(u)$ is readily seen to satisfy

$$\Psi_M(u) = 0 \quad \text{for } 0 \leqslant u \leqslant 2,$$

$$\Psi_M(u) = u - 2 \quad \text{for } 2 \leqslant u \leqslant 5,$$

$$\Psi_M(u) = 3 \quad \text{for } 5 \leqslant u \leqslant 7,$$

$$\Psi_M(u) = 3 + (u - 7) \quad \text{for } 7 \leqslant u \leqslant 10, \text{ etc.}$$

Now for $(\alpha,\beta) = (0,0), (\alpha,\beta)(1,1)^T = \Psi_M(1)$ and $(\alpha,\beta)(1,2)^T = \Psi_M(2)$.
For $(\alpha,\beta) = (-\frac{6}{5}, \frac{3}{5}), (\alpha,\beta)(1,2)^T = \Psi_M(2)$ and $(\alpha,\beta)(1,7)^T = \Psi_M(7)$.
For $(\alpha,\beta) = (-4,1), (\alpha,\beta)(1,7)^T = \Psi_M(7)$ and $(\alpha,\beta)(1,8)^T = \Psi_M(8)$.
So three facet-defining inequalities

$$y_1 + y_2 + y_3 + y_4 - 2x_1 - 2x_2 - 2x_3 - 2x_4 \leqslant 9,$$

$$y_1 + y_2 + y_3 + y_4 - 2x_1 - 2x_2 - 2x_3 - 2x_4 + \tfrac{3}{5}y_5 - \tfrac{6}{5}x_5 \leqslant 9$$

and

$$y_1 + y_2 + y_3 + y_4 - 2x_1 - 2x_2 - 2x_3 - 2x_4 + y_5 - 4x_5 \leqslant 9$$

are obtained for $\mathrm{conv}(X')$. In Fig. 1 the function $\Psi = \lambda\Psi_M$ is shown.

In theory, "exact" sequential lifting can be applied to derive valid inequalities for any kind of mixed integer set. However, in practice, this approach is only useful to generate valid inequalities for sets for which one can associate a lifting function that can be evaluated efficiently.

Lifting is applied in the context of set packing problems to obtain facets from odd-hole inequalities [84], see Section 2.4. Other uses of sequential lifting can be found in [29] where the lifting of continuous and integer variables is used to extend the class of lifted cover inequalities to a mixed knapsack set with general integer variables. In [72,73] lifting is used to define (lifted) feasible set inequalities for an integer set defined by multiple integer knapsack constraints, see Section 2.3.

Sequential lifting is not the only way of computing lifting coefficients. We now discuss a general approach in which an "a priori" characterization is used to compute lifting coefficients.

2.2.2. Sequence independent lifting and superadditivity

Returning to the $0-1$ knapsack set K, we show how to evaluate lifting coefficients $\{\pi_j\}_{j \in N \setminus M}$ when we want to lift all variables in $N \setminus M$ *simultaneously*.

Because the function Φ_M may decrease as more variables are lifted in, taking $\{\Phi_M(a_j)\}_{j \in N \setminus M}$ as lifting coefficients does not in general lead to a valid inequality for K. Therefore to obtain a "sequence independent lifting", we have to find a function $\Psi : \mathbb{R} \to \mathbb{R}$ with $\Psi(u) \leqslant \Phi_M(u)$ so that

$$\sum_{j \in N \setminus M} \Psi(a_j)x_j + \sum_{j \in M} \pi_j x_j \leqslant \pi_0 \qquad (21)$$

is valid for K. In the next proposition we characterize such a function Ψ. We first introduce a definition.

Definition 2.9. A function $F : \mathbb{R} \to \mathbb{R}$ is superadditive on $\mathbb{R}$ if $F(d_1)+F(d_2) \leqslant F(d_1+d_2)$ for all $d_1, d_2 \in \mathbb{R}$.

Proposition 2.10. *Sequence independent lifting* [52,104]. *Let* $\Psi : \mathbb{R} \to \mathbb{R}$ *be a function. If* (i) $\Psi(u) \leqslant \Phi_M(u)$ *for all* $u \in \mathbb{R}$ *and* (ii) $\Psi(u)$ *is superadditive on* $\mathbb{R}$, *then inequality* (21) *is valid for* K.

Condition (ii) is quite restrictive. However, by considering the lifting of variables whose coefficients in the knapsack constraint take particular values, one can relax assumption (ii). In particular, if we suppose that all coefficients a_j are positive, condition (ii) becomes $\Psi(u)$ is superadditive on $\mathbb{R}_+$. We now illustrate this idea by deriving particular lifted cover inequalities using a superadditive function.

Consider a $0-1$ knapsack set K in which $a_j > 0$ for all $j \in N$. If $C \subseteq N$ is a minimal cover, the cover inequality

$$\sum_{j \in C} x_j \leqslant |C| - 1$$

is valid for $K_C = K \cap \{x: x_j = 0, j \in N \setminus C\}$. The lifting function here is

$$\Phi_C(u) = \min \left\{ |C| - 1 - \sum_{j \in C} x_j \,\middle|\, \sum_{j \in C} a_j x_j \leqslant b - u, x \in \{0,1\}^{|C|} \right\}.$$

Suppose $C = \{1,\ldots,r\}$ and $a_j \geqslant a_{j+1}$ for all $j \in \{1,\ldots,r-1\}$. Let $A_j = \sum_{t=1}^{j} a_t$ and let $A_0 = 0$. The function

$$\Psi(u) = \begin{cases} j & \text{if } A_j \leqslant u \leqslant A_{j+1} - \lambda \text{ for } j = 0,\ldots,r-1, \\ j + [u - A_j]/\lambda & \text{if } A_j - \lambda \leqslant u \leqslant A_j \quad \text{for } j = 1,\ldots,r-1, \\ r + [u - A_r]/\lambda & \text{if } A_r - \lambda \leqslant u \end{cases}$$

is dominated by $\Phi_C(u)$ and is superadditive on $\mathbb{R}_+$. Therefore

$$\sum_{j \in N \setminus C} \Psi(a_j) x_j + \sum_{j \in C} x_j \leqslant |C| - 1 \tag{22}$$

is valid for K.

Example 2.7 (continued). The inequality (22) associated to $C = \{1,2,3,4\}$ is

$$x_1 + x_2 + x_3 + x_4 + \tfrac{1}{3}x_5 + \tfrac{4}{3}x_6 \leqslant 3.$$

The functions $\Phi_C(u)$ and $\Psi(u)$ are shown in Fig. 1.

Again sequence independent lifting can be extended to the lifting of valid inequalities for more general mixed integer sets [52]. In [53], simultaneous lifting of pairs of variables (included in the same variable upper bound constraint) is studied. Sequence independent lifted flow cover inequalities are obtained. In some of the cases studied there, the lifting function itself is shown to be superadditive. In [71], classes of facet-defining inequalities for the mixed knapsack set are obtained using the superadditivity of the lifting function first on $\mathbb{R}_+$ and then on $\mathbb{R}_-$, i.e., first lifting variables with positive coefficients, and then those with negative coefficients.

Other uses of lifting can be found in the literature. In [10,11], lift-and-project cuts are generated in the space of the fractional variables. The cutting planes are then lifted in the full space of variables. Lifting in this approach plays a central role because it reduces the computational effort required to generate lift-and-project cuts. A similar idea is used in [3] where cutting planes for the symmetric travelling salesman problem are generated from a polytope obtained by projection onto a small subset of the original variables.

2.3. Knapsacks and feasible set inequalities

Section 1.1 showed a way to derive an elementary inequality by forbidding an infeasible subset of items of a 0–1 knapsack set. We now investigate a way of defining valid inequalities for the 0–1 knapsack set starting with a feasible set and again using sequential lifting. This yields a generalization of the cover inequalities.

Consider again the 0–1 knapsack set

$$K = \left\{ x \in \{0,1\}^N : \sum_{j \in N} a_j x_j \leqslant b \right\}$$

with $a_j > 0$ for $j \in N$.

Let $T \subseteq N$ be a feasible set, i.e., $\sum_{j \in T} a_j \leqslant b$ and $w : T \to \mathbb{Z}_+ \setminus \{0\}$ a weighting of the items in T. We denote the slack by $r = b - \sum_{j \in T} a_j \geqslant 0$. Clearly, the inequality $\sum_{i \in T} w_i x_i \leqslant w(T)$ is valid for $K \cap \{x : x_i = 0 \text{ for } i \in N \setminus T\}$. Then we carry out sequential lifting as in the previous section.

Proposition 2.11 (Weismantel [102]). *If T is a feasible set and $w : T \to \mathbb{Z}_+ \setminus \{0\}$, the inequality*

$$\sum_{i \in T} w_i x_i + \sum_{j \in N \setminus T} \pi_j x_j \leqslant w(T)$$

is valid for K, where $(\mu_1, \ldots, \mu_{n-|T|})$ is a permutation of $N \setminus T$, Φ_T is the lifting function (20) *with $\pi_j = w_j$ for $j \in T$ and $\pi_0 = w(T)$, and $\pi_{\mu_i} = \Phi_{T \cup \{\mu_1, \ldots, \mu_{i-1}\}} (a_{\mu_i})$.*

We observe that if $w_i = 1$ for all $i \in T$, then $\pi_0 = |T|$ and

$$\Phi_T(u) = \min \left\{ |S| : S \subseteq T, \sum_{j \in S} a_j \geqslant u - r \right\}.$$

It follows immediately in this case that $\Phi_T(u) = 0$ for $0 \leqslant u \leqslant r$, and thus $\pi_j = 0$ whenever $j \in N \setminus T$ and $a_j \leqslant r$.

Example 2.12. Consider the knapsack polytope conv(K) defined as the convex hull of all 0–1 vectors that satisfy the constraint

$$3x_1 + 4x_2 + 6x_3 + 7x_4 + 9x_5 + 18x_6 \leqslant 21.$$

Taking the feasible set $T = \{1, 2, 3, 4\}$, we obtain a slack $r = 1$. Choosing the permutation (5,6), we obtain coefficients $\pi_5 = 2$ and $\pi_6 = 3$. The resulting feasible set inequality

$$x_1 + x_2 + x_3 + x_4 + 2x_5 + 3x_6 \leqslant 4$$

defines a facet of conv(K).

Feasible set inequalities associated with a set T and weights $w_i = 1$ for all $i \in T$ subsume the family of lifted cover and $(1,k)$-configuration inequalities. Specifically, a set $T \cup \{z\} \subseteq N$ with $\sum_{i \in T} a_i \leqslant b$ is called a $(1,k)$-configuration, if every k-element subset of T together with the element z forms a minimal cover. This configuration gives rise to a valid inequality for K,

$$\sum_{i \in T} x_i + (|T| - k + 1)x_z \leqslant |T|.$$

It is a characteristic of feasible set inequalities that lifting coefficients can be computed in polynomial time under modest assumptions on the weights w_i of the items $i \in T$, see [102]. Indeed, the exact lifting coefficient of an item either equals a certain lower bound or equals this lower bound plus one. This generalizes an earlier result where this property was shown to hold for the lifting of minimal cover inequalities [14].

Theorem 2.13. *For $i \in N \setminus T$ with $a_i > r$, the coefficient π_i in any feasible set inequality associated with T and the weights $w_i = 1$ for all $i \in T$ satisfies*

$$\Phi_T(a_i) - 1 \leqslant \pi_i \leqslant \Phi_T(a_i).$$

In fact, Theorem 2.13 extends to more general families of feasible set inequalities where the coefficients of the items in the feasible set are not restricted to the value one, see [102]. Another extension of feasible set inequalities in [72,73] applies to general integer programs.

2.4. 0–1 matrices and valid inequalities

Integer and mixed integer programs often contain some constraints with only 0–1 coefficients. In addition, many preprocessors for integer programs automatically generate logical inequalities of the form $x_i + x_j \leqslant 1, x_i \leqslant x_j$, cover inequalities, etc. This naturally leads to the study of integer programs with 0–1 matrices.

The study of such problems, and in particular the set packing and covering problems, plays a prominent role in combinatorial optimization. These problems are among the most studied with a beautiful theory involving topics such as perfect, ideal, or balanced matrices, perfect graphs, the theory of blocking and anti-blocking polyhedra, independence systems and semidefinite programming.

The focus of this section is on a (partial) description of the associated polyhedra by means of inequalities. Assuming that relaxations of various integer programs yield set packing/covering problems, knowledge about these polyhedra can be used to strengthen the formulation of the original problem.

Definition 2.14. Let $A \in \{0,1\}^{m \times n}$ be a 0–1 matrix and $c \in \mathbb{R}^n$. The 0–1 integer programs

$$\max \{c^\mathrm{T} x \colon Ax \leqslant \mathbf{1}, x \in \{0,1\}^n\}, \tag{23}$$

$$\min \{c^\mathrm{T} x \colon Ax \geqslant \mathbf{1}, x \in \{0,1\}^n\} \tag{24}$$

are called the *set packing* and *set covering problems*, respectively.

Each column j of A can be viewed as the incidence vector of a subset F_j of the ground set $\{1, \ldots, m\}$, i.e., $F_j := \{i \in \{1, \ldots, m\} \colon A_{ij} = 1\}$. With this interpretation, the set packing problem consists of finding a collection of sets from $F_1, \ldots, F_n$ that are mutually disjoint and maximal with respect to the objective function c. Analogously, the covering problem aims at finding a collection of subsets whose union yields the ground set and is minimal with respect to c.

2.4.1. The set packing polytope

Feasible solutions of the set packing problem have a nice graph theoretic interpretation. Introduce a node for each column index of A and an edge (i,j) between two nodes i and j if their corresponding columns have a common non-zero entry in some row. The resulting graph, denoted by $G(A)$, is called (column) intersection graph. Obviously, every feasible 0–1 vector x satisfying $Ax \leqslant \mathbf{1}$ is the incidence vector of a stable set ($U \subseteq V$ is a stable set if $i,j \in U$ implies $(i,j) \notin E$) in the graph $G(A)$. Conversely, the incidence vector of any stable set in $G(A)$ is a feasible solution of the set packing problem $Ax \leqslant \mathbf{1}$. So a study of stable sets in graphs is equivalent to a study of the set packing problem.

Now consider some 0–1 matrix A and denote by

$$P(A) = \mathrm{conv}\{x \in \{0,1\}^N : Ax \leqslant \mathbf{1}\}$$

the set packing polytope. Let $G = (V,E)$ be the intersection graph $G(A)$. From our previous discussion it follows that $P(A) = \mathrm{conv}\{x \in \{0,1\}^n : x_i + x_j \leqslant 1, (i,j) \in E\}$, where the latter is an integer programming formulation of the stable set problem in G. In other words, with two matrices A and A' one may associate the same set packing polytope if and only if their corresponding intersection graphs coincide. It is therefore customary to study $P(A)$ via the graph G and denote the set packing polytope and the stable set polytope, respectively, by $P(G)$.

The following observations about $P(G)$ are immediate:

(i) $P(G)$ is full dimensional.
(ii) $P(G)$ is down monotone, i.e., $x \in P(G)$ implies $y \in P(G)$ for all $0 \leqslant y \leqslant x$. All non-trivial facets of $P(G)$ have non-negative coefficients.
(iii) The non-negativity constraints $x_j \geqslant 0$ induce facets of $P(G)$.

It is also well known that the edge and non-negativity constraints suffice to describe $P(G)$ if and only if G is bipartite (i.e., there is a partition (V_1, V_2) of the nodes such that every edge has one endpoint in V_1 and the other in V_2).

Non-bipartite graphs contain odd cycles. Odd cycles give rise to new valid inequalities that cannot be derived as linear combinations of the edge inequalities.

Proposition 2.15 (Padberg [84]). *Let $C \subseteq E$ be a cycle of odd cardinality in G. The odd cycle inequality*

$$\sum_{i \in V(C)} x_i \leqslant \frac{|V(C)| - 1}{2}$$

is valid for $P(G)$. It defines a facet of $P((V(C)), E(V(C)))$ if and only if C is an odd hole, i.e., a cycle without chords.

Odd cycle inequalities can be separated in polynomial time using the algorithm of Lemma 9.1.11 in [50] based on shortest paths. Graphs $G = (V,E)$ for which $P(G)$ is completely described by the edge inequalities $x_i + x_j \leqslant 1$ for $(i,j) \in E$ and the odd cycle inequalities are called *t-perfect*. This notion was introduced in [31] and includes series parallel and bipartite graphs.

Another important class of valid inequalities for the stable set polytope are clique inequalities.

Proposition 2.16 (Fulkerson [44], Padberg [84]). *Let* $(C, E(C))$ *be a clique in G. The inequality*

$$\sum_{i \in C} x_i \leq 1$$

is valid for $P(G)$. *It defines a facet of* $P(G)$ *if and only if* $(C, E(C))$ *is maximal with respect to node-inclusion.*

Graphs $G = (V, E)$ for which $P(G)$ is completely described by the clique inequalities are called *perfect*, a notion going back to Berge [19].

Unlike the class of odd cycle inequalities, the separation problem for the class of clique inequalities is NP-hard, see Theorem 9.2.9 in [50]. Surprisingly, however, there exists a larger class of inequalities, called *orthonormal representation inequalities* (see Proposition 3.5), that includes the clique inequalities and that can be separated in polynomial time. See Section 2.3 for a further discussion. Besides cycle, clique and OR-inequalities, there are many other inequalities known for the stable set polytope. Among these are blossom, odd antihole, wheel, antiweb and web, wedge inequalities and many more. Reference [23] gives a survey on these inequalities including a discussion on their separability.

2.4.2. The independence system polytope

Independence systems provide a framework in combinatorial optimization that generalizes among others the feasible sets of knapsack and set packing problems. To see this, let N be a finite ground set. A collection $\mathscr{I}$ of subsets of N is an *independence system* if it is closed under taking subsets, i.e.,

$$F \in \mathscr{I} \text{ and } G \subseteq F \text{ implies } G \in \mathscr{I}.$$

Associated with an independence system is a second system $\mathscr{C}$ of subsets of N. $\mathscr{C}$ is called the system of *circuits*. It includes all subsets of N of minimal cardinality that do not belong to $\mathscr{I}$.

From the definition of an independence system it is clear that, for instance, the set of all feasible points in a 0–1 knapsack set forms an independence system, and the minimal covers are the circuits. Also the set of stable sets in a graph forms an independence system. Here the cardinality of each circuit is two, and the circuits are precisely the edges of the graph.

More generally, let $A \in \mathbb{R}_+^{m \times n}$ be a non-negative matrix. The set of all 0–1 solutions satisfying $Ax \leq b$ for $b \in \mathbb{R}^m$ forms an independence system $\mathscr{I}$ on the ground set $N = \{1, \ldots, n\}$. Let

$$P_{\mathscr{I}} := \text{conv}\{x \in \{0, 1\}^n : Ax \leq b\}.$$

$P_{\mathscr{I}}$ is called an independence system polyhedron. The following fact about the facet-defining inequalities of $P_{\mathscr{I}}$ is immediate.

Proposition 2.17. *Let $c^T x \leqslant \gamma$ be a facet-defining inequality that is not a positive multiple of one of the non-negativity constraints $-x_i \leqslant 0$. Then c is a non-negative vector and $\gamma > 0$.*

Observe that for the set packing problem Proposition 2.17 was stated in (ii) in Section 2.4.1. An easy example of a valid inequality for the polyhedron of a general independence system is the *circuit constraint*.

Proposition 2.18. *Let $\mathscr{I}$ be an independence system and let $C \subseteq N$ be a circuit. The inequality*

$$\sum_{i \in C} x_i \leqslant |C| - 1$$

is valid for $P_{\mathscr{I}}$.

In fact the problem of finding a maximum weight set in an independence system can be formulated as the integer program

$$\max \left\{ c^T x \colon \sum_{i \in C} x_i \leqslant |C| - 1 \text{ for all } C \in \mathscr{C}, \ x \in \{0,1\}^N \right\}.$$

Except for special cases, a circuit constraint does not necessarily define a facet of the associated independence system polyhedron. Recall that this applies in particular to the stable set problem for which clique constraints subsume the edge constraints. This motivates the following definition.

Definition 2.19. For $T \subseteq N$, the inequality

$$\sum_{i \in T} x_i \leqslant r(T) := \max\{|S| \colon S \subseteq T, S \in \mathscr{I}\}$$

is called a *rank inequality*, since the right-hand side reflects the maximal cardinality of an independence set with support in T.

Calculating the rank of a set is typically a difficult problem. For instance for the stable set problem, the rank inequality for an arbitrary graph G takes the form

$$\sum_{i \in V} x_i \leqslant \alpha(G),$$

where $\alpha(G)$ is the size of a maximum stable set in G, and it is NP-hard to calculate its value.

If $\mathscr{I}$ is an arbitrary independence system, then one cannot expect to derive a system of inequalities that describes $P_{\mathscr{I}}$. This motivates the search for a partial description. A natural starting point is again the stable set polyhedron. Specifically, we can think of an odd cycle on $\{1,\ldots,2k+1\}$ as a set of adjacent pairs $e_i = (i, i+1) \bmod 2k+1$ for $i = 1,\ldots,2k+1$ such that at most one item can be chosen from each pair.

Generalizing, we now consider a set $\{1,\ldots,n\}$ and the set of adjacent t-tuples $N^i = \{i, i+1,\ldots,i+t-1\} \bmod n$ for $i = 1,\ldots,n$. For $q \leqslant t$, the set consisting of all sets containing at most $q-1$ elements from each set N^i is an independence system, known as an *antiweb*, denoted $\mathscr{A}\mathscr{W}(n,t,q)$. Thus

$$\mathscr{A}\mathscr{W}(n,t,q) := \{I \subseteq N : |I \cap N^j| \leqslant q - 1 \text{ for all } j = 1,\ldots,n\}.$$

For example the antiweb $\mathscr{A}\mathscr{W}(5,3,3)$ is the set of subsets represented by the feasible incidence vectors of the 0–1 integer program with constraints

$$\begin{pmatrix} 1 & 1 & 1 & 0 & 0 \\ 0 & 1 & 1 & 1 & 0 \\ 0 & 0 & 1 & 1 & 1 \\ 1 & 0 & 0 & 1 & 1 \\ 1 & 1 & 0 & 0 & 1 \end{pmatrix} x \leqslant \begin{pmatrix} 2 \\ 2 \\ 2 \\ 2 \\ 2 \end{pmatrix}.$$

The set $\mathscr{C}$ of all circuits of $\mathscr{A}\mathscr{W}(n,t,q)$ is equal to

$$\mathscr{C} := \{C \subseteq N : |C| = q, C \subseteq N^j \text{ for some } j \in \{1,\ldots,n\}\}.$$

An antiweb gives rise to a valid inequality for the associated independence system polyhedron $P_{\mathscr{I}}$. In the example of $\mathscr{A}\mathscr{W}(5,3,3)$, the inequality reads $\sum_{i \in N} x_i \leqslant 3$. More generally, one obtains

Proposition 2.20. *Let $\mathscr{A}\mathscr{W}(n,t,q)$ be an antiweb and $P_{\mathscr{I}}$ the associated polyhedron. The inequality $\sum_{i \in N} x_i \leqslant \lfloor n(q-1)/t \rfloor$, called an antiweb inequality, is valid for $P_{\mathscr{I}}$.*

Proof. The sum of all constraints $\sum_{i \in N^j} x_i \leqslant q-1$ for $j = 0,\ldots,n-1$ reads $\sum_{i \in N} t x_i \leqslant n(q-1)$. Therefore, the antiweb inequality coincides with the Chvátal–Gomory cutting plane $\sum_{i \in N} x_i \leqslant \lfloor n(q-1)/t \rfloor$ that is valid for $P_{\mathscr{I}}$. □

No polynomial time algorithms are known for the antiweb inequalities. For an antiweb $\mathscr{A}\mathscr{W}(n,t,q)$ the associated inequality always defines a facet if $n = t$. Hence we may assume that $n > t$. In this case a necessary condition for the antiweb inequality to define a facet of $P_{\mathscr{I}}$ is that t is not a divisor of $n(q-1)$. This condition is also sufficient. This condition, the definition of an antiweb and Proposition 2.20 are taken from Laurent [64]. The antiweb inequality in Laurent's paper extends, in particular, the generalized odd holes and antiholes of [42]. It also includes generalized cliques that were introduced in [77]. There are various other families of inequalities known for the independence system that we refrain from discussing here in detail.

Very special independence systems in which the rank inequalities and non-negativity constraints suffice to describe the convex hull $P_{\mathscr{I}}$ include matroids, see [40]. A generalization of the result to the intersection of two matroids can be found in [41].

2.4.3. The set covering polytope

The feasible solutions of the set covering problem

$$\{x \in \{0,1\}^n : Ax \geqslant \mathbf{1}\}$$

are in one-to-one correspondence with the independent sets of the system $\mathscr{I}$

$$\left\{ \bar{x} \in \{0,1\}^n : \sum_{j \in C} \bar{x}_j \leqslant |C| - 1 \text{ for } C \in \mathscr{C} \right\},$$

when the rows of A correspond to the incidence vectors of circuits $C \in \mathscr{C}$ and $\bar{x}_j = 1 - x_j$ for $j \in N = \{1, \ldots, n\}$.

Note that the antiweb inequality has an equivalent counterpart for the set covering polytope that is derived by complementing every binary variable. In fact the (q, t) roses of [93] are precisely Laurent's antiweb inequalities, see also [80]. Further inequalities for the set covering polytope have been derived, see [23] for a survey, but again all separation algorithms known are of heuristic nature.

3. Extensions

So far we have tried to introduce various ways to derive cutting planes for integer and mixed integer programs of potential computational value. There are many further extensions that are algorithmically promising and worth further exploration. Below we discuss three such topics: the idea of mixing MIR inequalities, the approach of constructing discrete relaxations of integer programs, and the use of semidefinite programming for separation issues.

3.1. Mixing MIR inequalities

Consider the mixed integer set

$$X = \{(x, s) \in \mathbb{Z}^{|P|} \times \mathbb{R}_+ : s + Cx_i \geqslant b_i, i \in P\}, \text{ for } P = \{1, \ldots, p\}.$$

Let $\mu_i = \lceil b_i / C \rceil$ and $r_i = b_i - (\mu_i - 1)C$. We assume that the constraints defining X are ordered in such a way that $r_i \leqslant r_{i+1}$.

The MIR inequality associated with each constraint $i \in P$ of X is

$$s \geqslant r_i(\mu_i - x_i).$$

By "mixing" these inequalities, a new inequality is obtained.

Proposition 3.1 (Günlük and Pochet [56]). *Taking $r_0 = 0$, the inequality*

$$s \geqslant \sum_{i \in P} (r_i - r_{i-1})(\mu_i - x_i)$$

is valid for X.

We illustrate the mixing procedure on two examples.

Example 3.2. Consider an instance of a discrete constant capacity lot-sizing problem,

$$X = \{(x,s) \in \{0,1\}^3 \times \mathbb{R}_+^4 : s_{i-1} + Cx_i = b_i + s_i, i \in \{1,2,3\}\},$$

where $C = 10$, $b_1 = 6$, $b_2 = 7$ and $b_3 = 8$. Eliminating variables s_1, s_2 and s_3, we obtain the inequalities,

$$s_0 + 10x_1 + 10x_2 + 10x_3 \geqslant 21,$$
$$s_0 + 10x_1 + 10x_2 \geqslant 13,$$
$$s_0 + 10x_1 \geqslant 6,$$

to which we can associate the MIR inequalities

$$s_0 \geqslant 3 - x_1 - x_2 - x_3,$$
$$s_0 \geqslant 3(2 - x_1 - x_2),$$
$$s_0 \geqslant 6(1 - x_1).$$

Applying Proposition 3.1, we obtain the mixed MIR inequality

$$S_0 \geqslant (3 - x_1 - x_2 - x_3) + 2(2 - x_1 - x_2) + 3(1 - x_1).$$

In [56] it is shown that every (k,l,S,I) inequality for the constant capacity lot-sizing problem can be obtained by mixing MIR inequalities. These inequalities suffice to solve the constant capacity lot-sizing problem by linear programming when the objective function satisfies the Wagner–Whitin assumption [89]. See Section 3.2 for a more extensive discussion of inequalities for lot-sizing problems.

Mixing can also be used to derive valid inequalities for general integer programs.

Example 3.3. Consider the following integer set

$$X = \{x \in \mathbb{Z}_+^5 : x_1 + 3x_2 + 10x_4 \geqslant 25, x_1 + 2x_3 + 10x_5 \geqslant 37\}.$$

Defining $s = x_1 + 3x_2 + 2x_3$, the two constraints defining set X can be relaxed to give a set

$$X' = \{(x_4, x_5, s) \in \mathbb{Z}_+^2 \times \mathbb{R}_+^1 : s + 10x_4 \geqslant 25, s + 10x_5 \geqslant 37\}.$$

Applying Proposition 3.1 to X', we obtain the mixed MIR inequality

$$s \geqslant 5(3 - x_4) + 2(4 - x_5)$$

or equivalently

$$x_1 + 3x_2 + 2x_3 \geqslant 5(3 - x_4) + 2(4 - x_5)$$

a valid inequality for X.

Other examples of application of the mixing idea can be found in [56].

3.2. Set packing relaxations

In the introduction it was mentioned that knowledge about the set packing polytope can be used to strengthen certain integer programming formulations. Below we show by example how, by introducing additional variables, it is possible to derive a set packing relaxation, generate one or more valid inequalities, and then project back into the original space of variables. We then give a formal description of the approach.

Example 3.4. Let P_I be the convex hull of all 0–1 vectors that satisfy the system of inequalities

$$5x_1 + 5x_2 + 7x_3 + 2x_4 \leqslant 18,$$
$$8x_3 + x_4 + 6x_5 + 5x_6 \leqslant 19,$$
$$7x_1 + 2x_2 + 7x_5 + x_6 \leqslant 16.$$

Define variables $w_1 = x_1 x_2$, $w_2 = x_3 x_4$ and $w_3 = x_5 x_6$, so that $x_1, x_2 \geqslant w_1$, $x_1 + x_2 - 1 \leqslant w_1$, etc. From the first constraint we have that $10w_1 + 9w_2 \leqslant 18$, $w_1, w_2 \in \{0, 1\}$ from which we obtain the valid cover inequality $w_1 + w_2 \leqslant 1$. Similarly, from the second and third constraints, we obtain $w_2 + w_3 \leqslant 1$ and $w_1 + w_3 \leqslant 1$. Now the odd cycle (or clique) inequality $w_1 + w_2 + w_3 \leqslant 1$ is valid, leading finally to a valid inequality in the original variables $(x_1 + x_2 - 1) + (x_3 + x_4 - 1) + (x_5 + x_6 - 1) \leqslant 1$ or $\sum_{i=1}^{6} x_i \leqslant 4$ which is valid for P_I.

In general consider a 0–1 integer program $\max\{c^T x : Ax \leqslant b, x \in \{0, 1\}^n\}$ and let $P_I = \mathrm{conv}\{x \in \mathbb{Z}^n : Ax \leqslant b, 0 \leqslant x \leqslant 1\}$. We define a set of affine functions $f_i : \mathbb{R}^n \mapsto \mathbb{R}$ for $i = 1, \ldots, M$ with the property that $f_i(x) \leqslant 1$ and $f_i(x) \in \mathbb{Z}$ for $x \in P_I \cap \mathbb{Z}^N$. We define a graph G, called the *conflict graph*, by introducing a node for each of these M affine functions and edges (i, j) if $f_i(x) + f_j(x) \leqslant 1$ for all $x \in P_I$. Now it is readily seen that any valid inequality for the stable set polytope $P(G)$ associated with the conflict graph G yields a valid inequality for P_I.

Natural affine functions that come up are $f_i(x) = x_j$ or $f_i(x) = 1 - x_j$. These are the ones that are generally used in mixed integer programming solvers, see, for instance, [5,36,61]. In the above example we have used the affine functions $f_1(x) = x_1 + x_2 - 1$, $f_2(x) = x_3 + x_4 - 1$, $f_3(x) = x_5 + x_6 - 1$.

More complicated affine functions have been used in Borndörfer and Weismantel [25]. It is shown that various inequalities known for certain combinatorial optimization problems can be interpreted as inequalities from a set packing relaxation. For instance, it turns out that two-chorded cycle inequalities for the clique partitioning problem are odd cycle inequalities of an appropriate set packing relaxation, and that a large class of Möbius ladder inequalities and fence inequalities for the acyclic subdigraph problem are cycle and clique inequalities, respectively, of suitable set packing relaxations.

3.3. Polynomial separation algorithms via matrix cuts

Coming back to our earlier discussions on the stable set polytope, we indicated that there are polynomial time separation algorithms for various classes of valid inequalities, but that a polynomial time separation algorithm cannot be expected for the family of clique constraints. More striking is the fact that clique constraints can be generalized, and that this larger family can be separated in polynomial time. This result is one of the most appealing applications of semidefinite programming in combinatorial optimization, see [50,66].

Let $G=(V,E)$ be a graph with $|V|=n$ and $P(G)$ the associated stable set polyhedron. By P we denote the fractional stable set polytope. For $s \in \mathbb{Z}_+$ a sequence of vectors of unit length, $v^1, \ldots, v^n \in \mathbb{R}^s$, $||v^i||=1$, $i=1,\ldots,n$ is called an *orthonormal representation* of G if $(i,j) \notin E$ implies that $(v^i)^T v^j = 0$.

An orthonormal representation of G and a vector of unit length $c \in \mathbb{R}^s, ||c||=1$ lead to a valid inequality for the stable set polyhedron.

Proposition 3.5. *Let $v^1, \ldots, v^n \in \mathbb{R}^s$ be an orthonormal representation of G and $c \in \mathbb{R}^s$, $||c||=1$. The inequality*

$$\sum_{i \in V} (c^T v^i)^2 x_i \leqslant 1,$$

called an orthonormal representation (OR)-inequality, is valid for $P(G)$, and every clique inequality is an OR-inequality.

Proof. Let χ^S be the incidence vector of a stable set S in G. Then $(v^i)^T v^j = 0$ for all $i,j \in S$, $i \neq j$. We can express c as $c = \sum_{j \in S} \lambda_j v^j + \tilde{c}$ with $\lambda \in \mathbb{R}^S$ and $\tilde{c}$ in the orthogonal complement of the linear space induced by the vectors v^j, $j \in S$. Then

$$\sum_{i \in V} (c^T v^i)^2 \chi_i^S = \sum_{i \in S} (c^T v^i)^2 = \sum_{i \in S} \lambda_i^2 \leqslant 1,$$

because $||c||=1$.

If Q is a clique in G we may set $v^i = c = e^1 \in \mathbb{R}^n$ for all $i \in Q$, and $v^j = e^j$ for all $j \notin Q$. The corresponding orthonormal representation constraint is precisely the clique constraint $\sum_{i \in Q} x_i \leqslant 1$. □

In the following we denote

$$TH(G) = \{x \in \mathbb{R}^n_+ : x \text{ satisfies all OR-constraints}\}.$$

$TH(G)$ is a convex set that is a relaxation of $P(G)$. It is polyhedral if and only if G is perfect, see [50]. However, even when G is not perfect, one can optimize linear functions over $TH(G)$ in polynomial time. This in turn means that we can separate over $TH(G)$ in polynomial time, and thus satisfy all the OR-inequalities.

To get an impression why this is true, we indicate below how $TH(G)$ can be characterized via positive semidefinite matrices. This result is due to Lovász and Schrijver

[66]. Let

$$H(G) = \{Y \in \mathbb{R}^{V \cup \{v_0\}} \times \mathbb{R}^{V \cup \{v_0\}}:$$

Y symmetric,

$Y_{ii} = Y_{i0} \ \forall i \in V,$

$Y_{ij} = 0 \ \forall (i,j) \in E,$

Y positive semidefinite,

$e_0^{\mathrm{T}} Y e_0 = 1$

$\}.$

Theorem 3.6.

$$TH(G) = \{Ye^0: Y \in H(G)\}.$$

It follows, for instance, from the theory of interior point algorithms that, subject to certain conditions, linear functions can be optimized over the cone of symmetric positive semidefinite matrices subject to linear constraints in polynomial time to within a specified error. Since the constraints in Theorem 3.6 are linear in the space of $(n+1) \times (n+1)$ matrices and the conditions are satisfied, this applies to $TH(G)$.

In fact, $TH(G)$ is the projection of a semidefinite relaxation of the stable set problem. Notice that for any incidence vector x of a stable set we have that

$$x_i + x_j \leqslant x_0 \quad \forall (i,j) \in E \text{ with } x_0 = 1.$$

Therefore, the symmetric $(n+1) \times (n+1)$ matrix

$$\tilde{X} = \begin{bmatrix} x_0^2 & x_0 x^{\mathrm{T}} \\ x_0 x & x x^{\mathrm{T}} \end{bmatrix}$$

satisfies the condition that
(a) $\tilde{X}_{ij} = 0$ for all $(i,j) \in E$.
(b) $\tilde{X}_{00} = 1$.
(c) $\tilde{X}_{ii} = \tilde{X}_{i0}$ for all $i \in V$.
(d) $v^{\mathrm{T}} \tilde{X} v \geqslant 0$ for all $v \in \mathbb{R}^{n+1}$, i.e., $\tilde{X}$ is positive semidefinite.

Neglecting the condition that the $n \times n$ submatrix of $\tilde{X}$ is of the form xx^{T}, we end up with a relaxation of the stable set problem in the space of the symmetric $(n+1) \times (n+1)$ matrices. Projecting back to the space of x-variables (using the standard lift-and-project approach) yields precisely $TH(G)$. Important is the fact that $TH(G)$ can be strengthened by using further information in quadratic space about the matrices associated with stable sets and projecting back to the space of x-variables. This follows from the work of Lovász and Schrijver [66] on matrix cuts. We also refer to [50,65]. The conditions to be encountered in the quadratic space come from multiplying each constraint of the fractional stable set problem in the original space by

x_i and by $(1 - x_i)$, replacing the quadratic terms by the corresponding matrix variable and requiring that $x_i^2 = x_i$.

Theorem 3.7. *Let*

$$T(G) = \{Ye^0 \colon Y \in H(G)$$

$$u^{\mathrm{T}} Y e_i \geqslant 0 \ \forall u \in \mathrm{cone}(\{1\} \times P)', \ i = 1, \ldots, n$$

$$u^{\mathrm{T}} Y(e^0 - e_i) \geqslant 0 \ \forall u \in \mathrm{cone}(\{1\} \times P)', \ i = 1, \ldots, n$$

$$\},$$

where P denotes the fractional stable set polytope and $\mathrm{cone}(\{1\} \times P)'$ denotes the polar of the set $\mathrm{cone}(\{1\} \times P)$. Then the following is true:

$$T(G) \subseteq \{x \geqslant 0 \colon x \ satisfies \ all \ edge \ constraints,$$

$$x \ satisfies \ all \ OR\text{-}constraints,$$

$$x \ satisfies \ all \ odd \ hole \ constraints,$$

$$x \ satisfies \ all \ odd \ antihole \ constraints,$$

$$x \ satisfies \ all \ odd \ wheel \ constraints$$

$$\}.$$

4. Valid inequalities for some structured MIPs

Here, we look briefly at four problem areas that provide a large variety of applications: fixed charge network design, production planning, facility location and electricity generator scheduling. As many of the ideas for generating inequalities for network problems can be used in the other areas, we start with network design.

4.1. Fixed charge network design

Traditionally, single commodity fixed charge network problems arose in designing transport, water and electricity networks. In the last 10 years the design of telecommunication networks and VLSI have provided perhaps the bulk of applications in this area—these include both single commodity problems, such as the construction of two or multiply connected networks, and multicommodity problems which arise because messages/communications between two nodes A and B are distinct from messages being sent from C to D.

Below, we concentrate mainly on single commodity problems because the majority of valid inequalities can be explained in this simpler context. We present a variety of different ways to derive inequalities. In particular, we first look at the simplest single node model considering different variants, uncapacitated and capacitated, and with 0–1 or integer variables as appropriate. The same single node inequalities are then used when several nodes S are combined to form a macro-node, but the difficulty is now how to choose the set S.

We then present four classes of inequalities that use more of the network structure, such as the sparsity of the network, ways to combine different dicut inequalities, or submodularity.

Finally, we briefly touch on multicommodity problems with a single source and sink for each commodity. We look at a basic single arc model with both divisible and indivisible flows, and then again at how to choose a good macro-node set S on which to generate a dicut or other inequality.

4.1.1. Single commodity problems

We consider a basic single commodity fixed charge network flow problem consisting of a digraph $D = (V, A)$ and a vector $b \in \mathbb{R}^n$ with $\sum_{i=1}^n b_i = 0$, where $n = |V|$, $T = \{i \in V: b_i > 0\}$ is the set of demand nodes or *terminals*, and $U = \{i \in V: b_i < 0\}$ is the set of *sources*. Given unit flow costs p_{ij} and fixed arc capacity installation costs f_{ij} for an amount C_{ij} of capacity on arc $(i, j) \in A$, the problem is to find a feasible flow minimizing the sum of the flow and capacity installation costs. Much of the literature has been devoted to the special case of this problem without flow costs—special cases are the Steiner tree problem, or the problem of designing a two-connected network of minimum cost, etc. [48,54].

Below we use the notation $\bar{S} = V \setminus S$, $V^-(i) = \{j \in V: (j,i) \in A\}$, $V^+(i) = \{j \in V: (i,j) \in A\}$, and $\delta(S, \bar{S}) = \{(i, j) \in A: i \in S, j \in \bar{S}\}$.

Letting y_{ij} denote the flow in arc $(i, j) \in A$ and x_{ij} the number of times the capacity C_{ij} is installed, we obtain the natural formulation

$$\min \sum_{(i,j) \in A} p_{ij} y_{ij} + \sum_{(i,j) \in A} f_{ij} x_{ij}, \tag{25}$$

$$\sum_{j \in V^-(i)} y_{ji} - \sum_{j \in V^+(i)} y_{ij} = b_i \quad \text{for } i \in V, \tag{26}$$

$$0 \leqslant y_{ij} \leqslant C_{ij} x_{ij} \quad \text{for } (i,j) \in A, \tag{27}$$

$$x_{ij} \in \mathbb{Z}_+^1 \quad \text{for } (i,j) \in A. \tag{28}$$

Here (26) are flow conservation constraints and (27) are variable upper bound capacity constraints. We will denote the feasible region (26)–(28) by X^{FC}. In practice one also encounters many variants such as

(i) $x_{ij} \in \{0, 1\}$ in place of (28),
(ii) $C_{ij} = C$ and also possibly $f_{ij} = f$ for all $(i,j) \in A$ when standard equipment is installed throughout the network,
(iii) Capacity C_{ij}^0 already exists on certain arcs, and two or more different types of capacity can be installed, so we have $0 \leqslant y_{ij} \leqslant C_{ij}^0 + C^1 x_{ij}^1 + C^2 x_{ij}^2$ in place of (27),
(iv) Capacity is undirected, so we have $0 \leqslant y_{ij} + y_{ji} \leqslant C_e x_e$ in place of (27), where e represents the edge (i, j).

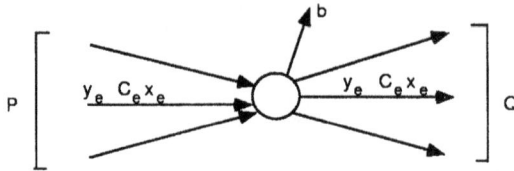

Fig. 2. Single node flow set.

4.1.2. Single node inequalities

If we just consider the flow conservation constraint (26) for node i along with the associated bounds on the flows (27), we obtain the situation shown in Fig. 2 and the corresponding single node flow set

$$X^{SN} = \left\{ (x, y) \in \mathbb{Z}_+^{p+q} \times \mathbb{R}_+^{p+q} : \sum_{e \in P} y_e - \sum_{e \in Q} y_e = b, \ y_e \leqslant C_e x_e \text{ for } e \in P \cup Q \right\}$$

with $p = |P|$ and $q = |Q|$, and its relaxation

$$X_>^{SN} = \left\{ (x, y) \in \mathbb{Z}_+^{p} \times \mathbb{R}_+^{p} : \sum_{e \in P} y_e \geqslant b, \ y_e \leqslant C_e x_e \text{ for } e \in P \right\}.$$

The uncapacitated case. If the capacities are so large that the flow on each arc is unrestricted, x_e can be restricted to be a 0–1 variable for all arcs $e \in A$. Now points in $X_>^{SN}$ satisfy $\sum_{e \in P} y_e \geqslant b$, $C \sum_{e \in P} x_e \geqslant b$, and thus if $b > 0$, the *cut* inequality

$$\sum_{e \in P} x_e \geqslant 1$$

is valid for $X_>^{SN}$. Note that if $b < 0$, a similar inequality is obtained with Q in place of P.

More generally, if F is a subset of the arcs in P, feasible points in $X_>^{SN}$ satisfy $\sum_{e \in P \setminus F} y_e + C \sum_{e \in F} x_e \geqslant b$ leading to the *mixed cut* inequality

$$\sum_{e \in P \setminus F} y_e + b \sum_{e \in F} x_e \geqslant b.$$

The constant capacity case—integer batches. For simplicity, we assume that the capacities C_e and demands b are integer. When $b > 0$ and $C_e = C$ for all $e \in F$, the inequality $\sum_{e \in P \setminus F} y_e + C \sum_{e \in F} x_e \geqslant b$ leads to the *residual capacity* or MIR inequality (see Section 1.2) for $X_>^{SN}$

$$\sum_{e \in P \setminus F} y_e + r \sum_{e \in F} x_e \geqslant r\mu, \tag{29}$$

where $\mu = \lceil b/C \rceil$ and $r = b - (\mu - 1)C$.

For X^{SN}, with $G \subseteq Q$, the inequality takes the more general form

$$\sum_{e \in P \setminus F} y_e + r \sum_{e \in F} x_e \geqslant r\mu + \sum_{e \in G} [y_e - (C - r)x_e], \tag{30}$$

see [4].

The capacitated 0–1 *case.* Rewriting the simple flow cover inequalities for single node flow sets that have been described in Section 1.1, we first present valid inequalities for $X_{\geq}^{SN} \cap \{(x, y) \in \{0, 1\}^p \times \mathbb{R}_+^p\}$. For F a cover, $(\sum_{e \in F} C_e - b = \lambda > 0)$, we obtain

$$\sum_{e \in P \backslash F} y_e + \sum_{e \in F} (C_e - \lambda)^+ x_e \geq \sum_{e \in F} (C_e - \lambda)^+.$$

Generalizing to include outflows, the basic inequality obtained for $X^{SN} \cap \{(x, y) \in \{0, 1\}^{p+q} \times \mathbb{R}_+^{p+q}\}$ is

$$\sum_{e \in P \backslash F_1} y_e + \sum_{e \in F_1} (C_e - \lambda)^+ x_e \geq \sum_{e \in F_1} (C_e - \lambda)^+ + \sum_{e \in F_2} (y_e - C_e) + \sum_{e \in L_2} (y_e - \lambda x_e),$$

where $F_1 \subseteq P, F_2, L_2 \subseteq Q, F_2 \cap L_2 = \emptyset$ and $\sum_{e \in F_1} C_e - \sum_{e \in F_2} C_e - b = \lambda > 0$.

In the constant capacity case, the inequalities for the 0–1 case take the same form as (29) and (30), and are known to describe the convex hull of solutions, see [86,4].

More general capacity constraints. Suppose that the constraints

$$y_e \leq C_e^0 + C_e^1 x_e^1 + C_e^2 x_e^2$$

describe the potential capacities. Feasible points now satisfy $\sum_{e \in P \backslash F} y_e + C^1 \sum_{e \in F} x_e^1 + C^2 \sum_{e \in F} x_e^2 \geq b - \sum_{e \in F} C_e^0$. Now assuming $b - \sum_{e \in F} C_e^0 > 0$, and divisible capacities (i.e., C^1 divides C^2), which is often the case in telecommunications applications, extensions of the residual capacity inequalities have been proposed in [20,67], and these have been generalized to handle an arbitrary number of divisible capacities in [90].

4.1.3. Aggregate node inequalities

By summing the flow conservation constraints (26) for $i \in S$, we obtain the set X^S:

$$\sum_{e \in \delta(\bar{S}, S)} y_e - \sum_{e \in \delta(S, \bar{S})} y_e = \sum_{i \in S} b_i, \tag{31}$$

$$0 \leq y_e \leq C_e x_e, \quad x_e \in \{0, 1\} \quad \text{for } e \in \delta(\bar{S}, S) \cup \delta(S, \bar{S}) \tag{32}$$

which is precisely in the form of the single node flow set X^{SN}. Thus, if $\sum_{i \in S} b_i > 0$, all the inequalities presented above can be generalized to the set X^S. In particular in the uncapacitated case we obtain the *dicut* inequality

$$\sum_{e \in \delta(\bar{S}, S)} x_e \geq 1$$

and if F is a subset of $\delta(\bar{S}, S)$, the *mixed dicut* inequality

$$\sum_{e \in \delta(\bar{S}, S) \backslash F} y_e + \left(\sum_{i \in S} b_i \right) \sum_{e \in F} x_e \geq \sum_{i \in S} b_i.$$

There is now however a major question to be answered before we can make use of these inequalities. How should the set S of nodes be chosen, given the huge number of possibilities?

The separation problem for dicut inequalities. Formally, we wish to solve the problem: given a solution (x^*, y^*) satisfying the linear programming relaxation of (25)–(28), does there exist a non-empty subset $S \subset V$ with $\sum_{i \in S} b_i > 0$ and $\sum_{e \in \delta(\bar{S}, S)} x_e^* < 1$?

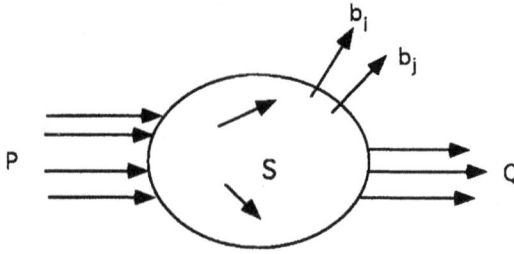

Fig. 3. Aggregate node set.

Special dicut inequalities: *maximum flow*. First we restrict the choice of subsets S. Remember the notation that $T = \{i: b_i > 0\}$ and $U = \{i: b_i < 0\}$. Let $\mathscr{S} = \{S \subset V: S \cap U = \emptyset, \ S \cap T \neq \emptyset\}$. Now if $S \in \mathscr{S}$, we are sure that $\sum_{i \in S} b_i > 0$. The separation problem then reduces to $|T|$ maximum flow problems.

Specifically, choose $s \in U$ and $t \in T$. Let ζ_t be the value of a maximum s–t flow in the digraph $D = (V, A)$ with capacities $h_{ij} = \infty$ if $i, j \in U$ and $h_{ij} = x_{ij}^*$ otherwise. If $\zeta_t \geqslant 1$, there is no violated dicut inequality with $s \in S$ and $t \in T$. Otherwise if $\zeta_t < 1$, the resulting minimal s–t cut gives a violated dicut inequality.

Note that for single source problems with $|U| = 1$, all dicuts of interest are included in this procedure.

All dicut inequalities: quadratic 0–1 Knapsack. To model the general case, let $z_j = 1$ if $j \in S$ and $z_j = 0$ otherwise. The resulting separation problem can now be written as

$$\zeta = \min \sum_{(i,j) \in A} x_{ij}^*(1 - z_i)z_j,$$

$$\sum_{j \in V} b_j z_j > 0,$$

$$z_j \in \{0, 1\} \quad \text{for } j \in V.$$

If S is the set minimizing ζ, a violated dicut inequality has been found if $\zeta < 1$, and in any case we can look at the single node flow set associated with S for other violated inequalities.

4.1.4. Inequalities using structure

Uncapacitated: inflow–outflow inequalities. When all arcs are present in an uncapacitated network, flow entering the network can reach any other node. However, when the network is sparse, this is no longer true. Specifically, consider the subgraph induced by the node set S as shown in Fig. 3. We will now take into account the internal structure of $D_S = (S, A_S)$. Write $P = \delta(\bar{S}, S)$ and $Q = \delta(S, \bar{S})$. Also let $R \subseteq A_S$ be a subset of the arcs in S. For an entering arc $e \in P$, let $S_e = \{i \in S: b_i > 0 \text{ and there exists a dipath in } D_{S,R} = (S, A_S \setminus R) \text{ from the head of arc } e \text{ to node } i\}$ and $\alpha_e = \sum_{i \in S_e} b_i$.

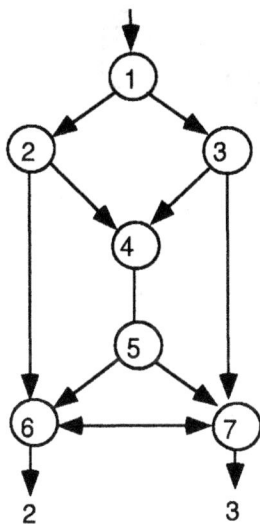

Fig. 4. Network for multicut inequality.

The *inflow–outflow inequality*

$$\sum_{e \in F} \alpha_e x_e + \sum_{e \in (P \setminus F) \cup R} y_e \geq \sum_{i \in S} b_i$$

is valid for any $F \subseteq P$.

Uncapacitated: multi-dicut inequalities. Rather than use just a single dicut inequality, here we show how to use several dicuts simultaneously. Suppose that for each $t \in T$, a family of dicuts $\{\delta(\bar{S}_t^k, S_t^k)\}_{k=1}^{K_t}$ is given with $t \in S_t^k$ and $S_t^k \cap U = \emptyset$ for all k and t. Also take $F_t^k \subseteq \delta(\bar{S}_t^k, S_t^k)$. The following *multi-dicut inequality*:

$$\sum_{e \in A} \max_{t \in T} \alpha_e(t) y_e + \sum_{e \in A} \sum_{t \in T} \beta_e(t) d_t x_e \geq \sum_{t \in T} K_t d_t$$

is shown to be valid in [91], where, for $e \in A$,
$\alpha_e(t)$ is the number of arc sets $\{F_t^k\}_{k=1}^{K_t}$ containing e, and
$\beta_e(t)$ is the number of arc sets $\{\delta(\bar{S}_t^k, S_t^k) \setminus F_t^k\}_{k=1}^{K_t}$ containing e.

Example 4.1. Consider the network shown in Fig. 4 with $T = \{6, 7\}$, $d_6 = 2$ and $d_7 = 3$. Taking $K_6 = K_7 = 2$, $S_6^1 = \{2567\}$, $S_6^2 = S_7^1 = \{3567\}$, $S_7^2 = \{567\}$, $F_6^1 = \{(37)\}$, $F_6^2 = F_7^1 = \{(26)\}$, $F_7^2 = \{(26), (37)\}$, we have $\alpha_{26}(1) = \alpha_{37}(1) = \alpha_{37}(2) = 1$, $\alpha_{26}(2) = 2$, and we obtain the multi-dicut inequality

$$y_{37} + 2y_{26} + 2x_{12} + 10x_{45} + 5x_{13} \geq 10.$$

0–1 Capacitated: submodular inequalities. An important, but rare structural property, in discrete optimization problems, is submodularity, which is some discrete form of non-increasing returns. Specifically, $f : \mathcal{P}(N) \to R$ is *submodular* if $f(A) + f(B) \geq$

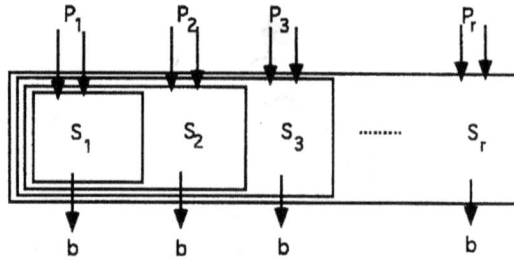

Fig. 5. Embedded node sets.

$f(A \cap B) + f(A \cup B)$ for all $A, B \subseteq N$. Not surprisingly, this structure is reflected in a family of valid inequalities. Consider again Fig. 3. For $F \subseteq P$, let $v(F)$ be the maximum flow that can enter D_S through the arcs of F, and leave via the demand nodes in S with $b_i > 0$. It can be shown that v is submodular. Define $\rho_j(T) = v(T \cup \{j\}) - v(T)$, and let $\{1, 2, \ldots, p\}$ be a chosen ordering of the elements of P. The following *submodular* inequality:

$$\sum_{j \in P} y_j \leq v(F) + \sum_{j \in P \setminus F} \rho_j(F \cup \{j+1, \ldots, p\}) x_j$$

$$- \sum_{j \in F} \rho_j(F \cap \{j+1, \ldots, p\} \cup \{1, \ldots, j-1\})(1 - x_j) + \sum_{e \in Q} y_e$$

is valid, see [109].

Capacitated: dynamic inequalities. Here, we use the idea of mixing to combine cut inequalities from different aggregate node sets, which can be viewed as generalizing the use of sparsity in the input–output inequalities. Suppose we have node sets $S_1 \subset S_2 \subset \cdots \subset S_t$, and entering arcs $P_1, \ldots, P_t$ as shown in Fig. 5. Let $Q_{pq} = \{(i, j) \in A: i \in S_p, j \in S_q\}$. Considering the sets $S_1, S_2, \ldots, S_t$ in turn, the inequalities based on the inflow to S_k being at least equal to the demand give

$$\sum_{i=1}^{k} \sum_{e \in P_i} C_e x_e + \sum_{p,q: p > q \geq k} \sum_{e \in Q_{pq}} y_e \geq \sum_{i \in S_k} b_i \tag{33}$$

for $k = 1, \ldots, t$.

With constant capacities, the mixing theorem can be applied to give inequalities of the form

$$\sum_{p,q: p > q} \sum_{e \in Q_{pq}} y_e \geq r_{[1]}(\mu_{[1]} - X_{[1]}) + \cdots + (r_{[t]} - r_{[t-1]})(\mu_{[t]} - X_{[t]}),$$

where $\mu_k = \lceil \sum_{i \in S_k} b_i / C \rceil$, $r_k = \sum_{i \in S_k} b_i - (\mu_k - 1)C$, $\{[1], \ldots, [t]\}$ is a permutation of $\{1, \ldots, t\}$ with $r_{[1]} \leq \cdots \leq r_{[t]}$, and $X_{[k]} = \sum_{i=1}^{[k]} \sum_{e \in P_i} x_e$.

Examples of such inequalities are given below both for lot-sizing and for facility location problems.

4.1.5. Multicommodity problems

In multicommodity problems feasible flows have to be determined for each of $k = 1, \ldots, K$ commodities satisfying demands b_i^k at each node $i \in V$, where the commodities share arc capacity. This can be formulated as

$$\min \sum_{(i,j) \in A} \sum_k p^k y_{(ij)}^k + \sum_{(i,j) \in A} \sum_k f^k x_{(ij)}^k, \tag{34}$$

$$N y^k = b^k \quad \text{for } k = 1, \ldots, K, \tag{35}$$

$$0 \leqslant \sum_k y_{ij}^k \leqslant C_{ij} x_{ij} \quad \text{for } (i,j) \in A, \tag{36}$$

$$x_{ij} \in Z^1 \quad \text{for } (i,j) \in A, \tag{37}$$

where N is the node-arc incidence matrix of D. In many instances each commodity k has a single source i^k and a single sink j^k, in which case we write $b_{j_k}^k = d_k$, $b_{i_k}^k = -d_k$ and $b_i^k = 0$ otherwise. From now on we limit our attention to this case. We also consider the network loading problem in which x_{ij} is integer rather than 0–1.

4.1.6. Single arc inequalities

Multiple routes: Consider flow in a single arc $(i,j) \in A$. Let y_k be the flow of commodity k in this arc, and x the associated capacity variable. The resulting set is

$$X^{SA} = \left\{ (x, y) \in Z_+^1 \times R_+^K : \sum_{k=1}^K y_k \leqslant Cx, y_k \leqslant d_k \text{ for } k = 1, \ldots, K \right\}.$$

Taking an arbitrary set $K' \subseteq \{1, \ldots, K\}$ of commodities and setting $w = \sum_{k \in K'} y_k$, we have that $w \leqslant Cx$ and $w \leqslant \sum_{k \in K'} d_k$ leading to the *arc residual capacity inequality*

$$\sum_{k \in K'} y_k \leqslant \sum_{k \in K'} d_k - r'(\mu' - x),$$

where $\mu' = \lceil \sum_{k \in K'} d_k / C \rceil$ and $r' = \sum_{k \in K'} d_k - (\mu' - 1)C$. It is shown in [67] that this family of inequalities completely describes the convex hull of X^{SA}.

Mono-routing. When each commodity must flow on a single path, the flow of commodity k in arc (i,j) is either 0 or d_k, and so we obtain the knapsack set

$$X^{SAM} = \left\{ (x_0, x) \in Z_+^1 \times \{0,1\}^K : \sum_k d_k x_k \leqslant C x_0 \right\}.$$

Valid inequalities for a more general model with capacities of the form $x_1 + C x_2$ have been derived in [26]. See also [98].

4.1.7. Multinode inequalities

If we choose a commodity k and a set $S \subset V$ with $i^k \in \bar{S}$ and $j^k \in S$, flow conservation for commodity k gives

$$\sum_{e \in \delta(\bar{S}, S)} y_e^k - \sum_{e \in \delta(S, \bar{S})} y_e^k = d_k.$$

One can first check for a violated dicut inequality by finding a maximum (i^k, j^k) flow with capacities $\min\{d_k, C_e\} x_e^*$ on the arcs.

More generally, with a constant capacity C and a subset K' of commodities, we have that

$$\sum_{k \in K'} \sum_{e \in \delta(\bar{S},S)} y_e^k \geq \sum_{k \in K': i^k \notin S, j^k \in S} d_k,$$

which after introduction of the capacity constraints gives

$$\sum_{e \in \delta(\bar{S},S)} x_e \geq \frac{\sum_{k: i^k \notin S, j^k \in S} d_k}{C}$$

and then applying Gomory integer rounding gives

$$\sum_{e \in \delta(\bar{S},S)} x_e \geq \left\lceil \frac{\sum_{k: i^k \notin S, j^k \in S} d_k}{C} \right\rceil.$$

Consider now the relaxed version of these inequalities without the round up of the right-hand side term and with $C = 1$. They are automatically satisfied by a point x^* if there exists a y such that (x^*, y) satisfies the linear programming relaxation of (34)–(37). More precisely such points satisfy the metric inequalities

$$\sum_e \mu_e x_e \geq \sum_k \pi_k d_k,$$

where $\mu \in \mathbb{R}_+^{|E|}$ are arbitrary edge lengths, and π_k is the corresponding length of a shortest path from i^k to j^k, see [59,82]. Note that if $\mu_e = 1$ for $e \in \delta(\bar{S},S)$, the relaxed inequality above is obtained as a special case.

However, separation for the special case is a max dicut problem, which is NP-hard. Specifically, it suffices to put a weight $-y_e^*$ on each arc of D, and a weight d_k/C on the arcs (i^k, j^k) for $k = 1, \ldots, K$, and find a maximum dicut. This separation procedure has been used in [15] in a model with edge capacities and no variable flow costs.

4.2. Lot-sizing

A single-item lot-sizing problem is a very special case of a fixed charge network flow problem, see Fig. 6.

The basic single-item lot-sizing problem is typically formulated as

$$\min \sum_t p_t y_t + \sum_t h_t s_t + \sum_t f_t x_t, \tag{38}$$

$$s_{t-1} + y_t = d_t + s_t \quad \text{for } t = 1, \ldots, n, \tag{39}$$

$$y_t \leq C_t x_t \quad \text{for } t = 1, \ldots, n, \tag{40}$$

$$s_t, y_t \geq 0, \ x_t \in \{0,1\} \quad \text{for } t = 1, \ldots, n. \tag{41}$$

Here d_t is the demand, p_t, h_t, f_t are the variable production, storage and fixed setup costs, and C_t is the maximum amount that can be produced in period t. y_t, s_t are continuous variables denoting the production and end-stock in period t, and x_t is a

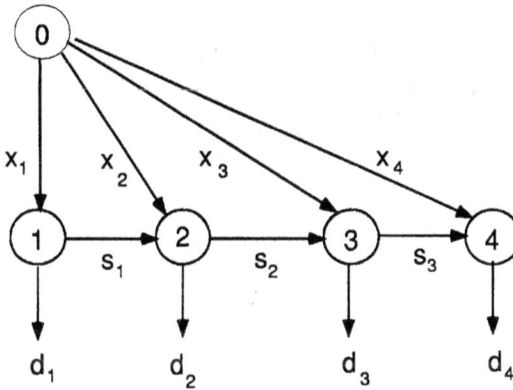

Fig. 6. Network for lot-sizing.

0–1 setup variable indicating whether the machine can produce in period t. Thus $y_t > 0$ only if $x_t = 1$. Constraints (39) are flow balance constraints, and (40) are capacity constraints linking the production and setup variables.

Much is known about the polyhedral structure of different variants of this problem. We will see below that all the valid inequalities can be derived using procedures that we have seen earlier either for general 0–1 MIPs in Sections 1.2 and 2.1, or for fixed charge network problems. Later in this section we will also introduce a natural way to derive valid inequalities for problems with start-ups. Let $d_{kt} = \sum_{j=k}^{t} d_j$.

Uncapacitated lot-sizing. Let X^{ULS} denote the set of feasible solutions of (39)–(41), where again we assume that C is very large and does not limit the amount produced in any period. Aggregating the flow balance constraints (39) for $t = k, \ldots, l$, and choosing a subset $S \subseteq \{k, \ldots, l\}$, of periods, leads to the relaxation

$$s_{k-1} + \sum_{j \notin S, k \leqslant j \leqslant t} y_j + C \sum_{j \in S, j \leqslant t} x_j \geqslant d_{kt} \qquad (42)$$

leading to the MIR inequalities

$$s_{k-1} + \sum_{j \notin S, k \leqslant j \leqslant t} y_j \geqslant d_{kt} \left(1 - \sum_{j \in S, j \leqslant t} x_j \right) \qquad (43)$$

and by the mixing procedure (Section 2.1) to the valid inequalities

$$s_{k-1} + \sum_{j \notin S, k \leqslant j \leqslant t} y_j \geqslant \sum_{j \in S} d_j \left(1 - \sum_{t \in S, k \leqslant t \leqslant j} x_t \right). \qquad (44)$$

These inequalities completely describe the convex hull of X^{ULS} [16].

Constant capacity lot-sizing. An identical approach leads to a large number of facet-defining inequalities when $C_t = C$ for $t = 1, \ldots, n$. First from (42) we obtain the MIR inequality

$$s_{k-1} + \sum_{j \notin S, k \leqslant j \leqslant t} y_j \geqslant r_{kt}\left(\mu_{kt} - \sum_{j \in S, j \leqslant t} x_j\right),$$

where $\mu_{kt} = \lceil d_{kt}/C \rceil$ and $r_{kt} = d_{kt} - (\mu_{kt} - 1)C$.

Now if the r_{kt} are placed in non-decreasing order, and written $r_{[1]} \leqslant r_{[2]} \cdots \leqslant r_{[q]}$, and $\mu_{[i]}$ and $X_{[i]}^S$ are the corresponding terms for μ and $\sum_j x_j$, the mixing procedure gives

$$s_{k-1} + \sum_{j \notin S, k \leqslant j \leqslant l} y_j \geqslant r_{[1]}(\mu_{[1]} - X_{[1]}^S) + (r_{[2]} - r_{[1]})(\mu_{[2]} - X_{[2]}^S)$$

$$+ \cdots + (r_{[q]} - r_{[q-1]})(\mu_{[q]} - X_{[q]}^S).$$

An example of this inequality has been shown in Example 3.2.

Varying capacity lot-sizing. Inequality (42) with varying capacities gives, setting $s' = s_{k-1} + \sum_{j \notin S, k \leqslant j \leqslant t} y_j$, the relaxation

$$s' + \sum_{j \in S, j \leqslant t} C_t x_j \geqslant d_{kt}, \quad s' \geqslant 0, x_j \in \{0,1\} \quad \text{for } j \in S,$$

for which mixed knapsack inequalities can be generated, see Section 1.1. Alternatively, aggregation of the flow balance constraints gives the inequality $\sum_{j=l}^{l} y_j \leqslant d_{kt} + s_l$, the bounds give us $y_j \leqslant C_j x_j$, and the uncapacitated inequality (43) gives $y_j \leqslant d_{jl} x_j + s_l$. Setting $s_l = 0$ temporarily, we have a single node flow set:

$$\left\{ (x,y) \in \{0,1\}^{k-l+1} \times \mathbb{R}_+^{k-l+1} : \sum_{j=k}^{l} y_j \leqslant d_{kl}, y_j \leqslant \min[C_j, d_{jl}]x_j \text{ for } j=k, \ldots, l \right\}.$$

Now it suffices to add the term $(+s_l)$ to the right-hand side of any flow cover inequality to have a valid inequality for X^{ULS}, see [91].

4.2.1. Modelling start-ups

If $x_1, x_2, \ldots, x_n \in Z_+^n$ denote the number of machines set-up in periods $1, \ldots, n$, it is often important to know the number $\max[x_t - x_{t-1}, 0]$ of machines that start-up in period t. If z_t is a variable representing the number of start-ups, we use the constraints

$$z_t \geqslant x_t - x_{t-1}, \quad z_t \geqslant 0$$

to get an upper bound on the number of start-ups, and

$$z_t \leqslant x_t, \quad z_t \leqslant u_t - x_{t-1}$$

to try to make the upper bound tight, where u_t is an upper bound on x_t. This provides an exact formulation if $x_t, x_{t-1} \in \{0,1\}$, but it is not tight otherwise.

Observation 4.2. Let $\chi_{kt} = \max\{x_k, \ldots, x_t\}$, then $x_k + z_{k+1} + \cdots + z_t \geqslant \chi_{kt}$.

Lot-sizing with start-ups. Let z_t be defined as above to take value 1 if and only if $x_t = 1$ and 0, and let χ_{kl} denote the maximum of $(x_k, \ldots, x_l)$. The uncapacitated inequality (44) says essentially that the stock at the end of period $k - 1$ contains the demand d_t if there is no production in periods $k, \ldots, t$, or in other words if $\chi_{kt} = 0$. This gives the valid inequality $s_{k-1} \geq \sum_{t=k}^{l} d_t(1 - \chi_{kt})$, or using Observation 4.2

$$s_{k-1} \geq \sum_{t=k}^{l} d_t(1 - x_k - z_{k+1} - \cdots - z_t).$$

In the constant capacity case, either $\chi_{j,l} = 0$ and

$$s_{k-1} + C\left(\sum_{i=k}^{j-1} x_i + \chi_{j,l}\right) \geq d_{kl} = d_{k,j-1} + d_{jl},$$

or $\chi_{j,l} = 1$ and so

$$s_{k-1} + C\left(\sum_{i=k}^{j-1} x_i + \chi_{j,l}\right) \geq d_{k,j-1} + C.$$

Thus, all feasible solutions to (42) satisfy $s_{k-1} + C(\sum_{i=k}^{j-1} x_i + \chi_{j,l}) \geq d_{k,j-1} + \min[C, d_{jl}]$, and from this we obtain the valid MIR inequality

$$s_{k-1} \geq \tilde{r}_{kl}\left(\tilde{\mu}_{kl} - \sum_{i=k}^{j-1} x_i - x_j - z_{j+1} - \cdots - z_l\right),$$

where $\tilde{d}_{kl} = d_{k,j-1} + \min[C, d_{jl}]$, $\tilde{\mu}_{kl} = \lceil \tilde{d}_{kl}/C \rceil$ and $\tilde{r}_{kl} = \tilde{d}_{kl} - (\tilde{\mu}_{kl} - 1)C$. Now varying l and using mixing, one can obtain the *left extended klSI inequalities* from [32].

4.3. Facility location problems

The capacitated facility location problem is also a special case of the fixed charge network flow problem. One particularity is that the fixed costs are incurred on opening nodes (locations) rather than arcs. We show that both flow cover and dynamic inequalities can be specialized for the special structure of this problem. A more combinatorial class of inequalities, a generalization of inequalities from the uncapacitated case, is also presented.

The feasible region is typically described as follows:

$$\sum_{j \in N} y_{ij} = a_i \quad \text{for } i \in M, \tag{45}$$

$$\sum_{i \in M} y_{ij} \leq C_j x_j \quad \text{for } j \in N, \tag{46}$$

$$0 \leq y_{ij} \leq \min[a_i, C_j] x_j \quad \text{for } i \in M, \ j \in N, \tag{47}$$

$$x_j \in \{0, 1\} \quad \text{for } j \in N, \tag{48}$$

where y_{ij} is the amount shipped from location j to client i, and $x_j = 1$ indicates that location j is in use.

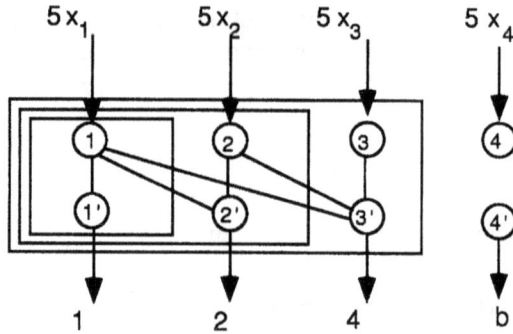

Fig. 7. Dynamic location set.

Letting $v_j = \Sigma_{i \in M} y_{ij}$ and summing up all the demand constraints (45) leads to a single node flow set X described by

$$\left\{ (v,x) \in \mathbb{R}^n_+ \times \{0,1\}^n : \sum_{j \in N} v_j = \sum_{i \in M} a_i, \ 0 \leqslant v_j \leqslant C_j x_j \text{ for } j \in N \right\}$$

for which knapsack and flow cover inequalities can be generated.

Next we consider the internal structure of the underlying digraph. Consider a subset $K \subseteq M$ of clients, a subset $J \subseteq N$ of locations, and for each $j \in J$ a possibly smaller subset $K_j \subseteq K$ of clients. Restricted to this subset, the *effective capacity* of location j is $\bar{C}_j = \min[C_j, \sum_{i \in K_j} a_i]$. Now we obtain a modified flow cover set based on the new variable $\tilde{v}_j = \sum_{i \in K_j} y_{ij}$, namely the set

$$X^{\text{EC}} := \left\{ (\tilde{v},x) \in \mathbb{R}^{|J|}_+ \times \{0,1\}^{|J|} : \sum_{j \in J} \tilde{v}_j \leqslant \sum_{i \in K} a_i, \tilde{v}_j \leqslant \bar{C}_j x_j \text{ for } j \in J \right\}.$$

Specifically, if J is a cover with excess $\lambda = \sum_{j \in J} \bar{C}_j - \sum_{i \in K} a_i > 0$, then we obtain the *effective capacity flow cover inequality*

$$\sum_{j \in J} \sum_{i \in K_j} y_{ij} + \sum_{j \in J} (\bar{C}_j - \lambda)^+ (1 - x_j) \leqslant \sum_{i \in K} a_i.$$

Submodular inequalities can also be defined for this model leading to very similar inequalities. The separation problem for the effective capacity and submodular inequalities involves a choice of the sets, J, K and K_j, and is necessarily heuristic, see [1].

Dynamic inequalities. When $K_r \subseteq K_{r-1} \cdots \subseteq K_1$, we can use the embedded set structure to obtain dynamic inequalities, see Section 3.1.

Example 4.3. Consider a problem with four clients and four locations as shown in Fig. 7.

Specifically, we have $J = \{1,2,3\}, K_1 = \{1',2',3'\}, K_2 = \{2',3'\}$ and $K_3 = \{3'\}$. This corresponds to an embedded node set with $S_1 = \{1,1'\}, S_2 = \{1,2,1',2'\}$,

$S_1 = \{1, 2, 1', 2', 3, 3'\}$ giving the surrogate capacity constraints

$$v_{21} + v_{31} \qquad + v_{41} \qquad\qquad\qquad +5x_1 \qquad\qquad\qquad \geqslant 1$$
$$+v_{31} + v_{32} + v_{41} + v_{42} \qquad +5x_1 + 5x_2 \qquad\qquad \geqslant 3$$
$$v_{41} + v_{42} + v_{43} + 5x_1 + 5x_2 + 5x_3 \geqslant 7$$

leading first to the standard MIR inequalities and then the dynamic inequality

$$v_{21} + v_{31} + v_{32} + v_{41} + v_{42} + v_{43} \geqslant 1(1 - x_1) + (2 - 1)(2 - x_1 - x_2 - x_3)$$
$$+ (3 - 2)(1 - x_1 - x_2).$$

Combinatorial inequalities. With the same structure J, K and K_j of locations and clients, let β be the minimum number of locations required to serve all the clients in K if location j is restricted to serving clients in K_j. Then it is shown in [2] that

$$\sum_{j \in J} \sum_{j \in K_j} \frac{1}{a_i} y_{ij} - \sum_{j \in J} x_j \leqslant |K| - \beta$$

is valid.

4.4. Unit commitment problems

The unit commitment problem (the problem of scheduling electricity generators to satisfy hourly demands for a day or a week) is not a fixed charge network flow problem. However, its formulation as a mixed integer program contains several constraints and variables that have been encountered in this chapter for which cuts can be generated, such as single node flow models and start-up variables linking the generators between time periods. A typical formulation involves the following variables:

x_t^i is the number of generators of type i functioning at period (hour) t (often each generator is distinct, and this is a 0–1 variable)

z_t^i is the increase in the number of generators of type i active in period t

y_t^i is the amount of electricity produced by generators of type i in period t, and as basic constraints

$$\sum_i y_t^i = d_t \quad \text{for all } t, \tag{49}$$

$$l^i x_t^i \leqslant y_t^i \leqslant C^i x_t^i \quad \text{for all } i, t, \tag{50}$$

$$z_t^i \geqslant x_t^i - x_{t-1}^i \quad \text{for all } i, t, \tag{51}$$

$$z_t^i \leqslant x_t^i \quad \text{for all } i, t, \tag{52}$$

$$y_t^i \geqslant 0, \quad x_t^i \leqslant u^i \quad \text{for all } i, t, \tag{53}$$

$$x_t^i, z_t^i \in Z_+^1 \quad \text{for all } i, t. \tag{54}$$

Typical models also contain ramping and reserve constraints, see [92]. Constraints (49), (50), (53), (54) lead to single node flow sets, or continuous knapsack sets on which various inequalities presented in Section 1 can be generated. In contrast to lot-sizing models, the flow balance constraints are not linked over time, as electricity cannot be stocked. However, the start-up variables provide a certain link between periods. Specifically, if we aggregate (49) for periods $t = k, \ldots, l$ and use (50), we obtain $\sum_{t=k}^{l} \sum_i C^i x_t^i \geq d_{kl}$. Letting $\chi_{kl}^i = \max\{x_k^i, \ldots, x_l^i\}$ and (I_1, I_2) be a partition of the generator set, we obtain

$$\sum_{i \in I_1} C^i \sum_{t=k}^{l} x_t^i + \sum_{i \in I_2} (k - l + 1) C^i \chi_{kl}^i \geq d_{kl}$$

with $x_t^i, \chi_{kl}^i \in Z_+^1$. Deriving valid inequalities for such knapsack sets, and then using Observation 4.2 to replace χ_{kl}^i by its upper bound $x_k^i + z_{k+1}^i + \cdots + z_l^i$ leads to new valid inequalities.

Example 4.4 (Marchand [68]). Consider two generator types and two periods with $C^1 = 4, C^2 = 5, d_1 = 12, d_2 = 13$ and $u^1 = u^2 = 4$. Taking $I_1 = \{2\}$ and $I_2 = \{1\}$, we obtain the set

$$5x_1^2 + 5x_2^2 + 8\chi_{12}^1 \geq 25, \quad 0 \leq x_1^2, x_2^2, \chi_{12}^1 \leq 4 \text{ and integer.}$$

A valid inequality for this set is

$$\chi_{12}^1 + x_1^2 + x_2^2 \geq 4.$$

Now using $x_1^1 + z_2^1 \geq \chi_{12}^1$, we obtain the valid inequality

$$x_1^1 + z_2^1 + x_1^2 + x_2^2 \geq 4.$$

This inequality cuts off the extreme point solution $x_1^1 = 3$, $x_2^1 = 3$, $z_2^1 = 0$ and $x_1^2 = 0$, $x_2^2 = \frac{1}{5}$, $z_2^2 = \frac{1}{5}$.

5. Note on computation with cutting planes

Several of the families of valid inequalities described above have been incorporated into branch-and-bound systems in the last fifteen years. If cuts are only added at the top node, we speak of a *cut-and-branch* system, while if cuts are added at other nodes in the enumeration tree, it is a *branch-and-cut* system. Introductions to branch-and-cut can be found in [62,107]. For a survey on branch-and-cut systems for combinatorial optimization problems, see [27,63].

5.1. General mixed integer programming systems

In [36] lifted cover inequalities for 0–1 knapsack inequalities were first incorporated in a cut-and-branch system for 0–1 integer programs. Later flow cover inequalities and an uncapacitated version of the dynamic inequalities on paths were included in MPSARX [101], a cut-and-branch system for MIPs. MINTO [75] was the first branch-and-cut system for MIPs incorporating lifted cover and flow cover inequalities, and more recently lifted cover inequalities for knapsack constraints with generalized upper bound constraints. Computational testing of lifted cover and flow cover inequalities has been reported in [54,53]. More recent systems include SIP [72,73] which also generates feasible set inequalities, and BC-OPT [34] that includes integer knapsack inequalities and recently also MIR inequalities. Taking a different approach, MIPO [11] is a branch-and-cut system for MIPs based on lift-and-project inequalities, where the importance of finding the right balance between cutting and branching is clearly demonstrated. With this system it has also been shown that Gomory mixed integer cuts can be used effectively [12].

Two of the commercial systems, CPLEX and XPRESS, have recently started incorporating lifted cover inequalities, flow cover and MIR inequalities into their systems. For those interested in testing new cuts, etc., a library of mixed integer programming test instances is available [22]. ABACUS [97] is a branch-and-cut framework more specifically suited to combinatorial optimization problems.

5.2. Packing and covering

Most set packing or covering inequalities are used in connection with the solution of set partitioning problems, for instance, [58] exploits clique and cycle inequalities, [23] uses aggregated cycle inequalities in addition. There seem to be virtually no efficient separation algorithms for set covering problems. To the best of our knowledge the only exceptions are the cutting planes from conditional bounds by [9], a class of k-projection inequalities by [81], and the mentioned aggregated cycle inequalities by [24], which also apply to set covering. A cutting plane algorithm for set packing problems has been developed in [76]. Note also that clique inequalities are used in many general mixed integer programming systems [35,36,75].

5.3. Network design problems

There is little specialized computational work on single commodity network design problems. However, the cutting planes in the general systems cited above significantly improve performance on some instances. In contrast there has been considerable work on multicommodity problems arising from telecommunications networks. Among others single arc sets [67] and MIR inequalities [26] have been used, and both heuristics [20,21], total enumeration [20] and max cut [15] have been used to generate good cut sets. See also [6,37,38,55].

5.4. Lot-sizing, facility location and other structured MIPs

A variety of multiitem and multilevel lot-sizing problems have been solved using the cutting planes described above, see [32,88,18]. A variety of problem instances are available.[3,4]

---

[3]Lot-sizing instances available at http://www.eng.auburn.edu/~gaoyubo.

[4](http://www.core.ucl.ac.be/wolsey/Lotsizeli.htm), a library of lot-sizing instances.

Some computation on capacitated facility location problems is presented in [1]. The library [17] contains a variety of instances.

Several instances in MIPLIB3.0 are unit commitment instances. For these and other electricity generation applications [74], using knapsack and MIR inequalities significantly improves solution performance [70].

Acknowledgements

We are grateful to K. Aardal for a careful reading of the text.

References

[1] K. Aardal, Capacitated facility location: separation algorithms and computational experience, Math. Programming 81 (1998) 149–175.

[2] K. Aardal, Y. Pochet, L.A. Wolsey, Capacitated facility location: valid inequalities and facets, Math. Oper. Res. 20 (1995) 562–582.

[3] D. Applegate, R.E. Bixby, V. Chvátal, W. Cook, Project-and-lift (a paradigm for finding cuts), Draft, Aussois, March, 1998.

[4] A. Atamturk, On network design cut-set polyhedra, Draft, Department of Industrial Engineering and Operations Research, U.C. at Berkeley, August, 1999.

[5] A. Atamturk, G.L. Nemhauser, M.W.P. Savelsbergh, Conflict graphs in integer programming, Technical Report LEC 98-03, Georgia Institute of Technology, 1998.

[6] A. Balakrishnan, T.L. Magnanti, J. Sokol, Y. Wang, Modeling and solving the single facility line restoration problem, Technical report, MIT, Operations Research Center, 1998. Available at http://web.mit.edu/yiwang/www.

[7] E. Balas, Facets of the knapsack polytope, Math. Programming 8 (1975) 146–164.

[8] E. Balas, Disjunctive programming, Ann. Discrete Math. 5 (1979) 3–51.

[9] E. Balas, A. Ho, Set covering algorithms using cutting planes, heuristics, and subgradient optimization: a computational study, Math. Programming 12 (1980) 37–60.

[10] E. Balas, S. Ceria, G. Cornuéjols, A lift-and-project cutting plane algorithm for mixed 0–1 programs, Math. Programming 58 (1993) 295–324.

[11] E. Balas, S. Ceria, G. Cornuéjols, Mixed 0–1 programming by lift-and-project in a branch-and-cut framework, Management Sci. 42 (1996) 1229–1246.

[12] E. Balas, S. Ceria, G. Cornuéjols, N. Natraj, Gomory cuts revisited, Oper. Res. Lett. 19 (1996) 1–9.

[13] E. Balas, E. Zemel, Lifting and complementing yields all the facets of positive zero-one programming polytopes, Math. Programming (1984) 13–24, Proceedings of the International Congress, Rio de Janeiro, 1981.

[14] E. Balas, E. Zemel, Facets of the knapsack polytope from minimal covers, SIAM J. Appl. Math. 34 (1978) 119–148.

[15] F. Barahona, Network design using cut inequalities, SIAM J Optim. 6 (1996) 823–837.

[16] I. Barany, T.J. Van Roy, L.A. Wolsey, Strong formulations for multi-item capacitated lot-sizing, Management Sci. 30 (1984) 1255–1261.

[17] J.E. Beasley, OR-Library: distributing test problems by electronic mail, J. Oper. Res. Soc. 41 (1990) 1069–1072, available at ⟨http://mscmga.ms.ic.ac.uk/info.html⟩.

[18] G. Belvaux, L.A. Wolsey, BC-PROD: A specialized branch-and-cut system for lot-sizing problems, Management Sci. 46 (2000) 724–738.

[19] C. Berge, Färbung von Graphen, deren sämtliche bzw. deren ungerade Kreise starr sind (Zusammenfassung), in Wissenschaftliche Zeitschrift, Mathematisch-Naturwissenschaftliche Reihe. Martin-Luther-Universität Halle-Wittenberg, 1961.

[20] D. Bienstock, O. Günlük, Capacitated network design—polyhedral structure and computation, ORSA J. Comput. 8 (1996) 243–259.

[21] D. Bienstock, S. Chopra, O. Günlük, C.-Y. Tsai, Minimum cost capacity installation for multicommodity network flows, Math. Programming 81 (1998) 177–199.

[22] R.E. Bixby, S. Ceria, C.M. McZeal, M.W.P. Savelsbergh, An updated mixed integer programming library: MIPLIB 3.0, text and problems available at http://www.caam.rice.edu/~bixby/miplib/miplib.html

[23] R. Borndörfer, Aspects of set packing, partitioning, and covering, Ph.D. Thesis, Technische Universität Berlin, 1998.

[24] R. Borndörfer, R. Weismantel, Relations among some combinatorial programs, Technical Report Preprint SC 97-54, Konrad-Zuse-Zentrum für Informationstechnik Berlin, 1997.

[25] R. Borndörfer, R. Weismantel, Set packing relaxations of some integer programs, Technical Report Preprint SC 97-30, Konrad-Zuse Zentrum für Informationstechnik Berlin, 1997.

[26] B. Brockmüller, O. Günlük, L.A. Wolsey, designing private line networks—polyhedral analysis and computation, Core Discussion Paper 9647, Université Catholique de Louvain, 1996, revised March, 1998.

[27] A. Caprara, M. Fischetti, Branch-and-cut algorithms, in: M. Dell'Amico, F. Maffioli, S. Martello (Eds.), Annotated Bibliographies in Combinatorial Optimization, Wiley, Chichester, 1997, pp. 45–63.

[28] A. Caprara, M. Fischetti, $\{0, \frac{1}{2}\}$-Chvátal–Gomory cuts, Math. Programming 74 (1996) 221–236.

[29] S. Ceria, C. Cordier, H. Marchand, L.A. Wolsey, Cutting planes for integer programs with general integer variables, Math. Programming 81 (1998) 201–214.

[30] V. Chvátal, Edmonds polytopes and a hierarchy of combinatorial problems, Discrete Math. 4 (1973) 305–337.

[31] V. Chvátal, On certain polytopes associated with graphs, J. Combin. Theory B 18 (1975) 305–337.

[32] M. Constantino, A cutting plane approach to capacitated lot-sizing with start-up costs, Math. Programming 75 (1996) 353–376.

[33] W. Cook, R. Kannan, A. Schrijver, Chvátal closures for mixed integer programming problems, Math. Programming 47 (1990) 155–174.

[34] C. Cordier, H. Marchand, R. Laundy, L.A. Wolsey, bc-opt: A branch-and-cut code for mixed integer programs, bc-opt: a branch-and-cut code for mixed integer programs, Math. Programming 86 (1999) 335–354.

[35] CPLEX, Using the CPLEX callable library, ILOG CPLEX Division, 889 Alder Avenue, Suite 200, Incline Village, NV 89451, USA, Information available at URL http://www.cplex.com (1998).

[36] H. Crowder, E. Johnson, M.W. Padberg, Solving large-scale zero-one linear programming problems, Oper. Res. 31 (1983) 803–834.

[37] G. Dahl, A. Martin, M. Stoer, Routing through virtual paths in layered telecommunication networks, Research Note N78/95, Telenor Research and Development, Kjeller, Norway, 1995, Oper. Res. 49 (1999) 693–702.

[38] G. Dahl, M. Stoer, A cutting plane algorithm for multicommodity survivable network design problems, INFORMS J. Comput. 10 (1998) 1–11.

[39] I.R. de Farias, E.L. Johnson, G.L. Nemhauser, Facets of the complementarity knapsack polytope, Technical Report LEC-98-08, Georgia Institute of Technology, 1998.

[40] J. Edmonds, Matroids and the greedy algorithm, Math. Programming 1 (1971) 127–136.

[41] J. Edmonds, Matroid intersection, Ann. Discrete Math. 4 (1979) 39–49.

[42] R.Euler, M. Jünger, G. Reinelt, Generalizations of odd cycles and anti-cycles and their relation to independence system polyhedra, Math. Oper. Res. 12 (1987) 451–462.

[43] C.E. Ferreira, A. Martin, R. Weismantel, Solving multiple knapsack problems by cutting planes, SIAM J. Optim. 6 (1996) 858–877.

[44] D.R. Fulkerson, Blocking and anti-blocking pairs of polyhedra, Math. Programming 1 (1971) 168–194.

[45] M.X. Goemans, The Steiner tree polytope and related polyhedra, Math. Programming 63 (1994) 157–183.

[46] R.E. Gomory, Outline of an algorithm for integer solutions to linear programs, Bull. Amer. Soc. 64 (1958) 275–278.

[47] R.E. Gomory, An algorithm for the mixed integer problem, Technical Report RM-2597, The RAND Cooperation, 1960.

[48] R.E. Gomory, Solving linear programming problems in integers, in: R. Bellman, M. Hall (Eds.), Combinatorial Analysis, Proceedings of Symposia in Applied Mathematics 10, Providence, RI, 1960.

[49] R.E. Gomory, An algorithm for integer solutions to linear programming, in: R.L. Graves, P. Wolfe (Eds.), Recent Advances in Mathematical Programming, McGraw-Hill, New York, 1969, pp. 269–302.

[50] M. Grötschel, L. Lovász, A. Schrijver, Geometric Algorithms and Combinatorial Optimization, Springer, Berlin, 1988.

[51] M. Grötschel, C.L. Monma, M. Stoer, Design of survivable networks, in: M.O. Ball et al. (Eds.), Network Models, Handbooks in OR and MS 7, Elsevier, Amsterdam, 1995 (chapter 10).

[52] Z. Gu, G.L. Nemhauser, M.W.P. Savelsbergh, Sequence independent lifting in mixed integer programming, J. Combin. Optim. 4 (2000) 109–129.

[53] Z. Gu, G.L. Nemhauser, M.W.P. Savelsbergh, Lifted flow cover inequalities for mixed 0–1 integer programs, Math. Programming A 85 (1999) 436–467.

[54] Z. Gu, G.L. Nemhauser, M.W.P. Savelsbergh, Lifted cover inequalities for 0–1 integer programs: computation, INFORMS J. Comput. 10 (1998) 427–437.

[55] O. Günlük, A branch-and-cut algorithm for capacitated network design, Technical report, Cornell University, 1966.

[56] O. Günlük, Y. Pochet, Mixing mixed-integer inequalities, Math. Programming (2001) 429–458.

[57] P.L. Hammer, E.L. Johnson, U.N. Peled, Facets of regular 0–1 polytopes, Math. Programming 8 (1975) 179–206.

[58] K.L. Hoffman, M.W. Padberg, Solving airline crew-scheduling problems by branch-and-cut, Management Sci. 39 (1993) 657–682.

[59] M. Iri, On an extension of the maximum-flow minimum-cut theorem to multi-commodity flows, J. Oper. Res. Soc. Japan 13 (1970/71) 129–135.

[60] R.G. Jeroslow, Cutting plane theory: disjunctive methods, Ann. Discrete Math. 1 (1977) 293–330.

[61] E.L. Johnson, M.W. Padberg, Degree-two inequalities, clique facets, and biperfect graphs, Ann. Discrete Math. 16 (1982) 169–187.

[62] E.L. Johnson, G.L. Nemhauser, M.W.P. Savelsbergh, Progress in linear programming based branch-and-bound algorithms: an exposition, INFORMS J. Comput. 12 (2000) 2–23.

[63] M. Jünger, G. Reinelt, S. Thienel, Practical problem solving with cutting plane algorithms in combinatorial optimization, in: W. Cook, L. Lovász, P. Seymour (Eds.), Combinatorial Optimization, DIMACS Series in Discrete Mathematics and Computer Science, AMS, Providence, RI, 1995, pp. 11–152.

[64] M. Laurent, A generalization of antiwebs to independence systems and their canonical facets, Math. Programming 45 (1989) 97–108.

[65] L. Lovász, On the Shannon capacity of a graph, IEEE Trans. Inform. Theory 25 (1979) 1–7.

[66] L. Lovász, A. Schrijver, Cones of matrices and set-functions and 0–1 optimization, SIAM J. Optim. 1 (1991) 166–190.

[67] T.L. Magnanti, P. Mirchandani, R. Vachani, Modelling and solving the two-facility network loading problem, Oper. Res. 43 (1995) 142–157.

[68] H. Marchand, Etude d'un problème d'optimisation lié à la gestion d'un parc électrique, Engineering Thesis, Faculté des Sciences Appliquées, 1994.

[69] H. Marchand, A polyhedral study of the mixed knapsack set and its use to solve mixed integer programs, Ph.D. Thesis, Université Catholique de Louvain, Louvain-la-Neuve, Belgium, 1998.

[70] H. Marchand, L.A. Wolsey, Aggregation and mixed integer rounding to solve MIPs, CORE DP9839, Université Catholique de Louvain, Louvain-la-Neuve, Belgium, 1998, Oper. Res., to appear.

[71] H. Marchand, L.A. Wolsey, The 0–1 knapsack problem with a single continuous variable, Math. Programming 85 (1999) 15–33.

[72] A. Martin, Integer programs with block structure, Habilitations-Schrift Technische Universität Berlin, 1998. Available at ⟨ftp://ftp.zib.de/pub/zib-publications/reports/SC-99-03.ps⟩.

[73] A. Martin, R. Weismantel, The intersection of knapsack polyhedra and extensions, in: R.E. Bixby, E.A. Boyd, R.Z. Rios-Mercado (Eds.), Lecture Notes in Computer Science, Vol. 1412, Springer, Berlin, 1998, pp. 243–256.

[74] MEMIPS, Model enhanced solution methods for integer programming software, Esprit Project 20118, Public Report Reference DR1.1.10, 1997.

[75] G.L. Nemhauser, M.W.P. Savelsbergh, G.C. Sigismondi, MINTO, a mixed INTeger optimizer, Oper. Res. Lett. 15 (1994) 47–58.

[76] G.L. Nemhauser, G. Sigismondi, A strong cutting plane/branch-and-bound algorithm for node packing, J. Oper. Res. Soc. 43 (1992) 443–457.

[77] G.L. Nemhauser, L.E. Trotter, J., Properties of vertex packing and independence system polyhedra, Math. Programming 6 (1974) 48–61.

[78] G.L. Nemhauser, L.A. Wolsey, Integer and Combinatorial Optimization, Wiley, New York, 1988.

[79] G.L. Nemhauser, L.A. Wolsey, A recursive procedure to generate all cuts for 0–1 mixed integer programs, Math. Programming 46 (1990) 379–390.

[80] P. Nobili, A. Sassano, Facets and lifting procedures for the set covering polytope, Math. Programming 45 (1989) 111–137.

[81] P. Nobili, A. Sassano, A separation routine for the set covering polytope, in: E. Balas, G. Cornuéjols, R. Kannan (Eds.), Integer Programming and Combinatorial Optimization, Proceedings of the 2nd IPCO Conference, 1992, pp. 201–219.

[82] K. Onaga, O. Kakusho, On feasibility conditions of multicommodity flows in networks, IEEE Trans. Circuit Theory 18 (1971) 425–429.

[83] J.H. Owen, S. Mehrotra, Math. Programming 89 (2001) 437–448.

[84] M.W. Padberg, On the facial structure of set packing polyhedra, Math. Programming 5 (1973) 199–215.

[85] M.W. Padberg, A note on zero-one programming, Oper. Res. 23 (1975) 833–837.

[86] M.W. Padberg, T.J. Van Roy, L.A. Wolsey, Valid linear inequalities for fixed charge problems, Oper. Res. 33 (1985) 842–861.

[87] Y. Pochet, Valid inequalities and separation for capacitated economic lot-sizing, Oper. Res. Lett. 7 (1988) 109–116.

[88] Y. Pochet, L.A. Wolsey, Solving multi-item lot-sizing problems using strong cutting planes, Management Sci. 37 (1991) 53–67.

[89] Y. Pochet, L.A. Wolsey, Lot-sizing with constant batches: formulation and valid inequalities, Math. Oper. Res. 18 (1993) 767–785.

[90] Y. Pochet, L.A. Wolsey, Integer knapsacks and flow covers with divisible coefficients: polyhedra, optimization and separation, Discrete Appl. Math. 59 (1995) 57–74.

[91] R. Rardin, L.A. Wolsey, Valid inequalities and projecting the multicommodity extended formulation for uncapacitated fixed charge network flow problems, European J. Oper. Res. 71 (1993) 95–109.

[92] A. Renaud, Daily generation management at electricité de France: from planning towards real time, IEEE Trans. Automat. Control 38 (1993) 1080–1093.

[93] A. Sassano, On the facial structure of the set covering polytope, Math. Programming 44 (1989) 181–202.

[94] A. Schrijver, On cutting planes, Ann. Discrete Math. 9 (1980) 291–296.

[95] A. Schrijver, Theory of Linear and Integer Programming, Wiley, Chichester, 1986.

[96] H. Sherali, W. Adams, A hierarchy of relaxations between the continuous and convex hull representations for zero-one programming problems, SIAM J. Discrete Math. 3 (1990) 411–430.

[97] S. Thienel, ABACUS—a branch-and-cut system, Doctoral Thesis, Universität zu Köln, 1995.

[98] R. van de Leensel, Models and algorithms for telecommunications network design, Proefschrift, Universiteit Maastricht, 1999.

[99] R. van de Leensel, C.P.M. van Hoesel, J.J. van de Klundert, Lifting valid inequalities for the precedence constrained knapsack problem, Math. Programming A 86 (1999) 161–185.

[100] T.J. Van Roy, L.A. Wolsey, Valid inequalities for mixed 0–1 programs, Discrete Appl. Math. 4 (1986) 199–213.

[101] T.J. Van Roy, L.A. Wolsey, Solving mixed 0–1 problems by automatic reformulation, Oper. Res. 35 (1987) 45–57.

[102] R. Weismantel, On the 0/1 knapsack polytope, Math. Programming 77 (1997) 49–68.

[103] L.A. Wolsey, Faces for a linear inequality in 0–1 variables, Math. Programming 8 (1975) 165–178.

[104] L.A. Wolsey, Valid inequalities and superadditivity for 0/1 integer programs, Math. Oper. Res. 2 (1977) 66–77.

[105] L.A. Wolsey, Submodularity and valid inequalities in capacitated fixed charge networks, Oper. Res. Lett. 8 (1989) 119–124.

[106] L.A. Wolsey, Valid inequalities for 0–1 knapsacks and MIPS with generalized upper bound constraints, Discrete Appl. Math. 29 (1990) 251–261.

[107] L.A. Wolsey, Integer Programming, Wiley, New York, 1998.

[108] F. Eisenbrand, On the membership problem for the elementary closure of a polyhedron, Combinatorica 12 (1999) 27–37.

For further reading

A. Balakrishnan, T.L. Magnanti, R.T. Wong, A decomposition algorithm for local access telecommunications network expansion planning, Oper. Res. 43 (1995) 58--76.

W. Cook, L. Lov{a} sz, P. Seymour, Combinatorial Optimization, DI-MACS Series in Discrete Mathematics and Computer Science, AMS, Providence, RI, 1995.

B. Gavish, K. Altinkemer, Backbone network design tools with economic tradeoffs, ORSA J. Comput. 2 (1990) 58--76.

T.I. Magnanti, P. Mirchandani, Shortest paths, single origin-destination network design and associated polyhedra, Networks 23 (1993) 103--121.

Discrete Applied Mathematics 123 (2002) 447–472

DISCRETE
APPLIED
MATHEMATICS

Graph connectivity and its augmentation: applications of MA orderings

Hiroshi Nagamochi[a],*, Toshihide Ibaraki[b]

[a] *Department of Information and Computer Sciences, Toyohashi University of Technology, Hibarigaoka, Tenpaku, Toyohashi 441-8580, Japan*
[b] *Department of Applied Mathematics and Physics, Kyoto University, Sakyo, Kyoto 606-8501, Japan*

Received 4 October 1999; received in revised form 26 June 2000; accepted 30 October 2000

Abstract

This paper surveys how the maximum adjacency (MA) ordering of the vertices in a graph can be used to solve various graph problems. We first explain that the minimum cut problem can be solved efficiently by utilizing the MA ordering. The idea is then extended to a fundamental operation of a graph, edge splitting. Based on this, the edge-connectivity augmentation problem for a given k (and also for the entire range of k) can be solved efficiently by making use of the MA ordering, where it is asked to add the smallest number of new edges to a given graph so that its edge-connectivity is increased to k. Other related topics are also surveyed. © 2002 Elsevier Science B.V. All rights reserved.

Keywords: Graphs; MA ordering; Minimum cuts; Edge-connectivity; Graph augmentation; Polynomial time algorithms

1. Introduction

This paper surveys how the maximum adjacency (MA) ordering of vertices of a graph can be used to solve various graph problems. Let an undirected multigraph (i.e., an undirected graph with integer edge weights) be given, which has n vertices. An ordering $v_1, v_2, \ldots, v_n$ of vertices is called an *MA ordering* if an arbitrary vertex is chosen as v_1, and after choosing the first i vertices $v_1, \ldots, v_i$, the $(i+1)$th vertex v_{i+1} is chosen from the vertices u that have the largest number of edges between $\{v_1, \ldots, v_i\}$ and u. The ordering was proposed in [20,69] to compute a sparse k-edge (resp., k-vertex)

---

* Corresponding author.
E-mail addresses: naga@ics.tut.ac.jp (H. Nagamochi), ibaraki@amp.i.kyoto-u.ac.jp (T. Ibaraki).

0166-218X/02/$ - see front matter © 2002 Elsevier Science B.V. All rights reserved.
PII: S 0 1 6 6 - 2 1 8 X (0 1) 0 0 3 4 9 - 3

connected spanning subgraph in a given k-edge (resp., k-vertex) connected graph. It was called a legal ordering in [20], and the name "MA ordering" was coined by Matula [67]. The MA ordering is identical with the maximum cardinality ordering which was discovered by Tarjan and Yannakakis [87] to test the chordality of graphs. But in an MA ordering, we also label the edges in order to decompose the graph into spanning forests (see Section 2.2).

An important property of an MA ordering is that it identifies a minimum cut between some two vertices, which are specified by the ordering. Based on this, we can solve the minimum cut problem of a graph in $O(mn + n^2 \log n)$ time [70,76], where n and m are the numbers of vertices and edges, respectively. This is an improvement over the conventional minimum cut algorithms which execute the computation of maximum flows n times [1,37]. The idea is then extended to a fundamental operation of a graph, edge splitting [71,78]. Based on the new edge splitting algorithm, the edge-connectivity augmentation problem can also be solved efficiently. The edge-connectivity augmentation problem asks to add to a given graph the smallest number of new edges so that the edge-connectivity of the resulting graph is increased to a target k. This problem has important applications such as the network construction problem [84], the rigidity problem in grid frameworks [3,27], the data security problem [29,56] and the rectangular dual graph problem in floor-planning [88].

The edge-connectivity and vertex-connectivity augmentation problems were first studied in 1976 by Eswaran and Tarjan [15] and Plesnil [82], and both problems were shown to be polynomially solvable for $k = 2$. For general k, Watanabe and Nakamura [89] established in 1987 a min–max theorem for the edge-connectivity augmentation problem, based on which they gave an $O(k^2(kn + m)n^4)$ time algorithm. Afterwards, Frank [17] gave a unified approach to various edge-connectivity augmentation problems by making use of the edge-splitting theorems of Lovász [62,63] and Mader [64,65]. Then Nagamochi and Ibaraki [75] proposed an $O((nm + n^2 \log n) \log n)$ time algorithm for the edge-connectivity augmentation problem for a given target k, by combining their minimum cut algorithm and the approach of Frank. If the graph under consideration is weighted by real numbers, this algorithm can be further simplified and can be extended to solve the edge-connectivity augmentation problem for the entire range of targets k in $O(nm + n^2 \log n)$ time [75].

This paper first defines the MA ordering and reviews its properties in Section 2. Then Section 3 describes the new minimum cut algorithm and its application to edge splitting. In Sections 4 and 5, we show how to extend the minimum cut algorithm to solve the edge-connectivity augmentation problem for integer weighted and real-weighted graphs, respectively. Recent results on some other augmentation problems are briefly surveyed in Section 6.

2. Preliminaries

2.1. Definitions

Let $G = (V, E)$ be an undirected graph with a vertex set V and an edge set E. The vertex set and the edge set of a graph G may be denoted by $V[G]$ and $E[G]$,

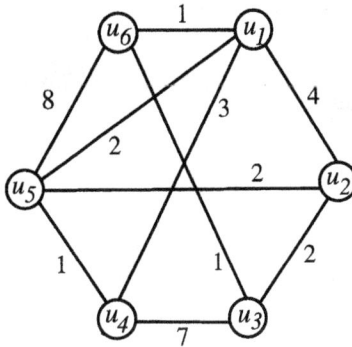

Fig. 1. An integer-weighted graph $G = (V, E)$.

respectively. A singleton set $\{x\}$ is sometimes written as x. The notation $\subseteq$ (resp., $\subset$) denotes the set inclusion (resp., proper set inclusion). For two nonempty and disjoint subsets $X, Y \subseteq V$, let $E_G(X, Y)$ denote the set of edges between X and Y, and $d_G(X, Y)$ denote its cardinality $|E_G(X, Y)|$. In particular, they are also written as $E_G(X)$ and $d_G(X)$, respectively, if $Y = V - X$. If $a = (u, v)$ is an edge of G, then u (resp., v) is a neighbor of v (resp., u) and u, v are said to be the end vertices of a. Edges with the same pair of end vertices are called *multiple edges*. A graph is called a *multigraph* if it is allowed to have multiple edges; *simple* otherwise. We denote

$$n = |V|, \quad e = |E|, \quad m = |\{(x, y) \mid d_G(x, y) \geq 1\}|.$$

The input size of a multigraph $G = (V, E)$ can be measured by n and e. However, it can also be represented by an edge-weighted graph having integer multiplicity $d_G(u, v)$ as the weight of edge (u, v). In this case, the input size if $O(n + m)$ (under the assumption that the logarithm of the maximum weight is constant), and this measure will be used in the rest of this paper. For example, Fig. 1 shows an integer weighted graph, which can be viewed as a multigraph having the multiplicity equal to its edge weight. Throughout this paper, we understand that a graph is a multigraph, unless otherwise specified, except that, in Section 5, we deal with a graph whose edges are weighted by real numbers. Such a graph will be called a *real weighted graph* to distinguish it from an *integer weighted graph*.

We call $E_G(X)$ (or X), satisfying $\emptyset \neq X \subset V$, a *cut*, and define its value by $d_G(X)$. A cut X separates vertices u and v if $u \in X$ and $v \notin X$ (or $u \notin X$ and $v \in X$). It is known that the cut function d_G always satisfies the following *submodular* inequality:

$$d_G(X) + d_G(Y) \geq d_G(X \cap Y) + d_G(X \cup Y), \tag{1}$$

where $d_G(\emptyset) = 0$ is assumed for convenience. A graph is called k-*edge-connected* if the value of any cut is at least k. The minimum value among all the cuts that separate two vertices u and v is called *the local edge-connectivity* between u and v, and is denoted by $\lambda_G(u,v)$. By the well-known theorem of Menger, $\lambda_G(u,v)$ is equal to the maximum number of edge disjoint paths between u and v in G (e.g., [1,37]). The minimum value among all the cuts in G is called its *edge-connectivity*. Finally, for a subset $X \subseteq V$, define the *inner edge-connectivity* of X by

$$\lambda_G(X) = \min\{d_G(X') \mid \emptyset \neq X' \subset X\}.$$

In particular, $\lambda_G(V)$ is equivalent to the edge-connectivity of G.

For a connected graph $G=(V,E)$, a subset $S \subset V$ is called a *vertex-cut* if $G-S$ has at least two connected components, and G is called *k-vertex-connected* if $|V| \geq k+1$ and there is no vertex-cut S with size $k-1$. The maximum number of vertex-disjoint paths from u to v is called *the local vertex-connectivity* between u and v, and is denoted by $\kappa_G(u,v)$ (if u and v are not adjacent, then $\kappa_G(u,v)$ is equal to the minimum size of vertex-cuts separating u and v).

2.2. MA ordering

For any pair of vertices $u,v \in V$ in a graph G, we can compute the local edge-connectivity $\lambda_G(u,v)$ by using the conventional maximum flow algorithm (e.g., [1,37,84]). However, the local edge-connectivity $\lambda_G(u,v)$ for some pair $u,v \in V$ (which are specified by the algorithm) can be computed by a significantly simpler method. An ordering $v_1,v_2,\ldots,v_n$ of vertices in G is called an *MA ordering* if it satisfies

$$d_G(\{v_1,v_2,\ldots,v_i\},v_{i+1}) \geq d_G(\{v_1,v_2,\ldots,v_i\},v_j), \quad 1 \leq i < j \leq n.$$

Such an ordering can be found by choosing an arbitrary vertex v_1, and choosing a vertex $u \in V - \{v_1,\ldots,v_i\}$ that has the largest number of edges between $\{v_1,\ldots,v_i\}$ and u as the $(i+1)$th vertex v_{i+1} after choosing the first i vertices $v_1,\ldots,v_i$. For example, an MA ordering of the graph G in Fig. 1 is obtained as $v_1=u_1, v_2=u_2, v_3=u_5, v_4=u_6, v_5=u_4$ and $v_6=u_3$ (see Fig. 2).

By using the data structure of Fibonacci heap [23], an MA ordering starting from an arbitrarily chosen vertex v_1 can be obtained in $O(n+e)$ or in $O(m+n\log n)$ time [70]. The following property of an MA ordering is the starting point of the rest of development.

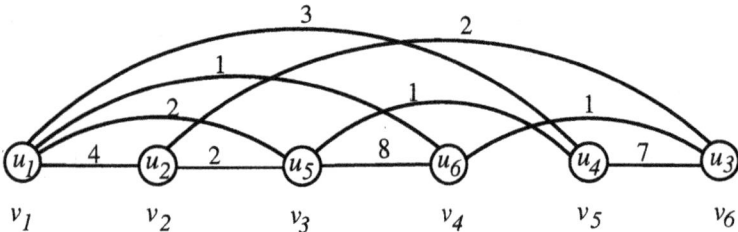

Fig. 2. An MA ordering for the graph G in Fig. 1.

Theorem 2.1 (Nagamochi and Ibaraki [70]). *For a graph $G \doteq (V, E)$, let $x = v_{n-1}$ and v_n be the last two vertices in an MA ordering. Then* (i) $\lambda_G(x, v_n) = d_G(v_n)$. (ii) $\kappa_G(x, v_n) = d_G(v_n)$ *if G is simple and unweighted.*

(The above results are originally proved for the case where x is the vertex v_p with the largest index p which is adjacent to v_n. However, this case implies the theorem because adding an edge between v_{n-1} and v_n preserves the MA ordering.)

Theorem 2.1(i) says that $X = \{v_n\}$ is a minimum cut that separates v_{n-1} and v_n. We first outline below a proof of this theorem when G is a multigraph before giving a short proof for G with edges weighted by real numbers.

It is trivial from the definition to see $\lambda_G(v_{n-1}, v_n) \leqslant d_G(v_n)$.

To show the converse, $\lambda_G(v_{n-1}, v_n) \geqslant d_G(v_n)$, we consider the following decomposition of the edge set E of G. First choose an arbitrary maximal forest $F_1 \subseteq E$ in G, then choose a maximal spanning forest $F_2 \subseteq E - F_1$ in $G - F_1$, where $G - F_1$ is a brief notation of the graph $(V, E - F_1)$. Similarly, let F_i be a maximal spanning forest in $G - (F_1 \cup \cdots \cup F_{i-1})$ for $i = 3, 4, \ldots$. An MA ordering actually provides such a decomposition. For each $i = 2, \ldots, n$, consider the set $E_G(\{v_1, \ldots, v_{i-1}\}, v_i)$ of edges between $\{v_1, \ldots, v_{i-1}\}$ and v_i, and let $e_{i,k} \in E_G(\{v_1, \ldots, v_{i-1}\}, v_i)$ be the edge that appears as the kth edge when the edges in $E_G(\{v_1, \ldots, v_{i-1}\}, v_i)$ are arranged in the order $e_{i,1} = (v_{j_1}, v_i), e_{i,2} = (v_{j2}, v_i), \ldots, e_{i,p} = (v_{j_p}, v_i)$, where $1 \leqslant j_1 \leqslant j_2 \leqslant \cdots \leqslant j_p$ holds. By letting

$$F_k = \{e_{2,k}, e_{3,k}, \ldots, e_{n,k}\}, \quad k = 1, 2, \ldots, |E| \tag{2}$$

(some of $e_{i,k}$ may be void), we have a partition $(F_1, \ldots, F_{|E|})$ of E. Then it is not difficult to see from the definition of an MA ordering that (V, F_i) is a maximal spanning forest in $G - (F_1 \cup F_2 \cup \cdots \cup F_{i-1})$. Notice that exactly one edge $e_{n,k}$ from each F_k, $k = 1, \ldots, d_G(v_n)$ is incident to the last vertex v_n. Now suppose that there is an edge between v_{n-1}, v_n (otherwise we can add such an edge without destroying the MA ordering). Then v_{n-1} and v_n are connected in $F_{d_G(v_n)}$ by edge $e_{n,d_G(v_n)} = (v_{n-1}, v_n)$. Thus, by the maximality of the forests (as edge sets), v_{n-1} and v_n are connected (by a path) in each F_k with $k < d_G(v_n)$. Hence, there are at least $d_G(v_n)$ edge-disjoint paths between v_{n-1} and v_n, implying $\lambda_G(v_{n-1}, v_n) \geqslant d_G(v_n)$.

For example, we have such spanning forests $F_1, \ldots, F_{10}$ in the MA ordering in Fig. 2, where

$$F_1 = \{(u_2, u_1), (u_5, u_1), (u_6, u_1), (u_4, u_1), (u_3, u_2)\},$$
$$F_2 = \{(u_2, u_1), (u_5, u_1), (u_6, u_5), (u_4, u_1), (u_3, u_2)\},$$
$$F_3 = \{(u_2, u_1), (u_5, u_2), (u_6, u_5), (u_4, u_1), (u_3, u_6)\},$$
$$F_4 = \{(u_2, u_1), (u_5, u_2), (u_6, u_5), (u_4, u_5), (u_3, u_4)\},$$
$$F_5 = \{(u_6, u_5), (u_3, u_4)\},$$
$$F_6 = \{(u_6, u_5), (u_3, u_4)\},$$
$$F_7 = \{(u_6, u_5), (u_3, u_4)\},$$
$$F_8 = \{(u_6, u_5), (u_3, u_4)\},$$
$$F_9 = \{(u_6, u_5), (u_3, u_4)\},$$
$$F_{10} = \{(u_3, u_4)\}.$$

An MA ordering has the following hierarchical structure. For the above forests $F_1, F_2, \ldots$, consider the spanning subgraph $G_k = (V, F_1 \cup F_2 \cup \cdots \cup F_k)$, $k = 1, 2, \ldots$. We easily see that the MA ordering $v_1, \ldots, v_n$ for G remains to be an MA ordering for all G_k. Therefore, it holds $\lambda_{G_k}(v_{n-1}, v_n) = d_{G_k}(v_n) = k$ for $1 \leqslant k \leqslant d_G(v_N)$.

Although the above proof works when G is a multigraph or edge weighted by integral (or rational) numbers, it is shown [70] that Theorem 2.1(i) remains valid even for real edge weights using some technical argument by approximating real numbers with rational numbers. The correctness of this theorem has been proved by several researchers [19,24,70,76,77,86]. To complete the proof of Theorem 2.1(i), we describe a simple proof for the case of real weights, which is due to Frank [19] (the proof is also found in [12]). We proceed by an induction on the numbers of vertices. The theorem is true for $|V| = 2$. Let $n = |V|$ for a given graph $G = (V, E)$, and assume that the theorem holds for all graphs which have $n' (< n)$ vertices. Consider an MA ordering $v_1, v_2, \ldots, v_{n-2}, v_{n-1}, v_n$ in G. As observed in the above, we can assume that the last two vertices are not adjacent, and $d_G(\{v_1, \ldots, v_{n-2}\}, v_{n-1}) = d_G(v_{n-1})$ and $d_G(\{v_1, \ldots, v_{n-2}\}, v_n) = d_G(v_n)$ holds. Notice that the ordering $v_1, v_2, \ldots, v_{n-2}, v_{n-1}$ is an MA ordering in the graph $G - v_n$ obtained from G by deleting v_n, and hence by inductive hypothesis $\lambda_{G-v_n}(v_{n-2}, v_{n-1}) = d_{G-v_n}(v_{n-1})(= d_G(v_{n-1}))$. Similarly, we have $\lambda_{G-v_{n-1}}(v_{n-2}, v_n) = d_{G-v_{n-1}}(v_n)(= d_G(v_n))$ since $v_1, v_2, \ldots, v_{n-2}, v_n$ is an MA ordering in $G - v_{n-1}$. Therefore, we obtain $\lambda_G(v_{n-1}, v_n) \geqslant \min\{\lambda_{G-v_n}(v_{n-2}, v_{n-1}), \lambda_{G-v_{n-1}}(v_{n-2}, v_n)\}$ $= \min\{d_G(v_{n-1}), d_G(v_n)\} = d_G(v_n)$ (by the choice of v_{n-1} in the MA ordering). This proves Theorem 2.1(i).

In fact, we can actually compute the maximum flow with flow value $d_G(v_n)$ between v_{n-1} and v_n in time complexity $O(m \log n)$ by using the hierarchical structure of MA orderings [77]. Based on the fact that all minimum cuts separating v_{n-1} and v_n can be obtained in linear time from the maximum flow between them [81], all minimum cuts with value $\lambda(G)$ and the corresponding *cactus structure* (introduced by Dinits et al. [13]), which is a compact representation of all minimum cuts in G, can be computed in $O(nm \log n)$ time without relying on the conventional maximum flow algorithm [79].

An MA ordering is also used to find a sparse spanning subgraph of a given graph while preserving the vertex and edge-connectivities of the original graph G [20,69].

Theorem 2.2. *For an unweighted multigraph* $G = (V, E)$, *let a set of forests* $F_1, F_2, \ldots$, $F_{|E|}$ *be the partition of E obtained from an MA ordering by* (2), *where* $F_i = F_{i+1} = \cdots = F_{|E|} = \emptyset$ *possibly holds for some i. Let* $G_k = (V, F_1 \cup F_2 \cup \cdots \cup F_k)$, *for* $k = 1, 2, \ldots, |E|$. *Then each G_k has at most $k(|V| - 1)$ edges and satisfies*
(i) $\lambda_{G_k}(u, v) \geqslant \min\{\lambda_G(u, v), k\}$ *for all* $u, v \in V$,
(ii) $\kappa_{G_k}(u, v) \geqslant \min\{\kappa_G(u, v), k\}$ *for all* $u, v \in V$ *if G is simple.*

Since the above decomposition of G into forests $F_1, \ldots, F_{|E|}$ can be found in $O(m + n \log n)$ time, such G_k is widely used as a fast preprocessing for sparsifying a given graph G, in order to reduce the time complexity of many graph connectivity algorithms (see [25,28,31,58,66] for its applications). A graph search for finding such partition of E in Theorem 2.2 is studied by Cheriyan et al. [10].

3. Computing a minimum cut and an edge splitting

3.1. A minimum cut algorithm

Based on Theorem 2.1(i), we can compute a minimum cut of a given graph $G=(V,E)$ as follows [70]:

Algorithm MIN-CUT
Input: A graph $G=(V,E)$.
Output: A minimum cut X in G.
Step 1: Let $G_1:=G$ and $i:=1$.
Step 2: **while** $i < n$ **do**

> Compute the local edge-connectivity $\lambda_{G_i}(u_i,v_i)=d_G(v_i)$ for the last two vertices $u_i, v_i \in G_i$, in an MA ordering of G_i (where v_i is assumed to be the last vertex), and contract vertices u_i, v_i into a single vertex, denoting the resulting graph by G_{i+1}. Let $i:=i+1$.
> **end** /* while */

Step 3: Find $i=i^*$ that minimizes $\lambda_{G_i}(u_i,v_i)$ among all $i=1,2,\ldots,n-1$. Then output the set of vertices X contracted to v_{i^*} before obtaining G_{i^*}.

It is not difficult to see that

$$\lambda(G) = \min\{\lambda_{G_i}(u_i,v_i) \mid i=1,2,\ldots,n-1\} \tag{3}$$

holds, because, for each i, either a minimum cut separating u_i and v_i in G_i is also a minimum cut of G_i, or a minimum cut of G_i is given as a minimum cut of G_{i+1}. If the $i=i^*$ that attains the minimum of (3) is identified, then a minimum cut $X \subset V$ of G is obtained as the set of all the vertices contracted into the vertex v_{i^*} (i.e., $d_{G_{i^*}}(v_{i^*})=d_G(X)$ holds). The running time of this minimum cut algorithm is $O(n(m+n\log n))$.

Theorem 3.1. *For a given graph $G=(V,E)$, algorithm MIN-CUT outputs a minimum cut of G in $O(nm+n^2\log n)$ time.*

In this decade, there has been a significant progress in the study of how to compute a minimum cut of a graph, from both practical and theoretical view points. Beside the above MIN-CUT, the following new algorithms may be worth mentioning.
(a) For a digraph with root s, we can consider a minimum cut that has a minimum number of arcs from X to $V-X$ over all X with $s\in X \subset V$. Such a minimum cut X can be found in $O(\lambda m \log(n^2/m))$ time due to Gabow's matroidal approach [25], where λ is the value of a minimum cut. This algorithm can also be used to compute a minimum cut in an undirected graph.
(b) For a real weighted digraph with root s, a minimum cut can be computed in $O(mn\log(n^2/m))$ time, due to Hao and Orlin's maximum flow algorithm [30]. Similarly to (a), this algorithm can also be used as an $O(mn\log(n^2/m))$ time minimum cut algorithm for an undirected graph.
(c) For a weighted undirected graph, all minimum cuts can be computed in $\tilde{O}(n^2)$ time by a randomized algorithm by Karger and Stein [60,61]. The running time

was reduced to almost linear by Karger [57]. Also an NC algorithm was found for this problem [59].

Practical performance of these algorithms (including the above MIN-CUT) has been extensively and systematically studied in [8]. According to this report, a heuristic proposed by Padberg and Rinaldi [80] for reducing a given graph is effective in practice. The algorithm by Hao and Orlin, followed by MIN-CUT, is practically most efficient for many benchmark graphs.

Recently, extensions of algorithm MIN-CUT to the minimization of the class of symmetric submodular functions (and a slightly wider class of functions, called posi-modular and sub-modular functions) were discussed in [72,73,83,85].

Now suppose that, given a graph $G' = (V', E')$, a designated vertex $s \in V'$ and $k \geq 0$, we want to test whether the inner edge-connectivity satisfies $\lambda_{G'}(V' - s) \geq k$ or not. This can be done by slightly modifying MIN-CUT. The key point here is to compute an MA ordering starting from $v_1 = s$.

Algorithm CONTRACT

Input: A graph $G' = (V', E')$, a designated vertex $s \in V'$ and a real $k \geq 0$.
Output: Yes if $\lambda_{G'}(V' - s) \geq k$ holds; otherwise No.
Step 1: If there is a vertex $v \in V' - s$ with $d_{G'}(v) < k$, then halt after outputting No.
Step 2: $H := G'$;
   while $|V[H]| \geq 4$ **do**
      Find a pair of vertices $v, w \in V[H] - s$ such that $\lambda_H(v, w) \geq k$ (by
      applying an MA ordering). Then contract them into a single vertex x^*,
      and denote the resulting graph also by H. Let $X^* \subset V'$ denote the set of
      all vertices contracted into x^* so far. If $d_H(x^*) < k$, then halt after
      outputting No (since $d_H(x^*) = d_{G'}(X^*) < k$ is detected).
   end /* while */
Step 3: /* H has at most three vertices */ Halt after outputting Yes.

The correctness of Step 1 is immediate. In Step 2, before starting the while loop in each iteration, we see by induction that graph H satisfies $d_H(u) \geq k$ for all $u \in V[H] - s$. Hence a pair of $v, w \in V[H] - s$ with $\lambda_H(v, w) \geq k$ can be found as the last two vertices in an MA ordering with $v_1 = s$ (we start from $v_1 = s$ to avoid the case in which vertex s is used as one of the contracted vertices). By theorem 2.1(i), it holds $\lambda_H(v, w) = d_H(w) \geq k$. Since this tells that no cut with value less than k separates v and w, all the cuts with values less than k (if such cuts exist) remain in H after contracting v and w. By induction, this shows that CONTRACT runs correctly. Its running time is $O(nm + n^2 \log n)$.

3.2. Edge-splitting theorem

The operation of an *edge splitting* in a graph $G' = (V', E')$ at $s \in V'$ is to replace two edges (u, s) and (s, w) incident to a vertex s with a single edge (u, w). For a real weighted graph $G' = (V', E')$, the definition is extended as follows. For two vertices $u, v \in V' - s$ (possibly $u = v$) and a nonnegative real $\delta \leq \min\{d_{G'}(s, u), d_{G'}(s, v)\}$, we decrease weights $d_{G'}(s, u)$ and $d_{G'}(s, v)$, respectively, by δ, and increase weight

$d_{G'}(u,v)$ by δ (after introducing a new edge (u,v) of zero weight if there was no edge (u,v)). After splitting by δ, the value of a cut in G' either remains unchanged or decreases by 2δ. Clearly the original edge-splitting for an integer weighted graph can be regarded as an extended edge-splitting with $\delta = 1$.

Let the given graph $G' = (V', E')$ satisfy $\lambda_{G'}(V' - s) \geqslant k$. Then splitting edges (s, u) and (s, v) by δ is called (k, s)-feasible if the resulting graph G'' also satisfies $\lambda_{G''}(V' - s) \geqslant k$. The next theorem is due to Lovász (see also [17] for a different proof).

Theorem 3.2 (Lovász [62,63]). (a) *Let $G'=(V', E')$ be a graph, $s \in V'$ be a designated vertex with even $d_{G'}(s)$, and k be an integer with $2 \leqslant k \leqslant \lambda_{G'}(V' - s)$. Then for any neighbor u of s, there are a neighbor v of s and an integer $\delta \geqslant 1$ such that splitting (s, u) and (s, v) by δ is (k, s)-feasible.*

(b) *Let $G'=(V', E')$ be a real weighted graph, $s \in V'$ be a designated vertex, and k be a real number with $0 \leqslant k \leqslant \lambda_{G'}(V' - s)$. Then for any neighbor u of s, there are a neighbor v of s and a $\delta > 0$ such that splitting (s, u) and (s, v) by δ is (k, s)-feasible, and, after this splitting, the pair (s, u) and (s, v) is no longer splittable.*

A sequence of splitting of edges incident to a designated vertex s is called a *complete* splitting if no edge is incident to s after the splittings. By repeatedly applying Theorem 3.2(a), we can obtain a complete (k, s)-feasible splitting at a vertex s with even $d_{G'}(s)$. It is shown in [71,78] that such a complete (k, s)-feasible splitting can be obtained in $O((nm + n^2 \log n) \log n)$ time by applying a modification of algorithm CONTRACT (which will be called AUGMENT, and will be described in Section 4.2) $O(\log n)$ times. The description of these edge-splitting algorithms will be given in Section 4.2.

There is the corresponding edge-splitting theorem in a digraph [65], where splitting two directed edges (u, s) and (s, v), one has head s and the other has tail s, means replacing them with a single directed edge (u, v) which tail u and head v. Other extensions of these edge splitting theorems have been studied in [2–4,14,27,40,46,52,64,68].

4. Edge-connectivity augmentation problem for a target k

In this section, we first describe the approach by Frank for solving the edge-connectivity augmentation problem, and then explain in the second subsection how to implement his algorithm to make it run efficiently. This is based on the efficient edge splitting algorithm to be described in the last subsection.

4.1. Edge-connectivity augmentation

Given a graph G and an integer $k \geqslant 2$, called a *target*, we consider the problem of finding the smallest number of edges to be added to G to obtain a k-edge-connected graph, where we allow multiple edges in the resulting graph. A family $\mathcal{X}$ of mutually disjoint subsets of V is called a *subpartition* of V. The next min–max result, which was a cornerstone in the study of connectivity augmentation problems, is due to Watanabe and Nakamura [89].

Theorem 4.1 (Watanabe and Nakamura [89]). *For a graph $G=(V,E)$ and an integer $k \geqslant \max\{\lambda_G(V)+1,2\}$, let*

$$\alpha_k(G) = \max \sum_{X \in \mathscr{X}} (k - d_G(X)), \tag{4}$$

where the maximum is taken over all nonempty subpartitions $\mathscr{X}$ of V. Then the minimum number of edges to be added to make G k-edge-connected is equal to $\lceil \alpha_k(G)/2 \rceil$.

Now we prove this theorem by following the argument of Frank [17], which also provides an efficient algorithm for the edge-connectivity augmentation problem. First, let a subpartition $\mathscr{X}$ realize the maximum of (4) (here $k - d_G(X) > 0$ can be assumed for all $X \in \mathscr{X}$, since any cut X with $k-d_G(X) \leqslant 0$ can be discarded from $\mathscr{X}$). Then $\lceil \alpha_k(G)/2 \rceil$ edges are necessary to made G k-edge-connected, because each $k - d_G(X)$ in the summation (4) shows the deficiency for the cut X with respect to k, and adding one edge can reduce the deficiency of at most two distinct cuts in $\mathscr{X}$ by 1, respectively.

Thus, the crucial part is to prove that $\lceil \alpha_k(G)/2 \rceil$ edges are sufficient to make G k-edge-connected. For this, we follow the algorithmic proof by Frank [17], which is based on Lovász's edge-splitting theorem (Theorem 3.1).

Algorithm INCREASE

Input: A graph $G = (V,E)$ and an integer $k \geqslant \max\{\lambda_G(V)+1,2\}$.

Output: A k-edge connected graph G_k^* optimally augmented from G.

Step 1: Add to $G = (V,E)$ a new vertex s together with edges of integer weights between s and some vertices in V so that the resulting graph $G_k' = (V' = V \cup \{s\}, E')$ satisfies the next conditions (i) and (ii):

    Optimality conditions:

      (i) $\lambda_{G_k'}(V'-s) \geqslant k$ (i.e., $\lambda_{G_k'}(x,y) \geqslant k$ for all $x,y \in V$).

      (ii) The weight $d_{G_k'}(s,v)$ of each edge (s,v) incident to s is minimal in the sense that decreasing weight $d_{G_k'}(s,v)$ by any amount $\varepsilon > 0$ violates (i).

    If the degree $d_{G_k'}(s)$ is odd, then choose an arbitrary vertex $v_1 \in V$ and increase the weight of edge (s,v_1) by 1.

Step 2: Let G_k' be the graph obtained in Step 1. Now $d_{G_k'}(s)$ is even, and it holds $2 \leqslant k \leqslant \lambda_{G_k'}(V'-s)$. By Lovász's theorem, there is a complete (k,s)-feasible edge-splitting at s in G_k'. Then output the graph G_k^* obtained by the complete splitting, ignoring the isolated vertex s.

We first note that it is easy to find a set of weighted edges incident to s that satisfies the above optimality conditions. For example, after adding an edge of weight k between s and each vertex in V, decrease each of their as long as (i) holds. However, we shall present in the next subsection a more systematic efficient method.

To execute Step 2, recall that Lovász's theorem says that the output graph G_k^* is k-edge connected, and we can view the split edges as the edges added to the original graph G. We now show that the output graph G_k^* is in fact optimally augment from G.

The sum of the weights of added edges in G_k^* is the half of the original degree of vertex s in G' (or the degree plus one if it is odd) after Step 1; i.e., $\lceil \frac{1}{2} d_{G_k'}(s) \rceil$. Thus, to prove the optimality of G_k^*, it suffices to show that $d_{G_k'}(s) \leqslant \alpha_k(G)$. For this, we introduce some terminology. In a graph $G' = (V \cup \{s\}, E')$, a cut $X \subset V$ is called (k,s)-semi-critical if

$$d_{G'}(s, X) > 0, \quad k \leqslant d_{G'}(X) \leqslant k + 1 \quad \text{and} \quad \lambda_{G'}(X) \geqslant k.$$

A (k,s)-semi-critical cut $X \subset V$ is called (k,s)-critical if $d_{G'}(X) = k$. A family $\mathcal{X} = \{X_1, X_2, \ldots, X_p\}$ of mutually disjoint subsets $X_i \subset V$ is called a *subpartition* (possibly $\mathcal{X} = \emptyset$). If every neighbor of s belongs to a subset $X_i \in \mathcal{X}$ (i.e., $\sum_{i=1}^{p} d_G(s, X_i) = d_G(s)$), then $\mathcal{X}$ is called a *covering* subpartition. A subpartition $\mathcal{X}$ is called (k,s)-*critical* (resp., (k,s)-*semi-critical*) if every $X_i \in \mathcal{X}$ is (k,s)-critical (resp., (k,s)-semi-critical) or $\mathcal{X} = \emptyset$.

Consider the graph G_k' after Step 1 of INCREASE, in which any neighbor v of s is contained in some cut $X \subset V$ with $d_{G_k'}(X) = k$ by optimality condition (ii). Such a cut X is (k,s)-critical, since G_k' satisfies $\lambda_{G_k'}(X) \geqslant k$ by optimality condition (i). Thus, any neighbor v of s is contained in a (k,s)-critical cut $X_v \subset V$. We then choose a minimal family $\mathcal{X}' \subseteq \{X_v \mid v \text{ is a neighbor of } s \text{ in } G_k'\}$ under the constraint that $\mathcal{X}' = \{X_1, \ldots, X_p\}$ is covering. If $X_i \cap X_j \neq \emptyset$ holds for some $X_i, X_j \in \mathcal{X}'$, then we modify $\mathcal{X}'$ by replacing X_i and X_j with $X_i - X_j$ and $X_j - X_i$. We see that, after this modification, $\mathcal{X}'$ remains covering since properties $d_{G_k'}(X_i - X_j) = d_{G_k'}(X_j - X_i) = k$ and $d_{G_k'}(s, X_i \cap X_j) = 0$ follow from the optimality condition (i) and the submodularity (1) of cut function $d_{G_k'}$ (note that $d_{G_k'}(X_i) = d_{G_k'}(X_j) = k$ by assumption, $d_{G_k'}(X_i - X_j), d_{G_k'}(X_j - X_i) \geqslant k$ by optimality condition (i), and $d_{G_k'}(X_i) + d_{G_k'}(X_j) \geqslant d_{G_k'}(X_i - X_j) + d_{G_k'}(X_j - X_i) + 2d_{G_k'}(X_i \cap X_j, V' - (X_i \cup X_j))$ by (1).) By repeatedly modifying $\mathcal{X}'$ until it contains only mutually disjoint cuts, we obtain a (k,s)-critical covering subpartition $\mathcal{X}'$ until it contains only mutually disjoint cuts, we obtain a (k,s)-critical covering subpartition $\mathcal{X}'$ in G_k'. Denote the resulting $\mathcal{X}'$ by $\mathcal{X}$. Now, optimality condition (ii) is rewritten as the following condition:

Optimality condition

(ii') G_k' has a (k,s)-critical covering subpartition $\mathcal{X}$.

For each cut $X \in \mathcal{X}$ in (ii'), $d_{G_k'}(s, X)$ represents the deficiency for cut X with respect to k in G. Thus we have

$$d_{G_k'}(s) = \sum_{X \in \mathcal{X}} d_{G_k'}(s, X) = \sum_{X \in \mathcal{X}} (k - d_G(X)) \leqslant \alpha_k(G).$$

This proves the optimality of G_k^* and hence the correctness of Theorem 4.1.

4.2. Efficient implementation of INCREASE

Now we consider how to execute INCREASE efficiently. For this, we first show that Step 1 of INCREASE can be implemented by using a modification of algorithm

CONTRACT. Let $G' = (V \cup \{s\}, E)$ be the graph obtained from G by adding a vertex s, and apply CONTRACT to G' with parameter k. CONTRACT halts whenever it finds a cut X with value less than k. In this case, we modify CONTRACT so that it continues the execution after increasing the value of the detected cut X up to k by adding weighted edges between s and X. More precisely, in Step 1 of CONTRACT, we add to G' an edge (s, v) with weight $d_{G'}(s, v) = k - d_{G'}(v)$ if $d_{G'}(v) < k$ holds. In Step 2, if $d_H(x^*) < k$ holds, then we choose an arbitrary vertex $x' \in X^*$ and increase the weight of edge (s, x') (resp., (s, x^*)) by $k - d_H(x^*)$ in G' (resp., in H).

Let us call the modified algorithm AUGMENT, which is described as follows:

Algorithm AUGMENT
Input: A graph $G' = (V, E)$ and a real $k \geqslant 0$.
Output: A graph G'_k which satisfies the optimality conditions (i) and (ii$'$), and
a (k, s)-critical covering subpartition $\mathcal{X}$ of G'_k.
Step 1: Let $U = \{u_1, u_2, \ldots, u_p\}$ be the set of vertices $u_i \in V$ such that $d_G(u_i) < k$;
$\qquad V' = V \cup \{s\}$; $E' := E \cup \{(s, u_1), \ldots, (s, u_p)\}$;
$\qquad$ **for** each $u_i \in U$ **do**
$\qquad\qquad d_{G'}(s, u_i) := k - d_G(u_i)$
$\qquad$ **end**; /* for */
$\qquad$ Let $G' = (V', E')$ be the resulting edge-weighted graph;
$\qquad \mathcal{X} := \{\{u_1\}, \{u_2\}, \ldots, \{u_p\}\}$; /* possibly $\mathcal{X} = \emptyset$ */
Step 2: $H := G'$;
$\qquad$ **while** $|V[H]| \geqslant 4$ **do** /* $d_H(u) \geqslant k$ holds for all vertices $u \in V[H] - s$ */
$\qquad\qquad$ Find two vertices $v, w \in V[H] - s$ such that $\lambda_H(v, w) \geqslant k$;
$\qquad\qquad$ /* Such v, w can be obtained by applying an MA ordering */
$\qquad\qquad$ Contract v and w in H into a single vertex x^*, and let H be the resulting
$\qquad\qquad$ graph; **if** $d_H(x^*) < k$ **then**
$\qquad\qquad\qquad$ Let $X^*(\subseteq V' - s)$ denote the set of vertices that have been contracted
$\qquad\qquad\qquad$ into x^* so far;
$\qquad\qquad\qquad$ Choose an arbitrary vertex $u \in X^*$, and let $d_{G'}(s, u) := d_{G'}(s, u) + k - $
$\qquad\qquad\qquad\qquad d_H(x^*)$ (after letting $E' := E' \cup \{(s, u)\}$ and $d_{G'}(s, u) := 0$, if
$\qquad\qquad\qquad (s, u) \notin E[G']$); Let G' be the resulting graph;
$\qquad\qquad\qquad$ Let H denote the graph obtained from H by setting $d_H(s, x^*) := $
$\qquad\qquad\qquad\qquad d_H(s, x^*) + k - d_H(x^*)$ (after creating edge (s, x^*) in H and letting
$\qquad\qquad\qquad\qquad c_H(s, x^*) := 0$, if $(s, x^*) \notin E[H]$);
$\qquad\qquad\qquad \mathcal{X} := \mathcal{X} \cup \{X^*\}$, after discarding all X' with $X' \subset X^*$ from $\mathcal{X}$
$\qquad\qquad$ **end**; /* if */
$\qquad$ **end** /* while */
Step 3: Output $G'_k := G'$ and $\mathcal{X}$.

Let G'_k be the graph output by AUGMENT. Then G'_k satisfies $\lambda_{G'}(V' - s) = \lambda_{G'_k}(V) \geqslant k$ (i.e., optimality condition (i)). It is not difficult to see that if a cut X^* with value less than k is found and its value is increased up to k during Step 2 of AUGMENT, then the cut X^* becomes (k, s)-critical in G'. Thus, each neighbor of s is always contained in some (k, s)-critical cut in G' during AUGMENT. This implies that optimality condition (ii$'$) holds in the output graph G'_k. AUGMENT runs in the same time

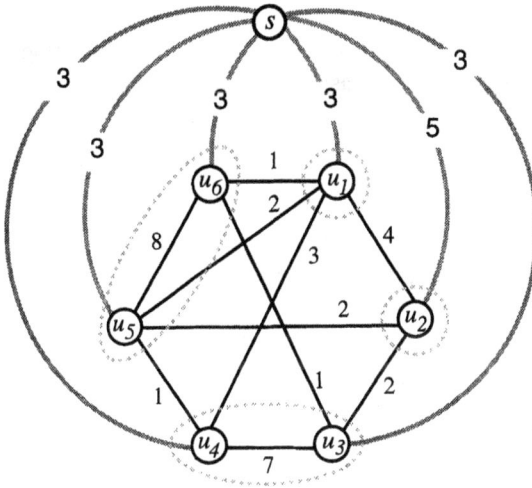

Fig. 3. The graph G'_k obtained from G of Fig. 1 by applying AUGMENT for $k = 13$, where subsets of vertices circled by broken lines form a (k,s)-critical covering subpartition $\mathscr{X}$.

complexity $O(nm + n^2 \log n)$ as that of CONTRACT. As an example, Fig. 3 shows the graph G'_k obtained from the graph G in Fig. 1 by applying AUGMENT for target $k = 13$. A (k,s)-critical covering subpartition in G'_k is given by $\mathscr{X} = \{\{u_1\}, \{u_2\}, \{u_3, u_4\}, \{u_5, u_6\}\}$.

Now we turn to Step 2 of INCREASE. This step can be executed in $O((nm + n^2 \log n) \log n)$ time by using an algorithm [71,78] for finding a complete (k,s)-feasible splitting at s. A description of this edge splitting algorithm will be given in the next subsection (as it uses algorithm AUGMENT of this subsection).

It is also interesting to note that, after Step 1 of INCREASE, we already know the optimal value (i.e., the minimum sum of weights of edges to be added to G') to obtain a k-edge-connected graph.

4.3. Edge-splitting algorithm

In this subsection, we sketch the edge-splitting algorithms proposed in [71,78]. Let $G' = (V' = V \cup \{s\}, E')$ and k satisfy Theorem 3.2 (i.e., $\lambda_{G'}(V' - s) \geq k$). Then we split arbitrarily all edges incident to s to obtain a complete splitting (which is not necessarily (k,s)-feasible). Let G^I denote the resulting graph and let B be the set of edges created by the complete splitting.

Suppose that Step 2 of algorithm AUGMENT is applied to the resulting graph $G' := G^I$ and the k (where $\mathscr{X}$ is set to be empty) to test whether the above complete splitting at s is (k,s)-feasible or not (i.e., $\lambda_{G^I}(V' - s) \geq k$ holds or not). If $\lambda_{G^I}(V' - s) \geq k$, then Step 2 of AUGMENT outputs $\mathscr{X} = \emptyset$ without finding any cut $X \subset V' - s$ with its value less than k. In this case, we conclude that the complete splitting is (k,s)-feasible.

On the other hand, if $\lambda_{G^I}(V' - s) < k$, then a cut $X^* \subset V' - s$ with $d_{G^I}(X^*) < k$ is detected in the while-loop of Step 2. For this cut X^*, the subgraph $G^I[X^*]$ induced from G^I by X^* must contain at least $\lceil \frac{1}{2}(k - d_{G^I}(X^*)) \rceil$ edges in B. This is because the original assumption $\lambda_{G'}(V' - s) \geqslant k$ implies that splitting δ pairs of edges (s, u) and (s, v) in G' decreases cut value $d_{G'}(X^*) \geqslant k$ by 2δ only when $\{u, v\} \subseteq X^*$ holds. We now say that an edge (u, v) in a graph is *hooked up* at a vertex s if we replace the edge (u, v) with two new edges (s, u) and (s, v). We then increase the cut value of X^* at least to k by hooking up $\lceil \frac{1}{2}(k - d_{G'}(X^*)) \rceil$ edges in $B \cap E(G^I[X^*])$ (instead of increasing $d_{G'}(s, u)$ up to k for some $u \in X^*$, as in the original AUGMENT). This modified iteration of the while-loop is repeated while $|V[H]| \geqslant 4$ holds.

Let G^{II} be the graph obtained from G^I by hooking up all the chosen edges for all the cuts X^* with $d_{G^I}(X^*) < k$ found in the above process. As a byproduct of this computation, a (k, s)-semi-critical covering subpartition $\mathcal{Y}$ in G^{II} can be found by retaining all the detected cuts X^* and choosing all the maximal ones among them. The detail is described as follows:

Algorithm HOOK-UP

Input: A graph $G^I = (V' = V \cup \{s\}, E^I)$, a designated vertex $s \in V$ with $d_{G^I}(s) = 0$, and a set $B \subseteq E^I$.

Output: A graph G^{II} obtained from G^I by hooking up some edges in B, where G^{II} satisfies $\lambda_{G^{II}}(V' - s) \geqslant k$, and a (k, s)-semi-critical covering subpartition $\mathcal{Y}$ in G^{II}.

Step 1: $H := G^I$; $\mathcal{Y} := \emptyset$; $B' := \emptyset$; /* B' denotes the set of edges to be hooked up. */

Step 2: **while** $|V[H]| \geqslant 4$ **do**

Find vertices $v, w \in V[H] - s$ with $\lambda_H(v, w) = d_H(w) \geqslant k$;
/* Such v, w can be found by an MA ordering. */

Contract v and w into a single vertex x^* and let H be the resulting graph;

if $d_H(x^*) < k$ **then**

Let $X^* \subseteq V' - s$ be the set of all vertices contracted so far into x^*;

Choose a set $\Delta B \subseteq B$ of arbitrary $\lceil \frac{1}{2}(k - d_{G^I}(X^*)) \rceil$ edges in $B \cap E[G^I[X^*]]$;
$B := B - \Delta B$; $B' := B' \cup \Delta B$;

Let G^I denote the graph obtained by hooking up these edges in ΔB at s in G^I;

Let H denote the graph obtained by adding new $2\lceil \frac{1}{2}(k - d_{G^I}(X^*)) \rceil$ edges between s and x^* in H;

$\mathcal{Y} := \mathcal{Y} \cup \{X^*\}$, after discarding from $\mathcal{Y}$ all $X' \in \mathcal{Y}$ such that $X' \subset X^*$;

end; /* if */

end /* while */

Step 3: Output $G^{II} := G^I$ and $\mathcal{Y}$.

For simplicity, we assume for a moment that k is an even integer. We consider how to find a complete (k, s)-feasible splitting in the output graph G^{II} based on the information of the output (k, s)-semi-critical covering subpartition $\mathcal{Y}$. Clearly, splitting two edges (s, u) and (s, v) is not (k, s)-feasible if u and v belong to the same subset $X \in \mathcal{Y}$ since $d_{G^{II}}(X) \leqslant k+1$. We call splitting edges (s, u) and (s, v) *$\mathcal{Y}$-astride* if $u \in X$ and $v \in X'$ hold for distinct $X, X' \in \mathcal{Y}$. It is not difficult to see that G^{II} always admits a complete $\mathcal{Y}$-astride edge-splitting and such splitting can be easily found.

Let $\mathscr{Y}_1 := \mathscr{Y}$ and $G_1^{II} := G^{II}$ initially. Then we split all edges incident to s in G_1^{II} by a complete $\mathscr{Y}_1$-astride edge-splitting, and apply algorithm HOOK-UP to test whether this complete edge-splitting is (k, s)-feasible or not in the resulting graph $G_2^{II} := G^{II}$. HOOK-UP obtains a new (k, s)-semi-critical covering subpartition $\mathscr{Y}_2$ if the edge-splitting is not (k, s)-feasible. We repeat this process until, for some $i > 1$, the complete $\mathscr{Y}_i$-astride edge-splitting becomes (k, s)-feasible in G_i^{II}. In this iteration, we can prove that $|\mathscr{Y}_{i+1}| \leqslant \frac{1}{2}|\mathscr{Y}_i|$ and $|\mathscr{Y}_i| \neq 1$ hold for all $i \geqslant 1$ [78]. Therefore, $\mathscr{Y}_i$ becomes empty after $O(\log n)$ iterations, and a complete (k, s)-feasible edge-splitting of the original graph $G' = (V' = V \cup \{s\}, E')$ can be obtained. The entire running time is $O((nm + n^2 \log n) \log n)$.

For an odd integer k, we have to be more careful to find a complete $\mathscr{Y}_i$-astride edge-splitting in G_i^{II}, which satisfies a certain condition to guarantee the property $|\mathscr{Y}_{i+1}| \leqslant \alpha|\mathscr{Y}_i|$ for some constant $\alpha < 1$ [71]. Another article [78] handles the case of odd k in a slightly different way. In the given graph $G' = (V' = V \cup \{s\}, E')$, it first finds a maximal set $E_k \subset E'$ of edges incident to s such that $G' - E_k$ satisfies $\lambda_{G'-E_k}(V' - s) \geqslant k - 1$ and $d_{G'-E_k}(s)$ is even. (Such E_k can be computed by AUG-MENT for integer $k - 1$ after removing all edges incident to s). Then by applying the above algorithm for the even integer $k - 1$, we can find a complete $(k - 1, s)$-feasible edge-splitting. Now the problem is to increase the edge-connectivity $k - 1$ of the result-ing graph G^* to k by adding some edges which can be created by splitting appropriate edges in E_k. This problem can be solved in linear time after constructing the cactus structure of G^*, which provides us the system of all cuts with value $k - 1$.

Theorem 4.2. *For a given graph $G' = (V \cup s, E')$ and an integer k satisfying $2 \leqslant k \leqslant \lambda_{G'}(V' - s)$, a complete (k, s)-feasible edge splitting can be obtained in $O((nm + n^2 \log n) \log n)$ time.*

Therefore, by the argument in Section 4.2, we have the next result.

Theorem 4.3. *For a given graph $G = (V, E)$ and an integer $k \geqslant \max\{\lambda_G(V) + 1, 2\}$, algorithm INCREASE optimally augments G into G_k^* so that the output G_k^* is k-edge-connected. INCREASE runs in $O((nm + n^2 \log n) \log n)$ time.*

5. Edge-connectivity augmentation for the entire range of targets

In this section, we consider a real weighted graph $G = (V, E)$, where edges are weighted by *nonnegative reals*. As we allow edges of zero weights, G can be assumed to be a complete graph. All the definitions used so far are also generalized to consider this model; e.g., the cut value $d_G(X)$ is defined to be the weight sum of the edges between X and $V - X$, and the edge-connectivity is the minimum value of the cuts in G. For such a real weighted graph G, we are permitted to increase an edge weight by an arbitrary nonnegative real, and the goal is to minimize the sum of weights to be increased in order to make G k-edge-connected, where the target k is a given nonneg-ative real number. We call this problem the *fractional version* of the edge-connectivity augmentation problem. It is not difficult to see that the problem can be formulated as a linear program, and hence can be solved in polynomial time. However, we present in the sequel of a graph theoretic algorithm, which is more efficient.

For a given target $k \geq 0$, let us denote by $\Lambda_G(k)$ the optimal value of the problem (i.e., the sum of added weights), and by $G^*(k)$ an optimally augmented graph. $\Lambda_G(k)$ is called the *edge-connectivity augmentation function*. We shall show that not only $\Lambda_G(k)$ for all $k \geq 0$ but also $G^*(k)$ for all $k \geq 0$ can be computed in $O(nm + n^2 \log n)$ time by using a compact representation of optimally augmented graphs.

First note that, based on the fractional version of Lovász's theorem (i.e Theorem 3.2(b)), it can be shown that, for a real weighted graph $G' = (V' = V \cup s, E')$ and a real $k \leq \lambda_{G'}(V' - s)$, there exists a complete (k,s)-feasible edge-splitting at s, where size $\delta > 0$ in each edge-splitting is not necessarily integer. Based on this, the fractional version can also be solved by algorithm INCREASE (where we need not round up an odd degree of s to an even degree at the end of Step 1). As remarked at the end of Section 4.2, the optimal value for a given target k is obtained after Step 1 of INCREASE. In other words, it holds $\Lambda_G(k) = d_{G'_k}(s)/2$ for the graph G'_k obtained after Step 1.

5.1. Edge-connectivity augmentation function

From the fact that the fractional version of the edge-connectivity augmentation problem is formulated as a linear program, it is clear that edge-connectivity augmentation function $\Lambda_G(k)$ is piecewise linear, convex and increasing over $k \in [0, +\infty)$.

For example, the graph in Fig. 1 has the edge-connectivity augmentation function of Fig. 4. In what follows, we show that the function Λ_G can be identified in $O(mn + n^2 \log n)$ time. We already know that, for a fixed target k, $\Lambda_G(k)$ can be computed in

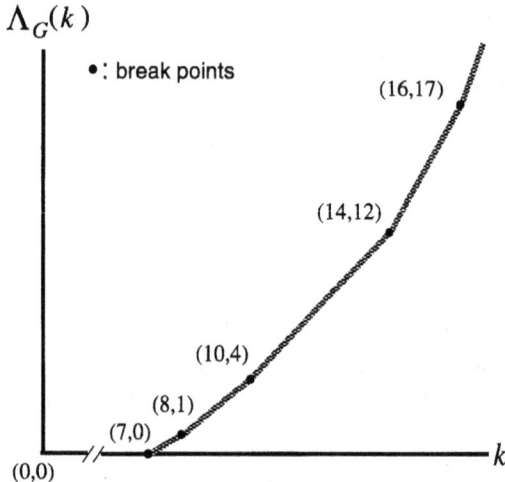

Fig. 4. The edge-connectivity augmentation function of graph G in Fig. 1.

$O(mn + n^2 \log n)$ time by AUGMENT (i.e., Step 1 of INCREASE). To compute $\Lambda_G(k)$ over the entire range $k \geqslant 0$, we try to execute AUGMENT simultaneously for all k. To realize this computation in finite time and space, we introduce the *ranged graph*.

For two reals a and b with $a < b$, the interval $[a, b]$ is called a *range*, and its *size* $\pi([a, b])$ is defined as $b - a$. Let $R = \{[a_1, b_1], [a_2, b_2], \ldots, [a_t, b_t]\}$ be a set of ranges. The size of R, denoted by $\pi(R)$, is defined as the sum of all range sizes in R:

$$\pi(R) = (b_1 - a_1) + (b_2 - a_2) + \cdots + (b_t - a_t),$$

where $\pi(\emptyset)$ is defined to be 0. For a given real h, the *upper h-truncation* of a range $[a, b]$ is defined by

$$[a, b]|^h = \begin{cases} [a, \min\{b, h\}] & \text{if } a < h, \\ \emptyset & \text{otherwise.} \end{cases}$$

Based on this, the upper h-truncation of a set R of ranges is defined by

$$R|^h = \{[a_i, b_i]^h \neq \emptyset \mid [a_i, b_i] \in R\}.$$

For example, we have $\{[1, 3], [2, 5], [4, 7]\}|^3 = \{[1, 3], [2, 3]\}$.

Now we modify algorithm AUGMENT for a given target k (described in Section 4.2) in order to deal with all targets k. Let us define the ranged graph $\mathscr{R}(G) = (V \cup \{s\}, E \cup E_s, R)$ for a given weighted graph $G = (V, E)$ as follows. Let $E_s = \{(s, v) \mid v \in V\}$ be the set of edges between s and all vertices in V. In $\mathscr{R}(G)$, let each edge in E have the same weight as in G, but let each edge $(s, v) \in E_s$ have a set R_v of ranges (i.e., $R = \{R_v \mid v \in V\}$). This R_v is initially set to be $R_v = \{[d_G(v), +\infty]\}$ and will be changed during execution of AUGMENT. Given a real $h \geqslant 0$, we define the weight of each edge $(s, v) \in E_s$ by $\pi(R_v|^h)$ (i.e., the sum of sizes of the upper h-truncated ranges in R_v). The resulting real weighted graph is denoted by $\mathscr{R}(G)|^h$. In other words, the ranged graph $\mathscr{R}(G)$ represents the real weighted graphs $\mathscr{R}(G)|^h$ for all reals $h \geqslant 0$.

To execute Step 1 of AUGMENT for all targets k, we construct the ranged graph $\mathscr{R}(G)$ having a set $R_v = \{[d_G(v), +\infty]\}$ of ranges for each $v \in V$. By definition, graph $\mathscr{R}(G)|^k$ is the same as the weighted graph G' obtained after Step 1 of the original AUGMENT for a target k. In Step 2 of AUGMENT for all targets, it is important to choose a pair of vertices v and w for contraction so that this pair v and w is commonly used for all targets k; otherwise, we cannot maintain the process of contractions for all targets in a single ranged graph. Fortunately, the existence of such a pair is guaranteed by the hierarchical structure of MA orderings described in Section 2.2. After the contraction of v and w, there is a way of updating the range sets R_v, $v \in V$ so that $\mathscr{R}(G)|^k$ for any $k \geqslant 0$ is always equivalent to the graph G'_k computed by AUGMENT for a single target k. However we omit the details (see [75]). The final ranged graph $\mathscr{R}(G)$ is totally optimal in the following sense:

Total optimality condition: Let $\mathscr{R}(G)$ be the ranged graph for a weighted graph G. If the weighted graph $G'_k = \mathscr{R}(G)|^k$ satisfies the optimality condition (i) and (ii') for all nonnegative reals k, then $\mathscr{R}(G)$ is called a *totally optimal ranged graph*.

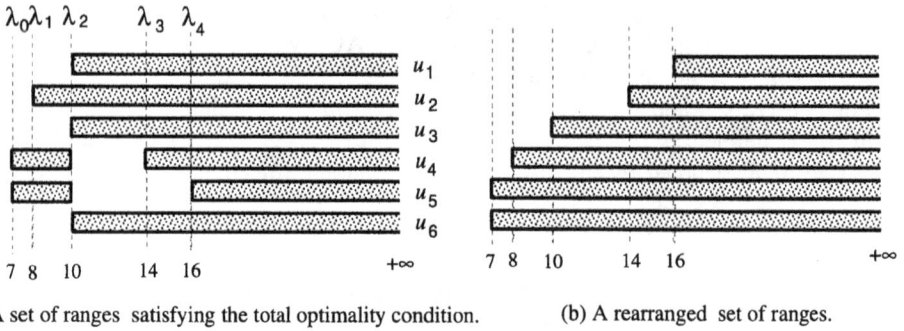

(a) A set of ranges satisfying the total optimality condition.   (b) A rearranged set of ranges.

Fig. 5. A totally optimal ranged graph $\mathscr{R}(G)$ for the graph G in Fig. 1: (a) A set of ranges satisfying the total optimality condition. (b) A rearranged set of ranges.

We can also prove that, for the final ranged graph $\mathscr{R}(G)$ computed by AUGMENT for all targets k, the total number of ranges $\sum_{v \in V} |R_v|$ is bounded by $10n \log_2 n$, where $|R_v|$ denotes the number of ranges in R_v. Given a totally optimal ranged graph $\mathscr{R}(G)$, we can obtain the optimal value $\Lambda_G(k)$ for a target k by

$$\Lambda_G(k) = d_{G'_k}(s)/2 = \sum_{v \in V} \pi(R_v|^k)/2. \tag{5}$$

For example, Fig. 5(a) shows the range sets R_{u_i} of vertices u_i in a totally optimal ranged graph $\mathscr{R}(G)$ obtained for the graph G in Fig. 1. Fig. 5(b) then shows the range sets obtained from those in Fig. 5(a) by moving some part of ranges to other ranges while keeping the property (5). The edge-connectivity function in Fig. 4 is then immediately obtained from Fig. 5(b). In general, the range sets of a totally optimal ranged graph can be transformed in this way into n ranges [75]. From this, we see that the number of break points in the edge-connectivity augmentation function is at most n.

Theorem 5.1. *For a real weighted graph $G=(V,E)$, its edge-connectivity augmentation function $\Lambda_G(k)$ for all targets $k \geqslant 0$ can be computed in $O(nm + n^2 \log n)$ time.*

5.2. Optimal solutions for all targets

We turn to the problem of computing optimally augmented graphs $G^*(k)$ for all $k \geqslant 0$. Again, we do not have to execute Step 2 of INCREASE for all targets k separately (actually it already seems difficult to execute Step 2 directly on the ranged graphs). We show that $G^*(k)$ can be obtained from the totally optimal ranged graph $\mathscr{R}(G)$ computed by AUGMENT of Section 5.1. First let $R = \{R_v \mid v \in V\}$ be the range sets of $\mathscr{R}(G)$, and let λ_i, $i = 0, 1, \ldots, q$ be the set of all the distinct end points appearing in ranges R_v, $v \in V$, where $\lambda_0 < \lambda_1 < \cdots < \lambda_q \, (= +\infty)$ is assumed. For example, Fig. 5(a) gives

$$\lambda_0 = 7, \quad \lambda_1 = 8, \quad \lambda_2 = 10, \quad \lambda_3 = 14, \quad \lambda_4 = 16 \quad \text{and} \quad \lambda_5 = +\infty.$$

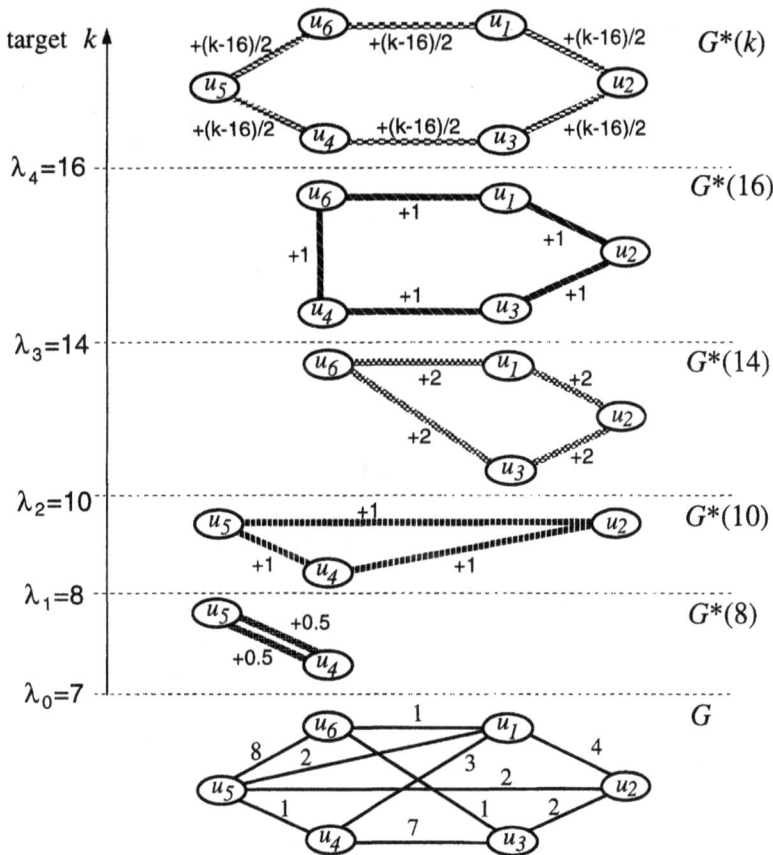

Fig. 6. A representation of optimal solutions $G^*(k)$ for all targets $k \in [0, +\infty)$.

Next consider q ranges $[\lambda_0, \lambda_1], [\lambda_1, \lambda_2], \ldots, [\lambda_{q-1}, \lambda_q]$, and let X_i be the set of vertices v such that one of the ranges in R_v contains $[\lambda_{i-1}, \lambda_i]$. Construct an arbitrary cycle C_i which visits all vertices in X_i. For Fig. 5(a), such cycles are chosen as $C_1 = (u_4, u_5)$, $C_2 = (u_2, u_4, u_5)$, $C_3 = (u_1, u_2, u_3, u_6)$, $C_4 = (u_1, u_2, u_3, u_4, u_6)$, $C_5 = (u_1, u_2, u_3, u_4, u_5, u_6)$. See Fig. 6.

We are now ready to construct an optimally augmented graph $G^*(k)$ for any target $k \geqslant 0$. Let i_k be the maximum index such that $\lambda_{i_k} < k$. We increase the edge weights by $(\lambda_i - \lambda_{i-1})/2$ along each cycle C_i for $i = 1, 2, \ldots, i_k$, and by $(k - \lambda_{i_k})/2$ along C_{i_k+1}. Obviously, the sum of the increased weights is equal to $\Lambda_G(k)$. It is not trivial to see that the resulting graph is in fact k-edge-connected; but can be shown from the total optimality condition [75]. From an analysis of the algorithm, we can also observe that the least number q of such cycles $C_1, \ldots, C_q$ is at most $6n + 4n \log_2 n$ [75].

Theorem 5.2 (Nagamochi and Ibaraki [75]). *For a real weighted graph* $G = (V, E)$, *the above* $\lambda_0 < \lambda_1 < \cdots < \lambda_q$ *and* $C_1, C_2, \ldots, C_q$ *(which can provide optimally augmented* *graphs $G^*(k)$ for all targets $k \geqslant 0$) can be obtained in* $O(nm + n^2 \log n)$ *time, where* $q \leqslant 6n + 4n \log_2 n$ *holds.*

Notice that optimally augmented graphs $G^*(k)$ constructed in this way have the monotone structure such that, for $k < k'$, $G^*(k')$ is obtained in increasing the weights of some edges in $G^*(k)$. Such monotone structures of optimal solutions are also observed in some other edge-connectivity augmentation problems [9].

6. Other augmentation problems

In this section, we summarize the results on augmentation problems of other types, including problems of increasing vertex-connectivity.

The problems of increasing the edge- or vertex-connectivity by adding the minimum number of edges are extensively surveyed by Frank [18], and here we give only a brief summary, where we first describe some results covered by the survey with the following four categories and then show some recent results:

Edge-connectivity augmentation of undirected graphs: After the edge-connectivity augmentation algorithm for an undirected graph proposed by Watanabe and Nakamura [89], the edge-splitting theorem was first used to solve the same problem by Cai and Sun [7]. Frank [17] then refined it by using the theorem of Lovász [63]. Moreover he showed that a more general augmentation problem can be solved polynomially in an undirected graph [17]. The local edge-connectivity augmentation problem asks to find a minimum set F of new edges to be added to a given undirected graph G such that, for each pair of vertices u and v, the resulting local edge-connectivity $\lambda_{G+F}(u,v)$ becomes larger than or equal to the target value $r(u,v)$ prescribed for each pair of $u, v \in V$. He proved that the problem can be solved in $O(n^3 m \log(n^2/m))$ time by applying Mader's edge-splitting theorem [64], a generalization of Lovász's theorem, which preserves the local edge-connectivity in an undirected graph.

Edge-connectivity augmentation of digraphs: A digraph is called k-*edge-connected* if it remains strongly connected by removal of any $(k-1)$ edges. The edge-connectivity augmentation problem in digraphs asks to find a minimum set of new directed edges to be added to a given digraph such that the augmented digraph becomes k-edge-connected for a prescribed target $k \geqslant 1$. Frank [17] pointed out that the edge-connectivity augmentation problem in a digraph can be solved in $O(n^3 m \log(n^2/m))$ time by Mader's edge-splitting theorem [65] in digraphs, which preserves the edge-connectivity in a digraph. In [17], the edge-connectivity augmentation problem in digraphs (undirected graphs) is still polynomially solvable even if lower and upper bounds are imposed on degree of each vertex.

One may introduce the local edge-connectivity augmentation problem in digraphs, whose undirected graph version is successfully solved by an edge-splitting theorem. However, there does not exist the corresponding edge-splitting theorem that preserves the local edge-connectivity in a digraph. In fact, the problem of increasing the local edge-connectivity (or the local vertex-connectivity) in a directed graph is NP-hard even if the target values satisfy $r(u,v) \in \{0,1\}$ for all pairs $u, v \in V$ [17].

Given a digraph $G = (V, E)$ with two specified subsets $S, T \subset V$ (which are not necessarily disjoint), Frank and Jordàn [21] first studied the problem of finding a minimum number of new edges directed from S to T to make G k-*edge-connected from* S *to* T (a digraph is called k-edge-connected from S to T if it has k edge disjoint directed paths from every vertex $s \in S$ to every vertex $t \in T$). They gave a min–max formula for the problem, based on which a polynomial time algorithm is obtained (where the algorithm is not combinatorial but rely on the ellipsoid method).

Vertex-connectivity augmentation of undirected graphs: The vertex-connectivity augmentation problem in undirected graphs is to increase the vertex-connectivity of a given undirected graph G to a target k by adding a minimum number of new edges. The problem is polynomially solvable for $k = 2, 3$ and 4, due to [15,35,36,82,90], and [32,33], respectively. For a general $k \geqslant 5$, it is not known whether the vertex-connectivity augmentation problem is NP-hard or not, even if a given graph is $(k-1)$-vertex-connected. As an approximation solution, Jordàn [49] proved that a solution with its absolute error from the optimal value being at most $k - 3$ can be found in $O(n^5)$ time. The problem of increasing the local-vertex-connectivity to prescribed target values $r(u, v)$ by a minimum number of edges is shown to be NP-hard in general. Jordàn [48] proved the NP-hardness of the problem in the case where a given graph $G = (V, E)$ is $(n/2)$-vertex-connected and there is a subset $S \subset V$ such that $r(u, v) = (n + 2) + 1$ for all $u, v \in S$ and $r(u, v) = 0$ otherwise. However, the complexity of the problem is not known if target values $r(u, v)$, $u, v \in V$ are independent of n (e.g., the case of $r(u, v) \in \{0, 2\}$).

Vertex-connectivity augmentation of digraphs: A digraph is called k-*vertex-connected* if it has at least $(k + 1)$ vertices and remains strongly connected by removal of any $(k - 1)$ vertices. The vertex-connectivity augmentation problem in digraphs is shown to be polynomially solvable by Frank and Jordàn [21]. This result is based on a min–max formula for the problem. They found that the minimum number of new directed edges to make a given digraph $G = (V, E)$ k-vertex-connected is given by the maximum of $\sum_i (k - |V - (A_i \cup B_i)|)$ over all families $\{(A_1, B_1), \ldots, (A_p, B_p)\}$ of pairs of disjoint subsets $A_i, B_i \subseteq V$ such that G has no directed edge from A_i to B_i for each i and $A_i \cap A_j = \emptyset$ or $B_i \cap B_j = \emptyset$ for each $i < j$. The algorithm obtained from this theorem again relies on the ellipsoid method. Afterwards, Frank and Jordàn [22] observed that the vertex-connectivity augmentation problem can be directly reduced to the problem of increasing edge-connectivity from S to T in digraphs. However, it remains open to design combinatorial polynomial time algorithms for solving these two problems.

Let us summarize some recent results obtained after the survey by Frank [18] was done.

As already observed in Theorem 4.3, the time complexity for solving the edge-connectivity augmentation problem in undirected graphs is reduced to $O((mn + n \log n) \log n)$ [71,78]. For this problem, efficient randomized algorithms are also proposed [5,6], among which the algorithm by Benczúr and Karger [6] runs in $O(n^2)$ time. By characterizing all graphs G'_k satisfying the optimality conditions (i)–(ii) in Section 4.1, Nagamochi and Ibaraki [74] showed that an optimal solution F that minimizes the number of vertices incident to F over all optimal solutions can be found in $O((mn + n \log n) \log)$ time.

Gabow [26] improved both the running times of the local edge-connectivity augmentation algorithm of undirected graphs and the edge-connectivity augmentation algorithm of digraphs in [17] to $O(n^2 m \log(n^2/m))$.

As to the vertex-connectivity augmentation problem, Jordàn reduced the absolute error of his algorithm to $\lceil (k-1)/2 \rceil$ [50]. By investigating structure of shredders (which are defined as vertex-cuts removal of which creates more than two components), Cheriyan and Thurimella [11] improves the time complexity of Jordàn's algorithm [49] to $O(\min\{k, \sqrt{n}\}k^2 n^2 + kn^2 \log n)$ time. Very recently, Ishii and Nagamochi [39] obtained an approximation algorithm for the problem without assuming that a given graph is $(k-1)$-vertex-connected for a target k. Given an ℓ-vertex-connected graph, they proved that a solution with absolute error $(k-\ell)(k-1) + \max\{0, (k-\ell-1)(\ell-3) - 1\}(=O((k-\ell)k))$ can be found in $O((k-\ell)(k^2 n^2 + k^3 n^{3/2}))$ time (see also [47] for a slightly better error bound).

The problem of increasing both edge- and vertex-connectivities in a given graph has been studied in [34,40–44]. Hsu and Kao [34] first treated the problem of augmenting the edge- and vertex-connectivities simultaneously, and presented a linear time algorithm for the problem of augmenting an undirected graph $G = (V, E)$ with two specified vertex sets $X, Y \subseteq V$ by adding a minimum number of edges such that the local vertex-connectivity (resp., local edge-connectivity) between every two vertices in X (resp., in Y) becomes at least 2. Afterwards, Ishii et al. considered the problem of augmenting a multigraph $G = (V, E)$ with two integers ℓ and k by adding a minimum number of edges such that G becomes ℓ-edge-connected and k-vertex-connected. They showed polynomial time algorithms for $k = 2$ [40,45] and for a fixed ℓ and $k = 3$ [42] (when a given graph is 2-vertex-connected) and [44] (for an arbitrary graph). For general ℓ and k, they also gave a polynomial time approximation algorithm which produces a solution whose size is at most $\max\{\ell + 1, 2k - 4\}$ over the optimum if a given graph is $(k-1)$-vertex-connected [43] (see [38] for the series of results by Ishii et al.).

For the edge-connectivity augmentation problem in an undirected graph, several types of restrictions on how to add edges have been studied (other than lower and upper bounds on degrees of vertices [17]). Jordán [51] proved that, if the given graph is simple and edges must be added without creating multiple edges, then the problem can be solved in polynomial time for a fixed target k, although the problem is shown to be NP-hard for general k. Bang-Jensen et al. [4] showed that if a partition $\{V_1, \ldots, V_p\}$ of V is given as a constraint such that only edges connecting distinct subsets V_i and V_j can be added to G, then the problem can be solved in polynomial time. The problem of augmenting a connected planar graph to a 2-vertex-connected planar graph is shown to be NP-hard [53,55], and a 5/3-approximation algorithm is proposed by Fialko and Mutzel [16]. If a given graph G is restricted to be outerplanar, then the following problems are polynomially solvable: Augment G to a 2-edge-connected (resp., 2-vertex-connected) planar graph (Kant [53,54]), and for an even k or $k = 3$, augment G to a k-edge-connected planar graph (Nagamochi and Eades [68]).

7. Concluding remarks

In this article, we started with the definition of an MA ordering of vertices in a graph, and then surveyed some of the new algorithms for solving the minimum cut problem and the edge-connectivity augmentation problem, as applications of MA orderings. Triggered by the work of Frank [17] that unified various approaches from the view point of edge-splitting, connectivity augmentation problems have been intensively studied in this decade. Additional constraints such as simplicitly or planarity of a graph have also been taken into account. In these developments, min–max type theorems played an important role in solving the corresponding connectivity augmentation problems exactly. Even if the min–max type theorem holds only approximately in the sense that there remains a small gap between the minimum and maximum objective functions, the problems are sometimes solvable by characterizing the graph structure that yields such a gap (see [4,3,27]). These results may lead to new frameworks for solving other types of combinatorial optimization problems related to connectivity and edge-splitting.

Acknowledgements

Finally, this research was partially supported by the Scientific Grant-in-Aid by the Ministry of Education, Science, Sports and Culture of Japan. The authors would like to thank an anonymous referee for his/her helpful comments.

References

[1] R.K. Ahuja, T.L. Magnanti, J.B. Orlin, Network Flows: Theory, Algorithms, and Applications, Prentice-Hall, Englewood Cliffs, NJ, 1993.

[2] J. Bang-Jensen, A. Frank, B. Jackson, Preserving and increasing local edge-connectivity in mixed graphs, SIAM J. Discrete Math. 8 (1995) 155–178.

[3] J. Bang-Jensen, H.N. Gabow, T. Jordán, Z. Szigeti, Edge-connectivity augmentation with partition constraints, SIAM J. Discrete Math. 12 (1999) 160–207.

[4] J. Bang-Jensen, T. Jordán, Edge-connectivity augmentation preserving simplicity, SIAM J. Discrete Math. 11 (1998) 603–623.

[5] A.A. Benczúr, Augmenting undirected connectivity in $\tilde{O}(n^3)$ time, Proceedings of 26th ACM Symposium on Theory of Computing, 1994, pp. 658–667.

[6] A.A. Benczúr, D.R. Karger, Augmenting undirected edge connectivity in $\tilde{O}(n^2)$ time, Proceedings of 9th Annual ACM-SIAM Symposium on Discrete Algorithms, 1998, pp. 500–519.

[7] G.-R. Cai, Y.-G. Sun, The minimum augmentation of any graph to k-edge-connected graph, Networks 19 (1989) 151–172.

[8] C.S. Chekuri, A.V. Goldberg, D.R. Karger, M.S. Levine, C. Stein, Experimental study of minimum cut algorithms, Proceedings of 8th Annual ACM-SIAM Symposium on Discrete Algorithm, 1997, pp. 324–333.

[9] E. Cheng, T. Jordán, Successive edge-connectivity augmentation problems, Math. Program. Ser. B 84 (1999) 577–594.

[10] J. Cheriyan, M.-Y. Kao, R. Thurimella, Scan-first search and sparse certificates: an improved parallel algorithm for k-vertex connectivity, SIAM J. Comput. 22 (1993) 157–174.

[11] J. Cheriyan, R. Thurimella, Fast algorithms for k-shredders and k-node connectivity augmentation, J. Algorithms 33 (1999) 15–50.

[12] W.J. Cook, W.H. Cunningham, W.R. Pulleyblank, A. Schrijver, Combin. Optimization, Wiley-Interscience, Wiley, Inc., New York, 1998.

[13] E.A. Dinits, A.V. Karzanov, M.V. Lomonosov, On the structure of a family of minimal weighted cuts in a graph, in: A.A. Fridman (Ed.), Studies in Discrete Optimization, Nauka, Moscow, 1976, pp. 290 –306 (in Russian).

[14] S. Enni, A note on mixed graphs and directed splitting off, J. Graph Theory 27 (1998) 213–221.

[15] K.P. Eswaran, R.E. Tarjan, Augmentation problems, SIAM J. Comput. 5 (1976) 653–665.

[16] S. Fialko, P. Mutzel, A new approximation algorithm for the planar augmentation problem, Proceedings of 9th Annual ACM-SIAM Symposium on Discrete Algorithms, 1998, pp. 260–269.

[17] A. Frank, Augmenting graphs to meet edge-connectivity requirements, SIAM J. Discrete Math. 5 (1992) 25–53.

[18] A. Frank, Connectivity augmentation problems in network design, in: J.R. Birge, K.G. Murty (Ed.), Mathematical Programming: State of the Art 1994, The University of Michigan, Ann Arbor, MI 1994, pp. 34–63.

[19] A. Frank, On the edge-connectivity algorithm of Nagamochi and Ibaraki, Laboratoire Artemis, IMAG, Université J. Fourier, Grenoble, March 1994.

[20] A. Frank, T. Ibaraki, H. Nagamochi, On sparse subgraphs preserving connectivity properties, J. Graph Theory 17 (1993) 275–281.

[21] A. Frank, T. Jordán, Minimal edge-coverings of pairs of sets, J. Combin. Theory Ser. B 65 (1995) 73–110.

[22] A. Frank, T. Jordán, Directed vertex-connectivity augmentation, Math. Program. Ser. B 84 (1999) 537–554.

[23] M.L. Fredman, R.E. Tarjan, Fibonacci heaps and their uses in improved network optimization algorithms, J. ACM 34 (1987) 596–615.

[24] S. Fujishige, Another simple proof of the validity of Nagamochi and Ibaraki's min-cut algorithm and Queyranne's extension to symmetric submodular function minimization, J. Oper. Res. Soc. Japan 41 (1998) 626–628.

[25] H.N. Gabow, A matroid approach to finding edge connectivity and packing arborescences, Proceedings of 23rd ACM Symposium on Theory of Computing, New Orleans, Louisiana, 1991, pp. 112–122.

[26] H.N. Gabow, Efficient splitting off algorithms for graphs, Proceedings of 26th ACM Symposium on Theory of Computing, 1994, 696–705.

[27] H.N. Gabow, T. Jordán, How to make a square grid framework with cables rigid, SIAM J. Comput. 30 (2000) 649–680.

[28] A.V. Goldberg, S. Rao, Flows in undirected unit capacity network, Proceedings of 38th Annual IEEE Symposium on Foundations of Computer Science, 1997, pp. 32–35.

[29] D. Gusfield, Optimal mixed graph augmentation, SIAM J. Comput. 16 (1987) 599–612.

[30] J. Hao, J.B. Orlin, A faster algorithm for finding the minimum cut in a directed graph, J. Algorithms 17 (1994) 424–446.

[31] M.R. Henzinger, S. Rao, H.N. Gabow, Computing vertex connectivity: new bounds from old techniques, J. Algorithms 34 (2000) 222–250.

[32] T. Hsu, On four-connecting a triconnected graph, J. Algorithms 35 (2000) 202–234.

[33] T. Hsu, Undirected vertex-connectivity structure and smallest four-vertex connectivity augmentation, Lecture Notes in Computer Science 1004, Sixth International Symposium on Algorithms and Computation, Springer, Berlin, 1995, pp. 274–283.

[34] T. Hsu, M. Kao, A unifying augmentation algorithm for two-edge-connectivity and biconnectivity, J. Combin. Optim. 2 (1998) 237–256.

[35] T. Hsu, V. Ramachandran, A linear time algorithm for triconnectivity augmentation, Proceedings of 32nd IEEE Symposium on Foundations of Computer Science, 1991, pp. 548–559.

[36] T. Hsu, V. Ramachandran, Finding a smallest augmentation to biconnect a graph, SIAM J. Comput. 22 (1993) 889–912.

[37] T.C. Hu, Integer Programming and Network Flows, Addison-Wesley, Reading, MA, 1969.

[38] T. Ishii, Studies on multigraph connectivity augmentation problems, Ph.D. Thesis, Dept. of Applied Mathematics and Physics, Kyoto University, Kyoto, Japan, 2000.

[39] T. Ishii, H. Nagamochi, On the minimum augmentation of an ℓ-connected graph to a k-connected graph, Lecture Notes in Computer Science, 1851, Springer, Berlin, Seventh Biennial Scandinavian Workshop on Algorithm Theory, Bergen, Norway, July 5–7, 2000, pp. 286–299.

[40] T. Ishii, H. Nagamochi, T. Ibaraki, Augmenting edge and vertex connectivities simultaneously, Lecture Notes in Computer Science, 1350, Springer, Berlin, Eighth International Symposium on Algorithms and Computation, 1997, pp. 102–111.

[41] T. Ishii, H. Nagamochi, T. Ibaraki, Optimal augmentation of a biconnected graph to a k-edge-connected and triconnected graph, Proceedings of 9th Annual ACM-SIAM Symposium on Discrete Algorithms, 1998, pp. 280–289.

[42] T. Ishii, H. Nagamochi, T. Ibaraki, k-edge and 3-vertex connectivity augmentation in an arbitrary multigraph, Lecture Notes in Computer Science, vol. 1533, Springer, Berlin, in: K.-Y. Chwa, O.H. Ibarra (Eds.), Algorithms and Computation, Ninth International Symposium on Algorithms and Computation, 1998, pp. 159–168.

[43] T. Ishii, H. Nagamochi, T. Ibaraki, Augmenting a $(k-1)$-vertex-connected multigraph to an ℓ-edge-connected and k-vertex-connected multigraph, Lecture Notes in Computer Science, 1643, Springer, Berlin, Seventh Annual European Symposium on Algorithms, 1999, pp. 414–425.

[44] T. Ishii, H. Nagamochi, T. Ibaraki, Optimal augmentation of a 2-vertex-connected multigraph to a k-edge-connected and 3-vertex-connected multigraph, J. Combin. Optim. 4 (2000) 35–78.

[45] T. Ishii, H. Nagamochi, T. Ibaraki, Multigraph augmentation under biconnectivity and general edge-connectivity requirements, Networks 37 (2001) 144–155.

[46] B. Jackson, Some remarks on arc-connectivity, vertex splitting, and orientation in graphs and digraphs, J. Graph Theory 12 (1988) 429–436.

[47] B. Jackson, T. Jordán, A near optimal algorithm for vertex connectivity augmentation, in: Proc. 11th Annual Int. Symp. on Algorithms and Computation ISAAC'00, Lecture Notes in Computer Science, Vol. 1969, Springer, Berlin, 2000, pp. 326–337.

[48] T. Jordán, Connectivity augmentation problem in graphs, Ph.D. Thesis, Department of Computer Science, Eötvös University, Budapest, Hungary, 1994.

[49] T. Jordán, On the optimal vertex-connectivity augmentation, J. Combin. Theory Ser. B 63 (1995) 8–20.

[50] T. Jordán, A note on the vertex-connectivity augmentation problem, J. Combin. Theory Ser. B 71 (1997) 294–301.

[51] T. Jordán, Two NP-complete augmentation problems, Odense University Preprints #8, 1997.

[52] T. Jordán, Edge-splitting problems with demands, Lecture Notes in Computer Science, 1610, Springer, Berlin, Seventh Conference on Integer Programming and Combinatorial Optimization, 1999, pp. 273–288.

[53] G. Kant, Algorithms for Drawing Planar Graphs, Ph.D. Thesis, Dept. Of Computer Science, Utrecht University, 1993.

[54] G. Kant, Augmenting outerplanar graphs, J. Algorithms 21 (1996) 1–25.

[55] G. Kant, H.L. Bodlaender, Planar graph augmentation problems, Lecture Notes in Computer Science, vol. 621, Springer, Berlin, 1992, pp. 258–271.

[56] M. Kao, Data security equals graph connectivity, SIAM J. Discrete Math. 9 (1996) 87–100.

[57] D.R. Karger, Minimum cuts in near-linear time, Proceedings of 28th ACM Symposium on Theory of Computing, 1996, pp. 56–63.

[58] D.R. Karger, M.S. Levine, Finding maximum flows in undirected graphs seems easier than bipartite matching, Proceedings of 30th ACM Symposium on Theory of Computing, 1998, pp. 69–78.

[59] D.R. Karger, R. Motwani, An NC algorithm for minimum cuts, SIAM J. Comput. 26 (1997) 255–272.

[60] D.R. Karger, C. Stein, An $\tilde{O}(n^2)$ algorithm for minimum cuts, Proceedings of 25th ACM Symposium on Theory of Computing, 1993, pp. 757–765.

[61] D.R. Karger, C. Stein, A new approach to the minimum cut problems, J. ACM 43 (1996) 601–640.

[62] L. Lovász, Conference on Graph Theory, Prague, 1974.

[63] L. Lovász, Combinatorial Problems and Exercises, North-Holland, Amsterdam, 1979.

[64] W. Mader, A reduction method for edge-connectivity in graphs, Ann. Discrete Math. 3 (1978) 145–164.

[65] W. Mader, Konstruktion aller n-fach kantenzusammenhängenden Digraphen, European J. Combin. 3 (1982) 63–67.

[66] D.W. Matula, A linear time $2 + \varepsilon$ approximation algorithm for edge connectivity, Proceedings of 4th Annual ACM-SIAM Symposium on Discrete Algorithms, 1993, pp. 500–504.

[67] D.W. Matula, Two results on search and edge connectivity, Graph Algorithms and Applications, Dagtuhl-Seminar-Report, vol. 145, 1996, p. 16.

[68] H. Nagamochi, P. Eades, Edge-splitting and edge-connectivity augmentation in planar graphs, Lecture Notes in Computer Science, 1412, Springer, Berlin, in: R.E. Bixby, E.A. Boyd, R.Z. Ríos-Mercado (Eds.), Sixth Conference on Integer Programming and Combinatorial Optimization, 1998, pp. 96–111.

[69] H. Nagamochi, T. Ibaraki, A linear-time algorithm for finding a sparse k-connected spanning subgraph of a k-connected graph, Algorithmica 7 (1992) 583–596.

[70] H. Nagamochi, T. Ibaraki, Computing edge-connectivity in multigraphs and capacitated graphs, SIAM J. Discrete Mathematics 5 (1992) 54–66.

[71] H. Nagamochi, T. Ibaraki, Deterministic $\tilde{O}(nm)$ time edge-splitting in undirected graphs, J. Combin. Optim. 1 (1997) 5–46.

[72] H. Nagamochi, T. Ibaraki, A note on minimizing submodular functions, Inform. Process. Lett. 67 (1998) 239–244.

[73] H. Nagamochi, T. Ibaraki, Polyhedral structure of submodular and posi-modular systems, Lecture Notes in Computer Science, 1533, in: K.-Y. Chwa, O.H. Ibarra (Eds.), Springer, Berlin, Algorithms and Computation, Ninth International Symposium on Algorithms and Computations, 1998, pp. 169–178.

[74] H. Nagamochi, T. Ibaraki, Augmenting edge-connectivity over the entire range in $\tilde{O}(nm)$ time, J. Algorithms 30 (1999) 253–301.

[75] H. Nagamochi, T. Ibaraki, Polyhedral structure of submodular and posi-modular systems, Discrete Appl. Math. 107 (2000) 165–189.

[76] H. Nagamochi, T. Ono, T. Ibaraki, Implementing an efficient minimum capacity cut algorithm, Math. Program. 67 (1994) 325–341.

[77] H. Nagamochi, T. Ishii, T. Ibaraki, A simple and constructive proof of a minimum cut algorithm, Inst. Electron. Inform. Comm. Eng. Trans. Fundamentals E82-A (1999) 2231–2236.

[78] H. Nagamochi, S. Nakamura, T. Ibaraki, A simplified $\tilde{O}(nm)$ time edge-splitting algorithm in undirected graphs, Algorithmica 26 (2000) 56–67.

[79] H. Nagamochi, Y. Nakao, T. Ibaraki, A fast algorithm for cactus representations of minimum cuts, J. Japan Soc. Ind. Appl. Math. 17 (2000) 245–264.

[80] M. Padberg, G. Rinaldi, An efficient algorithm for the minimum capacity cut problem, Math. Program. 47 (1990) 19–36.

[81] J.C. Picard, M. Queyranne, On the structure of all minimum cuts in a network and applications, Math. Program. Study 13 (1980) 8–16.

[82] J. Plesnik, Minimum block containing a given graph, Archiv der Mathmatik, XXVII 1976, Fasc. 6, pp. 668–672.

[83] M. Queyranne, Minimizing symmetric submodular functions, Math. Program. 82 (1998) 3–12.

[84] S. Raghavan, T.L. Magnanti, Network Connectivity, in: M. Dell'Amico, F. Maffioli, S. Martello (Eds.), Annotated Bibliographies in Combinatorial Optimization, Wiley, New York, 1997.

[85] R. Rizzi, On minimizing symmetric set functions, Combinatorica 20 (2000) 445–450.

[86] M. Stoer, F. Wagner, A simple min-cut algorithm, J. ACM 44 (1997) 585–591.

[87] R.E. Tarjan, M. Yannakakis, Simple linear-time algorithms to test chordality of graphs, test acyclicity of hypergraphs, and selectively reduce acyclic hypergraphs, SIAM J. Comput. 13 (1984) 566–579.

[88] S. Tsukiyama, K. Koike, I. Shirakawa, An algorithm to eliminate all complex triangles in a maximal planar graph for use in VLSI floor-plan, Proceedings of ISCAS'86, 1986, pp. 321–324.

[89] T. Watanabe, A. Nakamura, Edge-connectivity augmentation problems, J. Comput. System Sci. 35 (1987) 96–144.

[90] T. Watanabe, A. Nakamura, A minimum 3-connectivity augmentation of a graph, J. Comput. System Sci. 46 (1993) 91–128.

![N·H logo] ELSEVIER

Discrete Applied Mathematics 123 (2002) 473–485

DISCRETE
APPLIED
MATHEMATICS

Applications of combinatorics to statics—rigidity of grids

Norbert Radics, András Recski*

Department of Computer Science and Information Theory, Budapest University of Technology and Economics, H-1521 Budapest, Hungary

Received 5 January 2000; received in revised form 30 November 2000; accepted 22 January 2001

Abstract

The infinitesimal rigidity (or briefly rigidity) of a bar-and-joint framework (in any dimension) can be formulated as a rank condition of the so-called rigidity matrix. If there are n joints in the framework then the size of this matrix is $O(n)$, so the time complexity of determining its rank is $O(n^3)$. But in special cases we can work with graph and matroid theoretical models from which very fast and effective algorithms can be obtained. At first the case of planar square grids will be presented where they can be made rigid with diagonal rods and cables in the squares, and with long rods and cables which may be placed between any two joints of the grid. Then we will consider the one- and multi-story buildings, and finally some other results and algorithms. © 2002 Elsevier Science B.V. All rights reserved.

1. Planar square grids with diagonal and long rods

Let us consider a $k \times l$ square grid which consists of rigid rods and rotatable joints (in the grid points). The motions of such a planar square grid framework have a very simple description. Since the opposite rods of a square will be parallel after any planar motions, all the deformations can be described with the rotations of the rows and columns of the grid. The extra diagonal rod in Fig. 1 ensures $x_2 = y_1$ (that is, the rotations of row 2 and column 1 are identical, preserving the shape of the square in their intersection).

We can define a bipartite graph where the vertices of the graph correspond to the rows and columns, respectively, and there is an edge between two vertices if and only if there is a diagonal rod in the intersection of the corresponding row and column,

---

* Corresponding author.
  E-mail address: recski@cs.bme.hu (A. Recski).

0166-218X/02/$ - see front matter © 2002 Elsevier Science B.V. All rights reserved.
PII: S 0166-218X(01)00350-X

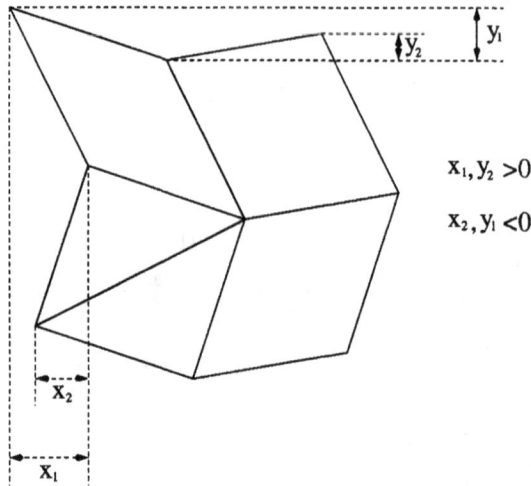

$$x_1, y_2 > 0$$
$$x_2, y_1 < 0$$

Fig. 1.

ensuring that the deformations of the corresponding row and column must be equal. This graph theoretical model of square grid frameworks was given by Bolker and Crapo [3]. They proved that a planar square grid framework will be rigid if and only if the corresponding bipartite graph is connected. In this case all the x_i and y_j quantities are equal and hence any motion is necessarily a congruent motion (a rotation) of the whole framework. As a corollary one can deduce that the minimal rigid systems of diagonal rods in a $k \times l$ square grid consist of $k + l - 1$ diagonals (forming a spanning tree in the graph) and the rigidity of the system can be recognized in $O(k + l)$ time (we only have to check the connectivity of the corresponding graph). Observe that using the rigidity matrix (see Abstract) we should need $O((k + l)^6)$ time.

The reader can immediately observe the difference between a "real" deformation (like at the right-hand side of Fig. 4) and an "infinitesimal" one (like in Fig. 8c). Throughout, rigidity will mean that even infinitesimal deformations are excluded. Most authors call this concept infinitesimal rigidity.

The natural generalization of this problem is if we use long rods in the square grid which can be placed between any two joints of the grid. The graph theoretical model does not work in this case. Long rods, parallel to the rows or columns, have no effect to the infinitesimal motions of the grid so they are ignored. The effect of a general long rod can be described with a linear equation [7]. These equations have very simple structure: they have the rotations of the rows and columns as variables and the coefficients are 0 and ± 1. In Fig. 2 we show an example, but we have to emphasize that the possible motions to be prevented by the long rod are infinitesimal only. Of course "short" diagonal rods are special long rods, so if we have a $k \times l$ square grid with s pieces of long rods we have to consider the system of equations with $k + l$ variables and s equations.

The planar frameworks are rigid if and only if they have only the trivial congruent motions in the plane, where the rotations of rows and columns are the same. In the

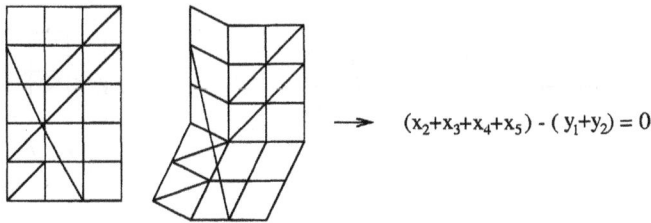

$$(x_2+x_3+x_4+x_5) - (y_1+y_2) = 0$$

Fig. 2.

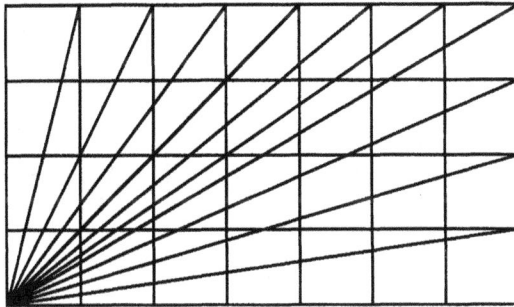

Fig. 3.

$(k + l)$-dimensional linear space all of the hyperplanes corresponding to the equations contain this one-dimensional subspace of the trivial solutions. To obtain the line of the trivial solutions as the intersection of the hyperplanes we need at least $k + l - 1$ hyperplanes.

Theorem 2.1 (Gáspár et al. [7]). *The square grid framework with long rods is infinitesimally rigid if and only if the corresponding system of equations has only a one-dimensional set of solutions (the congruent motions). So it requires at least $k + l - 1$ long rods to make the square grid rigid and so many rods are sufficient (see Fig. 3).*

2. Planar square grids with diagonal and long cables

Physically realizable rods, unfortunately, are less reliable against compression than tension. If we want to model the physically constructible frameworks and wish to permit only tension in the diagonals, we have to introduce the concept of tensegrity frameworks. Here we can use three kind of elements between joints: rods (which are rigid both under tension and compression), cables (which are reliable against only tension) and struts (which are reliable against only compression). Since a diagonal cable and a diagonal strut in the opposite (that is, perpendicular) position

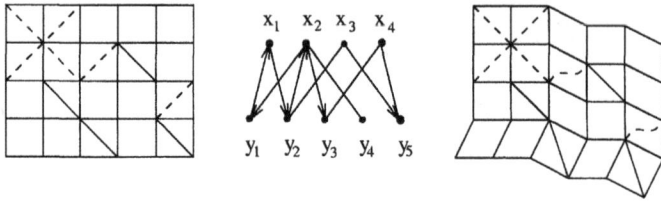

Fig. 4.

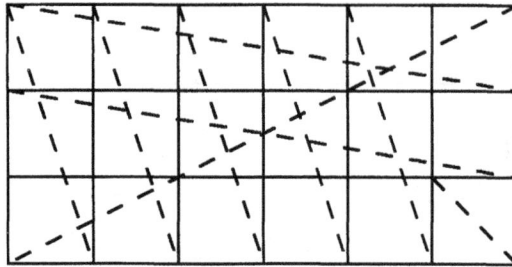

Fig. 5.

in a square framework have the same effect, we may disregard struts in our model. In what follows, rods and cables will be drawn by continuous and by broken lines, respectively.

If we use diagonal cables in the squares the problem will be similarly very simple. A diagonal cable can prevent the deformation of the square only in one direction which means that an inequality will hold between the rotations of the corresponding row and column. Since the effect of the cable depends on its position we have to use directed edges in the graph to indicate the direction of the cable (and also the inequality). Baglivo and Graver [1] showed that the square grid framework with diagonal cables is rigid if and only if the corresponding digraph is strongly connected. (Undirected edges, indicating diagonal rods, are considered as pairs of oppositely oriented directed edges.) For example, there is no directed edge from $\{y_3, y_5, x_4\}$ to the other vertices, hence the tensegrity framework of Fig. 4 is nonrigid. So we need at least $2\max(k,l)$ diagonal cables to make the $k \times l$ grid rigid.

To describe the effect of long cables we need linear inequalities because a cable can prevent motions in one direction only.

Theorem 3.1 (Gáspár et al. [7]). *The square grid framework with long cables is infinitesimally rigid if and only if the corresponding system of inequalities has only a one-dimensional set of solutions* (*the congruent motions*). *So it requires at least $k+1$ long cables to make the $k \times l$ square grid rigid and so many cables are sufficient* (*see* Fig. 5).

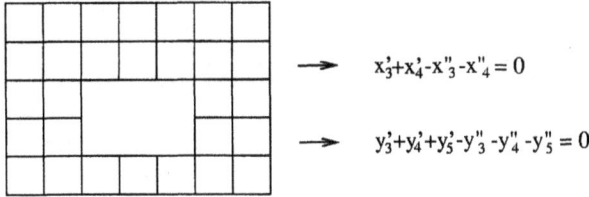

$$x'_3 + x'_4 - x''_3 - x''_4 = 0$$

$$y'_3 + y'_4 + y'_5 - y''_3 - y''_4 - y''_5 = 0$$

Fig. 6.

We can observe that using long rods instead of "short" diagonal rods the required number $k + l - 1$ of rods did not change. But in the case of cables this number decreased, sometimes significantly: the required number of "short" diagonal cables (in a $k \times l$ square grid) was $2 \max(k, l)$, while $k + l$ long cables are sufficient. For example in Fig. 5 this number is 9 instead of 12.

3. Planar square grids with holes

So far the bases of the frameworks were complete rectangular parts of the infinite square grid. It is easy to see that using "convex" square grids, where each row and column consist of one connected sequence of squares, the previous theorems remain valid without any changes, only the number of possible rods and cables will decrease. If the square grid is not "convex" but there is no hole in it then we have to increase the number of variables introducing different variables for the independent segments of rows and columns. So the previous theorems will hold if we substitute the number of row and column segments for k and l, respectively. But what is the situation if we have hole(s) in the grid? It was shown in [8] that each hole forces two more equations, one among the rotations of the row segments and one among the column segments, respectively (see Fig. 6).

This observation implies that at least

$$(\#\{\text{row and column segments}\}) - 2(\#\{\text{holes}\}) - 1$$

pieces of rods, or at least one more of cables, are required to infinitesimally rigidify the square grid framework with holes. For example the framework of Fig. 7a is nonrigid (an infinitesimal deformation is indicated in Fig. 7c), but changing the direction of a cable (Fig. 7b) it becomes rigid.

4. One-story buildings with diagonal and long rods

A one-story building is a square grid whose joints are connected to the ground via joints by rods of uniform length. The first observation about one-story buildings was that a diagonal rod in a vertical wall prevents the motions of the wall along its plane, so putting diagonal rods into two intersecting vertical walls the vertical rod in the

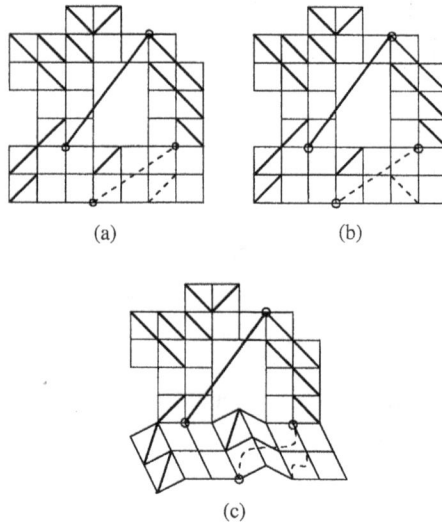

(a)   (b)

(c)

Fig. 7.

intersection will be fixed. If we put four diagonal rods into the four external vertical walls then the problem can be reduced to the planar problem of a square grid with its four corners fixed to the plane. That is, the sum of the rotations of the rows and the sum of the rotations of the columns must be equal to zero.

$$x_1 + x_2 + \cdots + x_k = 0, \qquad y_1 + y_2 + \cdots + y_l = 0. \tag{1}$$

These equations result that in the case of $k \times l$-sized one-story buildings the required number of diagonal rods is $k + l - 2$ (in addition to the four rods in the vertical walls).

Theorem 5.1 (Crapo [5]). *The framework of a one-story building which has rods in the external vertical walls is made infinitesimally rigid by certain diagonal rods of the ceiling if and only if the corresponding bipartite graph is either connected or is an asymmetric 2-component graph. Asymmetric means that*

$$\begin{vmatrix} |V_1 \cap A| & |V_1 \cap B| \\ |V_2 \cap A| & |V_2 \cap B| \end{vmatrix} \neq 0,$$

where V_1 and V_2 are the vertex sets of the two connected components of G, while A and B are the two subsets of the original bipartition of the bipartite graph.

For example Fig. 8 shows two square grids and the corresponding 2-component forests. The ratios of rows and columns in the components of the first graph are 3:3 and 3:6, respectively, the graph is asymmetric, so the first framework is rigid. In the second graph these ratios are 2:3 and 4:6 which are the same, the second graph is symmetric and we can see a deformation of the framework in Fig. 8c. (It is easy to see that these systems make the framework rigid if and only if, after a suitable

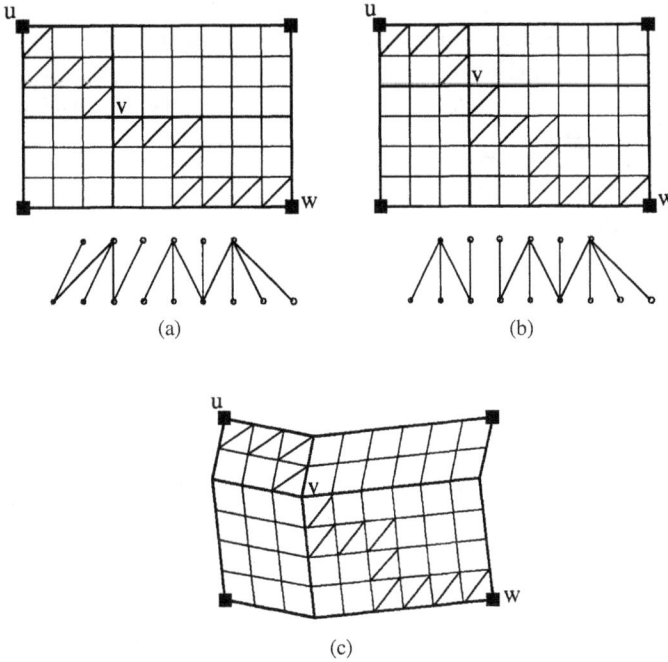

Fig. 8.

permutation among the rows and after another one among the columns to make the grid "block-diagonal", the three special points of the grid u, v and w, determined by the two-components of the forest, are not collinear.) One can prove that the asymmetric 2-component forests of a graph form the base set of a matroid, this result can be generalized to non-bipartite graphs as well [14].

In the general case of one-story buildings it is easy to see that at least three vertical walls, not all parallel, must contain diagonal rods. In such a good situation we have to consider only the horizontal motions of the joints because preventing the horizontal motions the framework will be rigid (the joints cannot move vertically). In this case the effect of a diagonal rod in a vertical wall can be described with an equation where the variables are the rotations of rows and columns and two further auxiliary variables:

Theorem 5.2 (Radics [13]). *Let us consider a $(k \times l)$-sized one-story building with some diagonal rods of certain (horizontal or vertical) squares. The building is infinitesimally rigid if and only if the rank of the coefficient matrix of the corresponding system of equations is $k + l + 2$. Hence this is the minimum number of diagonal rods to make the building rigid.*

The minimal rigid systems of diagonal rods form the base set of a matroid (the coefficient matrix of the equation system is a representation). In the simplest case, where $k = l = 1$ this matroid is isomorphic to the cycle matroid of the graph C_5 (cycle of length 5). In general the matroids have the following property:

Theorem 5.3 (Radics [13]). *The matroid of a $(k \times l)$-sized one-story building is binary if and only if $k = l = 1$.*

Making the one-story building rigid with long rods is solved only in the special case mentioned above, when we put diagonal rods into the external vertical walls. In this case, besides the linear equations as the effect of long rods, we have to put the two additional equations (1) into the system. The one-story building will be rigid if and only if the intersection of the hyperplanes corresponding to the equations of the long rods and the pinned points consists of the origin only.

Theorem 5.4 (Gáspár et al. [7]). *The one-story building based on a $k \times l$ square grid is made infinitesimally rigid by (neither horizontal nor vertical) long rods if and only if the solution of the corresponding system of linear equations is unique: the zero vector.*

So we need $k + l - 2$ long rods to make the building rigid (of course in addition to the rods in the four vertical walls).

5. One-story buildings with diagonal and long cables

The general problem of the one-story buildings with diagonal cables seems to be much more difficult than the problem with diagonal rods (see the previous section). Only the special case of the building with four external vertical walls braced with diagonal rods was solved, but the minimal rigid systems of diagonal cables have various structures.

The first observation is that the difference between the required numbers of diagonal cables or rods is not so big as it was in the planar case ($2 \max(k, l)$ versus $k + l - 1$).

Theorem 6.1 (Chakravarty et al. [4]). *Let $k, l \geqslant 2$ and $k + l \geqslant 5$. Then at least $k + l - 1$ diagonal cables are required to make the one-story building (with braced vertical walls) infinitesimally rigid, and that number will always do.*

Let us presume that the bipartite graph $G(A, B)$ of the system of cables is connected (where A and B are the vertex classes corresponding to the rows and columns, respectively). Then there is a necessary and sufficient condition for the graph to make the one-story building rigid:

Theorem 6.2 (Recski and Schwärzler [16]). *Let the graph $G(A, B)$ of the cables be connected. Then this system of cables makes the one-story building infinitesimally rigid if and only if*

$$|N^*(X)| \cdot k > |X| \cdot l$$

for all proper subsets X of A or

$$|N^*(Y)| \cdot l > |Y| \cdot k$$

for all proper subsets Y of B, where $N^*(Z)$ denotes the set of vertices in the other vertex class which can be reached with directed paths from Z.

There is an interesting question implied by this theorem: what kind of cable systems can occur as minimal rigid cable systems. The first special case is when we prescribe that all the cables must be parallel. Then we have a simple necessary and sufficient condition:

Theorem 6.3 (Recski [15]). *Consider a system of $k+l-1$ diagonal cables in the $k \times l$ square grid where the corners are pinned down, and suppose that all the diagonals are parallel. Let $G(A,B)$ be the corresponding bipartite graph. Then the system makes the grid infinitesimally rigid if and only if $|N(X)| > (l/k)|X|$ holds for every proper subset X of A, where $N(X)$ denotes the set of those vertices of B which are adjacent to at least one vertex of X.*

Recall that a bipartite graph, with bipartition subsets A, B of cardinalities k and l, respectively, has a perfect-matching if and only if $k = l$ and $|N(X)| \geq |X|$ holds for every subset X of A. This theorem of Hall has a strengthening which is in complete formal analogue of the condition of Theorem 6.3, namely that every edge of a connected bipartite graph is contained in some perfect matching if and only if $k = l$ and $|N(X)| > |X|$ holds for every proper subset X of A (see [9]).

On the other hand, the graph of a minimal system, since the required number of cables is $k + l - 1$, can be a directed tree. The natural questions are: which trees can also be the graphs of a minimal system, and are there any other possibilities as well? The answer to the second question is in the affirmative (i.e. the graph need not be a tree) but there is only one exception:

Theorem 6.4 (Recski [15]). *Consider a system of $k+l-1$ diagonal cables in the $k \times l$ square grid where the corners are pinned down. Suppose the corresponding graph is not a tree. Then the system of cables makes the grid infinitesimally rigid if and only if $k - l = \pm 1$ and the corresponding graph consists of an isolated vertex and a directed circuit with $2 \min(k, l)$ vertices (like in Fig. 9).*

As we saw the graph of most of the rigid minimal systems of cables is a directed tree. Let us consider the reverse problem: if we have an undirected tree as the graph of the cables (we have information only about the squares in which the cables are placed but the positions of the cables are unknown) is there a good orientation of the edges (and that of the cables in the squares) which makes the one-story building rigid? There is a simple characterization of these "rigid" trees. Call an edge e of the tree F *critical*, if $F - \{e\}$ is a symmetric 2-component forest (cf. Theorem 5.1).

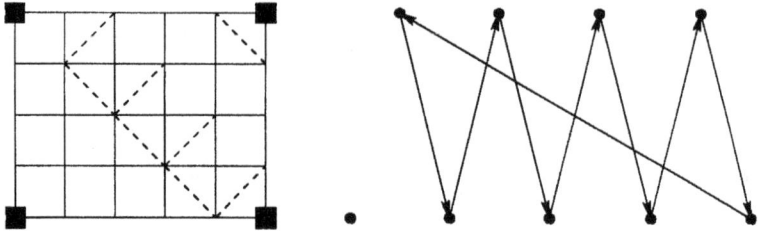

Fig. 9.

Theorem 6.5 (Recski and Schwärzler [16]). *F has a rigid orientation if and only if F has no critical edges. The rigid orientation, if it exists, is essentially (up to inversion of the whole orientation) unique.*

The necessity of the condition is obvious: the remark about the collinearity of the points u, v and w following Theorem 5.1 shows that if $F - \{e\}$ is symmetric then a single cable cannot prevent a deformation like that of Fig. 8c.

In [16] we can find an algorithm which provides a good orientation if it exists: Let V_1, V_2 denote the vertex sets of the two connected components of $F - \{e\}$. The edge e has tail in V_i and head in V_j if and only if $k_i l_j > k_j l_i$, where k_i denotes the number of vertices in V_i which correspond to rows of the grid, and l_i denotes the number of vertices in V_i which correspond to columns.

6. Results in 3-dimension

The 3-dimensional problem of cubic grids is much more difficult than the planar problem, we have no such effective methods in the space as the graph theoretic model was in the plane. But the problem is very interesting so partial results have been already published about rigidity of special cubic grids [10,13] or general observations about d-cube grids [11]. However, the problem of the t-story buildings with diagonal rods, as the simplest 3-dimensional case is solved.

The description of the one-story building can easily be generalized to the case of higher buildings. It is easy to see that in each floor at least three vertical walls, not all parallel, must contain diagonal rods, hence we have to consider only the horizontal motions of the floors. Describing the effect of a diagonal rod we will obtain similar linear equations as in the case of one-story buildings [13]. In this system of equations the number of variables is $t(k + l + 2)$ (in the case of a $k \times l$-sized t-story building), while the number of equations is equal to the number of diagonal rods.

Theorem 7.1 (Radics [13]). *A $k \times l$-sized t-story building is infinitesimally rigid if and only if there are at least three vertical walls braced in each floor and these are not all parallel, and the system of equations obtained from the rods has the zero vector as the only solution.*

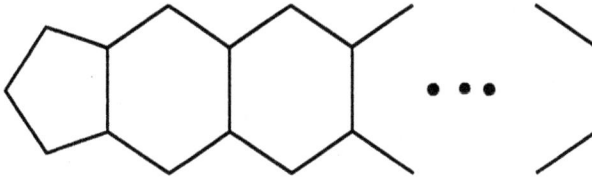

Fig. 10.

This means that we need at least $t(k+l+2)$ diagonal rods to make a $(k \times l)$-sized t-story building rigid.

Similarly to the one-story case we can define the matroid of a t-story building. Of course this matroid will be a representable matroid but from Theorem 5.3 it is obvious that such a matroid can be graphic only if $k = l = 1$ (that is, for a $1 \times 1 \times t$ building). However, in this case the structure of these matroids is very simple:

Theorem 7.2 (Radics [13]). *The matroid of a $1 \times 1 \times t$ building is always graphic and the corresponding graph is a "chain" of a pentagon (corresponding to the first floor) and $t - 1$ pieces of hexagons, like in* Fig. 10.

7. Other grid-like structures

It is easy to see that all the results in Sections 2–4 are almost the same if we have a planar grid of parallelograms [17]. The only changes will arise when we construct the linear equations or inequalities because the coefficients depend on the size of the parallelograms.

The graph theoretical method can be applied for other grid-like structures as well, see [12] for the Archimedian semiregular grids in the plane. The semiregular grids (33344) and (33434) are isomorphic to square grids with diagonal rods of certain squares (see Fig. 11) so we can use the graph theoretical method of square grids to make these semiregular grids rigid. But if there are hexagons or larger faces in the grid then this method can work only with special further assumptions about the motions of the rods.

8. Algorithmic aspects

As it was mentioned, using the rigidity matrix the number of operations required to determine the rigidity of a framework is proportional to the cube of the number of joints, hence can be bounded from above by $c \times n$ where $n = k + l$ ($k + l + t$ in Section 6) and c, q are constants. But using the new results all the above problems require less operations. The decrease is significant in every case except for the grid with holes: if we have a large number of holes, the new method decreases only the coefficient c in the bound for the number of operations, as compared to the original method. In all the other cases the exponent q is reduced in the upper bound.

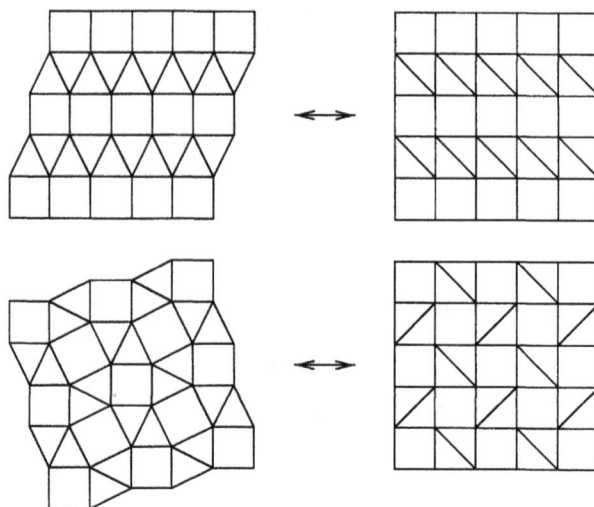

Fig. 11.

In all cases the set of minimal rigid systems form the base set of a matroid. It means that the greedy algorithm can work in these cases, so we can make these types of frameworks rigid even if we have special requirements about placing the rods or cables. For example, if we wish to find a minimum system of diagonals to make a system rigid then the "user" may specify priorities for certain diagonals.

Another problem is how can one extend a given set of diagonal rods or cables in a planar square gird to obtain a rigid system or to increase the reliability of the system (if it is not only rigid but also remains rigid if one diagonal is "broken"). Ref. [2] contains a linear time algorithm for the connectivity augmentation problem for graphs with special requirements. A special case of this problem—augmenting the connectivity of bipartite graphs while preserving bipartiteness—gives a linear time algorithm for our problem. The analogous problem concerning strong connectivity of digraphs was solved in [6]. Their algorithm solves, still in linear time, the problem of completing square grid framework with diagonal cables.

Acknowledgements

Grants No. OTKA 29772 and 30122 of the Hungarian National Science Foundation and grant No. FKFP 409/1997 of the Hungarian Ministry of Education are gratefully acknowledged. Part of the research has been performed while the second author was with the Forschungsinstitut für Diskrete Mathematik, Rheinische Friedrich-Wilhelms-Universität, Bonn, Germany, partly supported by the Alexander-von-Humboldt Foundation.

References

[1] J.A. Baglivo, J.E. Graver, Incidence and Symmetry in Design and Architecture, Cambridge University Press, Cambridge, 1983.

[2] J. Bang-Jensen, H.N. Gabow, T. Jordán, Z. Szigeti, Edge-connectivity augmentation with partition constraints, Proceedings of the Ninth Annual ACM-SIAM Symposium on Discrete Algorithms, 1998, pp. 306–315.

[3] E.D. Bolker, H. Crapo, Bracing rectangular frameworks, I. SIAM J. Appl. Math. 36 (1979) 473–490.

[4] N. Chakravarty, G. Holman, S. McGuinness, A. Recski, One-story buildings as tensegrity frameworks, Structural Topology 12 (1986) 11–18.

[5] H. Crapo, More on the bracing of one-story buildings, Environ. Planning B 4 (1977) 153–156.

[6] H.N. Gabow, T. Jordán, How to make a square grid framework with cables rigid, Proceedings of the 10th Annual ACM-SIAM Symposium on Discrete Algorithms, 1999, pp. 356–365.

[7] Zs. Gáspár, N. Radics, A. Recski, Square grids with long diagonals, Optim. Methods Software 10 (1998) 217–231.

[8] Zs. Gáspár, N. Radics, A. Recski, Rigidity of square grids with holes, Comput. Assisted Mech. Eng. Sci. 6 (1999) 329–335.

[9] G. Hetyei, On covering by 2 × 1 rectangles, Pécsi Tanárképző Főisk. Közl. 8 (1964) 351–368 (in Hungarian).

[10] Gy. Nagy, Diagonal bracing of special cube grid, Acta Tech. Acad. Sci. Hungar. 106 (3–4) (1994) 265–273.

[11] Gy. Nagy, The rigidity of special d cube grids, Ann. Univ. Sci. Budapest 39 (1996) 107–112.

[12] Gy. Nagy, Diagonal bracing of regular planar grids, Alkalmaz. Mat. Lapok 18 (1998) 101–109.

[13] N. Radics, Rigidity of t-story buildings, Proceedings of the 1st Japanese–Hungarian Symposium on Discrete Mathematics and Its Applications, 1999, pp. 181–187.

[14] A. Recski, Elementary strong maps of graphic matroids, Graphs and Combin. 3 (1987), 379–382 (Part 1);
A. Recski, Elementary strong maps of graphic matroids, Graphs and Combin. 10 (1994) 205–206 (Part 2).

[15] A. Recski, One-story buildings as tensegrity framework II, Structural Topology 17 (1991) 43–52.

[16] A. Recski, W. Schwärzler, One-story buildings as tensegrity frameworks III, Discrete Appl. Math. 39 (1992) 137–146.

[17] W. Whiteley, private communication, June 1990.

For further reading

Gy. Nagy, How to brace an annex building? Proceedings of the Second International Conference on Graphs and Mechanics, 1999, pp. 35–36.

A. Recski, Applications of combinatorics to statics—a survey, Rend. Circ. Mat. Palermo 3 (2) (1984) 237–247.

A. Recski, Applications of combinatorics to statics—a second survey, Discrete Math. 108 (1992) 183–188.

ELSEVIER

Discrete Applied Mathematics 123 (2002) 487–512

DISCRETE
APPLIED
MATHEMATICS

Models, relaxations and exact approaches for the capacitated vehicle routing problem

Paolo Toth, Daniele Vigo*

Università di Bologna, Dipartimento di Elettronica, Informatica e Sistemistica, Viale Risorgimento 2, 40136 Bologna, Italy

Received 28 September 1999; accepted 2 October 2000

Abstract

In this paper we review the exact algorithms based on the branch and bound approach proposed in the last years for the solution of the basic version of the vehicle routing problem (VRP), where only the vehicle capacity constraints are considered. These algorithms have considerably increased the size of VRPs that can be solved with respect to earlier approaches. Moreover, at least for the case in which the cost matrix is asymmetric, branch and bound algorithms still represent the state-of-the-art with respect to the exact solution. Computational results comparing the performance of different relaxations and algorithms on a set of benchmark instances are presented. We conclude by examining possible future directions of research in this field. © 2002 Elsevier Science B.V. All rights reserved.

Keywords: Vehicle routing problem; Exact algorithms; Branch and bound; Relaxations

1. Introduction

The vehicle routing problem (VRP) is one of the most studied among the combinatorial optimization problems, due both to its practical relevance and to its considerable difficulty.

The VRP is concerned with the determination of the optimal routes used by a fleet of vehicles, based at one or more depots, to serve a set of customers. Many additional requirements and operational constraints are imposed on the route construction in practical applications of the VRP. For example, the service may involve both deliveries and collections, the load along each route must not exceed the given capacity of the

---

* Corresponding author.
E-mail address: dvigo@deis.unibo.it (D. Vigo).

0166-218X/02/$ - see front matter © 2002 Elsevier Science B.V. All rights reserved.
PII: S0166-218X(01)00351-1

vehicles, the total length of each route must not be greater than a prescribed limit, the service of the customers must occur within given time windows, the fleet may contain heterogeneous vehicles, precedence relations may exist between the customers, the customer demands may not be completely known in advance, the service of a customer may be split among different vehicles, and some problem characteristics, as the demands or the travel times, may vary dynamically.

We consider the static and deterministic basic version of the problem, known as the *capacitated* VRP (CVRP). In the CVRP all the customers correspond to deliveries, the demands are deterministic, known in advance and may not be split, the vehicles are identical and are based at a single central depot, only the capacity restrictions for the vehicles are imposed, and the objective is to minimize the total cost (i.e., the number of routes and/or their length or travel time) needed to serve all the customers. Generally, the travel cost between each pair of customer locations is the same in both directions, i.e., the resulting cost matrix is *symmetric*, whereas in some applications, as the distribution in urban areas with one-way directions imposed on the roads, the cost matrix is *asymmetric*.

The CVRP has been extensively studied since the early sixties and in the last years many new heuristic and exact approaches were presented. The largest problems which can be consistently solved by the most effective exact algorithms proposed so far contain about 50 customers, whereas larger instances may be solved only in particular cases. So instances with hundreds of customers, as those arising in practical applications, may only be tackled with heuristic methods.

The CVRP extends the well-known *Traveling Salesman Problem* (TSP), calling for the determination of the circuit with associated minimum cost, visiting exactly once a given set of points. Therefore, many exact approaches for the CVRP were inherited from the huge and successful work done for the exact solution of the TSP.

Laporte and Nobert [32] presented an extensive survey which was entirely devoted to exact methods for the VRP and gave a complete and detailed analysis of the state of the art up to the late eighties. The aim of the present work is to provide an update of that survey, describing the algorithms recently proposed for the exact solution of CVRP both for the case with symmetric and asymmetric cost matrices. Up to the end of the last decade the most effective exact approaches for the CVRP were mainly branch and bound algorithms using basic relaxations, as the assignment problem and the shortest spanning tree. Recently, more sophisticated bounds were proposed, as those based on Lagrangian relaxations or on the additive approach, which increased the size of the problems that can be solved to optimality by branch and bound. Moreover, following the success obtained by branch and cut methods for the TSP, encouraging results were obtained by using these algorithms for the CVRP.

In this work we treat separately problems with symmetric and asymmetric cost matrices. In fact, although the symmetric problems are special cases of the asymmetric ones, the latter were much less studied in the literature and the exact methods developed for them have in general a poor performance when applied to symmetric instances. Analogously, not all the approaches proposed for symmetric problems may be directly adapted to solve also asymmetric ones. In the following, we will denote with SCVRP and ACVRP the symmetric and asymmetric CVRP, respectively. Moreover, when the explicit distinction between the two versions is not needed, we simply use CVRP.

Other surveys covering exact algorithms, but often mainly devoted to heuristic methods, were presented by Christofides et al. [10], Magnanti [37], Bodin et al. [4], Christofides [7], Laporte [29], Fisher [22], Toth and Vigo [45] and Golden et al. [25]. An annotated bibliography was recently proposed by Laporte [30], whereas an extensive bibliography was presented by Laporte and Osman [34]. A book on the subject was edited by Golden and Assad [24].

The work is organized as follows. In Section 2 we give a detailed description of CVRP as a graph theoretic problem, and introduce the corresponding notation. In Section 3 we consider the more general case of ACVRP, where the cost matrix is asymmetric, illustrating the branch and bound algorithms proposed by Laporte et al. [31] and by Fischetti et al. [18]. In Section 4 we examine the exact methods proposed for the more widely studied SCVRP. In particular, we discuss the basic relaxations based on K-tree and b-matching and their strengthening in a Lagrangian fashion proposed by Fisher [20] and Miller [40], respectively. We also briefly discuss the set-partitioning based relaxations used by Hadjcostantinou et al. [26]. Computational results comparing the performance of different relaxations and algorithms on a set of benchmark instances are presented. Finally, in Section 5, we draw some conclusions and discuss future directions of research.

The information about the performance, expressed in Mflops, of the computers used for testing the algorithms presented are taken (when available) from Dongarra [15]. Moreover, all the computational results reported in this paper are performed by using well-known test instances from the literature. As proposed in Vigo [48], the instances are identified with a name whose first character denotes the problem type (A, E and S for asymmetric, Euclidean and other symmetric problems, respectively), then the name includes the number of vertices, depot included, and the number of available vehicles, and the last letter indicates the source of the instance. For example, E051-05e is the famous 50 customers problem Euclidean described in Christofides and Eilon [8].

2. Problem definition and notation

The CVRP may be defined as the following graph theoretic problem. Let $G = (V, A)$ be a complete graph where $V = \{0, \ldots, n\}$ is the vertex set and A is the arc set. Vertices $j = 1, \ldots, n$ correspond to the customers, each with a known nonnegative *demand*, d_j, to be delivered, whereas vertex 0 corresponds to the depot (with a fictitious demand $d_0 = 0$). Given a customer set $S \subseteq V$, let $d(S) = \sum_{j \in S} d_j$ denote the total demand of the set.

A nonnegative *cost*, c_{ij} is associated with each arc $(i, j) \in A$ and represents the *travel cost* spent to go from vertex i to vertex j. Generally, the use of the loop arcs, (i, i), is not allowed and this is imposed by defining $c_{ii} = +\infty$ for all $i \in V$. If the cost matrix is asymmetric, A is a set of directed arcs and the corresponding problem is called *asymmetric* CVRP (ACVRP). Otherwise, i.e., when $c_{ij} = c_{ji}$ for all $i, j \in V$, the problem is called *symmetric* CVRP (SCVRP) and the arc set A is often replaced by a set of undirected edges, E. In the following, we denote the undirected edge set of graph G by A when edges are indicated by means of their endpoints $(i,j), i,j \in V$, and by E when edges are indicated through a single index e. Given a vertex set $S \subset V$, let $\delta(S)$ and $\sigma(S)$ denote the set of edges $e \in E$ (or arcs $(i,j) \in A$) which have only one or both endpoints in S, respectively. As usual, when single vertices $i \in V$ are considered, we write $\delta(i)$ rather than $\delta(\{i\})$.

Graph G is generally assumed to be *complete* (i.e., it includes the arcs connecting all the vertex pairs, possibly with the exception of loops) since this simplifies the notation. If this is not the case, a complete graph may be easily obtained by assigning an infinite cost value to nonexisting arcs.

In several practical situations the cost matrix satisfies the *triangle inequality*, $c_{ik} + c_{kj} \geqslant c_{ij}$ for all $i,j,k \in V$. In this case, it is not convenient to deviate from the direct link between two vertices i and j. The respect of the triangle inequality is sometimes required by the algorithms for CVRP. In such case, if the original instance does not satisfy the triangle inequality, and equivalent instance may be obtained in an immediate way by adding a suitably large positive quantity M to the cost of each arc. However, the drastic distortion of the metric induced by this operation may produce very bad solutions with respect to the original costs, mainly for what concerns the effectiveness of heuristic algorithms. If G is strongly connected but not complete, it is possible to obtain a complete graph where the cost of each arc (i,j) is defined as the cost of the shortest path from i to j, computed on the original graph. Note that in this case the complete graph satisfies the triangle inequality, therefore this may be seen also as a method for "triangularizing" complete graphs. Moreover, in some instances the vertices are associated with points of the plane with given coordinates and the cost c_{ij}, for all the arcs $(i,j) \in A$, is defined as the Euclidean distance between the two points corresponding to vertices i and j. In this case the cost matrix is symmetric and satisfies the triangle inequality, and the resulting problem is often called *Euclidean* CVRP. Observe that the frequently performed rounding to the nearest integer of the real-valued Euclidean arc costs may cause a violation of the triangular inequality, whereas this does not happen if the costs are rounded up.

A set of K identical vehicles, each with capacity C, is available at the depot. Each vehicle may perform at most one route, and we assume that K is not smaller than $K_{\min}$, where $K_{\min}$ is the minimum number of vehicles needed to serve all the customers. The value of $K_{\min}$ may be determined by solving the *bin packing problem* (BPP) associated with the CVRP, calling for the determination of the minimum number of bins, each with capacity C, required to load all the n items, each with nonnegative weight $d_j, j = 1,\ldots,n$. In spite of the fact that BPP is NP-hard in the strong sense, instances with hundreds of items can be optimally solved very effectively (see, e.g., Martello and Toth [38]). In the following, given a set $S \subseteq V \setminus \{0\}$, we denote by $\gamma(S)$ the minimum number of vehicles needed to serve all the customers in S, i.e., the optimal solution value of the BPP with item set S. Note that $\gamma(V \setminus \{0\}) = K_{\min}$. Often $\gamma(S)$ is replaced by the so-called *continuous* lower bound for BPP: $\lceil d(S)/C \rceil$. Moreover, to ensure feasibility we assume that $d_j \leqslant C$ for each $j = 1,\ldots,n$.

The CVRP consists of finding a collection of K simple *circuits* (corresponding to vehicle routes) with minimum cost, defined as the sum of the costs of the arcs belonging to the circuits, and such that:

(i) each circuit visits vertex 0, i.e., the depot vertex;
(ii) each vertex $j \in V \setminus \{0\}$ is visited by exactly one circuit;
(iii) the sum of the demand of the vertices visited by a circuit does not exceed the vehicle capacity, C.

Several variants of the basic versions of CVRP have been considered in the literature. First of all, when the number K of available vehicles is greater than K_{min}, it may be possible to leave some vehicle unused, thus requiring to determine *at most K* circuits. In this case, fixed costs are often associated with the use of the vehicles. This may be included in the CVRP by adding the constant value representing the fixed cost associated with the use of a vehicle, to the cost of the arcs leaving the depot.

In practical situations the additional objective requiring the minimization of the number of used circuits (i.e., vehicles) is frequently present. Normally, the algorithms proposed in the literature do not consider this objective explicitly, however, depending on the characteristics of the algorithm used, there are different ways to take it into account. When the algorithm allows for the determination of solutions using a number of circuits smaller than K, this objective may be easily included by adding a large constant value to the cost of the arcs leaving the depot. Thus, the optimal solution first minimizes the number of arcs leaving the depot (hence the number of circuits), then the cost of the other used arcs. If, as normally happens, the algorithm determines only solutions using all the K available vehicles, there are two possibilities. The first one is to compute K_{min} by solving the BPP associated with CVRP, and then to apply the algorithm with $K = K_{min}$. The second possibility is to define an extended instance with a complete graph $\bar{G} = (\bar{V}, \bar{A})$ obtained from G by adding $K - K_{min}$ dummy vertices to V, each with demand $d_j = 0$. Let $W = \{n + 1, \ldots, n + K - K_{min}\}$ be the set of these dummy vertices, the cost $\bar{c}_{ij}$ of the arcs $(i, j) \in \bar{A}$ is defined as

$$
\bar{c}_{ij} := \begin{cases}
c_{ij} & \text{for } i, j \in V; \\
0 & \text{for } i = 0, \ j \in W; \\
0 & \text{for } i \in W, \ j = 0; \\
c_{0j} & \text{for } i \in W, \ j \in V \setminus \{0\}; \\
M & \text{for } i \in V \setminus \{0\}, \ j \in W; \\
M & \text{for } i \in W, \ j \in W;
\end{cases} \tag{1}
$$

where M is a very large positive number. The optimal solution of the CVRP computed on the extended instance may contain "empty" routes made up by single dummy vertices. Note that by adding a large constant to $\bar{c}_{0j}$, $j \in W$, the number of empty routes is maximized, i.e., the number of used vehicles is minimized.

Note that, even in the case for which the triangle inequality holds, the minimization of the number of used circuits does not correspond, in general, to the minimization of the total cost of the circuits. On the other hand, solutions forced to use exactly K circuits (with $K > K_{min}$) do not lead, in general, to the minimum total cost.

The CVRP is known to be NP-hard (in the strong sense), and generalizes the well-known Traveling Salesman Problem, arising when $C \geqslant d(V)$ and $K = K_{min} = 1$.

Therefore, all the relaxations proposed for the TSP are valid for the CVRP. As already mentioned, the CVRP is also related to the bin packing problem.

3. The asymmetric CVRP

In this section we examine the CVRP with asymmetric cost matrix (ACVRP). Two different basic modeling approaches have been proposed for the VRP in the literature. The models of the first type, known as *vehicle flow formulations*, use integer variables, associated with each arc or edge of the graph, which count the number of times that the arc or edge is traversed by a vehicle. These are the most frequently used models for the basic versions of VRP. The linear programming relaxation of vehicle flow models can be very weak when the capacity constraints are tight.

The models of the second type have an exponential number of binary variables, each associated with a different feasible circuit. The VRP is then formulated as a *set partitioning problem* (SPP) calling for the determination of a collection of circuits with minimum cost, which serves each customer once. The corresponding linear programming relaxation is typically much tighter than in the previous models. Note, however, that these models generally require dealing with a very large number of variables. An example of SPP-based model for SCVRP is given in Section 4.4.

The integer linear programming model we describe for ACVRP is a two-index vehicle flow formulation which uses $O(n^2)$ binary variables x, to indicate if a vehicle traverses or not an arc in the optimal solution. In other words, variable x_{ij} takes value 1 if arc $(i,j) \in A$ belongs to the optimal solution, and value 0 otherwise.

$$(\text{VRP1}) \quad \min \sum_{i \in V} \sum_{j \in V} c_{ij} x_{ij} \tag{2}$$

$$\text{s.t.} \quad \sum_{i \in V} x_{ij} = 1 \quad \text{for all } j \in V \setminus \{0\} \tag{3}$$

$$\sum_{j \in V} x_{ij} = 1 \quad \text{for all } i \in V \setminus \{0\} \tag{4}$$

$$\sum_{i \in V} x_{i0} = K \tag{5}$$

$$\sum_{j \in V} x_{0j} = K \tag{6}$$

$$\sum_{i \notin S} \sum_{j \in S} x_{ij} \geqslant \gamma(S) \quad \text{for all } S \subseteq V \setminus \{0\}, \ S \neq \emptyset \tag{7}$$

$$x_{ij} \in \{0, 1\} \quad \text{for all } i, j \in V. \tag{8}$$

The *indegree* and *outdegree* constraints (3) and (4) impose that exactly one arc enters and leaves each vertex associated with a customer, respectively. Analogously, constraints (5) and (6) impose the degree requirements for the depot vertex. Note that one arbitrary constraint among the $2|V|$ constraints (3)–(6) is actually implied by the remaining $2|V|-1$ ones, hence it can be removed. The so-called *capacity-cut* constraints (7) impose both the connectivity of the solution and the vehicle capacity requirements. In fact, they stipulate that each cut $(V \setminus S, S)$ defined by a vertex set S is crossed by a number of arcs not smaller than $\gamma(S)$ (minimum number of vehicles needed to serve set S). The capacity-cut constraints remain valid also if $\gamma(S)$ is replaced by the continuous lower bound for BPP (see, e.g., [12]).

Observe that, when $|S|=1$ or $S=V\setminus\{0\}$ the capacity-cut constraints (7) are weakened versions of the corresponding degree constraints (3)–(6). Note also that, because of the degree constraints (3)–(6), we have

$$\sum_{i\notin S}\sum_{j\in S} x_{ij} = \sum_{i\in S}\sum_{j\notin S} x_{ij} \quad \text{for all } S \subseteq V \setminus \{0\}, \ S \neq \emptyset, \tag{9}$$

in other words, each cut $(V \setminus S, S)$ is crossed in both directions the same number of times. From (9) we may also re-state (7) as

$$\sum_{i\notin S}\sum_{j\in S} x_{ij} \geqslant \gamma(V \setminus S) \quad \text{for all } S \subset V, \ \{0\} \in S. \tag{10}$$

An alternative formulation may be obtained by transforming the capacity-cut constraints (7), by means of the degree constraints (3)–(6), into the well-known *generalized subtour elimination* constraints (GSEC):

$$\sum_{i\in S}\sum_{j\in S} x_{ij} \leqslant |S| - \gamma(S) \quad \text{for all } S \subseteq V \setminus \{0\}, \ S \neq \emptyset, \tag{11}$$

which impose that at least $\gamma(S)$ arcs leave each vertex set S.

Both families of constraints (7) and (11) have a cardinality growing exponentially with n. A possible way to partially overcome this drawback is to consider only a limited subset of these constraints. This can be done by relaxing them in a Lagrangian fashion as done in [20] and in [40] (see Section 4.3) or by explicitly including them in the linear programming relaxation as done in branch and cut approaches.

Alternatively, an equivalent family of constraints with polynomial cardinality may be obtained by considering the subtour elimination constraints proposed for the TSP by Miller et al. [39], and extending them to CVRP (see, e.g., [10] and [28]):

$$u_i - u_j + Cx_{ij} \leqslant C - d_j \quad \text{for all } i,j \in V \setminus \{0\}, \ i \neq j, \ \text{s.t. } d_i + d_j \leqslant C, \tag{12}$$

$$d_i \leqslant u_i \leqslant C \quad \text{for all } i \in V \setminus \{0\}, \tag{13}$$

where u_i, $i \in V \setminus \{0\}$, is an additional continuous variable representing the load of the vehicle after visiting customer i. It is easy to see that constraints (12)–(13) impose the capacity requirements of CVRP. In fact, when $x_{ij} = 0$ the constraint is not binding since $u_i \leqslant C$ and $u_j \geqslant d_j$, whereas when $x_{ij} = 1$ they impose that $u_j \geqslant u_i + d_j$. These constraints may be strengthened by lifting some coefficients as illustrated by Desrochers and Laporte [13].

Two exact algorithms, both based on the branch and bound approach were proposed for ACVRP so far. The algorithm described by Laporte et al. [31] uses a lower bound based on the *Assignment Problem* (AP) relaxation of ACVRP. The algorithm proposed by Fischetti et al. [18] combines, according to the so-called *additive approach*, the AP lower bound with a lower bound based on disjunction and one based on a min-cost flow relaxation. These bounds are briefly described in this section.

Other bounds for the ACVRP may be derived by generalizing the methods proposed for the symmetric case. For example, Fisher [20] proposed a way to extend to ACVRP the bounds based on K-tree he derived for the SCVRP (described in Sections 4.1 and 4.3). In this extension the Lagrangian problem calls for the determination of an undirected K-tree on the undirected graph obtained by replacing each pair of directed arcs (i,j) and (j,i) with a single edge (i,j) with cost $c'_{ij} = \min\{c_{ij}, c_{ji}\}$. No computational testing for this bound was presented in Fisher [20]. Possibly better bounds may be obtained by explicitly considering the asymmetry of the problem, i.e., by using K-arborescences rather than K-trees and by strengthening the bound in a Lagrangian fashion as proposed by Fisher for the CVRP (see [43,44] for an application to the capacitated shortest spanning arborescence problem, and to the VRP with backhauls, respectively).

3.1. The assignment lower bound

Carpaneto and Toth [6], and Laporte et al. [31] proposed to relax model VRP1 by dropping the capacity-cut constraints (7). The resulting relaxation, i.e. (2)–(6) and (8), is a *transportation problem* (TP), calling for a min-cost collection of circuits of G visiting once all the vertices in $V \setminus \{0\}$, and K times vertex 0. This solution can be infeasible for ACVRP since:

(i) the total customer demand on a circuit can exceed the vehicle capacity;
(ii) there may exist circuits not visiting vertex 0.

The solution of TP requires $O(n^3)$ time through a transportation algorithm. In practice, it is more effective to transform the problem into an *assignment problem* (AP) defined on the extended complete digraph $G' = (V', A')$, where $V' := V \cup W'$ and $W' = \{n + 1, \ldots, n + K - 1\}$ contains $K - 1$ additional copies of vertex 0, and the cost c'_{ij} of each arc in A' is defined as follows:

$$c'_{ij} := \begin{cases} c_{ij} & \text{for } i, j \in V \setminus \{0\}; \\ c_{i0} & \text{for } i \in V \setminus \{0\}, \ j \in W'; \\ c_{0j} & \text{for } i \in W', \ j \in V \setminus \{0\}; \\ \lambda & \text{for } i, j \in W'; \end{cases} \tag{14}$$

where $\lambda = M \gg 1$. After this transformation, constraint (5) may be replaced by K constraints of type (3), one for each copy of the depot. Analogously, constraint (6) may be replaced by K constraints of type (4). This extension was originally proposed by Lenstra and Rinnooy Kan [35], to transform into an ordinary TSP the m-TSP, which calls for the determination of a collection of m circuits visiting m times a distinguished vertex (i.e., the depot) and once all the remaining vertices. Observe that by defining λ in a different way we obtain an alternative transformation, with respect to that presented in Section 2, to obtain solutions using less than K vehicles. In particular, defining $\lambda = 0$ leads to the determination of the min-cost set of *at most* K routes, whereas defining $\lambda = -M$ leads to the determination of the min-cost set of $K_{\min}$ routes.

3.2. The bounds based on arborescences

In analogy with what is done for the SCVRP (see Section 4.1) another basic relaxation is that based on the solution of degree-constrained spanning arborescences. This relaxation may be obtained from model VRP1 by:

(i) removing the outdegree constraints (4) for all the customer vertices;
(ii) weakening the capacity-cut constraints (7) so as to impose only the connectivity of the solution, i.e. by replacing the right-hand side with 1.

The resulting relaxed problem, called K-shortest spanning arborescence problem (KSSA) is defined by

$$(\text{KSSA}) \quad \min \sum_{i \in V} \sum_{j \in V} c_{ij} x_{ij} \tag{15}$$

$$\text{s.t.} \quad \sum_{i \in V} x_{ij} = 1 \quad \text{for all } j \in V \setminus \{0\}, \tag{16}$$

$$\sum_{i \in V} x_{i0} = K, \tag{17}$$

$$\sum_{j \in V} x_{0j} = K, \tag{18}$$

$$\sum_{i \notin S} \sum_{j \in S} x_{ij} \geqslant 1 \quad \text{for all } S \subseteq V \setminus \{0\}, \ S \neq \emptyset, \tag{19}$$

$$x_{ij} \in \{0, 1\} \quad \text{for all } i, j \in V. \tag{20}$$

The KSSA can be effectively solved by considering two separate subproblems:

(i) the determination of a min-cost spanning arborescence with outdegree K at the depot vertex, defined by (15), (16), (18)–(20), with variables x_{ij} for $i \in V$, $j \in V \setminus \{0\}$, and
(ii) the determination of a set of K min-cost arcs entering the depot, defined by (15), (17), and (20), with variables x_{0i} for $i \in V$.

The KSSA can be determined in $O(n^2)$ since the first subproblem can be solved in $O(n^2)$ time (see, [23,43]), while the second subproblem clearly requires $O(n)$ time.

The above described lower bound was never used within branch and bound algorithms and the preliminary computational results discussed in Section 3.5 show that its quality is generally poor and inferior to that of the AP lower bound. However, it should be mentioned that for a problem closely related to the CVRP, as the VRP with backhauls, Toth and Vigo [44] successfully used a Lagrangian relaxation based on the solution of KSSAs, solving problems with up to 100 customers.

3.3. The disjunctive lower bound

The following two bounds were proposed by Fischetti et al. [18]. The first bound is based on a disjunction on infeasible arc subsets, whereas the second bound is based on a min-cost flow relaxation.

A given arc subset $B \subset A$ is called *infeasible* if no feasible solution to ACVRP can use all its arcs, i.e., when

$$\sum_{(a,b)\in B} x_{ab} \leqslant |B| - 1 \tag{21}$$

is a valid inequality for ACVRP. For any given (minimal) infeasible arc subset $B \subset A$, the following logical disjunction holds for each $x \in F$, where F is the set of all the ACVRP feasible solutions:

$$\bigvee_{(a,b)\in B} (x \in Q^{ab} := \{x \in \mathfrak{R}^A : x_{ab} = 0\}). \tag{22}$$

Then $|B|$ restricted problems are defined, each denoted as RP^{ab} and including the additional condition $x_{ab} = 0$ imposed for a different $(a, b) \in B$. For each RP^{ab}, a valid lower bound, ϑ^{ab}, is computed through the AP relaxation of the previous section (with $c_{ab} := +\infty$ to impose $x_{ab} = 0$). The disjunctive bound

$$L_D := \min\{\vartheta^{ab} : (a, b) \in B\}, \tag{23}$$

clearly dominates the AP lower bound, L_{AP}, since $\vartheta^{ab} \geqslant L_{AP}$ for all $(a, b) \in B$.

A possible way to determine infeasible arc subsets B is the following. First solve the AP relaxation with no additional constraints, and store the corresponding optimal solution $(x_{ij}^* : i, j \in V)$. If x^* is feasible for ACVRP, then clearly the lower bound L_{AP} cannot be improved. Otherwise, try to improve it by using a disjunction on a suitable infeasible arc subset B. Note that imposing $x_{ab} = 0$ for any $(a, b) \in A$ such that $x_{ab}^* = 0$ would produce $\vartheta^{ab} = L_{AP}$, hence a disjunctive bound $L_D = L_{AP}$. Therefore, B is chosen as a subset of $A^* := \{(i, j) \in A : x_{ij}^* = 1\}$, if any, corresponding to one of the following cases:

 (i) a circuit which is disconnected from the depot vertex,
 (ii) a path such that the total demand of the associated customer vertices exceeds C,
 (iii) a feasible circuit which leaves uncovered a set of customers, S, whose total demand cannot be served by the remaining $K-1$ vehicles, i.e., such that $\gamma(S) > K-1$.

Different choices of the infeasible arc subset B lead to different lower bounds. Therefore, Fischetti, Toth and Vigo [18] used an overall bounding procedure, called ADD_DISJ, based on the *additive approach* which considers, in sequence, different infeasible arc subsets so as to produce a possibly better overall lower bound.

The additive approach was proposed by Fischetti and Toth [16] and allows for the combination of different lower bounding procedures, each exploiting different substructures of the considered problem. When applied to a minimization problem, each procedure returns a lower bound ρ and a *residual cost matrix*, $\tilde{c}$, such that:

$$\tilde{c} \geqslant 0$$

$$\rho + \tilde{c}x \leqslant cx \quad \text{for all } x \in F.$$

The entries of $\tilde{c}$ represent lower bounds on the increment of the optimal solution value if the corresponding arc is imposed in the solution. The different bounding procedures are applied in sequence, and each of them uses as input costs the residual cost matrix given as output by the previous procedure (obviously, the first procedure starts with the original cost matrix). The overall additive lower bound is given by the sum of the lower bounds obtained by each procedure. It can be easily shown that if the lower bounding procedures are based on linear programming relaxations, as those previously described for ACVRP, the reduced costs are valid residual costs. For further details see [18,17].

Procedure ADD_DISJ starts by solving the AP relaxation with no additional constraints, and defines the initial lower bound LB as the optimal AP solution value, and the arc set A^* as the arcs used in the optimal AP solution. Then iteratively an infeasible subset B, if any, is chosen from A^* and used for the computation of the disjunctive lower bound, returning a lower bound L_D and the corresponding residual cost matrix. The current LB is increased by L_D, the set A^* is updated by removing from it all the arcs whose corresponding variables are not equal to 1 in the current optimal solution of the disjunctive bound. The process is iterated until A^* does not contain further infeasible arc subsets. Procedure ADD_DISJ can be implemented, through parametric techniques, so as to have an overall time complexity equal to $O(n^4)$.

3.4. The lower bound based on min-cost flow

Let $\{S_1, \ldots, S_m\}$ be a given partition of V with $0 \in S_1$, and define

$$A_1 := \bigcup_{h=1}^{m} \{(i,j) \in A : i, j \in S_h\}$$

$$A_2 := A \setminus A_1.$$

In other words, A is partitioned into $\{A_1, A_2\}$, where A_1 contains the arcs "internal" to the subsets S_h, and A_2 those connecting vertices belonging to different S_h's.

In the following, a lower bound L_P based on projection is described. The bound is given by $L_P := \vartheta_1 + \vartheta_2$, where $\vartheta_t, t = 1, 2$, is a lower bound on $\sum (c_{ij} : (i,j) \in A^* \cap A_t)$ for every (optimal) ACVRP solution $A^* \subset A$.

The contribution to L_P of the arcs internal to the given subsets S_h is initially neglected, i.e., ϑ_1 is set equal to 0. The rationale of this choice is clarified later. As to ϑ_2, this is computed by solving the following linear programming relaxation, called R1, obtained from model VRP1 by
  (i) weakening degree equations (3)–(6) into inequalities, to take into account the removal of the arcs in A_1;

(ii) imposing the capacity-cut constraints (7) and (10) only for the m subsets S_h's .
The model of R1 is

$$(\text{R1}) \quad \vartheta_2 = \min \sum_{(i,j) \in A_2} c_{ij} x_{ij} \tag{24}$$

$$\text{s.t.} \quad \sum_{i \in V : (i,j) \in A_2} x_{ij} \leqslant \begin{cases} 1 & \text{for all } j \in V \setminus \{0\}; \\ K & \text{for } j = 0; \end{cases} \tag{25}$$

$$\sum_{j \in V : (i,j) \in A_2} x_{ij} \leqslant \begin{cases} 1 & \text{for all } i \in V \setminus \{0\}; \\ K & \text{for } i = 0; \end{cases} \tag{26}$$

$$\sum_{i \notin S_h} \sum_{j \in S_h} x_{ij} = \sum_{i \in S_h} \sum_{j \notin S_h} x_{ij} \geqslant \begin{cases} \gamma(V \setminus S_h) & \text{for } h = 1; \\ \gamma(S_h) & \text{for } h = 2, \ldots, m; \end{cases} \tag{27}$$

$$x_{ij} \in \{0, 1\} \quad \text{for all } (i,j) \in A_2. \tag{28}$$

This model can be solved efficiently, since it can be viewed as an instance of a min-cost flow problem on an auxiliary layered network. The network contains $2(n + m + 2)$ vertices, namely:

- two vertices, say i^+ and i^-, for all $i \in V$;
- two vertices, say a_h and b_h, for all $h = 1, \ldots, m$;
- a source vertex, s, and a sink vertex, t.

The arcs in the network, and the associated capacities and costs, are:

- for all $(i,j) \in A_2$: arc (i^+, j^-) with cost c_{ij} and capacity 1;
- for all $h = 1, \ldots, m$: arcs (a_h, i^+) and (i^-, b_h) for all $i \in S_h$, with cost 0 and capacity 1 (if $i \neq 0$) or K (if $i = 0$);
- for all $h = 1, \ldots, m$: arc (a_h, b_h) with cost 0 and capacity $|S_h| - \gamma(S_h)$ (if $h \neq 1$) or $|S_1| + K - 1 - \gamma(V \setminus S_1)$ (if $h = 1$);
- for all $h = 1, \ldots, m$: arcs (s, a_h) and (b_h, t), both with cost 0 and capacity $|S_h|$ (if $h \neq 1$) or $|S_1| + K - 1$ (if $h = 1$).

It can be easily seen that finding the min-cost s-t flow of value $n + K$ on this network actually solves relaxation R1. The worst-case time complexity for the computation of ϑ_2, and of the corresponding residual costs, is $O(n^3)$ by using a specialized algorithm based on successive shortest path computations.

Different choices of the vertex partition $\{S_1, \ldots, S_m\}$ lead to different lower bounds. Note that choosing $S_h = \{h\}$ for all $h \in V$, produces a relaxation R1 that coincides with the AP relaxation of Section 3.1. When, on the other hand, non-singleton S_h's are present, relaxation R1 is capable of taking into account the associated capacity-cut constraints (that are, instead, neglected by AP), while loosing a possible contribution to the lower bound of the arcs inside S_h (which belong to A_1), and weakening the degree constraints of the vertices in S_h. Fischetti et al. [18] used, in sequence, different partitions obtaining an overall additive procedure, called ADD_FLOW.

The procedure is initialized with the partition $S_h = \{h\}$ for all $h = 1, \ldots, m = n$ (i.e., with the AP relaxation). At each iteration of the additive scheme, relaxation R1 is solved, the current lower bound is increased, and the current costs are reduced accordingly. Then a convenient collection of subsets $S_{h_1}, \ldots, S_{h_r}$ (with $r \geqslant 2$) belonging to the current partition is selected and the subsets are replaced with their union, say S^*. The choice of this collection is made so as to produce an *infeasible* set S^*, i.e., a vertex set whose associated capacity-cut constraint is violated by the solution of the current relaxation R1. This hopefully produces an increase of the additive lower bound in the next iteration. The additive scheme ends when either $m = 1$, or no infeasible S^* is detected.

Procedure ADD_FLOW takes $O(n^4)$ time and the resulting additive lower bound clearly dominates the AP bound, which is used to initialize it. On the other hand no dominance relation exists between ADD_FLOW and procedure ADD_DISJ of the previous section. Therefore, Fischetti, Toth and Vigo proposed to apply ADD_DISJ and ADD_FLOW in sequence, again in an additive fashion. To reduce the average overall computing time, procedure ADD_FLOW was stopped when no increase of the current lower bound LB was observed for five consecutive iterations.

3.5. Branch and bound algorithms for the ACVRP

We now briefly describe the main ingredients of the branch and bound algorithms used for the exact solution of the ACVRP proposed by Laporte et al. [31] and by Fischetti et al. [18]. The two algorithms have the same basic structure, derived from that of the algorithm for the asymmetric TSP described in Carpaneto and Toth [6] and based on that proposed in Bellmore and Malone [3]: the first one uses as lower bound the AP relaxation of Section 3.1, whereas the second uses the two additive bounding procedures described in Sections 3.3 and 3.4.

The algorithms adopt a *best-bound-first* search strategy, i.e., branching is always executed on the pending node of the branch-decision tree with the smallest lower bound value. This rule allows for the minimization of the number of subproblems solved at the expense of larger memory usage, and computationally proved to be more effective than the *depth-first* strategy, where the branching node is selected according to a last-in-first-out rule.

The branching rules used by both algorithms are related to the *subtour elimination* scheme used for the asymmetric TSP, and handle the relaxed constraints imposing the connectivity and the capacity requirements of the feasible ACVRP solutions. At a node v of the branch-decision tree, let I_v and F_v contains the arcs imposed and forbidden in the current solution, respectively.

Given the set A^* of arcs corresponding to the optimal solution of the current relaxation, a non-imposed arc subset $B := \{(a_1, b_1), (a_2, b_2), \ldots, (a_h, b_h)\} \subset A^*$ on which to branch is chosen.

Fischetti et al. defined B by considering the subset of A^* with the minimum number of non-imposed arcs, defining a path or a circuit which is infeasible according to the conditions of Section 3.3. Note that since the additive bounding procedure alters the objective function of the problem, an optimal solution of the relaxed problem which is feasible for ACVRP is not necessarily optimal for it. Therefore, if A^* defines a feasible ACVRP solution, set B is chosen as the feasible circuit through vertex 0 with the minimum number of non-imposed arcs. Then $h = |B|$ descendant nodes are generated. The subproblem associated with node v_i, $i = 1, \ldots, h$ is defined by excluding the ith arc of B and by imposing the arcs up to $i - 1$:

$$I_{v_i} := I_v \cup \{(a_1, b_1), \ldots, (a_{i-1}, b_{i-1})\},$$

$$F_{v_i} := F_v \cup \{(a_i, b_i)\},$$

where $I_{v_1} := I_v$.

Laporte et al. defined B as an infeasible subtour according to the conditions of Section 3.1, and used a more complex branching rule in which at each descendant node at most r arcs of B are simultaneously excluded, where $r := \lceil d(S)/C \rceil$ and S is the set of vertices spanned by B. In this case, since at most $\binom{|B|}{r}$ descendant nodes may be generated, the set B is chosen as the one minimizing $\binom{|B|}{r}$.

The performance of the branch and bound algorithms is enhanced by means of several additional procedures performing variable fixing, feasibility checks and dominance tests. The Fischetti et al. algorithm (FTV) at each node of the branch-decision tree uses a heuristic algorithm proposed by Vigo [47] which starts from the infeasible solution associated with the current relaxation and tries to obtain a feasible solution through an insertion procedure and a post-optimization phase based on arc exchanges.

Laporte et al. used their algorithm (LMN) to solve, on a VAX 11/780 computer (0.14 Mflops), test instances where demands d_j and costs c_{ij} were randomly generated from a uniform distribution in [0, 100], and rounded to the nearest integer. The vehicle capacity was defined as

$$C := (1 - \alpha) \max_{j \in V} \{d_j\} + \alpha d(V),$$

where α is a real parameter chosen in [0,1]. The number of available vehicles was defined as $K = K_{\min}$, and computed by using the trivial BPP lower bound. Note that larger values of α produce larger C, and hence smaller K (when $\alpha = 1$, ACVRP reduces to the asymmetric TSP, since $K = 1$). No monotone correlation between α and the *average percentage load* of a vehicle, defined as $100\, d(V)/(KC)$, can instead be inferred. Laporte et al. considered $\alpha = 0.25$, 0.50 and 0.75, producing $K = 4, 2$ and 2, respectively.

For each pair (n, α), five instances were generated and algorithm LMN was run by imposing a limit on the total available memory. The LMN algorithm was able to solve instances with up to 90 vertices if $\alpha \geq 0.50$ (i.e., with $K \leq 2$), although for $n \geq 70$ and $\alpha < 1$, only half or less of the instances were actually solved. With $\alpha = 0.25$ only the instances with 10 vertices and one of those with 20 vertices were solved. The computing times for the most difficult instances solved were almost 6000 s, whereas no statistics were reported for the non-solved instances. The algorithm was also tested on instances of the same type but with $K = K_{\min} + 2$ or $K = K_{\min} + 4$. These problems resulted to be much easier than the previous ones: algorithm LMN was able to solve instances with up to 260 vertices. Finally, randomly generated Euclidean instances were considered and, as expected, algorithm LMN obtained poor results, being able to solve only some of the problems with two vehicles and up to 30 vertices.

Fischetti et al. tested their algorithm FTV on the same randomly generated instances used for LMN with $K = K_{\min}$. Algorithm FTV was able to solve all the instances with

Table 1
Percentage ratios of different ACVRP lower bounds with respect to the optimal solution value on real-world instances

Problem	n	K	(%) AP	(%) KSSA	(%) ADD
A034-02v	33	2	85.8	78.7	90.1
A036-03v	35	3	90.9	75.2	93.2
A039-03v	38	3	93.8	77.6	96.7
A045-03v	44	3	93.4	75.6	95.7
A048-03v	47	3	93.6	79.0	97.2
A056-03v	55	3	88.5	75.4	94.3
A065-03v	64	3	92.6	75.6	95.5
A071-03v	70	3	91.7	79.3	94.6
			91.3	77.1	94.6

up to 300 vertices and up to 4 vehicles, within 1000 CPU seconds on a DECstation 5000/240 (5.3 Mflops). For $n = 90$, LMN solved one instance (out of 5) with $\alpha = 0.50$ and two instances with $\alpha = 0.75$, requiring average CPU times of 5787 and 1162 s, respectively. For the same values of n and α, FTV solved all the instances within CPU times of 15 and 1 s, respectively. On these instances the additive lower bound considerably improved the AP value.

Algorithm FTV was also tested on a class of more realistic instances where the cost matrices were obtained from those of the previous class by "triangularizing" the costs, i.e., by replacing each c_{ij} with the cost of the shortest path from i to j. The number of vehicles K and the average percentage vehicle load, say r, were fixed and the vehicle capacity was defined as $C := \lceil 100d(V)/(rK) \rceil$. Instances of this type with up to 300 vertices, 8 vehicles and with r equal to 80 and 90 were solved, those with $n \geqslant 150$ being easier than the smaller ones. Algorithm FTV was finally used to solve eight real-world instances with up to 70 vertices and 3 vehicles, coming from pharmaceutical and herbalists' product delivery in the center of an urban area with several one-way restrictions imposed on the roads. These instances resulted to be more difficult than the randomly generated ones: the computing time and the number of nodes were higher than those required for analogous random instances. The maximum CPU time required by FTV to solve the instances was about 30 minutes. Table 1 report the percentage ratios of the different lower bounds described in Sections 3.1–3.4, with respect to the optimal solution value, when applied to these real-world instances. In particular the table contains the ratios corresponding to AP, KSSA, and the overall additive bound (ADD). The average gap, over the eight instances, of the additive bound with respect to the optimal solution value was about 5.4% (that of AP being 8.7% and that of KSSA 22.9%) whereas on random instances the gap was normally much smaller (1–2% for the additive bound and 2–5% for the AP).

We recently applied algorithm FTV to some Euclidean SCVRP instances from the literature. The results we obtained show that SCVRP instances with up to 25–30 vertices may be consistently solved by this algorithm (see Table 3). Moreover, the largest instance we solved includes 47 customers.

4. The symmetric CVRP

In this section we examine the branch and bound algorithms for the symmetric version of CVRP proposed by Fisher [20] and Miller [40]. We first give a general model for SCVRP and describe the basic relaxations based on spanning trees and on b-matching. The strengthening of these basic relaxations in a Lagrangian fashion is then discussed and the overall branch and bound algorithms are described. The exact algorithm proposed by Hadjconstantinou, Christofides and Mingozzi [26], will be also briefly presented.

In the following we assume that single-customer routes are allowed.

The model we consider is obtained, as proposed in Laporte et al. [33], by adapting to SCVRP the two-index vehicle flow formulation VRP1 of ACVRP. To this end it should be noted that in SCVRP the routes are not oriented (i.e., the customers along a route may be visited indifferently clockwise or counter-clockwise). Therefore, it is not necessary to know in which direction edges are covered by the vehicles, and for each undirected edge $e \in E$ one integer variable x_e is used to indicate how many times the edge is covered in the optimal solution. In particular, if $e \notin \delta(0)$ then $x_e \in \{0, 1\}$, whereas if $e \in \delta(0)$ then $x_e \in \{0, 1, 2\}$. The case $x_e = 2$ indicates that the endpoint customer of edge e, say j, is contained into the single-customer route $0 \to j \to 0$. The model reads:

$$(\text{VRP2}) \quad \min \sum_{e \in E} c_e x_e \tag{29}$$

$$\text{s.t.} \quad \sum_{e \in \delta(i)} x_e = 2 \quad \text{for all } i \in V \setminus \{0\}, \tag{30}$$

$$\sum_{e \in \delta(0)} x_e = 2K, \tag{31}$$

$$\sum_{e \in \delta(S)} x_e \geqslant 2\gamma(S) \quad \text{for all } S \subseteq V \setminus \{0\}, S \neq \emptyset, \tag{32}$$

$$x_e \in \{0, 1, 2\} \quad \text{for all } e \in \delta(0), \tag{33}$$

$$x_e \in \{0, 1\} \quad \text{for all } e \notin \delta(0). \tag{34}$$

The *degree* constraints (30) and (31) impose that exactly two arcs are incident to each vertex associated with a customer, and $2K$ arcs are incident to the depot vertex, respectively. The capacity-cut constraints (32), where $\gamma(S)$ may be replaced by the trivial BPP lower bound, impose both the connectivity of the solution and the vehicle capacity requirements, by forcing that a sufficient number of arcs enter each subset of vertices. Also in this case, due to (30), these constraints may be rewritten as the generalized subtour elimination constraints (GSECs):

$$\sum_{e \in \sigma(S)} x_e \leqslant |S| - \gamma(S) \quad \text{for all } S \subseteq V \setminus \{0\}, S \neq \emptyset. \tag{35}$$

In addition, subtour elimination constraints as those proposed by Miller et al. [39] for the TSP may be easily extended to SCVRP (see also Section 3).

4.1. The lower bounds based on trees

Different relaxations based on spanning trees were presented for SCVRP by extending the well-known 1-tree relaxation proposed by Held and Karp [27] for the symmetric TSP.

Christofides et al. [11] proposed a branch and bound algorithm based on the *k-degree center tree* (*k*-DCT) relaxation of SCVRP. Given *k*, with $K \leqslant k \leqslant 2K$, the *k*-DCT is a min-cost spanning tree on *G* with degree *k* at the depot vertex. Then, *K* least cost arcs not in the tree are added, $2K - k$ of which are incident to the depot, and the remaining $k - K$ are not incident to it. The bound was tightened by using Lagrangian penalties associated with the degree constraints. The branch and bound algorithm was able to solve problems from the literature with up to 25 vertices within 244 s on a CDC 7600 (2 Mflops).

Another tree-based relaxation was presented in [20], and requires the determination of a *K-tree*, defined as a min-cost set of $n + K$ edges spanning the graph. The approach used by Fisher is based on formulation VRP2 with the additional assumption that single-customer routes are not allowed (by imposing $x_e \in \{0, 1\}$ for $e \in \delta(0)$). However, as he observed, in some cases this assumption is not constraining. In fact, customer *j* can be served alone in a route if and only if on the remaining $K - 1$ vehicles there is enough space to load the demand of the other customers, i.e., if $\gamma(V \setminus \{j\}) \leqslant K - 1$. By replacing $\gamma(\cdot)$ with the trivial BPP lower bound we may re-state the above condition as

$$d_j \geqslant C_{\min} = d(V) - (K - 1)C. \tag{36}$$

If, given a CVRP instance, condition (36) is satisfied by no $j \in V$, then in any feasible solution no customer may be served alone in a route (hence the constraint preventing it is superfluous). We checked the above condition on 65 SCVRP instances from the literature and it was satisfied, i.e., single-customer routes cannot be used, in 29 of these instances.

Fisher modeled the SCVRP as the problem of determining a *K*-tree with degree equal to $2K$ at the depot vertex, with additional constraints imposing: (i) the vehicle capacity requirements, and (ii) that the degree of each customer vertex must be equal to 2. These additional constraints are relaxed in a Lagrangian fashion, thus obtaining as Lagrangian problem the determination of a *K*-tree with degree $2K$ at the depot, which can be computed in $O(n^3)$ time (see [21]). This degree constrained *K*-tree relaxation may be easily obtained by considering formulation VRP2 and:

(i) removing the degree constraints (30);
(ii) weakening the capacity-cut constraints (32) into connectivity constraints by replacing the right-hand side with 1.

It can be easily seen that a *K*-tree solution may be infeasible for SCVRP because some vertices have degree different than two. Moreover, the demand associated with the branches leaving the depot may exceed the vehicle capacity.

Table 2
Percentage ratios of different basic SCVRP lower bounds with respect to the best known solution value of Euclidean instances

Problem	n	K	% b-matching	% K-tree[a]	% KSSA	% AP	% ADD
E045-04f	44	4	71.4	62.6	62.2	57.4	70.3
E051-05e	50	5	87.9	84.9	79.4	80.9	87.5
E072-04f	71	4	80.9	77.7	72.0	69.8	77.9
E076-10e	75	10	76.7	76.2	69.2	71.0	76.1
E101-08e	100	8	86.4	81.5	77.5	80.7	86.1
E101-10c	100	10	70.3	77.6	72.2	66.5	69.6
E135-07f	134	7	63.4	59.2	57.5	47.5	60.3
E151-12c	150	12	80.5	78.4	73.6	68.6	77.6
E200-16c	199	16	72.4	74.1	66.4	64.6	72.2
			76.7	74.7	70.0	67.4	75.5

[a] Single-customer routes not allowed.

4.2. The lower bound based on matching

The b-matching is a natural relaxation for SCVRP and is the counterpart for the symmetric version of the assignment relaxation for ACVRP described in Section 3.1. However, only recently this relaxation came on the scene, due to the work by Miller [40], after the development of efficient codes for the b-matching problem (see, e.g. [41]).

The b-matching relaxation of CVRP may be obtained by considering model VRP2 and removing the capacity-cut constraints (32). The resulting relaxed problem requires the determination of a subset of arcs covering all the vertices and such that the degree of each customer vertex is equal to two, while the degree of the depot is equal to $2K$. It can be noted that a b-matching solution may be infeasible for SCVRP since: (i) some connected components (i.e., subtours) may be disconnected from the depot, and (ii) the demand associated with a subtour may exceed the vehicle capacity. As for the AP relaxation for ACVRP, it is possible to obtain an equivalent 2-matching relaxation for SCVRP by adding $K - 1$ copies of the depot and by imposing $x_e \in \{0, 1\}$ for $e \in \delta(0)$ (see, Section 3.1 for further details, and Pekny and Miller [42] for an effective 2-matching algorithm). As described in the next section, in [40] some of the GSECs (35) are relaxed in a Lagrangian fashion, obtaining as Lagrangian problem the determination of a min-cost b-matching.

4.3. The Lagrangian lower bounds

The relaxations of SCVRP presented in the previous section have in general a poor quality. Table 2 reports the average percentage ratios of the basic lower bounds corresponding to the degree constrained K-tree and the b-matching with respect to the optimal or the best known solution value, for a set of widely used Euclidean SCVRP instances from the literature. The K-tree values are those reported in [20] who used real-valued cost matrices. The best known solution values that we used to compute the ratios are those reported in Toth and Vigo [46] which were obtained by using real-valued cost matrices. The b-matching values were computed with Cplex 6.0 ILP solver. The table also reports the ratios of the AP, KSSA and of the overall additive lower bound (ADD) of Fischetti et al. [18]. All these values have been computed by using integer cost matrices where the arc cost is defined as the real cost multiplied by 10,000 and rounded to the nearest integer. The final value is then scaled down by dividing it by 10,000. It should be recalled that the problem solved by Fisher in [20] was slightly different from what we defined as CVRP, since the single-customer routes were not allowed. As a consequence the K-tree values computed by Fisher may by slightly larger than those which could be obtained in the case in which single-customer routes are allowed.

By observing Table 2 it can be noted that none of the basic relaxations reaches a quality sufficient to solve moderate size problems. As an example, we used the Fischetti et al. code for ACVRP based on the additive bound ADD: the largest SCVRP instance which it was able to solve included 47 customers (problem E048-04y not included in the tables), and some problems with 25–30 customers were not solved to optimality.

Therefore, to obtain better bounds both Fisher [20] and Miller [40] strengthened the basic relaxations by dualizing, in a Lagrangian fashion, some of the relaxed constraints. In particular, Fisher included in the objective function the degree constraints (30) and some of the capacity-cut constraints (32), whereas Miller included some of the GSECs (35). It should be also remembered that Fisher did not allow single-customer routes. As in related problems, good values for the Lagrangian multipliers associated with the relaxed constraints are determined by using a standard subgradient optimization procedure (see, e.g., [19]).

The main difficulty associated with these Lagrangian relaxations is represented by the exponential cardinality of the set of relaxed constraints (i.e., the capacity-cuts and the GSECs) which does not allow for the explicit inclusion of all of them in the objective function. To this end, both authors proposed to include only a limited family $\mathscr{F}$ of relaxed capacity-cut or GSEC constraints and to iteratively add to the Lagrangian relaxation the constraints which are violated by the current solution of the Lagrangian problem. In particular, at each iteration of the subgradient optimization procedure, the arcs incident to the depot in the current Lagrangian solution are removed. Violated constraints (i.e., capacity-cuts or GSECs, depending on the approach), if any, are detected by examining the connected components obtained in this way. This detection routine is exact. In other words, if a constraint associated with, say, vertex set S, is violated by the current Lagrangian solution, then there is a connected component of that solution spanning all the vertices in S.

The new constraints are added to the Lagrangian problem, i.e., to $\mathscr{F}$, with an associated multiplier and the process is iterated until no violated constraint is detected (hence the Lagrangian solution is feasible) or a prefixed number of subgradient iterations has been executed. Slack constraints are periodically purged from $\mathscr{F}$.

Fisher [20] initialized $\mathscr{F}$ with an explicit set of constraints containing the customer subsets nested around $K+3$ *seed* customers. The seeds were chosen as the K customers farthest from the depot in the routes corresponding to an initial feasible solution, with the addition of the three customers maximally distant from the depot and the other

Table 3
Comparison of percentage ratios of the basic and improved lower bounds for SCVRP with respect to different test instances. Instances marked with an asterisk were solved to optimality by the corresponding branch and bound code

Problem	n	K	K-tree[a]		b-matching[b]		HCM[c]	ADD[c]
			% LB	% Lagr.	% LB	% Lagr.	% LB	% LB
S007-02a	6	2				100.0*		73.7*
S013-04d	12	4				96.8*		71.0*
E016-05m	15	5					97.6*	85.9*
E021-04m	20	4					100.0*	84.6*
E022-04g	21	4			90.1	99.7*		82.7*
E023-03g	22	3			96.5	100.0*		93.9*
E026-08m	25	8					100.0*	77.4
E030-03g	29	3			71.7	95.3*		—
S031-07w	30	7				96.0*		—
E031-09h	30	9					97.9*	72.8
E033-03n	32	3			86.5	98.9*		—
E036-11h	35	11					99.5*	77.1
E041-14h	40	14					98.9*	73.0
E045-04f	44	4	62.6	99.6*				70.3
E051-05e	50	5	84.9	96.7	92.9	96.9*	98.5*	87.5
E072-04f	71	4	77.7	98.3*				77.9
E076-10e	75	10	76.2	90.5			97.6	76.1
E101-08e	100	8	81.5	95.1			95.9	86.1
E101-10c	100	10	77.6	99.8*				69.6
E135-07f	134	7	59.2	97.4				60.3
E151-12c	150	12	78.4	90.7			97.2	77.6
E200-16b	199	16	74.1	84.7				72.2

[a]Real-valued costs and single-customer routes not allowed.
[b]Rounded integer costs.
[c]Real costs multiplied by 10 000 and rounded to the nearest integer.

seeds. For each seed, 60 sets were generated by including customers according to increasing distances from the seed. After 50 subgradient iterations, new sets were added to $\mathscr{F}$ by identifying violated capacity-cuts in the current Lagrangian solution as previously explained. The step size used in the subgradient optimization method was initially set to 2 and reduced by a factor of 0.75 if the lower bound was not improved in the last 30 iterations. The number of iterations of the subgradient optimization procedure performed at the root node of the branch and bound algorithm ranged between 2000 and 3000. The overall Lagrangian bound considerably improved the basic K-tree relaxation. Table 3 reports the percentage ratios of the K-tree and of the Lagrangian bound. We used the K-tree and Lagrangian bound values computed by Fisher [20] by using real-valued cost matrices and not allowing single-customers routes, and we compared the bounds with respect to the optimal or the best known solution values determined by using real-valued cost matrices. Over the nine instances considered by Fisher, the average ratio of the K-tree is 74.4% while that of the Lagrangian bound is 94.8%.

Miller [40] initialized $\mathscr{F}$ as the empty set and at each iteration of the subgradient procedure detected violated GSECs and additional constraints belonging to the following two classes. The first type of constraints is given by additional GSECs which were added when the current Lagrangian solution $\bar{x}$ contains two or more overloaded routes. The customer set of these new GSECs is the union of the sets $S_1, \ldots, S_k$ associated with the GSECs violated by $\bar{x}$. This increases the possibility that arcs connecting customers belonging to the overloaded routes to those in sets $S_1, \ldots, S_k$ are selected by the b-matching solution. The second type of constraints was added when $\bar{x}$ contained routes which were *underloaded*, i.e., routes whose associated load was smaller than the minimum vehicle load $C_{\min}$ defined by (36). In this case for each such set S, with $0 \in S$, a constraint of the form

$$\sum_{e \in \sigma(S)} x_e \leqslant |S| - 1, \tag{37}$$

which breaks the current underloaded route in $\bar{x}$, was added to $\mathscr{F}$. The procedure was iterated until no improvement was obtained since 50 subgradient iterations. The step size was modified in an adaptive way every five subgradient iterations to produce a slight oscillation in lower bound values during the progress of the subgradient procedure. If the lower bound was monotonically increasing, the step size was increased by 50%; if the oscillation of the lower bound value was greater than 2%, the step size was reduced by 20%, and when the oscillation was smaller than 0.5% it was increased by 10%.

As can be seen from Table 3, the final Lagrangian bound of Miller is considerably tight, being on average 98% of the optimal solution value for the eight problems with $n \leqslant 50$ solved by Miller by using integer rounded cost matrices. The author also communicated us some values of the pure b-matching relaxation which are reported in Table 3 (the corresponding ratio is on average 87.5%).

4.4. Bounds based on set partitioning formulation

Hadjcostantinou et al. [26] proposed a branch and bound algorithm where the lower bound is computed by heuristically solving the dual of the linear programming relaxation of the set partitioning formulation of the CVRP.

The *set partitioning* (SP) formulation of the VRP was originally proposed by Balinsky and Quandt [2] and uses a possibly exponential number of binary variables, each associated with the different feasible circuit of G. More specifically, let $\mathscr{H} = \{H_1, \ldots, H_M\}$ denote the collection of all the circuits of G each corresponding to a feasible route, with $M = |\mathscr{H}|$. Each circuit H_j has an associated optimal cost c_j, and let a_{ij} be a binary coefficient which takes value 1 if and only if vertex i is covered (i.e., visited) by route H_j. The binary variable $x_j, j = 1, \ldots, M$, is equal to 1 if and only if circuit H_j is selected in the optimal solution. The model is:

$$(\text{VRP3}) \quad \min \sum_{j=1}^{M} c_j x_j \tag{38}$$

$$\text{s.t.} \quad \sum_{j=1}^{M} a_{ij} x_j = 1 \quad i \in V \setminus \{0\} \tag{39}$$

$$\sum_{j=1}^{M} x_j = K \tag{40}$$

$$x_j \in \{0,1\} \quad j = 1,\ldots,M. \tag{41}$$

Constraints (39) impose that each customer i is covered by exactly one of the selected circuits, and (40) requires that K circuits are selected.

This is a very general model which may easily take into account several constraints as, for example, time windows, since route feasibility is implicitly considered in the definition of set $\mathscr{H}$. Agarwal et al. [1] proposed an exact algorithm for CVRP based on set partitioning approach, whereas several successful applications of this technique to tightly constrained VRPs are reported in Desrosiers et al. [14]. Moreover, the linear programming relaxation of this formulation is typically very tight (see also Bramel and Simchi-Levi [5] which give a detailed probabilistic analysis of the quality of the linear programming relaxation of the set partitioning formulation).

Hadjcostantinou et al. [26] proposed to obtain a valid lower bound to SCVRP by considering the dual of the linear relaxation of model VRP3:

$$(\text{DVRP3}) \quad \max K\pi_0 + \sum_{i=1}^{n} \pi_i \tag{42}$$

$$\text{s.t.} \quad \pi_0 + \sum_{i \in H_j} \pi_i \leqslant c_j \quad j = 1,\ldots,M, \tag{43}$$

$$\pi_i \quad \text{unrestricted } i = 0,\ldots,n. \tag{44}$$

Where π_i, $i = 1,\ldots,n$ are the dual variables associated with the partitioning constraints (39) and π_0 is that associated with constraint (40). It is clear that any feasible solution to problem DVRP3 provides a valid lower bound to SCVRP. Hadjcostantinou et al. [26] determined the heuristic dual solutions by combining two relaxations of the original problem: the q-path relaxation proposed in Christofides et al. [11], and the k-shortest path relaxation proposed in Christofides and Mingozzi [9]. The proposed approach was able to solve randomly generated Euclidean instances with up to 30 vertices and instances proposed in the literature with up to 50 vertices, within a time limit of 12 h on a Silicon Grapics Indigo R4000 (12 Mflops). The percentage ratio of the overall bound (HCM) are reported in Table 3.

4.5. Branching schemes and overall algorithms

Many branching schemes were used for SCVRP and almost all are extensions of those used for the TSP.

The first scheme we consider, proposed in [11], is known as *branching on arcs* and proceeds by extending partial paths, starting from the depot and finishing at a given vertex. At each node of the branch-decision tree an arc (i,j) is selected to extend the current partial path and two descendant nodes are generated: the first node is associated with the inclusion of the selected arc in the solution (i.e., $x_{ij} = 1$), while in the second node the arc is excluded (i.e., $x_{ij} = 0$).

Miller [40] used this branching scheme, where the arc to branch with is selected by examining the solution obtained by the Lagrangian relaxation based on b-matching described in Sections 4.3. When a partial path is present in the current subproblem ending, say, with vertex v, the arc (v, h) belonging to the current Lagrangian solution is selected. If the current subproblem does not contain a partially fixed path, e.g., at the root node or when a route has been closed by the last imposed arc, the arc connecting the depot with the unrouted customer j with the largest demand is selected for branching. In this case a third descendant node is also created, by imposing $x_{0j} = 2$, i.e., by considering, if feasible, the route containing only customer j. The resulting branch and bound algorithm was applied to Euclidean SCVRP instances from the literature, where the edge costs are computed as the Euclidean distances between the customers and rounded to the nearest integer. The algorithm was able to solve problems with up to 50 customers within $15,000$ s on a Sun Sparc 2 (4 Mflops).

Fisher [20] used a mixed scheme where branching on arcs is used when no partial path is present in the current subproblem. In this case the currently unserved customer i with the largest demand is chosen and the arc (i, j) is used for branching, where j is the unserved customer closest to i. At the node where arc (i, j) is excluded from the solution, branching on arcs is again used, whereas at the second node the scheme known as *branching on customers* is used. One of the two ending customers, say v, of the currently imposed sequence of customers is chosen, and branching is performed by enumerating the customers which may be appended to that end of the sequence. A subset T of currently unserved customers is selected, e.g., that including the unserved customers closest to v, and $|T| + 1$ nodes are generated. Each of the first $|T|$ nodes corresponds to the inclusion in the solution of a different customer $j \in T$, while in the last node all the arcs $(v, j), j \in T$ are excluded.

The performance of this branching scheme may be enhanced by means of a dominance test proposed by Christofides et al. [11]. A node of the branch-decision tree where a partial sequence of customers $v, \ldots, w$ is fixed, can be fathomed if there exists a lower cost ordering of the customers in the sequence starting with v and ending with w. The improved ordering may be heuristically determined, e.g., by means of exchange procedures as those proposed in Lin and Kernighan [36].

The mixed branching scheme with the described dominance rule was used by Fisher to attempt the solution of Euclidean SCVRP instances with real distances and about 100 customers, but proved unsuccessful. In fact, Fisher observed that in instances where many small clusters of close customers exist (as in the case of several instances from the literature) any solution in which these customers are served contiguously in the same route have almost the same cost. Thus, when the sequence of these customers have to be determined through branching, unless an extremely tight bound is used, it would be very difficult to fathom many of the resulting nodes. Therefore, in [20] an alternative branching scheme is proposed, aiming at exploiting macro-properties of the optimal solution whose violation would have a large impact on the cost, thus allowing the fathoming of the corresponding nodes. To this end a subset T of currently unserved customers is selected and two descendant nodes are created: at the first node the additional constraint $\sum_{e \in \delta(T)} x_e = 2\lceil d(T)/C \rceil$ is added to the current problem, while at the second node the constraint $\sum_{e \in \delta(T)} x_e \geqslant 2 \lceil d(T)/C \rceil + 2$ is imposed. Some ways of identifying suitable subsets as well as additional dominance rules are described in [20]. This second branch and bound algorithm was successfully applied to some Euclidean SCVRP instances with real distances and with no single customer route allowed. The largest solved instance included 100 customers and was solved within less than 60,000 s on a small Apollo Domain 3000 computer (0.071 Mflops). Note, however, that several Euclidean instances from the literature were not solved to optimality.

5. Conclusions

In this paper we reviewed the most important branch and bound algorithms proposed during the last decade for the capacitated vehicle routing problem with either symmetric or asymmetric cost matrix. The progress made with these algorithms with respect to those of the previous generation is considerable: the dimension of the largest instances solved has been increased from about 25 to more than 100 customers. However, the CVRP is still far from being a closed chapter in the combinatorial optimization book. In fact some Euclidean problems from the literature with 75 customers are still unsolved and, in our opinion, the size of the problems which may be actually solved in a systematic way by the present approaches is limited to few tenths of customers.

Several possible directions of research are still almost uncovered, e.g., Dantzig–Wolfe decomposition based approaches (also known as *branch and price* approaches), but also a more deep investigation and understanding of the capabilities of the available techniques is strongly needed. As an example, we may mention that a direct computational evaluation and comparison of the effectiveness of the algorithms presented in this paper for the symmetric case is not possible. In fact, as illustrated in Table 3, each author either considered a slightly different problem (e.g., in [20] single customers routes were not allowed, whereas Miller [40] allowed them) or solved a completely different set of instances. The only instance which has been tackled by almost all the authors we considered is the 50 customers Euclidean problem described in Christofides and Eilon [8]. However, for this instance Fisher [20] used a real-valued cost matrix with Euclidean distances, Miller [40] used an integer cost matrix with Euclidean distances rounded to the nearest integer. As to Hadjconstantinou et al. [26], they used a hybrid solution where the integer cost of each arc is defined as the Euclidean distance between its endpoints multiplied by 10^4 and then rounded to nearest integer. Another research issue which may lead to interesting results is represented by the adaptation to the symmetric CVRP of the exact approaches developed for the asymmetric case, and vice versa.

Acknowledgements

This work has been supported by Ministero dell'Università e della Ricerca Scientifica e Tecnologica (MURST), and by Consiglio Nazionale delle Ricerche (CNR), Italy.

The computational experiments have been executed at the Laboratory of Operations Research of the University of Bologna (Lab.O.R.).

References

[1] Y. Agarwal, K. Mathur, H.M. Salkin, A set-partitioning-based exact algorithm for the vehicle routing problem, Networks 19 (1989) 731–749.
[2] M. Balinski, R. Quandt, On an integer program for a delivery problem, Oper. Res. 12 (1964) 300–304.
[3] M. Bellmore, J.C. Malone, Pathology of travelling salesman subtour-elimination algorithms, Oper. Res. 19 (1971) 278–307.
[4] L. Bodin, B.L. Golden, A.A. Assad, M.O. Ball, Routing and scheduling of vehicles and crews, the state of the art, Comput. Oper. Res. 10 (2) (1983) 63–212.
[5] J. Bramel, D. Simchi-Levi, On the effectiveness of the set partitioning formulation for the vehicle routing problem, Oper. Res. 45 (1997) 295–301.
[6] G.Carpaneto, P. Toth, Some new branching and bounding criteria for the asymmetric traveling salesman problem, Manage. Sci. 26 (1980) 736–743.
[7] N. Christofides, Vehicle routing, in: E.L. Lawler, J.K. Lenstra, A.H.G. Rinnooy Kan, D.B. Shmoys (Eds.), The Traveling Salesman Problem, Wiley, Chichester, 1985, pp. 431–448.
[8] N. Christofides, S. Eilon, An algorithm for the vehicle dispatching problem, Oper. Res. Quart. 20 (1969) 309–318.
[9] N. Christofides, A. Mingozzi, Vehicle routing: practical and algorithmic aspects, in: C.F.H. van Rijn (Ed.), Logistics: Where Ends Have to Meet, Pergamon Press, Oxford, 1989, pp. 30–48.
[10] N. Christofides, A. Mingozzi, P. Toth, The vehicle routing problem, in: N. Christofides, A. Mingozzi, P. Toth, C. Sandi (Eds.), Combinatorial Optimization, Wiley, Chichester, 1979, pp. 315–338.
[11] N. Christofides, A. Mingozzi, P. Toth, Exact algorithms for the vehicle routing problem based on the spanning tree and shortest path relaxations, Math. Programming 20 (1981) 255–282.
[12] G. Cornuéjols, F. Harche, Polyhedral study of the capacitated vehicle routing problem, Math. Programming 60 (1) (1993) 21–52.
[13] M. Desrochers, G. Laporte, Improvements and extensions to the Miller-Tucker-Zemlin subtour elimination constraints, Oper. Res. Lett. 10 (1991) 27–36.
[14] J. Desrosiers, Y. Dumas, M.M. Solomon, F. Soumis, Time constrained routing and scheduling, in: M.O. Ball, T.L. Magnanti, C.L. Monma, G.L. Nemhauser (Eds.), Network Routing, Handbooks in Operations Research and Management Science, North-Holland, Amsterdam, 1995, pp. 35–139.
[15] J.J. Dongarra, Performance of various computers using standard linear equations software, Technical Report CS-89-85, University of Tennessee, Knoxville, 1996.
[16] M. Fischetti, P. Toth, An additive bounding procedure for combinatorial optimization problems, Oper. Res. 37 (1989) 319–328.
[17] M. Fischetti, P. Toth, An additive bounding procedure for the asymmetric travelling salesman problem, Math. Programming 53 (1992) 173–197.
[18] M. Fischetti, P. Toth, D. Vigo, A branch-and-bound algorithm for the capacitated vehicle routing problem on directed graphs, Oper. Res. 42 (5) (1994) 846–859.
[19] M.L. Fisher, An application oriented guide to Lagrangian relaxation, Interfaces 15 (1985) 10–21.
[20] M.L. Fisher, Optimal solution of vehicle routing problems using minimum k-trees, Oper. Res. 42 (1994) 626–642.
[21] M.L. Fisher, A polynomial algorithm for the degree constrained k-tree problem, Oper. Res. 42 (4) (1994) 776–780.
[22] M.L. Fisher, Vehicle routing, in: M.O. Ball, T.L. Magnanti, C.L. Monma, G.L. Nemhauser (Eds.), Network Routing, Handbooks in Operations Research and Management Science, North-Holland, Amsterdam, 1995, pp. 1–33.
[23] H.N. Gabow, R.E. Tarjan, Efficient algorithms for a family of matroid intersection problems, J. Algorithms 5 (1984) 80–131.
[24] B.L. Golden, A.A. Assad, Vehicle Routing: Methods and Studies, North-Holland, Amsterdam, 1988.

[25] B.L. Golden, E.A. Wasil, J.P. Kelly, I. Chao, The impact of metaheuristics on solving the vehicle routing problem: algorithms, problem sets, and computational results, in: T.G. Crainic, G. Laporte (Eds.), Fleet Management and Logistic, Kluwer Academic Publisher, Boston (MA), 1998, pp. 33–56.

[26] E. Hadjiconstantinou, N. Christofides, A. Mingozzi, A new exact algorithm for the vehicle routing problem based on q-Paths and k-Shortest paths relaxations, Ann. Oper. Res. 61 (1995) 21–43.

[27] M. Held, R.M. Karp, The traveling salesman problem and minimum spanning tress: Part II, Math. Programming 1 (1971) 6–25.

[28] R.V. Kulkarni, P.V. Bhave, Integer programming formulations of vehicle routing problems, European J. Oper. Res. 20 (1985) 58–67.

[29] G. Laporte, The vehicle routing problem: An overview of exact and approximate algorithms, European J. Oper. Res. 59 (1992) 345–358.

[30] G. Laporte, Vehicle routing, in: M. Dell'Amico, F. Maffioli, S. Martello (Eds.), Annotated Bibliographies in Combinatorial Optimization, Wiley, Chichester, 1997.

[31] G. Laporte, H. Mercure, Y. Nobert, An exact algorithm for the asymmetrical capacitated vehicle routing problem, Networks 16 (1986) 33–46.

[32] G. Laporte, Y. Nobert, Exact algorithms for the vehicle routing problem, Ann. Discrete Math. 31 (1987) 147–184.

[33] G. Laporte, Y. Nobert, M. Desrochers, Optimal routing under capacity and distance restrictions, Oper. Res. 33 (1985) 1050–1073.

[34] G. Laporte, I.H. Osman, Routing problems: a bibliography, Ann. Oper. Res. 61 (1995) 227–262.

[35] J.K. Lenstra, A.H.G. Rinnooy Kan, Some simple applications of the traveling salesman problem, Oper. Res. Quart. 26 (1975) 717–734.

[36] S. Lin, B.W. Kernighan, An effective heuristic algorithm for the travelling salesman problem, Oper. Res. 21 (1973) 498–516.

[37] T.L. Magnanti, Combinatorial optimization and vehicle fleet planning: Perspectives and prospects, Networks 11 (1981) 179–214.

[38] S. Martello, P. Toth, Knapsack Problems: Algorithms and Computer Implementations, Wiley, Chichester, 1990.

[39] C.E. Miller, A.W. Tucker, R.A. Zemlin, Integer programming formulations and traveling salesman problems, ACM 7 (1960) 326–329.

[40] D.L. Miller, A matching based exact algorithm for capacitated vehicle routing problems, ORSA J. Comput. 7 (1) (1995) 1–9.

[41] D.L. Miller, J.F. Pekny, A staged primal-dual algorithm for perfect b-matching with edge capacities, ORSA J. Comput. 7 (1995) 298–320.

[42] J.F. Pekny, D.L. Miller, A staged primal-dual algorithm for finding a minimum cost perfect two-matching in an undirected graph, ORSA J. Comput. 6 (1994) 68–81.

[43] P. Toth, D. Vigo, An exact algorithm for the capacitated shortest spanning arborescence, Ann. Oper. Res. 61 (1995) 121–142.

[44] P. Toth, D. Vigo, An exact algorithm for the vehicle routing problem with backhauls, Transport. Sci. 31 (1997) 372–385.

[45] P. Toth, D. Vigo, Exact solution of the vehicle routing problem, in: T.G. Crainic, G. Laporte (Eds.), Fleet Management and Logistic, Kluwer Academic Publisher, Boston (MA), 1998, pp. 1–31.

[46] P. Toth, D. Vigo, The granular tabu search (and its application to the vehicle routing problem). Technical Report OR/98/9, D.E.I.S. - Università di Bologna, 1998.

[47] D. Vigo, A heuristic algorithm for the asymmetric capacitated vehicle routing problem, European J. Oper. Res. 89 (1996) 108–126.

[48] D. Vigo, VRPLIB: a vehicle routing problem library, Technical Report OR/99/9, D.E.I.S., Università di Bologna, 1999.

Discrete Applied Mathematics 123 (2002) 513–577

DISCRETE
APPLIED
MATHEMATICS

ELSEVIER

Semidefinite programming for discrete optimization and matrix completion problems ☆

Henry Wolkowicz[*][1], Miguel F. Anjos[2]

Department of Combinatorics & Optimization, Faculty of Mathematics, University of Waterloo, Waterloo, Ontario, Canada N2L 3G1

Abstract

Semidefinite programming (SDP) is currently one of the most active areas of research in optimization. SDP has attracted researchers from a wide variety of areas because of its theoretical and numerical elegance as well as its wide applicability. In this paper we present a survey of two major areas of application for SDP, namely discrete optimization and matrix completion problems.

In the first part of this paper we present a recipe for finding SDP relaxations based on adding redundant constraints and using Lagrangian relaxation. We illustrate this with several examples. We first show that many relaxations for the max-cut problem (MC) are equivalent to both the Lagrangian and the well-known SDP relaxation. We then apply the recipe to obtain new strengthened SDP relaxations for MC as well as known SDP relaxations for several other hard discrete optimization problems.

In the second part of this paper we discuss two completion problems, the positive semidefinite matrix completion problem and the Euclidean distance matrix completion problem. We present some theoretical results on the existence of such completions and then proceed to the application of SDP to find approximate completions. We conclude this paper with a new application of SDP to find approximate matrix completions for large and sparse instances of Euclidean distance matrices. © 2002 Elsevier Science B.V. All rights reserved.

Keywords: Semidefinite programming; Discrete optimization; Lagrangian relaxation; Max-cut problem; Euclidean distance matrix; Matrix completion problem

☆ Survey article for the Proceedings of Discrete Optimization '99 where some of these results were presented as a plenary address. This paper is available by anonymous ftp at orion.math.uwaterloo.ca in directory pub/henry/reports or with URL.

* Corresponding author. Tel.: +1-519-888-4567x5589.

E-mail addresses: hwolkowi@orion.math.uwaterloo.ca (H. Wolkowicz), anjos@stanfordalumni.org (M. F. Anjos).

URL: http://orion.math.uwaterloo.ca/~hwolkowi/henry/reports/ABSTRACTS.html

[1] Research supported by The Natural Sciences and Engineering Research Council of Canada.

[2] Research supported by an FCAR Doctoral Research Scholarship.

1. Introduction

There have been many survey articles written in the last few years on semidefinite programming (SDP) and its applicability to discrete optimization and matrix completion problems [2,5,36–38,61,76,77,105] This highlights the fact that SDP is currently one of the most active areas of research in optimization. In this paper we survey in depth two application areas where SDP research has recently made significant contributions. Several new results are also included.

The first part of the paper is based on the premise that Lagrangian relaxation is "best". By this we mean that good tractable bounds can always be obtained using Lagrangian relaxation. Since the SDP relaxation is equivalent to the Lagrangian relaxation, we explore approaches to obtain tight SDP relaxations for discrete optimization by applying a recipe for finding SDP relaxations using Lagrangian duality. We begin by considering the max-cut problem (MC) in Section 2.2. We first present several different relaxations of MC that are equivalent to the SDP relaxation, including the Lagrangian relaxation, the relaxation over a sphere, the relaxation over a box, and the eigenvalue relaxation. This illustrates our theme on the strength of the Lagrangian relaxation. The question of which relaxation is most appropriate in practice for a given instance of MC remains open. Section 2.5 contains an overview of the main algorithms that have been proposed to compute the SDP bound, and Section 2.6 presents an overview of the known qualitative results about the quality of the SDP bound. We then proceed in Section 2.7 to derive new strengthened SDP relaxations for MC. To obtain these relaxations we apply the recipe for finding SDP relaxations presented in [100]. This recipe can be summarized as: add as many redundant quadratic constraints as possible; take the Lagrangian dual of the Lagrangian dual; remove redundant constraints and project the feasible set of the resulting SDP to guarantee strict feasibility. We also show several interesting properties for the tighter of the new SDP relaxations, denoted SDP3. In particular, we prove that SDP3 is a *strict* improvement on the addition of all the triangle inequalities to the well-known SDP relaxation, denoted SDP1 [7–9]. This shows that SDP3 often improves on SDP1 whenever the latter is not optimal. In Section 3 we discuss the application of Lagrangian relaxation to general quadratically constrained quadratic problems and in Section 4 we present applications of the recipe to other discrete optimization problems, including the graph partitioning, quadratic assignment, max-clique and max-stable-set problems.

The second part of the paper presents several SDP algorithms for the positive semidefinite and the Euclidean distance matrix completion problems. The algorithms are shown to be efficient for large sparse problems. Section 5.1 presents some theoretical existence results for completions based on chordality. This follows the work in [41]. An approach to solving large sparse completion problems based on approximate completions [56] is outlined in Section 5.2. In Section 5.3 a similar approach for Euclidean distance matrix completions [1] is presented. However, the latter does not take advantage of sparsity and has difficulty solving large sparse problems. We conclude in Section 5.4 with a new characterization of Euclidean distance matrices from which we derive an algorithm that successfully exploits sparsity [3].

1.1. Notation and preliminaries

We let $\mathcal{S}^n$ denote the space of $n \times n$ symmetric matrices. This space has dimension $t(n):=n(n+1)/2$ and is endowed with the trace inner product $\langle A, B \rangle = \text{trace } AB$. We let $A \circ B$ denote the Hadamard (elementwise) matrix product and $A \succcurlyeq 0$ denote the Löwner partial order on $\mathcal{S}^n$, i.e. for $A \in \mathcal{S}^n$, $A \succcurlyeq 0$ if and only if A is positive semidefinite. We denote by $\mathcal{P}$ the cone of positive semidefinite matrices.

We also work with matrices in the space $\mathcal{S}^{t(n)+1}$. For given $Y \in \mathcal{S}^{t(n)+1}$, we index the rows and columns of Y by $0, 1, \ldots, t(n)$. We will be particularly interested in the vector x obtained from the first (0th) row (or column) of Y with the first element dropped. Thus, in our notation, $x = Y_{0, 1:t(n)}$.

We let e denote the vector of ones and $E = ee^T$ the matrix of ones; their dimensions will be clear from the context. We also let e_i denote the ith unit vector and define the elementary matrices $E_{ij}:=\frac{1}{\sqrt{2}}(e_i e_j^T + e_j e_i^T)$, if $i \neq j$, $E_{ii}:=\frac{1}{2}(e_i e_j^T + e_j e_i^T)$. For any vector $v \in \mathfrak{R}^n$, we let $\|v\|:=\sqrt{v^T v}$ denote the ℓ_2 norm of v.

We use operator notation and operator adjoints. The adjoint of the linear operator $\mathcal{A}$ is denoted $\mathcal{A}^*$ and satisfies (by definition)

$$\langle \mathcal{A}x, y \rangle = \langle x, \mathcal{A}^* y \rangle, \quad \forall x, y.$$

Given a matrix $S \in \mathcal{S}^n$, we now define several useful operators. The operator diag(S) returns a vector with the entries on the diagonal of S. Given $v \in \mathfrak{R}^n$, the operator Diag(v) returns an $n \times n$ diagonal matrix with the vector v on the diagonal. It is straightforward to check that Diag is the adjoint operator of diag. We use both Diag(v) and Diag v provided the meaning is clear, and the same convention applies to diag and all the other operators. The symmetric vectorizing operator svec satisfies $s = \text{svec}(S) \in \mathfrak{R}^{t(n)}$ where s is formed column-wise from S and the strictly lower triangular part of S is ignored. Its inverse is the operator sMat, so $S = \text{sMat}(s)$ if and only if $s = \text{svec}(S)$. Note that the adjoint of svec is not sMat but svec$^* = $ hMat with hMat(s) being the operator that forms a symmetric matrix from s like sMat but also multiplies the off-diagonal terms by $\frac{1}{2}$ in the process. Similarly, the adjoint of sMat is the operator dsvec which acts like svec except that the off-diagonal elements are multiplied by 2.

For notational convenience, we also define the symmetrizing diagonal vector operator

$$\text{sdiag}(x):=\text{diag}(\text{sMat}(x))$$

and the vectorizing symmetric vector operator

$$\text{vsMat}(x):=\text{vec}(\text{sMat}(x)),$$

where vec(S) returns the n^2-dimensional vector formed columnwise from S like svec but with the complete columns of the matrix S. Note that the adjoint of vsMat is

$$\text{vsMat}^*(x) = \text{dsvec}[\tfrac{1}{2}(\text{Mat}(x) + \text{Mat}(x)^T)].$$

Let us summarize here some frequently used operators in this paper:

$$\text{diag}^* = \text{Diag},$$
$$\text{svec}^* = \text{hMat},$$

$$\text{svec}^{-1} = \text{sMat},$$
$$\text{dsvec}^* = \text{sMat},$$
$$\text{vsMat}^* = \text{dsvec}[\tfrac{1}{2}(\text{Mat}(\cdot) + \text{Mat}(\cdot)^{\text{T}})].$$

We will frequently use the following relationships between matrices and vectors:

$$X \cong vv^{\text{T}} \cong \text{sMat}(x) \in \mathscr{S}^n \quad \text{and} \quad Y \cong \begin{pmatrix} y_0 \\ x \end{pmatrix} (y_0 \quad x^{\text{T}}) \in \mathscr{S}^{t(n)+1}, \quad y_0 \in \mathfrak{R}.$$

2. The max-cut problem

We begin our presentation with the study of one of the simplest NP-hard problems, albeit one for which SDP has been successful. The max-cut problem (MC) is a discrete optimization problem on undirected graphs with weights on the edges. Given such a graph, the problem consists in finding a partition of the set of nodes into two parts, which we call shores, to maximize the sum of the weights on the edges that are cut by the partition (we say that an edge is cut if it has exactly one end on each shore of the partition). In this paper we shall assume that the graph in question is complete (if not, non-existing edges can be added with zero weight to complete the graph without changing the problem) and we require no restriction on the type of edge weights (so, in particular, negative edge weights are permitted).

Following [86], we can formulate MC as follows. Let the given graph G have node set $\{1,\ldots,n\}$ and let it be described by its weighted adjacency matrix $A = (w_{ij})$. Let $L:=\text{Diag}(Ae) - A$ denote the Laplacian matrix associated with the graph, where the linear operator Diag returns a diagonal matrix with diagonal formed from the vector given as its argument, and e denotes the vector of all ones. Let us also define the set $\mathscr{F}_n:=\{\pm 1\}^n$, and let the vector $v \in \mathscr{F}_n$ represent any cut in the graph via the interpretation that the sets $\{i: v_i = +1\}$ and $\{i: v_i = -1\}$ form a partition of the node set of the graph. Then we can formulate MC as

$$\text{(MC1)} \qquad \begin{aligned} \mu^* &:= \max \ \tfrac{1}{4}v^{\text{T}}Lv \\ &\text{s.t.} \ \ v \in \mathscr{F}_n, \end{aligned} \qquad (2.1)$$

where here and throughout this paper μ^* denotes the optimal value of the MC problem.

It is straightforward to check that

$$\frac{1}{4}v^{\text{T}}Lv = \sum_{i<j} w_{ij} \left(\frac{1 - v_i v_j}{2} \right)$$

and that the term multiplying w_{ij} in the sum equals one if the edge (i,j) is cut, and zero otherwise. Analogous quadratic terms having this property will be used in our formulations of MC.

We can view MC1 as the problem of maximizing a homogeneous quadratic function of v over the set $\mathscr{F}_n$. We show that this problem is equivalent to problem MCQ below which has a more general objective function. This equivalence shows that all the results about MC1 also extend to MCQ. Furthermore, the formulation MCQ will help us derive relaxations for MC in Sections 2.2 and 2.3.

Let us therefore consider the quadratic objective function

$$q_0(v) := v^{\mathrm{T}} Q v - 2 c^{\mathrm{T}} v$$

(the meaning of the subscript 0 will become clear at the beginning of Section 2.2) and the corresponding ± 1-constrained quadratic problem MCQ:

$$(\text{MCQ}) \quad \max_{v \in \mathscr{F}_n} q_0(v). \tag{2.2}$$

Clearly, MC1 corresponds to the choice $Q = \frac{1}{4} L$ and $c = 0$. Conversely, we can homogenize the problem MCQ by increasing the dimension by one. Indeed, given $q_0(v)$, define the $(n+1) \times (n+1)$ matrix Q^c obtained by adding a 0th dimension to Q and placing the vector c in the new row and column, so that

$$Q^c := \begin{bmatrix} 0 & -c^{\mathrm{T}} \\ -c & Q \end{bmatrix}. \tag{2.3}$$

If we consider the variable $\bar{v} = \binom{v_0}{v} \in \mathscr{F}_{n+1}$ and the new quadratic form

$$q_0^c(\bar{v}) := \bar{v}^{\mathrm{T}} Q^c \bar{v} = v^{\mathrm{T}} Q v - 2 v_0 (c^{\mathrm{T}} v),$$

then we get an equivalent MC problem.

2.1. Higher-dimensional embeddings of MC

We can express the feasible set $\mathscr{F}_n$ in several different ways by appropriately embedding all its points in spaces of varying dimensions. In this section we take a geometrical view of several such embeddings and the respective formulations of MC. Relaxations of these formulations will be considered in the remainder of Section 2.
1. If we define

$$\mathscr{F}_n(1) := \{ v \in \mathfrak{R}^n \colon |v_i| = 1, \ i = 1, \dots, n \}, \tag{2.4}$$

then clearly $\mathscr{F}_n(1) = \mathscr{F}_n$ and the formulation MC1 corresponds to optimizing $\frac{1}{4} v^{\mathrm{T}} L v$ over $\mathscr{F}_n(1)$.
For later reference, we note here that $\mathscr{F}_n(1)$ is the set of extreme points of the unit hypercube in $\mathfrak{R}^n$ (the ℓ_∞ norm unit ball). Furthermore, all the points $v \in \mathscr{F}_n(1)$ satisfy the constraints

$$\|v_i\|_2 = 1 \quad \forall i, \qquad |v_i v_j| = |v_i^{\mathrm{T}} v_j| = 1 \quad \forall i < j.$$

We have deliberately added the transpose, even though the variables v_i are all scalars, to emphasize the similarity with the next embeddings.
2. For any given positive integer p, we can lift each of the variables v_i from a scalar to a vector of length p by defining

$$\mathscr{F}_n(p) := \{ V \in \mathfrak{R}^{n \times p} \colon V = [v_1, \dots, v_n]^{\mathrm{T}}, \ \|v_i\|_2 = 1 \ \forall i, \ |v_i^{\mathrm{T}} v_j| = 1 \ \forall i < j \}. \tag{2.5}$$

Note that if $p = 1$ then we simply recover $\mathscr{F}_n(1)$. For $p > 1$ the constraints on the inner products restrict the cosines of the angles between any two vectors v_i and v_j to equal ± 1. Hence for $i = 2,\ldots,n$, either $v_i = v_1$ or $-v_1$, and we can obtain a cut by choosing the sets $\{i: v_i^{\mathrm{T}} v_1 = +1\}$ and $\{i: v_i^{\mathrm{T}} v_1 = -1\}$ as the shores. Thus, the objective function may be written as

$$\sum_{i<j} w_{ij} \left(\frac{1 - v_i^{\mathrm{T}} v_j}{2} \right)$$

or, in terms of the Laplacian, as

$$\tfrac{1}{4} \operatorname{trace} V^{\mathrm{T}} L V.$$

We have thus derived our second formulation of MC:

$$\mu^* = \max \tfrac{1}{4} \operatorname{trace} V^{\mathrm{T}} L V$$

$$\text{s.t. } v_i \in \mathfrak{R}^p, \ \|v_i\|_2 = 1 \quad \forall i,$$

(MC2)

$$|v_i^{\mathrm{T}} v_j| = 1 \quad \forall i < j, \qquad (2.6)$$

$$V = [v_1 \ldots v_n]^{\mathrm{T}}.$$

The commutativity of the arguments inside the trace means that

$$\tfrac{1}{4} \operatorname{trace} V^{\mathrm{T}} L V = \operatorname{trace}(\tfrac{1}{4} L) V V^{\mathrm{T}}$$

and this observation leads us to an embedding of MC into $\mathscr{S}^n$, the space of symmetric $n \times n$ matrices, by rewriting MC2 in terms of the variable $X \in \mathscr{S}^n$ which is defined by

$$X_{ij} := v_i^{\mathrm{T}} v_j, \quad \text{or equivalently,} \quad X := V V^{\mathrm{T}}.$$

Then the constraint $\|v_i\|_2 = 1$ is equivalent to $\operatorname{diag}(X) = e$ and

$$|v_i^{\mathrm{T}} v_j| = 1 \iff |X_{ij}| = 1, \quad \forall i < j.$$

Finally, $X = V V^{\mathrm{T}} \iff X \succcurlyeq 0$, and our third formulation of MC is

$$\mu^* = \max \tfrac{1}{4} \operatorname{trace} L X \ \left(= \tfrac{1}{2} \sum_{i<j} w_{ij}(1 - X_{ij}) \right)$$

(MC3) $$\text{s.t. } \operatorname{diag}(X) = e, \qquad (2.7)$$

$$|X_{ij}| = 1, \quad \forall i < j,$$

$$X \succcurlyeq 0.$$

Note that although each X_{ij} can be interpreted as the cosine of the angle between some vectors v_i and v_j, the length p of these vectors does not appear explicitly in the formulation MC3.

Having derived MC3, we can obtain yet another formulation by applying the following theorem:

Theorem 2.1 (Anjos and Wolkowicz [9, Theorem 3.2]). *Let X be an $n \times n$ symmetric matrix. Then*

$$X \succcurlyeq 0, X \in \{\pm 1\}^{n \times n} \quad \text{if and only if} \quad X = xx^{\mathrm{T}}, \quad \text{for some } x \in \{\pm 1\}^n.$$

Thus, we can replace the ± 1 constraint on the elements of X by the requirement that the rank of X be equal to one. Hence, we obtain our fourth formulation of MC:

$$\mu^* = \max \tfrac{1}{4} \operatorname{trace} LX$$

(MC4)
$$\text{s.t. } \operatorname{diag}(X) = e,$$

$$\operatorname{rank}(X) = 1, \tag{2.8}$$

$$X \succcurlyeq 0, \ X \in \mathscr{S}^n.$$

3. We now introduce an embedding of MC in a space of even higher dimension. This embedding is interesting because of its connection to the strengthened SDP relaxations that we present in Section 2.7.

For any given positive integer q, let us define the set

$$\mathscr{F}_{t(n)+1}(q) := \{ U \in \mathfrak{R}^{(t(n)+1) \times q} \colon U = [u_0, u_1, \dots, u_{t(n)}]^{\mathrm{T}},$$

$$\|u_i\|_2 = 1 \ \forall i = 0, 1, \dots, t(n),$$

$$|u_0^{\mathrm{T}} u_i| = 1 \ \forall i = 1, \dots, t(n),$$

$$\operatorname{sMat}(u_0^{\mathrm{T}} u_1, \dots, u_0^{\mathrm{T}} u_{t(n)}) \succcurlyeq 0 \}, \tag{2.9}$$

where $t(i) = i(i+1)/2$. The constraints

$$|u_0^{\mathrm{T}} u_i| = 1 \ \forall i \quad \text{and} \quad \operatorname{sMat}(u_0^{\mathrm{T}} u_1, \dots, u_0^{\mathrm{T}} u_{t(n)}) \succcurlyeq 0$$

imply (by Theorem 2.1) that the matrix $X = \operatorname{sMat}(u_0^{\mathrm{T}} u_1, \dots, u_0^{\mathrm{T}} u_{t(n)})$ has rank equal to one. By analogy with the previous embedding in $\mathscr{S}^n$, we can therefore write the following interpretation:

$$u_0^{\mathrm{T}} u_{t(j-1)+i} = X_{ij} = v_i^{\mathrm{T}} v_j, \quad \forall 1 \le i \le j \le n. \tag{2.10}$$

This means that we can think of the cosines of the angles between u_0, the first row of U, and every other row $u_{t(j-1)+i}$, $1 \le i \le j \le n$, as being equal to the cosines of the angles between the vectors v_i and v_j (corresponding to the indices i and j) in the previous embedding. We can thus write down the objective function in terms of the entries in the first row of U:

$$\sum_{i<j} w_{ij} \left(\frac{1 - u_0^{\mathrm{T}} u_{t(j-1)+i}}{2} \right).$$

Let us now define the matrix

$$H_L := \begin{pmatrix} 0 & \tfrac{1}{2} \operatorname{dsvec}(L)^{\mathrm{T}} \\ \tfrac{1}{2} \operatorname{dsvec}(L) & 0 \end{pmatrix}.$$

Then, since

$$\sum_{i<j} w_{ij} \left(\frac{1 - u_0^T u_{t(j-1)+i}}{2} \right) = \frac{1}{4} \text{trace}\, H_L U U^T = \frac{1}{4} \text{trace}\, U^T H_L U,$$

we can write down our fifth formulation of MC:

$$\mu^* = \max \tfrac{1}{4} \text{trace}\, U^T H_L U$$

$$\text{s.t.}\quad u_i \in \mathfrak{R}^q,\ \|u_i\|_2 = 1 \quad \forall i,$$

(MC5) $$|u_0^T u_i| = 1 \quad \forall i = 1, \dots, t(n),$$ (2.11)

$$\text{sMat}(u_0^T u_1, \dots, u_0^T u_{t(n)}) \succcurlyeq 0,$$

$$U = [u_0 \dots u_{t(n)}]^T.$$

As for the remaining entries of U, we can interpret them as

$$u_{t(j-1)+i}^T u_{t(l-1)+k} = (v_i^T v_j)(v_k^T v_l), \quad \forall 1 \leqslant i \leqslant j \leqslant n,\ \forall 1 \leqslant k \leqslant l \leqslant n.$$

(2.12)

This interpretation is particularly interesting if we use again the analogy with the previous embedding as in (2.10). If $X = VV^T$ with $V \in \mathscr{F}_n(1)$ (so V is a column vector) then the elements of X always satisfy the equation $X = (1/n)X^2$, i.e. each entry of X is equal to the average of n products of entries of X. Using interpretations (2.10) and (2.12), this is equivalent to the constraint that for each $k = 1, \dots, t(n)$, $u_0^T u_k$ be equal to the average of n specific elements $u_i^T u_j$ with $i, j \geqslant 1$. For the verification of the equation relating X and X^2 and a much detailed discussion of these interpretations, see Section 2.7.

We have thus embedded the feasible set $\mathscr{F}_n$ of MC in several different spaces and obtained corresponding formulations for MC. We now illustrate in the next three sections what we mean when we claim that the Lagrangian relaxation is "best". First, we introduce in Sections 2.2 and 2.3 a variety of (seemingly different) tractable relaxations obtained from these formulations. Then in Section 2.4 we present the Lagrangian relaxation and Theorem 2.3, which states that (surprisingly) the (upper) bounds on μ^* yielded by these relaxations, i.e. their optimal values, are all equal to the optimal value of the Lagrangian relaxation.

2.2. Relaxations for MC using $v_i \in \mathfrak{R}$

Let us begin our study of relaxations for MC by considering the embedding $\mathscr{F}_n(1)$ of the MC variables and the problem MCQ. We have already argued that MCQ is equivalent to MC. Before we continue, we show why it is helpful to allow a more general quadratic objective in this section.

Consider the formulation MC1:

$$\mu^* := \max v^T Q v$$

$$\text{s.t.}\quad v \in \mathscr{F}_n(1),$$

where $Q = \frac{1}{4}L$, and recall that $\mathscr{F}_n(1)$ is the set of extreme points of the unit hypercube in $\mathfrak{R}^n$. One obvious relaxation is to optimize $v^\mathrm{T} Q v$ over the entire hypercube. If we do so, the resulting relaxation falls into one of the following two cases:

1. If $Q \preccurlyeq 0$, i.e. Q is negative semidefinite, then the maximum over the hypercube is always equal to zero and is attained at the origin.
2. If Q is not negative semidefinite then (at least) one eigenvalue of Q is positive and Pardalos and Vavasis [98] showed that in this case the maximization of $v^\mathrm{T} Q v$ over the hypercube is NP-hard. So the relaxation is no more tractable than the original problem.

Clearly we do not obtain a useful relaxation in either case.

By considering instead the problem MCQ, the objective function $q_0(v) = v^\mathrm{T} Q v - 2 c^\mathrm{T} v$ has a linear term and this allows us to consider perturbations of $q_0(v)$ of the form

$$q_u(v) : = v^\mathrm{T} (Q + \mathrm{Diag}(u)) v - 2 c^\mathrm{T} v - u^\mathrm{T} e \tag{2.13}$$

with $u \in \mathfrak{R}^n$. It is important to note that if $v \in \mathscr{F}_n(1)$, then $v_i^2 = 1$ $\forall i$ and therefore

$$q_u(v) = q_0(v) \quad \forall v \in \mathscr{F}_n(1), \ \forall u \in \mathfrak{R}^n.$$

Hence,

$$\mu^* = \max \ q_u(v)$$
$$\text{s.t.} \ \ v \in \mathscr{F}_n(1),$$
$$u \in \mathfrak{R}^n.$$

We now show how these perturbations help.

2.2.1. The trivial relaxation in $\mathfrak{R}^n$

For given $u \in \mathfrak{R}^n$ let us maximize the perturbed objective function without any of the constraints on v, i.e. let us consider the function

$$f_0(u) := \max_v q_u(v).$$

For any choice of u, this function gives us an upper bound on μ^*, since $\mu^* \leqslant f_0(u)$. Hence, minimizing $f_0(u)$ over all $u \in \mathfrak{R}^n$ gives us a (trivial) relaxation of MC:

$$\mu^* \leqslant B_0 := \min_u f_0(u). \tag{2.14}$$

Remark 2.2. Note that $f_0(u)$ can take on the value $+\infty$. In particular, this will happen whenever the matrix $Q + \mathrm{Diag}(u)$ has at least one positive eigenvalue, since a quadratic function is unbounded above if the Hessian is not negative semidefinite. (In fact, a quadratic function is bounded above if and only if the Hessian is negative semidefinite and the stationarity equation is consistent. A proof of this well-known fact is given, for example, in [79, Lemma 3.6].) However, since (2.14) is a min–max problem, we can add the (hidden) semidefinite constraint $Q + \mathrm{Diag}(u) \preccurlyeq 0$ without changing the value of the bound B_0. The resulting problem is tractable since it consists of minimizing a convex function over a convex set. The trivial relaxation is thus equivalent to

$$B_0 = \min_{Q + \mathrm{Diag}(u) \preccurlyeq 0} f_0(u). \tag{2.15}$$

Furthermore, let us define the set

$$S := \{u: u^T e = 0, \; Q + \mathrm{Diag}(u) \preccurlyeq 0\}.$$

Provided that $S \neq \emptyset$, it is shown in [101] that the optimality conditions for min–max problems imply that

$$B_0 = \min_{u^T e=0} f_0(u) = \min_{\substack{Q + \mathrm{Diag}(u) \preccurlyeq 0 \\ u^T e = 0}} f_0(u).$$

2.2.2. The trust-region (spherical) relaxation in $\mathfrak{R}^n$

Next, let us relax the feasible set $\mathscr{F}_n(1)$ to the sphere in $\mathfrak{R}^n$ of radius $\sqrt{n}$ and centered at the origin. (Note that all the points of $\mathscr{F}_n(1)$ are contained in this sphere.) If we define the function

$$f_1(u) := \max_{\|v\|^2 = n} q_u(v), \tag{2.16}$$

then $\mu^* \leqslant f_1(u)$ for all u. This maximization problem is a trust-region subproblem and is tractable since its dual is a concave maximization problem over an interval [107,110,120]. Therefore we obtain the (tractable) trust-region relaxation:

$$\mu^* \leqslant B_1 := \min_u f_1(u). \tag{2.17}$$

2.2.3. The box relaxation in $\mathfrak{R}^n$

Alternatively, we can replace the spherical constraint with the box constraint or ℓ_∞ norm constraint (all the points of $\mathscr{F}_n(1)$ also lie in this unit box) and consider the function

$$\mu^* \leqslant f_2(u) := \max_{|v_i| \leqslant 1} q_u(v). \tag{2.18}$$

Since the maximization of a non-convex quadratic over the box constraint is NP-hard [98], we must add the hidden semidefinite constraint to make the calculation of $f_2(u)$ tractable and obtain the box relaxation:

$$\mu^* \leqslant B_2 := \min_{Q + \mathrm{Diag}(u) \preccurlyeq 0} f_2(u). \tag{2.19}$$

It is worth mentioning that it is precisely the addition of the hidden semidefinite constraint (to make the box relaxation tractable) that makes the bound B_2 equal to all the other bounds we are currently presenting (see Theorem 2.3).

2.2.4. The eigenvalue relaxation in $\mathfrak{R}^{n+1}$

We showed at the beginning of Section 2 how the problem MCQ can be homogenized at the price of increasing the dimension by 1. This homogenization yields three more bounds B_0^c, B_1^c and B_2^c via the same derivations used to obtain the bounds B_0, B_1 and B_2.

Given Q and c, recall the $(n + 1) \times (n + 1)$-matrix

$$Q^c = \begin{bmatrix} 0 & -c^{\mathrm{T}} \\ -c & Q \end{bmatrix} \tag{2.20}$$

and the vector $\bar{v} = \binom{v_0}{v}$. By analogy with the previous relaxations, we define

$$q_u^c(\bar{v}) := \bar{v}^{\mathrm{T}}(Q^c + \mathrm{Diag}(u))\bar{v} - u^{\mathrm{T}}e \tag{2.21}$$

and the functions $f_i^c(u)$, $i = 0, 1, 2$. Note that if v_0, the first component of $\bar{v}$, equals ± 1 then $q_u^c(\bar{v}) = q_u(v)$.

For brevity, we discuss only the relaxation B_1^c analogous to the trust-region relaxation. This particular relaxation is interesting because it turns out to be equivalent to an eigenvalue bound for μ^*. Indeed, since

$$f_1^c(u) := \max_{\|\bar{v}\|^2 = n+1} q_u^c(\bar{v}),$$

it follows from the Courant–Fisher Theorem (e.g. [54, Theorem 4.2.11]) that

$$f_1^c(u) = (n + 1)\lambda_{\max}(Q^c + \mathrm{Diag}(u)) - u^{\mathrm{T}}e,$$

where $\lambda_{\max}(\cdot)$ denotes the maximum eigenvalue of the matrix argument. Hence $f_1^c(u)$ is tractable and the (tractable) eigenvalue bound is

$$\mu^* \leqslant B_1^c := \min_u f_1^c(u). \tag{2.22}$$

2.3. Matrix relaxations for MC using $v_i \in \mathfrak{R}^p$

We now introduce relaxations arising from the formulations of MC using the feasible set $\mathscr{F}_n(p)$ with $p > 1$.

2.3.1. The Goemans–Williamson relaxation
This relaxation is obtained by considering the formulation MC2 and removing the (hard) ± 1 constraint on the inner products $v_i^{\mathrm{T}}v_j$. The resulting relaxation gives us the bound B_3:

$$B_3 := \max \frac{1}{2} \sum_{i<j} w_{ij}(1 - v_i^{\mathrm{T}}v_j)$$
$$\text{s.t. } \|v_i\|_2 = 1, \quad \forall i. \tag{2.23}$$

We note that in their well-known qualitative analysis of the SDP relaxation (the next relaxation we present), Goemans and Williamson [39] proved and used the fact that this relaxation and the SDP relaxation are equivalent. (For more details on their qualitative analysis, see Section 2.6.)

2.3.2. The semidefinite relaxation
This relaxation can be derived in (at least) two different ways. One way is to relax the formulation MC3 by removing the ± 1 constraint on the elements of X. The result

is the semidefinite programming problem

$$B_4 := \max \ \text{trace} \, QX$$

(SDP1)   s.t. $\text{diag}(X) = e,$ (2.24)

$$X \succcurlyeq 0,$$

where $Q = \frac{1}{4}L$. SDP1 is a convex programming problem and is therefore tractable [92].

Alternatively, this relaxation can be obtained from the formulation MC1 using the fact that the trace is commutative:

$$v^{\mathrm{T}} Q v = \text{trace} \, v^{\mathrm{T}} Q v = \text{trace} \, Q v v^{\mathrm{T}}$$

and that for $v \in \mathscr{F}_n$, $X_{ij} = v_i v_j$ defines a symmetric, rank-one, positive semidefinite matrix X with diagonal elements 1. Therefore, we can lift the problem MC1 into the (higher dimensional) space $\mathscr{S}^n$ of symmetric matrices. This is an alternative way to derive the formulation MC4:

$$\mu^* = \max \ \text{trace} \, QX$$

s.t. $\text{diag}(X) = e,$

(MC4) $\text{rank}(X) = 1,$ (2.25)

$$X \succcurlyeq 0, \quad X \in \mathscr{S}^n.$$

Removing the rank-one constraint from MC4 yields the SDP1 relaxation and the bound B_4.

2.4. Strength of the Lagrangian relaxation

Consider the problem MCQ and replace the constraint $v \in \mathscr{F}_n$ with the equivalent constraints $v_i^2 = 1$, $\forall i$. The result is yet another formulation of MC which we refer to as MC$_\mathrm{E}$:

(MC$_\mathrm{E}$) $\mu^* = \max \ q_0(v)$ (2.26)

s.t. $v_i^2 = 1, \quad i = 1, \ldots, n.$

It is straightforward to check that the Lagrangian dual of the problem MC$_\mathrm{E}$ is

$$\min_u \max_v q_0(v) + \sum_i u_i(v_i^2 - 1)$$

and that it yields precisely our first bound B_0.

It is shown in [101,100] that all the above relaxations and bounds for MC are equivalent to the Lagrangian dual of MC$_\mathrm{E}$. The strong duality result for the trust-region subproblem [110] is the key for proving the following theorem:

Theorem 2.3. *All the bounds for MCQ discussed above are equal to the optimal value of the Lagrangian dual of the equivalent problem MC$_\mathrm{E}$.*

Hence our theme about the strength of the Lagrangian relaxation.

The application of Lagrangian relaxation to obtain quadratic bounds has been extensively studied and used in the literature, for example in [69] and more recently in [70]. The latter calls the Lagrangian relaxation the "best convex bound". Discussions on Lagrangian relaxation for non-convex problems also appear in [29]. More references are given throughout this paper.

2.5. Computing the bounds

While it is true that all the relaxations we have presented so far yield the same bound, it is not necessarily true that all are equally efficient when it comes to computing bounds for MC. Since the qualitative analysis of Goemans and Williamson (see Section 2.6), a lot of research work has focused on the semidefinite relaxation SDP1. For this reason, and since all the bounds we have presented so far are equivalent to the SDP1 bound, we shall change our notation at this point and from now on denote the optimal value of SDP1 by v_1^* (our subsequent SDP relaxations will be similarly indexed). It is also for that reason that this Section mostly focuses on algorithms for computing the bound v_1^*. Nonetheless, we believe that it is still unclear at this time which are the best relaxations to use.

2.5.1. Computing the semidefinite programming bound

From a theoretical point of view, given a semidefinite programming problem, we can find in polynomial time an approximate solution to within any (fixed) accuracy using interior-point methods. This follows from the seminal work of Nesterov and Nemirovskii much of which is summarized in [92]. They also implemented the first interior-point method for SDP in [91]. Independently, Alizadeh extended interior-point polynomial-time algorithms from linear programming to SDP and studied applications to discrete optimization [4,5]. Non-smooth optimization methods for solving semidefinite programming problems have also been proposed (see e.g. [46]).

Before we proceed let us observe that $X = I$ is a strictly positive definite feasible point for SDP1 (usually refered to as a Slater point) and therefore strong duality holds, i.e. both SDP1 and its dual DSDP1:

$$v_1^* := \min \ e^T y$$

$$\text{(DSDP1)} \qquad \text{s.t.} \ \ Z = \text{Diag}(y) - Q, \tag{2.27}$$

$$Z \succcurlyeq 0, \quad y \in \mathfrak{R}^n,$$

have the same optimal value v_1^*. Hence to compute the bound it suffices to solve either one of these SDPs.

Most efficient interior-point methods available for solving SDPs (e.g. [48,45,113,6,93, 31,88,87,20,111]) are primal–dual methods that require solving a dense Newton system of dimension equal to the number of constraints. The solution of this system is then used as a search direction. Typically, some form of line search is performed and it requires a few Cholesky factorizations of the matrix variables concerned to ensure the positive semidefinite constraints are not violated by the next iterate. Although current research is exploring ways to exploit sparsity in this framework (see e.g. [32,33]), most interior-point approaches are still very slow when applied to large ($n \geqslant 1000$) instances of SDP1. (Practical applications typically have at least a few thousand variables.)

One important weakness of interior-point methods is that the matrix variables are usually dense even when the matrix Q and the linear constraints of the SDP are sparse and structured, as is the case for SDP1. Several researchers have therefore proposed alternative approaches to evaluate the bound v_1^* which seek to exploit the structure of SDP1. We summarize here several promising approaches in this direction.

2.5.2. Solving the primal problem SDP1

A successful approach in this direction was introduced by Homer and Peinado [53] and improved on by Burer and Monteiro [21].

These algorithms can be interpreted as projected gradient methods applied to a constrained non-linear reformulation of SDP1. More specifically, Homer and Peinado use the fact that the constraint $X \succcurlyeq 0$, $X \in \mathscr{S}^n$ is equivalently formulated as $X = VV^{\mathrm{T}}$, $V \in \mathfrak{R}^{n \times n}$ (recall the connections between formulation MC2 with $p = n$ and formulation MC3). Burer and Monteiro improve on the efficiency of this approach by observing further that V can be restricted to be a lower triangular matrix, and hence simplify the computations involved in each iteration of the projected gradient method. We refer the reader to the above references for more details.

2.5.3. Solving the dual problem DSDP1

Another alternative to interior-point methods is the use of bundle methods for min–max eigenvalue optimization. As seen in Section 2.2.4, the MC problem is equivalent to the min–max eigenvalue problem

$$\min \ e^{\mathrm{T}} y + n\lambda_{\max}(Q - \mathrm{Diag}(y))$$
$$\text{s.t.} \qquad y \in \mathfrak{R}^n. \tag{2.28}$$

Helmberg and Rendl [47] develop a suitable bundle method for solving this problem and report numerical results for relaxations of MC instances with up to $n = 3000$ nodes. A detailed survey of their work and related results appears in [46]. The min–max eigenvalue approach for more general SDPs is discussed in Section 3.1.

Finally, back in the realm of interior-point methods, Benson et al. [17] derived and implemented an efficient and promising potential-reduction affine scaling algorithm to solve DSDP1. This polynomial-time algorithm generates the Newton system very quickly by virtue of the special structure of the n linear constraints of SDP1. This approach is further improved by Choi and Ye [22] via the use of a preconditioned conjugate gradient method to accelerate the generation of an approximate solution for the Newton system. Computational results for problems of dimension up to $n = 14\,000$ are reported in [22].

2.5.4. Other relaxations

We conclude this section by recalling that there has been very little numerical experimentation with the other relaxations we have presented even though, for example, fast and efficient quadratic programming algorithms are available and could be used to compute the bound B_2 from the box relaxation. Furthermore, it is possible that other relaxations could be better numerically in certain circumstances and therefore that the choice of tractable bound to use should dependent on the particular instance of the problem. We believe that more research is needed in this direction.

2.6. Qualitative analysis of the bounds

Several interesting results on the quality of the bound v_1^*, and hence (by Theorem 2.3) on the quality of the Lagrangian relaxation, have been published in recent years. We have already mentioned the celebrated proof of Goemans and Williamson that, under the assumption that all the edge weights are non-negative, the SDP bound always satisfies

$$\mu^* \geqslant \alpha v_1^*, \tag{2.29}$$

where $\alpha = \min_{0 \leqslant \theta \leqslant \pi} (2/\pi)\theta/(1 - \cos\theta) \approx 0.87856$. This immediately implies that $v_1^* \leqslant 1.14\mu^*$, i.e. v_1^* is guaranteed to overestimate μ^* by at most 14%.

Alternatively, we can state this result as follows. Let us define the quantity μ_* as the optimal value of the problem:

$$\mu_* := \min \ v^{\mathrm{T}}Qv$$
$$\text{s.t.} \ \ v \in \mathscr{F}_n, \tag{2.30}$$

where $Q = \frac{1}{4}L$. (Note that $\mu_* = 0$ in the absence of negative edge weights.) Goemans and Williamson [39] proved that

$$\frac{\mu^* - v_1^*}{\mu^* - \mu_*} \leqslant (1 - \alpha) \approx 0.1214. \tag{2.31}$$

Nesterov [90] proved that without any assumption on the matrix Q, the following result holds:

$$\frac{\mu^* - v_1^*}{\mu^* - \mu_*} \leqslant \frac{4}{7}. \tag{2.32}$$

This line of analysis is extended further in [121,94].

We now proceed to illustrating the application of Lagrangian relaxation to obtain tighter bounds for the MC problem.

2.7. Strengthened SDP relaxations for MC

The results in Sections 2.2–2.4 may give the impression that we have the tightest possible tractable bound for MC. It turns out that this is not the case because adding redundant quadratic constraints to the MC formulations before applying Lagrangian relaxation makes it possible to obtain stronger bounds. In fact, the addition of redundant quadratic of the type that we use here was shown in [12,11] to guarantee strong duality for certain problems where duality gaps can exist.

The process we employ to obtain a strengthened SDP relaxation for MC is an illustration of the recipe to find SDP relaxations presented in [100]. This process also illustrates the power of using the Lagrangian relaxation to derive SDP relaxations. The

recipe is roughly the following:
- Add redundant constraints to the MC formulation.
- Take the Lagrangian dual of the Lagrangian dual to obtain the SDP relaxation.
- Finally, remove all the redundant constraints in the SDP relaxation.

The first step of the recipe asks that we add redundant constraints to the MC formulation. In order to apply the recipe effectively, we shall make a particular choice of formulation from among the various formulations of MC presented in Section 2.1. Indeed, if we restrict ourselves to formulating MC over the feasible set $\mathscr{F}_n(1)$, then it is not clear what redundant constraints one can add. However, when MC is formulated over $\mathscr{F}_n(p)$, there are many constraints that can be added. Let us therefore recall the formulation MC3:

$$\mu^* = \max \ \text{trace} \, QX$$

$$\text{(MC3)} \qquad \begin{aligned} \text{s.t.} \quad & \text{diag}(X) = e, \\ & |X_{ij}| = 1, \quad \forall i < j, \\ & X \succcurlyeq 0, \end{aligned}$$

where $Q = \frac{1}{4}L$.

First we may consider adding linear constraints. Among the many linear inequalities we may add are the well-known triangle inequalities that define the metric polytope M_n [45,47,48]:

$$M_n := \{X \in \mathscr{S}^n: \ \text{diag}(X) = e, \ \text{and}$$

$$X_{ij} + X_{ik} + X_{jk} \geqslant -1, \ X_{ij} - X_{ik} - X_{jk} \geqslant -1,$$

$$-X_{ij} + X_{ik} - X_{jk} \geqslant -1, \ -X_{ij} - X_{ik} + X_{jk} \geqslant -1,$$

$$\forall 1 \leqslant i < j < k \leqslant n\}.$$

These inequalities model the easy observation that for any three mutually connected nodes of the graph, only two or none of the edges may be cut. There are $4\binom{n}{3}$ such inequalities, which is a rather large number of constraints to add to the SDP, and it is not the case that adding a certain subset of triangle inequalities will improve every instance of MC. Instead of adding these constraints to MC3, we will instead add certain quadratic constraints (see below) that are closely related to these inequalities.

Beyond the addition of linear constraints, the addition of redundant quadratic constraints can be particularly effective, as was already mentioned above. In fact, the appropriate choice of quadratic constraints will play an important role in our derivation of tighter bounds for MC.

Several interesting choices of quadratic constraints are available. One obvious possibility is to formulate the constraints $|X_{ij}| = 1$ in MC3 as quadratic constraints using the Hadamard product:

$$X \circ X = E,$$

where $E \in \mathscr{S}^n$ is the matrix of all ones. Let us, in fact, replace the absolute value constraints with these quadratic constraints and obtain the formulation:

$$
\begin{aligned}
\mu^* = \max \ &\text{trace}\, QX \\
\text{s.t.} \quad &\text{diag}\, X = e, \\
&X \circ X = E, \\
&X \succcurlyeq 0.
\end{aligned}
\tag{2.33}
$$

Michel Goemans[3] recently suggested the following very interesting set of quadratic constraints:

$$
X_{ij} = X_{ik}X_{kj}, \quad \forall 1 \leqslant i,j,k \leqslant n.
\tag{2.34}
$$

One interpretation for these constraints arises from the alternative derivation of MC4 in Section 2.3.2. If $X_{ij} = v_i v_j$ and $v_k^2 = 1$ for $k = 1,\ldots,n$ then

$$
X_{ij} = v_i v_j = v_i v_k^2 v_j = v_i v_k \cdot v_k v_j = X_{ik} \cdot X_{kj}.
$$

There is also a connection between these constraints and the triangle inequalities in the definition of the metric polytope above. This connection is used in the proof of Theorem 2.12.

For reasons that will be clear later, we do not add these constraints exactly as we have stated them. We shall instead add a weaker form of these constraints by virtue of the observation that

$$
(X^2)_{ij} = \sum_{k=1}^{n} X_{ik}X_{kj}.
$$

If, according to Eq. (2.34), each of the elements in the sum on the right equals X_{ij}, then $(X^2)_{ij} = nX_{ij}$ or equivalently

$$
X^2 = nX.
$$

This very useful quadratic constraint can alternatively be obtained by considering the formulation MC2 with $p = 1$ (or formulation MC4) and observing that if $X = vv^{\mathrm{T}}$, $v \in \{\pm 1\}^n$, then

$$
X^2 = (vv^{\mathrm{T}})(vv^{\mathrm{T}}) = (v^{\mathrm{T}}v)vv^{\mathrm{T}} = nX.
$$

Therefore, we can add the redundant quadratic constraint $X^2 - nX = 0$ and obtain the formulation:

$$
\begin{aligned}
\mu^* = \max \ &\text{trace}\, QX \\
\text{s.t.} \quad &\text{diag}\, X = e, \\
&X \circ X = E, \\
&X^2 - nX = 0, \\
&X \succcurlyeq 0.
\end{aligned}
\tag{2.35}
$$

---

[3] Presented at The 4th International Conference on High Performance Optimization Techniques, June 1999, Rotterdam, Netherlands.

The constraint $X^2 - nX = 0$ will play a central role in the rest of this section. In fact, we shall use it right away to argue that we can drop the constraint $X \succcurlyeq 0$ from our formulation (2.35) of MC. Indeed, because we can simultaneously diagonalize X and X^2, the constraint $X^2 - nX = 0$ implies that the eigenvalues of X must satisfy the equation $\lambda^2 - n\lambda = 0$. Therefore, the only possible eigenvalues for X are 0 and n and we conclude that $X \succcurlyeq 0$ holds. (Let us note here that, by virtue of Lemma 2.8, we incur no loss by removing this constraint before proceeding.)

Hence after the first step of the recipe we have the following formulation of MC:

$$\mu^* = \max \ \text{trace} \ QX,$$

$$\text{(MC6)} \qquad \begin{aligned} \text{s.t.} \quad & \text{diag}(X) = e, \\ & X \circ X = E, \\ & X^2 - nX = 0. \end{aligned} \qquad\qquad (2.36)$$

The next step in the recipe is to form the Lagrangian dual of MC6 and then the dual of the dual. Before we construct the Lagrangian dual, we must pay special attention to the linear constraints $\text{diag}(s\text{Mat}(x)) = e$ in order to avoid increasing the duality gap when we go to the dual. The following simple example illustrates what may happen.

Example 2.4. Consider the problem

$$\max \ x^2$$

$$\text{s.t.} \quad x = 0.$$

Obviously the optimal value is 0. However, the Lagrangian dual has optimal value

$$\inf_{\lambda} \max_{x} x^2 + \lambda x = +\infty,$$

so we have introduced a duality gap by lifting the linear constraint as it is. However, if we first replace the linear constraint by $x^2 = 0$ then the Lagrangian dual yields

$$\inf_{\lambda} \max_{x} x^2 + \lambda x^2 = 0.$$

Hence squaring the linear constraint eliminates the duality gap.

It is perhaps surprising that the trick illustrated in the example works in general in our framework, i.e. replacing the constraints $Ax = b$ by $||Ax - b||^2 = 0$ before taking the dual ensures that $Ax = b$ holds in the dual. More precisely

Theorem 2.5 (Poljak et al. [100, Theorem 9]). *Let* $K \subset \mathfrak{R}^n$ *be a finite set, let* $q(x) = x^{\mathsf{T}} Q x - 2 c^{\mathsf{T}} x$, $A \in \mathfrak{R}^{m \times n}$ *and* $b \in \mathfrak{R}^n$. *Then there exists* $\bar{\lambda} \in \mathfrak{R}$ *such that*

$$\max_{x \in K} \{q(x) : ||Ax - b||^2 = 0\} = \max_{x \in K} \{q(x) - \lambda ||Ax - b||^2 \ \forall \lambda \geqslant \bar{\lambda}\}.$$

Hence

$$\max_{x \in K} \{q(x) : ||Ax - b||^2 = 0\} = \min_{\lambda} \max_{x \in K} q(x) - \lambda ||Ax - b||^2$$

and strong duality holds.

In fact, the proof of the theorem shows that the quadratic penalty function is exact in the case that K is a finite set; thus by changing linear constraints to the norm squared constraint before lifting them we are ensuring that they hold after taking the dual [100]. This observation sheds some light on the success of SDP relaxation in discrete optimization. Finally, let us note that the effectiveness of this approach to lift the linear constraints can also be argued via the use of an augmented Lagrangian, i.e. the exactness can be obtained in this alternate way [79].

Let us now return to the application of the recipe. To reduce the number of variables by taking advantage of the symmetry in the problem, let us rewrite MC6 using the variable $x \in \mathfrak{R}^{t(n)}$ such that $x = \mathrm{svec}(X)$:

$$\mu^* = \max \ \mathrm{trace}\, Q\, \mathrm{sMat}(x)$$

$$\text{s.t.} \quad \mathrm{diag}(\mathrm{sMat}(x)) = e,$$

$$\mathrm{sMat}(x) \circ \mathrm{sMat}(x) = E,$$

$$(\mathrm{sMat}(x))^2 - n\,\mathrm{sMat}(x) = 0,$$

$$x \in \mathfrak{R}^{t(n)}.$$

Replacing the linear constraint by the norm constraint and homogenizing the problem using the scalar variable y_0, we have

$$\mu^* = \max \ \mathrm{trace}(Q\, \mathrm{sMat}(x))y_0$$

$$\text{s.t.} \quad \mathrm{sdiag}(x)^{\mathrm{T}} \mathrm{sdiag}(x) - 2e^{\mathrm{T}} \mathrm{sdiag}(x)y_0 + n = 0,$$

$$E - \mathrm{sMat}(x) \circ \mathrm{sMat}(x) = 0,$$

$$\mathrm{sMat}(x)^2 - n\,\mathrm{sMat}(x)y_0 = 0, \tag{2.37}$$

$$1 - y_0^2 = 0,$$

$$x \in \mathfrak{R}^{t(n)}, \quad y_0 \in \mathfrak{R}.$$

Note that this problem is equivalent to the previous formulation since we can change x to $-x$ if $y_0 = -1$.

We now write down the Lagrangian dual of 2.37 using Lagrange multipliers $w, t \in \mathfrak{R}$ and $T, S \in \mathscr{S}^n$:

$$\mu^* \leqslant v_2^* := \min_{t,w,T,S} \max_{x,y_0} \mathrm{trace}(Q\, \mathrm{sMat}(x))y_0$$

$$+ w(\mathrm{sdiag}(x)^{\mathrm{T}} \mathrm{sdiag}(x) - 2e^{\mathrm{T}} \mathrm{sdiag}(x)y_0 + n)$$

$$+ \mathrm{trace}\, T(E - \mathrm{sMat}(x) \circ \mathrm{sMat}(x))$$

$$+ \mathrm{trace}\, S((\mathrm{sMat}(x))^2 - n\,\mathrm{sMat}(x)y_0) + t(1 - y_0^2). \tag{2.38}$$

The inner maximization of the above relaxation is an unconstrained pure quadratic maximization whose optimal value is $+\infty$ unless the Hessian is negative semidefinite in which case $x = 0$, $y_0 = 0$ is optimal. Therefore let us calculate the Hessian.

Using trace Q sMat$(x) = x^T$ dsvec(Q), and pulling out a 2 (for convenience later), we can express H_Q, the constant part (without Lagrange multipliers) of the Hessian as

$$2H_Q := 2 \begin{pmatrix} 0 & \frac{1}{2}\text{dsvec}(Q)^T \\ \frac{1}{2}\text{dsvec}(Q) & 0 \end{pmatrix}. \tag{2.39}$$

For notational convenience, we let $\mathscr{H}(w, T, S, t)$ denote the *negative* of the non-constant part of the Hessian, and we split it into four linear operators with the factor 2:

$$2\mathscr{H}(w, T, S, t) := 2\mathscr{H}_1(w) + 2\mathscr{H}_2(T) + 2\mathscr{H}_3(S) + 2\mathscr{H}_4(t)$$

$$= 2w \begin{pmatrix} 0 & (\text{dsvec Diag } e)^T \\ (\text{dsvec Diag } e) & -\text{sdiag}^*\text{sdiag} \end{pmatrix}$$

$$+ 2 \begin{pmatrix} 0 & 0 \\ 0 & \text{dsvec}(T \circ \text{sMat}) \end{pmatrix}$$

$$+ 2 \begin{pmatrix} 0 & \frac{n}{2}\text{dsvec}(S)^T \\ \frac{n}{2}\text{dsvec}(S) & (\text{Mat vsMat})^* S(\text{Mat vsMat}) \end{pmatrix}$$

$$+ 2t \begin{pmatrix} 1 & 0 \\ 0 & 0 \end{pmatrix}. \tag{2.40}$$

We can cancel the 2 in (2.40) and (2.39) and get the (equivalent to the Lagrangian dual) semidefinite program DSDP2:

$$\text{(DSDP2)} \quad \begin{aligned} v_2^* = \min \ & nw + \text{trace } ET + \text{trace } 0S + t \\ \text{s.t.} \ & \mathscr{H}(w, T, S, t) \succcurlyeq H_Q. \end{aligned} \tag{2.41}$$

If we take T sufficiently positive definite and t sufficiently large, then we can guarantee Slater's constraint qualification. Therefore, the dual of DSDP2 has the same optimal value v_2^* and it provides a strengthened SDP relaxation of MC:

$$\text{(SDP2)} \quad \begin{aligned} v_2^* = \max \ & \text{trace } H_Q Y \\ \text{s.t.} \ & \mathscr{H}_1^*(Y) = n, \\ & \mathscr{H}_2^*(Y) = E, \\ & \mathscr{H}_3^*(Y) = 0, \\ & \mathscr{H}_4^*(Y) = 1, \\ & Y \succcurlyeq 0, \quad Y \in \mathscr{S}^{t(n)+1}. \end{aligned} \tag{2.42}$$

To help define the adjoint operators we partition Y as

$$Y = \begin{pmatrix} Y_{00} & x^T \\ x & \bar{Y} \end{pmatrix}, \quad \bar{Y} \in \mathscr{S}^{t(n)}.$$

It is straightforward to check that

$$\mathscr{H}_2^*(Y) = \text{sMat diag}(\bar{Y}) \quad \text{and} \quad \mathscr{H}_4^*(Y) = Y_{00},$$

so the constraints $\mathscr{H}_2^*(Y) = E$ and $\mathscr{H}_4^*(Y) = 1$ are equivalent to $\text{diag}(Y) = e$. Also, $\mathscr{H}_1^*(Y)$ is twice the sum of the elements in the first row of Y corresponding to the positions of the diagonal of $\text{sMat}(x)$ minus the sum of the same elements in the diagonal of $\bar{Y}$, i.e.

$$\mathscr{H}_1^*(Y) = 2\text{svec}(I_n)^{\mathrm{T}}x - \text{trace Diag}(\text{svec}(I_n))\bar{Y}.$$

The constraint $\mathscr{H}_1^*(Y) = n$ requires that $Y_{0,t(i)} = 1, \forall i = 1,\ldots,n$, as shown in the proof of Lemma 2.7 below.

Finally, to find $\mathscr{H}_3^*(Y)$, recall that by definition,

$$\langle \mathscr{H}_3(S), Y \rangle = n\,\text{dsvec}(S)^{\mathrm{T}}x - \langle(\text{Mat vsMat})^*S(\text{Mat vsMat}), \bar{Y}\rangle.$$

taking adjoints,

$$\langle S, \mathscr{H}_3^*(Y)\rangle = \text{trace } Sn\,\text{sMat}(x) - \langle S, (\text{Mat vsMat})\bar{Y}(\text{Mat vsMat})^*\rangle$$

$$= \langle S, n\,\text{sMat}(x) - (\text{Mat vsMat})\bar{Y}(\text{Mat vsMat})^*\rangle.$$

Note that $(\text{Mat vsMat})^* = \text{vsMat}^*\text{vec}$ is essentially (and in the symmetric case reduces to) sMat^* except that it acts on possibly non-symmetric matrices. Hence,

$$\mathscr{H}_3^*(Y) = n\,\text{sMat}(x) - (\text{Mat vsMat})\bar{Y}(\text{Mat vsMat})^*. \tag{2.43}$$

Equivalently, $\mathscr{H}_3^*(Y)$ consists of the sums in SDP2 below. The constraint $\mathscr{H}_3^*(Y) = 0$ is key to showing that for Y feasible for SDP2, $\text{sMat}(x)$ is always positive semidefinite (and in fact feasible for SDP1).

The end result as an SDP with linear constraints. The last step of the recipe consists of removing the redundant constraints in this SDP. This is usually done using the structure of the problem. The result after deleting redundant constraints is the following SDP relaxation of MC (see [7,9] for details):

$$v_2^* = \max \text{ trace } H_Q Y$$

$$\text{s.t.} \quad \text{diag}(Y) = e,$$

$$\text{(SDP2)} \qquad Y_{0,t(i)} = 1, \quad i = 1,\ldots,n, \tag{2.44}$$

$$Y_{0,T(i,j)} = \frac{1}{n}\sum_{k=1}^{n} Y_{T(i,k),T(k,j)}, \quad \forall i,j \text{ s.t. } 1 \leqslant i < j \leqslant n,$$

$$Y \succcurlyeq 0, \quad Y \in \mathscr{S}^{t(n)+1},$$

where

$$T(i,j) := \begin{cases} t(j-1)+i & \text{if } i \leqslant j, \\ t(i-1)+j & \text{otherwise.} \end{cases} \tag{2.45}$$

Remark 2.6. The indices for the linear constraints in SDP2 may be thought of as the entries of a matrix T constructed in the following way. Expanding the relationship $X = \text{sMat}(x)$ we have

$$X = \begin{pmatrix} x_1 & x_2 & x_4 & \cdots \\ x_2 & x_3 & x_5 & \cdots \\ & \cdots & & x_{t(n)} \end{pmatrix}.$$

Let us now keep only the indices of the entries of x and thereby define the matrix T:

$$T = \begin{pmatrix} 1 & 2 & 4 & \cdots \\ 2 & 3 & 5 & \cdots \\ & \cdots & & t(n) \end{pmatrix}.$$

In fact, there is still some redundancy in the constraints of SDP2 as we now show that Slater's constraint qualification does not hold.

Lemma 2.7. *If Y is feasible for SDP2, then Y is singular.*

Proof. Let Y be feasible for SDP2. The constraints $\mathscr{H}_2^*(Y) = E$ and $\mathscr{H}_4^*(Y) = 1$ together imply that $\text{diag}(Y) = e$. The constraint $\mathscr{H}_1^*(Y) = n$ can be written as

$$2\text{svec}(I_n)^{\mathrm{T}} x - \text{trace Diag}(\text{svec}(I_n))\bar{Y} = n$$

with

$$Y = \begin{pmatrix} 1 & x^{\mathrm{T}} \\ x & \bar{Y} \end{pmatrix}.$$

Since $\text{diag}(Y) = e$, $\text{trace Diag}(\text{svec}(I_n))\bar{Y} = n$ and so $\text{svec}(I_n)^{\mathrm{T}} x = n$, or equivalently $\sum_{i=1}^{n} Y_{0,t(i)} = n$. Now $Y \succcurlyeq 0$ implies every principal minor of Y is non-negative, so $|Y_{0,t(i)}| \leqslant 1$ must hold (again because $\text{diag}(Y) = e$). So $\sum_{i=1}^{n} Y_{0,t(i)} = n \Rightarrow Y_{0,t(i)} = 1$, $i = 1,\ldots,n$. Hence each of the 2×2 principal minors obtained from the subsets of rows and columns $\{0, t(i)\}$, $i = 1,\ldots,n$ equals zero. Hence Y is not positive definite. $\square$

This result makes it possible to further reduce the number of constraints in SDP2 by projecting the problem onto the positive semidefinite cone of dimension $t(n-1) + 1$. This is done in detail in [7,9].

2.7.1. Properties of the strengthened relaxation

We now state and prove some of the interesting properties of the relaxation SDP2.

One surprising result is that the matrix obtained by applying sMat to the first row of a feasible Y is positive semidefinite, even though this nonlinear constraint was not explicitly included in the formulation MC6.

Lemma 2.8. *Suppose that Y is feasible for SDP2. Then*

$$\mathrm{sMat}(Y_{0,1:t(n)}) \succcurlyeq 0$$

and so is feasible for SDP1.

Proof. For Y feasible for SDP2, write

$$Y = \begin{pmatrix} 1 & x^{\mathrm{T}} \\ x & \bar{Y} \end{pmatrix}$$

with $x = Y_{0,1:t(n)}$. Note that $\bar{Y}$ is a principal submatrix of Y and therefore $\bar{Y} \succcurlyeq 0$. By (2.43), the constraint $\mathscr{H}_3^*(Y) = 0$ is equivalent to

$$\mathrm{sMat}(x) = \frac{1}{n} (\mathrm{Mat\ vsMat}) \bar{Y} (\mathrm{Mat\ vsMat})^*$$

and thus $\mathrm{sMat}(x)$ is a congruence of the positive semidefinite matrix $\bar{Y}$. The result follows. $\square$

We now prove that the relaxation SDP2 is a strengthening of SDP1.

Theorem 2.9. *The optimal values satisfy*

$$v_2^* \leqslant v^*. \tag{2.46}$$

Proof. Suppose that

$$Y^* = \begin{pmatrix} 1 & x^{*\mathrm{T}} \\ x^* & \bar{Y}^* \end{pmatrix}$$

solves SDP2. From Lemma 2.8, it is clear that $\mathrm{sMat}(x^*)$ is feasible for SDP1. Therefore,

$$
\begin{aligned}
v_2^* &= \mathrm{trace}\, H_Q Y^* \\
&= (\mathrm{dsvec}\, Q)^{\mathrm{T}} x^* \\
&= \mathrm{trace}\, Q\, \mathrm{sMat}(x^*) \\
&\leqslant v^*. \quad \square
\end{aligned}
$$

2.7.2. A further strengthening of the relaxation SDP2

We now examine SDP2 more closely and show how an even tighter relaxation can be obtained. It may be helpful to the reader at this point to reexamine the formulation MC5 since both SDP2 and the upcoming relaxation SDP3 have connections to that formulation.

Let us begin by recalling the alternative derivation of MC4 in Section 2.3.2 and the rank-one matrices $X = vv^{\mathrm{T}}$, $v \in \{\pm 1\}^n$. We know that these matrices X have all their entries equal to ± 1. Hence, the corresponding matrices Y feasible for SDP2 have all their entries in the first row and column equal to ± 1. Looking back to the formulation MC5 this statement corresponds to the (hard) constraints $|u_0^{\mathrm{T}} u_i| = 1$ for all i.

Now let us consider the following constraints of SDP2:

$$Y_{0,T(i,j)} = \frac{1}{n}\sum_{k=1}^{n} Y_{T(i,k),T(k,j)}, \quad \forall 1 \leqslant i < j \leqslant n \tag{2.47}$$

for

$$Y = \begin{pmatrix} 1 & x^{\mathrm{T}} \\ x & \bar{Y} \end{pmatrix} \quad \text{and} \quad x = \mathrm{svec}(vv^{\mathrm{T}}).$$

The entry $Y_{0,T(i,j)}$ is in the first row of Y and therefore it is equal to 1 in magnitude. The corresponding constraint in (2.47) says that it must be equal to the average of n specific entries in the block $\bar{Y}$. But each of these n entries has magnitude at most 1, and hence for equality to hold, they must all have magnitude equal to 1, and in fact they must all equal $Y_{0,T(i,j)}$.

Let us state this observation in a different way. If Y and X are both rank-one, then the block $\bar{Y} = xx^{\mathrm{T}}$ and $Y_{T(i,k),T(k,j)} = x_{T(i,k)}x_{T(k,j)} = v_i v_k \cdot v_k v_j$. But if $v_k^2 = 1$, then $Y_{T(i,k),T(k,j)} = v_i v_j = X_{ij} = Y_{0,T(i,j)}$.

There is yet another interpretation for this observation. Recall that we obtained the quadratic constraint $X^2 = nX$ by appropriately adding up (and thereby weakening) the quadratic constraints (2.34). What we are observing here is that constraints (2.47) consist of sums that originate in this weakening of constraints (2.34). Hence we can now "undo" these sums and retrieve a linearized version of constraints (2.34) in terms of the entries of the matrix variable Y.

This discussion leads us to define the relaxation SDP3 as

$$v_3^* = \max \ \mathrm{trace}\, H_Q Z$$

$$\text{s.t.} \quad \mathrm{diag}(Z) = e,$$

(SDP3) $\qquad\qquad Z_{0,t(i)} = 1, \quad i = 1, \ldots, n,$ $\hfill (2.48)$

$$Z_{0,T(i,j)} = Z_{T(i,k),T(k,j)}, \quad \forall k, \ \forall 1 \leqslant i < j \leqslant n,$$

$$Z \succcurlyeq 0, \quad Z \in \mathcal{S}^{t(n)+1}.$$

We proceed to prove that SDP3 is a *strict* improvement on the addition of all the triangle inequalities to SDP1. First, let us define

$$F_n := \{X \in \mathcal{S}: X = \mathrm{sMat}(Z_{0,1:t(n)}), \ Z \text{ feasible for SDP3}\}.$$

Since the feasible set of SDP3 is convex and compact, and since F_n is the image of that feasible set under a linear transformation, it follows that F_n is also convex and compact.

For completeness, we begin by proving that SDP3 is indeed a relaxation of MC. This is not guaranteed a priori since SDP3 is a strengthening of SDP2.

Lemma 2.10. $C_n \subseteq F_n.$

Proof. Consider an extreme point of C_n, $X = vv^{\mathrm{T}}$, $v \in \{\pm 1\}^n$. Let

$$x = \mathrm{svec}(X) \quad \text{and} \quad Z = \begin{pmatrix} 1 \\ x \end{pmatrix}\begin{pmatrix} 1 \\ x \end{pmatrix}^{\mathrm{T}}.$$

We show that Z is feasible for SDP3.

Clearly $Z \succcurlyeq 0$ and $Z_{0,0} = 1$. Since $x_{T(i,j)} = v_i v_j$, for $1 \leqslant i \leqslant j \leqslant n$,

$$Z_{T(i,j), T(i,j)} = (x_{T(i,j)})^2 = v_i^2 v_j^2 = 1.$$

Therefore $\operatorname{diag}(Z) = e$. Also, $Z_{0,t(i)} = Z_{0,T(i,i)} = x_{T(i,i)} = v_i^2 = 1$. Finally, for $1 \leqslant i < j \leqslant n$,

$$Z_{T(i,k), T(k,j)} = x_{T(i,k)} x_{T(k,j)}$$

$$= v_i v_k v_k v_j$$

$$= v_i v_j$$

$$= x_{T(i,j)}$$

$$= Z_{0, T(i,j)}.$$

Hence, each $X = vv^{\mathrm{T}}$, $v \in \{\pm 1\}^n$ has a corresponding Z feasible for SDP3, and so $X \in F_n$. Since C_n and F_n are convex, we are done. $\square$

Clearly, every Z feasible for SDP3 is feasible for SDP2. Therefore, by Lemma 2.8 above, we have the inclusion:

Corollary 2.11. $F_n \subseteq \mathscr{E}_n$.

Using Lemma 2.10, we observe that $\mu^* \leqslant v_3^* \leqslant v_2^* \leqslant v_1^*$. Furthermore, the strengthening result of Theorem 2.9 also holds for SDP3.

We now exploit the fact that there is a strong connection between the quadratic constraints (2.34) and the triangle inequalities to prove the next theorem.

Theorem 2.12. $F_n \subseteq M_n$.

Proof. Suppose $X \in F_n$, then $X = \mathrm{sMat}(Z_{0,1:t(n)})$ for some Z feasible for SDP3. Since $Z_{0,t(i)} = 1 \ \forall i$, it follows that $\operatorname{diag}(X) = e$ holds.

Given i, j, k such that $1 \leqslant i < j < k \leqslant n$, let $Z_{i,j,k}$ denote the 4×4 principal minor of Z corresponding to the indices $0, T(i,j), T(i,k), T(j,k)$. Let $a = X_{ij} = Z_{0,T(i,j)}$, $b = X_{ik} = Z_{0,T(i,k)}$, $c = X_{jk} = Z_{0,T(j,k)}$. Then

$$Z_{i,j,k} = \begin{pmatrix} 1 & a & b & c \\ a & 1 & c & b \\ b & c & 1 & a \\ c & b & a & 1 \end{pmatrix},$$

since $\operatorname{diag}(Z) = e$ and $Z_{0,T(i,j)} = Z_{T(i,k),T(k,j)}$, $Z_{0,T(i,k)} = Z_{T(i,j),T(j,k)}$ and $Z_{0,T(j,k)} = Z_{T(j,i),T(i,k)}$ all hold for Z feasible for SDP3. Now

$$
Z_{i,j,k} \succcurlyeq 0 \Leftrightarrow \begin{pmatrix} 1 & c & b \\ c & 1 & a \\ b & a & 1 \end{pmatrix} - \begin{pmatrix} a \\ b \\ c \end{pmatrix} (a \ \ b \ \ c) \succcurlyeq 0
$$

$$
\Leftrightarrow \begin{pmatrix} 1 - a^2 & c - ab & b - ac \\ c - ab & 1 - b^2 & a - bc \\ b - ac & a - bc & 1 - c^2 \end{pmatrix} \succcurlyeq 0
$$

$$
\Rightarrow e^{\mathrm{T}} \begin{pmatrix} 1 - a^2 & c - ab & b - ac \\ c - ab & 1 - b^2 & a - bc \\ b - ac & a - bc & 1 - c^2 \end{pmatrix} e \geqslant 0.
$$

Hence,

$$
Z_{i,j,k} \succcurlyeq 0 \Rightarrow 3 - (a + b + c)^2 + 2(a + b + c) \geqslant 0
$$

$$
\Leftrightarrow \gamma^2 - 2\gamma - 3 \leqslant 0, \quad \text{where } \gamma := a + b + c
$$

$$
\Leftrightarrow (\gamma - 3)(\gamma + 1) \leqslant 0
$$

$$
\Leftrightarrow -1 \leqslant \gamma \leqslant 3
$$

$$
\Rightarrow a + b + c \geqslant -1.
$$

Therefore, $X_{ij} + X_{ik} + X_{jk} \geqslant -1$ holds for X.

Because multiplication of row and column i of $Z_{i,j,k}$ by -1 will not affect the positive semidefiniteness of $Z_{i,j,k}$, if we multiply the two rows and two columns of $Z_{i,j,k}$ with indices $T(i,k)$ and $T(j,k)$ and apply the same argument to the resulting matrix, we obtain the inequality $X_{ij} - X_{ik} - X_{jk} \geqslant -1$. Similarly, the inequalities $-X_{ij} + X_{ik} - X_{jk} \geqslant -1$ and $-X_{ij} - X_{ik} + X_{jk} \geqslant -1$ also hold. □

We have thus proved the following:

Corollary 2.13. $C_n \subseteq F_n \subseteq \mathscr{E}_n \cap M_n$.

Appropriate examples are provided in [7,9] to prove the following strict improvement result for SDP3:

Theorem 2.14. $C_n \subsetneqq F_n \subsetneqq \mathscr{E}_n \cap M_n$ for $n \geqslant 5$.

Table 1
Numerical comparison of all MC relaxations for small test problems

Graph	μ^*	SDP1 bound	SDP2 bound	M_n bound	$\mathscr{E}_n \cap M_n$ bound	SDP3 bound
C_5	4	4.5225	4.2889	4.0000	4.0000	4.0000
		$\rho = 0.8845$	$\rho = 0.9326$	$\rho = 1.0000$	$\rho = 1.0000$	$\rho = 1.0000$
		R.E.: 13.06%	R.E.: 7.22%	R.E.: 0%	R.E.: 0%	R.E.: 0%
$K_5 \setminus e$	6	6.2500	6.1160	6.0000	6.0000	6.0000
		$\rho = 0.9600$	$\rho = 0.9810$	$\rho = 1.0000$	$\rho = 1.0000$	$\rho = 1.0000$
		R.E.: 4.17%	R.E.: 1.93%	R.E.: 0%	R.E.: 0%	R.E.: 0%
K_5	6	6.2500	6.2500	6.6667	6.2500	6.2500
		$\rho = 0.9600$	$\rho = 0.9600$	$\rho = 0.9000$	$\rho = 0.9600$	$\rho = 0.9600$
		R.E.: 4.17%	R.E.: 4.17%	R.E.: 11.11%	R.E.: 4.17%	R.E.: 4.17%
Given by $A(G)$	9.28	9.6040	9.4056	9.3867	9.2961	9.2800
		$\rho = 0.9663$	$\rho = 0.9866$	$\rho = 0.9886$	$\rho = 0.9983$	$\rho = 1.0000$
		R.E.: 3.49%	R.E.: 1.35%	R.E.: 1.15%	R.E.: 0.17%	R.E.: 0%
AW_9^2	12	13.5	12.9827	12.8571	12.6114	12.4967
		$\rho = 0.8889$	$\rho = 0.9243$	$\rho = 0.9333$	$\rho = 0.9515$	$\rho = 0.9603$
		R.E.: 12.50%	R.E.: 8.19%	R.E.: 7.14%	R.E.: 5.10%	R.E.: 4.14%
Pet.	12	12.5	12.3781	12.0000	12.0000	12.0000
		$\rho = 0.9600$	$\rho = 0.9695$	$\rho = 1.0000$	$\rho = 1.0000$	$\rho = 1.0000$
		R.E.: 4.17%	R.E.: 3.15%	R.E.: 0%	R.E.: 0%	R.E.: 0%
Rand. gen.	88	90.3919	89.5733	89.3333	88.0029	88.0000
		$\rho = 0.9735$	$\rho = 0.9824$	$\rho = 0.9851$	$\rho = 1.0000$	$\rho = 1.0000$
		R.E.: 2.72%	R.E.: 1.79%	R.E.: 1.52%	R.E.: $3.3E - 5$	R.E.: $9.9E - 7$

2.7.3. Numerical results

The relaxations SDP1, SDP2 and SDP3 were compared for several interesting problems using the software package SDPPACK (version 0.9 Beta) [6]. For completeness we also solved the linear relaxation over the metric polytope:

$$\max \ \text{trace}\, QX$$

$$\text{s.t.} \quad X \in M_n.$$

This relaxation is easily formulated as an LP and we solved it using the Matlab solver LINPROG. The results are summarized in Table 1. The value ρ equals the value of the optimal cut divided by the bound, and R.E. denotes the relative error with respect to the optimal cut.

The test problems in Table 1 are as follows:
1. The first line of results corresponds to solving the three SDP relaxations for a 5-cycle with unit edge-weights.
2. The second line corresponds to the complete graph on 5 vertices with unit edge-weights on all edges except one, which is assigned weight zero.
3. The third line corresponds to the complete graph on 5 vertices with unit edge-weights. In this example, none of the four SDP relaxations attains the MC optimal value,

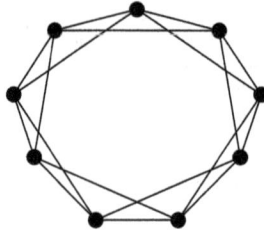

Fig. 1. Antiweb AW_9^2.

and in fact they are not distinguishable. Only the linear relaxation M_n gives a noticeably weaker bound.

4. The fourth line corresponds to the graph defined by the weighted adjacency matrix

$$A(G) = \begin{pmatrix} 0 & 1.52 & 1.52 & 1.52 & 0.16 \\ 1.52 & 0 & 1.60 & 1.60 & 1.52 \\ 1.52 & 1.60 & 0 & 1.60 & 1.52 \\ 1.52 & 1.60 & 1.60 & 0 & 1.52 \\ 0.16 & 1.52 & 1.52 & 1.52 & 0 \end{pmatrix}.$$

This problem is interesting because it shows a significant difference between SDP3 and all the other relaxations; in this case, SDP3 is the only relaxation that attains the MC optimal value.

5. The fifth line corresponds to the graph in Fig. 1 with unit edge weights. This graph is the antiweb AW_9^2 and it is the hardest example [4] that the authors know for the relaxation $\mathcal{E}_n \cap M_n$. It is interesting that SDP3 performs better on this example than on the K_5 with unit edge weights.

6. The last two lines correspond to slightly larger graphs. The graph on 10 vertices is the Petersen graph with unit edge-weights. The graph on 12 vertices is a randomly generated graph that gives slightly different results for each relaxation (the exact description of the graph is in [7,9]).

In Table 1, a relative error equal to zero means that the relative error was below 10^{-11}, the value of the smallest default stopping criteria used by SDPPACK.

We conclude by pointing out that solving the relaxations SDP2 and SDP3 using an interior-point method becomes very time consuming and requires large amounts of memory even for moderate values of n. Nonetheless, their constraints are very sparse and have a special structure therefore it is hoped that research efforts like those mentioned in Section 2.5, or perhaps even entirely new approaches, will allow these relaxations to be solved efficiently for larger values of n.

---

[4] We thank Franz Rendl for suggesting this interesting example.

3. SDP and Lagrangian relaxation for Q^2P's

We now move on to illustrate the Lagrangian relaxation approach for general quadratically constrained quadratic problems (Q^2P). In this section we briefly outline the approach for the general Q^2P and specific instances are considered in some detail in Section 4. This general quadratic problem is also studied in e.g. [30,67,66,68,115,103, 73,71,83,16]. The more general polynomial optimization problem is considered in [74] which presents a relaxation very similar to SDP3 but motivated by result results in the theory of moments and positive polynomials.

The quadratic problem we consider is the following Q^2P:

$$q^* := \max\ q_0(x) : = x^T Q_0 x + 2g_0^T x + \alpha_0$$

$$(Q^2P_x) \qquad \text{s.t.} \qquad q_k(x) : = x^T Q_k x + 2g_k^T x + \alpha_k \leqslant 0, \qquad (3.49)$$

$$k \in \mathscr{I} := \{1, \ldots, m\},$$

$$x \in \mathfrak{R}^n,$$

where the matrices $Q_k \neq 0$, $k = 0, \ldots, m$, are symmetric.

Let us define

$$P_k := \begin{bmatrix} \alpha_k & g_k^T \\ g_k & Q_k \end{bmatrix} \qquad (3.50)$$

and (by abuse of notation)

$$q_k(y) := y^T P_k y, \quad k = 0, 1, \ldots, m.$$

Using the technique for proving the equivalence of MC and MCQ at the beginning of Section 2, we obtain a homogenized formulation of Q^2P_x in terms of the new variable y and we denote it Q^2P_y:

$$q^* = \max\ q_0(y)$$

$$\text{s.t.} \qquad q_k(y) \leqslant 0, \quad k \in \mathscr{I},$$

$$(Q^2P_y) \qquad y_0^2 = 1, \qquad (3.51)$$

$$y = \begin{pmatrix} y_0 \\ x \end{pmatrix} \in \mathfrak{R}^{n+1}.$$

If $y_0 = 1$ is optimal for Q^2P_y, then y is optimal for Q^2P_x and if $y_0 = -1$ is optimal for Q^2P_y, then $-y$ is optimal for Q^2P_x. Hence the optimal values of Q^2P_x and Q^2P_y are equal.

The Lagrangian relaxation of the homogenized problem Q^2P_y provides a simpler path for obtaining the SDP relaxation. Indeed, the Lagrangian of Q^2P_y is

$$L(y, \mu, \lambda) := y^T P_0 y - \mu(y_0^2 - 1) + \sum_{k \in \mathscr{I}} \lambda_k y^T P_k y$$

and therefore the Lagrangian relaxation of Q^2P_y is

$$(DQ^2P_y) \quad d^* := \min_{\mu} \ \max_{y} \ y^T P_0 y - \mu(y_0^2 - 1) + \sum_{k \in \mathscr{I}} \lambda_k y^T P_k y.$$
$$\lambda \geqslant 0$$

Note that

$$d^* = \min_{\lambda \geqslant 0} \min_{\mu} \max_{y} \ y^T P_0 y - \mu(y_0^2 - 1) + \sum_{k \in \mathscr{I}} \lambda_k y^T P_k y$$

$$= \min_{\lambda \geqslant 0} \max_{y_0^2 = 1} \ y^T P_0 y + \sum_{k \in \mathscr{I}} \lambda_k y^T P_k y$$

by the strong duality of the trust-region subproblem [110]. Therefore,

$$(DQ^2P_x) \quad d^* = \min_{\lambda \geqslant 0} \max_{x} \ q_0(x) + \sum_{k \in \mathscr{I}} \lambda_k q_k(x)$$

and we have shown the equivalence of the dual values for the problems in x and in y. (This is similar to the approaches in [119,109].)

By weak duality, we have

$$d^* \geqslant q^* = \max_{y} \ \min_{\mu} \ y^T P_0 y - \mu(y_0^2 - 1) + \sum_{k \in \mathscr{I}} \lambda_k y^T P_k y.$$
$$\lambda \geqslant 0$$

If we can find the optimal values μ^* and λ^* for the dual variables, then we obtain a single quadratic function whose maximal value is an upper bound on q^*:

$$q^* \leqslant d^* = \max_{y} \ y^T P_0 y - \mu^*(y_0^2 - 1) + \sum_{k \in \mathscr{I}} \lambda_k^* y^T P_k y. \tag{3.52}$$

Furthermore, the Lagrangian $L(y, \mu, \lambda)$ is a quadratic function of y and therefore we can add the following hidden semidefinite constraint to the outer minimization of DQ^2P_y:

$$P_0 - \mu E_{00} + \sum_{k \in \mathscr{I}} \lambda_k P_k \preccurlyeq 0, \quad \lambda \geqslant 0, \tag{3.53}$$

where E_{00} is the zero matrix with 1 in the top left corner. The maximum of the maximization subproblem is attained for $y = 0$ and thus the dual problem DQ^2P_y is equivalent to the SDP

$$d^* = \min \ \mu$$

$$\text{s.t.} \qquad \mu E_{00} - \sum_{k \in \mathscr{I}} \lambda_k P_k \succcurlyeq P_0.$$
$$\lambda \geqslant 0.$$

One important observation is that a greater number of quadratic constraints $q_k(y)$ means that we obtain a stronger dual. This is equivalent to our earlier claim that adding redundant quadratic constraints strengthens the SDP relaxation. An excellent illustration of the effectiveness of this strategy is presented in Section 4.4 where this approach achieves strong duality.

Another approach is presented in detail in Kojima and Tunçel [67,66]. For problems that also have linear equality constraints the notion of copositivity can be used to strengthen the SDP relaxation [102]. However, the result is not a tractable relaxation in general.

3.1. Solving SDPs arising from Q^2Ps

There are many existing packages for solving SDPs in the public domain (see e.g. Christoph Helmberg's SDP web page

```
http://www.zib.de/helmberg/semidef.html
```

or The Handbook of Semidefinite Programming [118].) However, we already alluded in Section 2.5 to the limitations of many algorithms when it comes to solving large SDPs. We outlined in that section several research directions that seek to exploit the sparsity and structure of SDP1 and/or DSDP1. For more general SDPs Kojima et al [65,32,33] have made promising advances and Borchers [20,19] exploits the BLAS routines. Nonetheless, the question of efficiently exploiting sparsity is still very much an open question.

We briefly outline one approach that may help exploit structure and sparsity for SDPs in discrete optimization. Recall from Section 2 that the SDP relaxation SDP1 gives the same bound for MC as the eigenvalue bound (2.22). In fact, this equivalence of the bounds holds for any SDP for which trace X is constant over all feasible matrices X [47,46]. (Note that many SDPs that arise in applications satisfy this property.) In particular, the constant trace condition holds for all SDPs that arise from problems which have a bounded feasible set. We can see this by homogenizing Q^2P as in (3.51) and then adding the redundant constraint $||y||^2 \leq K$ with K sufficiently large. Now the identity I is in the range of the linear operator $\mathscr{A}^*$ and this is precisely equivalent to the constant trace condition. Therefore, Q^2Ps with bounded feasible set can all be phrased as min–max eigenvalue problems for which the inherent structure and sparsity can be exploited. Clearly all 0,1 or ±1 problems satisfy this boundedness condition and in particular the graph partitioning and the quadratic assignment problem that we study in the next section fall into this class.

4. Specific instances of SDP relaxations

We now study in some detail four specific problems and show how to apply the recipe for SDP relaxations. In each case we derive a min–max eigenvalue problem from the Lagrangian dual of an appropriately chosen quadratically constrained problem. The dual of this min–max eigenvalue problem then provides an SDP relaxation for the original problem. Adding redundant quadratic constraints at the start helps in reducing the duality gap. Once we obtain the SDP relaxation, any remaining redundancy in the constraints is eliminated if we ensure that the linear constraints have full row rank and that Slater's condition holds. This illustrates again the strength of this Lagrangian approach.

4.1. The graph partitioning problem

Let $G = (V, E)$ be an undirected graph as in the description of the MC problem. The graph partitioning problem is the problem of partitioning the node set V into k disjoint subsets of specified sizes to minimize the total weight of the edges connecting nodes in distinct subsets of the partition. Let $A = (a_{ij})$ be the weighted adjacency matrix of G, i.e.

$$a_{ij} = \begin{cases} w_{ij}, & ij \in E, \\ 0 & \text{otherwise.} \end{cases}$$

The graph partitioning problem can be described by the following 0,1 quadratic problem [106]:

$$
\begin{aligned}
w(E_{\text{uncut}}) &= \max \ \tfrac{1}{2} \operatorname{trace} X^t A X \\
\text{(GP)} \quad \text{s.t.} \quad & X e_k = e_n, \\
& X^\mathrm{T} e_n = m, \\
& X_{ij} \in \{0, 1\}, \quad \forall ij,
\end{aligned}
$$

where e_k is the vector of ones of appropriate size and m is the vector of ordered set sizes

$$m_1 \geqslant \cdots \geqslant m_k \geqslant 1 \quad \text{and} \quad k < n.$$

The columns of the 0,1 $n \times k$ matrices X are the indicator vectors for the sets. If we replace the 0,1 constraints by quadratic constraints and the linear constraints taking their norm squared, we obtain the equivalent problem:

$$
\begin{aligned}
w(E_{\text{uncut}}) &= \max \ \tfrac{1}{2} \operatorname{trace} X^t A X \\
\text{s.t.} \quad & \|X e_k - e_n\|^2 + \|X^\mathrm{T} e_n - m\|^2 = 0, \\
& X_{ij}^2 - X_{ij} = 0, \quad \forall ij.
\end{aligned}
$$

The Lagrangian relaxation yields the following bound:

$$
B_{\text{GP}} := \min_{\alpha, W} \max_{X} \operatorname{trace}[\tfrac{1}{2} X^\mathrm{T} A X + \alpha(e_k e_k^\mathrm{T} X^\mathrm{T} X + X^\mathrm{T} e_n e_n^\mathrm{T} X) + W^\mathrm{T}(X \circ X)
$$

$$
- 2\alpha(e_k e_n^\mathrm{T} X + m e_n^\mathrm{T} X) - W^\mathrm{T} X] + \alpha \left(n + \sum_i m_i^2 \right). \tag{4.54}
$$

We can now homogenize the problem by adding a variable x:

$$B_{GP} := \min_{\alpha,W} \max_{\substack{X \\ x^2=1}} \operatorname{trace}[\tfrac{1}{2}X^T A X + \alpha(e_k e_k^T X^T X + X^T e_n e_n^T X) + W^T(X \circ X)$$

$$+ x(-2\alpha(e_k e_n^T X + m e_n^T X) - W^T X)] + \alpha\left(n + \sum_i m_i^2\right).$$

We now lift the variable x into the Lagrangian to get a min–max eigenvalue problem:

$$B_{GP} := \min_{\alpha,W,\delta} \max_{X,x} \operatorname{trace}[\tfrac{1}{2}X^T A X + \alpha(e_k e_k^T X^T X + X^T e_n e_n^T X) + W^T(X \circ X) + \delta x^2$$

$$+ x(-2\alpha(e_k e_n^T X + m e_n^T X) - W^T X)] + \alpha\left(n + \sum_i m_i^2\right) - \delta.$$

The above has a hidden semidefinite constraint:

$$\min \ \alpha\left(n + \sum_i m_i^2\right) - \delta \tag{4.55}$$

$$\text{s.t.} \quad L_A + \operatorname{Arrow}(\delta, \operatorname{vec}(W)) + \alpha L_\alpha \preccurlyeq 0,$$

where we define the matrices

$$L_A := \begin{bmatrix} 0 & 0 \\ 0 & \tfrac{1}{2}I \otimes A \end{bmatrix},$$

$$v = \operatorname{vec} e_n m^T, \tag{4.56}$$

$$L_\alpha := \begin{bmatrix} 0 & -(e+v)^T \\ -(e+v) & (e_k e_k^T I \otimes I + I \otimes e_n e_n^T) \end{bmatrix} \tag{4.57}$$

and the linear operator

$$\operatorname{Arrow}(\delta, \operatorname{vec}(W)) := \begin{bmatrix} \delta & -\tfrac{1}{2}(\operatorname{vec}(W))^T \\ -\tfrac{1}{2}(\operatorname{vec}(W)) & \operatorname{Diag}(\operatorname{vec}(W)) \end{bmatrix}. \tag{4.58}$$

The dual problem yields the semidefinite relaxation of (GP):

$$\max \ \operatorname{trace} L_A Y$$

$$\text{s.t.} \quad \operatorname{diag}(Y) = (1, Y_{0,1:n})^T,$$

$$\operatorname{trace} Y L_\alpha = 0, \tag{4.59}$$

$$Y \succcurlyeq 0.$$

4.2. The quadratic assignment problem

While MC can be considered the simplest of the NP-hard problems, the quadratic assignment problem (QAP) can be considered the hardest. This is an area where $n = 30$

is a large-scale problem. We shall use the trace formulation of the QAP where the variable X is a permutation matrix, i.e. X is a 0,1 matrix and all row and column sums are equal to one. The formulation is

$$\mu^* := \max \ q(X) = \text{trace}(AXB - 2C)X^{\mathrm{T}}$$

$$(\text{QAP}_E) \qquad \text{s.t.} \qquad Xe = e,$$

$$X^{\mathrm{T}}e = e, \qquad\qquad\qquad (4.60)$$

$$X_{ij} \in \{0, 1\} \quad \forall i, j.$$

(See [97] for applications and other formulations of the QAP.)

Let us apply the recipe. We first add redundant quadratic constraints to the model. Since the set of permutation matrices is equal to the intersection of the set of orthogonal matrices with the 0,1 matrices [122], we can add both of the following (equivalent) definitions of orthogonality: $XX^{\mathrm{T}} = I$ and $X^{\mathrm{T}}X = I$. The recipe also requires the application of Theorem 2.5 to the linear constraints before taking the Lagrangian dual. We thus obtain the following formulation for the QAP:

$$\mu^* := \min \ \text{trace} \, AXBX^{\mathrm{T}} - 2CX^{\mathrm{T}}$$

$$\text{s.t.} \qquad XX^{\mathrm{T}} = I,$$

$$X^{\mathrm{T}}X = I,$$

$$(\text{QAP}_E) \qquad ||Xe - e||^2 = 0,$$

$$||X^{\mathrm{T}}e - e||^2 = 0,$$

$$X_{ij}^2 - X_{ij} = 0, \quad \forall i, j.$$

Other relaxations and bounds can be obtained by adding redundant constraints such as trace $XX^{\mathrm{T}} = n$ or $0 \leqslant X_{ij} \leqslant 1$, $\forall i, j$.

It turns out that the squared linear constraints are eliminated by the projection later so we can add them together without any loss: $||Xe - e||^2 + ||X^{\mathrm{T}}e - e||^2 = 0$. We first add the 0,1 and row–column sum constraints to the objective function using Lagrange multipliers W_{ij} and u_0, respectively:

$$\mu_\mathcal{O} = \min_{XX^{\mathrm{T}} = X^{\mathrm{T}}X = I} \ \max_{W, u_0} \left\{ \text{trace} \, AXBX^{\mathrm{T}} - 2CX^{\mathrm{T}} + \sum_{ij} W_{ij}(X_{ij}^2 - X_{ij}) \right.$$

$$\left. + u_0(||Xe - e||^2 + ||X^{\mathrm{T}}e - e||^2) \right\}. \qquad (4.61)$$

Interchanging min and max yields

$$\mu_\mathcal{O} \geqslant \mu_\mathcal{L} := \max_{W, u_0} \ \min_{XX^{\mathrm{T}} = X^{\mathrm{T}}X = I} \left\{ \text{trace} \, AXBX^{\mathrm{T}} - 2CX^{\mathrm{T}} + \sum_{ij} W_{ij}(X_{ij}^2 - X_{ij}) \right.$$

$$\left. + u_0(||Xe - e||^2 + ||X^{\mathrm{T}}e - e||^2) \right\}. \qquad (4.62)$$

We now homogenize the objective function using the constrained scalar variable x_0 and increasing the dimension of the problem by 1. This simplifies the transition to an

SDP:

$$\mu_{\mathcal{O}} \geq \mu_{\mathscr{L}} = \max_{W, u_0} \min_{XX^{\mathrm{T}} = X^{\mathrm{T}}X = I, x_0^2 = 1} \{ \mathrm{trace}[AXBX^{\mathrm{T}} + W(X \circ X)^{\mathrm{T}}$$

$$+ u_0(\|Xe\|^2 + \|X^{\mathrm{T}}e\|^2) - x_0(2C + W)X^{\mathrm{T}}]$$

$$- 2x_0 u_0 e^{\mathrm{T}}(X + X^{\mathrm{T}})e + 2n u_0 x_0^2 \}. \tag{4.63}$$

Introducing a Lagrange multiplier w_0 for the constraint on x_0 and Lagrange multipliers S_b for $XX^{\mathrm{T}} = I$ and S_o for $X^{\mathrm{T}}X = I$ we get the lower bound μ_R:

$$\mu_{\mathcal{O}} \geq \mu_{\mathscr{L}} \geq \mu_R := \max_{W, S_b, S_o, u_0, w_0} \min_{X, x_0} \{ \mathrm{trace}[AXBX^{\mathrm{T}} + u_0(\|Xe\|^2 + \|X^{\mathrm{T}}e\|^2)$$

$$+ W(X \circ X)^{\mathrm{T}} + w_0 x_0^2 + S_b XX^{\mathrm{T}} + S_o X^{\mathrm{T}}X]$$

$$- \mathrm{trace}\, x_0(2C + W)X^{\mathrm{T}} - 2x_0 u_0 e^{\mathrm{T}}(X + X^{\mathrm{T}})e$$

$$- w_0 - \mathrm{trace}\, S_b - \mathrm{trace}\, S_o + 2n u_0 x_0^2 \}. \tag{4.64}$$

Note that we grouped the quadratic, linear, and constant terms together in (4.64). Now we define $x := \mathrm{vec}(X)$, $y^{\mathrm{T}} := (x_0, x^{\mathrm{T}})$ and $w^{\mathrm{T}} := (w_0, \mathrm{vec}(W)^{\mathrm{T}})$ to obtain

$$\mu_R = \max_{W, S_b, S_o, u_0} \min_{y} \{ y^{\mathrm{T}}[L_Q + \mathrm{Arrow}(w) + \mathrm{B}^0 \mathrm{Diag}(S_b)$$

$$+ \mathrm{O}^0 \mathrm{Diag}(S_o) + u_0 D]y - w_0 - \mathrm{trace}\, S_b - \mathrm{trace}\, S_o \}, \tag{4.65}$$

where L_Q is as above and we used the linear operators

$$\mathrm{Arrow}(w) := \begin{bmatrix} w_0 & -\frac{1}{2}w_{1:n^2}^{\mathrm{T}} \\ -\frac{1}{2}w_{1:n^2} & \mathrm{Diag}(w_{1:n^2}) \end{bmatrix}, \tag{4.66}$$

$$\mathrm{B}^0 \mathrm{Diag}(S) := \begin{bmatrix} 0 & 0 \\ 0 & I \otimes S_b \end{bmatrix}, \tag{4.67}$$

$$\mathrm{O}^0 \mathrm{Diag}(S) := \begin{bmatrix} 0 & 0 \\ 0 & S_o \otimes I \end{bmatrix}, \tag{4.68}$$

and

$$D := \begin{bmatrix} n & -e^{\mathrm{T}} \otimes e^{\mathrm{T}} \\ -e \otimes e & I \otimes E \end{bmatrix} + \begin{bmatrix} n & -e^{\mathrm{T}} \otimes e^{\mathrm{T}} \\ -e \otimes e & E \otimes I \end{bmatrix}.$$

There is a hidden semidefinite constraint in (4.65): the inner minimization problem is bounded below only if the Hessian of the quadratic form is positive semidefinite. And in that case the quadratic form has minimum value 0. Hence we have the equivalent SDP:

$$(\mathrm{D}_{\mathcal{O}}) \quad \begin{aligned} & \max \quad -w_0 - \mathrm{trace}\, S_b - \mathrm{trace}\, S_o \\ & \text{s.t.} \quad L_Q + \mathrm{Arrow}(w) + \mathrm{B}^0 \mathrm{Diag}(S_b) + \mathrm{O}^0 \mathrm{Diag}(S_o) + u_0 D \succeq 0. \end{aligned}$$

We now obtain our desired SDP relaxation of $(QAP_{\mathcal{O}})$ as the Lagrangian dual of $(D_{\mathcal{O}})$. We introduce the $(n^2 + 1) \times (n^2 + 1)$ dual matrix variable $Y \succcurlyeq 0$ and derive the dual problem to the SDP $(D_{\mathcal{O}})$.

$$\min \; \text{trace} \, L_Q Y$$

$$(SDP_{\mathcal{O}}) \quad \text{s.t.} \quad b^0 \text{diag}(Y) = I, \quad o^0 \text{diag}(Y) = I, \tag{4.69}$$

$$\text{arrow}(Y) = e_0, \quad \text{trace} \, DY = 0,$$

$$Y \succcurlyeq 0,$$

where the *arrow operator*, acting on the $(n^2 + 1) \times (n^2 + 1)$ matrix Y, is the adjoint operator to $\text{Arrow}(\cdot)$ and is defined by

$$\text{arrow}(Y) := \text{diag}(Y) - (0, (Y_{0,1:n^2})^T), \tag{4.70}$$

i.e. the arrow constraint guarantees that the diagonal and the first (0th) row (or column) are identical.

The *block-0-diagonal operator* and *off-0-diagonal operator* acting on Y are defined by

$$b^0 \text{diag}(Y) := \sum_{k=1}^{n} Y_{(k,\cdot),(k,\cdot)} \tag{4.71}$$

and

$$o^0 \text{diag}(Y) := \sum_{k=1}^{n} Y_{(\cdot,k),(\cdot,k)}. \tag{4.72}$$

These are the adjoint operators of $B^0 \text{Diag}(\cdot)$ and $O^0 \text{Diag}(\cdot)$, respectively. The block-0-diagonal operator guarantees that the sum of the diagonal blocks equals the identity. The off-0- diagonal operator guarantees that the trace of each diagonal block is 1, while the trace of the off-diagonal blocks is 0. These constraints come from the orthogonality constraints, $XX^T = I$ and $X^T X = I$, respectively.

We have expressed the orthogonality constraints with both $XX^T = I$ and $X^T X = I$. It is interesting to note that this redundancy adds extra constraints into the relaxation which are *not* redundant. These constraints reduce the size of the feasible set and so tighten the bounds.

Proposition 4.1. *Suppose that Y is feasible for the SDP relaxation* (4.69). *Then Y is singular.*

Proof. Note that $D \neq 0$ is positive semidefinite. Therefore Y has to be singular to satisfy the constraint $\text{trace} \, DY = 0$. $\square$

This means that the feasible set of the primal problem $(SDP_{\mathcal{O}})$ has no interior. It is not difficult to find an interior-point for the dual $(D_{\mathcal{O}})$, which means that Slater's constraint qualification (strict feasibility) holds for $(D_{\mathcal{O}})$. Therefore $(SDP_{\mathcal{O}})$ is attained and there is no duality gap in theory, for the usual primal–dual pair. However if Slater's constraint qualification fails, then this is not the proper dual, since perturbations in the right-hand side will not result in the dual value. This is because we cannot stay exactly feasible, since the interior is empty (see [104]). In fact, we may never attain the supremum of $(D_\mathcal{O})$, which may cause instability when implementing any kind of interior-point method. Since Slater's constraint qualification fails for the primal, the set of optimal solutions of the dual is an unbounded set and an interior-point method may never converge. Therefore, we have to express the feasible set of $(SDP_\mathcal{O})$ in some lower dimensional space. We study this below when we project the problem onto a face of the semidefinite cone.

However, if we add the rank-one condition, then the relaxation is exact.

Theorem 4.2. *Suppose that* Y *is restricted to be rank-one in* $(SDP_\mathcal{O})$, *i.e.*

$$Y = \begin{pmatrix} 1 \\ x \end{pmatrix} (1 x^T), \text{ for some } x \in \mathfrak{R}^{n^2}. \text{ Then the optimal solution of } (SDP_\mathcal{O}) \text{ provides}$$

the permutation matrix $X = \text{Mat}(x)$ *that solves the QAP.*

Proof. The arrow constraint in $(SDP_\mathcal{O})$ guarantees that the diagonal of Y is 0 or 1. The 0-diagonal and assignment constraint now guarantee that $\text{Mat}(x)$ is a permutation matrix. Therefore, the optimization is over the permutation matrices and so the optimum of QAP is obtained. □

We now devote our attention to homogenization since that results in a min–max eigenvalue problem and an equivalent SDP. We have seen that we can homogenize by increasing the dimension of the problem by 1. We first add the 0,1 constraints to the objective function using Lagrange multipliers W_{ij}.

$$\min_{W} \max_{XX^T=I} \text{trace}(AXB - 2C)X^T + \sum_{ij} W_{ij}(X_{ij}^2 - X_{ij}). \tag{4.73}$$

We now homogenize the objective function by multiplying by a constrained scalar x.

$$\min_{W} \max_{XX^T=I, x^2=1} \text{trace}[AXBX^T + W(X \circ X)^T - x(2C + W)X^T]. \tag{4.74}$$

We can now use Lagrange multipliers to get a parameterized min–max eigenvalue problem in dimension $n^2 + 1$. We get the following bound. The parameters are: the symmetric $n \times n$ matrix $\Lambda = \Lambda^T$, the general $n \times n$ matrix W and the scalar α.

$$B_{QAP} := \min_{\Lambda, W, \alpha} \max_{X} \text{trace}[AXBX^T + \Lambda XX^T + W^T(X \circ X) + \alpha x^2$$

$$-x(2C + W)X^T] - \alpha - \text{trace} \Lambda. \tag{4.75}$$

We have grouped the quadratic, original linear, and constant terms together. The hidden semidefinite constraint now yields an SDP:

$$\min -\text{trace} \Lambda - \alpha$$

$$\text{s.t.} \quad L_Q + \text{Arrow}(\alpha, \text{vec}(W)) + B^0\text{Diag}(\Lambda) \preccurlyeq 0, \tag{4.76}$$

where we define the matrix

$$L_Q := \begin{bmatrix} 0 & -\text{vec}(C)^{\mathrm{T}} \\ -\text{vec}(C) & B \otimes A \end{bmatrix} \tag{4.77}$$

and the linear operators

$$\text{Arrow}(\alpha, \text{vec}(W)) := \begin{bmatrix} \alpha & -\tfrac{1}{2}\text{vec}(W)^{\mathrm{T}} \\ -\tfrac{1}{2}\text{vec}(W) & \text{Diag}(\text{vec}(W)) \end{bmatrix}, \tag{4.78}$$

$$B^0\text{Diag}(\Lambda) := \begin{bmatrix} 0 & 0 \\ 0 & I \otimes \Lambda \end{bmatrix}. \tag{4.79}$$

We can now introduce the $(n^2+1) \times (n^2+1)$ dual variable matrix $Y \succcurlyeq 0$ and derive the dual problem to this min–max eigenvalue problem, i.e.

$$\max_{Y \succcurlyeq 0} \min_{\Lambda, W, \alpha} - \text{trace}\, \Lambda - \alpha + \text{trace}\, Y(L_Q + \text{Arrow}(\alpha, \text{vec}(W)) + B^0\text{Diag}(\Lambda)).$$

The inner minimization problem is unconstrained and linear in the variables. Therefore, after reorganizing the variables, we can differentiate to get the dual problem to this dual problem, or the semidefinite relaxation to the original QAP. (Recall that $Y_{i,j:k}$ refers to the ith row and columns j to k of the matrix Y; and $b^0\text{diag}(Y)$ is the block diagonal sum of Y which ignores the first row.) The derivatives with respect to α and W yields the first constraint and the derivative with respect to Λ yields the second constraint in the following problem. Equivalently, the constraints are the adjoints of the linear operators Arrow and B^0Diag:

$$\max\ \text{trace}\, L_Q Y$$

$$\text{s.t.}\quad \text{diag}(Y) = (1, Y_{0,1:n^2})^{\mathrm{T}},$$
$$b^0\text{diag}(Y) = I, \tag{4.80}$$
$$Y \succcurlyeq 0.$$

Another primal–dual pair can be obtained using a trust-region subproblem as the inner maximization problem, rather than homogenizing to an eigenvalue problem. This is done by adding the redundant trust-region constraint $\text{trace}\, XX^{\mathrm{T}} = n$. As mentioned above, we can also add the redundant constraint

$$\|Xe - e\|^2 + \|X^{\mathrm{T}}e - e\|^2 = 0.$$

A primal–dual interior-point method based on these types of dual pairs of problems, such as (4.80) and (4.76), is tested and studied in [122].

4.3. The max-clique and max-stable-set problems

Consider again the undirected graph $G = (E, V)$ defined above. The max-clique problem consists in finding the largest connected subgraph. We let $\omega(G)$ denote the size of the largest clique in G. A stable set is a subset of nodes of V such that no two nodes are adjacent. We denote the size of the largest stable set in $\bar{G}$, the complement of G, by $\alpha(\bar{G})$. Clearly,

$$\alpha(\bar{G}) = \omega(G).$$

Bounds for these problems and relationships to the theta function, or Lovász number of the graph, are described in the expository paper of Knuth [64] (see also [108]).

In this section we show that the Lovász bound on $\omega(G)$ can be alternatively obtained from two distinct 0,1 problems (4.81) and (4.84) by Lagrangian relaxations. Let A be the incidence matrix of the graph, i.e. $A = (a_{ij})$ with $a_{ij} = 1$ if $ij \in E$ and 0 otherwise. If x is the indicator vector for the largest clique in G of size k, A then $x^T(I + A)x/x^Tx = k^2/k = k$. A quadratic formulation of the max-clique problem is the following 0,1 quadratic problem:

$$\omega(G) = \max \frac{x^T(I+A)x}{x^Tx}$$

$$\text{s.t.} \quad x_ix_j = 0, \quad \text{if } ij \notin E, \ i \neq j, \tag{4.81}$$

$$x_i \in \{0,1\}, \ \forall i.$$

Therefore, a quadratic relaxation of the max-clique problem is the following quadratic constrained problem:

$$\omega(G) \leqslant \omega_1^* := \max x^T(I + A)x$$

$$\text{s.t.} \quad x_ix_j = 0, \quad \text{if } ij \notin E, \ i \neq j, \tag{4.82}$$

$$x^Tx = 1.$$

The Lagrangian relaxation for this problem is the perturbed min–max eigenvalue problem and the equivalent SDP:

$$\omega_1^* \leqslant \min_{\substack{W_{ij}=0, \text{ if } ij \in E, \text{ or } i=j}} \lambda_{\max}(I + A + W) - \alpha x^Tx + \alpha$$

$$= \min_{w, \alpha} \max_x x^T(I + A)x + \sum_{ij \notin E, \ i \neq j} w_{ij}x_ix_j - \alpha x^Tx + \alpha$$

$$= \min_{\substack{I+A+W \preceq \alpha I \\ W_{ij}=0, \text{ if } ij \in E, \text{ or } i=j}} \alpha,$$

i.e. minimize the max eigenvalue over perturbations in the off-diagonal elements corresponding to disjoint nodes. This bound is equal to the Lovász theta function on the complementary graph:

$$\vartheta(\bar{G}) = \min_{A \in \mathscr{A}} \lambda_{\max}(A), \tag{4.83}$$

where

$$\mathscr{A} = \{A: A \text{ symmetric } n \times n \text{ matrix with } A_{ij} = 1, \text{ if } ij \in E, \text{ or } i = j\}.$$

By considering the (optimal) indicator vector for the largest clique, we see that the following 0,1 quadratic problem describes exactly the max-clique problem. Note that if node i is not in the largest clique, then necessarily, $x_i x_j = 0$ for some j with node j in the clique, i.e. necessarily $x_i = 0$ in the indicator vector.

$$\omega(G) = \max \ x^{\mathrm{T}} x$$

$$\text{s.t.} \quad x_i x_j = 0, \quad \text{if } ij \notin E, \ i \neq j, \tag{4.84}$$

$$x_i^2 - x_i = 0, \quad \forall i.$$

The Lagrangian relaxation yields the bound

$$B_{\text{clique}} := \min_{W, \lambda} \max_{x} \ x^{\mathrm{T}} x + \sum_{ij \notin E, \ i \neq j} w_{ij} x_i x_j + \sum_i \lambda_i (x_i^2 - x_i).$$

We let W be an $n \times n$ matrix with zeros in positions where $ij \in E$. We can homogenize by adding the constraint $y^2 = 1$ and then lifting it into the Lagrangian:

$$\min_{\alpha, W, \lambda} \max_{x, y} \ x^{\mathrm{T}} x + \sum_{ij \notin E} w_{ij} x_i x_j + \sum_i \lambda_i x_i^2 + \alpha y^2 - y \sum_i \lambda_i x_i - \alpha.$$

We now exploit the hidden semidefinite constraint to obtain the SDP:

$$B_{\text{clique}} = \min_{W, \lambda, \alpha} \ -\alpha$$

$$\text{s.t.} \quad L_A + L_W(W) + \text{Arrow}(\alpha, \lambda) \preccurlyeq 0, \tag{4.85}$$

$$W_{ij} = 0, \ \forall ij \in E, \text{ or } i = j,$$

where the matrix

$$L_A := \begin{bmatrix} 0 & 0 \\ 0 & I \end{bmatrix} \tag{4.86}$$

and the linear operators

$$L_W(W) := \begin{bmatrix} 0 & 0 \\ 0 & W \end{bmatrix}, \tag{4.87}$$

$$\text{Arrow}(\alpha, \lambda) := \begin{bmatrix} \alpha & -\tfrac{1}{2}\lambda^{\mathrm{T}} \\ -\tfrac{1}{2}\lambda & \text{Diag}(\lambda) \end{bmatrix}. \tag{4.88}$$

The dual of the above min–max eigenvalue problem yields the semidefinite relaxation for the max-clique problem with $Y \in \mathscr{S}_{n+1}$:

$$\max \ \text{trace} \ L_A Y$$

$$\text{s.t.} \quad \text{diag}(Y) = (1, Y_{0,1:n})^{\mathrm{T}},$$

$$Y_{ij} = 0, \ \forall ij \notin E, \tag{4.89}$$

$$Y \succcurlyeq 0.$$

The equivalence of bounds (4.83) and (4.89) was shown in Lemma 2.17 of [81]. Consider problem (4.81) with an additional redundant constraint

$$x_i x_j \geqslant 0 \quad \text{for } ij \in E, \tag{4.90}$$

that is

$$\omega(G) = \max \ \frac{x^{\mathrm{T}}(I+A)x}{x^{\mathrm{T}}x}$$

$$\text{s.t.} \quad x_i x_j = 0 \quad \text{if } ij \notin E, \ i \neq j,$$

$$x_i x_j \geq 0 \quad \text{if } ij \in E,$$

$$x_i \in \{0,1\}, \ \forall i.$$

(4.91)

A quadratic relaxation of the max-clique problem is the following quadratically constrained problem:

$$\omega(G) \leq \omega_1^* := \max \ x^{\mathrm{T}}(I+A)x$$

$$\text{s.t.} \quad x_i x_j = 0 \quad \text{if } ij \notin E, \ i \neq j,$$

$$x_i x_j \geq 0 \quad \text{if } ij \in E,$$

$$x^{\mathrm{T}}x = 1.$$

(4.92)

The Lagrangian relaxation for this problem is equal to Schrijver's improvement [108] of the theta function on the complementary graph:

$$\vartheta'(\bar{G}) = \min_{A \in \mathscr{A}'} \ \lambda_{\max}(A),$$

where

$$\mathscr{A}' = \{A: A \text{ symmetric } n \times n \text{ matrix with } A_{ij} \geq 1, \text{ if } ij \in E, \text{ or } i = j\}.$$

Haemers [43] constructed graphs where $\vartheta'(\bar{G})$ is strictly smaller than $\vartheta(\bar{G})$.

Analogously, it is possible to modify problem (4.84) by adding constraint (4.90).

4.4. Orthogonally constrained problems: achieving zero duality gaps

As a final illustration of the strength of Lagrangian relaxation and the power of adding appropriate redundant quadratic constraints we consider the orthonormal-type constraints:

$$X^{\mathrm{T}}X = I, \quad X \in \mathscr{M}_{m,n}.$$

(This set is sometimes known as the Stiefel manifold. Applications and algorithms for optimization over orthonormal sets of matrices are discussed in [26].) We also consider the trust-region-type constraint

$$X^{\mathrm{T}}X \leq I, \quad X \in \mathscr{M}_{m,n}.$$

We follow the approach in [10–12] and show that if $m = n$ then strong duality holds for certain (non-convex) quadratic problems defined over orthonormal matrices after adding some quadratic redundant constraints. Because of the similarity of the orthonormality constraint to the (vector) norm constraint $x^{\mathrm{T}}x = 1$, the results of this section can be viewed as a matrix generalization of the strong duality result for the well-known Rayleigh quotient problem [99].

Let A and B be $n \times n$ symmetric matrices, and consider the orthonormal constrained homogeneous problem:

$$(\text{QQP}_O) \quad \mu^O := \begin{array}{c} \min \ \text{trace}\, AXBX^T \\ \text{s.t.} \ \ XX^T = I. \end{array} \tag{4.92}$$

This problem can be solved exactly using Lagrange multipliers [42] or the classical Hoffman-Wielandt inequality [18].

Proposition 4.3. *Suppose that the orthogonal diagonalizations of A, B are $A = V\Sigma V^T$ and $B = U\Lambda U^T$, respectively, where the eigenvalues in Σ are ordered non-increasing, and the eigenvalues in Λ are ordered non-decreasing. Then the optimal value of QQP_O is $\mu^O = \text{trace}\,\Sigma\Lambda$, and the optimal solution is obtained using the orthogonal matrices that yield the diagonalizations, i.e. $X^* = VU^T$.*

The Lagrangian dual of QQP_O is

$$\max_{S=S^T} \min_X \text{trace}\, AXBX^T - \text{trace}\, S(XX^T - I). \tag{4.93}$$

However, there can be a non-zero duality gap for the Lagrangian dual, see [122] for an example. The inner minimization in the dual problem (4.93) is an unconstrained quadratic minimization in the variables $\text{vec}(X)$, with hidden constraint on the Hessian

$$B \otimes A - I \otimes S \succeq 0.$$

The first-order stationarity conditions are equivalent to $AXB = SX$ or $AXBX^T = S$. One can easily construct examples where the semidefinite condition and the stationarity conditions are in conflict and thus a duality gap occurs. In order to close the duality gap, we need a larger class of quadratic functions.

Note that in QQP_O the constraints $XX^T = I$ and $X^TX = I$ are equivalent. Adding the redundant constraints $X^TX = I$, we arrive at

$$\text{QQP}_{OO} \quad \mu^O := \begin{array}{c} \min \ \text{trace}\, AXBX^T \\ \text{s.t.} \ \ XX^T = I, \quad X^TX = I. \end{array}$$

Using symmetric matrices S and T to relax the constraints $XX^T = I$ and $X^TX = I$, respectively, we obtain a dual problem

$$\begin{array}{c} \max \ \text{trace}\, S + \text{trace}\, T \\ \text{DQQP}_{OO} \quad \mu^O \geqslant \mu^D := \text{s.t.} \ \ (I \otimes S) + (T \otimes I) \preccurlyeq (B \otimes A), \\ S = S^T, \quad T = T^T. \end{array}$$

Theorem 4.4. *Strong duality holds for QQP_{OO} and $DQQP_{OO}$, i.e., $\mu^D = \mu^O$ and both primal and dual are attained.*

A further relaxation of the above orthogonal relaxation is the trust-region relaxation studied in [63]:

$$\mu^*_{\text{QAPT}}:= \min \ \text{trace} \, AXBX^{\text{T}}$$

$$\text{s.t.} \ \ XX^{\text{T}} \leqslant I.$$

The constraints are convex with respect to the Löwner partial order and so it is hoped that solving this problem would be useful. The set

$$\{X : W = XX^{\text{T}} \leqslant I\}$$

is studied in [96,28] and is useful in eigenvalue variational principles. Furthermore, problem (4.93) is visually similar to the trust-region subproblem so we would like to find a characterization of optimality.

We study the matrix trust-region relaxation of QAP:

$$\mu^*_{\text{SDPT}}= \min \ \text{trace} \, AXBX^{\text{T}}$$

$$\text{s.t.} \ \ XX^{\text{T}} \leqslant I.$$

The following generalization of the Hoffman–Wielandt inequality holds.

Theorem 4.5. *For any* $XX^{\text{T}} \leqslant I$, *we have*

$$\sum_{i=1}^{n} \min\{\lambda_i \mu_{n-i+1}, 0\} \leqslant \text{tr} \, AXBX^{\text{T}} \leqslant \sum_{i=1}^{n} \max\{\lambda_i \mu_i, 0\}$$

and the upper bound is attained if

$$X = P \, \text{Diag}(\varepsilon_1, \varepsilon_2, \ldots, \varepsilon_n) \, Q^{\text{T}}, \tag{4.93}$$

where

$$\epsilon_i = \begin{cases} 1, & \lambda_i \mu_i > 0, \\ \alpha \in [0,1], & \lambda_i \mu_i = 0, \\ 0, & \lambda_i \mu_i < 0. \end{cases} \tag{4.93}$$

The lower bound is attained if

$$X = P \, \text{Diag}(\varepsilon_1, \varepsilon_2, \ldots, \varepsilon_n) \, Q^{\text{T}}, \tag{4.93}$$

where

$$\epsilon_i = \begin{cases} 1, & \lambda_i \mu_{n-i+1} < 0, \\ \alpha \in [0,1], & \lambda_i \mu_{n-i+1} = 0, \\ 0, & \lambda_i \mu_{n-i+1} > 0. \end{cases} \tag{4.93}$$

The lower bound in the above theorem states that $\mu^*_{\text{SDPT}} = \sum_{i=1}^{n} [\lambda_i \mu_i]^-$. Since the theorem provides the feasible point of attainment, i.e. an upper bound for the relaxation problem, we will prove the theorem by proving another theorem that shows that the value μ^*_{SDPT} is also attained by a Lagrangian dual problem. Note that since XX^{T} and

X^TX have the same eigenvalues, $XX^T \preccurlyeq I$ if and only if $X^TX \preccurlyeq I$. Explicitly using both sets of constraints, as in [12], we obtain

QAPTR $\mu^*_{\mathrm{QAPT}} := \min \operatorname{trace} AXBX^T$

$$\text{s.t. } XX^T \preccurlyeq I, \quad X^TX \preccurlyeq I.$$

Next, we apply Lagrangian relaxation to QAPTR, using matrices $S \succcurlyeq 0$ and $T \succcurlyeq 0$ to relax the constraints $XX^T \preccurlyeq I$ and $X^TX \preccurlyeq I$, respectively. This results in the dual problem

DQAPTR $\mu^*_{\mathrm{QAPT}} \geqslant \mu^D_{\mathrm{QAPT}} := \max \ -\operatorname{trace} S - \operatorname{trace} T$

$$\text{s.t. } (B \otimes A) + (I \otimes S) + (T \otimes I) \succcurlyeq 0,$$

$$S \succcurlyeq 0, \quad T \succcurlyeq 0.$$

To prove that $\mu^*_{\mathrm{QAPT}} = \mu^D_{\mathrm{QAPT}}$ we will use the following simple result:

Lemma 4.6. *Let* $\lambda \in \mathfrak{R}^n$, $\lambda_1 \leqslant \lambda_2 \leqslant \cdots \leqslant \lambda_n$. *For* $\gamma \in \mathfrak{R}^n$ *consider the problem*

$$\min z_\pi := \sum_{i=1}^n [\lambda_i \gamma_{\pi(i)}]^-,$$

where $\pi(\cdot)$ *is a permutation of* $\{1,\ldots,n\}$. *Then the permutation that minimizes* z_π *satisfies* $\gamma_{\pi(1)} \geqslant \gamma_{\pi(2)} \geqslant \cdots \gamma_{\pi(n)}$.

Theorem 4.7. *Strong duality holds for QAPTR and DQAPTR:*

$$\mu^D_{\mathrm{QAPT}} = \mu^*_{\mathrm{QAPT}}$$

and both primal and dual optimal values are attained.

These results conclude the first part of the paper which illustrated the strength of the Lagrangian relaxation. We now proceed to our second application of SDP.

5. Matrix completion problems

Semidefinite programming problems arise in surprisingly many different areas of mathematics and engineering where they sometimes have different names. In engineering they are often referred to as linear matrix inequalities problems. In matrix theory, the class of problems called matrix completion problems is closely related to SDP. In this last section we study application of SDP to this class of problems.

A *symmetric partial matrix* is a symmetric matrix where certain entries are fixed or specified while the remaining entries are unspecified or free. The symmetric matrix completion problem endeavors to specify the free elements in such a way that the resulting matrix satisfies certain required properties. For example, the positive semidefinite matrix completion problem (PSDM) consists of finding a completion so that the resulting matrix is symmetric positive semidefinite, while the Euclidean distance matrix completion problem (EDM) seeks a completion that forms a Euclidean distance matrix (a precise definition of this class of matrices is given below).

In this section we show how successful SDP has been in solving matrix completion problems. We begin in Section 5.1 with theoretical existence results for completions based on chordality. This follows the work in [41]. We then present an efficient approach to solve PSDM completion problems [56]. This approach successfully solves large sparse problems. In Section 5.3 this approach is extended to the EDM completion problem (based on the work in [1]) but is shown to exhibit difficulties in the large sparse case. Hence, we conclude by presenting in Section 5.4 a new characterization of Euclidean distance matrices and new algorithms that efficiently solve large sparse problems.

5.1. Existence results

Both the PSDM and EDM completion problems have been extensively studied in the literature. Let us first phrase the completion problem using the graph of the matrix. Suppose that $\mathcal{G}(V, E)$ is a finite undirected graph. The edges of the graph correspond to fixed elements in the matrix, i.e. $A(\mathcal{G})$ is a $\mathcal{G}$-partial matrix

if a_{ij} is defined if and only if $\{i, j\} \in E$.

$A(\mathcal{G})$ is a $\mathcal{G}$-partial positive matrix if $a_{ij} = \overline{a_{ji}}$, $\forall \{i, j\} \in E$ and all existing principal minors are positive. With $\mathcal{J} = (V, \bar{E})$, $E \subset \bar{E}$ a $\mathcal{J}$-partial matrix. $B(\mathcal{J})$ extends the $\mathcal{G}$-partial matrix $A(\mathcal{G})$ if $b_{ij} = a_{ij}$, $\forall \{i, j\} \in E$, i.e. the missing (free) elements in the matrix are filled in.

$\mathcal{G}$ is positive completable if every $\mathcal{G}$-partial positive matrix can be extended to a positive definite matrix. With this definition we look at the pattern of fixed elements in the matrix rather than specific elements. The following is the key property to guarantee that a completion is possible.

Definition 5.1. $\mathcal{G}$ is chordal if there are no minimal cycles of length $\geqslant 4$ (every cycle of length $\geqslant 4$ has a chord).

Theorem 5.2 (Grone et al. [41]). *$\mathcal{G}$ is positive completable if and only if $\mathcal{G}$ is chordal.*

When a positive definite completion is possible, then the one of maximum determinant is unique and can be characterized.

Theorem 5.3 (Grone et al. [41]). *Let A be a partial symmetric matrix all of whose diagonal entries are specified, and suppose that A has a positive definite completion. Then, among all positive definite completions, there is a unique one with maximum determinant.*

The 1990 survey paper [55] presents many of the theoretical results for completion problems. Similar existence results are known for the EDM completion problem, see e.g. the comparison of the two problems [76], as well as the survey paper

Table 2
PSD completion data for dual-step-first method (20 problems per test)

Dim.	Toler.	H dens./infty.	A ⩾ 0	cond(A)	H ≻ 0	min/max	Iters
83	10^{-6}	0.007/0.001	No	235.1	No	24/29	25.5
85	10^{-5}	0.008/0.001	Yes	94.7	No	11/17	13.1
85	10^{-6}	0.0075/0.001	No	299.9	No	23/27	25.2
87	10^{-6}	0.006/0.001	Yes	74.2	Yes	14/19	16.9
89	10^{-6}	0.006/0.001	No	179.3	No	23/28	15.2
110	10^{-6}	0.007/0.001	Yes	172.3	Yes	15/20	17.8
155	10^{-6}	0.01/0	Yes	643.9	Yes	14/18	15.3
655	10^{-6}	0.017/0	Yes	1.4	No	13/16	14.
755	10^{-6}	0.002/0	Yes	1.5	No	14/17	15.

Table 3
PSD completion data for primal-step-first (20 problems per test)

Dim.	Toler.	H dens./infty.	A ⩾ 0	cond(A)	H ≻ 0	min/max	Iters
85	10^{-5}	0.0219/0.02	Yes	1374.5	No	16/23	18.9
95	10^{-5}	0.0206/0.02	Yes	2.7	No	8/14	11.1
95	10^{-6}	1/0.999	Yes	196.	Yes	14/18	16.8
145	10^{-6}	0.01/0.997	Yes	658.5	Yes	13/17	14.9

5.3. Approximate EDM completions

We now look at the EDM completion problem. We follow the successful approach above and use some known characterizations of EDMs. (The details can be found in [1].)

An $n \times n$ symmetric matrix $D = (d_{ij})$ with non-negative elements and zero diagonal is called a *pre-distance matrix* (or dissimilarity matrix). A pre-distance matrix such that there exists points $x^1, x^2, \ldots, x^n$ in $\Re^r$ with

$$d_{ij} = ||x^i - x^j||^2, \quad i, j = 1, 2, \ldots, n$$

is called a *(squared) Euclidean distance matrix* (EDM). The smallest value of r is called *the embedding dimension* of D. (r is always $\leqslant n - 1$.)

Given a partial symmetric matrix A with certain elements specified, the *Euclidean distance matrix completion problem* (EDMCP) consists in finding the unspecified elements of A that make A an EDM. In other words, we wish to determine the relative locations of points in Euclidean space, when we are only given a subset of the pairwise distances between the points.

There are surprisingly many applications for this problem, sometimes called the molecule problem. These applications include NMR data, determination of protein structure, surveying, satellite ranging, and molecular conformation; see e.g. the survey [24] and the discussion in [50] and the related papers [49,89,114,117,44].

We now consider the approximate EDMCP and follow the approach in [1], where the reader will find all the proofs and details omitted here. Let A be a pre-distance matrix and let H be an $n \times n$ symmetric matrix with non-negative elements (weights). Consider the objective function

$$f(D) := \|H \circ (A - D)\|_F^2,$$

where $\circ$ denotes *Hadamard product*. The *weighted, closest Euclidean distance matrix problem* is

$$(\text{CDM}_0) \qquad \begin{aligned} \mu^* &:= \min \ f(D) \\ &\text{s.t.} \ \ D \in \mathscr{E}, \end{aligned}$$

where $\mathscr{E}$ denotes the cone of EDMs.

5.3.1. EDM model

The cone of EDM is homeomorphic to a face of the cone of positive semidefinite matrices. This can be seen from the fact that a pre-distance matrix D is a EDM if and only if D is negative semidefinite on

$$M := \{x \in \mathfrak{R}^n : x^t e = 0\},$$

where e is the vector of all ones. (For these and other related results see the development in [1].) Now, define

$$V \text{ is } n \times (n-1) \text{ full column rank with } V^t e = 0. \tag{5.2}$$

Then

$$J := VV^\dagger = I - \frac{ee^t}{n} \tag{5.3}$$

is the orthogonal projection onto M, where $V^\dagger$ denotes the Moore–Penrose generalized inverse.

Define the *centered* and *hollow* subspaces

$$\mathscr{S}_C := \{B \in \mathscr{S}^n : Be = 0\},$$

$$\mathscr{S}_H := \{D \in \mathscr{S}^n : \text{diag}(D) = 0\}$$

and the two linear operators

$$\mathscr{K}(B) := \text{diag}(B)\, e^t + e\, \text{diag}(B)^t - 2B,$$

$$\mathscr{T}(D) := -\tfrac{1}{2} JDJ.$$

The operator $-2\mathscr{T}$ is an orthogonal projection onto $\mathscr{S}_C$.

Theorem 5.6. *The linear operators satisfy*

$$\mathscr{K}(\mathscr{S}_C) = \mathscr{S}_H,$$

$$\mathscr{T}(\mathscr{S}_H) = \mathscr{S}_C$$

and $\mathscr{K}_{|\mathscr{S}_C}$ and $\mathscr{T}_{|\mathscr{S}_H}$ are inverses of each other.

Lemma 5.7. *The hollow matrix* $D \in \mathcal{E}$ *if and only if*

$$v^{\mathrm{T}}e = 0 \;\Rightarrow\; v^{\mathrm{T}}Xv \leqslant 0.$$

From the above we see that a hollow matrix D is EDM if and only if it is negative semidefinite on the orthogonal complement of e, i.e. if and only if $B = \mathcal{T}(D) \succcurlyeq 0$ (positive semidefinite). Alternatively, D is EDM if and only if $D = \mathcal{K}(B)$, for some B with $Be = 0$ and $B \succcurlyeq 0$. In this case the embedding dimension r is given by the rank of B. Moreover if $B = XX^{\mathrm{t}}$, then the coordinates of the points $x^1, x^2, \ldots, x^n$ that generate D are given by the rows of X and, since $Be = 0$, it follows that the origin coincides with the centroid of these points.

The cone of EDMs, $\mathcal{E}$, has empty interior. This can cause problems for interior-point methods. We can correct this by projection and moving to a smaller dimensional space [1]; note that

$$V \cdot V : \mathcal{S}_{n-1} \to \mathcal{S}_n,$$

$$V \cdot V : \mathcal{P}_{n-1} \to \mathcal{P}_n.$$

Define the composite operators

$$\mathcal{K}_V(X) := \mathcal{K}(VXV^{\mathrm{t}})$$

and

$$\mathcal{T}_V(D) := V^{\dagger}\mathcal{T}(D)(V^{\dagger})^{\mathrm{t}} = -\tfrac{1}{2}V^{\dagger}D(V^{\dagger})^{\mathrm{t}}.$$

Lemma 5.8.

$$\mathcal{K}_V(\mathcal{S}_{n-1}) = \mathcal{S}_{\mathrm{H}},$$

$$\mathcal{T}_V(\mathcal{S}_{\mathrm{H}}) = \mathcal{S}_{n-1}$$

and $\mathcal{K}_V$ *and* $\mathcal{T}_V$ *are inverses of each other on these two spaces.*

Corollary 5.9.

$$\mathcal{K}_V(\mathcal{P}) = \mathcal{E},$$

$$\mathcal{T}_V(\mathcal{E}) = \mathcal{P}.$$

We can summarize the above and obtain the model used in [1]. (Re)Define the closest EDM problem:

$$f_0(X) := \|H \circ (A - \mathcal{K}_V(X))\|_F^2$$

$$= \|H \circ \mathcal{K}_V(B - X)\|_F^2,$$

where $B = \mathcal{T}_V(A)$ ($\mathcal{K}_V$ and $\mathcal{T}_V$ are both linear operators):

$$\mu_0^* := \min \; f_0(X)$$

(CDM$_0$)   s.t. $\mathcal{A}X = b,$

$$X \succcurlyeq 0.$$

The additional constraint using $\mathscr{A}: \mathscr{S}_{n-1} \to \mathfrak{R}^m$ could represent some of the fixed elements in the given matrix A.

Numerical tests for this model are given in [1]. The number of iterations are comparable to those for the semidefinite completion problem (Section 5.2), though the time per iteration was much higher, i.e. sparsity was not exploited efficiently.

5.4. Alternate EDM model for the large sparse case

The above model appears to be quite efficient for solving the EDM completion problem. It handles the lack of interiority and actually reduces the dimension of the problem. There is one major difference between this model CDM$_0$ and the one used in Section 5.2. That is, the operator $H\circ$ in the objective function is replaced by $H \circ \mathscr{K}_V$. This change allows one to reduce the dimension of the problem and obtain Slater's constraint qualification for both the primal and dual problems. However, one cannot exploit sparsity as one did in CDM. As is often the case in modelling, a model that appears to be simpler is often not more efficient in computations. We now outline a different approach that increases the dimension of the problem but can exploit sparsity. The details can be found in [3]. (Recall that e denotes the vector of ones.)

Lemma 5.10. *Let*

$$\mathscr{F}:=\{X \in \mathscr{S}^n: v^{\mathrm{T}}e = 0 \Rightarrow v^{\mathrm{T}}Xv \leqslant 0\},$$

$$\mathscr{F}_0:=\{X \in \mathscr{S}^n: X - \alpha ee^{\mathrm{t}} \leqslant 0, \text{ for some } \alpha \geqslant 0\},$$

$$\mathscr{F}_1:=\{X \in \mathscr{S}^n: X - \alpha ee^{\mathrm{t}} \leqslant 0, \forall \alpha \geqslant \bar{\alpha}, \text{ for some } \bar{\alpha} \geqslant 0\}.$$

Then

$$\mathrm{ri}(\mathscr{F}) \subset \mathscr{F}_0 = \mathscr{F}_1 \subset \mathscr{F} \subset \overline{\mathscr{F}_0}. \tag{5.4}$$

Proof. Suppose that $\bar{X} \in \mathrm{ri}(\mathscr{F})$ (i.e. $v^{\mathrm{T}}e = 0, v \neq 0 \Rightarrow v^{\mathrm{T}}\bar{X}v < 0$) but $\bar{X} \notin \mathscr{F}_0$. Then, for each $\alpha \geqslant 0$, there exists w_α with $\|w_\alpha\| = 1$, such that $w_\alpha \to \bar{w}$, as $\alpha \to \infty$ and

$$w_\alpha^{\mathrm{T}}(\bar{X} - \alpha ee^{\mathrm{t}})w_\alpha > 0, \quad \forall \alpha \geqslant 0,$$

i.e.

$$w_\alpha^{\mathrm{T}}\bar{X}w_\alpha > \alpha w_\alpha^{\mathrm{T}}ee^{\mathrm{t}}w_\alpha, \quad \forall \alpha \geqslant 0.$$

Since w_α converges and the left-hand side of the above inequality must be finite, this implies that $e^{\mathrm{t}}\bar{w} = \bar{w}^{\mathrm{T}}\bar{X}\bar{w} = 0$, a contradiction. Therefore, $\mathrm{ri}(\mathscr{F}) \subset \mathscr{F}_0$. That $\mathscr{F}_0 = \mathscr{F}_1$ is clear.

Now suppose that $\bar{X} - \alpha ee^{\mathrm{t}} \leqslant 0$, $\alpha \geqslant 0$. Let $v^{\mathrm{T}}e = 0$. Then $0 \geqslant v^{\mathrm{T}}(\bar{X} - \alpha ee^{\mathrm{t}})v = v^{\mathrm{T}}\bar{X}v$, i.e. $\mathscr{F}_0 \subset \mathscr{F}$. The final inclusion comes from the first and the fact that $\mathscr{F}$ is closed. □

Unfortunately, we cannot enforce equality in (5.4). This can be seen from the fact that $\mathscr{F}_0 = \mathscr{P} + \mathrm{span}\{ee^{\mathrm{t}}\} = \mathscr{P} + \mathrm{span}\,\mathscr{F}$, where $\mathscr{F} = \mathrm{cone}\{ee^{\mathrm{t}}\}$ is a face (actually a ray) of the positive semidefinite cone generated by ee^t. $\mathscr{P}$ and the sum of $\mathscr{P}$ and the span of a face is never closed, see [104, Lemma 2.2]. If we assume that X is hollow, then the same result holds. This is used in the algorithm for large problems.

Corollary 5.11. *Let*

$$\mathscr{E} := \{X \in \mathscr{S}_H : v^T e = 0 \ \Rightarrow \ v^T X v \leqslant 0\},$$

$$\mathscr{E}_0 := \{X \in \mathscr{S}_H : X - \alpha e e^t \leqslant 0 \ for \ some \ \alpha\},$$

$$\mathscr{E}_1 := \{X \in \mathscr{S}_H : X - \alpha e e^t \leqslant 0 \ \forall \alpha \geqslant \bar{\alpha}, \ for \ some \ \bar{\alpha}\}.$$

Then

$$\mathrm{ri}(\mathscr{E}) \subset \mathscr{E} = \mathscr{E} \subset \mathscr{E} \subset \bar{\mathscr{E}}. \tag{5.5}$$

Proof. The proof is similar to that in the above Lemma 5.10. We only include the details about the closure.

Suppose that $0 \neq X_k \in \mathscr{E}_0$, i.e. $\mathrm{diag}(X_k) = 0$, $X_k \leqslant \alpha_k E$, for some α_k; and, suppose that $X_k \to \bar{X}$. Since X_k is hollow it has exactly one positive eigenvalue and this must be smaller than α_k. However, since X_k converges to $\bar{X}$, we conclude that $\bar{X} \leqslant \lambda_{\max}(\bar{X})E$, where $\lambda_{\max}(\bar{X})$ is the largest eigenvalue of $\bar{X}$. $\square$

We can now use a different simplified objective function to obtain a new model. We let $E = ee^t$ and

$$f(P) := \|H \circ (A - P)\|_F^2$$

and

$$\mu^* := \min \ f(P)$$

(CDM)   s.t. $\mathscr{K}P = b$,

$$\alpha E - P \succcurlyeq 0,$$

where $\mathscr{K}$ is a linear operator. We assume that this linear equality constraint contains the constraint $\mathrm{diag}(P) = 0$, i.e. that P is a hollow matrix.

We now derive the dual problem for CDM. For $\Lambda \in \mathscr{S}^n$ and $y \in \mathfrak{R}^m$, let

$$L(P, \alpha, \Lambda, y) = f(P) + \langle y, b - KP \rangle - \mathrm{trace} \, \Lambda(\alpha E - P) \tag{5.6}$$

denote the *Lagrangian* of CDM. It is easy to see that the primal problem CDM is equivalent to

$$\mu^* = \min_{P, \alpha} \ \max_{y} \ L(P, \alpha, \Lambda, y). \tag{5.7}$$

$$\Lambda \succcurlyeq 0$$

We assume that the generalized Slater's constraint qualification,

$$\exists \alpha, P \quad \text{with} \ P - \alpha E \prec 0, \quad KP = b$$

holds for CDM.

Slater's condition implies that strong duality holds, i.e. this means

$$\mu^* = v^* := \max_{\substack{y \\ \Lambda \succcurlyeq 0}} \min_{P,\alpha} L(P, \alpha, \Lambda, y) \tag{5.8}$$

and v^* is attained for some $\Lambda \succcurlyeq 0$, y see e.g. [82]. The inner minimization of the convex, in P, Lagrangian is unconstrained and we can differentiate to get the equivalent problem

$$v^* = \max_{\substack{\nabla L(P,\alpha,\Lambda,y)=0 \\ \Lambda \succcurlyeq 0}} f(P) + \langle y, b - KP \rangle - \operatorname{trace} \Lambda(\alpha E - P). \tag{5.9}$$

We can now state the dual problem:

$$v^* := \max \ f(P) + \langle y, b - KP \rangle - \operatorname{trace} \Lambda(\alpha E - P)$$

(DCDM)   s.t. $\nabla_P f(P) - \mathscr{K}^* y + \Lambda = 0,$

$$-\operatorname{trace} \Lambda E = 0, \tag{5.10}$$

$$\Lambda \succcurlyeq 0.$$

The above pair of dual problems, CDM and DCDM, provide an optimality criteria in terms of feasibility and complementary slackness. This provides the basis for many algorithms including primal–dual interior-point algorithms. In particular, we see that the duality gap, in the case of primal and dual feasibility, is given by the difference of the primal and dual optimal values:

$$-\langle y, b - KP \rangle + \operatorname{trace} \Lambda(\alpha E - P) = \operatorname{trace} \Lambda(\alpha E - P). \tag{5.11}$$

Using the derivative $\nabla_P f(P) = 2H^{(2)} \circ (P - A)$, and primal–dual feasibility, we see that complementary slackness is given by

$$\operatorname{trace}(\alpha E - P)(-2H^{(2)} \circ (P - A) + K^* y) = 0. \tag{5.12}$$

Theorem 5.12. *The pair* $\bar{P} \succcurlyeq 0, \bar{\alpha}$ *and* $\bar{\Lambda} \succcurlyeq 0, \bar{y}$ *solve CDM and DCDM if and only if*

$K\bar{P} = b,$ *primal feasibility,*

$2H^{(2)} \circ (\bar{P} - A) - K^* \bar{y} - \bar{\Lambda} = 0,$ $-\operatorname{trace} \Lambda E = 0,$ *dual feasibility,*

$\operatorname{trace} \bar{\Lambda}(\bar{\alpha} E - \bar{P}) = 0,$ *compl. slack.*

The above yields an equation for the solution of CDM. (Recall that the primal feasibility constraint is assumed to include the fact that P is a hollow matrix.) However, we do not apply a Newton-type method directly to this equation but rather to a perturbed equation which allows us to stay interior to $\mathscr{P}$ and $\mathfrak{R}_+$. We note that though the generalized Slater's constraint qualification holds for the primal, it fails for the dual since $\Lambda \succ 0 \Rightarrow \text{trace } \Lambda E > 0$. Therefore, there is no duality gap between the optimal values, but numerical complications can arise. We address this later on.

5.4.1. Interior-point algorithms

We now present the interior-point algorithms for CDM. We present a dual-step-first algorithm. (A primal-step-first version can be similarly derived.) The difference in efficiency arises from the fact that the primal variable P does not change very much if few elements of A are free, while the dual variable Λ does not change very much if many elements of A are free.

Since we can increase the weights in H to try and fix certain elements of P, we restrict ourselves to the case where the only linear equality constraints are those that fix the diagonal at 0.

5.4.2. The log-barrier approach

We now derive a primal–dual interior-point method using the log-barrier approach, see e.g. [48]. This is an alternative way of deriving the optimality conditions in Theorem 5.12. The log-barrier problem for CDM is

$$\min_{\substack{\text{diag}(P)=0 \\ P \succ 0}} B_\mu(P) := f(P) - \mu \log \det(\alpha E - P),$$

where $\mu \downarrow 0$. For each $\mu > 0$ we take one Newton step toward minimizing the log-barrier function. The Lagrangian for this problem is

$$f(P) - y^t \operatorname{diag}(P) - \mu \log \det(\alpha E - P).$$

Therefore, we take one Newton step for solving the stationarity conditions

$$\nabla_P = 2H^{(2)} \circ (P - A) - \operatorname{Diag}(y) + \mu(\alpha E - P)^{-1} = 0,$$

$$\nabla_\alpha = -\mu \operatorname{trace} E(\alpha E - P)^{-1} = 0,$$

$$\operatorname{diag}(P) = 0. \tag{5.13}$$

After the substitution $-\mu(\alpha E - P)^{-1} = 2H^{(2)} \circ (P - A) - \operatorname{Diag}(y)$, the first two equations become the perturbed complementary slackness equations. The new optimality conditions are

$$(\alpha E - P)(-2H^{(2)} \circ (P - A) + \operatorname{Diag}(y)) = \mu I,$$

$$\operatorname{trace} E(2H^{(2)} \circ (P - A) - \operatorname{Diag}(y)) = 0,$$

$$\operatorname{diag}(P) = 0. \tag{5.14}$$

And, the estimate of the barrier parameter is

$$n\mu = \operatorname{trace}(\alpha E - P)(-2H^{(2)} \circ (P - A) + \operatorname{Diag}(y)). \tag{5.15}$$

The Newton direction is dependent on which of the Eqs. (5.13), (5.14) we choose to solve. Eq. (5.14) is shown to perform better in many applications. A discussion on various choices is given in [112]. (See also [72].) However we choose (5.13) below in order to exploit sparsity. The linearization to find the Newton direction is done below.

5.4.3. Primal–dual feasible algorithm—dual step first

The algorithm essentially solves for the step h, w and backtracks to ensure both primal and dual strict feasibility. This yields the primal-step-first algorithm since we only solve for the step h, w for changes in the primal variables P, α. We do need to evaluate the dual variable to update the barrier parameter μ using the perturbed complementarity condition.

Alternatively, we can work with dual step and perturbed complementary slackness. (We follow the approach in [48]. See also [85].) We keep primal feasibility, identify Λ

$$\Lambda = \mu(\alpha E - P)^{-1} \tag{5.16}$$

and replace Eqs. (5.13) and (5.14). This yields

$$\mathrm{diag}(P) = 0, \quad \text{primal feasibility,}$$

$$2H^{(2)} \circ (P - A) - \mathrm{Diag}(y) + \Lambda = 0, \quad -\mathrm{trace}\, \Lambda E = 0, \quad \text{dual feasibility,}$$

$$-(\alpha E - P) + \mu\Lambda^{-1} = 0, \quad \text{pert. compl. slack.} \tag{5.17}$$

Remark 5.13. Dual feasibility implies that trace $\Lambda E = 0$. Therefore,

$$\Lambda = V\hat{\Lambda}V^{\mathrm{t}}, \quad \hat{\Lambda} \succ 0,$$

where V is defined in (5.2). There are many choices for V. In particular, we can make a sparse choice, i.e. one with many zero elements. Therefore, in an interior-point approach we cannot maintain dual feasibility, e.g. during the algorithm trace $\Lambda E > 0$ with $= 0$ only in the limit.

Alternatively, we could eliminate the troublesome equation in the dual to obtain the following equivalent characterization of optimality:

$$\mathrm{diag}(P) = 0, \quad \text{primal feasibility,}$$

$$2H^{(2)} \circ (P - A) - \mathrm{Diag}(y) + V\Lambda V^{\mathrm{t}} = 0, \quad \text{dual feasibility,}$$

$$-(\alpha E - P) + \mu V\Lambda^{-1}V^{\mathrm{t}} = 0, \quad \text{pert. compl. slack.} \tag{5.18}$$

We apply Newton's method to solve (5.17). We let

h denote the step for P,

w denote the step for α,

l denote the step for Λ,

s denote the step for y.

(By abuse of notation, we use l as a matrix here and also as an index. The meaning is clear from the context.) We get

$$\mathrm{diag}(h) = -\mathrm{diag}(P), \tag{5.19}$$

i.e. the diagonal (linear) constraint will be satisfied if we take a full Newton step or if we start with the initial $\mathrm{diag}(P) = 0$. Therefore, we may as well start with $\mathrm{diag}(P) = 0$ and restrict $\mathrm{diag}(h) = 0$. Then linearization of the complementary slackness equation yields

$$-(\alpha + w)E + (P + h) + \mu \Lambda^{-1} - \mu \Lambda^{-1} l \Lambda^{-1} = 0$$

or

$$(\alpha + w)E - P - h = \mu \Lambda^{-1} - \mu \Lambda^{-1} l \Lambda^{-1}, \tag{5.20}$$

where $\mathrm{diag}(P) = \mathrm{diag}(h) = 0$. We get

$$h = -\mu \Lambda^{-1} + \mu \Lambda^{-1} l \Lambda^{-1} - P + (\alpha + w)E \tag{5.21}$$

and

$$l = \frac{1}{\mu} \Lambda \{ P + h - (\alpha + w)E \} \Lambda + \Lambda. \tag{5.22}$$

The linearization of the dual feasibility equations yields

$$2H^{(2)} \circ h - \mathrm{Diag}(s) + l = -(2H^{(2)} \circ (P - A) - \mathrm{Diag}(y) + \Lambda),$$

$$-\mathrm{trace}\, lE = \mathrm{trace}\, \Lambda E, \tag{5.23}$$

with $\mathrm{diag}(P) = \mathrm{diag}(h) = 0$. We assume that we start with an initial primal–dual feasible solution. However, we include the feasibility equation on the right-hand side of (5.23), because roundoff error can cause loss of feasibility. (Since Newton directions maintain linear equations, we could theoretically substitute for h in this linearization with the right-hand side being 0. We do however forcibly maintain a zero diagonal.)

We can eliminate the primal step h and dual step s and solve for the dual step l, w. From the linearization of the dual in (5.23) and the expression for h in (5.21),

$$-\mathrm{Diag}(s) + l = -2H^{(2)} \circ h - (2H^{(2)} \circ (P - A) - \mathrm{Diag}(y) + \Lambda)$$

$$= -2H^{(2)} \circ (-\mu \Lambda^{-1} + \mu \Lambda^{-1} l \Lambda^{-1} - P + (\alpha + w)E)$$

$$- (2H^{(2)} \circ (P - A) - \mathrm{Diag}(y) + \Lambda),$$

$$\mathrm{diag}(h) = \mathrm{diag}(-\mu \Lambda^{-1} + \mu \Lambda^{-1} l \Lambda^{-1} - P + (\alpha + w)E) = 0,$$

$$\mathrm{trace}(lE) = -\mathrm{trace}(\Lambda E). \tag{5.24}$$

Since we have the constraint $\mathrm{diag}(P) = 0$ in CDM, we can, without loss of generality, set the diagonal of the weight matrix H to zero, i.e. $\mathrm{diag}(H) = 0$. We can start with initial $\mathrm{diag}(P) = 0$ and $\mathrm{diag}(\Lambda) = y$. Therefore

$$s = \mathrm{diag}(l).$$

We can now eliminate s from the first equation.

$$- \operatorname{Diag diag}(l) + l = -2H^{(2)} \circ (-\mu \Lambda^{-1} + \mu \Lambda^{-1} l \Lambda^{-1} + (\alpha + w)E)$$

$$- (2H^{(2)} \circ (-\Lambda) - \operatorname{Diag}(y) + \Lambda) \tag{5.25}$$

and, assuming that $\operatorname{diag}(P) = 0$,

$$0 = \operatorname{diag}(-\mu \Lambda^{-1} + \mu \Lambda^{-1} l \Lambda^{-1} - P + (\alpha + w)E)$$

$$= \mu \operatorname{diag}(-\Lambda^{-1} + \Lambda^{-1} l \Lambda^{-1}) + (\alpha + w)e. \tag{5.26}$$

From this we already see that if Λ started sparse and H was similarly sparse, then Λ stays sparse and l is sparse.

We can now move the variables to the left and get the Newton equation

$$2H^{(2)} \circ (wE + \mu \Lambda^{-1} l \Lambda^{-1}) - \operatorname{Diag diag}(l) + l = 2H^{(2)} \circ \{\mu \Lambda^{-1} + \Lambda - \alpha E\}$$

$$+ \operatorname{Diag}(y) - \Lambda,$$

$$\operatorname{diag}(\mu \Lambda^{-1} l \Lambda^{-1}) + we = \operatorname{diag}(\mu \Lambda^{-1}) - \alpha e,$$

$$\operatorname{trace}(lE) = -\operatorname{trace}(\Lambda E). \tag{5.27}$$

This system is square, order $1 + t(n) = 1 + n(n+1)/2$, since we need only consider the strictly upper triangular part in the first equation and Λ, l are symmetric matrices.

We can now solve this system for l, set $s = \operatorname{diag}(l)$, $t = -\operatorname{trace}(\Lambda + l)E - \lambda$, and substitute to find h, w. We then take the primal–dual step and backtrack to ensure both primal and dual positive definiteness. Note that we cannot maintain dual positive definiteness if we maintain dual feasibility. However, we can maintain dual positive definiteness on the orthogonal complement of e, i.e. maintain $V^t \Lambda V \succ 0$.

Let nnz denote the number of nonzero, upper triangular, elements of H. We assume that the diagonal of H is zero and H is symmetric. Let F denote the $nnz + n \times 2$ matrix with row p denoting the indices of the pth nonzero, upper triangular, element of $H + I$ ordered by columns, i.e. for $p = 1, \ldots, nnz + n$,

$$\{(F_{p1}, F_{p2})_{p=1,\ldots,nnz+n}\} = \{ij : H_{ij} \neq 0, \ i \leqslant j, \text{ ordered by columns}\}. \tag{5.28}$$

Let δ_{ij} denote the *Kronecker delta function*, i.e. it is 1 if $i = j$ and 0 otherwise; δ_{ijkl} is 1 when all $i = j = k = l$ and 0 otherwise; $\delta_{(ij)(kl)}$ is 1 when $(ij) = (kl)$ and 0 otherwise. Let $E_{ij} = (e_i e_j^t + e_j e_i^t)/\sqrt{2}$ denote the ij unit matrix in $\mathscr{S}^n$, where $E_{ij} = (e_i e_j^t + e_j e_i^t)/2$ if $i = j$. (This set of matrices forms an orthonormal basis of $\mathscr{S}^n$.) Then $\operatorname{trace} EE_{ij} = \sqrt{2}$ (resp. 1) if $i \neq j$ (resp. $i = j$). From (5.27) the first $t(n)$ rows, with $w = 0$, $k \neq l$, and $k = l$, components of the left-hand side are, respectively,

$$k \neq l, \ i \neq j \ \text{LHS (5.27)} = \operatorname{trace} E_{kl}\{2H^{(2)} \circ (\mu \Lambda^{-1} E_{ij} \Lambda^{-1})$$

$$- \operatorname{Diag diag}(E_{ij}) + E_{ij}\}$$

$$= \mu \operatorname{trace}(e_k e_l^t + e_l e_k^t)(H^{(2)} \circ \Lambda^{-1}$$

$$(e_i e_j^t + e_j e_i^t)\Lambda^{-1}) + \delta_{(ij)(kl)}$$

$$= \mu[2e_i^t(H^{(2)} \circ \Lambda_{:,i}^{-1} \Lambda_{j:}^{-1})e_k$$
$$+ 2e_k^t(H^{(2)} \circ \Lambda_{:,i}^{-1} \Lambda_{j:}^{-1})e_l] + \delta_{(ij)(kl)},$$

$k \neq l, \ i \neq j$ LHS $(5.27) = 2\mu H_{kl}^{(2)}(\Lambda_{li}^{-1}\Lambda_{jk}^{-1} + \Lambda_{ki}^{-1}\Lambda_{jl}^{-1}) + \delta_{(ij)(kl)},$

$k \neq l, \ i = j$ LHS $(5.27) = \text{trace}\, E_{kl}\{2\mu H^{(2)} \circ [\Lambda^{-1} E_{jj} \Lambda^{-1}]$

$$- \text{Diag diag}(E_{jj}) + E_{jj}\}$$

$$= 2\sqrt{2}\mu\, \text{trace}\, e_k e_i^t (H^{(2)} \circ \Lambda^{-1} e_j e_j^t \Lambda^{-1})$$

$$= 2\sqrt{2}\mu H_{kl}^{(2)}(\Lambda_{lj}^{-1}\Lambda_{jk}^{-1}),$$

$k = l, \ i \neq j$ LHS $(5.27) = \sqrt{2}\mu \Lambda_{ki}^{-1}\Lambda_{jk}^{-1}, \quad k = 1,\dots,n,$

$k = l, \ i = j$ LHS $(5.27) = \mu \Lambda_{ki}^{-1}\Lambda_{ik}^{-1}, \quad k = 1,\dots,n.$ $\qquad (5.29)$

The last column of LHS, with the matrix $l = 0$ and $w = 1$, is

$w = 1, \ k \neq l$ LHS $(5.27) = \text{trace}(E_{kl}(2H^{(2)} \circ E)),$

$w = 1, \ k = l$ LHS $(5.27) = 1.$ $\qquad (5.30)$

While the last row of LHS is

$i \neq j$ LHS $(5.27) = \text{trace}(E_{ij}E)) = \sqrt{2},$

$i = j$ LHS $(5.27) = 1.$ $\qquad (5.31)$

Suppose that we represent the Newton system as

$$\text{sMat}[L(\text{svec}(l))] = \text{sMat}[\text{svec}(\text{RHS})], \qquad (5.32)$$

where $\text{svec}(S)$ denotes the vector formed from the non-zero elements of the columns of the upper triangular part of the symmetric matrix S, where the strict upper triangular part of S is multiplied by $\sqrt{2}$. This guarantees that trace $XY = \text{svec}(X)^t \text{svec}(Y)$, i.e. svec is an isometry; the operator sMat is its inverse, and RHS is the matrix on the right-hand side of (5.27). The system is order $nnz + n$. From (5.32), we can write the system as a matrix and column vector equation with matrix L and vector of unknowns $\text{svec}(l)$:

$$L(\text{svec}(l)) = \text{svec}(\text{RHS}). \qquad (5.33)$$

Then for

$$p = kl, \quad k \leq l, \qquad q = ij, \quad i \leq j,$$

the pq component of the matrix L is

$$
L_{pq} = \begin{cases}
2\mu H^{(2)}_{F_{p_2},F_{p_1}} (\Lambda^{-1}_{F_{p_2},F_{q_1}} \Lambda^{-1}_{F_{q_2},F_{p_1}} + \Lambda^{-1}_{F_{p_1},F_{q_1}} \Lambda^{-1}_{F_{q_2},F_{p_2}}) & \text{if } p \neq q, \ k \neq l, \\
& i \neq j, \\[4pt]
2\sqrt{2}\mu H^{(2)}_{F_{p_2},F_{p_1}} (\Lambda^{-1}_{F_{p_2},F_{q_2}} \Lambda^{-1}_{F_{q_2},F_{p_1}}) & \text{if } p \neq q, \ k \neq l, \\
& i = j, \\[4pt]
2\sqrt{2}\mu H^{(2)}_{F_{p_2},F_{p_1}} (\Lambda^{-1}_{F_{p_2},F_{q_2}} \Lambda^{-1}_{F_{q_2},F_{p_1}}) & \text{if } p = q, \ k \neq l, \\
& i = j, \\[4pt]
2\mu H^{(2)}_{F_{p_2},F_{p_1}} (\Lambda^{-1}_{F_{p_2},F_{q_1}} \Lambda^{-1}_{F_{q_2},F_{p_1}} + \Lambda^{-1}_{F_{p_1},F_{q_1}} \Lambda^{-1}_{F_{q_2},F_{p_2}}) + 1 & \text{if } p = q, \ k \neq l, \\
& i \neq j, \\[4pt]
\sqrt{2}\mu \Lambda^{-1}_{F_{p_1},F_{q_1}} \Lambda^{-1}_{F_{q_2},F_{p_1}} & \text{if } k = l, \ i \neq j, \\[4pt]
\mu \Lambda^{-1}_{F_{p_1},F_{q_1}} \Lambda^{-1}_{F_{q_1},F_{p_1}} & \text{if } k = l, \ i = j, \\[4pt]
2\sqrt{(2)} H^{(2)}_{F_{p_2},F_{p_1}} & \text{if } w = 1, \ k \neq l, \\[4pt]
1 & \text{if } w = 1, \ k = l.
\end{cases}
$$
(5.34)

The pth row can be calculated using the Hadamard product of pairs of columns of Λ^{-1},

$$
\Lambda^{-1}_{F_{p_2},F_{:,1}} \circ \Lambda^{-1}_{F_{p_1},F_{:,2}}. \tag{5.35}
$$

This allows for complete vectorization and simplifies the construction of the linear system, especially in the large sparse case.

The $p = kl$, $k \leqslant l$, and last row, component of the right-hand side of system (5.32) is

$$
\mathrm{RHS}_p = \begin{cases}
\sqrt{2}(2H^{(2)}_p \circ \{\mu\Lambda^{-1}_p + A_p - \alpha\} - \Lambda_p) & \text{if } k \neq l, \\
\mu\Lambda^{-1}_{kk} - \alpha & \text{if } k = l, \\
-\mathrm{trace}(\Lambda E) & \text{last row.}
\end{cases}
$$

The above provides a sparse system of linear equations for the search direction in a primal–dual interior-point algorithm. One would then take a step in this direction, backtrack to guarantee positive definiteness and then repeat the process with a new system, i.e. follow the standard paradigm for these algorithms.

5.4.4. Primal–dual feasible algorithm—primal step first

Alternatively, if many elements of H are sufficiently large, i.e. if we fix (or specify) many elements of A, then it is more efficient to eliminate l and solve for h first. The algorithm is similar to the dual-step-first one. The details can be found in [3].

6. Conclusion

In this paper we showed the strength of Lagrangian relaxation for obtaining semidefinite programming relaxations for several discrete optimization problems. We have presented a recipe for finding such relaxations based on adding redundant quadratic constraints and using Lagrangian duality and illustrated it with several examples, including the derivation of new strengthened SDP relaxations for MC. We also discussed the application of SDP to matrix completion problems. We showed how SDP can be used to find approximate positive semidefinite and Euclidean distance matrix completions and we concluded by presenting a new SDP algorithm which exploits sparsity and structure in large instances of Euclidean distance matrix completion problems.

References

[1] A. Alfakih, A. Khandani, H. Wolkowicz, Solving Euclidean distance matrix completion problems via semidefinite programming, Comput. Optim. Appl. 12(1–3) (1999) 13–30 (Computational optimization—a tribute to Olvi Mangasarian, Part I).

[2] A. Alfakih, H. Wolkowicz, Matrix completion problems, in: H. Wolkowicz, R. Saigal, L. Vandenberghe (Eds.), Handbook of semidefinite programming: Theory, Algorithms, and Applications, Kluwer Academic Publishers, Boston, MA, 2000, pp. 533–545.

[3] A. Alfakih, H. Wolkowicz, A new semidefinite programming model for large sparse Euclidean distance matrix completion problems, Technical Report CORR 2000-37, University of Waterloo, Waterloo, Canada, 2000, in progress.

[4] F. Alizadeh, Combinatorial optimization with interior point methods and semidefinite matrices, Ph.D. Thesis, University of Minnesota, 1991.

[5] F. Alizadeh, Interior point methods in semidefinite programming with applications to combinatorial optimization, SIAM J. Optim. 5 (1995) 13–51.

[6] F. Alizadeh, J.-P. Haeberly, M.V. Nayakkankuppam, M.L. Overton, S. Schmieta, SDPpack user's guide—version 0.9 Beta, Technical Report TR1997-737, Courant Institute of Mathematical Sciences, NYU, New York, NY, June 1997.

[7] M.F. Anjos, New convex relaxations for the maximum cut and VLSI layout problems, Ph.D. Thesis, University of Waterloo, 2001.

[8] M.F. Anjos, H. Wolkowicz, Geometry of semidefinite max-cut relaxations via matrix ranks, J. Combin. Optim., to appear.

[9] M.F. Anjos, H. Wolkowicz, A strengthened SDP relaxation via a second lifting for the Max-Cut problem, Discrete Appl. Math., to appear.

[10] K.M. Anstreicher, Eigenvalue bounds versus semidefinite relaxations for the quadratic assignment problem, Technical Report, University of Iowa, Iowa City, IA, 1999.

[11] K.M. Anstreicher, X. Chen, H. Wolkowicz, Y. Yuan, Strong duality for a trust-region type relaxation of the quadratic assignment problem, Linear Algebra Appl. 301 (1–3) (1999) 121–136.

[12] K.M. Anstreicher, H. Wolkowicz, On Lagrangian relaxation of quadratic matrix constraints, SIAM J. Matrix Anal. Appl. 22 (1) (2000) 41–55.

[13] M. Bakonyi, C.R. Johnson, The Euclidean distance matrix completion problem, SIAM J. Matrix Anal. Appl. 16 (2) (1995) 646–654.

[14] M. Bakonyi, G. Naevdal, On the matrix completion method for multidimensional moment problems, Acta Sci. Math. (Szeged) 64 (3–4) (1998) 547–558.

[15] W.W. Barrett, C.R. Johnson, M. Lundquist, Determinantal formulae for matrix completions associated with chordal graphs, Linear Algebra Appl. 121 (1989) 265–289 (Linear algebra and applications, Valencia, 1987).

[16] A. Barvinok, On convex properties of the quadratic image of the sphere, Technical Report, University of Michigan, 1999.

[17] S.J. Benson, Y. Ye, X. Zhang, Solving large-scale sparse semidefinite programs for combinatorial optimization, SIAM J. Optim. 10 (2) (2000) 443–461 (electronic).

[18] R. Bhatia, Perturbation Bounds for Matrix Eigenvalues, Pitman Research Notes in Mathematics Series, Vol. 162, Longman, New York, 1987.

[19] B. Borchers, CSDP, a C library for semidefinite programming, Optim. Methods Softw. 11/12 (1–4) (1999) 613–623 (Interior point methods).

[20] B. Borchers, SDPLIB 1.2, a library of semidefinite programming test problems, Optim. Methods Softw. 11(1) (1999) 683–690 (Interior point methods).

[21] S. Burer, R.D.C. Monteiro, A projected gradient algorithm for solving the maxcut SDP relaxation, Optimization Methods and Software 15 (2001) 175–200.

[22] C.C. Choi, Y. Ye, Application of semidefinite programming to circuit partitioning, Technical Report, The University of Iowa, Iowa City, IW, 1999.

[23] N. Cohen, J. Dancis, Maximal rank Hermitian completions of partially specified Hermitian matrices, Linear Algebra Appl. 244 (1996) 265–276.

[24] G.M. Crippen, T.F. Havel, Distance Geometry and Molecular Conformation, Wiley, New York, 1988.

[25] J. Dancis, Positive semidefinite completions of partial Hermitian matrices, Linear Algebra Appl. 175 (1992) 97–114.

[26] A. Edelman, T. Arias, S.T. Smith, The geometry of algorithms with orthogonality constraints. SIAM J. Matrix Anal. Appl. 20(2) (1999) 303–353 (electronic).

[27] S.M. Fallat, C.R. Johnson, R.L. Smith, The general totally positive matrix completion problem with few unspecified entries, Electron. J. Linear Algebra 7 (2000) 1–20 (electronic).

[28] P.A. Fillmore, J.P. Williams, Some convexity theorems for matrices, Glasgow Math. J. 10 (1971) 110–117.

[29] F. Forgó, Nonconvex Programming, Akadémiai Kiadó, Budapest, 1988.

[30] T. Fujie, M. Kojima, Semidefinite programming relaxation for nonconvex quadratic programs, J. Global Optim. 10 (4) (1997) 367–380.

[31] K. Fujisawa, M. Kojima, K. Nakata, SDPA semidefinite programming algorithm, Technical Report, Dept. of Information Sciences, Tokyo Institute of Technology, Tokyo, Japan, 1995.

[32] K. Fujisawa, M. Kojima, K. Nakata, Exploiting sparsity in primal–dual interior-point methods for semidefinite programming, Math. Programming B 79 (1997) 235–253.

[33] K. Fukuda, M. Kojima, K. Murota, K. Nakata, Exploiting sparsity in semidefinite programming via matrix completion I: general framework, Technical Report B-358, Dept. of Information Sciences, Tokyo Institute of Technology, Tokyo, Japan, 1999.

[34] J.F. Geelen, Maximum rank matrix completion, Linear Algebra Appl. 288 (1–3) (1999) 211–217.

[35] W. Glunt, T.L. Hayden, C.R. Johnson, P. Tarazaga, Positive definite completions and determinant maximization, Linear Algebra Appl. 288 (1–3) (1999) 1–10.

[36] M.X. Goemans, Semidefinite programming in combinatorial optimization, Math. Programming 79 (1997) 143–162.

[37] M.X. Goemans, Semidefinite programming and combinatorial optimization, Documenta Math. (Extra Volume ICM) 1998 (1998) 657–666 (Invited talk at the International Congress of Mathematicians, Berlin, 1998).

[38] M.X. Goemans, F. Rendl, Combinatorial optimization, in: H. Wolkowicz, R. Saigal, L. Vandenberghe (Eds.), Handbook of Semidefinite Programming: Theory, Algorithms, and Applications, Kluwer Academic Publishers, Boston, MA, 2000.

[39] M.X. Goemans, D.P. Williamson, Improved approximation algorithms for maximum cut and satisfiability problems using semidefinite programming, J. Assoc. Comput. Mach. 42 (6) (1995) 1115–1145.

[40] I. Gohberg, M.A. Kaashoek, F. Van schagen, Partially Specified Matrices and Operators: Classification, Completion, Applications, Birkhäuser, Verlag, Basel, 1995.

[41] B. Grone, C.R. Johnson, E. Marquês de Sa, H. Wolkowicz, Positive definite completions of partial Hermitian matrices, Linear Algebra Appl. 58 (1984) 109–124.

[42] S.W. Hadley, F. Rendl, H. Wolkowicz, A new lower bound via projection for the quadratic assignment problem, Math. Oper. Res. 17 (3) (1992) 727–739.

[43] W. Haemers, On some problems of Lovász concerning the Shannon capacity of graphs, IEEE Trans. Inform. Theory 25 (1979) 231–232.

[44] T.L. Hayden, R. Reams, J. Wells, Methods for constructing distance matrices and the inverse eigenvalue problem, Linear Algebra Appl. 295 (1–3) (1999) 97–112.

[45] C. Helmberg, An interior-point method for semidefinite programming and max-cut bounds, Ph.D. Thesis, Graz University of Technology, Austria, 1994.

[46] C. Helmberg, F. Oustry, Bundle methods to minimize the maximum eigenvalue function, in: H. Wolkowicz, R. Saigal, L. Vandenberghe (Eds.), Handbook of Semidefinite Programming: Theory, Algorithms, and Applications, Kluwer Academic Publishers, Boston, MA, 2000.

[47] C. Helmberg, F. Rendl, A spectral bundle method for semidefinite programming, SIAM J. Optim. 10 (3) (2000) 673–696.

[48] C. Helmberg, F. Rendl, R.J. Vanderbei, H. Wolkowicz, An interior-point method for semidefinite programming, SIAM J. Optim. 6 (2) (1996) 342–361.

[49] B. Hendrickson, The molecule problem: Determining conformation from pairwise distances, Ph.D. Thesis, Cornell University, Ithaca, New York, 1991.

[50] B. Hendrickson, The molecule problem: exploiting structure in global optimization, SIAM J. Optim. 5 (4) (1995) 835–857.

[51] L. Hogben, Completions of inverse M-matrix patterns, Linear Algebra Appl. 282 (1–3) (1998) 145–160.

[52] L. Hogben, Completions of M-matrix patterns, Linear Algebra Appl. 285 (1–3) (1998) 143–152.

[53] S. Homer, M. Peinado, In the performance of polynomial-time clique algorithms on very large graphs, Technical Report, Boston University, Boston, MA, 1994.

[54] R.A. Horn, C.R. Johnson, Matrix Analysis, Cambridge University Press, Cambridge, 1990 (Corrected reprint of the 1985 original).

[55] C.R. Johnson, Matrix completion problems: a survey, Matrix Theory and Applications, Phoenix, AZ, 1989, Amer. Math. Soc., Providence, RI, 1990, pp. 171–198.

[56] C.R. Johnson, B. Kroschel, H. Wolkowicz, An interior-point method for approximate positive semidefinite completions, Comput. Optim. Appl. 9 (2) (1998) 175–190.

[57] C.R. Johnson, B.K. Kroschel, The combinatorially symmetric P-matrix completion problem, Electron. J. Linear Algebra 1 (electronic) (1996) 59–63.

[58] C.R. Johnson, L. Rodman, Inertia possibilities for completions of partial Hermitian matrices, Linear Multilinear Algebra 16 (1–4) (1984) 179–195.

[59] C.R. Johnson, L. Rodman, Chordal inheritance principles and positive definite completions of partial matrices over function rings, Contributions to Operator Theory and its Applications, Mesa, AZ, 1987, Birkhäuser, Basel, 1988, pp. 107–127.

[60] C.R. Johnson, R.L. Smith, The symmetric inverse M-matrix completion problem, Linear Algebra Appl. 290 (1–3) (1999) 193–212.

[61] C.R. Johnson, P. Tarazaga, Connections between the real positive semidefinite and distance matrix completion problems (special issue honoring Miroslav Fiedler and Vlastimil Pták), Linear Algebra Appl. 223/224 (1995) 375–391.

[62] C.R. Johnson, P. Tarazaga, Approximate semidefinite matrices in a subspace, SIAM J. Matrix Anal. Appl. 19 (4) (1999) 861–871.

[63] S.E. Karisch, F. Rendl, H. Wolkowicz, Trust regions and relaxations for the quadratic assignment problem, Quadratic Assignment and Related Problems, New Brunswick, NJ, 1993, Amer. Math. Soc., Providence, RI, 1994, pp. 199–219.

[64] D.E. Knuth, The sandwich theorem, Electron. J. Combin. 1 (1994) 48pp.

[65] M. Kojima, S. Kojima, S. Hara, Linear algebra for semidefinite programming, Technical Report 1004, Dept. of Information Sciences, Tokyo Institute of Technology, Tokyo, Japan, 1997 (Linear matrix inequalities and positive semidefinite programming, Kyoto, 1996 (Japanese)).

[66] M. Kojima, L. Tunçel, Discretization and localization in successive convex relaxation methods for nonconvex quadratic optimization problems, Technical Report CORR98-34, Dept. of Combinatorics and Optimization, University of Waterloo, 1998.

[67] M. Kojima, L. Tunçel, Cones of matrices and successive convex relaxations of nonconvex sets, SIAM J. Optim. 10 (2000) 750–778.

[68] M. Kojima, L. Tunçel, On the finite convergence of successive convex relaxation methods, Technical Report CORR99-36, Dept. of Combinatorics and Optimization, University of Waterloo, 1999.

[69] F. Körner, A tight bound for the Boolean quadratic optimization problem and its use in a branch and bound algorithm, Optimization 19 (5) (1988) 711–721.

[70] F. Körner, Remarks on a difficult test problem for quadratic Boolean programming, Optimization 26 (1992) 355–357.

[71] S. Kruk, Semidefinite programming applied to nonlinear programming, Master's Thesis, University of Waterloo, 1996.

[72] S. Kruk, M. Muramatsu, F. Rendl, R.J. Vanderbei, H. Wolkowicz, The Gauss-Newton direction in linear and semidefinite programming, Optim. Methods Softw. 15 (1) (2001) 1–27.

[73] S. Kruk, H. Wolkowicz, sq^{2}p, sequential quadratic constrained quadratic programming, Advances in Nonlinear Programming, Beijing, 1996, Kluwer Acad. Publ., Dordrecht, 1998, pp. 177–204.

[74] J.B. Lasserre, Optimality conditions and lmi relaxations for 0–1 programs, Laas Research Report, LAAS-CNRS, Toulouse, France, 2000.

[75] M. Laurent, Cuts, matrix completions and graph rigidity, Math. Programming 79 (1997) 255–284.

[76] M. Laurent, A connection between positive semidefinite and Euclidean distance matrix completion problems, Linear Algebra Appl. 273 (1998) 9–22.

[77] M. Laurent, A tour d'horizon on positive semidefinite and Euclidean distance matrix completion problems, Topics in Semidefinite and Interior-Point Methods, The Fields Institute for Research in Mathematical Sciences, Communications Series, Vol. 18, American Mathematical Society, Providence, RI, 1998, pp. 51–76.

[78] M. Laurent, Polynomial instances of the positive semidefinite and Euclidean distance matrix completion problems, SIAM J. Matrix Anal. Appl. 22 (2000) 874–894.

[79] C. Lemaréchal, F. Oustry, Semidefinite relaxations and Lagrangian duality with application to combinatorial optimization, Technical Report, Institut National de Recherche en Informatique et en Automatique, INRIA, St Martin, France, 1999.

[80] M. Littman, D.F. Swayne, N. Dean, A. Buja, Visualizing the embedding of objects in Euclidean space, Technical Report, Bellcore, Morristown, NJ 07962-1910, 1999.

[81] L. Lovász, A. Schrijver, Cones of matrices and set-functions and 0–1 optimization, SIAM J. Optim. 1 (2) (1991) 166–190.

[82] D.G. Luenberger, Optimization by Vector Space Methods, Wiley, New York, 1969.

[83] Z-Q. Luo, S. Zhang, On the extension of Frank-Wolfe theorem, Technical Report, Erasmus University Rotterdam, The Netherlands, 1997.

[84] R. Mathias, Matrix completions, norms and Hadamard products, Proc. Amer. Math. Soc. 117 (4) (1993) 905–918.

[85] S. Mizuno, M.J. Todd, Y. Ye, On adaptive-step primal–dual interior-point algorithms for linear programming, Math. Oper. Res. 18 (4) (1993) 964–981.

[86] B. Mohar, S. Poljak, Eigenvalues in combinatorial optimization, Combinatorial Graph-Theoretical Problems in Linear Algebra, IMA Vol. 50, Springer, Berlin, 1993.

[87] R.D.C. Monteiro, Primal–dual path-following algorithms for semidefinite programming, SIAM J. Optim. 7 (3) (1997) 663–678.

[88] R.D.C. Monteiro, P.R. Zanjácomo, Implementation of primal–dual methods for semidefinite programming based on Monteiro and Tsuchiya Newton directions and their variants, Technical Report, Georgia Tech, Atlanta, GA, 1997.

[89] J.J. Moré, Z. Wu, Distance geometry optimization for protein structures, J. Global Optim. 15 (3) (1999) 219–234.

[90] Y.E. Nesterov, Quality of semidefinite relaxation for nonconvex quadratic optimization, Technical Report, CORE, Universite Catholique de Louvain, Belgium, 1997.

[91] Y.E. Nesterov, A.S. Nemirovski, Optimization over Positive Semidefinite Matrices: Mathematical Background and User's Manual, USSR Acad. Sci. Centr. Econ. & Math. Inst., 32 Krasikova St., Moscow 117418 USSR, 1990.

[92] Y.E. Nesterov, A.S. Nemirovski, Interior Point Polynomial Algorithms in Convex Programming, SIAM Publications, SIAM, Philadelphia, USA, 1994.

[93] Y.E. Nesterov, M.J. Todd, Self-scaled barriers and interior-point methods for convex programming, Math. Oper. Res. 22 (1) (1997) 1–42.

[94] Y.E. Nesterov, H. Wolkowicz, Y. Ye, Semidefinite programming relaxations of nonconvex quadratic optimization, in: H. Wolkowicz, R. Saigal, L. Vandenberghe (Eds.), Handbook of Semidefinite Programming: Theory, Algorithms, and Applications, Kluwer Academic Publishers, Boston, MA, 2000, pp. 361–420.

[95] M. Newman, Matrix completion theorems, Proc. Amer. Math. Soc. 94 (1) (1985) 39–45.

[96] M.L. Overton, R.S. Womersley, Optimality conditions and duality theory for minimizing sums of the largest eigenvalues of symmetric matrices, Math. Programming 62 (2, Ser. B) (1993) 321–357.

[97] P. Pardalos, F. Rendl, H. Wolkowicz, The quadratic assignment problem: a survey and recent developments. in: P. Pardalos, H. Wolkowicz, (Eds.), Quadratic Assignment and Related Problems, New Brunswick, NJ, 1993. Amer. Math. Soc., Providence, RI, 1994, pp. 1–42.

[98] P.M. Pardalos, S.A. Vavasis, Quadratic programming with one negative eigenvalue is NP-hard, J. Global Optim. 1 (1) (1991) 15–22.

[99] B.N. Parlett, The Symmetric Eigenvalue Problem, Society for Industrial and Applied Mathematics (SIAM), Philadelphia, PA, 1998 (Corrected reprint of the 1980 original).

[100] S. Poljak, F. Rendl, H. Wolkowicz, A recipe for semidefinite relaxation for $(0,1)$-quadratic programming, J. Global Optim. 7 (1) (1995) 51–73.

[101] S. Poljak, H. Wolkowicz, Convex relaxations of $(0,1)$-quadratic programming, Math. Oper. Res. 20 (3) (1995) 550–561.

[102] A.J. Quist, E. de Klerk, C. Roos, T. Terlaky, Copositive relaxation for general quadratic programming (special issue Celebrating the 60th Birthday of Professor Naum Shor), Optim. Methods Softw. 9 (1998) 185–208.

[103] M.V. Ramana, An algorithmic analysis of multiquadratic and semidefinite programming problems, Ph.D. Thesis, Johns Hopkins University, Baltimore, MD, 1993.

[104] M.V. Ramana, L. Tunçel, H. Wolkowicz, Strong duality for semidefinite programming, SIAM J. Optim. 7 (3) (1997) 641–662.

[105] F. Rendl, Semidefinite programming and combinatorial optimization, Appl. Numer. Math. 29 (1999) 255–281.

[106] F. Rendl, H. Wolkowicz, A projection technique for partitioning the nodes of a graph, Ann. Oper. Res. 58 (1995) 155–179 (Applied mathematical programming and modeling, II (APMOD 93), Budapest, 1993).

[107] F. Rendl, H. Wolkowicz, A semidefinite framework for trust region subproblems with applications to large scale minimization, Math. Programming 77 (2, Ser. B) (1997) 273–299.

[108] A. Schrijver, A comparison of the Delsarte and Lovász bounds, IEEE Trans. Inform. Theory IT-25 (1979) 425–429.

[109] N.Z. Shor, Quadratic optimization problems, Izv. Akad. Nauk SSSR Tekhn. Kibernet. 222 (1) (1987) 128–139, 222.

[110] R. Stern, H. Wolkowicz, Indefinite trust region subproblems and nonsymmetric eigenvalue perturbations, SIAM J. Optim. 5 (2) (1995) 286–313.

[111] J.F. Sturm, Using SeDuMi 1.02, a MATLAB toolbox for optimization over symmetric cones, Optim. Methods Softw. 11/12 (1–4) (1999) 625–653.

[112] M.J. Todd, A study of search directions in primal–dual interior-point methods for semidefinite programming, Optim. Methods Softw. 11& 12 (1999) 1–46.

[113] M.J. Todd, K.C. Toh, R.H. Tutuncu, A MATLAB software package for semidefinite programming, Technical Report, School of OR and IE, Cornell University, Ithaca, NY, 1996.

[114] M.W. Trosset, Applications of multidimensional scaling to molecular conformation, Comput. Sci. Statist. 29 (1998) 148–152.

[115] L. Tunçel, S. Xu, Complexity analyses of discretized successive convex relaxation methods, Technical Report CORR 99-37, Department of Combinatorics and Optimization, Waterloo, Ont., 1999.

[116] L. Vandenberghe, S. Boyd, S.-P. Wu, Determinant maximization with linear matrix inequality constraints, SIAM J. Matrix Anal. Appl. 19 (2) (1998) 499–533.

[117] C.P. Wells, An improved method for sampling of molecular conformation space, Ph.D. Thesis, University of Kentucky, 1995.

[118] H. Wolkowicz, R. Saigal, L. Vandenberghe (Eds.), Handbook of Semidefinite Programming: Theory, Algorithms, and Applications, Kluwer Academic Publishers, Boston, MA, 2000, xxvi+654 pages.

[119] V.A. Yakubovich, The S-procedure and duality theorems for nonconvex problems of quadratic programming, Vestnik Leningrad. Univ. 1973 (1) (1973) 81–87.

[120] Y. Ye, A new complexity result on minimization of a quadratic function with a sphere constraint, Recent Advances in Global Optimization, Princeton University Press, Princeton, 1992, pp. 19–31.

[121] Y. Ye, Approximating quadratic programming with bound and quadratic constraints, Math. Programming 84 (1999) 219–226.

[122] Q. Zhao, S.E. Karisch, F. Rendl, H. Wolkowicz, Semidefinite programming relaxations for the quadratic assignment problem, J. Combin. Optim. 2 (1) (1998) 71–109 (Semidefinite programming and interior-point approaches for combinatorial optimization problems, Toronto, ON, 1996).

ELSEVIER

Discrete Applied Mathematics 123 (2002) 579–580

DISCRETE
APPLIED
MATHEMATICS

Author Index
Volume 123 (2002)

PII: S0166-218X(02)00353-0

www.ingramcontent.com/pod-product-compliance
Lightning Source LLC
Chambersburg PA
CBHW050519190326
41458CB00005B/1597